Proceedings
"WASCOM 2001"
11th Conference on
Waves and Stability in Continuous Media

Proceedings
"WASCOM 2001"
11th Conference on
Waves and Stability in Continuous Media

Porto Ercole (Grosseto), Italy

3–9 June 2001

Editors

Roberto Monaco
Politecnico di Torino, Italy

Miriam Pandolfi Bianchi
Politecnico di Torino, Italy

Salvatore Rionero
Universitá di Napoli, Italy

World Scientific
Singapore • New Jersey • London • Hong Kong

Published by

World Scientific Publishing Co. Pte. Ltd.

P O Box 128, Farrer Road, Singapore 912805

USA office: Suite 1B, 1060 Main Street, River Edge, NJ 07661

UK office: 57 Shelton Street, Covent Garden, London WC2H 9HE

British Library Cataloguing-in-Publication Data
A catalogue record for this book is available from the British Library.

WAVES AND STABILITY IN CONTINUOUS MEDIA — WASCOM 2001

ISBN 981-23-8017-5

Printed in Singapore by World Scientific Printers (S) Pte Ltd

PREFACE

This volume contains the invited lectures and contributed papers presented at the XI International Conference on *Waves and Stability in Continuous Media* (WASCOM 2001) held June 3–9, 2001, in Porto Ercole (GR), Italy.

Ever since its initial edition organized in Catania 1981, the Conference aimed to bring together foreign and Italian researchers and scientists to discuss problems, promote collaborations and shape future directions for research in the field of stability and wave propagation in continuous media.

This cycle of conferences became a fixed meeting every two years: the further conferences have been held in Cosenza ('83), Bari ('85), Taormina ('87), Sorrento ('89), Acireale ('91), Bologna ('93), Palermo ('95), Monopoli ('97) and Vulcano ('99).

Every time the proceedings have been published, documenting the research work and progress in the area of waves and stability.

From a scientific point of view the success of this experience is confirmed by the fact that a remarkable group of italian researchers, from many different universities, has proposed several national projects in the field. The last project, entitled "Non Linear Mathematical Problems of Wave Propagation and Stability in Models of Continuous Media", co-ordinated by Prof. T. Ruggeri (Bologna), is the main proposer of the present conference.

The eleventh edition, the first of the third millennium, registered over 110 participants coming from more than 11 different countries. The topics covered by 29 main lectures and 52 short communications, within 10 sessions, were

- Discontinuity and shock waves
- Stability in Fluid Dynamics
- Small parameter problems
- Kinetic theories towards continuum models
- Non equilibrium thermodynamics
- Numerical applications

The Editors of the proceedings would like to thank the Scientific Committee who carefully suggested the invited lectures and selected the contributed papers, as well as the members of the Organizing Committee, coming from the Departments of Mathematics of the Universities of Napoli, Messina and Politecnico of Torino.

A special thank is addressed to all the participants to whom ultimately the success of the conference has to be ascribed.

Finally, the Editors are especially indebted to the institutions which have provided the financial support for publishing this book:

- Fondazione della Cassa di Risparmio di Torino (CRT)
- Provincia di Grosseto

Torino and Napoli, March 2002

THE EDITORS
Roberto Monaco
Miriam Pandolfi Bianchi
Salvatore Rionero

CONFERENCE DATA

WASCOM 2001
11th International Conference on
Waves and Stability in Continuous Media
Porto Ercole (GR), Italy, June 3–9, 2001

Scientific Committee

Chairmen: **R. Monaco** (Torino) and **S. Rionero** (Napoli)
G. Boillat (Clermont Ferrand), **C. Dafermos** (Providence)
D. Fusco (Messina), **A. Greco** (Palermo), **G. Mulone** (Catania)
I. Müller (Berlin), **C. Rogers** (Sydney), **T. Ruggeri** (Bologna)
B. Straughan (Durham), **C. Tebaldi** (Lecce)

Organizing Committee

Chairman: **M. Pandolfi Bianchi** (Torino)
F. Conforto (Messina), **M. Gentile** (Napoli), **A. Rossani** (Torino)
F. Salvarani (Torino)

Supported by

- PRIN 2000 "Non Linear Mathematical Problems of Wave Propagation and Stability in Models of Continuous Media"
- Gruppo Nazionale per la Fisica Matematica – INDAM
- Dipartimento di Matematica – Politecnico di Torino
- Regione Toscana
- Comune di Monte Argentario
- A.P.T. Grosseto
- Camera di Commercio Industria Artigianato e Agricoltura di Grosseto
- AIPE and Pro Loco Porto Ercole

TABLE OF CONTENTS

LINEAR STABILITY ANALYSIS OF A 1D FLOW IN A POROELASTIC MATERIAL UNDER DISTURBANCES WITH ADSORPTION

BETTINA ALBERS

Weierstrass Institute for Applied Analysis and Stochastics, Mohrenstr. 39, D-10117 Berlin, Germany

E-mail: albers@wias-berlin.de

In this paper we investigate the linear stability behavior of a 1D flow in a poroelastic material with respect to longitudinal as well as transversal disturbances with mass exchange. Fields are assumed to be a superposition of a stationary (nonuniform) 1D solution and of infinitesimal disturbances in the form of a one- or two-dimensional linear wave ansatz. We solve numerically the eigenvalue problem for the first step field equations using a finite-difference-scheme.

1 Introduction

In general flow instabilities arise due to at least two competing mechanisms. Conceivable are problems like a kinematic nonlinearity working against viscosity or gravity competing with a temperature gradient. In the case of our multicomponent model for porous systems, where a fluid flows through the channels of a skeleton, a kinematic nonlinearity is the counterpart to the permeability of the medium. Adsorption processes contribute in a nonlinear way to field equations and influence stability properties.

2 Base flow

We investigate the stability behaviour of a 1D flow in a channel of length l and width $2b$ under longitudinal and transversal perturbations with adsorption. The base flow (without adsorption) satisfies the following mass and momentum balances of the fluid

$$\frac{\partial \rho^F}{\partial t} + \frac{\partial \rho^F v_x^F}{\partial x} = 0, \qquad \rho^F \left(\frac{\partial v_x^F}{\partial t} + v_x^F \frac{\partial v_x^F}{\partial x} \right) = -\frac{\partial p^F}{\partial x} - \pi v_x^F . \tag{1}$$

Here, ρ^F is the mass density of the fluid component, v_x^F is the fluid velocity in x-direction and π is the bulk permeability coefficient. It describes the effective resistance of the skeleton to the flow of the fluid as well as the true viscosity of the fluid. The partial pressure in the fluid p^F is given by the following constitutive relation

$$p^F = p_0^F + \kappa \left(\rho^F - \rho_0^F \right), \tag{2}$$

where p_0^F and ρ_0^F are initial values of the pressure and the mass density in the fluid phase. κ denotes the compressibility. It is assumed that a deformed skeleton does not contribute to a dynamical disturbance.

The stationary form of (1) yields $\rho^F v_x^F = C_1 = \text{const.}$ and it remains to determine the unknown constants in the formal solution of $(1)_2$

$$\ln v_x^F + \frac{1}{2}\frac{\kappa}{v_x^{F2}} = \frac{\pi}{C_1}\left(C_2 - x\right), \tag{3}$$

where C_2 is the second integration constant. Boundary conditions for the base flow have the form:

$$\begin{aligned} -\left.\rho^F v_x^F\right|_{x=0} &= \alpha\left[\left.p^F\right|_{x=0} - n_E p_l\right], \\ \left.\rho^F v_x^F\right|_{x=l} &= \alpha\left[\left.p^F\right|_{x=l} - n_E p_r\right]. \end{aligned} \tag{4}$$

They are of third type and express the fact that the flow through the boundary of the body depends on the pressure difference of the partial pressure in the fluid $\left(p^F\right)$ and the external pressure which works on the fluid (on the left hand side p_l and on the right hand side p_r) as well as on the permeability of the surface which is denoted by α.

A comparison of the implicit solutions with and without the logarithmic part of (3) shows that we are allowed to skip the logarithmic part (i.e. the nonlinearity $\rho^F v_x^F \frac{\partial v_x^F}{\partial x}$ in $(1)_2$) so that we are able to find explicit solutions for the base flow.

Accounting for the fact that the pressure difference $p_d := n_E\left(p_r - p_l\right) \ll 2\kappa\rho_0^F$, and $p_0^F := n_E p_r$ we get the following base solution

$$v_x^F \cong \frac{\alpha p_d}{2\rho_0^F + \alpha\pi l}\frac{1}{\sqrt{1 - \dfrac{2\pi}{\kappa\rho_0^F}\dfrac{\alpha p_d}{2\rho_0^F + \alpha\pi l}x}}, \qquad \rho^F \cong \rho_0^F\sqrt{1 - \frac{2\pi}{\kappa\rho_0^F}\frac{\alpha p_d}{2\rho_0^F + \alpha\pi l}x}. \tag{5}$$

3 Adsorption/diffusion model

We consider a flow process in a porous medium in which there appears an exchange of mass between the skeleton and the liquid which is the mixture of a fluid carrier and an adsorbate and flows through the channels of the skeleton. In principle the model (see B. ALBERS [1], [2]) takes into account three components: the skeleton, the fluid and an adsorbate which flows with the same velocity as the fluid through the channels of the skeleton until it settles down on the inner surface of the porous body. As before we neglegt the dynamics of the skeleton.

Then the mass balance equation for the liquid and the concentration balance in 2D have the following form

$$\frac{\partial \rho^L}{\partial t} + \frac{\partial \rho^L v_x^F}{\partial x} + \frac{\partial \rho^L v_z^F}{\partial z} = \hat{\rho}^A, \tag{6}$$
$$\rho^L \left(\frac{\partial c}{\partial t} + v_x^F \frac{\partial c}{\partial x} + v_z^F \frac{\partial c}{\partial z} \right) = (1 - c)\, \hat{\rho}^A,$$

where ρ^L is the mass density of the liquid phases, i.e. the sum of the mass densities of the fluid and the adsorbate, and c denotes the concentration of the adsorbate in the fluid component. The common velocity of fluid and adsorbate has two components $\left(v_x^F, v_z^F\right)$. Finally, the intensity of the mass source is denoted by $\hat{\rho}^A$. Here it is assumed in the simplified version without changes of porosity and inner surface

$$\hat{\rho}^A = -\rho_{ad}^A \left\{ \left[\frac{cp^L}{p_0} (1 - \xi) - \xi \right] \frac{1}{\tau_{ad}} \right\}. \tag{7}$$

Its derivation is based on the classical LANGMUIR adsorption theory about *occupied* (ξ) and *bare* $(1 - \xi)$ *sites* (see [4]) on the internal surface f_{int} of the solid. $\rho_{ad}^A := \frac{m^A f_{int}}{V}$ where V is the representative elementary volume REV and m_A denotes the mass of adsorbate per unit of the internal surface area. p^L denotes the partial pressure in the liquid (fluid and adsorbate together), p_0 is a Langmuir reference pressure and τ_{ad} is the characteristic time of adsorption. The complete form of of the mass source is introduced in [2] and [1].

Due to the same velocity of fluid and adsorbate we need only one momentum balance together for these both components. E.g. in 2D we have

$$\rho^L \left[\frac{\partial v_x^F}{\partial t} + v_x^F \frac{\partial v_x^F}{\partial x} + v_z^F \frac{\partial v_x^F}{\partial z} \right] = -\kappa \frac{\partial \rho^L}{\partial x} - \pi v_x^F,$$
$$\rho^L \left[\frac{\partial v_z^F}{\partial t} + v_x^F \frac{\partial v_z^F}{\partial x} + v_z^F \frac{\partial v_z^F}{\partial z} \right] = -\kappa \frac{\partial \rho^L}{\partial z} - \pi v_z^F.$$

where also the linear constitutive relation for the pressure in the liquid phase p^L (2) has been used.

4 Linear stability analysis (1D and 2D)

4.1 Perturbation

We use a regular perturbation method restricted to zeroth and first order contributions for the analysis. This means that we expect the fields to be a superposition of the base solution (indicated by 0) and a small perturbation (indicated by 1)

$$\begin{aligned}
&\text{1D case} \quad \rho^L = \overset{0}{\rho}(x) + \overset{1}{\rho}(x,t)\,, \quad c = \overset{0}{c}(x) + \overset{1}{c}(x,t)\,, \\
&\qquad\qquad \xi = \overset{0}{\xi}(x) + \overset{1}{\xi}(x,t)\,, \quad v_x^F = \overset{0}{v}_x(x) + \overset{1}{v}_x(x,t)\,, \\
&\text{2D case} \quad \rho^L = \overset{0}{\rho}(x) + \overset{1}{\rho}(x,z,t)\,, \; c = \overset{0}{c}(x) + \overset{1}{c}(x,z,t)\,, \; \xi = \overset{0}{\xi}(x) + \overset{1}{\xi}(x,z,t)\,, \\
&\qquad\qquad v_x^F = \overset{0}{v}_x(x) + \overset{1}{v}_x(x,z,t)\,, \quad v_z^F = \overset{1}{v}_z(x,z,t)\,.
\end{aligned} \tag{8}$$

Additional to the quantities of the base flow without mass exchange we obtain the relation for the Langmuir isotherm

$$\overset{0}{\xi} = \frac{\dfrac{\overset{0}{c}\left[p_0^L + \kappa\left(\overset{0}{\rho} - \rho_0^L\right)\right]}{p_0}}{1 + \dfrac{\overset{0}{c}\left[p_0^L + \kappa\left(\overset{0}{\rho} - \rho_0^L\right)\right]}{p_0}} = \frac{\dfrac{\overset{0}{c}\overset{0}{p}{}^L}{p_0}}{1 + \dfrac{\overset{0}{c}\overset{0}{p}{}^L}{p_0}},$$

which, of course, varies with x. Simultaneously we have $\overset{0}{c} = \text{const.} = c_0$.

For the first step of perturbation we obtain the following set of equations

$$\frac{\partial \overset{1}{\rho}}{\partial t} + \overset{0}{\rho}\left(\frac{\partial \overset{1}{v}_x}{\partial x} + \frac{\partial \overset{1}{v}_z}{\partial z}\right) + \overset{1}{\rho}\frac{\partial \overset{0}{v}_x}{\partial x} + \overset{0}{v}_x\frac{\partial \overset{1}{\rho}}{\partial x} + \overset{1}{v}_x\frac{\partial \overset{0}{\rho}}{\partial x} = \overset{1}{\hat{\rho}}{}^A,$$

$$\overset{0}{\rho}\left(\frac{\partial \overset{1}{c}}{\partial t} + \overset{0}{v}_x\frac{\partial \overset{1}{c}}{\partial x}\right) = (1 - c_0)\,\overset{1}{\hat{\rho}}{}^A - \overset{1}{c}\frac{\rho_{ad}^A}{\tau_{ad}}\left[\frac{c_0\left[p_0^L + \kappa\left(\overset{0}{\rho} - \rho_0^L\right)\right]}{p_0}\left(1 - \overset{0}{\xi}\right) - \overset{0}{\xi}\right],$$

$$\frac{\partial \overset{1}{\xi}}{\partial t} = \left[\left(1 - \overset{0}{\xi}\right)\frac{c_0\kappa\overset{1}{\rho} + \overset{1}{c}\left[p_0^L + \kappa\left(\overset{0}{\rho} - \rho_0^L\right)\right]}{p_0} - c_0\overset{1}{\xi}\frac{p_0^L + \kappa\left(\overset{0}{\rho} - \rho_0^L\right)}{p_0} - \overset{1}{\xi}\right]\frac{1}{\tau_{ad}},$$

$$\overset{0}{\rho}\left[\frac{\partial \overset{1}{v}_x}{\partial t} + \overset{0}{v}_x\frac{\partial \overset{1}{v}_x}{\partial x} + \overset{1}{v}_x\frac{\partial \overset{0}{v}_x}{\partial x}\right] + \overset{1}{\rho}\overset{0}{v}_x\frac{\partial \overset{0}{v}_x}{\partial x} = -\kappa\frac{\partial \overset{1}{\rho}}{\partial x} - \pi\overset{1}{v}_x,$$

$$\overset{0}{\rho}\left[\frac{\partial \overset{1}{v}_z}{\partial t} + \overset{0}{v}_x\frac{\partial \overset{1}{v}_z}{\partial x}\right] = -\kappa\frac{\partial \overset{1}{\rho}}{\partial z} - \pi\overset{1}{v}_z,$$

with

$$\overset{1}{\hat{\rho}}{}^A = -\frac{\rho_{ad}^A}{\tau_{ad}}\left[\frac{c_0\kappa\overset{1}{\rho} + \overset{1}{c}\left[p_0^L + \kappa\left(\overset{0}{\rho} - \rho_0^L\right)\right]}{p_0}\left(1 - \overset{0}{\xi}\right) - \overset{1}{\xi}\left(1 + \frac{c_0\left[p_0^L + \kappa\left(\overset{0}{\rho} - \rho_0^L\right)\right]}{p_0}\right)\right].$$

4.2 Ansatz

The perturbations are expressed in terms of the following ansatz

1D case	2D case
$\overset{1}{\rho} = \bar{\rho}(x)\,e^{\omega t}, \quad \overset{1}{c} = \bar{c}(x)\,e^{\omega t},$	$\overset{1}{\rho} = \bar{\rho}(x)\,e^{\omega t}\cos kz, \quad \overset{1}{c} = \bar{c}(x)\,e^{\omega t}\cos kz,$
$\overset{1}{\xi} = \bar{\xi}(x)\,e^{\omega t}, \quad \overset{1}{v}_x = \bar{v}_x(x)\,e^{\omega t},$	$\overset{1}{\xi} = \bar{\xi}(x)\,e^{\omega t}\cos kz, \quad \overset{1}{v}_x = \bar{v}_x(x)\,e^{\omega t}\cos kz,$
	$\overset{1}{v}_z = \bar{v}_z(x)\,e^{\omega t}\sin kz,$

(9)

where $k := \frac{m\Pi}{b}$ is the real wave number of the disturbance and ω is the possibly complex frequency.

The boundary conditions have the form

$$\bar{v}_x\big|_{x=0} = -\frac{\alpha\kappa + \overset{0}{v}_x\big|_{x=0}}{\overset{0}{\rho}\big|_{x=0}}\,\bar{\rho}\big|_{x=0}, \qquad \bar{v}_z\big|_{z=b} = 0,$$

$$\bar{v}_x\big|_{x=l} = \frac{\alpha\kappa - \overset{0}{v}_x\big|_{x=l}}{\overset{0}{\rho}\big|_{x=l}}\,\bar{\rho}\big|_{x=l}, \qquad \bar{v}_z\big|_{z=-b} = 0.$$

5 Results

For calculations we use the following parameters

Length, width of channel l, b	1m	Initial concentration c_0	10^{-3}
Initial mass density ρ_0^L	$2.3 \cdot 10^2\ \frac{\text{kg}}{\text{m}^3}$	Langmuir pressure p_0	10kPa
Initial pressure p_0^L	23kPa	Compressibility κ	$2.25 \cdot 10^6\ \frac{\text{m}^2}{\text{s}^2}$
mass density of adsorbate on internal surface ρ_{ad}^A	$40\ \frac{\text{kg}}{\text{m}^3}$	Charact. time of adsorp. τ_{ad}	1s
		Proportionality factor ν	10
Permeability of solid π	$10^9\ \frac{\text{kg}}{\text{m}^3\text{s}}$	Surface permeability par. α	$4 \cdot 10^{-8}\ \frac{\text{s}}{\text{m}}$

Table 1: Typical model parameters for adsorption processes in soils

We solve the eigenvalue problem for ω numerically, using a second order finite difference scheme in a equidistant mesh (length of the body l divided into n parts of length h). The derivatives of disturbances are written as central differences for inner mesh points, or as asymmetric ones for the first, and the last point. For this linear eigenvalue problem we obtain for the 1D case $4n+2$ and for the 2D case $5n+3$ eigenvalues ω_i. The exponential form of the time ansatz yields that the base flow is stable if all real parts of ω_i are negative and unstable if at least one of the real parts is positive.

The results of the numerical investigation are given in Fig.1:

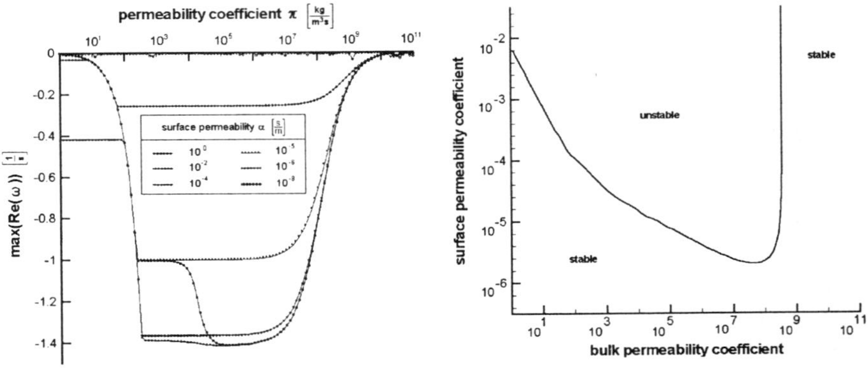

Fig 1: Relaxation of disturbance as a function of permeability π for different surface permeabilities α in 1D disturbance (left) and instability range on the parametric (π, α) plane in 2D disturbance (right)

6 Discussion

Similar to both longitudinal and transversal disturbances without mass exchange (see [3]) the base steady state flow is stable with respect to a longitudinal disturbance with mass exchange for any choice of parameters (α, π). The left part of Fig. 1 shows that the smaller the surface permeability parameter α the slower is the relaxation.

With respect to transversal disturbances the stability behavior is different (right hand side of Fig.1). Here exists a certain range of permeability parameters for which the base flow is unstable. This is the case for any α but very large π not shown anymore at the diagram. Most likely the model is not applicable in this range anyway. An essential range of instability appears for π lower than app. $10^{8.548}$ and surface permeability larger than app. $10^{-5.690}$. This range corresponds to higher velocities for the same pressure difference due to a lower resistance of the boundary. Simultaneously it corresponds to a lower "internal friction" πv_x^F in momentum balance equation due to lower values of permeability π.

References

1. ALBERS, B.: Makroskopische Beschreibung von Adsorptions-Diffusions-Vorgängen in porösen Körpern, PhD thesis TU Berlin, Logos-Verlag, Berlin, 2000.
2. ALBERS, B.: Coupling of Adsorption and Diffusion in Porous and Granular Materials. A 1-D Example of the Boundary Value Problem, *Arch.Appl.Mech.* **70** (2000) 7, 519-531.
3. ALBERS, B.; WILMAŃSKI, K.: Relaxation Properties of a 1D Flow through a Porous Material without and with Adsorption, WIAS-Preprint No. 707, 2001.
4. GREGG, S.J.; SING, K.S.W.: Adsorption, Surface Area and Porosity, Academic Press, London, 1982.
5. WILMAŃSKI, K.: A Thermodynamic Model of Compressible Porous Materials with the Balance Equation of Porosity, *Transport in Porous Media*, **32**, 21-47 (1998).

COMPRESSIBLE FLOWS IN TIME-DEPENDENT DOMAINS: ZERO MACH NUMBER LIMIT

G. ALÌ

Institute for Applied Mathematics, Consiglio Nazionale delle Ricerche, via P. Castellino 111, 80131 Napoli, Italy
E-mail: ali@iam.na.cnr.it

We perform a multiple time scale, single space scale analysis of a compressible fluid in a time-dependent domain, when the time variations of the boudary are small with respect to the acoustic velocity. We introduce an average operator with respect to the fast time. The averaged leading order variables satisfy modified incompressible equations, which are coupled to linear acoustic equations with respect to the fast time.

1 Introduction

The relationship between compressible and incompressible flows has been rigorously stated and proved for isentropic flows in unbounded domains by Klainerman and Majda at the beginning of the Eighties[6,7], and later extended to non-isentropic flows in bounded domains by Schochet[14]. In all these papers, the initial data are "prepared", that is, they are compatible with the limiting incompressible equations. When the data are not prepared, the limit is still valid but it is not uniform for times close to zero[2,3,4,5,13,15], since the initial time derivatives fail to be uniformly bounded.

Roughly speaking, it can be proved that the solution to the compressible Euler equations exists in a time interval independent of ϵ. Moreover, it converges to a solution of the corresponding incompressible Euler equations, whose initial data is the incompressible part of the original initial data.

The low Mach number theory has been applied to combustion, both in confined and unbounded domains[9,10], and to astrophysics, in the framework of magnetohydrodynamics, in order to justify the small density fluctuations, called pseudosound, in the solar wind incompressible model[11,16,17].

A rigorous justification of this theory, in the context of multiple scale expansion, was given by Klein[8], with application to a numerical scheme which is uniformly valid as the Mach number tends to zero. A rigorous account can be found in a recent paper[12].

In this note we perform a multiple time scale, single space scale analysis of a compressible fluid in a time-dependent domain, when the time variations of the boudary are small with respect to the acoustic velocity. This approach differs from Klein's theory, which makes use of two length scales and applies to unbounded (or "large") domain.

2 The incompressible limit of the Euler equations

The compressible Euler equations are

$$\begin{aligned} &\partial_t \rho + \nabla \cdot (\rho \mathbf{u}) = 0, \\ &\partial_t (\rho \mathbf{u}) + \nabla \cdot (\rho \mathbf{u} \otimes \mathbf{u} + pI) = 0, \\ &\partial_t E + \nabla \cdot [(E + p)\mathbf{u}] = 0, \end{aligned} \tag{1}$$

where the total energy is given by

$$E = \rho \left(e + \frac{1}{2} u^2 \right).$$

The internal energy, e, the pressure, p, and the density, ρ, are related to the entropy, s, and the temperature, T, by

$$de + pd\left(\frac{1}{\rho}\right) = Tds.$$

Using an equation of state of the form $s = s(\rho, p)$, we can replace the energy equation with an equation for the pressure,

$$\partial_t p + \mathbf{u} \cdot \nabla p + \gamma p \nabla \cdot \mathbf{u} = 0. \tag{2}$$

The adiabatic exponent γ is defined by

$$\gamma = \frac{\rho}{p} \left. \frac{\partial p}{\partial \rho} \right|_s \equiv \frac{\rho}{p} \frac{\partial s / \partial \rho}{\partial s / \partial p}.$$

In the following we will consider a politropic gas, for which γ is constant.

We introduce the rescaled variables

$$\begin{aligned} \rho' &= \frac{\rho}{\rho_m}, \quad \mathbf{u}' = \frac{\mathbf{u}}{u_m}, \quad p' = \frac{p}{p_m}, \\ \mathbf{x}' &= \frac{\mathbf{x}}{l_m}, \quad t' = \frac{t}{t_m} \quad u_m = \frac{l_m}{t_m}. \end{aligned}$$

Dropping the prime, we obtain

$$\begin{aligned} &\frac{1}{\rho}(\partial_t + \mathbf{u} \cdot \nabla)\rho + \nabla \cdot \mathbf{u} = 0, \\ &\rho(\partial_t + \mathbf{u} \cdot \nabla)\mathbf{u} + \frac{1}{\epsilon^2} \nabla p = 0, \\ &\frac{1}{\gamma p}(\partial_t + \mathbf{u} \cdot \nabla)p + \nabla \cdot \mathbf{u} = 0. \end{aligned} \tag{3}$$

The small parameter

$$\epsilon = \frac{u_m}{\sqrt{p_m/\rho_m}} \equiv \frac{M}{\sqrt{\gamma}},$$

is proportional to the Mach number $M = u_m/c_m$, where c_m is the typical sound speed,

$$c_m^2 = \left.\frac{\partial p}{\partial \rho}\right|_s (\rho_m, p_m) \equiv \frac{\gamma p_m}{\rho_m}.$$

Formally, as ϵ tends to 0, we have

$$\nabla p = 0, \tag{4}$$

which implies $p = p^{(0)}(t)$. The remaining equations become

$$\nabla \cdot \mathbf{u} = -\frac{1}{\gamma p^{(0)}} \frac{dp^{(0)}}{dt}, \tag{5}$$

$$\frac{1}{\rho}\frac{D\rho}{Dt} = \frac{1}{\gamma p^{(0)}} \frac{dp^{(0)}}{dt}, \tag{6}$$

with $\frac{D}{Dt} = \partial_t + \mathbf{u} \cdot \nabla$.

When these equations are considered in an unbounded domain $\Omega \subseteq \mathbf{R}^n$, it follows that the leading order pressure $p^{(0)}$ is constant and (5) reduces to the incompressibility condition.

To recover an equation for $\mathbf{u}$, we need to assume that the pressure has spatial variations at order ϵ^2. Then, we have

$$\rho \frac{D\mathbf{u}}{Dt} + \nabla p^* = 0. \tag{7}$$

An additional second order pressure term comes from fast acoustic wave generation, and it is coupled to the first order fluid velocity[6,7,9]. For this reason, a multiple scale analysis is needed for an effective description of the low Mach number limit.

3 Multiple scale expansion

A first multiple scale approach is due to Klein[8,12]. We assume that two different space scales are present, that is a reference length (l_{ref}) and an acoustic length ($l_{\text{ac}} = c_{\text{ref}} t_{\text{ref}}$). Here, t_{ref} is a characteristic time scale and l_{ac} is a characteristic length scale of the acoustic waves. Two space variables follow naturally,

$$\mathbf{x}' = \frac{\mathbf{x}}{l_{\text{ref}}}, \qquad \xi' = \frac{\mathbf{x}}{l_{\text{ac}}} \equiv \frac{l_{\text{ref}}}{l_{\text{ac}}} \mathbf{x}' \equiv \epsilon \mathbf{x}',$$

where $\mathbf{x}$ is the physical space variable. With this motivation, cropping the prime, the following expansion for $U = (\rho, \mathbf{u}, p)$ is considered:

$$U(x, t; \epsilon) = U^{(0)}(x, \xi, t) + \epsilon U^{(1)}(x, \xi, t) + \epsilon^2 U^{(2)}(x, \xi, t) + O(\epsilon^3). \qquad (8)$$

It follows that the pressure admits the expansion

$$p = p^{(0)}(t) + \epsilon p^{(1)}(\xi, t) + O(\epsilon^2), \qquad (9)$$

and the fluid velocity satisfies the constraint

$$\nabla \cdot \mathbf{u}^{(0)} = -\frac{1}{\gamma p^{(0)}} \frac{dp^{(0)}}{dt}. \qquad (10)$$

Moreover, we obtain the acoustic equation

$$\overline{\partial_t(\rho^{(0)}\mathbf{u}^{(0)})} + \nabla_\xi p^{(1)},$$
$$\partial_t p^{(1)} + \gamma p^{(0)} \overline{\nabla_\xi \cdot \mathbf{u}^{(0)}},$$

where the overline denotes spatial average with respect to $\mathbf{x}$. The details can be found in Meister[12].

We follow a different approach, which is more appropriate when we consider the Euler equations on a bounded, time-dependent domain $\Omega(t) \in \mathbf{R}^n$. We denote by Ω_0 the domain at initial time, and assume that there exists a family of continuous maps $\Phi_t : \mathbf{R}^n \to \mathbf{R}^n$, depending continously on the time t, such that

$$\Omega(t) = \Phi_t(\Omega_0) \qquad \text{for all } t.$$

The map Φ_t has a geometric meaning, and is neather related to the fluid motion nor to the Lagrangian variables.

The domain $\Omega(t)$ may represent, for instance, a fixed domain, a moving cylinder with fixed side boundary (piston problem) or a contracting-expanding sphere (star).

The velocity of the point $\mathbf{x} = \Phi_t(\mathbf{x}_0) \in \partial\Omega(t)$ is given by

$$\mathbf{v}_b(\mathbf{x}) = \partial_t \Phi_t(\mathbf{x}_0). \qquad (11)$$

We assume the boundary conditions

$$\nu \cdot \mathbf{u} = \nu \cdot \mathbf{v}_b, \qquad (12)$$

on $\partial\Omega(t)$, where $\nu(t)$ denotes the outward normal on the boundary.

In this particular case, two different time scales arise, a reference time ($t_{\text{ref}} = l_{\text{ref}}/u_{\text{ref}}$) and an "acoustic" time ($t_{\text{ac}} = l_{\text{ref}}/c_{\text{ref}}$). Here, again, l_{ref} is a characteristic length scale, while u_{ref} and c_{ref} are characteristic velocities, related to the boundary and to acoustic waves, respectively. Two time variables follow naturally,

$$t' = \frac{t}{t_{\text{ref}}}, \qquad \tau' = \frac{t}{t_{\text{ac}}} \equiv \left(\frac{u_{\text{ref}}}{c_{\text{ref}}}\right)^{-1} t' \equiv M^{-1}t',$$

where t is the physical time variable. Then (dropping the prime), we consider an expansion for $U = (\rho, \mathbf{u}, p)$ of the form

$$U(\mathbf{x}, t; \epsilon) = U^{(0)}(\mathbf{x}, t, \tau) + \epsilon U^{(1)}(\mathbf{x}, t, \tau) + \epsilon^2 U^{(2)}(\mathbf{x}, t, \tau) + O(\epsilon^3). \tag{13}$$

This expansion is uniformily valid in time if the following consistency condition holds,

$$\lim_{\tau\to\infty} \frac{1}{\tau} U^{(j)}(\mathbf{x}, t, \tau) = 0. \tag{14}$$

Also, for any appropriate function f, we introduce the average operator

$$\langle f \rangle (\mathbf{x}, t) = \lim_{T\to\infty} \frac{1}{T} \int_0^T f(\mathbf{x}, t, \tau')d\tau',$$

and split the function into its average $\langle f \rangle$ and its average free part,

$$\delta f = f(\mathbf{x}, t, \tau) - \langle f \rangle (\mathbf{x}, t).$$

It is possible to show[1] that a solution of (3), which admits the expansion (13), takes the form

$$\begin{aligned}
\rho(\mathbf{x}, t; \epsilon) &= \rho^{(0)}(\mathbf{x}, t) + \epsilon \left[\bar{\rho}^{(1)}(\mathbf{x}, t) + \delta\rho^{(1)}(\mathbf{x}, \tau; t)\right] + O(\epsilon^2), \\
\mathbf{u}(\mathbf{x}, t; \epsilon) &= \left[\bar{\mathbf{u}}^{(0)}(\mathbf{x}, t) + \delta\mathbf{u}^{(0)}(\mathbf{x}, \tau; t)\right] \\
&\quad + \epsilon \left[\bar{\mathbf{u}}^{(1)}(\mathbf{x}, t) + \delta\mathbf{u}^{(1)}(\mathbf{x}, \tau; t)\right] + O(\epsilon^2), \\
p(\mathbf{x}, t; \epsilon) &= p^{(0)}(t) + \epsilon \left[\delta p^{(1)}(\mathbf{x}, \tau; t)\right] \\
&\quad + \epsilon^2 \left[\bar{p}^{(2)}(\mathbf{x}, t) + \delta p^{(2)}(\mathbf{x}, \tau; t)\right] + O(\epsilon^3).
\end{aligned} \tag{15}$$

Here, the leading order pressure $p^{(0)}$ satisfies the adiabatic relation

$$p^{(0)} = C_0 |\Omega_t|^{-\gamma}, \tag{16}$$

where $|\Omega_t|$ is the measure of the domain Ω_t, and the average-free leading order velocity and first order pressure $(\delta\mathbf{u}^{(0)}, \delta p^{(1)})$ satisfy the acoustic system

$$\begin{aligned}
\partial_\tau \delta\mathbf{u}^{(0)} + \frac{1}{\rho^{(0)}} \nabla \delta p^{(1)} &= 0, \\
\partial_\tau \delta p^{(1)} + \gamma p^{(0)} \nabla \cdot \delta\mathbf{u}^{(0)} &= 0.
\end{aligned} \tag{17}$$

Moreover, $(\rho^{(0)}, \bar{\mathbf{u}}^{(0)})$ satisfy the system

$$\begin{aligned}
\frac{1}{\rho^{(0)}}\frac{\bar{D}\rho^{(0)}}{Dt} &= -\frac{1}{|\Omega_t|}\frac{d|\Omega_t|}{dt}, \\
\rho^{(0)}\frac{\bar{D}\bar{\mathbf{u}}^{(0)}}{Dt} + \nabla\bar{p}^{(2)} &= -\nabla\cdot\left(\rho^{(0)}\left\langle\delta\mathbf{u}^{(0)}\otimes\delta\mathbf{u}^{(0)}\right\rangle\right), \\
\nabla\cdot\bar{\mathbf{u}}^{(0)} &= \frac{1}{|\Omega_t|}\frac{d|\Omega_t|}{dt},
\end{aligned} \tag{18}$$

with $\bar{D}/Dt = \partial_t + \bar{\mathbf{u}}\cdot\nabla$.

The details of the derivation of these limit equations, and a preliminary discussion of the boundary and initial data induced by the data for the starting compressible equations (3) will be given in a forthcoming paper[1].

References

1. G. Alì, submitted to *SIAM J. Appl. Math.*
2. K. Asano, *Japan J. Appl. Math.* **4**, 455 (1987).
3. H. Isozaki, *Comm. Math. Phys.* **110**, 519 (1987).
4. H. Isozaki, *J. Reine Angew. Math.* **381**, 1 (1987).
5. H. Isozaki, *Osaka J. Math.* **26**, 399 (1989).
6. S. Klainerman, and A. Majda, *Comm. Pure Appl. Math.* **34**, 481 (1981).
7. S. Klainerman, and A. Majda, *Comm. Pure Appl. Math.* **35**, 629 (1982).
8. R. Klein, *J. Comp. Phys.* **121**, 213 (1995).
9. A. Majda, *Compressible Fluid Flow and Systems of Conservation Laws in Several Space Variables*, (Springer-Verlag, New York Berlin Heidelberg Tokio, 1984).
10. A. Majda, and J. Sethian, *Comb. Sci. Techn.* **42**, 185 (1985).
11. W. H. Matthaeus, and M. R. Brown, *Phys. Fluids* **31**, 3634 (1988).
12. A. Meister, *SIAM J. Appl. Math.* **60**, 256 (1999).
13. G. Métivier, and S. Schochet, *Arch. Rat. Mech. Anal.* **158**, 61 (2001).
14. S. Schochet, *Comm. Math. Phys.* **104**, 49 (1986).
15. S. Ukai, *J. Math. Kyoto Univ.* **26**, 323 (1986).
16. G. P. Zank, and W. H. Matthaeus, *Phys. Fluids A* **3**, 69 (1991).
17. G. P. Zank, and W. H. Matthaeus, *Phys. Fluids A* **5**, 257 (1993).

RECOVERING THE POTENTIAL IN THE SCHRÖDINGER EQUATION FROM THE N-D MAP

SERGEI AVDONIN

Department of Mathematical Sciences
University of Alaska
Fairbanks, AK 99775-6660

SUZANNE LENHART

Department of Mathematics
University of Tennessee
Knoxville, TN 37996–1300

VLADIMIR PROTOPOPESCU

Computer Science and Mathematics Division
Oak Ridge National Laboratory
Oak Ridge, TN 37831–6355

We consider the inverse problem of determining the potential in the one-dimensional Schrödinger equation from surface dynamical observations, namely from the range values of the Neumann to Dirichlet map. Dynamical boundary data have not been used in the inverse problem for the Schrödinger equation, since the traditional Gelfand-Levitan-Marchenko approach reconstructs the potential from spectral or scattering data. Here we show that one can completely recover the spectral data from the dynamical data. The construction of the spectral data uses a new result on spectral controllability for the Schrödinger equation, which we obtain by using the properties of exponential Riesz bases (nonharmonic Fourier series). From the spectral data, we solve the inverse problem by the Boundary Control method.

1 Introduction

Inverse spectral methods have been actively studied in the past years for their relevance in a variety of classical and quantum applications [1,12,19,21,22]. As most inverse problems, identification problems are ill conditioned. Since experimental observations are usually incomplete and/or affected by noise, the ill conditioning of the problem has important consequences. From this perspective, the major difference between classical and quantum situations is that the latter are always described by a linear evolution system. This important feature makes it possible to address inverse problems for quantum systems in a systematically and consistently *linear* manner, as opposed to recovery/identification methods based on optimization and control which, in general, result in nonlinear problems [8,21,22]. The framework that allows this linear implementation is the so-called Boundary Control (BC) method. The BC method has been recently proposed by Belishev[9,11] and applied to the wave, heat, and beam equations[2,3,4,7,10,16].

The BC method uses the connection between inverse problems of mathematical physics for distributed parameter systems on the one hand, and functional analysis, geometry, asymptotic methods, control theory, and system theory, on the other hand. This approach has several advantages over previous identification methods, namely: (i) it maintains linearity throughout the whole approach; (ii) it is applicable to a wide range of linear point and/or distributed systems and reconstruction situations; (iii) it can deal with identification of coefficients appearing in highest order terms, which other methods could not easily treat; (iv) it is, in principle, dimension-independent; and, finally, (v) it lends itself to straightforward algorithmic implementations and stable numerical schemes[2,3,4,7,9,10,11,16].

In this paper, we extend the BC method to the Schrödinger equation and determine the potential in the 1-d Schrödinger equation using boundary data, more precisely the Neumann-to-Dirichlet (ND) map. This extension is not trivial, insofar as it is based on new regularity and controllability results for the Schrödinger and wave equations with nonsmooth coefficients. The use of boundary data is new as well, since previous methods[1,13,15] rely on the knowledge of either spectral or scattering data.

In the BC approach, the original inverse problem, namely the recovery of the potential from the ND map is divided in two related but distinct inverse sub-problems. The first recovers the spectral data from the ND map of the Schrödinger equation. The second explicitly recovers the potential from an auxiliary wave equation with the same spectral data, via the BC method. The remaining of the paper is divided in two sections corresponding to these two subproblems. Specifically, in Section 2 we define the Schrödinger problem and the response operator, i.e. the time dependent ND map, we derive a priori regularity results, and prove controllability for the Schrödinger equation. Then we define the connecting operator and express it in terms of the ND map. The connecting operator has an obvious interpretation in terms of systems theory in that it makes the connection between the outer space of the controls and the inner space of the states. A variational algorithm is applied to the connecting operator to compute the spectral data for the Schrödinger equation thereby completing the solution of the first inverse sub-problem. In Section 3 we introduce an auxiliary wave equation that has precisely the same spectral data that we have computed. This simply means that the two equations (Schrödinger and wave) share the same time-independent part, i.e.

the same zero order (potential) term. We use the propagation of singularities for the wave equation and the BC method to find an explicit formula for the sought potential.

2 Recovering the Spectral Data from the Neumann to Dirichlet Map

Let T, $\ell > 0$ and let q be a real valued function from $\mathcal{H} := L^2(0, \ell)$. We consider a system described by the Schrödinger equation

$$i\frac{\partial u(x,t)}{\partial t} + \frac{\partial^2 u(x,t)}{\partial x^2} - q(x)u(x,t) = 0, \quad x \in (0,\ell)\,,\ t \in (0,T), \tag{1}$$

with the boundary conditions

$$\frac{\partial u(0,t)}{\partial x} = f(t)\,, \quad u(\ell,t) = 0\,, \quad t \in (0,T) \tag{2}$$

and zero initial condition

$$u(x,0) = 0. \tag{3}$$

Let us denote the solution of this initial boundary value problem by $u^f(x,t)$. We suppose that $f \in \mathcal{F}^T := L^2(0,T)$ and introduce in this space the response operator R^T, defined by:

$$(R^T f)(t) = u^f(0,t). \tag{4}$$

The dynamical inverse problem consists of recovering the unknown potential $q(x)$ from the given operator R^T which is the ND map for the problem (1)–(3). This is not standard, since in spectral inverse problem eigenvalues of the operator of the system and 'traces' of eigenfunctions are given. We prove that the dynamical inverse problem has a unique solution for any $T > 0$ and give a constructive method to find it. However, since most applications involve the Dirichlet-to-Neumann map (see e.g. Avdonin[2,3,4], Belishev[9,11]), we chose to extend the BC method to the Schrödinger equation for the ND map. The present results are stronger and more general than those from Horn[14], Lasiecka[18], Lebeau[20], Littman[23], Machtyngier[24] in that we consider a non-smooth potential. See Avdonin[6] for further details on the results given here.

We consider the solution of the initial boundary value problem (1)–(3) with $f \in \mathcal{F}^T$ in the generalized sense (in the sense of the theory of distributions) and use the Fourier method to obtain the sharp regularity results for the solution to this problem.

We introduce the operator $\mathcal{L}$, defined by

$$(\mathcal{L}\varphi)(x) = -\varphi''(x) + q(x)\varphi(x), \quad x \in (0,\ell), \tag{5}$$

with domain

$$\mathcal{D}(\mathcal{L}) = H^2(0,\ell) \tag{6}$$

and the operator $\mathcal{L}_0$ acting by the same rule (5) with domain

$$\mathcal{D}(\mathcal{L}_0) = \{\varphi \in H^2(0,\ell) : \varphi'(0) = \varphi(\ell) = 0\}.$$

The operator $\mathcal{L}_0$ is self adjoint in $\mathcal{H}$[26]. As an operator with compact resolvent, $\mathcal{L}_0$ has only discrete spectrum. Its eigenvalues λ_n and eigenfunctions $\varphi_n(x)$ satisfy the relations (see e.g. Pöschel[25])

$$\sqrt{\lambda_n} = \frac{\pi(n-1/2)}{\ell} + o(1), \quad n = 1, 2, \dots, \tag{7}$$

$$0 < \inf_{n\in\mathbb{N}} |\varphi_n(0)| \le \sup_{n\in\mathbb{N}} |\varphi_n(0)| < \infty \tag{8}$$

(we assume that $\|\varphi_n\|_{\mathcal{H}} = 1$). The set $\{\lambda_n, \kappa_n\}_{n\in\mathbb{N}}$ where $\kappa_n := \varphi_n(0)$ represents the *spectral data* for problem (1)–(3).

We look for the solution of the problem (1)–(3) in the form

$$u^f(x,t) = \sum_{n=1}^{\infty} a_n(t)\varphi_n(x). \tag{9}$$

Starting with

$$0 = \int_0^\ell \int_0^t (iu_t^f + u_{xx}^f - qu^f)\varphi_n(x)e^{-i\lambda_n(t-\tau)}\,d\tau\,dx$$

for $f \in C_0^\infty[0,T]$ and using integration by parts, (2), (3), and (9), we obtain

$$a_n(t) = -i\kappa_n \int_0^t f(\tau)e^{-i\lambda_n(t-\tau)}\,d\tau, \quad n \in \mathbb{N}. \tag{10}$$

These equalities are evidently true for $f \in \mathcal{F}^T$ as well.

Using (7) we can prove that the family $\mathcal{E} = \{e^{i\lambda_n t}\}$ forms an $\mathcal{L}$ - basis, i.e. a Riesz basis in the closure of its linear span in $\mathcal{F}^T$ for any $T > 0$. This means (see e.g. Avdonin[5] for the definition and properties of Riesz bases) that for some positive m and M and any finite sequence $\{\alpha_n\}$ we have

$$m\sum|\alpha_n|^2 \le \left\|\sum \alpha_n e^{i\lambda_n t}\right\|_{\mathcal{F}^T}^2 \le M\sum|\alpha_n|^2.$$

From (8)–(10) it follows now (see Avdonin[5] for details) that

$$\sum_{n=1}^{\infty} |a_n(\cdot)|^2 \in C[0,T]$$

and hence

$$u^f \in C([0,T];\mathcal{H}).$$

Moreover, the control operator U^T, defined by

$$U^T : \mathcal{F}^T \mapsto \mathcal{H},\ U^T f = u^f(\cdot,T),$$

is bounded.

Our approach to the identification problem is based on controllability of the system (1)–(3).

Proposition 1 *Let the potential $q \in \mathcal{H}$ be known. Then for any $n \in I\!N$, there exists a control $f_n \in \mathcal{F}_0^T := H_0^1(0,T)$ such that*

$$u^{f_n}(x,T) = \varphi_n(x). \tag{11}$$

This property is called *spectral controllability* (see e.g. Russell[27], Avdonin[5]). See Avdonin[6] for a new related exact controllability result.

Proof. Using representation (9) for u^f and formula (10) we need to prove the existence of $\{f_k\}_{k\in\mathbb{N}}$ in $\mathcal{F}_0^T$ such that

$$\delta_{nk} = -i\kappa_n \int_0^T f_k(t) e^{-i\lambda_n(T-t)}\,dt, \quad n \in \mathbb{N}, \tag{12}$$

where δ_{nk} is the Kronecker delta. Denote $f_k'(t) = g_k(t)$, integrate by parts in (14) and recall that $f_k(0) = f_k(T) = 0$. Then, finding such a sequence is equivalent to finding a family $\{g_k\}_{k\in\mathbb{N}}$ in $\mathcal{F}^T$ such that

$$f_k(t) = \int_0^t g_k(\tau)\,d\tau\,,$$

$$\delta_{nk} = \frac{\kappa_n}{\lambda_n} \int_0^T g_k(t) e^{-i\lambda_n(T-t)}\,dt, \quad n \in \mathbb{N}, \tag{13}$$

and

$$0 = \int_0^T g_k(t)\,dt\,. \tag{14}$$

Note that (13) came from substituting the representation of f_k in terms of g_k in (12) and integrating by parts. Conditions (13) and (14) express the biorthogonal property of the family $\{g_k\}_{k\in\mathbb{N}}$ with respect to the family $\{1\}\cup\{e^{i\lambda_n t}\}_{n\in\mathbb{N}}$ in $\mathcal{F}^T$. Based on the spectral properties (6), one can show[6] that the family $\{1\}\cup\{e^{i\lambda_n t}\}_{n\in\mathbb{N}}$ forms an $\mathcal{L}$-basis in $\mathcal{F}^T$; in turn, this implies the existence of such a biorthogonal family (see Avdonin[5]), which concludes the proof of Proposition 1.

Next, we introduce the connecting operator, C^T, which plays a central role in our approach to inverse problems. Indeed, C^T connects the (metrics of the) control space $\mathcal{F}^T$ and space of solutions $\mathcal{H}$ and can be explicitly expressed in terms of the response operator, R^T.

In the language of system theory, $\mathcal{F}^T$ is the outer space of the system (1)–(3), $\mathcal{H}$ is the inner space, and R^T is the input–output map. We shall define the connecting operator C^T through the bilinear form of the elements of the inner space, $\mathcal{H}$; however, its action can be expressed via the action of the operator R^T on the outer space, $\mathcal{F}^T$.

More precisely, we define the connecting operator $C^T : \mathcal{F}^T \to \mathcal{F}^T$ as

$$(C^T f, g)_{\mathcal{F}^T} := (u^f(\cdot,T), u^g(\cdot,T))_{\mathcal{H}}. \tag{15}$$

Here u^f and u^g are solutions of (1)–(3) corresponding to the boundary controls f and g. Since $u^f(\cdot,T) = U^T f$, we can write C^T in operator form:

$$C^T = (U^T)^* U^T. \tag{16}$$

The operator C^T is bounded in $\mathcal{F}^T$ since the operator U^T is bounded from $\mathcal{F}^T$ to $\mathcal{H}$.

To obtain an expression for C^T in terms of R^T, we use the equations (1)–(4) for $f,g\in C_0^\infty[0,T]$ and write:

$$\begin{aligned}
0 &= \int_0^\ell \int_0^T \left[\left(iu_t^f(x,t) + u_{xx}^f(x,t) - q(x)u^f(x,t)\right)\overline{u^g(x,t)}\right] dx\,dt \\
&= \int_0^\ell \int_0^T \left[u^f(x,t)\left(\overline{iu_t^g(x,t) + u_{xx}^g(x,t) - q(x)u^g(x,t)}\right)\right] dx\,dt \\
&+ i\int_0^\ell \left[u^f(x,t)\overline{u^g(x,t)}\right]_{t=0}^{t=T} dx \\
&+ \int_0^T \left[u_x^f(x,t)\overline{u^g(x,t)} - u^f(x,t)\overline{u_x^g(x,t)}\right]_{x=0}^{x=\ell} dt.
\end{aligned} \tag{17}$$

Finally, we have

$$i\left(u^f(\cdot,T), u^g(\cdot,T)\right)_{\mathcal{H}} = \left(f, R^T g\right)_{\mathcal{F}^T} - \left(R^T f, g\right)_{\mathcal{F}^T}. \tag{18}$$

Comparing (15) and (18) and using a standard density argument, we get:

$$C^T = i\left[R^T - (R^T)^*\right]. \tag{19}$$

This formula shows that C^T is selfadjoint, which is consistent with the representation (16).

We implement now for the Schrödinger equation the idea proposed in Belishev[10] for the heat equation (see also Avdonin[4]), namely we recover the spectral inverse data ($\{\lambda_n, \kappa_n\}$) from the dynamical inverse data (operator R^T) using the connecting operator C^T and the spectral controllability of the system (1)–(3).

From (15) and (1), we have for $f, g \in H_0^1(0,T) = \mathcal{F}_0^T$ the following equalities:

$$\begin{aligned}(C^T(if'), g)_{\mathcal{F}^T} &= \left(u^{if'}(\cdot,T), u^g(\cdot,T)\right)_{\mathcal{H}} = \left(iu_t^f(\cdot,T), u^g(\cdot,T)\right)_{\mathcal{H}} \\ &= \left(-u_{xx}^f(\cdot,T) + qu^f(\cdot,T), u^g(\cdot,T)\right)_{\mathcal{H}} = (\mathcal{L}u^f(\cdot,T), u^g(\cdot,T))_{\mathcal{H}} \\ &= (\mathcal{L}_0 u^f(\cdot,T), u^g(\cdot,T))_{\mathcal{H}}. \end{aligned} \tag{20}$$

In deriving (20), we used that $u^f(\cdot,T) \in \mathcal{D}(\mathcal{L}_0)$ for $f \in \mathcal{F}_0^T$. Indeed, integration by parts in (10) gives

$$\sum_{n=1}^{\infty} |a_n(T)\lambda_n|^2 < \infty .$$

Taking into account (9) it means that $\mathcal{L}_0 u^f(\cdot,T) \in \mathcal{H}$.

In Proposition 1 we proved that the system (1)–(3) is spectrally controllable, i.e. for any $n \in \mathbb{N}$, there exists a control $f_n \in \mathcal{F}_0^T$ such that $u^{f_n}(x,T) = \varphi_n(x)$.

Relations (15) and (20) imply

$$(C^T f_n, f_n)_{\mathcal{F}^T} = (\varphi_n, \varphi_n)_{\mathcal{H}} = 1 \ , \ (C^T(if_n'), f_n)_{\mathcal{F}^T} = \lambda_n . \tag{21}$$

By the definitions of R^T and f_n we have

$$(R^T f_n)(T) = u^{f_n}(0,T) = \varphi_n(0) = \kappa_n . \tag{22}$$

These relations allow us to find the spectral data, $\{\lambda_n, \kappa_n\}$, using the known operator R^T. We can do it in the following way.

The spectral analysis of the operator $\mathcal{L}_0$ may be realized with well known variational principle:

$$\begin{aligned} &\lambda_1 = \inf_{\varphi\in\mathcal{H},\, \|\varphi\|_{\mathcal{H}}^2=1} (\mathcal{L}_0\varphi, \varphi)_{\mathcal{H}} , \\ &\varphi_1 : \ (\mathcal{L}_0\varphi_1, \varphi_1) = \lambda_1 , \\ &\lambda_n = \inf_{\varphi\in\mathcal{H},\, \|\varphi\|_{\mathcal{H}}^2=1,\, (\varphi,\varphi_j)_{\mathcal{H}}=0,\, j=1,\dots,n-1} (\mathcal{L}_0\varphi, \varphi)_{\mathcal{H}} , \\ &\varphi_n : \ (\mathcal{L}_0\varphi_n, \varphi_n) = \lambda_n \ , \ n \in \mathbb{N} . \end{aligned}$$

Using relation (21) and the spectral controllability of system (1) – (3), we can realize this principle using the operator C^T instead of $\mathcal{L}_0$:

$$\lambda_1 = \inf\, (C^T(if'), f)_{\mathcal{F}^T}$$

where the infimum is taken over

$$f \in \mathcal{F}_0^T, \quad (C^T f, f)_{\mathcal{F}^T} = 1$$

and

$$f_1 : (C^T(if_1'), f_1)_{\mathcal{F}^T} = \lambda_1 \ .$$

Further,

$$\lambda_n = \inf\, (C^T(if'), f)_{\mathcal{F}^T} \ ,$$

where the infimum is taken over

$$f \in \mathcal{F}_0^T, \quad (C^T f, f)_{\mathcal{F}^T} = 1, \quad (C^T f, f_j)_{\mathcal{F}^T} = 0, \quad j = 1, \ldots, n-1 \, ,$$

and

$$f_n : \ (C^T(if_n'), f_n)_{\mathcal{F}^T} = \lambda_n \, , \ n \in \mathbb{N}.$$

Thus we find λ_n, f_n and, using (22), we can also find κ_n , $n \in \mathbb{N}$, and complete the recovery of the spectral data.

3 Recovering the Potential from the Spectral Data Using the Boundary Control Method

From the spectral data, $\{\lambda_n, \kappa_n\}$, (or their approximation), one could recover the potential by using the Gelfand–Levitan–Marchenko[1] and Krein[15] methods. Instead, we choose to solve the inverse problem using the extension of the BC method. Besides offering yet another alternative to identification methods based on control and optimization[8,21,22], the BC formalism is entirely linear and essentially independent on dimensionality. Indeed, using the BC approach, Belishev[12] presents a derivation the Gelfand-Levitan-type equations for the multidimensional wave equation. See particular applications of the BC method in multi-dimensions for the wave equation[9,11,16,17] and the heat equation[10]. The application of the BC method to solve the dynamical inverse problem for the system (1)–(3) relies on regularity and controllability results for a closely related boundary value problem for the wave equation. These results are new since, on the one hand, they are obtained for nonsmooth coefficients and, on the other hand, the time derivative of the solution instead of the solution itself is the desired target.

We consider, in connection with the Schrödinger system (1)–(3), the related auxiliary initial boundary value problem for wave equation:

$$\frac{\partial^2 w(x,t)}{\partial t^2} - \frac{\partial^2 w(x,t)}{\partial x^2} + q(x)w(x,t) = 0, \quad x \in (0,\ell),\ t \in (0,T), \tag{23}$$

with the boundary conditions

$$\frac{\partial w(0,t)}{\partial x} = f(t), \quad w(\ell,t) = 0,\ t \in (0,T), \tag{24}$$

and zero initial conditions

$$w(x,0) = \frac{\partial w}{\partial t}(x,0) = 0. \tag{25}$$

We denote the solution of (23)–(25) by w^f. Here and below we assume all functions are real.

To represent the solution w^f and study its properties, we consider a Goursat problem:

$$k_{tt} - k_{xx} + q(x)k = 0, \quad 0 < x < t < T\,, \tag{26}$$

$$k_x(0,t) = 0\,, \quad k(x,x) = \frac{1}{2}\int_0^x q(s)\,ds\,. \tag{27}$$

Using the standard successive approximation method, one can prove the following:

Proposition 2 *For $q \in \mathcal{H}$ problem (26), (27) has a unique solution such that $k \in H^1(\Omega^T)$, $\Omega^T := \{(x,t) : 0 < x < t < T \leq \ell\}$, $k_{tt} - k_{xx} \in L^2(\Omega^T)$ and equation (26) is satisfied almost everywhere.*

Similar proposition is proved e.g. in Tikhonov[28] for smooth q, but the method works as well for $q \in L^2(0,\ell)$.

The following proposition can be checked by direct calculations.

Proposition 3 *(i) If $f \in C^1[0,T]$ and $f(0) = 0$, the problem (23)–(25) has the unique solution $w = w^f(x,t)$,*

$$w^f(x,t) = \begin{cases} -F(t-x) + \int_x^t k(x,s)F(t-s)\,ds & \text{for } x < t \\ 0 & \text{for } x \geq t\,, \end{cases} \tag{28}$$

where $F(t) := \int_0^t f(s)\,ds$ and $w^f \in H^2(Q^T)$, $Q^T := (0,\ell) \times (0,T)$, equation (23) is satisfied almost everywhere, and the boundary and initial conditions are satisfied in a classical sense.

(ii) For $f \in \mathcal{F}^T$ the function $w^f(x,t)$ defined by (28) gives a generalized solution of problem (23)–(25) such that $w^f \in C([0,T];H^1(0,\ell))$; $w_t^f, w_x^f \in C([0,T];\mathcal{H})$.

Let

$$\mathcal{H}^T := \{f \in \mathcal{H} : \ \text{supp}\, f \subset [0,T]\}.$$

From Proposition 3 it also follows that the control operator $\mathcal{W}^T$,

$$\mathcal{W}^T : \mathcal{F}^T \mapsto \mathcal{H}^T, \ \mathcal{W}^T f = w_t^f(\cdot, T),$$

is bounded. We can prove a controllability result which shows that the operator $\mathcal{W}^T$ is boundedly invertible:

Proposition 4 *When the potential $q \in \mathcal{H}$ is known and $T \in (0,\ell]$, then for any function $z \in \mathcal{H}^T$, there exists a unique control $f \in \mathcal{F}^T$ such that*

$$w_t^f(x,T) = z(x) \ \ in\ \mathcal{H}^T. \tag{29}$$

Proof. According to (28), condition (29) is equivalent to the equation

$$z(x) = -f(T-x) + \int_x^T k(x,s) f(T-s)\, ds \ \ x \in (0,T). \tag{30}$$

This is the Volterra equation of the second kind with respect to f. Its solvability proves Proposition 4.

Next, we recover the potential $q(x)$ from the known spectral data, $\{\lambda_n, \kappa_n\}, n \in \mathbb{N}$.

Introduce another connecting operator $\mathcal{C}^T : \mathcal{F}^T \to \mathcal{F}^T$, but now for the wave equation (23)–(25):

$$\left(\mathcal{C}^T f, g\right)_{\mathcal{F}^T} = \left(w_t^f(\cdot,T), w_t^g(\cdot,T)\right)_{\mathcal{H}}.$$

The operator $\mathcal{C}^T$ is bounded and boundedly invertible, since $\mathcal{C}^T = \left(\mathcal{W}^T\right)^* \mathcal{W}^T$. Note that our definition of the connecting operator differs from the one used for the wave equation by Avdonin[2,3] and Belishev[11], since it is defined by the inner product of w_t instead of w. This definition is more convenient in problems with Neumann spectral data and dynamical problems involving ND maps. For the wave equation with dissipation, a bilinear form involving w_t has been used in Kurylev[17].

The solution $w^f(x,t)$ admits the (Fourier series) representation

$$w^f(x,t) = \sum_{n=1}^{\infty} b_n(t) \varphi_n(x),$$

wherefrom we obtain the equalities

$$b_n(t) = -\kappa_n \int_0^t f(\tau) \frac{\sin\sqrt{\lambda_n}(t-\tau)}{\sqrt{\lambda_n}}\, d\tau, \quad n \in \mathbb{N} \tag{31}$$

and

$$b_n'(t) = -\kappa_n \int_0^t f(\tau) \cos\sqrt{\lambda_n}(t-\tau)\, d\tau, \quad n \in \mathbb{N}. \tag{32}$$

From (32) we compute

$$\begin{aligned}(\mathcal{C}^T f, g)_{\mathcal{F}^T} = \sum_{n=1}^{\infty} \kappa_n^2 \int_0^T f(t) \cos\sqrt{\lambda_n}(T-t)\, dt \,\cdot \\ \cdot \int_0^T g(s) \cos\sqrt{\lambda_n}(T-s)\, ds,\end{aligned} \tag{33}$$

which shows that the spectral data completely determine the operator $\mathcal{C}^T$.

Let $y(x)$ be a solution to the boundary value problem

$$y''(x) - q(x)y(x) = 0,\ y(0) = 1,\ y'(0) = 0, x \in (0,\, \ell) \tag{34}$$

and let us find a control $p^T \in \mathcal{F}^T$ such that

$$w_t^{p^T}(x,T) = \begin{cases} y(x), & x \le T, \\ 0, & x > T. \end{cases} \tag{35}$$

Note that, since $q(x)$ is unknown, both $y(x)$ and $p^T(t)$ are unknown functions at this point. For any $g \in C_0^\infty[0,T]$, we have

$$\begin{aligned}(\mathcal{C}^T p^T, g)_{\mathcal{F}^T} &= \left(w_t^{p^T}(\cdot,T), w_t^g(\cdot,T)\right)_{\mathcal{H}} \\ &= \int_0^T y(x) w_t^g(x,T)\, dx \\ &= \int_0^T dt \int_0^T y(x) w_{tt}^g(x,t)\, dx \\ &= \int_0^T dt \int_0^T y(x) \left[w_{xx}^g(x,t) - q(x) w^g(x,t)\right] dx \\ &= -\int_0^T y(0) w_x^g(0,t)\, dt \\ &= -\int_0^T g(t)\, dt\end{aligned}$$

(we used that for $g \in C_0^\infty[0,T]$, the function w^g and its derivatives are equal to zero at $x = T$). Hence the function p^T satisfies the equation

$$\left(\mathcal{C}^T p^T\right)(t) = -1, \quad t \in [0,T].$$

Since $\mathcal{C}^T$ is boundedly invertible, this equation has a unique solution, $p^T \in \mathcal{F}^T$, for any $T \leq \ell$. Moreover, it can be proved that $p^T \in H^1(0,T)$.

Formula (28) implies that solution of the initial boundary value problem (23)–(25) satisfies the relation

$$w_t^f(x,t) = -f(t-x) + \int_x^t k(x,s) f(t-s)\, ds \text{ for } x < t\,.$$

Therefore,

$$w_t^f(t-0,t) = -f(+0)$$

provided the limit $f(+0)$ exists. Applying this propagation of singularities property for $f = p^T$, we obtain

$$w_t^{p^T}(T-0,T) = -p^T(+0).$$

Denote $p^T(+0)$ by $\mu(T)$. From (35), $w_t^{p^T}(T-0,T) = y(T)$, thus $\mu(T)$ is twice differentiable with respect to T, and (34) gives

$$q(T) = \frac{y''(T)}{y(T)} = \frac{\mu''(T)}{\mu(T)}.$$

By varying T in $(0,\ell)$, we obtain $q(\cdot)$ in that interval. Since the function $y(T)$ may have only a finite number of zeroes in $(0,\ell)$, this completes the solution of the identification problem.

Acknowledgments

S. A.'s research is supported in part by the Australian Research Council, grant # A00000723. S. L.'s and V. P.'s research is supported in part by the U.S. Department of Energy, Office of Basic Energy Sciences. The Oak Ridge National Laboratory is managed for the U.S. DOE by UT-Battelle, LLC, under contract No. DE-AC05-00OR22725. V. P. would like to thank Carole Holbrook for her expert typing and support.

References

1. M. J. Ablowitz, and H. Segur, *Solitons and the Inverse Scattering Transform* (SIAM, Philadelphia, 1981).
2. S. A. Avdonin, and M. I. Belishev, Boundary control and dynamical inverse problem for nonselfadjoint Sturm-Liouville operator (BC-method), *Control and Cybernetics* **25**, 429-440 (1996).
3. S. A. Avdonin, M. I. Belishev, and S. A. Ivanov, Boundary control and inverse matrix problem for the equation $u_{tt} - u_{xx} + V(x)u = 0$, *Math. USSR Sbornik* **7**, 287-310 (1992).
4. S. A. Avdonin, M. I. Belishev, and Yu. S. Rozhkov, The BC-method in the inverse problem for the heat equation, *J. Inv. Ill-Posed Problems* **5**, 1-14 (1997).
5. S. A. Avdonin, and S. A. Ivanov, *Families of Exponentials. The Method of Moments in Controllability Problems for Distributed Parameter Systems* (Cambridge University Press, New York, 1995).
6. S. A. Avdonin, S. Lenhart, and V. Protopopescu, Solving the dynamical inverse problem for the Schrödinger equation by the Boundary Control method, *Inverse Problems*, to appear.
7. S. A. Avdonin, N. G. Medhin, and T. L. Sheronova, Identification of a piecewise constant coefficient in the beam equation, *J. Comp. Appl. Math.* **114**, 11-21 (2000).
8. H. T. Banks, and K. Kunisch, *Estimation Techniques for Distributed Parameter Systems* (Birkhäuser, Boston, 1989).
9. M. I. Belishev, An approach to multidimensional inverse problems for the wave equation, *Dokl. AN SSSR* **297**, 524-527 (1987) (in Russian).
10. M. I. Belishev, Canonical model of a dynamical system with boundary control in inverse problem for the heat equation, *Algebra and Analysis* **7**, 3-32 (1995) (in Russian).
11. M. I. Belishev, Boundary control method in reconstruction of manifolds and metrics (the BC method), *Inverse Problems* **13**, R1-R45 (1997).
12. M. I. Belishev, Dynamical systems and boundary control: models and characterization of inverse data, *Inverse Problems* **17**, 659-682 (2001).
13. L. D. Faddeev, The inverse problem of quantum scattering theory, *Soviet J. Math.* **5**, 14-21 (1976).
14. M. A. Horn and W. Littman, Boundary control of a Schrödinger equation with nonconstant principal part, in *Control of PDEs and Applications*, E. Casas, ed., pp. 163-184. (Marcel Dekker, New York, 1995).
15. I. S. Kac and M. G. Krein, On the spectral function of the string, *Amer. Math. Soc. Transl.* Ser. 2 **103**, 19-102 (1974).

16. Ya. V. Kurylev, Multi-dimensional inverse boundary problems by BC-method: groups of transformations and uniqueness results, *Math. Comput. Modelling* **18**, 33-45 (1993).
17. Ya. V. Kurylev, and M. Lassas, Hyperbolic inverse problem with data on a part of the boundary, *AMS/IP Studies in Advanced Mathematics* **16**, 259-272 (2000).
18. I. Lasiecka, and R. Triggiani, Optimal regularity, exact controllability and uniform stabilization of Schrödinger equations with Dirichlet control, *Diff. and Int. Eqs.* **5**, 521-534 (1992).
19. P. D. Lax and R. S. Phillips, *Scattering Theory* (Academic Press, New York, 1967).
20. G. Lebeau, Controle de l'équation de Schrödinger, *J. Math. Pures Appl.* **71**, 267-277 (1992) (in French).
21. S. Lenhart, V. Protopopescu, and J. Yong, Optimal control of a reflection boundary coefficient in an acoustic wave equation, *Applicable Analysis* **69**, 179-188 (1998).
22. S.Lenhart, V. Protopopescu, and J. Yong, Identification of boundary shape and reflectivity in a wave equation by optimal control techniques, *Diff. and Int. Eqs.* **13**, 941-975 (2000).
23. W. Littman, and S. Taylor, Smoothing evolution equations and boundary control theory, *J. Analyse Math.* **59**, 117-126 (1992).
24. E. Machtyngier, Exact controllability for the Schrödinger equation, *SIAM J. Control & Optim.* **32**, 24-34 (1994).
25. J. Pöschel and E. Trubowitz, *Inverse Spectral Theory* (Academic Press, Orlando, 1987).
26. M. Reed, and B. Simon, *Methods of Modern Mathematical Physics*, vol. II (Academic Press, New York, 1975).
27. D. L. Russell, Controllability and stabilizability theory for linear partial differential equations, *SIAM Review* **20**, 639-739 (1978).
28. A. N. Tikhonov, and A. A. Samarskii, *Equations of Mathematical Physics* (Pergamon Press, New York, 1963).

MULTI-RESOLUTION ANALYSIS IN THE FRACTIONAL QUANTUM HALL EFFECT

FABIO BAGARELLO

Dip. di Matematica ed Applicazioni, Fac.Ingegneria, Università di Palermo, I - 90128 Palermo, Italy
E-mail: bagarell@unipa.it

We review some recent results concerning possible applications of wavelets and, more specifically, of multi-resolution analysis (MRA) to the problem of constructing the ground state of a two-dimensional electron gas in a strong magnetic field. We show how to produce wavelet-like orthonormal bases in the lowest Landau level (LLL). Further, we prove that any MRA produces a Slater determinant belonging to the LLL and viceversa.

1 Introduction

The role of wavelets in various applications of mathematics and to some physical problems like signal analysis is now completely established[1]. What is still not clear is the relevance of wavelets in quantum mechanics: in our knowledge, very few are the applications proposed in this field[2,3,4,5,6] and [7] among the others. This is the kind of applications we are interested in here.

The physical system to consider in the analysis of the fractional Quantum Hall effect (FQHE) is a (quasi)-planar gas of electrons in a strong magnetic field perpendicular to the plane and in a neutralizing uniform background. The first problem to tackle for discussing the static features of the FQHE is to find its ground state, and this is already a very hard problem. We will consider here only the Hartree-Fock approach, in which the (totally antisymmetric) ground state wave function is taken as a Slater determinant of one-electron wave functions (see, for instance, [8,9]). For that we have to select an adequate wave function for a single electron in the magnetic field. As it is well-known, the energy levels, the so-called *Landau levels*, are infinitely degenerate. Thus the problem arises of finding a good (orthogonal) basis in the corresponding Hilbert subspace, the so-called lowest Landau level (LLL).

It is a standard result [10] that the Hamiltonian of a single electron of our system, $H_0 = \frac{1}{2}(p_x - y/2)^2 + \frac{1}{2}(p_y + x/2)^2$ in suitable units, can be transformed into that of a harmonic oscillator by introducing the canonical variables

$$\widetilde{P} = p_x - y/2, \qquad \widetilde{Q} = p_y + x/2 : \tag{1}$$

$H_0 = \frac{1}{2}(\widetilde{P}^2 + \widetilde{Q}^2)$. The eigenstates of H_0 can be explicitly found, $\Phi_{lm}(x,y)$, and they correspond to the eigenvalues $E_{lm} \equiv E_l = l + 1/2$, $l, m = 0, 1, \ldots$.

Thus we see that the energy levels are all degenerate in m, so that the ground level (LLL) is spanned by the set $\{\Phi_{0m}(x,y)\}$, which forms an orthonormal basis for the LLL, but are not very well localized. Yet the physics of the problem requires that the wave functions be fairly well localized, in particular for approaching the low density limit of the celebrated Wigner crystal [9]. For this reason it is convenient to look for different bases.

An efficient method for constructing o.n. basis of the LLL has been discussed in some detail in [10] and [9], and we will use it here. The transformation (1) can be seen as a part of a canonical transformation from the variables x, y, p_x, p_y into the new ones $Q, P, \widetilde{Q}, \widetilde{P}$, where

$$P = p_y - x/2, \qquad Q = p_x + y/2. \tag{2}$$

These operators satisfy the following commutation relations: $[Q,P] = [\widetilde{Q},\widetilde{P}] = i$, all other commutators vanishing.

It is shown in [10] that a wave function in the (x,y)-space it is related to its $P\widetilde{P}$-expression by the formula

$$\Psi(x,y) = \frac{e^{ixy/2}}{2\pi} \int_{-\infty}^{\infty} dP \int_{-\infty}^{\infty} d\widetilde{P}\, e^{i(x\widetilde{P}+yP+P\widetilde{P})}\, \Psi(P,\widetilde{P}), \tag{3}$$

which becomes, when restricted to the LLL[9], $\Psi(P,\widetilde{P}) = \pi^{-1/4}\, e^{-\widetilde{P}^2/2} h(P)$,

$$\Psi^{\mathrm{L}}(x,y) = \frac{e^{ixy/2}}{\sqrt{2}\pi^{3/4}} \int_{-\infty}^{\infty} dP\, e^{iyP} e^{-(x+P)^2/2}\, h(P)\,. \tag{4}$$

In order to study the localization properties of the function $\Psi^{\mathrm{L}}(x,y)$, we have to analyze its asymptotic behavior. From (4), standard convolution arguments indicate that the behavior of $\Psi^{\mathrm{L}}(x,y)$ for large x is related to the behavior for large P of $h(P)$, while that for large y is related to the behavior for large values of Q of the Fourier transform $\widehat{h}(Q)$ of $h(P)$.

Now, since the transformation (1), (2) is canonical, it is clear that the transformed family $\{\Psi^{\mathrm{L}}_m(x,y) \in \mathcal{L}(\boldsymbol{R}^2)\}$ of any family $\{h_m(P) \in \mathcal{L}(\boldsymbol{R})\}$ is orthonormal in $\mathcal{L}(\boldsymbol{R}^2)$ if and only if $\{h_m(P)\}$ is orthonormal in $\mathcal{L}(\boldsymbol{R})$. Thus, the LLL basis problem boils down to finding suitable functions $h_m(P)$ that constitute a basis with the desirable properties. Following classical suggestions, we would like to find the best localization of the single electron basis compatible with the physics of the problem. Moreover, orthogonality of the basis makes life much easier, so that one would like to impose it too. The problem is that sharp localization and orthogonality are *not* compatible, as results from the Balian-Low and the Battle Theorems[1,11], so that the choice of the basis turns out to be a non trivial task.

Let us discuss some possible approaches, in all of which the basis functions $h_m(P)$ are generated from a single function $h(P)$. We begin with bases generated with magnetic translations.

A physically interesting possibility is to generate a basis by acting on a suitable function $\psi_o(x,y)$ with the so-called magnetic translation operators $T(\underline{a}_i)$ defined by

$$T(\underline{a}_i) \equiv \exp{(i\underline{\Pi}_c \cdot \underline{a}_i)}, \quad i = 1, 2, \tag{5}$$

where $\underline{\Pi}_c \equiv (Q, P)$ and $\mathbf{a}_i$ are the basis vectors of a suitable lattice Λ. The set obtained in this way is still in the LLL since the operators $T(\underline{a}_i)$ commute with H_0, in virtue of the commutation relations above. Moreover $[T(\underline{a}_1), T(\underline{a}_2)] = 0$ if the area of the lattice cell is such that $a_{1_x}a_{2_y} - a_{1_y}a_{2_x} = 2\pi$.

The authors of [9] have choosen $h(P) = f_0(P)$ and show that the corresponding wave function in the (x,y)-space is a Gaussian. Using magnetic translation operators $T(\underline{a}_i)$, they also construct a complete not o.n. set of functions for the LLL with Gaussian localization as in

$$\Psi_{m,n}(x,y) = T(\underline{a}_1)^m \, T(\underline{a}_2)^n \, \Psi(x,y), \qquad m, n \in \boldsymbol{Z}. \tag{6}$$

This lack of orthogonality follows from the Balian-Low theorem and any orthogonalization procedure necessarely destroys the exponential localization, again in agreement with the Balian-Low theorem.

As an alternative, one can try to use dilations and translations instead of the magnetic translations (5), that is, to turn to wavelets for constructing an orthonormal set in $\mathcal{L}(\boldsymbol{R})$ which, via (4), produces an orthonormal set for the LLL[2,3]. Here too, however, a no-go result concerning the localization of the single electron wave-function arises, namely, Battle's Theorem[11,6]. However, we may still expect that the use of wavelets can give a significative insight in the comprehension of the FQHE, as widely discussed in [6].

2 Magnetic translations and MRA

We refer to the original papers [2,3] for many examples of wavelet-like o.n. bases in the LLL, where one makes use of dilations and translations. Here we rather focus on more recent results [5], concerning a deeper connection between MRA and the FQHE. Whereas wavelets have been used in [2,3] for producing nicely localized orthonormal bases, here one establishes, via the so-called *kq*-representation (or Zak transform), a direct connection between FQHE states generated by magnetic translations and the coefficients of the low-pass filter m_o of an MRA. In what follows we will assume that the reader is familiar with the *kq*-representation[13,12].

Let us go back to the magnetic translations $T(\underline{a}_i)$ defined in (5), and let Λ be the lattice generated by $\underline{a}_1$ and $\underline{a}_2$. For simplicity, we take here for Λ a square lattice with unit cell of area 2π: $\underline{a}_1 = a(1,0)$, $\underline{a}_2 = a(0,1)$, $a^2 = 2\pi$. With this choice, (6) becomes $\Psi_{m,n}(x,y) = (-1)^{mn} e^{i\frac{1}{2}a(my-nx)}\Psi(x+ma, y+na)$. Taking now $\Psi(x,y)$ in the form (4), we obtain

$$\Psi_{m,n}(x,y) = \frac{e^{i\frac{xy}{2}+iamy}}{\sqrt{2}\pi^{3/4}} \int_{-\infty}^{\infty} dP\, e^{i(y+na)P-(x+ma+P)^2/2}\, h(P)\,. \tag{7}$$

We are interested now in finding conditions on $h(P)$ that guarantee the orthogonality of the resulting wave functions, viz. $S_{m,n} \equiv \langle \Psi_{0,0}, \Psi_{m,n} \rangle = \delta_{m,0}\,\delta_{n,0}$. Using (7), we find

$$S_{m,n} = \int_{-\infty}^{\infty} dp\, e^{inap}\, \overline{h(p+ma)}\, h(p). \tag{8}$$

Now we introduce the kq-representation[13]. The so called Zak transform Z of a function $H(P) \in \mathcal{L}^2(\boldsymbol{R})$ is defined as[12]

$$h(k,q) = (ZH)(k,q) = \frac{1}{\sqrt{a}} \sum_{n\in\boldsymbol{Z}} e^{-ikna} H(q+na), \tag{9}$$

and belongs to $\mathcal{L}^2(\diamond)$, where $\diamond = [0,a) \times [0,a)$ is the unit cell in the (k,q)-plane. It is known that Z^{-1} exists[12], and it defines a map from $\mathcal{L}^2(\diamond)$ to $\mathcal{L}^2(\boldsymbol{R})$.

It is convenient to split the integral (8) in an infinite sum of integrals restricted to $[ra,(r+1)a)$, $r \in \boldsymbol{Z}$, to use the inverse Zak transform Z^{-1} and the well-known equality $\sum_{l\in\boldsymbol{Z}} e^{ixl2\pi/c} = c\sum_{l\in\boldsymbol{Z}} \delta(x-cl)$, so to get $S_{m,n} = \int_\diamond dk\,dq\, e^{inaq-ikma}\, |h(k,q)|^2$. By the completeness of the set $\{e^{inaq-ikma},\ n,m \in \boldsymbol{Z}\}$ in the unit cell $\diamond$, we can conclude now that $S_{m,n} = \delta_{m,0}\,\delta_{n,0}$ if and only if $h(k,q)$ is a pure phase. This result holds for a filling factor $\nu = 1$.

It is easy to generalize this result to a filling factor $\nu = \frac{1}{2}$, using the following geometrical idea. A filling factor $\nu = 1$ corresponds to our square lattice Λ with all the sites occupied. Therefore, a $\nu = \frac{1}{2}$ 2-D electron gas can be associated to the same lattice, but only half occupied. If we require the orthonormality of the related set of single electron wave functions, it is enough to ask that $S_{m,2n} = \delta_{m,0}\,\delta_{n,0}$. Equation (8) and a little bit of algebra, together again with the completeness of the functions $\{e^{inaq-ikma},\ n,m \in \boldsymbol{Z}\}$ in $\diamond$, gives:

$$J_2(k,q) \equiv \left|h(k,\frac{q}{2})\right|^2 + \left|h(k,\frac{q+a}{2})\right|^2 = \frac{1}{\pi}, \qquad \text{for almost all } k,q \in \diamond. \tag{10}$$

The generalization to $\nu = \frac{1}{M}$, $M \in \boldsymbol{N}$, is straightforward. We simply require the orthonormality of the wave functions located at a (e.g. horizontal) distance of M sites: $S_{m,Mn} = \int_\diamond dk\, dq\, e^{iMnaq-ikma}\, |h(k,q)|^2 = \delta_{m,0}\,\delta_{n,0}$, and, proceeding as above, we deduce that $h(k,q)$ must satisfies the equality

$$J_M(k,q) \equiv \left|h(k,\frac{q}{M})\right|^2 + \left|h(k,\frac{q+a}{M})\right|^2 + \ldots + \left|h(k,\frac{q+(M-1)a}{M})\right|^2 = \frac{M}{2\pi}, \tag{11}$$

almost everywhere for $k, q \in \diamond$.

Using these results, we will now establish the correspondence between FQHE states and MRA. As it is well known an MRA of $\mathcal{L}(\boldsymbol{R})$ is determined by the sequence $\{h_n\}_{n\in\boldsymbol{Z}}$ of the Fourier coefficients of the 2π-periodic function $m_o(\omega)$, the low-pass filter of the MRA[1]. Now we use the sequence $\{h_n\}_{n\in\boldsymbol{Z}}$ to define the following function, which strongly resembles $m_o(\omega)$:

$$T_2(\omega) = \begin{cases} \frac{1}{\sqrt{a}} \sum_{l\in\boldsymbol{Z}} h_l e^{-il\omega a}, & \omega \in [0,a), \\ 0, & \text{otherwise.} \end{cases} \tag{12}$$

It is clear that $T_2(\omega)$ is square integrable and nonperiodic. In particular, by the sum rule for the coefficients of any MRA

$$\frac{1}{2} \sum_{n,l\in\boldsymbol{Z}} h_n\, \overline{h_l}\, e^{i(l-n)\omega} (1 + (-1)^{l+n}) = 1, \qquad \text{a.e.} \tag{13}$$

we deduce that $\|T_2\|_2^2 = \int_{\boldsymbol{R}} |T_2(\omega)|^2 d\omega = 1$. Therefore, the Zak transform of this function, (9), $t_2(k,q) = (ZT_2)(k,q)$, is well defined in $L^2(\diamond)$.

It is easy to check that $t_2(k,q)$ satisfies the orthonormality conditions (10). In fact, for $(k,q) \in \diamond$, we obtain $J_2(k,q) = |t_2(k,\frac{q}{2})|^2 + |t_2(k,\frac{q+a}{2})|^2 = \frac{1}{a^2} \sum_{l,s\in\boldsymbol{Z}} h_l\, \overline{h_s}\, e^{i(s-l)qa/2}\, (1 + (-1)^{l+s})$, which is equal to $1/\pi$ a.e. in $k, q \in \diamond$, by (13). This implies that $t_2(k,q)$ gives rise to a family of functions $\Psi_{m,n}(x,y)$ in the LLL, mutually orthonormal and corresponding to $\nu = 1/2$. The explicit form of these $\psi_{m,n}(x,y)$ is considered in [5], where we also compare these results with the ones obtained in [2].

The above procedure can be easily extended to fillings $\nu = \frac{1}{2L}$ [5], while the extension to an odd denominator is not so straightforward.

Conversely, it is easy to show also that a function $h(k,q)$ which produces an orthonormal set of translated functions in the LLL, also produces a sequence $\{h_n\}$ satisfying condition (13) and, therefore, generating an MRA. Let indeed $h(k,q)$ be a function in $L^2(\diamond)$, satisfying $|h(k,q/2)|^2 + |h(k,(q+a)/2)|^2 = \frac{1}{\pi}$, a.e. in $\diamond$. This means that, in configuration space, the corresponding set $\{\Psi_{m,n}(x,y)\}$ is orthonormal for $\nu = 1/2$. Define now $h_n(k) = \int_0^a dq\, e^{inaq}\, h(k,q)$, $\quad k \in [0,a)$.

Although $\overline{h_n(k)}$ is a function of k, it is straightforward to check that $\sum_{n\in\mathbf{Z}} h_n(k)\,\overline{h_{n+2l}(k)} = \delta_{l,0}$ [5]. This result shows that any orthonormal basis in the LLL for a filling factor $\nu = \frac{1}{2}$ produces an MRA of $\mathcal{L}(\boldsymbol{R})$ which, in general, depends on an external parameter $k \in [0, a)$. Again, the extension to a filling factor $\nu = \frac{1}{2L}$, $L \in \boldsymbol{N}$, is straightforward[5].

Summarizing, we have proved that there exists a **complete** equivalence between an MRA and an orthonormal set of single electron wave functions in the LLL (for $\nu = \frac{1}{2L}$, $L \in \boldsymbol{N}$).

Acknowledgments

F.B. aknowledges financial support by the Murst, within the project *Problemi Matematici Non Lineari di Propagazione e Stabilità nei Modelli del Continuo*, coordinated by Prof. T. Ruggeri.

References

1. I. Daubechies, *Ten Lectures on Wavelets.* (SIAM, Philadelphia, 1992); H.L. Resnikoff, R.O. Wells, jr., *Wavelet Analysis*, (Springer-Verlag, New York, 1998).
2. J-P. Antoine and F. Bagarello, J. Phys. A: Math. Gen. **27**, 2471–2481 (1994).
3. F. Bagarello, J. Phys. A: Math. Gen. **27**, 5583–5597 (1994).
4. F. Bagarello, J. Phys. A: Math. Gen. **29**, 565–576 (1996).
5. F. Bagarello, J. Math. Phys. **42**, 2001 (to appear).
6. J-P. Antoine and F. Bagarello, in *Wavelets in Physics, vol. 2*, ed. J.C. van den Berg, (Cambridge Univ. Press, Cambridge, to appear).
7. K. Cho, T.A. Arias, J.D. Joannopoulos, Phys. Rev. Lett. **71**, 1808 (1993); L. De Windt, P. Fisher, M. Defranceschi, J. Delhalle, J.G. Fripiat, J. Comp. Phys. **111**, 266 (1994); J-P. Antoine, Ph. Antoine, and B. Piraux, in *Wavelets in Physics*, ed. J.C. van den Berg (ed.), pp. 299–338, (Cambridge Univ. Press, Cambridge, 1999).
8. K. Maki and X. Zotos, Phys. Rev. B **28**, 4349–4356 (1983).
9. F. Bagarello, G. Morchio, and F. Strocchi, Phys. Rev. B **48**, 5306–5314 (1993).
10. I. Dana and J. Zak, Phys. Rev. B **28**, 811–820 (1983).
11. G. Battle, J. Math. Phys. **30**, 2195–2196 (1989).
12. J. Zak, J. Math. Phys. **39**, 4195–4200 (1998).
13. J. Zak, Solid State Physics **27**, 1–62 (1972).

Space homogeneous solutions of the linear Boltzmann equation for semiconductors: a semigroup approach

Jacek Banasiak
School of Mathematical and Statistical Sciences
University of Natal, Durban 4041 South Africa

Francesco Mugelli and Giovanni Frosali
Dipartimento di Matematica Applicata "G.Sansone"
Università di Firenze - I-50139 Firenze

1 Introduction

In a recent paper Majorana and Milazzo [7] considered the linear Boltzmann equation describing electron flow in a semiconductor. The electron-phonon interactions are simulated by a version of the $\delta-$like kernel, first mathematically investigated in ref. 8, that in the setting of this paper may be unbounded for large energies. The initial value problem for the linear Boltzmann equation reads as follows

$$\begin{cases} \dfrac{\partial f}{\partial t}(t, \mathbf{k}) = Kf(t, \mathbf{k}) - \nu(\mathbf{k}) f(t, \mathbf{k}) , \\[2ex] f(0, \mathbf{k}) = f_0(\mathbf{k}) , \end{cases} \tag{1.1}$$

with the gain collision operator K and the collision frequency ν defined by

$$Kf(t, \mathbf{k}) = \int_{\mathbb{R}^3} S(\mathbf{k}', \mathbf{k}) f(t, \mathbf{k}')\, d\mathbf{k}', \qquad \nu(\mathbf{k}) = \int_{\mathbb{R}^3} S(\mathbf{k}, \mathbf{k}')\, d\mathbf{k}' .$$

The kernel $S(\mathbf{k}, \mathbf{k}')$ which accounts for the scattering processes between electrons and the background takes the following form

$$\begin{aligned} S(\mathbf{k}, \mathbf{k}') = {} & \mathcal{G}(\mathbf{k}, \mathbf{k}')[(n_q + 1)\delta(\varepsilon(\mathbf{k}') - \varepsilon(\mathbf{k}) + \hbar\omega) + n_q \delta(\varepsilon(\mathbf{k}') - \varepsilon(\mathbf{k}) - \hbar\omega)] \\ & + \mathcal{G}_0(\mathbf{k}, \mathbf{k}')\delta(\varepsilon(\mathbf{k}') - \varepsilon(\mathbf{k})) . \end{aligned}$$

$\mathcal{G}(\mathbf{k}, \mathbf{k}')$ and $\mathcal{G}_0(\mathbf{k}, \mathbf{k}')$ denote symmetric continuous functions on $\mathbb{R}^3 \times \mathbb{R}^3$ characterizing inelastic collisions with optical phonons and elastic collisions with acoustic phonons and impurities, respectively. The constant positive parameter n_q is given by $n_q = \left[\exp\left(\frac{\hbar\omega}{k_B T_L}\right) - 1\right]^{-1}$, where $\hbar$ is the Planck constant divided by 2π, ω is the positive constant phonon frequency, k_B is the Boltzmann constant and T_L is the lattice temperature. $\delta(\varepsilon(\mathbf{k}))$ denotes the Dirac function composed with the electron energy function $\varepsilon(\mathbf{k})$:

$$\varepsilon(\mathbf{k}) = \frac{\hbar^2|\mathbf{k}|^2}{m^* + \sqrt{m^*(m^* + 2|\mathbf{k}|^2\alpha\hbar^2)}} \tag{1.2}$$

where m^* is the effective mass. Note that for $\alpha = 0$, eq. (1.2) gives the mass parabolic approximation.

In ref. 7 the Cauchy problem for eq. (1.1) is investigated in the Banach space of summable functions, without requiring the boundedness of the collision frequency, and the existence of the solution is established. The uniqueness of the solution can be recovered adding suitable conditions on the growth speed of the collision frequency $\nu(\mathbf{k})$ (ref. 7, Th. 7).

In this paper we shall study the initial value problem (1.1) by semigroups. Problems of this form fall into the general framework of the theory of substochastic semigroups (see ref. 1, 2, 4 and references therein). This approach gives almost immediately the existence of the semigroup, however the full characterization of the generator is usually obtained by some other means. This creates the possibility that the class of solutions obtained in this theory may not coincide with the solutions obtained by e.g. the approximation approach as in ref. 7. A more detailed analysis of the problem of multiple solutions is given in ref. 5.

Here we prove that the generator of the evolution semigroup is the closure of the physical operator, following the same lines of the results obtained by J.Banasiak for the Spiga model (that have similar mathematical structure to the Majorana model). For more details and discussions on the uniqueness and non-uniqueness see the papers quoted in the references.

2 Abstract semigroup approach

Consider the abstract formulation of the Cauchy problem for eq. (1.1):

$$\begin{cases} \dfrac{df}{dt} = Af + Bf\,, \\ f(0) = f_0\,, \end{cases} \tag{2.3}$$

where A and B are in general unbounded operators in a Banach space X and f is, say, a distribution of particles. For problems of this type the most suitable seems to be the method developed by Kato [6] for Kolmogorov's system. His results were extended to a more general situation and applied to kinetic theory by Voigt, Myiadera, van der Mee, Protopopescu, Desch, Mokhtar-Kharroubi,

Banasiak and many others. In the general case, one can only prove that the generator T of the semigroup $\{G(t), t \geq 0\}$ solving the Cauchy problem (2.3) is an extension of the operator $A + B$ and such a result is usually insufficient for applications. The reason for this is that the semigroup G solving eq. (2.3) should be a *transition (stochastic) semigroup*, i.e. one should have

$$||G(t)f|| = ||f||, \quad \forall t \geq 0, \ \forall f \geq 0;$$

this condition expresses the fact that the total number of particles in conserved through time. A sufficient condition for G to be stochastic is $T = \overline{A + B}$.

Three situations are possible: i) $T = A+B$ or ii) $T = \overline{A + B}, T \neq A+B$, whereby G is stochastic, and iii) T is a proper extension of $\overline{A + B}$, in which case G may be not stochastic. The total number of particles is preserved only if G is stochastic, so only in the first case we can claim that the obtained semigroup has physical relevance [4].

The following is an extension of the Kato-Voigt perturbation theorem that generalizes theorem by L. Arlotti [1,4].

Theorem 2.1 *Let A, B operators in $X = L^1(\Omega, \mu)$. Suppose*

i) $(A, \mathcal{D}(A))$ generates substochastic semigroup $\{G_A(t),\ t \geq 0\}$;

ii) $\mathcal{D}(B) \supseteq \mathcal{D}(A)$ and $Bf \geq 0$, $\forall f \in \mathcal{D}(B), f \geq 0$;

iii) $$\int_\Omega (Af + Bf)\, d\mu \leq 0\,, \quad \forall f \in \mathcal{D}(A), f \geq 0. \tag{2.4}$$

Then, there exists a smallest substochastic semigroup $\{G(t), t \geq 0\}$ generated by an extension T of $A + B$ and satisfying the integral equation

$$G(t)f = G_A(t)f + \int_0^t G(t-s)BG_A(s)f\, ds, \qquad \forall f \in \mathcal{D}(A),\ \forall t \geq 0\,.$$

It can also obtained by the Philips-Dyson expansion $G(t)f = \sum_{n=0}^{\infty} S_n(t)f$, $f \in X$ where $S_0(t)f = G_A(t)f$, $S_n(t)f = \int_0^t S_{n-1}(t-s)BG_A(s)f\, ds$.

Note that theorem 2.1 *does not* give any characterization of the domain of the generator T.

3 Existence of Solutions by Semigroup Theory

Let $X = L^1(\mathbb{R}^3)$. The multiplication operator by $-\nu(\mathbf{k})$ plays the role of A in the abstract problem (2.3), with domain $\mathcal{D}(A) = \{f \in X : \nu(\mathbf{k})f \in X\}$, under the assumption that $\nu(\mathbf{k})$ is non-negative and belonging to $L^1_{loc}(\mathbb{R}^3)$. The role of B is played by the positive gain integral operator with symmetric kernel $S(\mathbf{k},\mathbf{k}') \geq 0$, with domain $\mathcal{D}(B) = \{f \in X : Kf \in X\}$. The kernel is such that $\int_{\mathbb{R}^3} S(\mathbf{k},\mathbf{k}'))\,d\mathbf{k}' \notin L^1(\mathbb{R}^3)$ but is only $L^1_{loc}(\mathbb{R}^3)$.

Proposition 3.1

$$||Bf||_{L^1(\mathbb{R}^3)} \leq ||Af||_{L^1(\mathbb{R}^3)} \quad , \quad \forall f \in \mathcal{D}(A)\,. \tag{3.5}$$

Proof: Let $f \in L^1(\mathbb{R}^3) \cap \mathcal{D}(A)$. Evaluating the l.h.s of (3.5) and applying triangle inequality, the result follows immediately using the symmetry and the positiveness of the kernel S. ■

From Proposition 3.1 it follows that $\mathcal{D}(B) \supseteq \mathcal{D}(A)$. A generates the substochastic semigroup $G_A(t) = \exp\{-\nu(\mathbf{k})t\}$. Then it is easy to prove that

$$\int_{\mathbb{R}^3} (Af + Bf)\,d\mathbf{k} = 0, \quad \forall f \in \mathcal{D}(A), \quad f \geq 0, \tag{3.6}$$

thus we can apply Theorem 2.1, obtaining the existence of the semigroup of the evolution operator of our process. This semigroup, obtained by the Phillips-Dyson expansion, is substochastic and its generator is an extension of $A + B = -\nu I + K$, where I is the identity operator.

In the present context we use the method introduced by Arlotti [1] and adapted by Banasiak [3,5], which consists in extending suitably the domain of the collision operator. With this understanding, we have

Theorem 3.1 *If for any $f \in X, f \geq 0$, such that the expression $Kf(\mathbf{k})$ is finite almost everywhere and such that $-\nu f + Kf \in X$, we have*

$$\int_{\mathbb{R}^3} (-\nu(\mathbf{k})f(\mathbf{k}) + Kf(\mathbf{k}))\,d\mathbf{k} \geq 0, \tag{3.7}$$

then $T = \overline{-\nu I + K}$.

Before giving the characterization of T, we prove the following technical lemma

Lemma 3.1 *Let $B_n = \{(\varepsilon, \mathbf{u}) : 0 \leq \varepsilon < n\hbar\omega,\ \mathbf{u} \in S^2\}$. Then,*

$$\int_{B_n} (-\nu(\mathbf{k})f(\mathbf{k}) + (Kf)(\mathbf{k}))\,d\mathbf{k} = \tag{3.8}$$

$$= -\,n_q \int_{n\hbar\omega}^{(n+1)\hbar\omega} D(\varepsilon)D(\varepsilon - \hbar\omega) \int_{S^2\times S^2} \mathcal{G}(\varepsilon, \varepsilon - \hbar\omega, \mathbf{u}, \mathbf{u}')f(\varepsilon - \hbar\omega, \mathbf{u}')\,d\mathbf{u}'\,d\mathbf{u}\,d\varepsilon$$

$$+\ (n_q + 1) \int_{n\hbar\omega}^{(n+1)\hbar\omega} D(\varepsilon)D(\varepsilon - \hbar\omega) \int_{S^2\times S^2} \mathcal{G}(\varepsilon, \varepsilon - \hbar\omega, \mathbf{u}', \mathbf{u})f(\varepsilon, \mathbf{u}')\,d\mathbf{u}'\,d\mathbf{u}\,d\varepsilon\,.$$

Proof. Function $\varepsilon(\mathbf{k})$ is invertible on $\mathbb{R}^+$ if considered as function of $k = |\mathbf{k}|$: $k = \sqrt{2\alpha(\varepsilon + a\varepsilon^2)}$, where $a = m^*/\hbar^2$. From now on, we will express quantities in terms of the energy ε instead of $\mathbf{k}$, having $k\,dk = 2\alpha(1+2a\varepsilon)\,d\varepsilon$ and

$$\int_{\mathbb{R}^3} \cdot\, d\mathbf{k} = \int_0^\infty k^2\,dk \int_{S^2} \cdot\, d\mathbf{u} = \int_0^\infty D(\varepsilon)\,d\varepsilon \int_{S^2} \cdot\, d\mathbf{u}$$

where $\mathbf{k} = k\mathbf{u}$, $|\mathbf{u}| = 1$ and $D(\varepsilon) = (2\alpha)^{3/2}(1+2a\varepsilon)\sqrt{\varepsilon + a\varepsilon^2}$.

Rewriting the operators in terms of ε, a straight calculation gives:

$$\int_{B_n} (-\nu(\mathbf{k})f(\mathbf{k}) + (Kf)(\mathbf{k}))\,d\mathbf{k} =$$

$$= -(n_q+1)\int_{\hbar\omega}^{n\hbar\omega} D(\varepsilon-\hbar\omega)\int_{S^2\times S^2} \mathcal{G}(\varepsilon,\varepsilon-\hbar\omega,\mathbf{u},\mathbf{u}')D(\varepsilon)f(\varepsilon,\mathbf{u})\,d\mathbf{u}\,d\mathbf{u}'\,d\varepsilon$$

$$- n_q\int_0^{n\hbar\omega} D(\varepsilon+\hbar\omega)\int_{S^2\times S^2} \mathcal{G}(\varepsilon,\varepsilon+\hbar\omega,\mathbf{u},\mathbf{u}')D(\varepsilon)f(\varepsilon,\mathbf{u})\,d\mathbf{u}\,d\mathbf{u}'\,d\varepsilon$$

$$- \int_0^{n\hbar\omega} D^2(\varepsilon)\int_{S^2\times S^2} \mathcal{G}_0(\varepsilon,\varepsilon,\mathbf{u},\mathbf{u}')f(\varepsilon,\mathbf{u})\,d\mathbf{u}\,d\mathbf{u}'\,d\varepsilon \tag{3.9}$$

$$+ (n_q+1)\int_0^{n\hbar\omega} D(\varepsilon)\int_{S^2\times S^2} \mathcal{G}(\varepsilon,\varepsilon+\hbar\omega,\mathbf{u},\mathbf{u}')D(\varepsilon+\hbar\omega)f(\varepsilon+\hbar\omega,\mathbf{u}')\,d\mathbf{u}\,d\mathbf{u}'\,d\varepsilon$$

$$+ n_q\int_{\hbar\omega}^{n\hbar\omega} D(\varepsilon)\int_{S^2\times S^2} \mathcal{G}(\varepsilon,\varepsilon-\hbar\omega,\mathbf{u},\mathbf{u}')D(\varepsilon-\hbar\omega)f(\varepsilon-\hbar\omega,\mathbf{u}')\,d\mathbf{u}\,d\mathbf{u}'\,d\varepsilon$$

$$+ \int_0^{n\hbar\omega} D^2(\varepsilon)\int_{S^2\times S^2} \mathcal{G}_0(\varepsilon,\varepsilon,\mathbf{u},\mathbf{u}')f(\varepsilon,\mathbf{u}')\,d\mathbf{u}\,d\mathbf{u}'\,d\varepsilon\,. \tag{3.10}$$

We get immediately that (3.9) + (3.10) = 0, and, shifting ε variable in the other terms in a way to have $f(\varepsilon,\mathbf{u}')$ inside the integrals and using the symmetry of the kernel $\mathcal{G}$, we obtain the r.h.s. of (3.8). Hence the lemma is completely proved. ∎

Now we are able to characterize the generator T in the following way.

Theorem 3.2 *Assume that there exists $n_0 \in \mathbb{N}$ such that $\mathcal{G}$, as a function of ϵ, is a strictly positive function for $\epsilon > n_0$ and that there exists $q < 1$ such that for all $n \geq n_0$*

$$\sup_{B_{n+1}\setminus B_n} \mathcal{G}(\varepsilon,\varepsilon-\hbar\omega,\mathbf{u},\mathbf{u}') \leq q\frac{n_q+1}{n_q} \inf_{B_n\setminus B_{n-1}} \mathcal{G}(\varepsilon,\varepsilon-\hbar\omega,\mathbf{u}',\mathbf{u}), \tag{3.11}$$

then $T = \overline{-\nu I + K}$.

Proof. Following Lemma 3.1 we denote by b_n the right-hand side of (3.8). The proof of the characterization of the generator T relies on proving (3.7) of Theorem 3.1, that is, that the limit of the sequence b_n, that exists by the assumption $-\nu f + Kf \in X$, is non-negative. So, suppose on the contrary that $\lim_{n\to\infty} b_n < 0$. Then there exist n_0 and $b > 0$ such that $b_n < -b$ and the assumption (3.11) holds for all $n > n_0$. Let

$$G_n = \sup_{B_{n+1}\setminus B_n} \mathcal{G}(\varepsilon, \varepsilon - \hbar\omega, \mathbf{u}, \mathbf{u}'), \quad c_{n+1} = 4\pi \int_{S^2} \int_{n\hbar\omega}^{(n+1)\hbar\omega} D(\varepsilon) f(\varepsilon, \mathbf{u}')\, d\varepsilon d\mathbf{u}',$$

$$g_n = \inf_{B_{n+1}\setminus B_n} \mathcal{G}(\varepsilon, \varepsilon - \hbar\omega, \mathbf{u}', \mathbf{u}), \quad c_n = 4\pi \int_{S^2} \int_{n\hbar\omega}^{(n+1)\hbar\omega} D(\varepsilon - \hbar\omega) f(\varepsilon - \hbar\omega, \mathbf{u}')\, d\varepsilon d\mathbf{u}'.$$

Let $n = n_0 + k$. Then, for any $k \geq 0$, using the fact that D is an increasing function

$$\begin{aligned} b_n \geq & -n_q D((n_0 + k + 1)\hbar\omega) G_{n_0+k} c_{n_0+k} \\ & + (n_q + 1) D((n_0 + k - 1)\hbar\omega) g_{n_0+k} c_{n_0+k+1}. \end{aligned}$$

To shorten the notation we introduce $\beta_k = n_q D((n_0 + k + 1)\hbar\omega) G_{n_0+k}$, and $\alpha_k = (n_q + 1) D((n_0 + k - 2)\hbar\omega) g_{n_0+k-1}$. In the new notation, we have

$$0 > -b \geq b_n \geq -\beta_k c_{n_0+k} + \alpha_{k+1} c_{n_0+k+1}, \qquad \forall k \geq 0,$$

and we obtain the recurrence for c_{n_0+k+1}:

$$c_{n_0+k+1} \leq -\frac{b}{\alpha_{k+1}} + \frac{\beta_k}{\alpha_{k+1}} c_{n_0+k} \qquad \forall k \geq 0.$$

By induction, it is easy to show that

$$c_{n_0+k} \leq -bA_k + c_{n_0} B_k = B_k \left(-b \frac{A_k}{B_k} + c_{n_0} \right) \tag{3.12}$$

where

$$A_k = \frac{1}{\alpha_k} \sum_{l=1}^{k} \prod_{i=1}^{k-l} \frac{\beta_{k-i}}{\alpha_{k-i}}, \qquad B_k = \prod_{i=0}^{k-1} \frac{\beta_i}{\alpha_{i+1}}$$

and we put $\prod_{i=1}^{0} = 1$. Now it is easy to show that the sequence A_k/B_k is not divergent. In fact, if we assume $\lim_{k\to\infty} A_k/B_k = \infty$, it follows that $-b\frac{A_k}{B_k} + c_{n_0} \to -\infty$. Thus $c_{n_0+k} < 0$ for k sufficiently large; but this is impossible as by definition $c_{n_0+k} > 0$. After a simple manipulation we have

$$\frac{A_k}{B_k} = \sum_{l=0}^{k} \prod_{i=1}^{l} \frac{\alpha_i}{\beta_i},$$

hence the previous conclusion means that the series $\sum_{l=0}^{\infty} C_l = \sum_{l=0}^{\infty} \prod_{i=1}^{l} \frac{\alpha_i}{\beta_i}$ converges. The generic term of this series has the form

$$C_l = \frac{(n_q + 1) D((n_0 + l - 2)\hbar\omega) g_{n_0+l-1}}{n_q D((n_0 + l + 1)\hbar\omega) G_{n_0+l}} \cdot \ldots \cdot \frac{(n_q + 1) D((n_0 - 1)\hbar\omega) g_{n_0}}{n_q D((n_0 + 1)\hbar\omega) G_{n_0+1}}.$$

From the hypothesis there exists a $q < 1$ such that $G_{n_0+l} \leq q\frac{(n_q+1)}{n_q} g_{n_0+l-1}$ for l sufficiently large. Since $\lim_{l\to\infty} \frac{D((n_0+l-2)\hbar\omega)}{D((n_0+l+1)\hbar\omega)} = 1$, we can find $q' < 1$ such that for l sufficiently large,

$$\frac{(n_q+1)D((n_0+l-2)\hbar\omega)g_{n_0+l-1}}{n_q D((n_0+l+1)\hbar\omega)G_{n_0+l}} \geq q' > 1$$

which shows that the sum does not converge. Hence inequality (3.7) is proved and Theorem 3.1 gives the result. ■

In this paper we have studied the initial value problem (1.1) obtaining existence and uniqueness of the evolution substochastic semigroup by standard semigroup theory. In Theorem 3.1 we have proved a condition on the growth of the collision kernel $\mathcal{G}$, as function of the energy, sufficient to characterize the generator T by the closure of the physical operator $-\nu I + K$. Condition (3.11) implies that the collision kernel, as function of ϵ, belongs to the L^1 space with the weighted measure $p^\epsilon d\epsilon$, with $p < \frac{n_q}{q(n_q+1)}$. Conversely, we can observe that if the collision kernel is of a polynomial growth or behaves as p^ϵ with $p \leq \sqrt{\frac{n_q+1}{n_q}}$ for large ϵ, then condition (3.11) is satisfied.

Acknowledgements

This work was performed under the auspices of the *National Group for Mathematical Physics* (INdAM) and was partly supported by the *Italian Ministery of University (MURST* National Project "Mathematical Problems of Kinetic Theories", Cofin2000). Additional support from the *Italian Research National Council (CNR* Strategic Project "Modelli Matematici per Semiconduttori") is also gratefully acknowledged.

References

1. L. Arlotti, *Acta Applicandae Mathematicae* **23**, 129 (1991).
2. J. Banasiak, *Math. Mod. Meth. Appl. Sci.* **10**(2), 163 (2000).
3. J. Banasiak, *Math. Meth. in the Appl. Sci.* **23**, 1237 (2000).
4. J. Banasiak, *Taiwanese J. of Mathematics* **5**(1), 169 (2001).
5. J. Banasiak, *Transport Theory Stat. Phys.* (2001) (submitted).
6. T. Kato, *J. Math. Soc. Japan* **6**, 1 (1954).
7. A. Majorana and C. Milazzo, *J. Math. Anal. Appl.* **259**(2), 609 (2001).
8. A. Majorana, *Transport Theory Statist. Phys.* **20**, 261 (1991).

STATIONARY HEAT CONDUCTION IN A ROTATING FRAME

ELVIRA BARBERA

Department of Mathematics, University of Calabria, Cosenza, Italy

We consider a problem of stationary heat conduction in a monatomic gas between two co-axial cylinders which are at rest in a non-inertial frame. Instead of the usual Navier-Stokes-Fourier theory, we employ extended thermodynamics with 13 moments. Three results follow: There is a tangential heat flux between the cylinders, no rigid rotation of the heat conducting gas is possible and there is a normal pressure field in the axial direction.

1 Introduction

In the kinetic theory of gas and in extended thermodynamics[1] the set of balance equations for the first 13 moments in a non-inertial frame is[a]

$$
\begin{aligned}
&\frac{\partial F}{\partial t} + F^k_{;k} = 0, \\
&\frac{\partial F^i}{\partial t} + F^{ik}_{;k} - F i^i_0 - 2F^k W^i_k = 0, \\
&\frac{\partial F^{ij}}{\partial t} + F^{ijk}_{;k} - 2F^{(i}i^{j)}_0 - 4F^{k(i}W^{j)}_k = \frac{1}{\tau}(F_E^{<ij>} - F^{<ij>}), \\
&\frac{\partial F^{ji}_j}{\partial t} + F^{jik}_{j;k} - 3F^{(i}_j i^{j)}_0 - 6F^{k(i}_j W^{j)}_k = \frac{1}{\tau}(F^{ji}_{Ej} - F^{ji}_j),
\end{aligned}
\tag{1}
$$

where i^i_0 denotes the velocity-independent inertial accelerations and W^i_k is the matrix of the angular velocity of the frame. The right hand sides of (1) represent the productions due to collisions, simplified according to the BGK approach[2]. τ is a relaxation time of the order of magnitude of a mean time of free flight of the atoms of the gas.

It may be observed that all the equations, except the first one, contain inertial terms and these inertial contributions provide some unusual phenomena.

We use the system (1) in order to describe an heat conduction problem: We consider a monatomic gas between two co-axial cylinders and we heat the inner cylinder.

[a] Upper and lower indices represent contra– and covariant components respectively while the semicolon indicates covariant differentiation. Symmetrization is understood with respect to bracketed indices. Angular brackets denote trace-less tensors.

If the two cylinders are at rest in a inertial frame we expect an inward radial temperature gradient and an outward radial heat flux. However, if the cylinders are placed in a rotating frame with their axes parallel to the angular velocity of the frame, we expect a tangential component of the heat flux as well as a radial component in contrast to the customary N.S.F. theory.

This expectation is motivated by the following argument from the elementary kinetic theory of gases: In Figure 1 we focus the attention on a small volume element of the order of magnitude of a mean free path. Atoms from below will carry less energy upward than atoms from above will carry downward. Thus, if a pair of atoms passes through the middle layer, a downward transport of energy is effected, i.e. a heat flux in the radial direction. In a non-inertial frame the atoms do not fly in straight paths, rather their orbits are bent by the Coriolis force so that, along with the radial heat flux, there is a tangential one upon the passage of each pair of particles.

2 Field Equations

The moments $F^{i_1 i_2 \dots i_n}$ in (1) may be decomposed into internal parts, i.e. velocity independent parts, and other parts that depend on the velocity explicitly. This decomposition is dictated by the Galilean invariance of the system (1). It reads

$$\begin{aligned}
F &= \rho \\
F_i &= \rho v_i \\
F_{ij} &= \rho_{ij} + \rho v_i v_j \\
F_{ijk} &= \rho_{ijk} + 3\rho_{(ij}v_{k)} + \rho v_i v_j v_k \\
F_{ikll} &= \rho_{ikll} + 4\rho_{(ikl}v_{l)} + 6\rho_{(ik}v_l v_{l)} + \rho v_i v_k v^2 ,
\end{aligned} \tag{2}$$

where the ρ's are the internal moments. Some of them are easily interpreted. Thus ρ is the mass-density, ρ_{ij} is the pressure tensor with $\rho^i_i = 3p$, being p the pressure, and ρ^j_{ij} is the twice the heat flux.

Elimination of the **F**'s between (1) and (2) provides the set of balance equations for the internal moments. This system is not closed because of the occurrence of the fluxes $\rho^{<ij>k}$ and ρ^{ijk}_j. We close it by calculating these fluxes from the 13-moment Grad distribution function[3]

$$f_G = f_E \left(1 + \frac{\rho_{<ij>} C^i C^j}{2\frac{k}{m} p\theta} - \frac{\rho^j_{ij} C^i}{2\frac{k}{m} p\theta} \left(1 - \frac{g_{pq} C^p C^q}{5\frac{k}{m}\theta} \right) \right) . \tag{3}$$

f_E is the Maxwell distribution and θ is the kinetic temperature of the gas. C^i denotes the components of the velocity of the atoms in the local rest frame of the gas. With (3) the two internal moments $\rho^{<ij>k}$ and ρ_j^{ijk} result as

$$\rho^{<ij>k} = \frac{\rho_n^{in} g^{jk} + \rho_n^{jn} g^{ik} - \frac{2}{3}\rho_n^{kn} g^{ij}}{5}, \qquad \rho_j^{ijk} = \frac{5k}{m} p\theta g^{ik} + \frac{7k}{m}\theta\rho^{<ik>} . \tag{4}$$

We shall be interested in heat conduction in a gas between two co-axial cylinders. Therefore we use the cylindrical coordinates (r, ϑ, z). The bulk of the paper will refer to a non-inertial frame with a constant angular velocity ω and without translational motion with respect to an inertial frame.

Müller and Ruggeri[4] have used the equations (1), (2), (4) with $w = 0$ to calculate the stationary fields of the gas at rest in an inertial frame between two co-axial cylinders with inner and outer radii r_i and r_e. The gas was heated by the heat flux q_i^r at r_i and the temperature was kept at the value θ_e at r_e. From the 13 moments theory it follows that the temperature at the inner cylinder remains bounded as r_i tends to zero and there are normal components of the deviatoric stress $\rho^{<kl>}$.

2.1 A heat-conducting gas is incapable of a rigid rotation

In order to determine the non-inertial effect on the stationary heat conduction between the two cylinders, we use the system (1), (2), (4). For the heuristic argument about the tangential component of the heat flux presented in the introduction, we assume the tangential component $q^\vartheta(r)$ different from zero and we make the semi-inverse ansatz:

$$p = p(r), \qquad \mathbf{v} = 0, \qquad \theta = \theta(r),$$

$$\rho_j^{ij} = 2\begin{bmatrix} q^r(r) \\ q^\vartheta(r) \\ 0 \end{bmatrix}, \quad \rho^{<ij>} = \begin{bmatrix} \sigma^{rr}(r) & \sigma^{r\vartheta}(r) & 0 \\ \sigma^{r\vartheta}(r) & \sigma^{\vartheta\vartheta}(r) & 0 \\ 0 & 0 & \sigma^{zz}(r) \end{bmatrix}. \tag{5}$$

We find out that there is no non-trivial solution of the system for this case. Nor is there a non-trivial solution, if – for $\mathbf{v} = 0$ – we also set $\sigma^{r\vartheta} = 0$ or $\sigma^{r\vartheta}$ *and* q^ϑ equal to zero.

Therefore we conclude that the heat conducting gas cannot be at rest in a non-inertial frame. Or in other words: A heat conducting gas is incapable of a rigid rotation.

3 General planar solution

Since heat-conduction in a rigidly rotating gas is impossible, we generalise the semi-inverse ansatz (5) by assuming that the velocity field can have the form $\mathbf{v} = (v^r(r), v^\vartheta(r), 0)$.

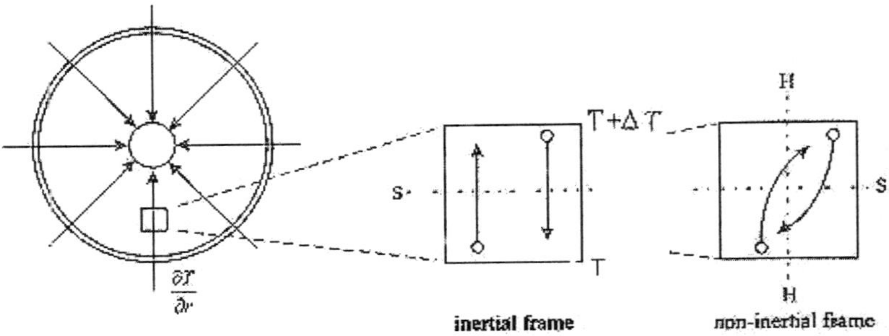

Figure 1. Radial and tangential components of the heat flux.

Actually the balance of mass quickly requires that v^r is zero. Thus only *one* new field, v^ϑ, is added to the fields (5). This modest addition, however, makes for a formidable system of coupled equations which cannot be solved analytically. The system is given in (6), where now we have employed physical components, that are characterised by the indices in square brackets.

$$\frac{d(p+\sigma[rr])}{dr} - \frac{\sigma[\vartheta\vartheta]-\sigma[rr]}{r} = \frac{p(r\omega+v[\vartheta])^2}{\frac{k}{m}r\theta},$$

$$\sigma[r\vartheta] = \frac{D}{r^2}, \qquad q[r] + \sigma[r\vartheta]\,v[\vartheta] = \frac{C}{r}$$

$$-\frac{5\sigma[rr]}{2\tau} = -10\omega\sigma[r\vartheta] + \frac{4}{3}\frac{dq[r]}{dr} - \frac{5\sigma[r\vartheta]}{3}\frac{dv[\vartheta]}{dr} - \frac{2}{3}\frac{q[r]}{r} - \frac{25}{3}\frac{\sigma[r\vartheta]v[\vartheta]}{r}$$

$$-\frac{5\sigma[r\vartheta]}{2\tau} = \frac{5(\sigma[rr]-\sigma[\vartheta\vartheta])(r\omega+v[\vartheta])}{r} + \frac{dq[\vartheta]}{dr} - \frac{q[\vartheta]}{r} + \frac{5(p+\sigma[rr])\left(\frac{dv[\vartheta]}{dr}-\frac{v[\vartheta]}{r}\right)}{2}$$

$$-\frac{5\sigma[\vartheta\vartheta]}{2\tau} = 10\omega\sigma[r\vartheta] + \frac{2q[r]}{r} + 4\sigma[r\vartheta]\frac{dv[\vartheta]}{dr} + \frac{6\sigma[r\vartheta]v[\vartheta]}{r}$$

$$-\frac{2q[r]}{\tau} = -4\omega q[\vartheta] - \frac{2\sigma[rr](r\omega+v[\vartheta])^2}{r} + \frac{k}{m}(5p+7\sigma[rr])\frac{d\theta}{dr} +$$

$$+2\frac{k}{m}\theta\left(\frac{d(\sigma[rr])}{dr} + \frac{\sigma[rr]-\sigma[\vartheta\vartheta]}{r}\right) + \frac{4q[\vartheta]}{5}\frac{dv[\vartheta]}{dr} - \frac{24}{5}\frac{q[\vartheta]v[\vartheta]}{r}$$

$$-\frac{2q[\vartheta]}{\tau} = \frac{4q[r](r\omega+v[\vartheta])}{r} - \frac{2\sigma[r\vartheta](r\omega+v[\vartheta])^2}{r} + \frac{7k}{m}\sigma[r\vartheta]\frac{d\theta}{dr} + \frac{14q[r]}{5}\left(\frac{dv[\vartheta]}{dr} - \frac{v[\vartheta]}{r}\right) \tag{6}$$

C and D are constants of integration.

This system will be solved after seven appropriate boundary values are

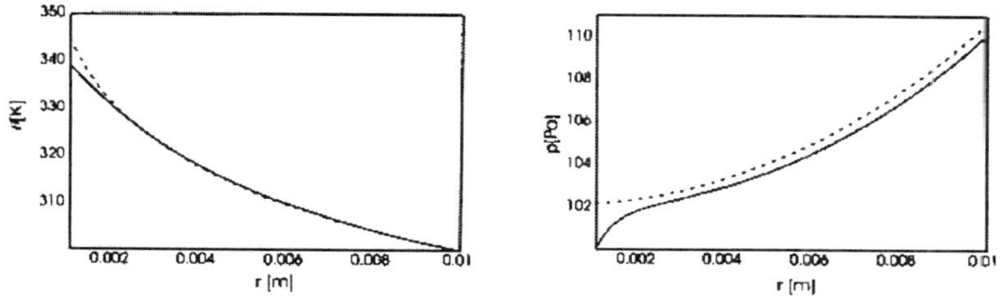

Figure 2. Temperature(left), pressure(right). Bold lines: 13 moments. Dashed lines: N.S.F.

chosen.

3.1 Boundary values

We fix $\tau = 10^{-5}s$, appropriate to a pressure of about 10^2 Pa, and let ω assume the large value $10^4 1/s$ in order to emphasize the rotational effects; the cylinders have radii $r_i = 10^{-3}m$ and $r_e = 10^{-2}m$.

We envisage both cylinders to be at rest in the non-inertial frame and the natural boundary condition for $v[\vartheta]$ at r_i and r_e is therefore the no-slip condition $v[\vartheta] = 0$. We apply at r_i a radial heat flux $q[r] = 10^4 W/m^2$ and we keep the temperature $\theta_e = 300K$ at r_e.

We assign at r_e the normal pressures $\{p + \sigma[rr]\}_e = \{p - \sigma[zz]\}_e = 110Pa$.

There is also a question mark about the assignability of the tangential heat flux $q[\vartheta]$ at r_e. This is a very serious difficulty and we do not know how to solve it. However there is an indication that the system in an unknown manner, – mysterious to us – knows which boundary value for $q[\vartheta]$ is permissible:

If we solve the system of equations for the previous set of boundary values *and for widely different values of* $q[\vartheta]_e$, the corresponding function $q[\vartheta]$ will sweep into essentially one single function within a very thin boundary layer, while the corresponding curves for $v[\vartheta]$ differ considerably with the data $q[\vartheta]_e$. If we choose $q[\vartheta]_e$ approximately equal to $-200W/m^2$ we eliminate the boundary layer. Incidentally this is also the value of $q[\vartheta]$ which we expect from the heuristic argument in the introduction.

While from the above it is suggestive to take the this value, there is not proof that we *have to* take it. This matter is currently under investigation. However, in the paper we proceed with this boundary value.

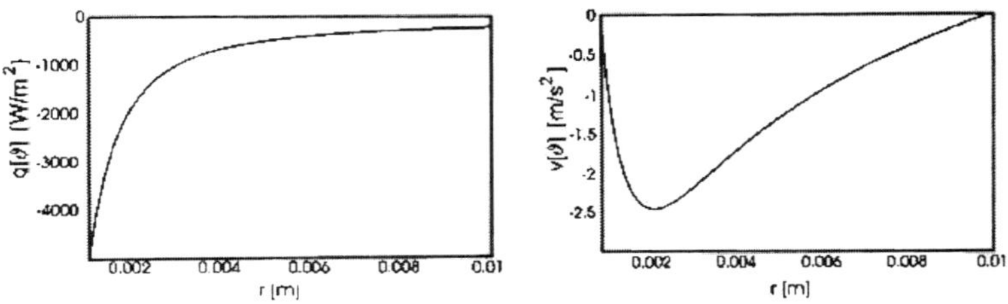

Figure 3. Tangential component of $q\,[\vartheta]$(left) and $v\,[\vartheta]$(right).

3.2 Solutions

Figures 2, 3, 4 represent the complete solution of the system (6) for the discussed boundary data:

Figure 2 shows the fields $\theta\,(r)$ and $p\,(r)$, the two functions that are also non-constant if we consider the gas as a Navier-Stokes-Fourier fluid. In order to exhibit the effect of 13 moments as opposed to Navier-Stokes-Fourier we have printed both in the graphs of Figure 2.

The next set of graphs, in Figure 3, represent two effects due to the inertial terms: non-zero functions for $q\,[\vartheta]$ and $v\,[\vartheta]$. We conclude that the heat flux has a tangential component, i.e. a component perpendicular to the temperature gradient. This phenomenon was already predicted by Müller[5] by an argument from the kinetic theory of gases. Also we conclude that there is a tangential component of the velocity, so that the gas does not rotate rigidly with the cylinders except at the boundaries, because there is not slip. To our knowledge this phenomenon has never been observed nor predicted from calculations.

The remaining results, Figure 4, concern the deviatoric components of the pressure tensor which all vanish in a Navier-Stokes-Fourier theory of the rotating heat-conducting gas. Here, in the 13 moment theory, they do not vanish. Note that there is no obvious relation between the shear pressure $\sigma\,[r\vartheta]$ and the velocity gradient $\frac{dv[\vartheta]}{dr}$.

We have unequal normal stresses, – much like in a non-Newtonian fluid, albeit for an entirely different reason. Of particular interest is the normal pressure $p + \sigma\,[zz]$ which is quite non-uniform mostly due to the centrifugal force which tends to drive the gas upwards at the outher cylinder. We conclude from this that our semi-inverse ansatz, which implies $v\,[z] = 0$, can only be realized by applying a stress field on the horizontal surfaces. Note that in the neighbourhood of the *inner cylinder* the normal pressure turns sharply up;

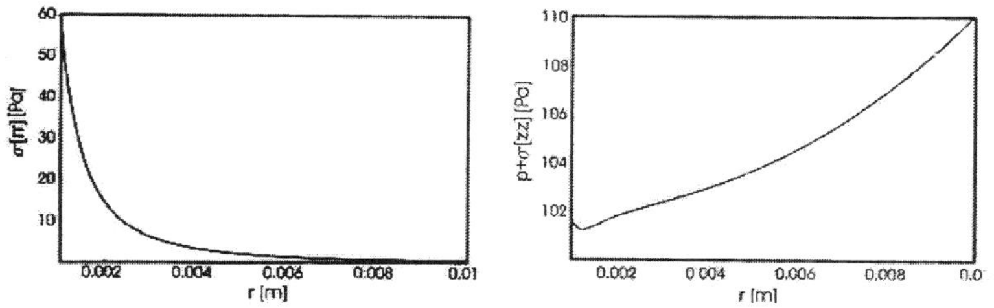

Figure 4. Normal stresses $\sigma\,[rr]$ (left) and $p+\sigma\,[zz]$ (right).

once again the analogy to non-Newtonian fluids, – and in particular to the Weissenberg effect, – catches the eye.

References

1. I. Müller and T. Ruggeri Rational Extended Thermodynamics (2nd ed.) Springer Tracts in Natural Philosophy **37**, New York (1998).
2. P. L. Bhatnagar, E.P. Gross, M.A. Krook, Phys.Rev. **94**, (1954).
3. H. Grad, Principles of the Kinetic Theory of Gases. Handbuch der Physik **XII**, Springer Heidelberg (1958).
4. I. Müller, T. Ruggeri, Stationary Heat Conduction in Radially Symmetric Situations – An Application of Extended Thermodynamics. Submitted.
5. I. Müller, Arc.Rat.Mech.An. **45**, (1972).

GRAD'S CLOSURE IN THE KINETIC THEORY OF A CHEMICALLY REACTING GAS

MARZIA BISI, MARIA GROPPI, GIAMPIERO SPIGA

Dipartimento di Matematica, Università di Parma
Via D'Azeglio 85, 43100 Parma, Italy

A Grad's closure strategy is applied to the moment equations relevant to a four species gas mixture undergoing a bimolecular chemical reaction. The resulting robust approximation is shown to satisfy an entropy inequality. Numerical results are presented, taking into account the chemical activation energies.

1 Introduction

The kinetic description of chemical reactions has stimulated an increasing interest in the last half century (see for instance the recent book [1] and the survey paper [2]). We will deal here with a simple model for a bimolecular chemical reaction [3], hoping to be able to extend our treatment to more complicated situations, like those considered in [4] and [5]. The problem has been already tackled by a semi–continuous approach [6] and by the reactive Euler equations [7]. Recently a Grad's 13-moment expansion has been used [8], restricted for simplicity to a kind of Maxwell molecules assumption.

In this paper we aim at proceeding further by using more realistic cross sections for the chemical reaction. We shall resort to the model proposed in [9], which accounts for the threshold effects in the exothermic reaction due to the activation energy for the chemical process. This determines also the cross section of the endothermic reaction by means of the principle of micro–reversibility. The whole algorithm is sketched in Section 2, where moments of the chemical collision integrals are cast in terms of incomplete gamma functions and it is shown that this approximation retains the dissipativity of the kinetic equations (a suitable functional is non-increasing following a solution). A sample of numerical results is presented in Section 3, in space homogeneous conditions and for randomly chosen (dimensionless) values of the physical quantities, in order to illustrate the main qualitative features of the model and the role played by the physical parameters.

2 Grad's equations

We shall not enter the details of the quite elaborate procedure, but rather refer the interested reader to [8]. It suffices to recall here that we are studying the bimolecular chemical reaction $A^1 + A^2 \rightleftharpoons A^3 + A^4$, where the direct reaction is conventionally assumed to be endothermic with a gap $\Delta E = E^3 + E^4 - E^1 - E^2 > 0$ in the energies of chemical link. A superscript $s = 1, 2, 3, 4$ is used to label species, μ^{sr} denotes the reduced mass for the pair (s, r), and all possible elastic collisions are allowed. Collision integrals for the transition $(s, r) \rightarrow (h, k)$ are expressed in terms of the differential cross section σ_{sr}^{hk}, and we may define for each species the 13 first power moments (density n^s, drift velocity $\mathbf{u}^s$, temperature T^s, viscosity tensor $\mathbf{p}^s$, heat flux $\mathbf{q}^s$) of the one particle distribution function f^s. The global distribution function is the four dimensional vector $\mathbf{f}$ of the f^s, and the global macroscopic fields are proper averages of the previous 52 moments. The Grad's approximation to f^s involves its first 13 moments (see eq.(16) of [8]), and is expressed in terms of the peculiar velocities $\mathbf{c}^s = \mathbf{v} - \mathbf{u}^s$, whereas $\mathbf{c}$ is used to mean $\mathbf{v} - \mathbf{u}$, where $\mathbf{u}$ is the global mass velocity. The governing equations are obtained by taking suitable moments of the kinetic equations and using the above approximation, along the lines of [8]. We make use of the same Maxwell-like elastic cross section, but we resort here to the isotropic cross section for exothermic reaction [9]

$$\sigma_{34}^{12}(g) = \frac{1}{4\pi} \frac{\nu_{34}^{12}}{g} \left(1 - \frac{\chi^2}{g^2}\right) U(g^2 - \chi^2) \tag{1}$$

where g is the relative speed, ν_{34}^{12} a positive constant, U the unit step function, and χ the threshold to be overcome in order to involve the chemical bonds in the interaction. The 52 partial differential equations corresponding to Grad's 13-moment expansion read as:

$$\frac{\partial n^s}{\partial t} + \frac{\partial (n^s u_i^s)}{\partial x_i} = \int_{\mathbb{R}^3} J^s[\mathbf{f}_M]\, d^3\mathbf{v} = Q^s, \tag{2}$$

$$\begin{aligned} m^s n^s \left(\frac{\partial u_i^s}{\partial t} + u_j^s \frac{\partial u_i^s}{\partial x_j}\right) + \frac{\partial (n^s K T^s)}{\partial x_i} + \frac{\partial p_{ij}^s}{\partial x_j} = \\ = n^s \sum_{r=1}^{4} \nu_1^{sr} \mu^{sr} n^r (u_i^r - u_i^s) - m^s (u_i^s - u_i)\, Q^s, \end{aligned} \tag{3}$$

$$\begin{aligned} &\frac{3}{2} n^s \left(\frac{\partial (KT^s)}{\partial t} + u_i^s \frac{\partial (KT^s)}{\partial x_i}\right) + n^s K T^s \frac{\partial u_i^s}{\partial x_i} + p_{ij}^s \frac{\partial u_j^s}{\partial x_i} + \frac{\partial q_i^s}{\partial x_i} = \\ &= 2 n^s \sum_{r=1}^{4} \nu_1^{sr} \frac{\mu^{sr}}{m^s + m^r} n^r \left[\frac{3}{2} K (T^r - T^s) + \frac{1}{2} m^r (\mathbf{u}^r - \mathbf{u}^s)^2\right] \\ &+ \frac{1}{2} m^s \int_{\mathbb{R}^3} c^2 J^s[\mathbf{f}_M]\, d^3\mathbf{v} + \left(\frac{1}{2} m^s (\mathbf{u}^s - \mathbf{u})^2 - \frac{3}{2} K T^s\right) Q^s, \end{aligned} \tag{4}$$

$$\frac{\partial p^s_{ij}}{\partial t}+\frac{\partial}{\partial x_k}(u^s_k\,p^s_{ij})+p^s_{ik}\frac{\partial u^s_j}{\partial x_k}+p^s_{jk}\frac{\partial u^s_i}{\partial x_k}-\frac{2}{3}\delta_{ij}p^s_{kl}\frac{\partial u^s_l}{\partial x_k}$$
$$+n^sKT^s\left(\frac{\partial u^s_i}{\partial x_j}+\frac{\partial u^s_j}{\partial x_i}-\frac{2}{3}\delta_{ij}\frac{\partial u^s_k}{\partial x_k}\right)+\frac{2}{5}\left(\frac{\partial q^s_i}{\partial x_j}+\frac{\partial q^s_j}{\partial x_i}-\frac{2}{3}\delta_{ij}\frac{\partial q^s_k}{\partial x_k}\right)=$$
$$=2\sum_{r=1}^{4}\frac{\nu_1^{sr}\mu^{sr}}{m^s+m^r}\Big\{n^sn^rm^r\Big[(u^r_i-u^s_i)(u^r_j-u^s_j)-\frac{1}{3}\delta_{ij}(\mathbf{u}^r-\mathbf{u}^s)^2\Big]$$
$$+n^sp^r_{ij}-n^rp^s_{ij}\Big\}-\sum_{r=1}^{4}\frac{3\,\nu_2^{sr}m^r}{2(m^s+m^r)^2}\Big\{m^sn^sm^rn^r\Big[(u^r_i-u^s_i)(u^r_j-u^s_j)-\frac{1}{3}\delta_{ij}(\mathbf{u}^r-\mathbf{u}^s)^2\Big]$$
$$+m^sn^sp^r_{ij}+m^rn^rp^s_{ij}\Big\}+m^s\Big[(u^s_i-u_i)(u^s_j-u_j)-\frac{1}{3}\delta_{ij}(\mathbf{u}^s-\mathbf{u})^2\Big]Q^s, \tag{5}$$

$$\frac{\partial q^s_i}{\partial t}+\frac{\partial}{\partial x_j}(u^s_j\,q^s_i)+\frac{7}{5}\,q^s_j\frac{\partial u^s_i}{\partial x_j}+\frac{2}{5}\,q^s_j\frac{\partial u^s_j}{\partial x_i}+\frac{2}{5}\,q^s_i\frac{\partial u^s_j}{\partial x_j}$$
$$+\frac{5}{2}\,n^sKT^s\frac{\partial}{\partial x_i}\left(\frac{KT^s}{m^s}\right)+\frac{5}{2}p^s_{ij}\frac{\partial}{\partial x_j}\left(\frac{KT^s}{m^s}\right)+n^sKT^s\frac{\partial}{\partial x_j}\left(\frac{p^s_{ij}}{m^sn^s}\right)-\frac{p^s_{ij}}{m^sn^s}\frac{\partial p^s_{kj}}{\partial x_k}=$$
$$=-\frac{5}{2}\frac{n^s}{m^s}KT^s\sum_{r=1}^{4}\nu_1^{sr}\mu^{sr}n^r(u^r_i-u^s_i)-\frac{1}{m^s}\,p^s_{ij}\sum_{r=1}^{4}\nu_1^{sr}\mu^{sr}n^r(u^r_j-u^s_j)$$
$$+\sum_{r=1}^{4}\frac{m^rn^r}{(m^s+m^r)^3}\Big\{\beta_1^{sr}q^s_i+\frac{m^sn^s}{m^rn^r}\beta_4^{sr}q^r_i+(u^r_j-u^s_j)\Big(\frac{1}{2}\,\beta_2^{sr}p^s_{ij}+\frac{m^sn^s}{m^rn^r}\beta_4^{sr}p^r_{ij}\Big)$$
$$+\frac{1}{2}(u^r_i-u^s_i)\Big[(\beta_2^{sr}+3\,\beta_3^{sr})\,n^sKT^s+5\,\frac{m^sn^s}{m^rn^r}\beta_4^{sr}\,n^rKT^r+m^sn^s\beta_4^{sr}(\mathbf{u}^r-\mathbf{u}^s)^2\Big]\Big\}$$
$$-\frac{5}{6}m^s(u^s_i-u_i)\int_{\mathbb{R}^3}c^2J^s[\mathbf{f}_M]\,d^3\mathbf{v}+(u^s_j-u_j)\Big[\frac{5}{2}\delta_{ij}KT^s+\frac{p^s_{ij}}{n^s}-\frac{1}{2}\delta_{ij}m^s(\mathbf{u}^s-\mathbf{u})^2\Big]Q^s. \tag{6}$$

The symbols which have not been defined are the same as in [8], and will not be repeated here. In the two remaining integrals of the chemical collision terms J^s the argument $\mathbf{f}_M$ means that such terms have to be evaluated, in our approximation, at the Maxwellians corresponding to elastic equilibrium, namely $M^s(\mathbf{v})=n^s(m^s/2\pi KT)^{\frac{3}{2}}\exp(-m^sc^2/2KT)$. We get, with $\lambda^1=\lambda^2=-\lambda^3=-\lambda^4=1$ and $M=m^1+m^2=m^3+m^4$

$$Q^s=-\lambda^s\left[n^1n^2\left(\frac{\mu^{34}}{\mu^{12}}\right)^{\frac{3}{2}}\mathrm{e}^{-\frac{\Delta E}{KT}}-n^3n^4\right]$$
$$\cdot\frac{2}{\sqrt{\pi}}\nu_{34}^{12}\left[\Gamma\left(\frac{3}{2},\frac{\mu^{34}\chi^2}{2KT}\right)-\frac{\mu^{34}\chi^2}{2KT}\Gamma\left(\frac{1}{2},\frac{\mu^{34}\chi^2}{2KT}\right)\right], \tag{7}$$

$$\int_{\mathbb{R}^3} c^2 J^s[\mathbf{f}_M]\, d^3\mathbf{v} =$$
$$= -\lambda^s \left[n^1 n^2 \left(\frac{\mu^{34}}{\mu^{12}}\right)^{\frac{3}{2}} \mathrm{e}^{-\frac{\Delta E}{KT}} - n^3 n^4 \right] \frac{2}{\sqrt{\pi}} \nu_{34}^{12} \left\{ 2KT \frac{M - m^s}{M\, m^s} \Gamma\left(\frac{5}{2}, \frac{\mu^{34}\chi^2}{2KT}\right) \right.$$
$$+ \left[\frac{3KT}{M} + \frac{M - m^s}{M\, m^s} \Big((1+\lambda^s)\Delta E - \mu^{34}\chi^2 \Big) \right] \Gamma\left(\frac{3}{2}, \frac{\mu^{34}\chi^2}{2KT}\right)$$
$$\left. - \left[\frac{3\mu^{34}\chi^2}{2M} + \frac{M - m^s}{M\, m^s} (1+\lambda^s) \frac{\Delta E}{2KT} \mu^{34}\chi^2 \right] \Gamma\left(\frac{1}{2}, \frac{\mu^{34}\chi^2}{2KT}\right) \right\}. \tag{8}$$

The robustness of the model equations (2)-(6) has been already discussed in [8], concerning exact fulfilment of the macroscopic conservation equations and correct projection of collision equilibria. Another preserved essential feature is irreversibility, since an entropy inequality holds. Consider, in space homogeneous conditions, the functional (essentially, the opposite of entropy)

$$H = -\frac{3}{2} n \log KT + \sum_{s=1}^{4} n^s \log \left[n^s / (m^s)^{\frac{3}{2}} \right] \tag{9}$$

where $n = \sum_{s=1}^{4} n^s$ is a first integral, while T can be expressed in terms of the T^s, $\mathbf{u}^s$, and n^s. Taking time derivative of H along a solution leads to

$$\dot{H} = -\frac{2}{\sqrt{\pi}} \nu_{34}^{12} \left[\Gamma\left(\frac{3}{2}, \frac{\mu^{34}\chi^2}{2KT}\right) - \frac{\mu^{34}\chi^2}{2KT} \Gamma\left(\frac{1}{2}, \frac{\mu^{34}\chi^2}{2KT}\right) \right]$$
$$\cdot\, n^3 n^4 \left[\frac{n^1 n^2}{n^3 n^4} \left(\frac{\mu^{34}}{\mu^{12}}\right)^{\frac{3}{2}} \mathrm{e}^{-\frac{\Delta E}{KT}} - 1 \right] \log \left(\frac{n^1 n^2}{n^3 n^4} \left(\frac{\mu^{34}}{\mu^{12}}\right)^{\frac{3}{2}} \mathrm{e}^{-\frac{\Delta E}{KT}} \right) \tag{10}$$

so that the familiar $(x-1)\log x$ function comes out, and $\dot{H}$ has a definite sign, which is negative since the square bracket containing the Γ functions is a monotonically decreasing function of χ, ranging from $\sqrt{\pi}/2$ (at $\chi = 0$) to zero (when $\chi \to +\infty$). Therefore $\dot{H} \leq 0$, with $\dot{H} = 0$ if and only if the argument of the logarithm is unity, which corresponds to the mass action law [8]

$$\frac{n^1 n^2}{n^3 n^4} = \left(\frac{\mu^{12}}{\mu^{34}}\right)^{\frac{3}{2}} \exp\left(\frac{\Delta E}{KT}\right). \tag{11}$$

3 Numerical results

We report in this Section on the response of the above model equations to varying input data. Calculations are restricted for simplicity to space homogeneous conditions, and illustrate specifically the effects of collisions. They correspond to a random choice of the physical parameters of the collision integrals, in dimensionless form with arbitrary units. For brevity, the numerical values are given only for the quantities which are varied in the simulations.

Figure 1 describes a reference case in which the energy gap is $\Delta E = 5000$, the activation threshold is $\chi = 10$, and masses of molecules are given by the string (58.5, 18, 40, 36.5). Collision parameters are chosen in such a way that the elastic relaxation time is much shorter than the chemical one, in agreement with our model. The first significant change with respect to previous calculations [8], due to use of cross section (1), is the occurrence of a quite long induction time followed by a strong and fast ignition of the chemical reaction [9], as shown by the plots of the densities n^s and of temperature T on the long time scale. During induction, on the shorter mechanical time scale, we observe convergence to the elastic collision equilibrium, with equalization of all velocities to their final common value, and equalization to zero of viscosity tensors and heat fluxes (not shown in the picture), whereas temperatures get also equalized on this short scale, but then undergo the variations implied by the reaction.

Figure 2 differs from the reference case only in that the threshold χ is decreased from 10 to 7. As expected, the lower threshold renders the exothermic process easier and then makes more energy available for an earlier and smoother transition of the number densities towards their final equilibrium values. The trend turns out indeed to be highly nonlinear, as shown by the first plots in the two Figures. The second plot shows another interesting feature in the evolution of temperatures. After equalization on the mechanical scale, they separate again at the start of a significant reaction rate. It is in fact physically clear, and quantitatively described by the balance equations (4), that the general collision contribution depends on the species via m^s and n^s, and it is nonzero when all temperatures and velocities are equal but chemical equilibrium is not established. This effect is present also in the reference case, but it can not be seen just because of graphical resolution.

Finally, Figure 3 differs from the reference case only for the values of masses, which are now (10.8814, 65.6186, 60, 16.5), and are such that M and μ^{12}/μ^{34} remain unchanged. Mass action law is left invariant, but we observe a dramatic reduction in the induction period. The reason is that now m^2 and m^3 are considerably larger than m^1 and m^4. In fact, at elastic equilibrium all species share the same temperature, so that their typical speeds are proportional to $(m^s)^{-\frac{1}{2}}$, and thus the typical values of the relative speed g in a (3,4) encounter are now much higher than in the reference case. Bearing (1) in mind, it is then obvious that the chemical reaction is much more easily trig-

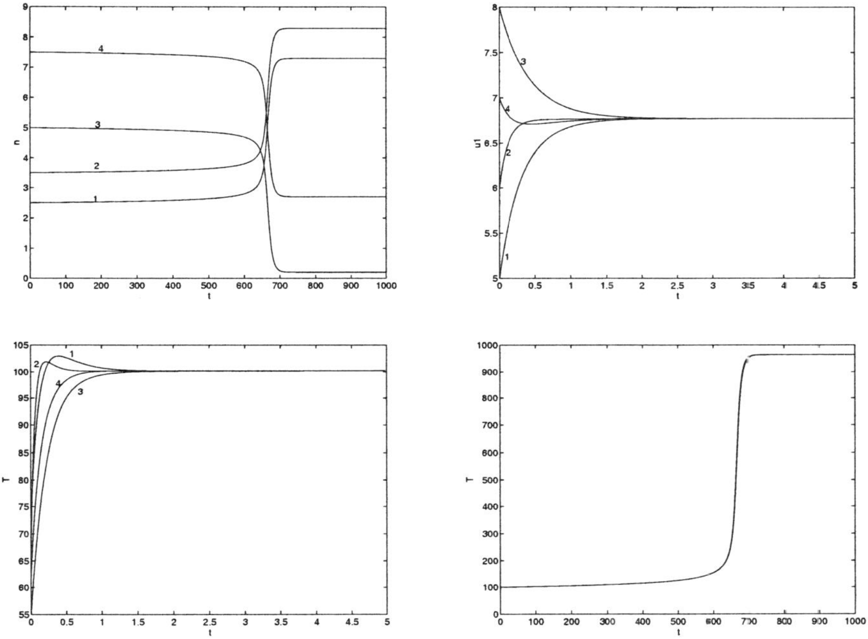

Figure 1.

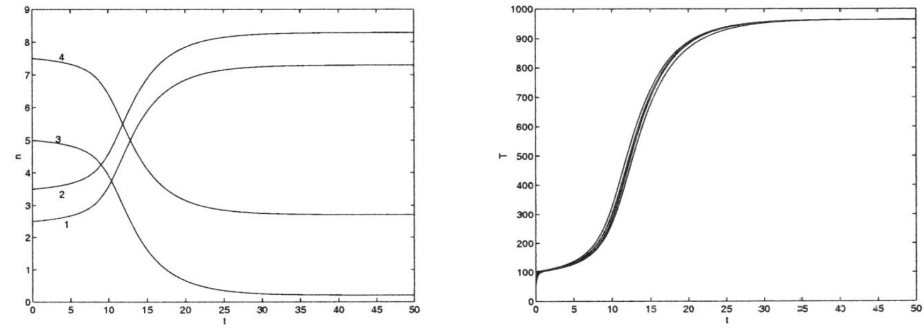

Figure 2.

gered in this case. The evolution for temperatures shows the same splitting effect at the chemical ignition already discussed.

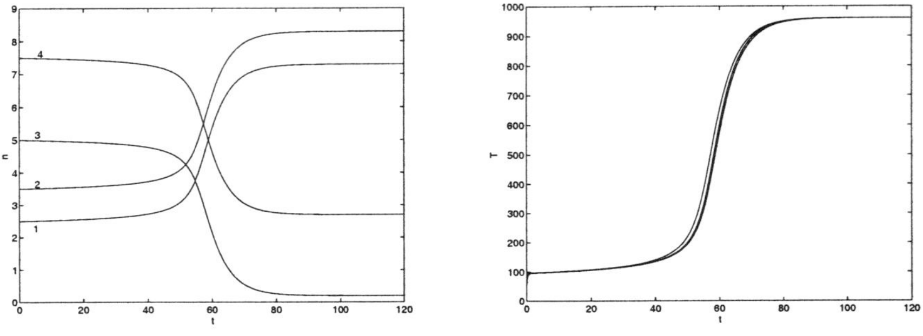

Figure 3.

Acknowledgements

This work was performed in the frame of the activities sponsored by MURST (Project "Mathematical Problems of Kinetic Theories"), by CNR, by GNFM, and by the University of Parma (Italy), and by the European TMR Network "Asymptotic Methods in Kinetic Theory". Enlightening discussions with F. Schürrer (Technical University of Graz) are also gratefully acknowledged.

References

1. C. Cercignani, *Rarefied Gas Dynamics. From Basic Concepts to Actual Calculations*, Cambridge University Press, Cambridge (U.K.), 2000.
2. W. Koller, F. Schürrer, *Riv. Mat. Univ. Parma* (6) **4*** (2001), 109–169.
3. A. Rossani, G. Spiga, *Phys. A* **272** (1999) 563–573.
4. M. Groppi, G. Spiga, *J. Math. Chem.* **26** (1999) 197-219.
5. M. Groppi, A. Rossani, G. Spiga, *J. Phys. A: Math. Gen.* **33** (2000) 8819-8833.
6. M. Groppi, W. Koller, *Z. Angew. Math. Phys.*, in press, preprint No. 263, Dept. Math., Univ. Parma (2001).
7. M. Groppi, G. Spiga, *Transport Theory Statist. Phys.* **30** (2001) 305-324.
8. M. Bisi, M. Groppi, G. Spiga, *Contin. Mech. Thermodyn.*, in press, preprint No. 262, Dept. Math., Univ. Parma (2001).
9. P. Griehsnig, F. Schürrer, G. Kügerl, in *Rarefied Gas Dynamics: Theory and Simulations*, B. D. Shizgal, D. P. Weaver Eds., Vol. 159, AIAA, Washington (1992) 581-589.

ELECTRON FLOW STABILITY IN BULK SILICON IN THE LIMIT OF SMALL ELECTRIC FIELD

A. M. BLOKHIN

Institute of Mathematics, Novosibirsk 630090, Russia

R. S. BUSHMANOV

Novosibirsk State University, Novosibirsk 630090, Russia

V. ROMANO

Dipartimento di Matematica e Informatica, Università di Catania, viale A. Doria, 6, 95125 Catania, Italy

E-mail: romano@dmi.unict.it, web: www.dmi.unict.it/~romano

Asymptotic stability of the homogeneous stationary solution to the hydrodynamical model of charge transport in semiconductors based on the maximum entropy principle (MEP)[1,2], is proved in the linear case in the limit of small electric fields.

1 Introduction

As known, in the hierarchy of the macroscopic models for the description of charge transport in semiconductors, beyond the drift-diffusion equations, one finds the hydrodynamical models which are obtained from the infinite set of moment equations of the Boltzmann Transport Equation (BTE) by a suitable truncation procedure. It is well-known too that moment systems require a closure assumption in order to lead to closed system of evolution equations. In [1,2] by using the maximum entropy ansatz for the closure, explicit constitutive relations for fluxes and production terms have been obtained, by including also the effects of nonparabolicity. For a complete review of the problem the interested reader is referred to [3].

In the present paper the stability properties of the solutions of such a model are investigated in the parabolic band approximation of the conduction band. In particular we will consider a one dimensional problem for bulk semiconductors, i.e. when the doping density is uniform, and prove that the stationary solutions are asymptotically stable for small electric fields.

2 Basic equations and formulation of the problem

The MEP model of semiconductors is represented by the following set of balance equations[1,2]

$$\frac{\partial n}{\partial t} + \frac{\partial (nV^i)}{\partial x^i} = 0, \tag{2.1}$$

$$\frac{\partial (nP^i)}{\partial t} + \frac{\partial (nU^{ij})}{\partial x^j} + neE^i = n\mathbb{C}_P^i, \; i = 1,2,3, \tag{2.2}$$

$$\frac{\partial (nW)}{\partial t} + \frac{\partial (nS^i)}{\partial x^i} + neV_iE^i = n\mathbb{C}_W, \tag{2.3}$$

$$\frac{\partial (nS^i)}{\partial t} + \frac{\partial (nF^{ij})}{\partial x^j} + neE_jG^{ij} = n\mathbb{C}_W^i, \; i = 1,2,3, \tag{2.4}$$

where n is the electron density, V^i are the components of the average electron velocity, $P^i = m^*V^i$ are the components of the average crystal momentum, m^* is the effective electron mass (for silicon $m^* = 0.32m_e$ with m_e mass of one electron in the vacuum), U^{ij} is the flow of crystal momentum, W is the average electron energy, e is the absolute value of the electron charge, E^i are the components of the electric field, $\mathbb{C}_P^i$ are the components of the production of the crystal momentum balance equations, S^i are the components of the energy flux, $\mathbb{C}_W$ is the production of the energy balance equation, F^{ij} is the flux of the energy flux, $\mathbb{C}_W^i$ are the components of the production of the energy flux balance equations. In the parabolic case $U^{ij} = \frac{2}{3}W\delta^{ij}$, $F^{ij} = \frac{9}{10}\frac{W^2}{m^*}\delta^{ij}$, $G^{ij} = \frac{5}{3}\frac{W}{m^*}\delta^{ij}$, $i = 1,2,3$. The productions can be also analytically expressed in terms of the Bessel functions of second kind. However the explicit expressions of $\mathbb{C}_P^i$, $\mathbb{C}_W$, $\mathbb{C}_W^i$ are not relevant for the aims of this article and therefore we do no report them.The interested reader may see [2].

Since the electric field is related to the electric potential Φ as $E^i = -\frac{\partial \Phi}{\partial x^i}, i = 1,2,3$, the system (2.1)–(2.4) is coupled with the Poisson equation

$$\epsilon \Delta \Phi = e(n - N) \tag{2.5}$$

where N is the doping density and ϵ is the dielectric constant.

Let us introduce the adimensional variables

$$R = \frac{n}{N+}, u^i = \frac{V^i}{\mathbb{C}_0}, E = \frac{W}{m^*\mathbb{C}_0^2}, q^i = \frac{S^i}{m^*\mathbb{C}_0^3}, \varphi = \frac{\Phi e}{m^*\mathbb{C}_0^2}, \tau = \frac{\mathbb{C}_0 t}{L}, \tilde{x}^i = \frac{x^i}{L}.$$

N^+ is a typical doping density, $\mathbb{C}_0 = \sqrt{\frac{K_B T_0}{m*}}$ is a sort of sound speed, T_0 is the lattice temperature, K_B is the Boltzmann constant, L is the length of the device. Further we will write again x^i instead of $\tilde{x}^i$.

For one dimensional problems the system of balance equations reads

$$U_\tau + \mathcal{B}U_x = F(Q,U), \quad \varepsilon^2 \varphi_{xx} = R - \rho. \tag{2.6}$$

$$\text{Here}\quad U=\begin{pmatrix}R\\ Ru\\ RE\\ Rq\end{pmatrix}=\begin{pmatrix}R\\ J\\ \mathcal{W}\\ \mathcal{N}\end{pmatrix},\quad \mathcal{B}=\begin{pmatrix}0&1&0&0\\ 0&0&\frac{2}{3}&0\\ 0&0&0&1\\ -\frac{10}{9}E^2&0&\frac{20}{9}E&0\end{pmatrix},$$

$$F=\begin{pmatrix}0\\ RQ+R\widetilde{\mathbb{C}}_P^1\\ JQ+R\widetilde{\mathbb{C}}_W\\ \frac{5}{3}\mathcal{W}Q+R\widetilde{\mathbb{C}}_W^1\end{pmatrix},\quad \begin{pmatrix}R\widetilde{\mathbb{C}}_P^1\\ R\widetilde{\mathbb{C}}_W^1\end{pmatrix}=\widetilde{\mathbb{C}}(E)\begin{pmatrix}J\\ \mathcal{N}\end{pmatrix},\widetilde{\mathbb{C}}=\begin{pmatrix}\tilde{c}_{11}&\tilde{c}_{12}\\ \tilde{c}_{21}&\tilde{c}_{22}\end{pmatrix}$$

$$Q=\varphi_x,\quad \rho=\frac{N}{N^+},\quad \beta=\frac{1}{\varepsilon^2}=\frac{e^2L^2N^+}{\epsilon m^*\mathbb{C}_0^2},$$

$$\widetilde{\mathbb{C}}_P^i=\mathbb{C}_P^i\frac{L}{n^*\mathbb{C}_0^2},\quad \widetilde{\mathbb{C}}_W=\mathbb{C}_W\frac{L}{n^*\mathbb{C}_0^3},\quad \widetilde{\mathbb{C}}_W^i=\mathbb{C}_W^i\frac{L}{n^*\mathbb{C}_0^4}$$

In this paper we will consider the semiconductor of infinite legth and moreover we assume an uniform doping density, i.e. $\rho(x)\equiv 1$, for each $x\in\mathbb{R}$. The coefficient $\widetilde{\mathbb{C}}_W$ must be considered as function of ζ, where $\zeta=\frac{3\xi}{4E}$, $\xi=\frac{\hbar\omega_{np}}{K_BT_0}$, $\hbar=\frac{h}{2\pi}$, with h the Planck constant and $\hbar\omega_{np}$ the optical phonon energy (for the explicit expression of the coefficients see [2] and for a formulation more usefull for the present problem see [5]).

The system (2.6) is hyperbolic in the physically relevant region of the phase space, as proved for the general case in [4]. As matter of fact the eigenvalues of the matrix $\mathcal{B}$ are $\lambda_{1,2}=\pm\left(\frac{10+2\sqrt{10}}{9}E\right)^{1/2}$, $\lambda_{3,4}=\pm\left(\frac{10-2\sqrt{10}}{9}E\right)^{1/2}$, provided that $E>0$.

The system of equations (2.6) has the stationary constant solution, that describes a uniform electron flow driven by a constant electric field,

$$\left.\begin{array}{l}R(\tau,x)=\widehat{R}=1,\quad u(\tau,x)=\hat{u},\quad E(\tau,x)=\widehat{E},\\ q(\tau,x)=\hat{q},\quad \varphi(\tau,x)=\hat{\varphi}=A+Bx,\quad Q(\tau,x)=\widehat{Q}=E,\end{array}\right\}\tag{2.7}$$

where $\hat{u}$ (>0), $\widehat{E}$ (>1), $\hat{q}$, A, B (>0) are constants and B represents the bias voltage across the device. $\hat{u}$, $\widehat{E}$, $\hat{q}$ are found from the next system of algebraic equations once the constant B (>0) is assigned

$$\widetilde{\mathbb{C}}_W(\hat{\zeta})+B\hat{u}=0,\quad \widetilde{\mathbb{C}}(\widehat{E})\begin{pmatrix}\hat{u}\\ \hat{q}\end{pmatrix}+B\begin{pmatrix}1\\ \frac{5}{3}\widehat{E}\end{pmatrix}=\begin{pmatrix}0\\ 0\end{pmatrix},\quad \text{with } \hat{\zeta}=\frac{3\xi}{4\widehat{E}}.\tag{2.8}$$

Note that for $B = 0$ the solution of the previous equation is the global thermodynamical equilibrium $\widehat{E} = \frac{3}{2}, \hat{u} = \hat{q} = 0$. Consequently for small B, that is in the limit of small electric fields, we may take $\widehat{E} = \frac{3}{2}, \quad \hat{u} = B\, f\left(\frac{3}{2}\right), \quad \hat{q} = B\, g\left(\frac{3}{2}\right)$ with good exactness.

Our problem is to investigate the linear stability of the solution given by the system (2.8).

By linearizing the original quasilinear problem (2.6) around the basic solution, we get the following linear system where the small perturbations of the sought values are denoted with same symbols of the unperturbed variables:

$$U_\tau + \widehat{\mathcal{B}} U_x = \mathcal{F}(Q, U), \tag{2.9}$$

$$\varepsilon^2 Q_s = R. \tag{2.10}$$

Here $\widehat{\mathcal{B}} = \mathcal{B}\left(\widehat{E}\right), \quad \mathcal{F} = \begin{pmatrix} 0 \\ Q + r_1 R + c_{11} J + w_1 \mathcal{W} + c_{12} \mathcal{N} \\ \hat{u} Q + r_2 R + BJ - \hat{\mu} \mathcal{W} \\ \frac{5}{2} Q - r_3 R + c_{21} J + w_2 \mathcal{W} + c_{22} \mathcal{N} \end{pmatrix},$

$$r_1 = B - \frac{3}{2} w_1, \quad w_1 = \hat{u} \tilde{c}'_{11} \left(\frac{3}{2}\right) + \hat{q} \tilde{c}'_{12} \left(\frac{3}{2}\right),$$

$$r_2 = \hat{\zeta} \widetilde{\mathbb{C}}'_W(\hat{\zeta}), \quad \hat{\mu} = \frac{\xi}{3} \widetilde{\mathbb{C}}'_W(\hat{\zeta}), \quad \widetilde{\mathbb{C}}'_W = \frac{d\widetilde{\mathbb{C}}_W}{d\zeta},$$

$$r_3 = \frac{3}{2} \left[\hat{u} \tilde{c}'_{21} \left(\frac{3}{2}\right) + \hat{q} \tilde{c}'_{22} \left(\frac{3}{2}\right) \right], \quad w_2 = \frac{5}{3} B + \frac{2}{3} r_3,$$

$$c_{ij} = \tilde{c}_{ij} \left(\frac{3}{2}\right), \quad \tilde{c}'_{ij} = \frac{d\tilde{c}_{ij}}{dE}, \ i, j = 1, 2, \quad \hat{\zeta} = \frac{\xi}{2}.$$

It is convenient to introduce a vector V such that

$$V = \begin{pmatrix} R \\ J \\ P \\ \Theta \end{pmatrix} = \begin{pmatrix} 1 & 0 & 0 & 0 \\ 0 & 1 & 0 & 0 \\ -1 & 0 & \frac{2}{3} & 0 \\ 0 & -\frac{5}{2} & 0 & 1 \end{pmatrix} U. \tag{2.11}$$

The system (2.9) in view of (2.11) transforms into

$$V_\tau + \widetilde{\mathcal{B}} V_x = \widetilde{F}, \tag{2.12}$$

where

$$\widetilde{\mathcal{B}} = \begin{pmatrix} 0 & 1 & 0 & 0 \\ 1 & 0 & 1 & 0 \\ 0 & \frac{2}{3} & 0 & \frac{2}{3} \\ 0 & 0 & \frac{5}{2} & 0 \end{pmatrix}, \quad \widetilde{F} = Q \begin{pmatrix} 0 \\ 1 \\ \frac{2}{3}\hat{u} \\ 0 \end{pmatrix} + \begin{pmatrix} 0 & 0 & 0 & 0 \\ B & \mu_{11} & \frac{3}{2}w_1 & \mu_{12} \\ 0 & \frac{2}{3}B & -\hat{\mu} & 0 \\ 0 & \mu_{21} & \frac{3}{2}\bar{w}_2 & \mu_{22} \end{pmatrix} V,$$

$$\mu_{11} = c_{11} + \frac{5}{2}c_{12}, \quad \mu_{12} = c_{12}, \quad \mu_{21} = c_{21} - \frac{5}{2}c_{11} + \frac{5}{2}\mu_{22},$$

$$\mu_{22} = c_{22} - \frac{5}{2}c_{12}, \quad \bar{w}_2 = w_2 - \frac{5}{2}w_1.$$

Let us denote with $V_0(x), Q_0(x)$ the initial data for (2.12), (2.10). Besides $R_0(x) = \varepsilon^2 Q_0'(x)$ in view of (2.10).

Note that we obtain from the first equation of the system (2.12) and the equation (2.10) the next relation

$$-\varepsilon^2 Q_\tau = J. \tag{2.13}$$

By taking into account (2.10), (2.13), we rewrite the system (2.12) as

$$\left.\begin{aligned} &Q_{\tau\tau} - Q_{xx} - \mathcal{P}_x + \beta Q + BQ_x - \mu_{11}Q_\tau + \tfrac{3}{2}w_1\mathcal{P} + \mu_{12}\vartheta = 0, \\ &\tfrac{3}{2}\mathcal{P}_\tau - Q_{\tau x} + \vartheta_x - \hat{u}\beta Q + BQ_\tau + \tfrac{3}{2}\hat{\mu}\mathcal{P} = 0, \\ &\tfrac{2}{5}\vartheta_\tau + \mathcal{P}_x + \tfrac{2}{5}\mu_{21}Q_\tau - \tfrac{3}{5}\bar{w}_2\mathcal{P} - \tfrac{2}{5}\mu_{22}\vartheta = 0. \end{aligned}\right\} \tag{2.14}$$

Here $\mathcal{P} = \beta P$, $\vartheta = \beta\Theta$.

3 Asymptotic stability

First we proceed to the derivation of some preliminary estimates. If we multiply the first equation of the system (2.14) by $2Q_\tau$, the second equation by $2\mathcal{P}$, the third equation by 2ϑ, sum them up and integrate it by x, one has

$$\frac{d}{d\tau}I_1(\tau) + 2J_1(\tau) = 0 \tag{3.1}$$

$$I_1(\tau) = \int\limits_{R^1} \left\{ Q_\tau^2(\tau,x) + Q_x^2(\tau,x) + \frac{3}{2}\mathcal{P}^2(\tau,x) + \frac{2}{5}\vartheta^2(\tau,x) + \beta Q^2(\tau,x) \right\} dx,$$

$$J_1(\tau) = \int\limits_{R^1} \Big\{ -\mu_{11}Q_\tau^2(\tau,x) + BQ_\tau(\tau,x)Q_x(\tau,x) + \bar{w}_1 Q_\tau(\tau,x)\mathcal{P}(\tau,x) +$$

$$+2\bar{\mu}_{12}Q_\tau(\tau,x)\vartheta(\tau,x)+\frac{3}{2}\hat{\mu}\mathcal{P}^2(\tau,x)-\hat{u}\beta Q(\tau,x)\mathcal{P}(\tau,x)-$$

$$-\frac{3}{5}\bar{w}_2\mathcal{P}(\tau,x)\vartheta(\tau,x)-\frac{2}{5}\mu_{22}\vartheta^2(\tau,x)\Big\}\,dx,$$

$$\bar{w}_1=\frac{3}{2}w_1+B,\quad \bar{\mu}_{12}=\frac{1}{2}\left(\mu_{12}+\frac{2}{5}\mu_{21}\right).$$

Similarly multiplying the first equation of system (2.14) by $2Q$ and integrating by x, we have:

$$\frac{d}{d\tau}I_2(\tau)+2J_2(\tau)=0, \tag{3.2}$$

$$I_2(\tau)=\int_{R^1}\left\{-\mu_{11}Q^2(\tau,x)+2Q(\tau,x)Q_\tau(\tau,x)\right\}dx,$$

$$J_2(\tau)=\int_{R^1}\Big\{\beta Q^2(\tau,x)+\frac{3}{2}w_1\mathcal{P}(\tau,x)Q(\tau,x)+$$

$$+\mu_{12}\vartheta(\tau,x)Q(\tau,x)-Q_\tau^2(\tau,x)+Q_x^2(\tau,x)+\mathcal{P}(\tau,x)Q_x(\tau,x)\Big\}\,dx.$$

Summing up (3.1), (3.2) and assuming that $|\vartheta|$, $|\mathcal{P}|$, $|Q|$, $|Q_x|$, $|Q_\tau|\to 0$ as $|x|\to\infty$, we obtain the expression:

$$\frac{d}{d\tau}I(\tau)+2J(\tau)=0. \tag{3.3}$$

The expression under the integral sign in $I(\tau)$ is a positive definite squared form of the variables Q_τ, Q_x, Q, $\mathcal{P}$, ϑ if the inequality

$$\beta-\mu_{11}-1>0 \tag{3.4}$$

holds. Similarly the expression under the integral sign in $J(\tau)$ is a positive defined squared form of the variables Q_τ, Q_x, Q, $\mathcal{P}$, ϑ if the inequalities

$$1+\mu_{11}<0,\quad \hat{\mu}>\tfrac{1}{6},\quad \tfrac{2}{5}\mu_{22}(1+\mu_{11})-\bar{\mu}_{12}^2+\tfrac{1}{4}\tfrac{\mu_{12}^2(1+\mu_{11})}{\beta}>0 \tag{3.5}$$

hold. The coefficient $\hat{\mu}$ comes from the explicit form of $\widetilde{\mathbb{C}}_W$ (see [5]).

The inequalities (3.5) have been obtained under the condition $B=0$. If B is sufficiently small and the inequalities (3.4), (3.5) are satisfied then the expressions under the integral sign in $I(\tau)$, $J(\tau)$ are still positive defined squared forms.

Therefore for sufficiently small B, that is in the limit of small electric fields, with the help of the Courant inequality for positive definite squared forms we can say that there exists a constant $M > 0$ such that $J(\tau) \geq MI(\tau)$ and (3.3) transforms into

$$\frac{d}{dt}I(\tau) + 2MI(\tau) \leq 0,$$

i.e.

$$I(\tau) \leq e^{-2M\tau} I(0) \quad \text{for all } \tau > 0. \tag{3.6}$$

Note, that the inequalities (3.4), (3.5) can be rewritten as (see [5])

$$\mu_{11} + 1 < 0, \quad \mu_{22} < 0, \quad \hat{\mu} > \tfrac{1}{6}, \quad \tfrac{2}{5}\mu_{22}(1 + \mu_{11}) - \bar{\mu}_{12}^2 > 0. \tag{3.7}$$

If we take for the parameters entering in the stability conditions the same values as in Ref.5, the inequalities (3.7) are satisfied and therefore the stationary solution in the linear approximation is asymptotically stable.

Acknowledgments

This work was supported by Russian Foundation for Basic Research (99 – 01 – 00486) and by INTAS, project "Conservation laws of mechanics of continua: waves and fronts", grant number 868. The author V. R. acknowledges the support from Italian CNR (program Agenzia2000, grant n. CNRG000DB7) and from TMR (program Asymptotic Methods in Kinetic Theory, grant n. ERBFMRXCT970157)

References

1. A. M. Anile and V. Romano, Cont. Mech. Thermodyn. **11** 307 (1999).
2. V. Romano, Cont. Mech. Thermodyn. **12** 31 (2000).
3. A. M. Anile and V. Romano, MECCANICA **35** 249 (2000).
4. V. Romano, Math. Meth. Appl. Sci. **24** 439 (2001).
5. A. M. Blokhin, R. S. Bushmanov and V. Romano, *Asymptotic stability of the equilibrium state for the hydrodynamical model of charge transport in semiconductors based on the maximum entropy principle*, preprint (2001).

END EFFECTS OF SAINT-VENANT'S TYPE IN MIXTURES OF THERMOELASTIC SOLIDS

F. BOFILL, M. DALMAU, R. QUINTANILLA
Dept. Matemàtica Aplicada 2. Universidad Politècnica de Catalunya, Colom, 11. Terrassa. Barcelona. Spain
E-mail: ramon@ma2.upc.es

This paper derives spatial decay bounds in a dynamical problem of mixtures of thermoelastic solids defined on a semi-infinite cylindrical region. Previous results for isothermal elastodynamics and for the parabolic heat equation lead us to suspect that the solution of the problem should tend to zero faster than a decaying exponential of the distance from the finite end of the cylinder. We prove that an energy expression is actually bounded above by a decaying exponential of a quadratic polynomial of the distance.

1 Introduction

An intensive study of the continuum theory of mixtures has been developed in recent years. In some theories for a mixture of materials, (fluids or gases) the constitutive independent variables are the displacement gradients and the relative velocities, and the spatial description is used. In the theory for a mixture of elastic materials using the Lagrangian description the independent variables are the relative displacement and the gradients displacement fields.

Many studies on Saint Venant's principle and spatial decay estimates have been developed since the decade of the sixties. First contributions were focused on elliptic [10,11,15], and parabolic [2,9] problems. Since the pioneering contribution of Flavin *et al.* [5] a growing attention has been given to hyperbolic problems. We may recall the contributions of Chirita *et al.* [4], and Bofill *et al.* [1]. The isothermal problem of elastic mixtures was studied in the reference [1]. We may also recall the surveys on these problems [6,7,8].

Chirita [3] considered recently an alternative way that allowed the study of the combination of hyperbolic and parabolic equations describing the linear problem of the thermoelasticity. He proved that a certain measure of the solutions decays exponentially with distance. The result were extend by Pompei and Scalia [12] in the case of mixtures. It is known that for the isothermal elastic case the displacement (in the transient problem) vanishes if the distance is large enough and the solutions of the transient problem for the heat equation decay as an exponential of the square of the distance. As systems of the thermoelasticity are combinations of hyperbolic and parabolic equations, one suspects that it is possible to obtain a faster rate of decay than an exponential of distance. It has been proved[13] that the decay is (at least) as the exponential of a polynomial of second order of the distance. The aim of this paper is to prove that the result obtained in [13] can be extended to the problem of the mixture of thermoelastic solids.

2 Notation and basic equations

Consider a body that occupies a prismatic semi-infinite cylinder B and select a rectangular system of coordinates such that the finite end of the cylinder lies in the (x_1, x_2)-coordinate plane and contains the origin. We denote by $D(z)$ the bounded cross section at a distance z from the finite plane end. We assume that it is a connected region for all $z \geq 0$. The boundary of the cross section is assumed sufficiently smooth to admit the application of the divergence theorem.

We assume that B is occupied by a chemically inert mixture of two constituents s_1, s_2 which are homogeneous and isotropic materials. Let $\rho_i, (i = 1, 2)$ be the mass density of the constituents. The equations of the dynamical problem of a linear mixture of thermoelastic solids are given by the equations of motion, the energy equation and the constitutive equations. In the absence of supply terms the evolution equations are

$$\rho_1 \ddot{u}_i - p_i = t_{ji,j}, \ \ \rho_2 \ddot{w}_i + p_i = s_{ji,j}, \ \ \rho_0 T_0 \dot{\eta} = q_{A,A}. \tag{1}$$

and for centrosymetric homogeneous and isotropic mixture the constitutive equations are

$$\begin{aligned} t_{ji} &= (\lambda + \nu)\delta_{ij} e_{rr} + 2(\mu + \zeta) e_{ji} + (\alpha + \nu)\delta_{ij} g_{ss} \\ &\quad + (2\kappa + \zeta) g_{ji} + (2\gamma + \zeta) g_{ij} - (\beta + m)\delta_{ij}\theta, \\ s_{ji} &= \nu \delta_{ij} e_{rr} + 2\zeta e_{ij} + \alpha \delta_{ij} g_{ss} + 2\kappa g_{ij} + 2\gamma g_{ji} - m\delta_{ij}\theta, \\ p_i &= \xi d_i, \ \ \rho_0 \eta = \beta e_{rr} + m g_{rr} + a\theta, \ \ q_i = k\theta_{,i}. \end{aligned} \tag{2}$$

Here t_{ij} and s_{ij} denote the partial stress tensors associated with constituents s_1, s_2 respectively, $(u_i), (w_i)$ are the displacement vector fields associated with constituents s_1, s_2 respectively, (p_i) is the internal body force, η is the entropy per unit mass of the mixture, $T_0 (> 0)$ is the constant temperature in the reference configuration, θ is the temperature measured from the reference configuration, (q_i) is the heat flux vector, $\rho_0 = \rho_1 + \rho_2$, $\alpha, \beta, \kappa, \gamma, \lambda, \mu, \nu, \zeta, \xi, m, a$ and k are constitutive constants and δ_{ij} is the Kronecker symbol. The tensors e_{ij}, g_{ij} and d_i are related to the displacements by the equations

$$e_{ij} = \frac{1}{2}(u_{i,j} + u_{j,i}), \ \ g_{ij} = (w_{i,j} + u_{j,i}), \ \ d_i = u_i - w_i. \tag{3}$$

From equations (1), (2) we may obtain the system of field equations that govern the dynamical problem of the mixture of the thermoelastic solids

$$\begin{aligned}\rho_1\ddot{u}_i &= \alpha_1 u_{i,jj}+\alpha_2 u_{j,ji}+\beta_1 w_{i,jj}+\beta_2 w_{j,ji}-\xi(u_i-w_i)-(\beta+m)\theta_{,i},\\ \rho_2\ddot{w}_i &= \beta_1 u_{i,jj}+\beta_2 u_{j,ji}+\gamma_1 w_{i,jj}+\gamma_2 w_{j,ji}+\xi(u_i-w_i)-m\theta_{,i},\\ aT_0\dot{\theta} &= -T_0(\beta+m)\dot{u}_{i,i}-T_0 m\dot{w}_{i,i}+k\theta_{,ii},\end{aligned} \tag{4}$$

where

$$\begin{aligned}&\alpha_1=\mu+2\kappa+2\zeta, \quad \alpha_2=\lambda+\mu+\alpha+2\nu+2\gamma+2\zeta,\\ &\beta_1=2\gamma+\zeta, \quad\quad \beta_2=\alpha+\nu+2\kappa+\zeta, \gamma_1=2\kappa, \gamma_2=2\gamma+\alpha.\end{aligned} \tag{5}$$

We denote

$$2W=2W_M+a\theta^2, \tag{6}$$

where

$$W_M=\frac{1}{2}\lambda e_{rr}e_{ss}+\mu e_{ij}e_{ij}+\frac{1}{2}\alpha g_{rr}g_{ss}+\kappa g_{ij}g_{ij}+\gamma g_{ji}g_{ij}+\nu g_{rr}e_{ss}+2\zeta g_{ij}e_{ij}+\frac{1}{2}d_i d_i.$$

We will obtain spatial decay estimates whenever the following conditions are satisfied:
(i) $\rho_1>0$, $\rho_2>0$, $a>0$, $T_0>0$,
(ii) There exists a positive constant K_1 such that the inequality

$$W_M\geq K_1[e_{ij}e_{ij}+g_{ij}g_{ij}+d_i d_i], \tag{7}$$

holds, for every tensor g_{ij}, symmetric tensor e_{ij} and every vector d_i .

Conditions (i) and (ii), imply the existence of two positive constants m_1, m_2 such that

$$t_{1i}t_{1i}\leq 2m_1 W, \quad s_{1i}s_{1i}\leq 2m_2 W. \tag{8}$$

We assume the boundary conditions, the asymptotic conditions and the homogeneous initial conditions

$$(t_{ji}+s_{ji})n_j=0, \quad u_i-w_i=0, \quad q_i n_i=0 \text{ on } \partial D(x_1), \quad x_1\in[0,\infty), \tag{9}$$

$$u_{i,t}, u_{i,K}, w_{i,t}, w_{i,K}, \theta \text{ and } \theta_{,K}\to 0 \text{ as } x_1\to\infty \text{ (uniformly as) } x_1^{-3}, \tag{10}$$

$$u_i(\mathbf{x},0)=\frac{\partial u_i}{\partial t}(\mathbf{x},0)=w_i(\mathbf{x},0)=\frac{\partial w_i}{\partial t}(\mathbf{x},0)=\theta(\mathbf{x},0)=0. \tag{11}$$

3 The decay estimate

In this section we obtain a decay estimate. If we define the function

$$E(z,t) = -\int_0^t \int_{D(z)} (t_{1i}u_{i,s} + s_{1i}w_{i,s} + T_0^{-1}q_1\theta)dads$$

$$= -\frac{1}{2}\int_0^t \int_{D(z)} \Big[(t_{1i}+s_{1i})(u_{i,s}+w_{i,s}) + (t_{1i}-s_{1i})(u_{i,s}-w_{i,s}) + 2T_0^{-1}q_1\theta\Big] dads,$$

the asymptotic conditions (10) imply that for finite time

$$\lim_{z\to\infty} E(z,t) = 0. \tag{12}$$

If we use the divergence theorem, the evolution equations, the boundary conditions, the asymptotic condition (12) and the initial conditions, we have:

$$E(z,t) = \frac{1}{2}\int_{B(z)} (\rho_1 u_{i,t}u_{i,t} + \rho_2 w_{i,t}w_{i,t} + 2W)dv + T_0^{-1}\int_0^t \int_{B(z)} k\theta_{,i}\theta_{,i}dvds. \tag{13}$$

where $B(z) = \{\mathbf{x} \in B, x_1 \geq z\}$. We now define the function

$$F(z,t) = \int_z^\infty E(\xi,t)d\xi. \tag{14}$$

From equations (13) and (14), it follows that

$$\frac{\partial F(z,t)}{\partial z} = -E(z,t) \tag{15}$$

and

$$\frac{\partial^2 F(z,t)}{\partial z^2} = \frac{1}{2}\int_{D(z)} (\rho_1 u_{i,t}u_{i,t} + \rho_2 w_{i,t}w_{i,t} + 2W)da + T_0^{-1}\int_0^t \int_{D(z)} k\theta_{,i}\theta_{,i}dads. \tag{16}$$

We also have that

$$\frac{\partial F(z,t)}{\partial t} = -\int_{B(z)} (t_{1i}u_{i,t} + s_{1i}w_{i,t} + T_0^{-1}q_1\theta)dv. \tag{17}$$

It is worth remarking that

$$\int_{B(z)} T_0^{-1}q_1\theta dv = \int_z^\infty \int_{D(\xi)} T_0^{-1}k\theta\theta_{,1}dad\xi = -\frac{1}{2T_0}\int_{D(z)} k\theta^2 da$$

Our next step is to estimate the time derivative of F in terms of the two first spatial derivatives of F. Using Holder's inequality, estimates (8) and the arithmetic-geometric mean inequality we have

$$|\int_{B(z)} t_{1i}u_{i,t}dv| \leq -\beta_3\frac{\partial F(z,t)}{\partial z}, \quad |\int_{B(z)} s_{i1}w_{i,t}dv| \leq -\beta_4\frac{\partial F(z,t)}{\partial z}, \tag{18}$$

where β_3 and β_4 can be easily obtained, and

$$\frac{1}{2T_0}\int_{D(z)} k\theta^2 da \leq \frac{k}{aT_0}\frac{\partial^2 F(\alpha,z,t)}{\partial z^2}. \tag{19}$$

From (18) and (19) we obtain the inequality

$$\frac{\partial F(z,t)}{\partial t} \leq -G\frac{\partial F(z,t)}{\partial z} + C_2\frac{\partial^2 F(z,t)}{\partial z^2}, \tag{20}$$

where $G = \beta_3 + \beta_4$, $C_2 = k/(aT_0)$.

It is known that the solutions of the inequality (20) satisfy (see [13,14]) the estimate

$$F(z,t) \leq \exp(\hat{a}z - \hat{\mu}t) \sup_{0\leq s\leq t} [\exp\hat{\mu}s\, F(0,s)]$$
$$\cdot \frac{z}{(4\pi C_2)^{1/2}}\int_0^t (t-s)^{-3/2}\exp -\frac{z^2}{4C_2(t-s)}ds,$$

for all $z \geq 0$, where $\hat{a} = (2C_2)^{-1}G$, and $\hat{\mu} = \hat{a}^2 C_2$.

Some manipulations (see [13,14]) allow us to obtain

$$F(z,t) \leq \frac{A(t)}{z}\exp(\hat{a}z - \frac{z^2}{4tC_2}), \tag{21}$$

where

$$A(t) = (4tC_2)^{1/2}\exp(-\hat{\mu}t)\sup_{0\leq s\leq t}\Big(\exp(\hat{\mu}s)F(0,s)\Big).$$

Thus, we have proved:

Theorem 3.1 . *Let (u_i, w_i, θ) be a solution of the initial boundary value problem determined by system (4), initial condition (11), boundary conditions (8) and the asymptotic condition (3). Then, the energy function $F(z,t)$ defined in (14) satisfies estimate (21).*

Thus, we see that the spatial behaviour becomes that of the solutions of parabolic equations.

Acknowledgments

This work is supported by the project (BFM2000-0809).

References

1. F. Bofill & R. Quintanilla, *Spatial estimates for dynamical problems in several elasticity theories*, Rich. Mat. XLVI (1997) 425-441.
2. B. A. Boley, *Some observations on Saint-Venant' s principle*, 3rd. US Natl. Cong. Appl. Mech. ASME New York 259–264, 1958.
3. S.Chirita, Saint-Venant's principle in linear thermoelasticity, *Jour. Thermal Stresses*, **18**, (1995) 485-496.
4. S.Chirita and R.Quintanilla, Saint-Venant's principle in linear elastodynamics, *J. of Elasticity*, **42**, (1996) 201-215.
5. J. N. Flavin, R.J. Knops andL. E. Payne, Energy bounds in dynamical problems for a semi-infinite elastic beam, *Elasticity: Mathematical methods and applications*, Ellis-Horwood Chichester 101–111, 1989.
6. C.O. Horgan, Recent developments concerning Saint-Venant' s Principle: An update, *Applied Mechanics Reviews*, **42**, (1989) 295-303.
7. C.O. Horgan, Recent developments concerning Saint-Venant' s Principle: A second update, *Applied Mechanics Reviews*, **49**, (1996) 101-111.
8. C.O. Horgan and J.K. Knowles, Recent developments concerning Saint-Venant' s Principle, *Advances in Applied Mechanics*, (Ed. J. W. Hutchinson & T.Y. Wu), 23 Academic Press, New York 179-269, 1983.
9. C.O. Horgan, L.E. Payne and L.T. Wheeler, Spatial decay estimates in transient heat conduction, *Quat. Appl. Mathematics*, **XLII**,(1984) 119-127.
10. O.A. Oleinick and G.A Yosifian, On the behavior at infinity of solutions of second order elliptic equations in domains with noncompact boundary, *Math.USSR Sbornik*, **40**, (1981) 527-548.
11. O.A. Oleinick and G.A Yosifian, On the asymptotic behaviour at infinity of solutions in linear elasticity, *Arch. Rational Mech. Anal.*, **78**, (1982) 29-53.
12. A. Pompei and A. Scalia, On the dynamic theory of mixtures of thermoelastic solids, *Jour. Thermal Stresses* **22**, (1999) 23-34.
13. R. Quintanilla, End effects in thermoelasticity, *Mathematical Methods in Applied Sciences* , **24**, (2001) 93-102.
14. R. Quintanilla, Damping of end effects in a thermoelastic theory, *Applied Mathematics Letters*, **14**, (2001) 137-141.
15. R. A. Toupin, Saint-Venant' s principle, *Arch. Rat. Mech. Anal.*, **18**, (1965) 83-96.

CHARACTERISTIC SHOCKS IN EXCEPTIONAL DIRECTIONS

G. BOILLAT & A. MURACCHINI

Department of Mathematics and Research Center of Applied Mathematics. University of Bologna, C.I.R.A.M. Via Saragozza 8, 40123 - Bologna, Italy. E-mail: boillat@riemann.ing.unibo.it - muracchi@ciram.ing.unibo.it

For a general hyperbolic conservative system in three-dimensional space variables the Lax condition of genuine non-linearity depends on the normal to the wave front. When it fails for some directions of the normal (*exceptional* directions) we study the possibility of existence of a characteristic shock and give its explicit expression. Conditions for the boundedness of the shock are given in the case of the Euler's variational equations.

1 Introduction

As usual in Mathematical Physics the system of N conservation laws

$$\partial_t u + \partial_i f^i(u) \equiv \partial_t u + A^i \partial_i u = 0, \quad A^i = \nabla f^i \qquad (i = 1, 2, 3)$$

admits an additional law

$$\partial_t h^o + \partial_i h^i = 0$$

where the density of entropy or energy h^o is a convex function of u. So the system can be written in a symmetric form (in the sense of Friedrichs[1]) which guarantees its hyperbolicity i.e. for any unit vector $\mathbf{n} = (n_i)$ the matrix $A_n \stackrel{\text{def}}{=} A^i n_i$ has a complete system of real eigenvectors $l(u, \mathbf{n})$, $d(u, \mathbf{n})$ satisfying

$$l(A_n - \lambda I) = 0, \qquad (A_n - \lambda I)d = 0; \quad ld = 1 \tag{1}$$

The eigenvalues $\lambda(u, \mathbf{n})$ are the *normal* wave velocities. Now, it is also useful to introduce the vector g defined as follows

$$g(A_n - \lambda I) = -\nabla\lambda + (\nabla\lambda d)l, \qquad gd = 0 \tag{2}$$

An important criterion has been introduced by Lax [2]. The scalar product

$$q(u, \mathbf{n}) \stackrel{\text{def}}{=} \nabla\lambda d, \qquad \nabla \stackrel{\text{def}}{=} (\partial/\partial u)_{\mathbf{n}=const.}$$

represents the coefficient of the nonlinear part in the propagation law of asymptotic [3] or discontinuity waves. When it is different from zero (for instance for a sound wave in a fluid) the wave is *genuinely nonlinear*. When it is identically equal to zero the wave is said to be *exceptional* or *linearly degenerated.*

This latter case occurs in many physical theories and also if λ is multiple [4] and is associated with the existence of *characteristic shocks* i.e. shocks which move with the velocity λ. In one-dimensional propagation $q(u)$ may vanish for some value of the field. This interesting case has been studied extensively together with its applications to rigid heat conductors at low temperature [5]. On the other hand in three dimensional propagation some conditions may be satisfied only for particular directions $\mathbf{n} = \mathbf{n}^*(u)$ depending on the field. For instance the crossing of velocities allows for such characteristic shocks whose explicit form[4,6,7] can be given in terms of the jump of the so-called *main field* $u' = \breve{\nabla} h^o$ (the breve denotes the transpose).
In this paper we study the possibility to have a characteristic shock moving in the *exceptional* direction $\mathbf{n}^*(u)$ when

$$q(u, \mathbf{n}^*(u)) = 0, \quad \forall u \tag{3}$$

An example of such a case is given by the scalar field deriving from the Lagrangian $L = L(g^{\alpha\beta}\partial_\alpha u\, \partial_\beta u/2)$ [4].

2 Directional characteristic shock

According to the above observations, λ is assumed to be a simple eigenvalue, the state $u(\tau, u_o)$ after the shock depends on a parameter τ, $u(0, u_o) = u_o$ is the state before the shock and satisfies the Rankine-Hugoniot equations

$$f_n(u) - \lambda u = f_n(u_o) - \lambda u_o, \quad \mathbf{n} = \mathbf{n}^*(u) = \mathbf{n}^*(u_o), \quad \lambda(u, \mathbf{n}^*) = \lambda(u_o, \mathbf{n}^*) \tag{4}$$

For any quantity $Q(u, \mathbf{n})$ we define

$$[Q] \stackrel{\text{def}}{=} Q(u, \mathbf{n}^*) - Q(u_o, \mathbf{n}^*)$$

and the total directional derivative

$$DQ(u, \mathbf{n}^*) = \nabla Q(u, \mathbf{n}^*) + (\partial Q(u, \mathbf{n}^*)/\partial n_i)\, D n_i{}^*$$

For instance, from (3), dropping the asterisk we can write

$$Dq = \nabla(\nabla\lambda d) + (\nabla\Lambda^i d + \nabla\lambda \partial d/\partial n^i) D n_i = 0 \tag{5}$$

where $\mathbf{\Lambda}$ is the *radial* velocity

$$\mathbf{\Lambda}(u, \mathbf{n}) \stackrel{\text{def}}{=} \lambda \mathbf{n} + \partial\lambda/\partial\mathbf{n} - (\mathbf{n} \cdot \partial\lambda/\partial\mathbf{n})\, \mathbf{n}$$

On the other hand if $Q(u, \mathbf{n}) = 0$, $(\forall u, \mathbf{n})$ then

$$\nabla Q(u, \mathbf{n}) = 0, \quad (\partial Q(u, \mathbf{n})/\partial n_i)\, \delta n^i = 0, \quad \mathbf{n}\, \delta\mathbf{n} = 0 \tag{6}$$

Differentiating the equations (4) with respect to the parameter τ it follows

$$\partial_\tau \mathbf{n} = D\mathbf{n}\partial_\tau u = 0, \quad \partial_\tau \lambda = D\lambda \partial_\tau u = 0, \quad (A_n - \lambda I)\partial_\tau u = 0$$

and so we can see that

$$\partial_\tau u \propto d, \; D\mathbf{n}d = 0$$

As a consequence, from (5), we get this compatibility condition

$$\nabla(\nabla\lambda d)d = 0, \quad \text{for } \mathbf{n} = \mathbf{n}^*(u) \tag{7}$$

Now following [4,7] we introduce the generating function of the shock

$$\eta = [h_n] - \lambda[h^o]$$

which is zero since [1]

$$\nabla h_n = \nabla h^o A_n, \quad \partial_\tau \eta = \breve{u}'(A_n - \lambda I)\partial_\tau u = 0$$

Differentiating η and (4) with respect to u_o yields, by some calculations,

$$D_o\eta = \breve{u}'(A_n - \lambda I)D_o u - \breve{u}'_o(A_{on} - \lambda_o I) + [h^i]D_o n_i - [h^o]D_o\lambda_o = 0$$

$$(A_n - \lambda I)D_o u - [u]D_o\lambda_o + [f^i]D_o n_i = A_{on} - \lambda_o I$$

The latter, inserted in the preceding relation, furnishes

$$(\breve{u}' - \breve{u}'_o)(A_{on} - \lambda_o I) + w\nabla_o\lambda_o - \varphi^i D_o n_i = 0 \tag{8}$$

where

$$\varphi^i = \breve{u}'[f^i] - [h^i] - w\Lambda^i_o$$

and

$$w = \breve{u}'[u] - [h^o] \tag{9}$$

Note that w, because of the convexity of $h^o(u)$, is positive when $u \neq u_o$. The second term in (8) can be expressed in terms of g_o remembering (2). To evaluate the last term in (8), we observe that the unit vector $\mathbf{n}^*$ constrained by (3) still has, in general, a degree of freedom. Differentiating (4) and $\eta = 0$ with respect to this parameter, keeping u_o constant, we obtain

$$([f^i] - \Lambda^i_o[u])\dot{n}_i + (A_n - \lambda I)\dot{u} = 0, \quad ([h^i] - \Lambda^i_o[h^o])\dot{n}_i + \breve{u}'(A_n - \lambda I)\dot{u} = 0$$

Eliminating $\dot{u}$ results in

$$\varphi^i \dot{n}_i = 0$$

Proceeding in the same way for (3) and $(1)_2$, we get

$$(\nabla\Lambda^i d)_o \dot{n}_i + \nabla\lambda_o \dot{d}_o = 0; \quad (A^i_o - \Lambda^i_o I) d_o \dot{n}_i + (A_{on} - \lambda_o I)\dot{d}_c = 0$$

Multiplying the second one by g_o, defined by (2), follows

$$V^i_o \dot{n}_i = 0, \quad V^i = gA^i d + \nabla\Lambda^i d$$

Since the vectors φ and $\mathbf{V}_o$ are both orthogonal to $\mathbf{n}$ and $\dot{\mathbf{n}}$, they are proportional

$$\varphi^i D_o n_i \propto V^i_o D_o n_i \tag{10}$$

It remains to calculate this last quantity.

3 Explicit form of the shock

Since the expression of the eigenvector holds for any $\mathbf{n}$ (in contradistinction to what happens for crossing eigenvalues where the multiplicity increases in the direction of $\mathbf{n}^*$) from (1) and (6) we have

$$(A^i - \Lambda^i I) d D n_i + (A_n - \lambda I)\,(\partial d/\partial n_i)\, D n_i = 0$$

Multiplying by g we obtain

$$(gA^i d) D n_i - \nabla\lambda\,(\partial d/\partial n_i)\, D n_i = 0$$

and therefore, from (5),

$$V^i D n_i = -\nabla(\nabla\lambda d) \tag{11}$$

This quantity can be expressed in another way. To this aim let us differentiate (2) in an arbitrary direction X, taking into account of (3) and (7)

$$\nabla g X(A_n - \lambda I) + g(\nabla A_n X - I\nabla\lambda X) = -\nabla\nabla\lambda X + \nabla(\nabla\lambda d) X l$$

and multiply it by d to obtain

$$g\nabla A_n d = -\nabla\nabla\lambda d + \nabla(\nabla\lambda d)$$

Now in the preceding equation let us set $X = d$ so obtaining

$$\nabla g d(A_n - \lambda I) + g\nabla A_n d = -\nabla\nabla\lambda d$$

by the condition (7) and therefore

$$\nabla g d(A_n - \lambda I) = -\nabla(\nabla\lambda d) \tag{12}$$

(For any $\mathbf{n}$, i.e. without the constraints (3) and (7), the additional quantity $Y = q(g + \nabla l d - \breve{d}\nabla\breve{l}) + (\nabla q d) l$ must be added to the right member of (12).)

Taking into account (2), (10), (11), the equation (8) gives the jump of the main field

$$\breve{u}' - \breve{u}'_o = \omega l_o + w g_o + \sigma(\nabla g d)_o \tag{13}$$

where the vectors are evaluated for u_o and $\mathbf{n}^*$ and the scalars ω (which can be taken equal to τ), w (given by (9)), σ are functions of τ and u_o.
The jump (13) differs from the usual expression [4] of the characteristic shocks by the last term which, unlike g_o, is not generally orthogonal to d_o as we shall see in the case of the Euler's variational equations.
First we need the expression of the derivative of l in the direction of d.
From (1)

$$\nabla l d(A_n - \lambda I) + l \nabla A_n d - (\nabla \lambda d) l = 0$$

$$(A_n - \lambda I)\nabla d + \nabla A_n d - d \nabla \lambda = 0 \quad \Rightarrow \quad l \nabla A_n d = \nabla \lambda$$

Hence

$$\nabla l d(A_n - \lambda I) = -\nabla \lambda + (\nabla \lambda d) l$$

and

$$\nabla l d = g + \gamma l$$

It is clear that γ depends on the norm of l (see also next paragraph). Suppose that the *k-th* component of g is not zero. The corresponding component of l cannot be zero either and can be taken equal to 1. This shows that, in this case, $\gamma \neq 0$. Conversely to a null component of l corresponds a null component of g.

4 Euler's variational equations

When the Euler's equations which derive from a Lagrangian $L(\partial_\alpha q^s, q^s)$ are reduced to a first order system they can be written [4]

$$H'(u')\partial_t u' + A'^i \partial_i u' = B' u', \quad H' = \nabla' u, \quad A'^i = A^i H'$$

where H' is positive definite, the matrices A'^i, constant and symmetric, satisfy the Duffin-Kemmer-Petiau relations while B' is constant and skew-symmetric. This means that the nonlinear character of the system is due only to the first term. As a consequence

$$l A'_n = \lambda \breve{d}, \quad l H' = \breve{d}, \quad l H' \breve{l} = 1 \tag{14}$$

from which follows

$$\nabla l d A'_n = \lambda \nabla \breve{d} d + \breve{d}(\nabla \lambda d) = g A'_n + \gamma \lambda \breve{d}, \quad \Rightarrow \lambda \nabla \breve{d} d \breve{g} = -\lambda \nabla g d \breve{d} = g A'_n \breve{g}$$

Thus the component of $\nabla g d$ on the vector l is

$$\nabla g d d = -\alpha$$

where $\alpha = g A'_n \breve{g}/\lambda$ is just the coefficient which appears in the expression of w of the usual characteristic shock. From (6), $l A'_n \breve{l} = \lambda$ and the norm (14) results in $l A'^i \breve{l} = \Lambda^i$. Differentiating in the direction of d follows

$$g A^i d + \gamma \Lambda^i = \nabla \Lambda^i d/2$$

which implies $\gamma = q/2\lambda$. Further $l \nabla' H' \breve{l}\breve{l} = -q/\lambda$ so that both these quantities vanish with q. It appears now that

$$V^i = -\gamma \Lambda^i + 3 \nabla \Lambda^i d/2$$

This shows, by (11), that the radial velocity is not continuous across the shock differing from the ordinary case $q(u, \mathbf{n}) \equiv 0$. (The additional quantity in (12) takes the explicit form $Y = (\nabla q d + q^2/2\lambda) l + q(2g - \nabla \lambda / 2\lambda)$.)
Now w is given by [4]

$$2\lambda w = [\breve{u}'] A'_n [u'] = \lambda \omega^2 + 2\sigma\omega\lambda \breve{d} \nabla \breve{g} d + w^2 g A'_n \breve{g} + 2 w \sigma g A'_n \nabla \breve{g} d + \sigma^2 (\nabla g d) A'_n (\nabla \breve{g} d)$$

where the coefficients are calculated for $u = u_o$ and $\mathbf{n} = \mathbf{n}^*(u_o)$. But

$$\nabla (g A'_n \breve{g}) d = 2 \nabla g d A'_n \breve{g} = \nabla(\lambda \alpha) d = \lambda \nabla \alpha d + \alpha \nabla \lambda d$$

Finally, setting $\beta = (\nabla g d) A'_n (\nabla \breve{g} d)/\lambda$, we have

$$\omega^2 + \alpha_o w^2 + \beta_o \sigma^2 - 2\alpha_o \omega \sigma + (\nabla \alpha d)_o w \sigma - 2w = 0$$

It follows that the shock will be bounded if this equation represents an ellipsoid in the (ω, w, σ)-space, i.e. if

$$\alpha_o > 0, \qquad \beta_o > \alpha_o^2 + (\nabla \alpha d)_o^2 / 4\alpha_o$$

5 Conclusion

If linear degeneracy (3) occurs in some directions $\mathbf{n}^*(u)$ a characteristic shock may exist if condition (7) is satisfied. The expression of the jump of the *main field* is then given by (13) and contains three vectors i.e. one more than the usual characteristic shock (corresponding to $q \equiv 0$ for any $\mathbf{n}$).

References

1. Friedrichs K. O., Lax P. D., Systems of conservation equations with a convex extension, Proc. Nat. Acad. Sci. U.S.A. *68* (1971) 1686-1688.
2. Lax P. D., The initial value problem for nonlinear hyperbolic equations in two independent variables, Ann. Math. Studies, Princeton, *33*, (1954) pp. 211-29; Hyperbolic systems of conservation laws II, Comm. Pure Appl. Math. *10* (1957) 537-566.
3. Choquet-Bruhat Y., Ondes asymptotiques pour un système d'équations aux dérivées partielles non linéaires, J. Maths. pures appl., *48* (1969) 117-158.
4. Boillat G., Nonlinear Hyperbolic Fields and Waves, In: *Recent Mathematical Methods in Nonlinear Wave Propagation*, CIME Course, T. Ruggeri, Ed., Lecture Notes in Mathematics, N° 1640, Springer Verlag, 1996.
5. Ruggeri T., Muracchini A., Seccia L., Continuum approach to phonon gas and shape changes of second sound via shock waves theory, Il Nuovo Cimento vol. *16*, N° 1 (1994) 15-44.
6. Boillat G., Muracchini A., Chocs caractéristiques de croisement, C.R. Acad. Sci. Paris, *310*, Série I, (1990) 229-232; Characteristic shocks of crossing velocities in Magnetohydrodynamics, NoDEA *3* (1996) 217-230.
7. Boillat G., Expression explicite des chocs caractéristiques de croisement, C. R. Acad. Sci. Paris, *312*, Série I, (1991) 653-656.

SCHRÖDINGER-LIKE MODEL FOR INTERBAND TUNNELING IN HETEROGENEOUS SEMICONDUCTOR DEVICES: A CURRENT ESTIMATE

Giovanni Borgioli and **Matteo Camprini**
Dipartimento di Elettronica e Telecomunicazioni
Via S. Marta, 3 - 50139 Firenze
Università di Firenze, Italy

1 Introduction

The aim of this paper is to develop and justify a simple model for the steady electronic current behaviour of a Resonant Interband Tunneling Diode (RITD)[9], with a Double-Barrier Quantum Well (DBQW) structure (see Fig.2). RITDs are very recent electronic devices working on tunneling effect between the conduction and the valence energy bands. We recall that the main characteristic of a tunnel diode is exhibiting regions of negative differential resistance in the $I(V)$ characteristic, i.e. in the current vs. voltage diagram. In this sense RITDs show a hugely interesting behaviour. Nanometric semiconductor devices like RITDs require appropriate models which take into account quantum processes. The literature on RITDs is still rather poor on the modeling side and we propose here the so-called Kane model [6,7] in order to evaluate the tunneling probability $T(E,V)$ for an incoming electron, whose total energy is E, with an external bias (applied voltage) V (Section 2). In Section 3 we propose a model for the steady current which depends on the evaluated tunneling probability and takes into account of the physical and the geometrical characteristics of the device. The numerical results obtained by such a model are compared to the laboratory data (see Fig. 3) with a satisfactory matching.

The specific diodes submitted to laboratory tests are Motorola K11A-M21 MBE840 and K11A-M21 MBE842 [4,5]. Their general structure is very similar: these devices are based on an heterogeneous multi-layer lattice-matched structure. The semiconductors used are $In_xGa_{1-x}As$ and $In_yAl_{1-y}As$ with suitable doping. The DBQW in the profile of the two bands energies (see Fig. 2) is achieved in the thinest inner layers, as it is shown in Fig. 1, where a schematic portrait of RITD is displayed.

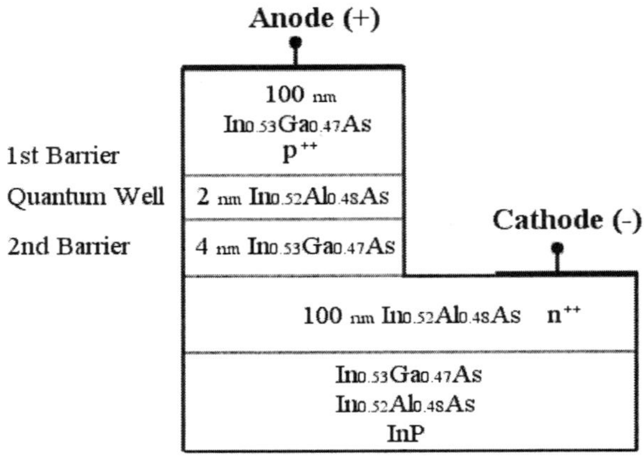

Fig. 1

2 The Kane Model

A simple model for interband tunneling is provided by a system of two Schrödinger–like equations coupled by a $\mathbf{k} \cdot \mathbf{P}$ term, known as the two–band Kane model,

$$\begin{cases} i\hbar \dfrac{\partial \psi_c}{\partial t} = \left[-\dfrac{\hbar^2}{2m_0} \dfrac{\partial^2}{\partial x^2} + E_c \right] \psi_c + \dfrac{\hbar}{i} P \dfrac{\partial \psi_v}{\partial x} \\ i\hbar \dfrac{\partial \psi_v}{\partial t} = \left[\dfrac{\hbar^2}{2m_0} \dfrac{\partial^2}{\partial x^2} + E_v \right] \psi_v + \dfrac{\hbar}{i} P \dfrac{\partial \psi_c}{\partial x} . \end{cases} \tag{1}$$

ψ_c and ψ_v are envelope functions [3] for the conduction and the valence band, m_0 is the "bare" electron mass; E_c and E_v are the conduction and valence energy edges, depending on x when the medium is heterogeneous. In (1) P is the momentum matrix element between the conduction and the valence band, responsable of a non zero probability of interband tunneling:

$$P = \sqrt{\frac{m_0 - m^*}{2m_0 m^*} E_g} \tag{2}$$

where m^* is the effective electron mass and $E_g = E_c - E_v$ is the gap between the energy of the conduction band and the energy of the valence band (see Fig. 2). All the above mentioned parameters are known by laboratory measures (with their dependence on x, in case of an heterogeneous medium), such that system (1) is explicitly solvable in the stationary case. Knowing envelope functions ψ_c and ψ_v permits to derive the tunneling probability coefficient and to deduce a rigorous formulation for the tunneling current which can be calculated numerically and compared with the laboratory results. In Fig.2, here below, we show the approximated behaviour of the bands energy profiles in the considered heterogeneous structure.

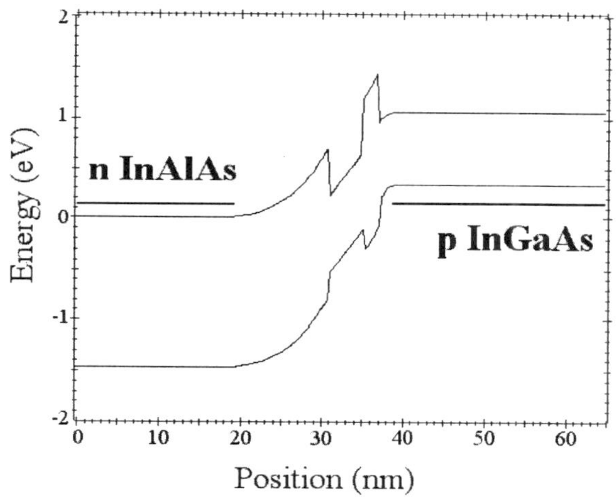

Fig. 2

The derivation of (1) from the "full" Schrödinger equation

$$i\hbar\frac{\partial\Psi}{\partial t} = H\Psi \tag{3}$$

and its physical justification are widely discussed in a previous paper [2] and we recall here the main assumptions adopted:
– the total energy E of the electron is constant (the model is conservative; no collisions or scattering phenomena are taken into account);
– the effective mass m^* is constant in each layer of the heterogeneous structure (i.e. m^* does not depend on E, it is constant inside each layer and changes abruptely on the interface between two different layers);
– the electric field inside the active region of the diode is supposed to be constant.
¿From the geometrical point of view, the main assumption, leading to a one-dimensional problem, is that the thickness (described by the x coordinate) of the considered device is much smaller than the other two dimensions, which

can be supposed to be infinitely extended in the $y - z$ plane and, hence, no boundary condition arise for y and z variables.

The crystal is supposed to be homogeneous in any $y - z$ plane and the external field or the impurity effect, etc. do not depend on the $y - z$ coordinates. These assumptions can be expressed by considering the Hamiltonian H as separated in a part depending only on x and in a part depending on the orthogonal plane variables y and z, such that the "full" wave function for (3) can be written as $\Psi(x, y, z, t) = \psi(x, t)e^{i(k_y y + k_z z)}$, where k_y and k_z are the y and z components of the wave vector $\mathbf{k}$. Moreover, the one-dimensional wave function $\psi(x, t)$ is expanded in the Bloch functions basis [2] at $\mathbf{k} = 0$ (where, in the reciprocal lattice [1,10], the conduction band energy attains a minimum, while the valence band energy exhibits a maximum) for the conduction and valence bands, i.e. $u_0^c(x)$ and $u_0^v(x)$, which can be supposed to depend only on the variable x [11], i.e. $\psi(x, t) = \psi_c(x, t)u_0^c(x) + \psi_v(x, t)u_0^v(x)$.

Looking for stationary solution to (1), we put $\psi_{c,v} = C_{c,v}\exp(i\eta(x))$. Introducing this kind of envelope function in the stationary version of (1) we obtain the so-called "dispersion relation" [9] which presents $\eta(x)$ as a solution of two different differential equations, depending on the position in the energy band diagram, i.e. if the electron is in the forbidden region D_f (the energy region between the two bands profiles) or in the allowed regions D_a (inside one of the considered bands):

$$\begin{cases} \dfrac{d}{dx}\eta(x) = \dfrac{i}{\hbar}\sqrt{\dfrac{m^*}{2E_g}}\sqrt{E_g^2 - 4\left(E - \dfrac{1}{2}(E_c + E_v)\right)^2} & \text{in } D_f \\[2ex] \dfrac{d}{dx}\eta(x) = \dfrac{1}{\hbar}\sqrt{\dfrac{m^*}{2E_g}}\sqrt{4\left(E - \dfrac{1}{2}(E_c + E_v)\right)^2 - E_g^2} & \text{in } D_a\,. \end{cases} \tag{4}$$

3 The Tunneling Probability and the Current Estimate

The tunneling probability is obtained from ψ_c and ψ_v by the following steps:
– identification of the tunneling path across the barrier. The path depends on the electron energy E and on the applied bias V, whose presence affects the barriers offset;
– conditions for the envelope functions at the so-called "turning points" (i.e. the points where an electron with energy E "crosses" the band energy profile, Fig. 2). In this case it is not possible to impose the usual condition of continuity for the envelope functions [4] and the problem is overcome imposing the WKB conditions [8], which permit the evaluation of the solution to (1) in

any point of the layers out of sufficiently small neighborhoods of the turning points.

By applying the WKB conditions to a DBQW system we derive the following expression of the tunneling probability, where the amplitudes $C_{c,v}$ of the envelope functions $\psi_{c,v}$ disappear by cancellation:

$$T(E,V) = \frac{4}{16\exp\left[2\left(\gamma_1+\gamma_2\right)\right]\cos^2(\vartheta) + 4\cosh^2\left(\gamma_2-\gamma_1\right)\sin^2(\vartheta)} ; \qquad (5)$$

where γ_1 and γ_2 represent the attenuations of the first and the second barrier respectively and ϑ is a phase term due to the crossing of the quantum well:

$$\gamma_1(E,V) = \eta\left(x_2^-\right) - \eta\left(x_1^+\right) \qquad (6)$$

$$\gamma_2(E,V) = \eta\left(x_4^-\right) - \eta\left(x_3^+\right) \qquad (7)$$

$$\vartheta(E,V) = \eta\left(x_3^-\right) - \eta\left(x_2^+\right) , \qquad (8)$$

where x_1 and x_2 are the turning points of the first barrier and x_3 and x_4 are the turning points of the second barrier and the $\pm$ exponents denote the incoming and outgoing directions, respectively. The resonance condition is reached when $\cos\vartheta = 1$; in this case the tunneling probability becomes:

$$T = T_{res} = \frac{1}{\cosh^2\left(\gamma_2-\gamma_1\right)} ; \qquad (9)$$

moreover, if $\gamma_1 = \gamma_2$ we obtain $T(E,V) = 1$.

The model we assume for the tunneling current I is the following [9]:

$$I(V) = \mathcal{I}\int_0^{+\infty}\int_{E_c^n}^{E_v^p}\left[f_c(E,V) - f_v(E,V)\right]T(E,V)\exp\left(-\frac{2E_\perp}{\overline{E}_\perp}\right)dE\,dE_\perp \qquad (10)$$

where $\mathcal{I} = A\frac{em^*}{2\pi^2\hbar^3}$ has the dimensions of a current over the square of an energy. A is the device section area, f_c and f_v are the Fermi electron distribution functions in the conduction and the valence band respectively. The electron energy on the orthogonal plane $y-z$ is $E_\perp = \frac{\hbar}{2m^*}\left(k_y^2+k_z^2\right)$ and the "mean" transversal energy is $\overline{E}_\perp = \frac{\sqrt{2}\hbar e\varepsilon}{\pi\sqrt{m^*E_g}}$, where ε is the electric field in the active region, known by V and the extension of the DBQW region. The current density at a fixed energy E between E_c^n (the minimum of the conduction band energy on the $n-$side) and E_v^p (the maximum of the valence band energy on the $p-$side) is proportional to the tunneling probability $T(E,V)$ and to the difference of the Fermi - Dirac distributions in the $n-$doped and the $p-$doped layers of the diode. This term takes into account the matching between the occupied states on the n side and the empty states on the p side. The exponential funcion of $E_\perp$ describes the effects of the electron energy in the orthogonal plane $y - z$.

The behaviour of the current under the applied bias is described in the following figure:

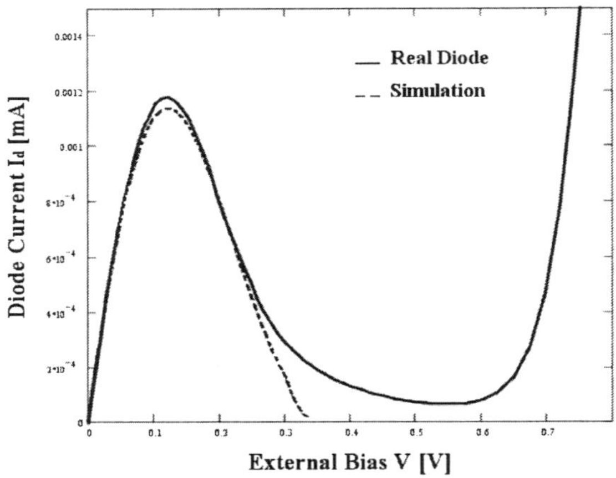

Fig. 3

In Fig. 3 there are plotted the numerical calculations of (5), performed by routine methods (Mathcad®), which are compared to an "average" of the laboratory measures. The real I/V characteristic is a mean value of the measured I/V plot of more diodes. Of course in this plot we have not only the tunneling current, but also the excess and thermoionic currents, which become predominant when the voltage V is higher than 0.3 V. These currents do not depend on tunneling phenomena and, therefore, cannot be described by the Kane model (1).

Acknowledgements. The authors are grateful to the Motorola Physical Science Research Lab (Tempe, Az.) for the technical support. This work was performed under the auspices of the *National Group for Mathematical Physics* of the *Istituto Nazionale di Alta Matematica* and was partly supported by the *Italian Ministery of University* (*MURST* National Project "Mathematical Problems of Kinetic Theories", Cofin2000). Additional support from the *Italian Research National Council* (*CNR* Strategic Project "Modelli Matematici per Semiconduttori") and from the University of Florence (Fondi di

Ricerca di Ateneo) is also gratefully acknowledged.

References

1. N.W. Ashcroft, N.D. Mermin, *Solid State Physics*, Harcourt Brace College Pub., Forth Worth (1976).
2. G. Borgioli, G. Frosali, P.F. Zweifel, "Wigner approach to the two-band Kane model for a tunneling diode", submitted to *Transp. Theor. Stat. Phys.* (2001).
3. M.G. Burt, *J. Phys. Condens. Matter* **4**, 6651-6690 (1992).
4. M. Camprini, "Modellistica del trasporto per diodi DBQW interbanda", Tesi di Laurea in Ingegneria Elettronica, Facoltà di Ingegneria, Università degli Studi di Firenze (2000).
5. A. Cidronali, G. Collodi, M. Deshpande, N. El-Zein, G. Manes, V. Nair, H. Goronkin, "A highly linear single balanced mixer based on heterojunction interband tunneling diode", IEEE *Transaction on Microwave Theory and Tchniques* (to appear in **vol. 4**, 12, December 2001).
6. E. C. Kane, *J. Phys. Chem. Solids*, **12**, 181–188 (1959).
7. E. O. Kane, "The $\mathbf{k} \cdot \mathbf{P}$ method", in *Semiconductors and Semimetals*, edited by R.K. Willardson and A.C. Bear, Vol. **1**, 75-100, Academic Press, New York (1966). Clarendon Press, Oxford (1999).
8. E.C. Kemble, *The Fundamental Principles of Quantum Mechanics with Elementary Applications*, Dover, New York (1937).
9. R.Q. Yang, M. Sweeny, D. Day and J.M. Xu, *IEEE Transactions on Electron Devices*, **38**(3) 442-446 (1991).
10. W.T. Wenckebach, *Essential of Semiconductor Physics*, J.Wiley & Sons, Chichester (1999).
11. S.R. White, L.J. Sham, *Phys. Rev. Letters*, **47**(12) 879-882, (1981).

INSTABILITIES IN COMPRESSIBLE FLUIDS

ANDREAS S. BORMANN

Linear stability analysis in combination with the Boussinesq approximation is usually used in order to determine the critical conditions for the onset of convection for the Rayleigh-Bénard problem. In this way a non-dimensional number, the Rayleigh number, is obtained, which reflects whether the fluid is at rest (stability) or in motion (instability). From a thermodynamic point of view the Boussinesq approximation has a shortcoming that contradicts thermodynamic stability. Therefore a linear stability analysis for the compressible Navier-Stokes-Fourier equations has been carried out numerically for rigid-rigid, rigid-free and free-free boundaries. The results show that the critical value of the Rayleigh number is not constant anymore as in the Boussinesq case, but that it depends on the thickness of the fluid layer.

1 Introduction and Motivation

The Rayleigh-Bénard Problem consists of a fluid layer that is bounded by two plates and where the lower plate is heated. If the temperature gradient in the fluid at rest exceeds a critical value, convection will appear and stability is lost. For the determination of the state of stability a non-dimensional parameter was introduced [1,2], that is called the Rayleigh number

$$Ra = \frac{g\alpha|\Delta T|d^3}{\kappa_c \nu}. \tag{1}$$

One way to determine the critical value of the Rayleigh number is to use a linear stability analysis [3], where, in order to simplify the equations for the linear stability analysis, usually the Boussinesq approximation is applied. According to this approximation the density of a fluid is assumed to depend on the temperature only, while the dependence on the pressure, i.e. the compressibility κ_T, is neglected.

The result of such a linear stability analysis is a constant value for the critical Rayleigh number and critical wave length k_{crit} that depends only on the boundary conditions.

rigid-rigid boundaries:	$Ra_{crit}^{Bouss\ I} = 1707.8$	$k_{crit}^{I} = 3.177$
rigid-free boundaries:	$Ra_{crit}^{Bouss\ II} = 1100.7$	$k_{crit}^{II} = 2.682$
free-free boundaries:	$Ra_{crit}^{Bouss\ III} = 657.5$	$k_{crit}^{III} = 2.221$

(2)

However, from a thermodynamic point of view the above described approximation is a little precarious because thermodynamic stability requires the inequality[4].

$$\alpha^2 \leqq \frac{\rho c_p}{T}\kappa_T. \tag{3}$$

So, if κ_T is neglected as it is in the Boussinesq approximation, then also the volume expansion α coefficient must vanish.

The majority of works about the Rayleigh-Bénard problem are based on the Boussinesq approximation and are therefore restricted to incompressible fluids. However, there are some works that deal with compressible fluids[5,6] and more recently by Carlès and Ugurtas [7] in the context of near-critical fluids. So far, all methods for the determination of the onset of convection are based on approximations of the field equations or simplifications concerning the solution. This fact and the above explained shortcoming of the Boussinesq approximation offered the motivation to recalculate the critical conditions for the stability threshold on a proper thermodynamic basis, i.e. without neglecting compressibility and without the use of similar simplifications.

2 The Rayleigh and the Schwarzschild Criteria

Consider the situation of a typical Rayleigh-Bénard arrangement as described in the first section, in which two solutions of the field equations compete; a static one and a convective one. The static solution is characterized by the density gradient

$$\left(\frac{\partial \rho}{\partial x_1}\right)_{static} = \alpha\rho_0\frac{T_0 - T_1}{d} - g\kappa_T\rho_0^2 \tag{4}$$

at the bottom of the layer. Convection starts at the bottom when a bubble of the liquid, or the gas, moves upwards into a region of lower pressure. The rise of the bubble by dx_1 may be considered as adiabatic so that the bubble changes its density by $d\rho$ with

$$\left(\frac{\partial \rho}{\partial x_1}\right)_{adiabatic} = \frac{g\alpha^2\rho_0 T_0}{c_p} - g\kappa_T\rho_0^2. \tag{5}$$

Rather obviously, the bubble sinks back to the bottom if the adiabatic gradient is larger than the static one. Thus, by this suggestive argument, the stability criterion reads

$$\text{stability prevails, if} \quad \frac{T_0 - T_1}{d} \leq \frac{g\alpha T_0}{c_p}, \tag{6}$$

provided $\alpha > 0$ holds. This is the stability criterion that is described by Jeffreys[8]. It takes no account of viscosity but it does allow for compressibility. The temperature gradient in (6) is also known as adiabatic temperature gradient (ATG) and is often attributed to Schwarzschild. According to this criterion, the critical temperature gradient for a non-viscous fluid reads

$$\left.\frac{T_0 - T_1}{d}\right|_{crit}^{ATG} = \frac{g\alpha T_0}{c_p}. \tag{7}$$

Note that, while the compressibility affects both, the density gradients (4) and (5), it drops out of the stability criterion (6). Note also that, if (6) is violated, instability does not necessarily occur, because the whole argument ignores the viscosity which has a stabilizing effect on the fluid.

Throughout the present work a Rayleigh number defined as

$$Ra = \frac{g\alpha c_p \rho_0^2 d^4}{\kappa\eta}\frac{T_0 - T_1}{d}. \tag{8}$$

Note that, for nearly incompressible fluids, (8) may be reduced to the canonical definition of the Rayleigh number as given in (1). Equation (7) may be used to replace $T_0 - T_1$ in (8). A critical Rayleigh number is thus obtained that represents the critical adiabatic temperature gradient in a nondimensional manner.

$$Ra_{crit}^{ATG} = \frac{\alpha^2 \rho_0^2}{\kappa\eta} T_0 g^2 d^4 \tag{9}$$

Thus, we are given two critical Rayleigh numbers, viz.

- $Ra_{crit}^{Bouss\ I,\ II,\ III}$ in (1,2), a constant value which ignores compressibility but accounts for viscosity and the effects of the boundaries, and
- Ra_{crit}^{ATG} in (9), which accounts for the compressibility but ignores viscosity.

It seems eminently reasonable to use these two critical Rayleigh numbers in an additive superposition in order to find the true critical Rayleigh number. Thus

$$Ra_{crit} = Ra_{crit}^{Bouss\ I,\ II,\ III} + Ra_{crit}^{ATG} \text{ or, by (2) and (9)}$$

$$Ra_{crit} = \begin{Bmatrix} 1707.8 \\ 1100.7 \\ 657.5 \end{Bmatrix} + \frac{\alpha^2 \rho_0^2}{\kappa\eta} T_0 g^2 d^4. \tag{10}$$

This relation, at least for rigid-rigid boundaries, is in fact the criterion for the onset of stability which is derived by Giterman & Shteinberg [6] from their approximation of the solution of the field equation. (10) is also confirmed by Carlès & Ugurtas[7] by using an asymptotic analysis of the compressible Navier-Stokes equations for near-critical fluids.

In the next section, a method is presented with which the field equation are solved numerically exact. Anticipating the results, it is only fair to say that the numerical exact solution will confirm the ansatz of Giterman & Shteinberg[6] and Carlès & Ugurtas[7].

3 Linear Stability Analysis

The constitutive equations for the heat flux and the stress tensor according to Fourier and Navier-Stokes are inserted into the complete 3D balance equations for mass, momentum and energy, i.e. without the use of the Boussinesq approximation, in order to form the field equations. The stationary solutions of the field variables density $\overline{\rho}$ and temperature $\overline{T}$ for the fluid at rest are calculated analytically from these field equations. A linearization of the balance equations is made in the usual way by.

$$\psi\left(x_1, x_2, x_3, t\right) = \overline{\psi}\left(x_1\right) + \widetilde{\psi}(x_1, x_2, x_3, t),\ \psi \in \{\rho, v_i, T\}\,, \tag{11}$$

where the perturbations are analysed with respect to normal modes

$$\widetilde{\psi}\left(x_1, x_2, x_3, t\right) = \widehat{\psi}\left(x_1\right) \exp\left(i\left(k_2 x_2 + k_3 x_3\right) + ct\right), \tag{12}$$

where k_i are positive real wave numbers and c is a complex constant that governs the temporal behavior of the disturbance. In this way a system of five differential equations for $\widehat{\rho}$, $\widehat{v}_i$ and $\widehat{T}$ is obtained.

In order to find a solution for the values of the critical Rayleigh number, the differential equations are discretized and the derivatives are linearized. Now boundary values for rigid and free boundaries are inserted. For rigid boundaries, all $\widehat{\psi}$ are set to zero at the first and last discretized point while for free boundaries, $\widehat{v}_1$, $\widehat{T}$ and the derivatives of the horizontal velocities $\frac{d\widehat{v}_2}{dx_1}$ and $\frac{d\widehat{v}_3}{dx_1}$ are set to zero at the boundary points. The resulting algebraic problem is transformed into an eigenvalue problem for c. From equation (12) it is obvious that the stability threshold is determined by

$$\max(\mathrm{Re}(c)) = 0, \tag{13}$$

because in that case a disturbance $\widetilde{\psi}$ does not grow in time. For a given layer thickness d and wavenumber $k = \sqrt{k_2^2 + k_3^2}$ the value of the wave number dependend critical Rayleigh number $\widetilde{Ra}_{crit}(d, k)$ is found by an iterative calculation of the eigenvalues c until condition (13) is satisfied. The global critical Rayleigh number is now obtained by

$$\min\{\widetilde{Ra}_{crit}(d,k),\ \forall k \in R\} =: Ra_{crit}(d) \tag{14}$$

The eigenvalues have been calculated by an implicitly restarted Arnoldi algorithm [9]. A more detailed explanation of the linear stability analysis and the numerical calculation of the critical Rayleigh number is presented in[10].

4 Results and Discussion

A stability analysis for compressible fluids was performed as outlined in the previous section for both, liquids and gases. The selected liquids were water, ether (DEE), and mercury. As a gas argon was chosen. All calculations refer to a temperature $T_1 = 298\,\mathrm{K}$ and a pressure of $p_1 = 1\,\mathrm{bar}$ for the liquids, respectively a pressure of 10 bar down to 0.001 bar for argon.

In the Figure 1 the results of the stability analysis for the selected liquids and the gas argon are presented for the different boundary conditions. The horizontal lines indicate the constant value for the critical Rayleigh number according to (2).

The numerical results show that the stability behavior of a the Rayleigh-Bénard problem is characterized by two parameters instead of only the constant Rayleigh number for the Boussinesq case. For a linear stability analysis for compressible fluids the value of the critical Rayleigh number becomes dependent on the thickness d of the fluid layer. For small layer thicknesses the critical Rayleigh number approximates asymptotically the constant value of the Rayleigh number of the Boussinesq case, while the critical Rayleigh number for large layer thicknesses is strongly dependent on the layer thickness.

This behaviour has been confirmed by experiments with near-critical fluids by Kogan, Murphy et al.[11] and may be understood by the fact that a large fluid layer exerts a non negligible pressure on the fluid layer at the bottom. The fluid becomes compressed and density is larger than in the Boussinesq case where a change of density due to pressure is neglected. In order to now achieve a critical density gradient for instability, the temperature gradient - and with it the thermal expansion - has to be increased beyond the one of an incompressible fluid. This implies that the Rayleigh number must increase with an increasing layer thickness.

The critical wave numbers for the different boundary cases resulted in the same values as calculated by use of the Boussinesq approximation[3] (see also (2)).

The simple Ansatz that is described by (10) agrees perfecty with the results of the numerical analysis and may therefore be used for all applications

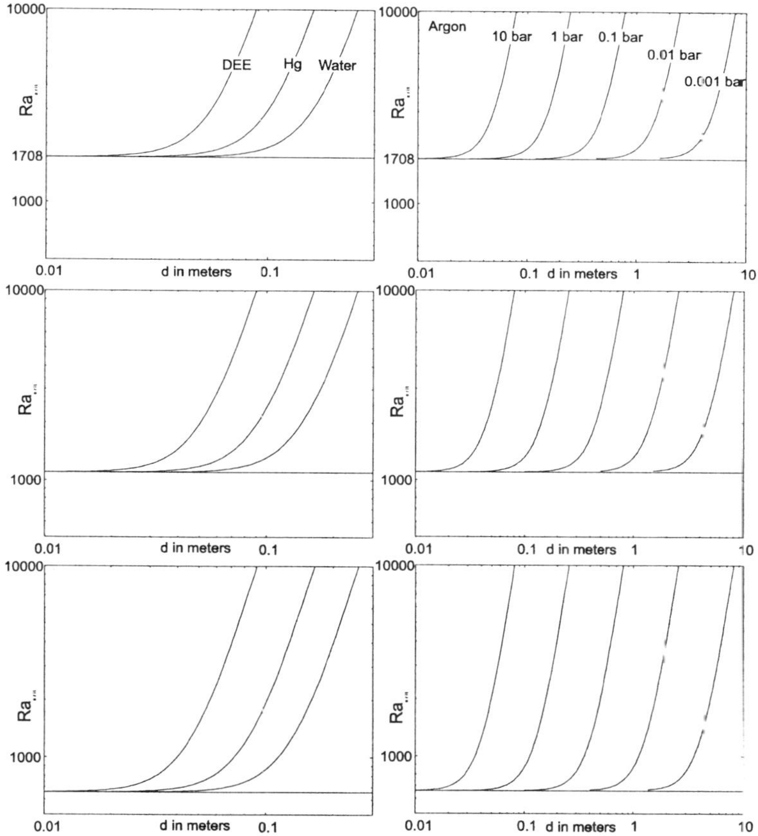

Figure 1. Crititcal Rayleigh number vs. layer thickness. Top row: rigid-rigid boundaries; middle row: rigid-free boundaries; bottom row: free-free boundaries.

where the Rayleigh number is involved.

A detailed comparison of the results of the present work with the literature is given in [10].

5 Summary

For the Rayleigh-Bénard problem a linear stability analysis has been performed numerically on the basis of the compressible, viscous and heat-conducting Navier-Stokes-Fourier equations in order to calculate the value of the critical Rayleigh number for compressible fluids. Unlike in previous works, the results were obtained without approximations of the field equations and without simplification of the solution. The analysis was carried out for liquids and gases and results are presented for a wide range of layer thicknesses. In contrast to the results obtained in the Boussinesq case, the value of the critical Rayleigh number for compressible fluids is not constant but depends on the layer thickness of the fluid layer. For small layer thickness the Rayleigh number approaches asymptotically the value of the Boussinesq case, while for a large fluid layer the critical Rayleigh number of a compressible fluid is greater than the value of the Boussinesq case. This behavior is due to the fact that a great layer thickness affects the density of the fluid even for nearly incompressible fluids.

Acknowledgments

This work was supported by the Deutsche Forschungsgemeinschaft (DFG) priority research program "Analysis and Numerics for Conservation Laws" under the project number MU 313/14-1.

References

1. Lord Rayleigh, Phil. Mag. **32,** 529 (1916).
2. H. Jeffreys, Phil. Mag. **2,** 833 (1926).
3. S. Chandrasekhar, Hydrodynamic and Hydromagnetic Stability, At the Clarendon Press, Oxford (1961).
4. I. Müller, Thermodynamics, Pitman Advanced Publishing Program, Boston, London, Melbourne (1985).
5. E. A. Spiegel, Astrophys. J. **141,3**, 1068 (1965).
6. M. Giterman, V. Shteinberg, J. Appl. Math. Mech. **34,2**, 305 (1970).
7. P. Carlès, B. Ugurtas, Physica D **126**, 69 (1999).
8. H. Jeffreys, Proc. Camb. Phil. Soc. **26,** 170 (1930).
9. R. B. Lehoucq, J. A. Scott, Technical report SAND97-2712J (1997).
10. A. S. Bormann, Cont. Mech. Thermodyn. **13,1**, 9 (2001).
11. A. B. Kogan, D. Murphy, H. Meyer, Phys. Rev. Let. **82,23**, 4635 (1999).

AN INVERSE PROBLEM IN FLUID DYNAMICS

GEORGE BOZIS
Department of Theoretical Mechanics, University of Thessaloniki, GR-540 06 Thessaloniki, Greece
E-mail: gbozis@ccf.auth.gr

FRANCESCO BORGHERO
Dipartimento di Matematica, Università di Cagliari, Via Ospedale 72, 09124 Cagliari, Italy
E-mail: borghero@unica.it

Given a monoparametric family of trajectories $f(x, y) = c$ in a two-dimensional perfect fluid in planar steady motion, we find two compatibility conditions among the function $f(x, y)$, the pressure $p(x, y)$, the density $\rho(x, y)$ and the components $X(x, y)$, $Y(x, y)$, of the body force acting on the fluid. These conditions serve to determine, apart from some arbitrary functions, the pressure and density of the fluid if, f, X, Y are known.

1 INTRODUCTION

One of the fundamental problems in Classical Mechanics is the following version of the inverse problem of Dynamics: *Given a monoparametric family of planar curves $f(x, y) = c$, find all possible force fields, conservative or not, which can give rise to these curves as orbits.* This problem has received much attention after Szebehely's work (1974) [1]. He derived a first order linear partial differential equation *giving all potential functions $V = V(x, y)$* which can generate the given family of trajectories $f(x, y) = c$, traced by a material point of unit mass with a preassigned dependence, $E = E(f)$, of the total energy E of the moving point on its trajectory.

A number of papers generalize Szebehely's equation from various viewpoints. For an almost complete list of the papers in this field see [2].

Simple as it may sound the study of this problem leads to *the understanding of how Geometry is linked to Dynamics.*

It is the aim of this paper to convey the above version of the inverse problem from the field of Newtonian or Lagrangian Mechanics to Fluid Dynamics; in other words to apply the method used in treating Newton's or Lagrange's equations to the Euler equations of motion of a *perfect fluid* [3].

The motivation to consider this problem hardly needs justification. In practice we do observe trajectories described by the particles of the fluid. So our problem is: *to find the compatibility conditions between all possible dis-*

tributions of density $\rho(x,y)$ and pressure $p(x,y)$ of a perfect two-dimensional fluid in planar steady motion which would allow for the creation of certain given (or observed) families $f(x,y)=c$ of streamlines. As for the body forces $X(x,y)$, $Y(x,y)$, one might think that these can either be disregarded or, if not, can be assumed known.

2 NOTATIONS

u,v : velocity components of a two-dimensional fluid; p : pressure; ρ : density; X, Y : body force components; $f(x,y)=c$: equation of a monoparametric family of streamlines in the fluid.
Subscripts denote partial differentiation and dots denote total derivatives with respect to time t.

Additional notations:

$$\gamma=\frac{f_y}{f_x},\quad \Gamma=\gamma\gamma_x-\gamma_y,\quad \kappa=\frac{1-\gamma^2}{\gamma},\quad \lambda=\frac{\Gamma_y-\gamma\Gamma_x}{\gamma\Gamma},\quad \mu=\lambda\gamma+\frac{3\Gamma}{\gamma},\tag{1}$$

$$\Lambda=\lambda+\frac{\gamma_x}{\gamma},\quad M=\mu+\frac{\gamma_y}{\gamma},\quad \alpha=\frac{X+\gamma Y}{\Gamma},\quad \beta=\frac{p_x+\gamma p_y}{\Gamma},\quad \varepsilon=\frac{\rho_y-\gamma\rho_x}{\rho}\tag{2}$$

$$\begin{aligned} b&=-p_{xx}+\kappa p_{xy}+p_{yy}-\lambda p_x-\mu p_y,\\ B&=-p_{xx}+\kappa p_{xy}+p_{yy}-\Lambda p_x-M p_y,\end{aligned}\tag{3}$$

$$\alpha^*=\alpha_y-\gamma\alpha_x,\quad \beta^*=\beta_y-\gamma\beta_x\tag{4}$$

$$\begin{aligned} \xi&=-X_x+\tfrac{1}{\gamma}X_y-\gamma Y_x+Y_y-\lambda X-\mu Y,\\ Z&=-X_x+\tfrac{1}{\gamma}X_y-\gamma Y_x+Y_y-\Lambda X-MY.\end{aligned}\tag{5}$$

In view of $(2)_{3,4}$, (4) and (5), it can be shown by direct calculations that

$$\alpha^*=2Y+\frac{\gamma}{\Gamma}\xi,\quad \beta^*=2p_y+\frac{\gamma}{\Gamma}b\tag{6}$$

$$b^*=b_y-\gamma b_x,\quad B^*=B_y-\gamma B_x,\quad \xi^*=\xi_y-\gamma\xi_x,\quad Z^*=Z_y-\gamma Z_x.\tag{7}$$

3 ANALYSIS OF THE PROBLEM

We consider a two-dimensional steady flow of a perfect fluid. In Euler variables the components of the velocity field **V** are $u = u(x, y)$, $v = v(x, y)$ and the components of the body force **F** acting on the fluid are $X(x, y)$, $Y(x, y)$. The pressure and the density are respectively $p(x, y)$ and $\rho(x, y)$.
In the same variables the Euler equations of motion are

$$\dot{u} = X - \frac{1}{\rho}p_x, \quad \dot{v} = Y - \frac{1}{\rho}p_y \tag{8}$$

and the continuity equation is

$$u\rho_x + v\rho_y + \rho(u_x + v_y) = 0. \tag{9}$$

No assumption is made in advance about any constitutive relation $\phi(p, \rho) = 0$ between the pressure and the density. In fact, at this point, it is not even clear if such an assumption can be made in advance. Suppose now that in the fluid and in the xy plane we observe a monoparametric family of streamlines

$$f(x, y) = c. \tag{10}$$

Since the motion is steady, streamlines and trajectories coincide, so what is actually observed is geometry. The question is: *if the body force* $\{X, Y\}$ *is known, what should the pressure* $p = p(x, y)$ *and the density* $\rho = \rho(x, y)$ *be in order for these functions to be compatible with the observed trajectories* (10) ?

To answer the above question we proceed as follows: from equation (10) differentiating with respect the time, we have

$$f_x u + f_y v = 0,$$

or in view of $(1)_{1,2}$

$$u = -\gamma v. \tag{11}$$

As the flow is in steady motion, the "slope function" $\gamma = -\frac{u}{v}$ depends merely on x, y. Actually this function represents the slope of the trajectories which are orthogonal to the given family (10). From the four equations (8), (9), and (11) we now eliminate the velocity components u, v. In fact the time t is also eliminated from the equations of motion (8). In so doing, we obtain two equations (equations (16) and (19) below) relating the "observed" function γ, the given body force $\{X, Y\}$ and the unknown pressure and density of the fluid. This elimination is effectuated as follows: we write the derivative $\dot{\gamma}$ in the following two ways

$$\frac{u\dot{v} - \dot{u}v}{v^2} = \gamma_x u + \gamma_y v. \tag{12}$$

From (12), in view of (8),(11) and $(2)_{3,4}$, we obtain

$$v^2 = \alpha - \frac{1}{\rho}\beta. \tag{13}$$

Replacing u from (11) by $u = -v\gamma$ in the ratio $\frac{\dot{u}}{\dot{v}} = \frac{u_x u + u_y v}{v_x u + v_y v}$ we find

$$\frac{\dot{u}}{\dot{v}} = -\gamma + \frac{2v^2\Gamma}{(v^2)_y - \gamma(v^2)_x} \tag{14}$$

with Γ given by $(1)_2$. We also take the ratio $\frac{\dot{u}}{\dot{v}} = \frac{\rho X - p_x}{\rho Y - p_y}$ from equations (8) and equate it to the ratio given by (14). So we find

$$\frac{\rho\alpha - \beta}{\rho Y - p_y} = \frac{2v^2}{(v^2)_y - \gamma(v^2)_x} \tag{15}$$

with α, β given by $(2)_{3,4}$. In view of (13) and (4), we calculate the expression $(v^2)_y - \gamma(v^2)_x = \alpha^* - \frac{1}{\rho}\beta^* + \frac{\beta}{\rho}\varepsilon$ with ε given by $(2)_5$ and α^*, β^* by (4). With $\rho\alpha - \beta = \rho v^2$, equation (15) is written as $(\alpha^* - 2Y)\rho = \beta^* - 2p_y - \beta\varepsilon$. So, in view of (6), we obtain our **first result**

Proposition : *Given a monoparametric family of trajectories $f(x, y) = c$ in a two-dimensional perfect fluid in steady motion, we have*

$$\Gamma\beta\varepsilon = \gamma(b - \rho\xi) \tag{16}$$

relating pressure and density to the known slope function $\gamma(x, y)$ and the given body force $\{X, Y\}$. It is a differential relation of the ***first order*** *for the density but of the* ***second order*** *for the pressure.*

Comment: A basic result in the framework of the inverse problem of Dynamics referring to the motion of one material point in the xy plane is the following: the components X, Y of an autonomous force field and the slope function γ of a family (10) generated by this field are related by the (differential) relation (Bozis, 1995, [2])

$$\xi = 0. \tag{17}$$

Equation (17) was derived by eliminating the time t between the two equations of motion $\ddot{x} = X(x, y)$, $\ddot{y} = Y(x, y)$. In our case, the second members of equations (8) also depend on x, y. So, as expected, if we replace X by $X - \frac{1}{\rho}p_x$ and Y by $Y - \frac{1}{\rho}p_y$ and apply (17), we obtain directly the relation (16).

We now turn attention to the continuity equation (9), which, for Fluid Dynamics, is an additional fact. Keeping the policy of expressing the velocity component u from (11), we write (9) in the form

$$v(\varepsilon - \gamma_x) + v_y - \gamma v_x = 0$$

or, multiplying through by $2v$,

$$2v^2(\varepsilon - \gamma_x) + (v^2)_y - \gamma(v^2)_x = 0. \tag{18}$$

Comparing (15) and (18), we obtain

$$(\rho\alpha - \beta)(\varepsilon - \gamma_x) + (\rho Y - p_y) = 0 \tag{19}$$

which is our **second result**, of the same nature with (16). It is a relation free of the velocity components. It is important to notice that, both in (16) and (19), the (first order) derivatives of the density ρ appear through the function ε, given by $(2)_5$. So, eliminating ε between (16) and (19) we obtain our **third result**, that is a quadratic in ρ, algebraic equation:

$$\alpha\xi\rho^2 - (\alpha b + \beta Z)\rho + \beta B = 0. \tag{20}$$

4 THE ROLE OF THE BODY FORCES

In the analysis of the previous section the body forces acting on the flow are considered to be known, be it zero or not. (In fact, in many instances the body forces are negligible). However the result expressed by formula (20) shows that *there arises an essential difference between the zero and nonzero force field.* Indeed:

Case a: For zero body forces ($X = Y = 0$) it is $\alpha = 0$, $\xi = Z = 0$ and equation (20) reduces to $B = 0$, i.e.

$$-p_{xx} + \kappa p_{xy} + p_{yy} = \Lambda p_x + M p_y. \tag{21}$$

Equation (21) relates pressure and trajectories. For material point Dynamics it reminds of the equation (Bozis, 1984, [4])

$$-V_{xx} + \kappa V_{xy} + V_{yy} = \lambda V_x + \mu V_y \tag{22}$$

relating potentials V and trajectories.

Since (21) is a second order PDE in p, to an observed family (10), there correspond as many possible pressure distributions $p = p(x, y)$ as two arbitrary functions allow. The example treated in section 5 refers to this case.

It is striking that, in this particular case ($X = Y = 0$), for a compatible pair (γ, p) the density has to be calculated by a PDE in ρ (e.g. equation (16)).

Case b: For nonzero forces, equation (20) offers two possible solutions-functions for the density $\rho = \rho(x, y)$, provided that a family of trajectories (10) is observed and also that, somehow, the pressure function $p = p(x, y)$ is known. In place of equation (21) (relating pressure to trajectories), we can find an analogous condition (now containing also the force components X, Y) by eliminating the density ρ between (20) and (16). The result is indeed lengthy and complicated and we do not give it here. We report only that the pertinent differential condition for the pressure $p(x, y)$ is of *the third order.*

5 AN EXAMPLE

We suppose (i) that the flow of our perfect fluid is in steady planar motion, (ii) that $X = Y = 0$, and (iii) that the family (10) of hyperbolas

$$f(x, y) = xy = c \tag{23}$$

is observed in the flow. From equations (1), $(2)_{1,2}$ we obtain the pertinent functions and, in view of (3), we write the equation (21). It reads

$$-p_{xx} + \frac{y^2 - x^2}{xy} p_{xy} + p_{yy} = -\frac{2}{x} p_x + \frac{2}{y} p_y. \tag{24}$$

To fix ideas, let us consider out of the totality of solutions of (24), the (simple) solution

$$p = \frac{1}{2}(x^2 + y^2). \tag{25}$$

With this pressure p and with $\gamma = \frac{x}{y}$, we express the function β (from $(2)_4$), b (from $(3)_1$), $\xi = 0$ (from $(5)_1$) and, with ε given by $(2)_5$, we write equation (16). It reads

$$x\rho_x - y\rho_y = 0 \tag{26}$$

with general solution $\rho = G(xy)$, G being an arbitrary function. The density is constant along each trajectory of the family (23).

Other possible solutions for p and ρ may be searched for possibly imposed boundary conditions. The idea is that, on the basis of observed trajectories, we obtain information regarding the rest of fluid dynamical functions.

References

1. V. Szebehely, *On the Determination of the Potential by Satellite Observations*, Rend. Sem. Fac. Scien. Univ. Cagliari, **XLIV**, 31, (1974).
2. G. Bozis, *The Inverse Problem of Dynamics: Basic Facts*, Special issue of *Inverse Problems*, **11**, 687-708, (1995).
3. J. Serrin, *Mathematical Principles of Classical Fluid Mechanics*, Encyclopaedia of Physics, ed. by S. Flugge, vol. **VIII/1**, Fluid Dynamics I, 125-350, Springer-Verlag, (1959).
4. G, Bozis, *Szebehely's Inverse Problem for Finite Material Concentrations*, Astron. Astroph., **134**, 360-364, (1984)

CONTINUUM MECHANICS AND DYNAMICAL PERMUTATIONS

Y. BRENIER
CNRS UMR 6621, Parc Valrose, Université de Nice,
06108 Nice, France
E-mail: brenier@math.unice.fr

Many models in Mathematical Physics and Continuum Mechanics describe the motion of a continuum of identical particles (or extended objects such as strings or membranes) that are subject to various kind of interactions (collisions, repulsive or attractive forces...). Some of them can be approximated by very simple dynamical systems involving permutations of the particles labels. Several examples will be discussed, including the simplest model of adhesion dynamics, linked to one-dimensional scalar conservation laws, the Euler equations of inviscid incompressible fluids, some models in Electrodynamics and Geophysics (such as the semi-geostrophic equations for atmospheric fronts) and, finally, isothermal gas dynamics equations through the concept of harmonic functions "up to rearrangement".

1 Permutations and sticky particles

1.1 Sticky particles

Let us consider a set of particles, labelled by $a \in \{1, ..., N\}$, of unit mass, moving along the real axis, with position $X(t, a)$ and velocity $X'(t, a)$ at time $t \in [0, T]$, where $T > 0$ is a fixed horizon time. These particles are supposed to freely stream until they collide. Then, they stick and their total momentum is preserved. Typically, as a binary collision occurs at time t^* between particle a and particle $\tilde{a}$, we get

$$X'(t^* + 0, a) = X'(t^* + 0, \tilde{a}) = \frac{1}{2}(X'(t^* - 0, a) + X'(t^* + 0, \tilde{a})).$$

Of course, such collisions dissipate energy. So, this model of "sticky particles" is probably the simplest model of particle interaction, combining free streaming, conservation of momentum and maximal dissipation of energy.

1.2 Velocity exchanching particles

Let us introduce a time step $h > 0$. Assume that all particles freely stream between $t = (n-1)h + 0$ and $t = nh - 0$, for all integer $n > 0$. Namely

$$X(t, a) = X((n-1)h, a) + (t - (n-1)h)X'((n-1)h + 0, a), \quad \forall t \in [(n-1)h, nh].$$

Then, they may cross and the order along the real axis of their positions may change accordingly. Generically, all positions $X(nh, a)$ are distinct from each other for different values of a. (We may slightly randomize them, otherwise.) Let us denote by $a \to \sigma_n(a)$ the unique permutation such that $X(nh, \sigma_n(a))$ is strictly increasing in a. Then, the velocities are exchanged accordingly. Namely,

$$X'(nh + 0, \sigma_n(a)) = X'((n-1)h + 0, \sigma_{n-1}(a)).$$

(In other words, velocities are assigned to each particle according to its order along the real axis.) Notice that there is no energy dissipation in this process. From the mathematical point of view, we have

Theorem 1.1 *Let us denote by $X(t, a)$ and $X_h(t, a)$ the particle positions, for $t \in [0, T]$ and $a = 1, ..., N$, corresponding respectively to the sticky particle dynamics and the velocity exchanging dynamics. Assume that at time $t = 0$,*

$$X_h(0, a) = X(0, a), \quad X_h'(0, a) = X'(0, a), \quad a = 1, ..., n.$$

Then there is a constant C depending only on T and $\max_a |X'(0, a)|$ such that

$$\sup_{0 \le t \le T} \frac{1}{N} \sum_{a=1}^{n} |X_h(t, a) - X(t, a)| \le C\sqrt{h}.$$

The proof easily follows from two earlier results. First, it has been established in[7] (in the spirit of Dafermos' polygonal approximations of scalar conservation laws [9]) that sticky particles dynamics correspond to general scalar conservation laws in one space dimension with nondecreasing initial data. Next, the velocity exchanging dynamics is nothing more than a discrete version, described in [4], of the "transport-collapse"method for scalar conservation laws [3]. This method is equivalent to the kinetic approach to scalar conservation laws by Giga and Miyakawa [10]. (A more recent variant is the Perthame-Tadmor "BGK model" for scalar conservation laws [12].) As h goes to zero, the transport-collapse approximation is known to converge to the "entropy solutions" of scalar conservation laws.

2 Incompressible inviscid fluids

2.1 Harmonic oscillators with exchange of centers

Consider the periodic cube $D = \mathbf{T}^d = (\mathbf{R}/\mathbf{Z})^d$. Dispose $N = n^d$ particles equally spaced in D at points $Y(1), ..., Y(N) \in D$. Let us call them "ions". Now, consider N other particles, called "electrons", with label $a = 1, ..., N$ and position $X(t, a) \in D$ at time t. A time step $h > 0$ is fixed. In each time interval $nh \leq t < (n+1)h$, there is a one-to-one pairing (i.e. a permutation) $a \to \sigma(a)$ between each electron a and each ion $Y(\sigma(a))$. A spring links each pair so that each electron oscillates around the corresponding ion with frequency proportional to ϵ^{-1} :

$$\epsilon^2 X''(t,a) + X(t,a) = Y(\sigma(a)) \tag{1}$$

During the time interval $nh < t < (n+1)h$, the total energy of the springs

$$E(t) = \frac{1}{2}\sum_{a=1}^{N}(|X'(t,a)|^2 + \epsilon^{-2} d(X(t,a), Y(\sigma(a)))^2), \tag{2}$$

(where $d(.,.)$ is the Euclidean distance on D) is conserved. At each discrete time $t = nh$, the permutation σ is updated and chosen to minimize, among all permutations, the potential energy

$$\frac{1}{2}\sum_{a=1}^{N}\epsilon^{-2} d(X(t,a), Y(\sigma(a)))^2. \tag{3}$$

(Notice that they may be several solutions, in which case we arbitrarily choose one of them.) So, σ is time dependent, piecewise constant, and denoted by $\sigma(t)$. Of course, we assume both positions and velocities of each particle to be continuous at each discrete time nh and we prescribe their values at time 0. This gives a complete description of the dynamical system. Notice that the total energy, defined by (2), is preserved on each interval $nh < t < (n+1)h$, and can only decay at each time nh, by definition of $\sigma(nh)$. So the total energy is a non-increasing function of time. The possible dissipation is due to the fact that $\sigma(t)$ is updated only at $t = nh$, and not continuously in time (in which case, the system would be formally conservative). It has been shown in [5] that this model of harmonic oscillators with exchange of centers describe incompressible inviscid fluids in the limit as ϵ, h, N^{-1} approach zero. More precisely

Theorem 2.1 *Let (v,p) a smooth solution to the Euler equations*

$$(\partial_t + v.\nabla)v + \nabla p = 0, \quad \nabla.v = 0,$$

where $v(t,x)$ and $p(t,x)$ are respectively the velocity and the pressure at $t \in [0,T]$, $x \in D$. Assume the following ordering to be true

$$N^{-1} \leq O(\epsilon^{8d}), \quad h \leq O(\epsilon^4)$$

and $X(0,a) = Y(a) + o(\epsilon)$. Then, if

$$X'(t,a) = v(0, X(t,a)) + o(1),$$

holds true at $t = 0$, it holds true also for all $t \in [0,T]$.

2.2 *A caricature of Coulomb interaction*

At least formally, in the limit $h = 0, N = \infty$, as $\epsilon > 0$ is fixed, the density field ρ and the velocity field v of the particles are subject to the "Euler-Monge-Ampère" system

$$(\partial_t + v.\nabla)v + \nabla p = 0,$$

$$\partial_t \rho + \nabla.(\rho v) = 0,$$

$$det(\delta_{ij} - \epsilon^2 \partial^2_{x_i\ x_j} p) = \rho.$$

As $\epsilon > 0$ goes to zero, we can expand the determinant and get

$$1 - \epsilon^2 \Delta p = \rho + O(\epsilon^4)$$

(with no error if $d = 1$). Neglecting the $O(\epsilon^4)$ term in the last equation leads to the (pressureless) Euler-Poisson system which describes the Coulomb interaction between a continuum of electrons and a uniform background of ions. So the model of harmonic oscillators with exchange of centers can be seen as a caricature of Coulomb interaction.

3 The semi-geostrophic equations

Hoskins' frontogenesis model and the related semi-geostrophic equations in atmospheric sciences [8], see also [2]) can be discretized in a way involving permutations. Following [1], at least in two space dimensions, we have a set of N particles moving in the periodic square $D = \mathbf{T}^2$ with positions $X(t, a) \in D$, for $a = 1, ..., N$, $t \in [0, T]$, according to the dynamical system

$$i.X'(t, a) + X(t, a) = Y(\sigma(t, a)) \tag{4}$$

where i is the rotation matrix of angle $\pi/2$ in the plane, and $Y(a), \sigma(t, a)$ are defined exactly in the same way as in the previous section. This (formally) Hamiltonian system differs from the one discussed in the previous section exactly as vortex point dynamics (as described in [11] for instance) differs from classical Coulomb dynamics.

4 Jointly vibrating strings

4.1 *The joint potential energy*

Let us consider a collection of vibrating strings parameterized by

$$s \in [0, S] \rightarrow X(t, s, a) \in \mathbf{R}^d,$$

and labelled by $a \in A = [0,1]$. The kinetic energy of each vibrating string at time t is evaluated individually

$$\int_0^S \frac{1}{2}|\partial_t X(t,s,a)|^2 ds$$

and then integrated in $a \in A$. The potential energy of X is not evaluated *individually* for each vibrating string by the usual formula

$$\int_0^S \frac{1}{2}|\partial_s X(t,s,a)|^2 ds,$$

but *jointly* by rearranging the labels of the strings for each fixed value of (t,s), in order to get the lowest possible energy. In other words, we consider all possible collections of fictitious vibrating strings

$$s \in [0,S] \to Y(t,s,a)$$

having the spatial density as X has, i.e. satisfying

$$\int_A \delta(x - Y(t,s,a))da = \rho(t,s,x), \tag{5}$$

where ρ is the nonnegative measure defined by

$$\rho(t,s,x) = \int_A \delta(x - X(t,s,a))da \tag{6}$$

for all continuous test function h and for all $s \in [0,S]$. Then, we define the joint potential energy to be the infimum of

$$\int_A \int_0^S \frac{1}{2}|\partial_s Y(t,s,a)|^2 dsda,$$

with respect to all Y satisfying (5). Then, it can be shown (see [6] for more details) that, assuming ρ to be a smooth positive function of t,s,x, an optimal choice for Y is obtained by solving the following ode in s (t and a being fixed)

$$\partial_s Y(t,s,a) = (\nabla\phi)(t,s,Y(t,s,a)), \quad Y(t,s=0,a) = X(t.0,a), \tag{7}$$

where $\phi(t,s,x)$ solves the elliptic equation in x (t and s just being parameters)

$$-\nabla.(\rho\nabla\phi) = \partial_s\rho. \tag{8}$$

(N.B. In order to have a precise statement, we need appropriate boundary conditions. The easiest way, adopted in[6], is to consider that the strings vibrate in the peridic cube $D = \mathbf{T}^d = (\mathbf{R}/\mathbf{Z})^d$. Then, the elliptic equation can be uniquely solved in ϕ with prescribed zero mean on D.)

4.2 The Dynamical equations

Dynamical equations for systems of vibrating string are obtained by varying the difference between the potential energy and the kinetic energy both integrated in time. In the case of an individual string, we would get the simple one-dimensional wave equation

$$\partial_{tt}X = \partial_{ss}X.$$

Here, because the potential energy is computed after optimizing in labels a, the dynamical equations for the jointly vibrating strings are far more complicated. They are obtained by varying

$$\int \frac{1}{2}(\eta|\partial_t X|^2 + |\partial_s Y|^2)dtdsda - \int (q(t,s,X(t,s,a)) - q(t,s,Y(t,s,a)))dtdsda, \tag{9}$$

where $\eta = -1$ and $q(t,s,x)$ is the Lagrange multiplier for constraint (5). We get

$$\eta\partial_{tt}X(t,s,a) = (-\nabla q)(t,s,X(t,s,a)), \tag{10}$$

$$\partial_{ss}Y(t,s,a) = (+\nabla q)(t,s,Y(t,s,a)). \tag{11}$$

Because of (7), $\nabla\phi$, denoted by e, must satisfy

$$\partial_s e + (e.\nabla)e = \nabla q, \tag{12}$$

which leads, after integrating in x, to the following expression for q (up to an irrelevant additive constant)

$$q = \partial_s\phi + \frac{1}{2}|\nabla\phi|^2.$$

Summarizing the equations we obtained for the jointly vibrating strings, we have

$$\eta\partial_{tt}X(t,s,a) = (-\nabla q)(t,s,X(t,s,a)), \quad \rho(t,s,x) = \int \delta(x - X(t,s,a))da, \tag{13}$$

$$\partial_s\rho + \nabla.(\rho\nabla\phi) = 0, \quad \partial_s\phi + \frac{1}{2}|\nabla\phi|^2 = q. \tag{14}$$

The second order ode (13) can be expressed as well in terms of the "phase density" $f(t,s,x,\xi) \geq 0$ associated to X and defined by

$$f(t,s,x,\xi) = \int \delta(x - X(t,s,a))\delta(\xi - \partial_t X(t,s,a))da, \tag{15}$$

for $\xi \in \mathbf{R}^d$. We get for f

$$\partial_t f + \xi.\nabla_x f - \eta \nabla q.\nabla_\xi f = 0, \quad \int f(t,s,x,d\xi) = \rho(t,s,x). \tag{16}$$

So, the jointly vibrating strings equations can be written either as the combination of (13), (14), in terms of (X, ρ, ϕ), or as the combination of (14), (16), in terms of (f, ρ, ϕ). Special solutions of the latter formulation can be found by assuming f and q to be time independent. Indeed, in such a case,

$$f(s,x,\xi) = F(s, \frac{|\xi|^2}{2} - q(s,x)), \tag{17}$$

provides a solution for (16), for any choice of $F(s,r)$ as a smooth nonnegative function with sufficient decay at $r = +\infty$. To solve the jointly vibrating string equations, we now need

$$\rho(s,x) = \int f(s,x,\xi)d\xi = R(s, q(s,x)), \tag{18}$$

where (using polar coordinates)

$$R(s,\sigma) = C_d \int_0^\infty F(s, \tau - \sigma)\tau^{d/2-1} d\tau, \tag{19}$$

C_d being the $d-1$ Hausdorff measure of the unit sphere in $\mathbf{R}^d$. In particular, as $F(s,r) = (2\pi)^{-d/2}\exp(-r)$, we get

$$\rho(s,x) = \exp(q(s,x)), \tag{20}$$

and

$$f(s,x,\xi) = \rho(s,x)(2\pi)^{-d/2}\exp(-\frac{|\xi|^2}{2}). \tag{21}$$

From this calculation, it is not hard to deduce

Theorem 4.1 *Let us consider a smooth solution $(f, \rho > 0, \phi)$ to the jointly vibrating string equations (16), (14). Assume that ρ does not depend on t and is not uniform in (s,x). Then f has the "thermalized" form*

$$f(s,x,\xi) = \rho(s,x)(2\pi\Theta)^{-d/2}\exp(-\frac{|\xi|^2}{2\Theta})$$

for some positive constant $\Theta > 0$, if and only if $(\rho, u = \nabla\phi)$ satisfies

$$\partial_s \rho + \nabla.(\rho u) = 0, \tag{22}$$

$$\partial_s u + (u.\nabla)u + \frac{\nabla p}{\rho} = 0, \tag{23}$$

$$p = \eta\rho\Theta. \tag{24}$$

Equations (22),(23) describe an isothermal gas, with density $\rho(s,x)$, velocity $u(t,x)$ with an unphysical negative temperature $\eta\Theta$, since $\eta = -1$. This yields a way to recover the physical isothermal gas dynamics equations just by using $\eta = +1$, instead of $\eta = -1$ in definition (9) for the Action principe! This amounts to consider no longer jointly vibrating strings but rather "harmonic functions up to rearrangement" in the following sense :

Definition 4.2 *A collection of functions $(t,x,a) \to X(t,s,a)$ is harmonic up to rearrangement if it minimizes*

$$\int |\partial_t X(t,s,a)|^2 dtdsda + \inf_Y \int |\partial_s Y(t,s,a)|^2 dtdsda, \tag{25}$$

with respect to local perturbations, where the infimum in Y is performed over all possible rearrangement Y of X with respect to $a \in A$, i.e. all Y such that

$$\int_A \delta(x - Y(t,s,a))da = \int_A \delta(x - X(t,s,a))da.$$

Then it can be established than smooth solutions of the isothermal gas dynamics equations generate non trivial examples of harmonic maps up to rearrangement $X(t,s,a)$ [6].

References

1. S. Baigent, J. Norbury, Phys. D **109**, 333 (1997).
2. J.-D. Benamou, Y. Brenier, SIAM J. Appl. Math. **58**, 1450 (1998).
3. Y. Brenier, J. Differential Equations **50**, 375 (1983).
4. Y. Brenier, SIAM J. Numer. Anal. **21**, 1013 (1984).
5. Y. Brenier, Comm. Math. Phys. **212**, 93 (2000).
6. Y. Brenier, Harmonicity up to rearrangement and isothermal gas dynamics, Preprint CNRS, UMR 6621, Université de Nice, 2001.
7. Y. Brenier, E. Grenier . SIAM J. Numer. Anal. **35**, 2317 (1998).
8. M.J.P Cullen, R.J. Purser, J. Atmospheric Sci. **46**, 2684 (1989).
9. C. Dafermos, J. Math. Anal. Appl. **38**, 33 (1972).
10. Y. Giga, T. Miyakawa, Duke Math. J. **50**, 505 (1983).
11. C. Marchioro, M. Pulvirenti, *Mathematical theory of incompressible nonviscous fluids*, Springer-Verlag, New York, 1994.
12. B. Perthame, E. Tadmor, Comm. Math. Phys. **136**, 501 (1991).

Entropy Principle and Non Controllable Data for the Moment Systems

F. BRINI, T. RUGGERI
Department of Mathematics and Research Center of Applied Mathematics - C.I.R.A.M. - University of Bologna
via Saragozza, 8 40123 Bologna, Italy
E-mail: brini@ciram.ing.unibo.it, ruggeri@ciram.ing.unibo.it

From the Boltzmann equations it is possible to derive truncated and closed finite moment systems through the procedures of moment theory and Extended Thermodynamics. In particular the closure of the system is provided by the entropy principle which is surely one of the main requirement of Extended Thermodynamics. For the sake of simplicity, the corresponding truncated distribution function is usually expanded with respect to the non equilibrium variables in the neighborhood of an equilibrium state. For the approximated system we have shown that in general the entropy principle requirement fails if all the non-equilibrium variables are of the same order of magnitude. This fact constitutes a selection criterion for the possible solutions and suggests a possible way to select the so called non-controllable boundary data. The results are also illustrated through some simple examples.

1 Introduction

It is well known that the entropy principle plays a central role both in Thermodynamics and in Extended Thermodynamics. In particular the procedures related to this second theory refer very often to the principle [1] in order to determine, for example, the physical constitutive equations for classical solutions or to select admissible weak solutions [2] or, furthermore, to rewrite the field equations in a symmetric hyperbolic form [3]. When an infinite hierarchy of moment equations is derived from the BOLTZMANN equation and truncated to a finite differential system, we can close it requiring the validity of the entropy principle [4]. Therefore it is evident that for such a truncated and closed full non linear moment system the principle and the hyperbolicity property are valid just by construction. In fact the equations obtained in this manner constitute a hard task both from the analytical and from the numerical point of view. Hence, at least if we consider processes not far from an equilibrium state, the truncated distribution function is commonly expanded in the neighborhood of the Maxwellian distribution [1]. So, we have now to introduce two indexes: one related to the truncation order, in particular to the number of moments m, the other associated to the order α of the approximation of the truncated distribution function and we have as many theories ET_m^α as many different couples

(α, m) we can consider [5]. Apparently it seems that you may choose the values of α and m independently from each others and since now the two indexes were always considered separately [1]. Let us go back to the question about the entropy principle and the ET_m^α theories. Due to the expansion of the distribution function in the neighborhood of an equilibrium state, the properties mentioned above remain valid only locally [1,5,6] and we expect that the entropy principle is still valid only if the non equilibrium field variables and their first derivatives are sufficiently small according to the required approximation order. In this paper we show through some simple stationary examples that the entropy principle fails inevitably, if all the non-equilibrium variables are of the same order of magnitude. In fact there exist often some first derivatives of the non equilibrium variables (*critical derivatives*) that are not small along all the solutions.

Therefore the entropy principle becomes an admissibility criterion for the solutions of any ET_m^α and hence for the choice of the Cauchy or boundary data. This observation is, of course, general and it should be taken into account in any application of a ET_m^α theory, but its contribution becomes particularly helpful when we are dealing with the so-called non-controllable data [5]. With the previous expression we refer to the Cauchy or boundary data for the moments of order higher than the 13, since it is not possible to associate to them a physical meaning and consequently to prescribe the corresponding values through physical or experimental considerations at a macroscopic level. In 1998 STRUCHTRUP and WEISS [7] proposed the idea that, at least in the stationary case, the non-controllable boundary data could be prescribed requiring that the L^∞ norm of the entropy production should be minimal (Minimax Entropy Principle). That is to say nature "chooses" the solution which is as close as possible to an equilibrium state. Unfortunately this elegant principle [8,9,10] presents several weak points and it seems not to work well. In particular it is easily verified that the solutions selected by the Minimax Entropy Principle are usually not entropic. The discovery of the so called critical derivatives has suggested to us a possible alternative way to prescribe the non controllable boundary data in stationary problems. In this paper we present some applications of this new criterion in order to investigate its validity and its consequences. It is still an open question if it is possible to choose the two indexes (α, m) in order to avoid the presence of the critical derivatives.

2 Moment systems and entropy principle

Following the kinetic theory, it is possible to describe the state of a rarefied gas through the introduction of the phase density $f(\mathbf{x}, t, \mathbf{c})$ that obeys the

BOLTZMANN equation

$$\partial_t f + c_i \,\partial_i f = Q \tag{1}$$

where Q and $\mathbf{c} \equiv (c_i)$ denote respectively the collisional term and the microscopic velocity. To (1) the moment procedure associates an infinite hierarchy of balance laws, usually called moment system

$$\partial_t \mathbf{u} + \partial_i \mathbf{F}^i = \mathbf{P} \tag{2}$$

where:

$$\begin{cases} \mathbf{u}(\mathbf{x},t) = \int f(\mathbf{x},t,\mathbf{c})\,\mathbf{\Phi}(\mathbf{c})\,dc; \\ \mathbf{F}^i(\mathbf{x},t) = \int f(\mathbf{x},t,\mathbf{c})\,c_i\mathbf{\Phi}(\mathbf{c})\,dc; \quad \mathbf{P}(\mathbf{x},t) = \int Q(\mathbf{x},t,\mathbf{c})\,\mathbf{\Phi}(\mathbf{c})\,dc; \end{cases} \tag{3}$$

and

$$\mathbf{\Phi} \equiv \begin{pmatrix} 1 \\ c_{k_1} \\ c_{k_1}c_{k_2} \\ \vdots \\ c_{k_1}c_{k_2}\cdots c_{k_j} \\ \vdots \end{pmatrix}, \qquad j = 1, 2, \ldots$$

The integrals - supposed convergent - are taken over the whole the $\mathbf{c}$-space, $dc = dc_1 dc_2 dc_3$. Such a system presents some important peculiarity. First of all the moments $\mathbf{u}$ (the densities), the fluxes $\mathbf{F}^i$, and the production terms $\mathbf{P}$ are tensors of increasing order and the flux in one block equation becomes the density in the next one. Also the first five productions P, P_{k_1}, P_{kk} vanish reflecting the conservation of mass, momentum and energy. Thus we have

$$\mathbf{u} \equiv \begin{pmatrix} u \\ u_{k_1} \\ u_{k_1k_2} \\ \vdots \\ u_{k_1k_2\cdots k_j} \\ \vdots \end{pmatrix}, \qquad \mathbf{F^i} \equiv \begin{pmatrix} u_i \\ u_{ik_1} \\ u_{ik_1k_2} \\ \vdots \\ u_{ik_1k_2\cdots k_j} \\ \vdots \end{pmatrix}, \qquad \mathbf{P} \equiv \begin{pmatrix} 0 \\ 0 \\ P_{<k_1k_2>} \\ \vdots \\ P_{k_1k_2\cdots k_j} \\ \vdots \end{pmatrix}, \tag{4}$$

where $P_{<ij>}$ denotes the deviatoric part of the tensor P_{ij}.

When system (2) is truncated to a finite number of equations, the closure problem arises naturally. Besides the classical GRAD closure, a different approach can be used following the Extended Thermodynamics procedures [1,4] which require in particular that all the solutions of the truncated system (2) up to the tensorial order n are also solutions of the entropy law

$$\partial_t h + \partial_i h^i = \Sigma \geq 0, \tag{5}$$

where h, h^i and Σ are respectively the entropy density, the entropy flux and the entropy production. For a non-degenerate gas we have:

$$h = -k \int f \ \log f \, dc, \quad h^i = -k \int f \ \log f \, c^i \, dc,$$

where k denotes the BOLTZMANN constant and the atomic mass is supposed to be equal one. We recall that for infinitely many equations the entropy principle corresponds to the H-Theorem, but if n is finite it becomes a constraint for the admissible truncated distribution function f.

We may require that all the solutions of (2) satisfy also the supplementary entropy law (5) with h supposed to be a concave function. Under such conditions the general theory of hyperbolic systems of balance laws (2) ensures the existence of a privileged field (*main field*)[1,3] $\mathbf{u}'$ that can be used to combine linearly (2) in order to obtain (5):

$$\mathbf{u}' \cdot \left(\partial_t \mathbf{u} + \partial_i \mathbf{F}^i - \mathbf{P}\right) \equiv \partial_t h + \partial_i h^i - \Sigma. \tag{6}$$

For the quasi-linearity, the identity (6) implies:

$$\mathbf{u}' \cdot d\mathbf{u} = dh; \ \ \mathbf{u}' \cdot d\mathbf{F}^i = dh^i; \ \ \mathbf{u}' \cdot \mathbf{P} = \Sigma \geq 0, \tag{7}$$

and it is possible to verify that there exist four potentials $h^{'}, h^{'i}$ (generators)

$$h' = \mathbf{u}' \cdot \mathbf{u} - h, \ h'^i = \mathbf{u}' \cdot \mathbf{F}^i - h^i, \tag{8}$$

such that

$$\mathbf{u} = \frac{\partial h'}{\partial \mathbf{u}'}, \ \mathbf{F}^i = \frac{\partial h'^i}{\partial \mathbf{u}'}. \tag{9}$$

In the case of moment systems, BOILLAT and RUGGERI have proved[4] that the conditions $(7)_{1,2}$ become a strong constraint for the truncated distribution function f. If f is supposed to depend on $\mathbf{x}, \mathbf{t}$ trough the moments, we have:

$$f = \exp(-1 - \chi/k), \tag{10}$$

where

$$\chi = \mathbf{u}'(\mathbf{x},t)\cdot\mathbf{\Phi}(\mathbf{c}). \tag{11}$$

Of course f depends on the truncation index n and it would be more correct to denote it by f_n. In general f_n is not a solution of the BOLTZMANN equation and it is conjectured that for $n \to \infty$, f_n converges in "some sense" to the distribution function f which is the solution of the BOLTZMANN equation.

In [4] it was verified that this procedure of closure is equivalent to the one obtained by the so-called *maximum entropy principle* [11].

We notice that f given by (10) solves the closure problem. In fact if we insert (10) into $(3)_1$ we have a one to one map between $\mathbf{u}'$ and $\mathbf{u}$. Through this map $\mathbf{u}'$ can be expressed as a function of $\mathbf{u}$ and inserting it into $(3)_2$ we obtain the vector fluxes $\mathbf{F^i}$ and the productions $\mathbf{P}$ as functions of $\mathbf{u}$.

In equilibrium all the components of the main field vanish, except the first five [1] and (10) reduces to the Maxwellian f_M. Therefore for processes close to equilibrium the distribution function may be approximated by the expansion

$$f = f_M\left(1-\frac{1}{k}\,\boldsymbol{\lambda}\cdot\mathbf{\Phi}(\mathbf{c})+\frac{1}{2k^2}\left(\boldsymbol{\lambda}\cdot\mathbf{\Phi}(\mathbf{c})\right)^2+\ldots+\frac{(-1)^\alpha}{\alpha!k^\alpha}\left(\boldsymbol{\lambda}\cdot\mathbf{\Phi}(\mathbf{c})\right)^\alpha\right), \tag{12}$$

where

$$\boldsymbol{\lambda} = \mathbf{u}' - \mathbf{u}'_E$$

denotes the non-equilibrium part of the main field. If only the linear terms are kept, the form of the resulting distribution function is substantially equivalent to the GRAD distribution. Hence two different approximations are usually made to the initial hierarchy: a truncation (for m moments) and an expansion of the truncated distribution function (of order α).

The non linear closure with the distribution function given by (10) suffers by some analytical problems that were firstly found out by JUNK and his coworkers [12] in particular concerning the domain $\mathcal{D}$ of invertibility between $\mathbf{u}$ and $\mathbf{u}'$ and the integrability of the moments (see also [13]). Instead, the GRAD-like distribution function (12) does not present this kind of problems. In fact, thanks to the Maxwellian function that dominates any polynomial, all the expressions for the moments are integrable. So, from now on, we will forget the original non linear problem that justifies in a formal sense the expansion (12) and we will consider only the ET_m^α theories obtained directly by the distribution function (12).

Due to the expansion of the truncated distribution function, for any ET_m^α theory the general statements of Extended Thermodynamics hold only in an approximate sense. For example the concavity of the entropy density and the hyperbolicity are now valid only in a neighborhood of an equilibrium state (*hyperbolicity domain* [1,6]). And we can expect that also the entropy principle is now satisfied only approximatively.

In fact, if the approximation in (12) is of order α, the field equations becomes also of order α in the non equilibrium variables, while the entropy density, the entropy flux and the entropy production are of order $\alpha+1$ due to the fact that these quantities play the role of the potentials (see (7) and (9)). Therefore instead of (7) we have:

$$\mathbf{u}'\cdot\partial_t\mathbf{u}-\partial_t h = O(\alpha+2);\quad \mathbf{u}'\cdot\partial_i\mathbf{F}^i-\partial_i h^i = O(\alpha+2);\quad \mathbf{u}'\cdot\mathbf{P}-\Sigma = O(\alpha+2). \quad (13)$$

In the first two expressions the $O(\alpha+2)$ indicates terms of order $\alpha+2$ provided that the non-equilibrium variables are sufficiently small together with their derivatives. In reality it is very easy to find cases in which we obtain an entropy law with a wrong entropy production or, it is the same, the entropy principle is not satisfied within the approximation order we have fixed.

3 Entropy principle violated and critical derivatives: one simple example

As anticipated in the previous sections, we have often entropy law with a wrong entropy production when we deal with ET_m^α theories. In those cases it is verified the presence of some spatial derivatives of non-equilibrium variables which do not contain terms of the desired order for every solutions. Therefore not all the possible boundary data give rise to stationary solutions which are entropic, in the sense that they satisfy the entropy principle within the "right" order of approximation. The derivatives that can contain terms of the wrong order are called *critical derivatives.* To better illustrate the problem let us consider one of the simplest example: the one-dimensional stationary heat conduction problem described by ET_{14}^1 for a fluid at rest ($\mathbf{v}=\mathbf{0}$) confined in the interval $x\in[0,L]$. The components of the field vector (moments) depending only on x are:

$$\mathbf{u}\equiv\left(u, u_1, u_{ll}, u_{<11>}, u_{1ll}, u_{llkk}\right)^T,$$

that, in the usual physical variables, can be rewritten as

$$\mathbf{u}\equiv\left(\rho, 0, 3p, -\sigma, 2q, \Delta+\frac{15p^2}{\rho}\right)^T,$$

where ρ, p, σ, q have the usual meaning of mass density, pressure, deviatoric stress $t_{<11>}$ and the x-component of the heat flux, while Δ denotes the non equilibrium part of u_{llkk}.

Using the procedure described in [5] for the closure method, the flux vector

$$\mathbf{F}^1 \equiv (u_1, u_{11}, u_{1ll}, u_{1<11>}, u_{11ll}, u_{1llkk})^T, \tag{14}$$

becomes:

$$\mathbf{F}^1 \equiv \left(0, p-\sigma, 2\,q, \frac{8\,q}{15}, 5\,k\,p\,T + \frac{\Delta}{3} - 7\,k\,T\,\sigma, 28\,k\,q\,T\right)^T, \tag{15}$$

where $T = p/k\rho$ denotes the absolute kinetic temperature.

For what concerns the production vector, we consider the most simple one obtained through the BGK approximation of the collisional term [14]. But, we observe that our conclusions remain qualitatively unchanged for different choice of the production vector:

$$\mathbf{P} = \frac{\mathbf{u_E} - \mathbf{u}}{\tau} \equiv \left(0, 0, 0, \frac{\sigma}{\tau}, -\frac{2\,q}{\tau}, -\frac{\Delta}{\tau}\right)^T;$$

(τ =const. having the role of a common relaxation time). Therefore the stationary solutions satisfies in $x \in [0, L]$ the ordinary differential system:

$$\begin{cases} \partial_x(p-\sigma) = 0 \\ \partial_x(2q) = 0 \\ \partial_x\left(\dfrac{8\,q}{15}\right) = \dfrac{\sigma}{\tau} \\ \partial_x\left(5\,k\,p\,T + \dfrac{\Delta}{3} - 7\,k\,T\,\sigma\right) = -\dfrac{2\,q}{\tau} \\ \partial_x\,(28\,k\,q\,T) = -\dfrac{\Delta}{\tau}. \end{cases} \tag{16}$$

The x component of the entropy flux h^1 and the entropy production Σ are given by:

$$h^1 = \frac{q}{T} - \frac{q\,\Delta}{15\,k\,p\,T^2} + \frac{2\,q\,\sigma}{5\,p\,T}; \tag{17}$$

$$\Sigma = \frac{2\,q^2}{5\,k\,p\,T^2\,\tau} + \frac{\Delta^2}{120\,k^2\,p\,T^3\,\tau} + \frac{3\,\sigma^2}{4\,p\,T\,\tau}. \tag{18}$$

From the field equations (16) we obtain

$$\sigma = 0; \quad q = \mathit{const}; \quad p = \mathit{const}.$$

It is also possible [5] to determine the main field until the second order of approximation and to evaluate the rest in the approximate entropy condition $(13)_2$:

$$R = \mathbf{u}'.\partial_x \mathbf{F}^1 - \partial_x h^1 = \left(\frac{154\, q^3}{75\, k\, p^2\, T^3} + \frac{77\, q\, \Delta^2}{900\, k^2\, p^2\, T^4} \right) \partial_x T + \frac{11\, q\, \Delta}{450\, k^2\, p^2\, T^3} \partial_x \Delta. \tag{19}$$

According to the theory, the second member of (19) has to be of order 3, provided that the non equilibrium variables q, Δ and the derivative $\partial_x \Delta$ are small at least of order one [a]. But from the field system (16), we obtain:

$$\partial_x \Delta = \frac{3\,(5p\Delta - 56q^2)}{28\, q\, \tau}, \quad \partial_x T = -\frac{\Delta}{28\, k\, q\, \tau} \tag{20}$$

and it is easily verified that $\partial_x \Delta$ is a *critical derivative* since it contains a first term that is of order zero if Δ and q are of the some order of magnitude!

If we put (20) into the rest we obtain:

$$R = \underline{\frac{11\, \Delta^2}{840\, k^2\, p\, T^3\, \tau}} - \frac{11\, q^2\, \Delta}{50\, k^2\, p^2\, T^3\, \tau} - \frac{11\, \Delta^3}{3600\, k^3\, p^2\, T^4\, \tau} \tag{21}$$

and the underlined first term is of order 2 instead of the required order 3. So, a general solution of the field equation $\partial_x \mathbf{F^1} = \mathbf{P}$, is not an entropic one.

4 Non Controllable Boundary Data and Entropic Solutions

In the introduction we have mentioned about the complex problem related to the choice of the non controllable data. On one hand the previous considerations about entropic and non entropic solutions must be obviously taken into account in order to select admissible solutions. On the other hand it is easily verified that the Minimax Entropy Principle prescribes boundary data which do not correspond in general to entropic solutions. So, at least in all the cases when the entropy principle is violated, it is natural to think about the requirement that the critical derivatives are somehow "kept" small along the solution. Even if we just think to the simplest case of a one-dimensional stationary problem in a limited domain, it is easy to understand that we are not allowed to "impose" any condition along the whole interval $x = [0, L]$. We may only prescribe boundary data! Therefore it is reasonable to conjecture

[a] In (19) it is not necessary to have the second order terms of $\mathbf{F^1}$ because $\mathbf{u}'_{\mathbf{E}} \cdot \partial_x \mathbf{F}^1$ is for any α and for any number of moments equal to $\frac{1}{T}\partial_x q$.

the following criterion, at least in order to analyze the consequences of our requirement about entropic solutions.

Criterion of vanishing critical derivative:

The non controllable data can be chosen in such a manner that the critical derivatives vanish at one boundary with eventually their successive derivatives until the number of conditions is equal to the number of non-controllable data.

This simple criterion guarantees that in a neighborhood of the prescribed boundary the critical derivatives remain small enough and the solution satisfies the entropy principle in the required approximation. Of course this is a stronger requirement with respect to the constraint about the order of magnitude of the non-controllable variables and it is not possible to prove in general that it corresponds to a necessary condition for the validity of the entropy principle. Nevertheless, it constitutes surely a sufficient condition and its application is particularly simple and straightforward. In order to test the validity of our conjecture we have considered several different ET_m^α theories for the planar case and also an example for radial symmetries. In the following subsection we will present and compare briefly the results.

4.1 Some applications of the criterion in the planar case

In this section we present some applications of the criterion of the vanishing critical derivatives for a one-dimensional planar heat conduction problem. We suppose that a rarefied gas at rest is confined between two planes which are kept at different temperature or, it is the same, on one plane the temperature is fixed, while on the other plane we assign a constant heat flux orthogonal to the planes. For the sake of simplicity we suppose that all the variables (moments) depend only on one spatial coordinate ($x \in [0, L]$). The following tables summarize the results for ET_{14}^1, ET_{14}^2, ET_{21}^1, ET_{26}^1, ET_{35}^1. In order to better understand the meaning of the symbols and the conclusions presented there, it is necessary to make some preliminary remarks. First of all we denote with G, Δ, Ω and F respectively the component $u_{<111>}$, the non equilibrium part of the fourteenth moment u_{llkk}, the component $u_{<11>kk}$ and the component $u_{<1111>}$. All these four quantities do not have a physical meaning and in principle it is necessary to prescribe at the plane $x = 0$ a boundary values for each of them (G_0, Δ_0, Ω_0, F_0). In reality in certain cases it is possible to express some of those quantities as a function of the other ones through an algebraic relation. This is the reason why for ET_{26}^1, and ET_{35}^1 we do not consider Ω_0 and F_0 as non controllable boundary data.

ET^{α}_{m}	Components of the field vector	Critical derivatives	Non-controllable boundary data	Criterion of the vanishing critical derivatives
ET^{1}_{14}	$\begin{pmatrix} \rho \\ v=0 \\ 3p \\ -\sigma \\ 2q \\ \Delta+\frac{15p^2}{\rho} \end{pmatrix}$	$\partial_x\Delta$	Δ_0	at the boundary $\partial_x\Delta = 0$ hence $\Delta_0 = \frac{56q^2}{5p}$
ET^{2}_{14}	$\begin{pmatrix} \rho \\ v=0 \\ 3p \\ -\sigma \\ 2q \\ \Delta+\frac{15p^2}{\rho} \end{pmatrix}$	$\partial_x\Delta,\quad \partial_x\sigma$	Δ_0 σ_0	at the boundary $\partial_x\Delta = 0$ $\partial_x\sigma = 0$ hence $\sigma_0 = 0$ $\Delta_0 = \frac{56q^2}{5p}$
ET^{1}_{21}	$\begin{pmatrix} \rho \\ v=0 \\ 3p \\ -\sigma \\ 2q \\ G \\ \Delta+\frac{15p^2}{\rho} \end{pmatrix}$	$\partial_x\Delta$	σ_0, G_0, Δ_0	at the boundary $\partial_x\Delta = 0$, $\partial^2_{xx}\Delta = 0$ $\partial^3_{xxx}\Delta = 0$ hence $\sigma_0 = 0$ $G_0 = 0$ $\Delta_0 = \frac{56q^2}{5p}$

ET_m^α	Components of the field vector	Critical derivatives	Non-controllable boundary data	Criterion of the vanishing critical derivatives
ET_{26}^1	$\begin{pmatrix} \rho \\ v=0 \\ 3p \\ -\sigma \\ 2q \\ G \\ \Delta+\frac{15p^2}{\rho} \\ \Omega \end{pmatrix}$	$\partial_x\sigma$ $\partial_x\Delta$	σ_0, G_0, Δ_0	at the boundary $\begin{cases} \partial_x\sigma=0 \\ \partial_x\Delta=0, \\ \partial_{xx}^2\sigma=0 \\ \partial_{xx}^2\Delta=0 \end{cases}$ hence $\begin{matrix} \sigma_0=0 \\ G_0=0 \\ \Delta_0=\frac{56q^2}{5p} \end{matrix}$
ET_{35}^1	$\begin{pmatrix} \rho \\ v=0 \\ 3p \\ -\sigma \\ 2q \\ G \\ \Delta+\frac{15p^2}{\rho} \\ \Omega \\ F \end{pmatrix}$	$\partial_x\sigma$ $\partial_x\Delta$	σ_0, G_0, Δ_0	at the boundary $\begin{cases} \partial_x\sigma=0 \\ \partial_x\Delta=0, \\ \partial_{xx}^2\sigma=0 \\ \partial_{xx}^2\Delta=0 \end{cases}$ hence $\begin{matrix} \sigma_0=0 \\ G_0=0 \\ \Delta_0=\frac{56q^2}{5p} \end{matrix}$

Following the procedures summarized in the tables and, in particular, prescribing the boundary data chosen by the criterion of the vanishing derivatives, we obtain for all the examples considered the same solution which is the one predicted by the NAVIER-STOKES-FOURIER model. The temperature depends linearly on x and the other physical quantities such as the pressure p, the first component of the heat flux q, the 11-component of the stress tensor $-\sigma$ remain constant. Two natural requirements about the choice of boundary data are trivially satisfied. In fact, these examples guarantee a sort of "continuity" with respect to the number of moment and to the order of expansion. Moreover it is evident that for such boundary data the critical derivatives remain equal to zero along the whole solution and thus the entropy principle is satisfied.

4.2 The criterion of the vanishing derivatives in the radial case: an example of application

When we thought about this criterion as an alternative way to prescribe non controllable data, it was natural to try to apply it in a radial symmetry problem. As a matter of fact, in the radial case the Minimax Entropy Principle seems to give unreasonable results [10] and so it constitutes certainly a good test. We have also to remark that the preliminary results [15] obtained by MÜLLER and RUGGERI for the comparison between ET^1_{13} and the NAVIER-STOKES-FOURIER model in radial symmetries have shown different behavior for the temperature. In this section we will consider the case of the cylindrical symmetry by analogy with the work cited above of MÜLLER and RUGGERI done for 13 moments. In order to select a boundary value for Δ through the entropy principle we have to consider the field equations and to check if there exists at least one critical derivative. In this case the field vector is given by

$$\mathbf{u} \equiv \left(u, u_i, u_{ll}, u_{<ij>}, u_{ill}, u_{llkk}\right)^T,$$

that in the usual physical variables can be rewritten as

$$\mathbf{u} \equiv \left(\rho, 0, 3p, -\sigma^{ij}, 2q^i, \Delta + \frac{15p^2}{\rho}\right)^T,$$

where now the components of the tensors are considered with respect to the cylindrical coordinates $(x_1, x_2, x_3) = (r, \theta, z)$. In particular we suppose that all the fields depend only on $x_1 = r$ and $q^1 = q$ is the unique non-vanishing component of the heat flux. Following the same procedure as in [15], we can rewrite the 14 moment system in cylindrical symmetry and conclude that

$$p = const, \quad q = \frac{c_1}{r}, \quad \sigma^{<11>} = -\frac{4}{5}\,\tau\,\frac{c_1}{r^2}, \quad \sigma^{<22>} = \frac{4}{5}\,\tau\,\frac{c_1}{r^4}, \quad \sigma^{<ij>} = 0 \text{ else}, \tag{22}$$

$$\begin{cases} \partial_r T = -\dfrac{\Delta}{28kq\tau}, \\[2ex] \partial_r \Delta = -\dfrac{3(56\,q^2 - 5\,p\,\Delta + 7\,\Delta\,\sigma^{<11>})}{28\,q\,\tau}. \end{cases} \tag{23}$$

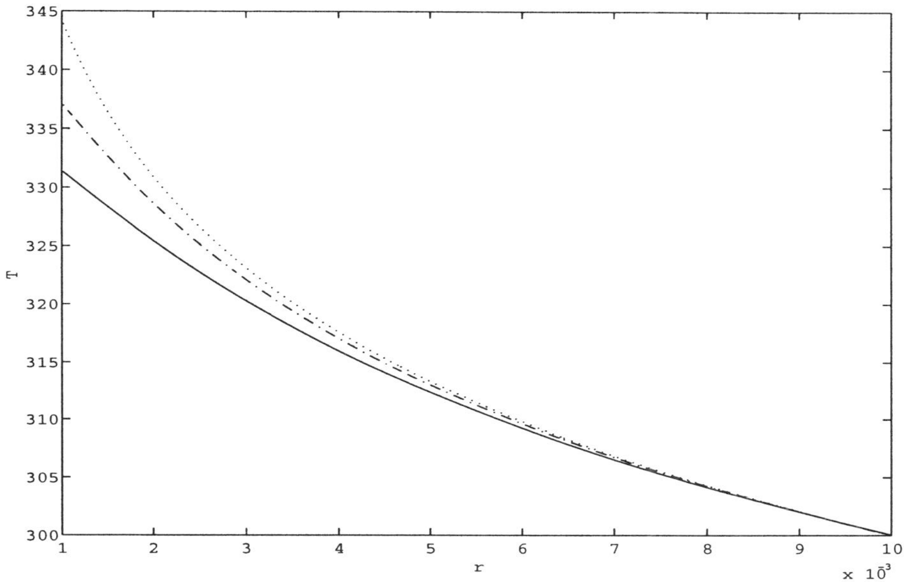

Figure 1: *Case of cylindrical symmetry - Comparison between the temperature obtained from the Navies Stokes Fourier model (dotted line), the one predicted by the* ET^1_{13} *theory (dashed-dotted line) and the temperature corresponding to the* ET^1_{14} *for the entropic solution (continuous line).*

if c_1 denotes a suitable constant. We focus now on the case of two concentric cylinders, as in [15]. On the surface of the internal one we suppose to assign the 1-component, q_{in} of the heat flux (in such a manner the value of c_1 is fixed), while on the external cylinder we prescribe the value of temperature, T_e, and pressure, p. There is still one boundary value that must be given in order to obtain the solution of (22), (23), but we do not have any other physical boundary data that can be imposed to the system. From (23) it is evident that $\partial_r \Delta$ constitutes a critical derivative. Thus we add to the previous boundary data the condition that, for example, on the external cylinder $\partial_r \Delta = 0$. This requirement implies that

$$\Delta_e = \frac{56 q_e^2}{5\,p \;-\; 7\,\sigma_e^{<11>}}.$$

For the previous system we have considered the numerical solution, referring to the boundary values chosen in [15] for the radius of the external and internal cylinders (respectively denoted by r_e and r_{in}), for T_e, q_{in} and p. In Figure 1 we show the behavior of the temperature (continuous line) as a function of r when the boundary value Δ_e is prescribed and we also compare the prediction for the temperature in ET^1_{14} with the ones for NAVIER-STOKES-FOURIER (briefly NSF) and ET^1_{13}. We remark that the expected temperature for a fixed r is less in ET^1_{14} than in ET^1_{13} and in NSF. In spite of the planar case, the critical derivative of Δ remains small only in a neighborhood of the external radius. Surely this effect is more evident if the difference between the external and the internal radius increases or, it is the same, if the KNUDSEN number increases and we are dealing with a more rarefied gas. We recall that more rarefied is the gas, larger is the number of moments needed to describe well the heat conduction phenomenon through Extended Thermodynamics. But higher is the number of moments we deal with, greater is the number of non-controllable boundary data. Increasing such a numbers, we have to impose the vanishing, together with the critical derivatives, of more and more successive derivatives. Consequently, we expect that for an increasing number of moments the critical derivatives tend to remain close to zero along all the interval and the solution becomes entropic also for large KNUDSEN number.

5 Acknowledgments

This paper was supported by fondi MURST Progetto di interesse Nazionale *Problemi Matematici Non Lineari di Propagazione e Stabilità nei Modelli del Continuo* Coordinatore T. Ruggeri, by the GNFM-INdAM, in part by Istituto Nazionale di Fisica Nucleare (INFN) (T.R.) and by Progetto Giovani Ricercatori dell'Università di Bologna and by GNFM-INdAM Piccolo Progetto *Metodi numerici per leggi di conservazione iperboliche con Applicazioni alla Termodinamica del non-equilibro e semiconduttori* (F.B.).

References

1. I. Müller and T. Ruggeri, *Rational Extended Thermodynamics*, Springer Tracts in Natural Philosophy, **37** (II Edition), Springer-Verlag, New York (1998).
2. C. Dafermos, *Hyperbolic Conservation Laws in Continuum Physics,* Springer-Verlag, Berlin (2001).
3. T. Ruggeri and A. Strumia, Ann. Inst. H. Poincaré , **34** A, 65 (1981).
4. G. Boillat and T. Ruggeri, Continuum Mech. Thermodyn. **9**, 205 (1997).
5. F. Brini and T. Ruggeri, *Entropy Principle for the Moment Systems of degree α associated to the Boltzmann Equation. Critical Derivatives and*

Non Controllable Boundary Data to appear on Continuum Mech. and Thermodyn (2002).

6. F. Brini, Continuum Mech. Thermodyn., **13**, 1-8 (2001)
7. H. Struchtrup and W. Weiss, Phys. Rev. Lett. **80** n. 23, 5048 (1998).
8. E. Barbera, I. Müller and M. Sugiyama, Meccanica, **34**, 103-113 (1999).
9. J. Au, I. Müller and T. Ruggeri, Continuum Mech. Thermodyn., **12**, 19-29 (2000).
10. F. Brini and T. Ruggeri, *Some critical remarks about the choice of the boundary data and the minimax entropy principle in Extended Thermodynamics* submitted to Accad. Naz. Lincei (2002).
11. W. Dreyer, J. Phys. A: Math. Gen. **20**, 6505 (1987).
12. M. Junk, J. Stat. Phys., **93**, 1143-1167 (1998); W. Dreyer, M. Junk and M. Kunik, WIAS-Preprint No 592, Berlin (2000).
13. T. Ruggeri, Proceedings of WASCOM 99, edited by V. Ciancio, A. Donato, F. Oliveri, S. Rionero –World Scientific, Singapore, 434-443, (2001). To appear in Lecture Notes of Wascom 99 – Vulcano – June 1999. World Scientific, Singapore (2001).
14. S.C. Chapman and T.G. Cowling, *The Mathematical Theory of Nonuniform Gases*, Cambridge University Press (1961). C. Cercignani, *Theory and Application of the Boltzmann Equation*, Scottish Academic Press (1975).
15. I. Müller and T. Ruggeri, *Stationary heat conduction in radially symmetric situations– an Application of Extended Thermodynamics*, to appear on the J. of Non-Newtonian Fluids (2002).

ON THE ONSET OF CONVECTION IN PRESENCE OF A BOUNDED NON UNIFORM TEMPERATURE GRADIENT

F. CAPONE, S. RIONERO

Department of Mathematics, University of Naples "Federico II",
Complesso Universitario Monte S. Angelo,
Via Cinzia, 80126 Naples - ITALY
E-mail: f.capone@unina.it; rionero@unina.it

1 Introduction

As it is well known, thermal convection problem has great relevance in many applications like - for example - climatology, the diffusion of pollution in the atmosphere and the diffusive convection in porous media. For this reason – in the past as nowadays – has attracted the attention of many authors [1, 2, 3, 11].

Recently this problem has been studied by many researches on assuming that the layer didn't be uniformly heated from below, and the presence of a spatially constant temperature gradient (in the horizontal directions) has been considered [4, 5, 6, 7, 8, 12]. Even if this is a more realistic situation it has a (serious) drawback: it implies, in fact, at large spatial distances an *infinite temperature on the horizontal planes.* The aim of this communication is to reconsider this problem, avoiding boundlessness temperature behaviour on the horizontal planes. To this end, we consider – on both the planes bounding the layer – a regular stationary spatially periodic temperature gradient.

The plan of the paper is the following. In Section 2 we introduce thermal convection problem for a fluid in an infinite horizontal layer with the aforesaid temperature gradient. Then we determine a motionless state with a bounded temperature profile in all directions. In Section 3 we perform the nonlinear stability analysis of the conduction solution and obtain a global (nonlinear) stability result with respect to the L^2-norm. Finally in Section 4, we consider the case of bi-dimensional perturbations to the velocity field and obtain an improvement of the energy stability threshold through the solution of a suitable variational problem.

2 Statement of the problem

Let us consider the dimensionless equations governing the motion of a newtonian, incompressible, non isothermal fluid in the horizontal layer $I\!R^2 \times [0, 1]$ and fluid motions periodic in the x and y directions. Denoting by Ω a periodicity cell, as model governing the motions we use suitably the Oberbeque-Boussinesq approximation [a], i.e.

$$\begin{cases} \mathbf{v}_{,t} + \mathbf{v}\cdot\nabla\mathbf{v} = -\nabla p + \Delta\mathbf{v} + R(T-T_0)\mathbf{k} \\ \nabla\cdot\mathbf{v} = 0 \\ Pr(T_{,t} + \mathbf{v}\cdot\nabla T) = \Delta T \end{cases} \tag{1}$$

in which $R^2 = \bar{\alpha}_0 \Delta T g d^3/k\nu$, $Pr = \nu/k$ are, respectively, the Rayleigh and the Prandtl numbers and ΔT is the temperature difference between the bounding planes at $x = 0$. To the system (1) we append the boundary conditions

$$T(z=0) = \alpha^\star \sin x + T_1 + 1, \qquad T(z=1) = \frac{\alpha^\star}{e}\sin x - T_1. \tag{2}$$

[a] In the Oberbeque-Boussinesq approximation the variation of density is neglected everywhere except in the buoyancy. From $\rho = \rho(T)$ it follows that for small temperature difference $T - T_0$, T_0 being a reference distribution of temperature, one obtains

$$\rho(T) = \rho(T_0) + \left(\frac{\partial\rho}{\partial T}\right)_{T=T_0}\cdot(T-T_0),$$

i.e., on setting $\alpha_0(T_0) = -\dfrac{1}{\rho(T_0)}\left(\dfrac{\partial\rho}{\partial T}\right)_{T=T_0}$

$$\rho(T) = \rho(T_0)[1-\alpha_0(T_0)(T-T_0)]. \tag{$*$}$$

In the present paper – letting T_0 be a non-constant reference temperature distribution – we take into account the density variations in the buoyancy, on the basis of

$$\bar{\rho}(T) = \bar{\rho}_0[1-\bar{\alpha}_0(T-T_0)] \tag{$**$}$$

with

$$\bar{\rho}_0 = \frac{1}{2}\left[\rho_m + \frac{\rho_M\alpha_M}{\bar{\alpha}_0}\right], \qquad \bar{\alpha}_0 = \min_\Omega \alpha_0[(T_0(x,y,z)],$$

where $\rho_m = \min_\Omega \rho[T_0(x,y,z)]$, $\rho_M = \max_\Omega \rho[T_0(x,y,z)]$, $\alpha_M = \max_\Omega \alpha_0[T_0(x,y,z)]$. It turns out that [10]

i) $1 - \bar{\alpha}_0(T-T_0) \geq 0$. In fact $(*)$ implies $1 \geq \alpha_0(T-T_0)$;

ii) in the case T_0 =constant, one obtains the usual Oberbeque-Boussinesq approximation;

iii) $|\bar{\rho}(T) - \rho(T)| \leq \dfrac{1}{2}\left[\dfrac{\rho_M\alpha_M}{\bar{\alpha}_0} - \rho_m\right](1 + \bar{\alpha}_0|T-T_0|)$;

iv) if one choose $\bar{\rho}_0 = \dfrac{1}{|\Omega|}\displaystyle\int_\Omega \rho(T_0)d\Omega$ it turns out that

$$\left|\int_\Omega (\bar{\rho}(T) - \rho(T))d\Omega\right| \leq |\Omega|\bar{\rho}_0\delta$$

where $\delta = \max\{\alpha_M\delta_2 - \bar{\alpha}_0\delta_1, \bar{\alpha}_0(\delta_2-\delta_1)\}$ with $\delta_1 = \max_\Omega|T-T_0|$ and $\delta_2 = \min_\Omega|T-T_0|$.

The boundary value problem (1)-(2) admits the following motionless state:

$$m_0 = \{\mathbf{v}_b = \mathbf{0};\ T_b(x,z) = \alpha^\star e^{-z}\sin x - z + T_1 + 1;\ p_b\}$$

in which the basic temperature field $T_b(x,z)$ exhibits a bounded periodic behaviour in the x-direction. In order to guarantee that the layer is heated from below it has to be $\alpha^* \in (0, e/(e-1))$. Let us notice that if one requires that $T_b(x,z)$ is a decreasing function of z in $[0,1]$, then $\alpha^* \in (0,1)$ [b].

On denoting by $\mathbf{u} = (u,v,w), \theta, \pi$, respectively, the perturbations to the velocity, temperature and pressure fields, the dimensionless equations governing the evolution of the perturbation fields in the strip $I\!R^2 \times [0,1]$ are

$$\begin{cases} \mathbf{u}_{,t} + \mathbf{u}\cdot\nabla\mathbf{u} = -\nabla\pi + \Delta\mathbf{u} + R\theta\mathbf{k} \\ \nabla\cdot\mathbf{u} = 0 \\ Pr(\theta_{,t} + \mathbf{u}\cdot\nabla\theta) = \Delta\theta + Rw - R\alpha^* e^{-z}\cos x\ u + R\alpha^* e^{-z}\sin x\ w. \end{cases} \tag{3}$$

To the previous system we append the boundary conditions:

$$w = \frac{\partial^2 w}{\partial z^2} = \Delta w = 0, \quad \frac{\partial u}{\partial z} = \frac{\partial v}{\partial z} = 0, \quad \theta = 0 \qquad \text{on } z = 0,1. \tag{4}$$

In the sequel we shall assume that the perturbed fields are periodic functions of x and y of periods 2π, $2\pi/a_y$, we shall denote by $\Omega = [0,2\pi]\times[0,2\pi/a_y]\times[0,1]$ the periodicity cell and by $<\cdot>$ and $\|\cdot\|$ respectively the integral and the L^2-norm on Ω. Finally, to be sure that m_0 is unique, we exclude any other rigid motion on requiring that $<u>=<v>=0$.

3 Nonlinear stability

Let us introduce the Liapunov functional

$$V(t) = \frac{1}{2}\parallel \mathbf{u}\parallel^2 + \frac{Pr}{2}\parallel\theta\parallel^2. \tag{5}$$

[b] On taking into account the boundary conditions (2), the layer is heated from below if $T(z=0) - T(z=1) = \alpha^*((e-1)/e)\sin x + 1 > 0$, i.e. $\alpha^* \in (0, e/(e-1))$. Concerning the decreasing of $T_b(x,z)$ in $[0,1]$, since $\frac{\partial T_b}{\partial z}(x,z) = -\alpha^* e^{-z}\sin x - 1$ one has that: if $\sin x > 0 \Longrightarrow \frac{\partial T_b}{\partial z}(x,z) < 0$; if $\sin x < 0$, since $\frac{\partial^2 T_b}{\partial z^2}(x,z) = \alpha^* e^{-z}\sin x \le 0\ \forall z \in [0,1]$, to have $\frac{\partial T_b}{\partial z}(x,z) < 0$ it is sufficient to choose $\alpha^*(>0)$ such that the maximum of $\frac{\partial T_b}{\partial z}(x,z)$ is negative, and it is certainly true if $\alpha^* \in (0,1)$.

On evaluating the time derivative of $V(t)$ along the solutions of (3)-(4) it turns out that

$$\frac{dV}{dt} = 2R < w\theta > - \parallel \nabla \mathbf{u} \parallel^2 - \parallel \nabla\theta \parallel^2 + \\ -R\alpha^* < e^{-z} \cos x \; u\theta > +R\alpha^* < e^{-z} \sin x \; w\theta > . \tag{6}$$

On applying the Cauchy, Poincaré and Wirtinger inequalities [2, 11] to the last two terms in (6), we obtain

$$R\alpha^* < e^{-z} \cos x \; u\theta > \leq \frac{R\alpha^*}{2\pi^2}(\parallel \nabla \mathbf{u} \parallel^2 + \parallel \nabla\theta \parallel^2) \tag{7}$$

$$R\alpha^* < e^{-z} \sin x \; w\theta > \leq \frac{R\alpha^*}{2\pi^2}(\parallel \nabla \mathbf{u} \parallel^2 + \parallel \nabla\theta \parallel^2). \tag{8}$$

On taking into account (7) and (8), from (6) it follows that

$$\frac{dV}{dt} \leq \left[\frac{(\pi^2 + \alpha^* R_B)R - \pi^2 R_B}{\pi^2 R_B}\right] (\parallel \nabla \mathbf{u} \parallel^2 + \parallel \nabla\theta \parallel^2), \tag{9}$$

where $R_B^2 = 657.511$ is the stability threshold of the Bénard problem in the case of stress-free boundary conditions (4) [1, 2, 3, 11].

By virtue of the Poincaré and Wirtinger inequalities [2, 11]: $\exists \gamma > 0 : \parallel \nabla \mathbf{u} \parallel^2 + \parallel \nabla\theta \parallel^2 \geq \gamma V$, hence from (9) one obtains the Theorem 1.

Theorem 1 – *Let*

$$R < R_E = \frac{\pi^2 R_B}{\pi^2 + \alpha^* R_B}, \tag{10}$$

then the asymptotic, exponential, global nonlinear stability of the conduction solution m_0 with respect to the V norm holds, according to the inequality

$$V(t) \leq V(0) \exp\left\{\gamma \frac{R - R_E}{R_E} t\right\}, \qquad t \geq 0. \tag{11}$$

The Table 1 contains the stability thresholds (10), versus α^*.

4 Nonlinear stability with respect to the bi-dimensional perturbations

Let us consider the bi-dimensional perturbations to the velocity field $\mathbf{u} = (0, v, w)$, and consider the Liapunov functional

$$V(t) = \frac{1}{2} \parallel \mathbf{u} \parallel^2 + \lambda \frac{Pr}{2} \parallel \theta \parallel^2, \tag{12}$$

Table 1: Stability thresholds (10), versus α^*

α^*	R_E^2
0.1	414.281
0.3	207.656
0.5	124.397

in which $\lambda = const. > 0$ is a suitable coupling parameter. Then (3)-(4) imply:

$$\frac{dV}{dt} = R(1+\lambda) < w\theta > + R\lambda\alpha^* < e^{-z} \sin x \, w\theta > - \| \nabla \mathbf{u} \|^2 - \lambda \| \nabla\theta \|^2 . \quad (13)$$

On introducing – by following the standard energy method – the nonlinear critical Rayleigh number [c]

$$\frac{1}{\tilde{R}_E} = \max_H \frac{(1+\lambda) < w\theta > + \lambda\alpha^* < e^{-z} \, w\theta >}{\| \nabla \mathbf{u} \|^2 + \lambda \| \nabla\theta \|^2}, \quad (14)$$

from (13), on setting $D = \| \nabla \mathbf{u} \|^2 + \lambda \| \nabla\theta \|^2$, we obtain:

$$\frac{dV}{dt} \leq \left[\frac{R}{\tilde{R}_E} - 1 \right] D. \quad (15)$$

Theorem 2 - *The global, asymptotic, exponential nonlinear stability of* m_0 *with respect to the bi-dimensional perturbations* $\mathbf{u} = (0, v, w)$ *in the* V*-norm is guarantee if*

$$R < \tilde{R}_E.$$

The values of $\tilde{R}_E$ obtained by solving numerically (14), for different values of α^*, are listed in Table 2. In this table are also labelled the values given by

$$R_E^* = \frac{2\pi^2 R_B}{2\pi^2 + \alpha^* R_B}, \quad (16)$$

which is the formula that one obtains from (8) (instead of (10)) in the bi-dimensional case.

[c] Let be $H = \{\mathbf{u}, \theta : \nabla \cdot \mathbf{u} = 0; \, \mathbf{u}, \theta$ periodic in the x and y directions, verifying (4), $D < \infty\}$ the class of the kinematically admissible perturbations. The maximum (14) exists by virtue of the Rionero's theorem [9].

Table 2: Stability thresholds versus α^* in the bi-dimensional case

α^*	R_E^{*2}	$\tilde{R}_E^2$
0.1	524.97	619.30
0.3	340.43	554.80
0.5	241.64	502.45

Acknowledgments

This work has been performed under the auspicies of the G.N.F.M. of I.N.D.A.M. and M.U.R.S.T. (P.R.I.N.): "Nonlinear mathematical problems of wave propagation and stability in continuous media".

References

1. S. Chandrasekhar in *Hydrodynamic and hydromagnetic stability*, (Dover, New York, 1961).
2. J.N. Flavin and S. Rionero in *Qualitative estimates for partial differential equations. An introduction*, (CRC Press, Boca Raton, Florida, 1996).
3. D.D. Joseph in *Stability of fluid motions I, II*, (Springer Verlag, New York, 1976).
4. D.M. Manole and J.L. Lange *J. Heat Mass Transfer* **37**, 2047 (1994)
5. D.A. Nield *J. Heat Mass Transfer* **34**, 87 (1991)
6. D.A. Nield *J. Heat Mass Transfer* **37**, 3021 (1994)
7. D.A. Nield *J. Heat Mass Transfer* **41**, 241 (1998)
8. Z. Qiao and P.N. Kaloni *J. Heat Mass Transfer* **41**, 2549 (1998)
9. S. Rionero *An. Mat. Pura Apl.* **78**, 339 (1968)
10. S. Rionero *A note on the Oberbeque-Boussinesq approximation in the Bénard problem* (to appear) (2002)
11. B. Straughan in *The energy method, stability and nonlinear convection*, (Springer Verlag, Appl. Math. Sci. Ser., New York, 1992).
12. J.E. Weber *J. Heat Mass Transfer* **17**, 241 (1974)

GLOBAL GEOMETRICAL SOLUTIONS FOR HAMILTON-JACOBI EQUATIONS OF EVOLUTION TYPE

F. CARDIN

Dip. di Matematica Pura ed Applicata, via Belzoni 7 - 35131 Padova - Italia
E-mail: cardin@math.unipd.it

We discuss the existence of global generating functions describing Lagrangian submanifolds connected with evolution problems for Hamilton-Jacobi equations. By using Viterbo's version of the Amann-Conley-Zehnder reduction, we compute, for generic (in a suitable sense) Hamiltonian functions and initial data, global space-time generating functions with finite parameters for geometric solutions of a H-J equations of evolution kind.

1 Introduction

1.1 The Hamilton-Jacobi problem

Many dynamical mechanical systems, sometimes infinite-dimensional evolution systems, like wave on Continuum Mechanics or Optics (WKB), are governed very often by a Hamilton-Jacobi equation.

Classical Cauchy Problem:
Searching for a C^1-function $S(x,t), \quad x \in R^n, \quad t \in [0,T]$,

$$\frac{\partial S}{\partial t}(x,t) + H\Big(x, \frac{\partial S}{\partial x}(x,t), t\Big) = 0, S(x,t)\big|_{t=0} = \sigma(x).$$

E.g., $S(x,t)$ represents the "phase" of a wave, etc. There are non-linearity obstructions (of H and σ) to existence of **global** classical solutions. A **many-valued** solutions theory arises, and its natural framework is Symplectic Geometry.

Let

$$S : R^n \times R \longrightarrow R, \qquad (x,t) \longmapsto S(x,t),$$

be a classical solution; by performing the so-called 'homogeneous version', we set

$$Q := R^n \times R : \ \ space - time, \quad q := (x,t), \quad (q,p) := (x,t;\xi,\tau),$$

$$\mathcal{H} : T^*Q \longrightarrow R \qquad \mathcal{H}(q,p) = \mathcal{H}(x,t;\xi,\tau) := \tau + H(x,\xi,t)$$

The H-J equation reads:

$$\mathcal{H}(q, \frac{\partial S}{\partial q}(q)) = \frac{\partial S}{\partial t} + H(x, \frac{\partial S}{\partial x}, t) = 0$$

and if $S : Q \to R$ is a **classical solution**, then the *image* of its differential

$$\Lambda := im(dS) = \Big\{(q,p) : p = \frac{\partial S}{\partial q}(q),\ q \in Q\Big\} \subset T^*Q,$$

solves H-J,

$$\mathcal{H}(x, \frac{\partial S}{\partial x}(x)) = 0,$$

geometrically:

$$\Lambda = im(dS) \subset \mathcal{H}^{-1}(0) \subset T^*Q$$

It is standard to see that Λ is a **transverse Lagrangian submanifold** of the symplectic manifold T^*Q inside $\mathcal{H}^{-1}(0)$.

• In order to generalize, the idea (after Maslov) is to relax transversality and looking for **general Lagrangian submanifolds** inside $\mathcal{H}^{-1}(0)$:

$$\Lambda \subset \mathcal{H}^{-1}(0)$$

A careful control of the **ramifications** of Λ will produce suitable "weak" solutions. The following theorem characterizes locally the Lagrangian submanifols.

Maslov-Hörmander Theorem: *Locally, every* Lagrangian submanifold, for some suitable generating function $S(x,u), u \in R^k$, is given by

$$\Lambda = \{(x,p) \in T^*R^{n+1} : \quad p_i = \frac{\partial S}{\partial x^i}(x,u), \quad 0 = \frac{\partial S}{\partial u^\alpha}(x,u)\},$$

with

$$rank\left(\frac{\partial^2 S}{\partial u_\alpha \partial x^i} \qquad \frac{\partial^2 S}{\partial u_\alpha \partial u_\beta}\right)\Bigg|_{\frac{\partial S}{\partial x^i}(x,u)=0} = k (= max).$$

1.2 Global Generating Functions

A question: Given a Lagrangian submanifold, geometrical solution of H-J

$$\Lambda \subset \mathcal{H}^{-1}(0) \subset T^*R^{n+1},$$

when does it admit a **global generating function**?

$$S : R^{n+1} \times R^k \longrightarrow R, \qquad (q,u) \longmapsto S(q,u)$$

It is important:
i) in order to construct **viscosity** solutions (Evans-Crandall-Lions) by some **inf-sup procedure** on the auxiliary parameters, Bardi&Evans[2], Bardi&Capuzzo-Dolcetta[1], Cardin[3].
ii) in order to construct the **min-max** solutions *of* Chaperon-Sikorav-Viterbo by means of Morse-Lusternik-Schnirelman techniques.
• For example, it is possible if the related Hamiltonian system is **global Liouville integrable**: a very hard hypothesis.
In this case, a **Complete Integral** there exists.
An example: $H = H(\xi)$,

$$\frac{\partial S}{\partial t}(x,t) + H(\frac{\partial S}{\partial x}(x,t)) = 0, \qquad S(x,t=0) = \sigma(x).$$

The **Complete Integral** is:

$$W(t,x;a) = -tH(a) + x \cdot a$$

and, if H is convex, the **viscosity solution** is given Bardi&Evans[2]

$$s(t,x) = \inf_{\chi \in R^n} \sup_{a \in R^n} [\underbrace{W(t,x;a) - W(0,\chi;a)}_{\mathcal{G}(t,x;a,\chi)} + \sigma(\chi)].$$

We see that the term in the formula above

$$\mathcal{G}(t,x;a,\chi) := W(t,x;a) - W(0,\chi;a)$$

can be interpreted as a sort of 'Propagator', or a geometrical Green-kernel.

2 The Reduction Theorem

Here, we sketch the following result (Cardin[4]):
By using suitably the *reduction* techniques by *Amann, Conley, Zehnder, and Viterbo* [a], we are able to make a **global generating function**

$$S(t,x;u), \quad t \in [0,T], \quad x \in R^n, \quad u \in R^k,$$

for the Lagrangian submanifold Λ —the geometrical solution— of a H-J evolution Cauchy problem, under the following *rather generic hypothesis*,

$$\sup_{t\in[0,T]} \sup_{z \in R^{2n}} |\nabla^2_{zz} H(t,z)| = C < +\infty \qquad (z := (x,\xi))$$

[a] they were searching periodic solutions of Hamiltonian systems.

We will build the global generating function for the **simpler** H-J problem starting, at $t=0$, from the **zero-section**:

$$\Lambda_0 = \{(x,0),\quad x\in R^n\},\quad in\ other\ words,\ for\ \sigma\equiv 0.$$

The more general H-J problem starting from an initial **exact Lagrangian manifold** (for non trivial $\sigma \neq$ const.):

$$\Lambda_\sigma = \{(x,\frac{\partial\sigma}{\partial x}(x)),\quad x\in R^n\}$$

will be solved

i) by using the simple *global canonical transformation*:

$$(x,t;\xi,\tau)\longmapsto(\bar{x}=x,\ \bar{t}=t;\ \bar{\xi}=\xi-\frac{\partial\sigma}{\partial x}(x),\ \bar{\tau}=\tau),$$

and then

ii) by solving the H-J problem related to the zero-section Λ_0 at $t=0$ for the *transformed* Hamiltonian K:

$$K(\bar{x},\ \bar{t},\bar{\xi}) := H(\bar{x},\bar{\xi}+\frac{\partial\sigma}{\partial x}(\bar{x}),\bar{t})\quad\Big(\text{hyp}:\ |\nabla^2_{zz}K(z,t)|<+\infty\Big).$$

Operatively, if the global generating function relative to the above Hamiltonian K and the zero section Λ_0 is

$$\bar{S}_\sigma(\bar{t},\bar{x},u),$$

then the global generating function relative to the above H and Λ_σ will given by, finally, $\forall t\in[0,T]$, $x\in R^n$ and $u\in R^k$,

$$S(t,x,u) := \sigma(x)+\bar{S}_\sigma(t,x,u).$$

First step: we start from (a version of) the Hamilton-Helmholtz variational principle, let consider the curves $\gamma(\cdot)$ belonging to

$$\Gamma=\Big\{\gamma(\cdot)=(\bar{x}(\cdot),\bar{\xi}(\cdot))\in H^{1,2}([0,T];R^{2n}),\ \bar{\xi}(0)=0\Big\}$$

(Sobolev $\Rightarrow\quad H^{1,2}\hookrightarrow\mathcal{C}^0$) we define the functional:

$$J:[0,T]\times\Gamma\to R,\qquad J(t;\gamma(\cdot))=\int_0^t[\bar{\xi}\cdot\dot{\bar{x}}-H(t',\bar{x},\bar{\xi})]dt'$$

Second step: A **representation** for Γ by the bijection $\hat{g}$,

$$\hat{g} : [0,T] \times R^n \times L^2\big([0,T], R^{2n}\big) \longrightarrow [0,T] \times \Gamma,$$

$$(t, x, \phi) \mapsto \hat{g}(t, x, \phi) = (t, \gamma(\cdot))$$

$$\gamma(s) = (x(s), \xi(s)) := \left(x - \int_s^t \phi_x(r)\, dr, \int_0^s \phi_\xi(r)\, dr \right).$$

The curve $\gamma(\cdot)$ is (i) starting from the 'zero section' of T^*R^n: $\xi(0) = 0$, it is such that (ii) $\dot{\gamma}(\cdot) = \phi(\cdot)$, and (iii) $x(t) = x$. *The new functional:*

$$W := J \circ \hat{g} : \ [0,T] \times R^n \times L^2\big([0,T], R^{2n}\big) \longrightarrow R$$

can be interpreted as a **global generating function** *with* **infinite** *auxiliary parameters* (that is, the derivative curves in L^2) for the following H-J problem: $W(t, x; \phi(\cdot))$:

$$\frac{\partial W}{\partial t}(t, x, \phi(\cdot)) + H\Big(x, \frac{\partial W}{\partial x}(t, x, \phi(\cdot)), t\Big) = 0, \qquad W(t, x, \phi(\cdot))\big|_{t=0} = 0,$$

$$\frac{DW}{D\phi(\cdot)}(t, x, \phi(\cdot)) = 0 \qquad Hamilton - Helmholtz\ Var.\ Pr.$$

Third step: **Fourier analysis**. For each **fixed** $m \in N$ let us to consider the projection maps on the basis $\{e^{i\frac{2\pi k}{T}s}\}_{k \in Z}$ of L^2,

$$P_m\phi(s) := \sum_{|k| \le m} \phi_k\, e^{i\frac{2\pi k}{T}s}, \qquad Q_m\phi(s) := \sum_{|k| > m} \phi_k\, e^{i\frac{2\pi k}{T}s}.$$

Clearly, $P_m L^2 \oplus Q_m L^2 = L^2$, and for $\phi \in L^2$ we will write $u := P_m\phi$ and $v := Q_m\phi$. $\Big[$Rem: Hamilton's o.d.e: $\dot{z} = E\nabla H(z,t)$, $z = (x,\xi)$.$\Big]$ We can prove that, (i) for m **large enough**, (ii) for fixed $(t,x) \in [0,T] \times R^n$ and $u \in P_m L^2$, and (iii) under the above hyp. $|\nabla^2_{zz} H(z,t)| = C < +\infty$, the following map

$$Q_m L^2 \longrightarrow Q_m L^2 \qquad v(\cdot) \longmapsto Q_m E\nabla H\Big(\cdot, \hat{g}(\cdot, t, x, u(\cdot) + v(\cdot))\Big)$$

is a **contraction map**. By Banach-Caccioppoli there exists one and only one **fixed point map** $f(t,x,u)$. This **fixed point map** f works in the following way:

$$f(t,x,u) = Q_m E\nabla H\Big(\hat{g}(t, x, u + f(t,x,u))\Big) \qquad (*)$$

Now, if we solve, in the unknowns $u \in P_m L^2$, the **finite** equation (algebraic, in a sense)

$$u = P_m E \nabla H\Big(\hat{g}\big(t, x, u + f(t, x, u)\big)\Big), \qquad (**)$$

then, by adding $(*)$ and $(**)$, we see that the curve

$$\gamma = \hat{g}(t, x, u + f(t, x, u))$$

solves Hamilton's system: $\dot{\gamma}(t) = E\nabla H\big(t, \gamma(t)\big)$.

Fourth (last) step: Finally, given the Hamiltonian

$$H = H(x, \xi, t) = H(z, t), \qquad |\nabla^2_{zz} H(z, t)| < +\infty,$$

and the H-J problem

$$\frac{\partial S}{\partial t}(x, t) + H\Big(x, \frac{\partial S}{\partial x}(x, t), t\Big) = 0, \qquad S(x, t = 0) \equiv 0,$$

the **global generating function** of the geometrical solution is

$$S(t, x, u) = W\big(t, x, \underbrace{u + f(t, x, u)}_{\phi(\cdot)}\big) = J \circ \hat{g}\big(t, x, u + f(t, x, u)\big)$$

$$\frac{\partial S}{\partial t}(t, x, u) + H\Big(x, \frac{\partial S}{\partial x}(t, x, u), t\Big) = 0,$$

$$S(t, x, u)\Big|_{t=0} = 0,$$

$$\frac{\partial S}{\partial u}(t, x, u) = 0.$$

References

1. M. Bardi, I. Capuzzo-Dolcetta, Optimal control and viscosity solutions of Hamilton-Jacobi-Bellman equations. *Systems & Control: Foundations & Applications.* Birkhäuser, xviii+570 pp. (1997).
2. M. Bardi, C. Evans, On Hopf's formulas for solutions of Hamilton-Jacobi equations, *Nonlinear Analysis, T. M. A.*, **8**, p. 1373-1389, (1984).
3. F. Cardin, On viscosity and geometric solutions for the Hamilton-Jacobi equations, *Nonlinear Analysis, T. M. A.*, **20**, n.6, p.713-719 (1993).
4. F. Cardin, The global finite structure of generic envelope loci for Hamilton-Jacobi equations, *Journal of Math. Physics* (2002).
5. C. Viterbo, Solutions of Hamilton-Jacobi equations and symplectic geometry *Sémin. Équ. Dériv. Partielles, École Polytech., Palaiseau* (1996).

HELMHOLTZ–BOLTZMANN THERMODYNAMICS ON CONFIGURATIONS SPACE

FRANCO CARDIN AND MARCO FAVRETTI

Dipartimento di Matematica Pura ed Applicata, Università di Padova
via Belzoni, 7 – 35131 Padova Italy
E-mail: cardin@math.unipd.it favretti@math.unipd.it

In a 1884 paper, Helmholtz showed that for a one–dimensional mechanical system with convex potential energy φ depending on a parameter V it is possible to define the temperature T, the pressure p and the entropy S verifying the Gibbs relation $TdS = de + pdV$ where $e =$ kin.en. $+ \varphi$. In the paper we propose an extension of the Helmholtz construction to general natural mechanical systems endowed with a fibration over the (possibly multidimensional) space of macroscopic parameters v. Moreover, we show that, for certain discrete mechanical systems with non convex potential energy, used as models for phase transitions in solids, the above introduced thermodynamic pressure $p = p(e, V)$ provides a single–valued macroscopic stress–strain relation.

1 Introduction

In a series of papers, Helmholtz[6 7], introduced the notion of temperature and pressure for a one–dimensional 'monocyclic' mechanical system with periodic orbits whose potential energy $\varphi(x, V)$ depends on a parameter V. For fixed values of the total energy e and V, the temperature T and pressure p are defined as the mean, on the interval between two points of inversion of the motion on a given orbit, of the kinetic energy K and of the quantity $\partial\varphi/\partial V$ respectively, with respect to the measure $d\mu = dx/\sqrt{e - \varphi(x, V)}$. It results, for these systems, that the one form $T^{-1}(de + pdV)$ is *exact.*

This paper is dedicated to a generalization, in each of its aspects, of this results and to the applications of the ensuing scheme to solids. We show that:

a) The theorem of Helmholtz can be generalized to n–dimensional natural mechanical systems, $(x \in \mathbf{R}^n)$,

b) one can develop a scheme with *multiple* macroscopic parameters, $(v \in \mathbf{R}^k$, with $k < n)$, where the v's are degrees of freedom of the system directly accessible to the observer, while the x–variables represent a microscopic state of the system whose description is outside the domain of the theory,

c) the projection $(x, v) \mapsto v$ from $\mathbf{R}^{n+k}$ to $\mathbf{R}^k$ can be substituted by a general fibration $g : \hat{Q} \longrightarrow \mathcal{V}$, where $\hat{Q}$ and $\mathcal{V}$ are manifolds.

Thermodynamics itself can be seen as a tool for reducing the complex description at the microscopic level of the many–particle system to few macroscopic parameters. In the Helmholtz scheme, this reduction task is accomplished by an integration which is a mean over the microscopic *state* variables. This procedure matches perfectly with the separation *in the configuration space* between microscopic and macroscopic degrees of freedom operated by the fibration g. Note that the thermodynamics developed later by Boltzmann and Gibbs deals with objects defined in the *phase space* of the system, and not in the configuration space, therefore it is not suited for our scheme.

Note also that the Helmholtz version of thermodynamics for mechanical systems need not to specify the inertial terms, that is the structure of kinetic energy. Interestingly, also, the probability measure $d\mu$, which is the time needed to the particle to travel a space interval dx, is the one associated to the classical system in the comparison between the classical and quantic description of the system.

Thermodynamics first gave a description of a many particle system at equilibrium. In the forthcoming paper Cardin–Favretti[2] we study the statics of a system formed by point masses free to move on a horizontal line and connected by nonlinear (i.e. with non convex potential energy) springs. One extremity of the chain is fixed on the line, $(x_1, \ldots, x_n)$ are the particle positions and $V = x_n$ is the length of the chain. For this system we show that the above introduced thermodynamic pressure does coincide with the reaction force on the controlled extremity of the chain (when all the particles are in equilibrium) computed through the equations of statics.

One of the major features of the thermodynamic construction, applied to this example, is that it provides a single–valued stress–strain relation $p = p(x_n)$.

2 Helmholtz thermodynamics of mechanical systems

Helmholtz [6 7] called *monocyclic* the (one–dimensional, with periodic orbits) mechanical systems with convex potential energy. See also Gallavotti [3 4 5] and the bibliography therein where the influence of Helmholtz' thermodynamic interpretation on Boltzmann' theory of Statistical Mechanics is analyzed.

Let us consider the differential equation $m\ddot{x} = f(x;V)$ describing a mechanical system where the force f depends on a parameter V which can be thought as a 'volume' in classical thermodynamics, and denote with

$$\varphi(x;V) := -\int_a^x f(\lambda;V)\, d\lambda, \qquad K(\dot{x}) := \frac{1}{2} m\, \dot{x}^2,$$

the potential and kinetic energy respectively. Now we introduce the following probability measure[a] defined on the Hill'region $\mathcal{H} := \{x \in \mathbf{R} : e - \varphi(x;V) > 0\}$

$$d\mu_{(e,V)}(x) = \frac{dx}{\sqrt{K\left(\dot{x}(x;e,V)\right)}} = \frac{dx}{\sqrt{e-\varphi(x;V)}}$$

and we compute the mean $T(e,V)$ of the kinetic energy with respect to $d\mu_{(e,V)}$:

$$T(e,V) := \langle K \rangle_{(e,V)} = \frac{\int_{x_1(e,V)}^{x_2(e,V)} K\left(\dot{x}(x;e,V)\right) d\mu}{\int_{x_1(e,V)}^{x_2(e,V)} d\mu}$$

This quantity can be formally interpreted as a *temperature* because it coincides with the mean of the kinetic energy over a period:

$$T(e,V) = \frac{\int_{x_1(e,V)}^{x_2(e,V)} \sqrt{K\left(\dot{x}(x;e,V)\right)}dx}{\int_{x_1(e,V)}^{x_2(e,V)} \frac{dx}{\sqrt{K\left(\dot{x}(x;e,V)\right)}}} = \frac{\int_0^{\tau/2} \sqrt{\frac{m}{2}}\,\dot{x}^2 dt}{\sqrt{\frac{2}{m}}\,\frac{\tau}{2}} = \frac{1}{\tau/2}\int_0^{\tau/2} K dt.$$

In the same line of thought, we define the *pressure* $p(e,V)$ as

$$p(e,V) := -\langle \frac{\partial \varphi}{\partial V} \rangle_{(e,V)}.$$

We shall see below the reason of this definition.

Theorem [Helmholtz,1884]. *For a one–dimensional, monocyclic mechanical system, the one–form*

$$\frac{1}{T}(de + pdV) \tag{1}$$

defined on $\{(e,V) \in \mathbf{R}^2,\ V > 0\}$ *is exact.*

The proof amounts to show that the above one–form has a primitive given by the *entropy* $S(e,V)$,

$$S(e,V) := 2\,\ln \int_{x_1(e,V)}^{x_2(e,V)} \sqrt{K\left(\dot{x}(x;e,V)\right)}\,dx = 2\,\ln \int_{x_1(e,V)}^{x_2(e,V)} \sqrt{e-\varphi(x;V)}\,dx.$$

The above Theorem provides simple formulae for computing the thermodynamic temperature and pressure given $S(e,V)$:

$$T(e,V) = \left(\frac{\partial S}{\partial e}\right)^{-1}, \qquad p(e,V) = \left(\frac{\partial S}{\partial V}\right)\left(\frac{\partial S}{\partial e}\right)^{-1}. \tag{2}$$

[a] The generalized integral of $d\mu$ over $\mathcal{H}$ is the time interval along the motion between two points of inversion of the motion. It is therefore always finite except for the value of e corresponding to the motion along a separatrix.

Remarks. 1) Note that for these simple systems, the ergodic hypothesis is satisfied with respect to the above introduced (spatial) probability measure $d\mu_{(e,V)}$.
2) Boltzmann[1], called *monade* the above probability measure and called it *horthode* —see Gallavotti[3] for an etymological analysis of Boltzmann' neologisms— in case the macroscopic quantities p, T defined through it satisfy Gibbs relation. Interestingly, this statistical picture was developed before Gibbs' theory of *ensembles*. Later, Boltzmann proved that the *microcanonical* and *canonical* probability measure (respectively $\mu(dpdx) = \text{cost}\ \delta(H - e)$ and $\mu(dpdx) = \text{cost}\ e^{-\beta H}$) are hortode.

3 Generalizations of Helmholtz' scheme

We realize an extension to n-dimensional natural mechanical systems of the above Theorem of Helmholtz. The thermodynamic description of a $n-$ dimensional mechanical system introduced below is based on probability measure $d\mu = \frac{dq}{\sqrt{e-\varphi(q,V)}}$ which has the same singularities of the *Jacobi metric* (in Hamiltonian form). Therefore, it could well be named "Thermodynamics of Jacobi".

Theorem. *Let us consider a natural mechanical system with potential energy depending on a macroscopic control parameter V:*

$$\varphi : \mathbf{R}^n \times \mathbf{R} \longrightarrow \mathbf{R}, \qquad (q, V) \longmapsto \varphi(q, V)$$

and the following probability measure

$$d\mu = d\mu_{(e,V)}(q) = \frac{dq}{\sqrt{e - \varphi(q, V)}}$$

defined on the compact set $\mathcal{H}(e, V) = \{q \in \mathbf{R}^n : e - \varphi(q, V) > 0\}$. Setting

$$p(e, V) := \frac{-\int_{\mathcal{H}(e,V)} \frac{\partial \varphi}{\partial V}(q, V) d\mu}{\int_{\mathcal{H}(e,V)} d\mu},$$

and

$$T(e, V) := \frac{\int_{\mathcal{H}(e,V)} [e - \varphi(q, V)]\ d\mu}{\int_{\mathcal{H}(e,V)} d\mu},$$

which have the formal interpretation of the system' pressure and temperature, the one-form in $\mathbf{R}^2$

$$\frac{de + p(e, V)\, dV}{T(e, V)}$$

is exact and one primitive for it is the (formal) entropy

$$S(e,V) = 2\ \ln \int_{\mathcal{H}(e,V)} \sqrt{e - \varphi(q,V)}\, dq.$$

Proof. The proof is an application of Fubini' Theorem. See Cardin–Favretti[2].

3.1 Multi–dimensional parameter space

Let us denote with $\hat{Q}$ the configuration manifold (collection of microscopic states) of a natural mechanical system and with $\hat{W} : \hat{Q} \longrightarrow \mathbf{R}$ its potential energy. Moreover, suppose that a fibration

$$g : \hat{Q} \to \mathcal{V}, \quad \hat{q}^i \mapsto v^L(\hat{q}^i)$$

is given. We can define the entropy of the reduced, macroscopic system, as

$$S : \mathbf{R} \times \mathcal{V} \longrightarrow \mathbf{R}$$

$$(e,v) \mapsto S(e,v) = 2\ \ln \int_{\mathcal{H}(e,v)} \sqrt{e - \hat{W}(\hat{q})}\ d\hat{q}$$

where

$$\mathcal{H}(e,v) = \{\hat{q} : e - \hat{W}(\hat{q}) > 0,\ g(\hat{q}) = v\}.$$

Note that $\mathcal{V}$ has dimension greater than one in the general case: see Pitteri[8] where the author develops a theory of Thermodynamics depending on multiple mechanical parameters.

3.2 An example with a non trivial fibration

We give an example of a non trivial fibration with multidimensional base space.

Denote with $(x_1^*, ..., x_n^*) \in \mathbf{R}^{3n}$ a stable equilibrium of the system of n particles $P_1, ..., P_n$ with potential energy $w(x_1, ..., x_n)$. Suppose that the number n of particles is sufficiently large to describe the deformations of the system from the equilibrium position $(x_1^*, ..., x_n^*)$ trough a *strain* matrix $F \in \mathrm{Lin}^+(3, \mathbf{R})$. The point is to find the matrix F that gives the best approximation of the real deformation $(x_1 - x_1^*, ..., x_n - x_n^*)$, therefore we are led to a classical least squares problem (Gauss problem): find $F(x_1, ..., x_n)$ such that

$$\mathcal{G}(F, x_1, ..., x_n) := \sum_{i=1}^{n} |F x_i^* - x_i|^2_{\mathbf{R}^3}$$

has a minimum. The problem has a unique solution: for a nondegenerate configuration $(x_1^*, ..., x_n^*)$, i.e. with $\det(x_i^* \otimes x_i^*) \neq 0$, the equation $\frac{\partial \mathcal{G}}{\partial F} = 0$ gives

$$F = \tilde{F}(x_1, ..., x_n) = (\sum_{j=1}^{n} x_j \otimes x_j^*)(\sum_{i=1}^{n} x_i^* \otimes x_i^*)^{-1}.$$

This solution introduces naturally a fibration between the description of the system at the microscopic and the macroscopic (mesoscale) level:

$$\mathbf{R}^{3n} \longrightarrow \mathrm{Lin}^+(3, \mathbf{R}), \quad (x_1, ..., x_n) \longmapsto \tilde{F}(x_1, ..., x_n)$$

with the related entropy

$$(e, F) \mapsto S(e, F) = 2 \ \ln \int_{\mathcal{H}} \sqrt{e - w(x_1, ..., x_n)} \, dx_1, ..., dx_n$$

where $\mathcal{H}(e, F) = \{(x_1, ..., x_n) \in \mathbf{R}^{3n} : \ e - w(x_1, ..., x_n) > 0, \ \tilde{F}(x_1, ..., x_n) = F \ \}$.

The above introduced fibration could be of interest when one looks for the stored energy density $w(F)$ of a continuum on the basis of a crystallographic model trough a Born Rule approach.

References

1. L. Boltzmann, Crelle's Journal 98, S.68–94 (1884), reproduced in F. Hasenöhrl *Wissenschaftliche Abhandlungen von Ludwig Boltzmann* Band III, p.122 (1909), reprinted by Chelsea Publ. Company New York, N.Y. (1968)
2. F. Cardin, M. Favretti, submitted
3. G. Gallavotti, Journal of Statistical Physics, **78**, 1571-1589, (1995). The paper is reprinted in Gallavotti[5], page 205.
4. G. Gallavotti, Documenta Mathematica · Extra Volume, International Congress of Mathematics, 1998.
5. G. Gallavotti, Meccanica Statistica, Quaderni del CNR, GNFM, pp.356, 1995.
6. H. Helmholtz, Crelle's Journal 97 S. 111–140 (1884), reproduced in "Wissenschaftliche Abhandlungen", vol. III, p.142-162 and p.179-202, Leipzig, 1895.
7. H. Helmholtz, Akademie der Wissenschaften zu Berlin, S. 159–177 (1884), reproduced in "Wissenschaftliche Abhandlungen", vol. III, p.163-172 and p. 173-178, Leipzig, 1895.
8. M. Pitteri, A.R.M.A. vol. 80, n.4, (1982), pp.333-385.

Variational Setting of Nonlinear Equilibrium Problem: Wedge Discontinuity Lines

S. Carillo
Dipartimento di Metodi e Modelli Matematici per le Scienze Applicate
University "La Sapienza"
Rome, Italy

A nonlinear equilibrium problem in elasticity is here studied. In particular, the model adopted consists of two elastic bodies which interact with each other: the first one represents the elastic body of interest and the latter the environment which is surrounding it. Accordingly, they are said to form a body-environment pair. The equilibrium problem is considered in the case when the interaction body-environment is *live*, that is the energy functional depends not only on the position and on the deformation which takes place at that position, but also on deformation gradients. In particular, when a material is of grade 1, the energy functional depends on the first deformation gradient and, when a material is of grade 2 it depends on the first deformation gradient and, in addition, on the second deformation gradient. Here, the environment is assumed to be a simple material, namely of grade 1, while the body immersed into it, of grade 1 or 2. In the two different cases,, respectively, the equilibrium conditions are written under the assumption that the body boundary is not a regular one, but is obtained as the union of two regular surfaces which intersect each other on a regular line. The latter, represents a *wedge* discontinuity line of the body boundary; that is, a line on which no outer normal unit vector is defined. The two different models, termed also First-Order and Second-Order Surface Interaction Potentials, in turn, are analyzed under this assumption on the body boundary. Thus, it is shown that further conditions need to be imposed. A comparison between the conditions in the two different cases is provided.

1 Introduction

Equilibrium problems in elasticity have been classically studied[2] on adoption of the model of *dead loads*, that is assuming that the external forces, which act on the elastic body, depend on the configuration of such a body only through its reference configuration. If a body immersed into a surrounding media is considered, such an approach, according to the results of Podio-Guidugli[3], is not satisfactory. To model *live interactions* between an elastic body and the surrounding environment, which is also thought as an elastic body, Podio-Guidugli and Vergara Caffarelli[4] introduced First-Order Surface Interaction Potentials. Indeed, in[4] a Surface Interaction Potential has been introduced which exhibits an explicit dependence on the first deformation gradient. The model has been subsequently reconsidered by Carillo, Podio-Guidugli and Vergara Caffarelli[1] who proposed a generalization on introduction of Second-Order Surface Interaction Potentials. The latter are those potentials whose corresponding density depends not only on the position on the body boundary, but also on its first as well as second deformation gradient: thus, Second-Order Surface Interaction Potentials are *more non local* than First-Order ones. The generalization from First-Order to Second-Order Surface Interaction Potentials allows to model further physical problems which could not be treated when only First-Order Surface Interaction Potentials were considered. A generalization to higher order has been briefly considered in [1], where, following the definition, given by Truesdell and Noll [5], a material is termed to be of grade N when its stress response depends on the first N deformation gradients. Accordingly, when N increases, the nonlocality of the stress response increases too.

In addition, to emphasize that the body and the environment interact with each other, they are termed to form a *body-environment pair of grade* (N, M), where N and M denote, in turn, the grade of the elastic body which represents the environment, and the grade of the body immersed into it. Here, body-environment pairs of grade $(1, 1)$ and $(1, 2)$, according to the results comprised in [4] and [1], are reconsidered. In both cases, the environment is supposed to be a simple elastic material, while, conversely, the elastic body is, respectively, assumed to be a simple material or a material of grade 2. These two cases, respectively, are also referred to as First-Order and Second-Order Surface Interaction Potentials. Here, specifically, the attention is focused on the effects of the presence of discontinuity lines on the boundary which separates the body and the environment from each other. The equilibrium problem under investigation refers to the particular case when the body boundary exhibits a *wedge discontinuity line* which will be defined in the opening Section 2.

Section 3 is devoted to briefly recall the variational setting of the equilibrium problem in the case of a body boundary with a wedge discontinuity line. Both body-environment pairs of grade $(1, 1)$ [4] and $(1, 2)$ [1], are reconsidered. The equilibrium conditions and, in addition, with respect to the case of a smooth boundary, further conditions are written; the latter are termed *Extra Line Conditions.*

2 Wedge Line Discontinuity: Geometrical Setting

In this Section, the model under investigation, consisting of a body immersed into a surrounding environment and in interaction with it, is described. Subsequently, a brief survey, which concerns some notational choices as well as some results which are needed to obtain the equilibrium conditions of interest, is given.

The body, whose equilibrium problem is going to be studied, is assumed to admit a reference configuration, $\Omega \subset \boldsymbol{R}^3$, which exhibits a line of discontinuity of the outer normal unit vector. The discontinuity line is termed to be a jump discontinuity when the following assumptions are adopted:

- the reference configuration of the body is represented by a closed set $\bar{\Omega} \subset \boldsymbol{R}^3$;
- the boundary of such set, $\partial\Omega$ is given by

$$\partial\Omega = \mathcal{S}^{(1)} \cup \mathcal{S}^{(2)} \cup \Gamma \quad ; \tag{1}$$

  where $\mathcal{S}^{(1)}$ and $\mathcal{S}^{(2)}$, for instance $\mathcal{S}^{(i)} \in C^4$, $\ i = 1, 2$, represent two *regular* surfaces, which, thus, at each point, admit a unit outer normal vector oriented from Ω, the reference configuration of the body, to the surrounding environment. These two unit vectors, are denoted, in turn, as $\mathbf{n}^{(1)}$ and $\mathbf{n}^{(2)}$;
- Γ is the regular closed curve representing the common boundary of $\mathcal{S}^{(1)}$ and $\mathcal{S}^{(2)}$, which are, both, supposed to admit a smooth prolongation through Γ. Hence, at each point of Γ, both $\mathbf{n}^{(1)}$, orthogonal to $\mathcal{S}^{(1)}$, and $\mathbf{n}^{(2)}$, orthogonal to $\mathcal{S}^{(2)}$, are defined;
- $\Gamma := \{$ discontinuity set of the normal vector $\mathbf{n}$ to the surface $\partial\Omega\}$: these discontinuities are *jump discontinuities* since, on Γ the two vectors $\mathbf{n}^{(1)}$ and $\mathbf{n}^{(2)}$ are required to satisfy the further condition

$$\mathbf{n}^{(1)} \times \mathbf{n}^{(2)} \neq 0 \tag{2}$$

  which guaranties that the line Γ represents a wedge, not a cuspidal line.

2.1 Notation and Background

Here, some symbols which are needed in later computations are recalled. According to Carillo, Podio-Guidugli and Vergara Caffarelli [1], the quantities of interest are decomposed into a *tangential* and a *normal* part to emphasize that the first one is parallel to the plane tangent to $\partial\Omega$ while the latter is orthogonal to it. Thus, given an arbitrary variation $\mathbf{v}$, then ${}^s\nabla\mathbf{v}$ denotes the tangential part of $\nabla\mathbf{v}$, ${}^s\mathrm{Div}\,\mathbf{v}$, the tangential part of $\mathrm{Div}\,\mathbf{v}$ and, in general, given any quantity of interest, say $\mathbf{A}$, then ${}^s\mathbf{A}$ denotes its tangential part. The details, here not reported, are comprised in Carillo, Podio-Guidugli and Vergara Caffarelli [1] where a wide introductory section is devoted to collect all those background results which are related to the equilibrium problem in the case of body-environment pairs of grade (M, N).

Furthermore, some classical results needed for later applications, are recalled. Let $\mathbf{a}$ and $\mathbf{A}$ denote, respectively, a vector and a tensor field over $\mathcal{S}^{(\alpha)}, \alpha = 1, 2$, then the *Surface-Divergence Identities* hold:

$$\int_{\mathcal{S}} {}^{s}\mathrm{Div}\,\mathbf{a}\,dA = -\int_{\mathcal{S}} 2H(\mathbf{a}\cdot\mathbf{n})\,dA + \int_{\Gamma} \mathbf{a}\cdot\mathbf{t}\times\mathbf{n}\,ds\,, \tag{3}$$

$$\int_{\mathcal{S}} {}^{s}\mathrm{Div}\,\mathbf{A}\,dA = -\int_{\mathcal{S}} 2H\mathbf{A}\mathbf{n}\,dA + \int_{\Gamma} \mathbf{A}(\mathbf{t}\times\mathbf{n})\,\mathbf{ds}\,. \tag{4}$$

where $H = {}^{s}\mathrm{Div}\,\mathbf{n}$, represents the *mean curvature* of $\mathcal{S}$, assumed to be a regular surface; here, in turn, $\mathcal{S} = \mathcal{S}^{(1)}$ or $\mathcal{S} = \mathcal{S}^{(2)}$, and Γ denotes the boundary common to the two regular surfaces. In addition, the unit vector $\mathbf{n}$, normal to $\mathcal{S}$, and the unit vector $\mathbf{t}$, tangent on Γ, are coerently oriented. In the case considered in this section, where $\partial\Omega = \mathcal{S}^{(1)} \cup \mathcal{S}^{(2)} \cup \Gamma$, $\mathcal{S}^{(1)}$ and $\mathcal{S}^{(2)}$ can be chosen so that the unit tangent vector $\mathbf{t}$ on Γ, can be represented as

$$\mathbf{t} = \frac{\mathbf{n}^{(1)} \times \mathbf{n}^{(2)}}{\left|\mathbf{n}^{(1)} \times \mathbf{n}^{(2)}\right|} \tag{5}$$

where $\times$ denotes the vector product and the orientation of $\mathbf{t}$ and $\mathbf{n}^{(1)}$ are coerently fixed.

A further consequence of the divergence theorem, applied to obtain the equilibrium conditions of interest, reads:

$$\int_{\mathcal{S}} \mathbf{A}\cdot{}^{\mathbf{s}}\nabla\mathbf{v} = -\int_{\mathcal{S}} ({}^{\mathbf{s}}\mathrm{Div}\,\mathbf{A} + \mathbf{2H}\,\mathbf{A}\mathbf{n})\cdot\mathbf{v}\,\mathbf{dA}\,. \tag{6}$$

3 Variational Formulation of the Elastic Equilibrium Problem

In this Section, the Equilibrium Problem related to the model under investigation, which consists of a body immersed into a surrounding environment and in interaction with it, is studied in the case when the reference configuration of the body, denoted by $\Omega \subset \boldsymbol{R}^3$, exhibits a line of discontinuity of the outer normal unit vector. The stationary condition reads

$$\delta\mathrm{E}\{f\}[\mathbf{v}] = \delta\Sigma\{f\}[\mathbf{v}] - \delta\mathrm{T}\{f\}[\mathbf{v}] = 0 \quad . \tag{1}$$

corresponding to all deformations f and all admissible variations $\mathbf{v}$, where Σ represents the elastic energy and T the Surface Interaction Potential. The environment is modeled by an elastic body of grade 1, also termed a simple material by Truesdell and Noll [5], then Σ assumes the same form in both the considered cases of body environment pairs of grade $(1,1)$ as well as $(1,2)$. Hence, first of all, the variation $\delta\Sigma$ is evaluated referring to the case when the boundary of the reference configuration of the body is $\mathcal{S}^{(1)} \cup \mathcal{S}^{(2)} \cup \Gamma$. Indeed, when the energy density per unit volume is denoted by $\sigma(x, F)$ at $x \in \Omega$ and $F \in \boldsymbol{R}^{3\times 3}$, then, given a deformation f whose gradient is $F := \nabla f$, the total stored energy is given by

$$\Sigma\{f\} = \int_{\Omega} \sigma(x, \nabla f(x)) \; dV \quad . \tag{2}$$

Hence, its variation is given by

$$\delta\Sigma\{f\}[\mathbf{v}] = \int_{\Omega} \sigma_F \cdot \nabla \mathbf{v} \, dV \tag{3}$$

which, on application of the divergence theorem, gives

$$\int_{\Omega} \sigma_F \cdot \nabla \mathbf{v} \, dV = -\int_{\Omega} \operatorname{Div} \sigma_F \cdot \mathbf{v} \, dV + \int_{\partial\Omega} \sigma_F \; \mathbf{n} \cdot \mathbf{v} \; dA \; . \tag{4}$$

where $\partial\Omega = \mathcal{S}^{(1)} \cup \mathcal{S}^{(2)} \cup \Gamma$. The integral on the right hand side of (4) is extended over $\mathcal{S}^{(1)} \cup \mathcal{S}^{(2)}$ since the line integral over Γ is zero. Furthermore, since no contribution on the volume integral comes from the variation of the surface potential T, (4) implies that the standard field equation in Ω, that is

$$\operatorname{Div} \sigma_F \, (x, \nabla f(x)) = 0 \quad , \quad x \in \Omega \tag{5}$$

is a necessary equilibrim condition. Conversely, the second term in (4), namely the surface contribution on the boundary of Ω, to the variation of Σ will be balanced with the variation of the Surface Interaction Potential T which is evaluated in the next two subsections referring to the two different cases of Body-Environment Pairs of grade $(1,1)$ and of grade $(1,2)$.

3.1 Body-Environment Pair of grade $(1,1)$

Here, the results obtained by Podio-Guidugli and Vergara Caffarelli [4], are summarized in the specific case when the body boundary $\partial\Omega$ exhibits one wedge discontinuity line. The assumption of a jump discontinuity of the normal vector $\mathbf{n}$ through the regular curve Γ suggests that the density τ, here denoted as $\tau^{(\alpha)}$ on $\mathcal{S}^{(\alpha)}$, $\alpha = 1, 2$, depends explicitly also on $\mathbf{n}$. Accordingly, in the case of body environment pairs of grade $(1,1)$ the surface potential density is assumed to be of the form

$$\tau^{(\alpha)} = \tau^{(\alpha)}(x, \mathbf{n}^{(\alpha)}, f, \nabla f) \quad , \quad on \ \mathcal{S}^{(\alpha)} \quad , \quad \alpha = 1, 2. \tag{6}$$

Consequently, on Γ, where the two different outer normal unit vectors $\mathbf{n}^{(1)}$ and $\mathbf{n}^{(2)}$ are both defined, both (6) $\alpha = 1, 2$ hold true both at the same time, where $\tau^{(\alpha)}$ denotes the prolongation of $\tau^{(\alpha)}$ defined over $\mathcal{S}^{(\alpha)} \cup \Gamma$.

Now, the variation of the Surface Potential is needed to write in explicit form the equilibrium condition (1). Such a variation, recalling the form of $\partial\Omega$, is given by

$$\delta \mathrm{T}\{f\}[\mathbf{v}] = \int_{\mathcal{S}^{(1)} \cup \mathcal{S}^{(2)}} \left(\tau_f^{(\alpha)} \cdot \mathbf{v} + \tau_F^{(\alpha)} \cdot \nabla \mathbf{v} \right) dA \tag{7}$$

which, when the divergence theorem is applied, omitting, here and throughout in the following, the superscript α [a], can be written in the form

$$\int_{\mathcal{S}^{(1)} \cup \mathcal{S}^{(2)}} \tau_F^{(\alpha)} \cdot \nabla \mathbf{v} \, dA = \int_{\mathcal{S}^{(1)} \cup \mathcal{S}^{(2)}} \left(-({}^s\mathrm{Div}\, \tau_F^{(\alpha)} + 2H\, \tau_F^{(\alpha)}[\mathbf{n}]) \cdot \mathbf{v} + \tau_{\mathbf{F}}^{(\alpha)}[\mathbf{n}] \cdot \partial_{\mathbf{n}} \mathbf{v} \right) \mathbf{dA} \,. \tag{8}$$

where, $\partial_{\mathbf{n}} \mathbf{v}$ denotes the derivative of $\mathbf{v}$ with respect to the outer normal vector[b]. Thus, the stationary condition (1), recalling (8) and (4), reads

$$(\partial_F \sigma)\mathbf{n}^{(\alpha)} = -\tau_F^{(\alpha)} + {}^s\mathrm{Div}\, \tau_F^{(\alpha)} \,, \quad \text{on} \quad \mathcal{S}^{(\alpha)} \quad , \quad \alpha = 1, 2; \tag{9}$$

The only tangency condition, obtained by Podio-Guidugli and Vergara Caffarelli [4], on each one of the regular surfaces $\mathcal{S}^{(1)}$ and $\mathcal{S}^{(2)}$, is given by

$$\tau_F^{(\alpha)} \left[\mathbf{n}^{(\alpha)} \right] = 0 \,, \quad \text{on} \quad \mathcal{S}^{(\alpha)} \quad , \quad \alpha = 1, 2 \tag{10}$$

In addition, when it is considered that the discontinuity line belongs to both the surfaces $\mathcal{S}^{(1)}$ and $\mathcal{S}^{(2)}$, the following Extra Line Condition

$$\tau_F{}^{(2)} \left[\mathbf{n}^{(1)} \right] + \tau_F{}^{(1)} \left[\mathbf{n}^{(2)} \right] = 0 \,, \quad \text{on} \ \Gamma \,. \tag{11}$$

is obtained.

3.2 Body-Environment Pair of grade (1, 2)

This subsection comprises the results obtained by Carillo, Podio-Guidugli and Vergara Caffarelli [1], in the particular case of a jump discontinuity of the normal vector $\mathbf{n}$ through the regular curve Γ, that is when the body boundary $\partial\Omega$ exhibits one wedge discontinuity line. Again, such an assumption suggests

---

[a] Hence, the vector $\mathbf{n}$, when it appears as the argument of any function, is understood to coincide with $\mathbf{n}^{(1)}$ on $\mathcal{S}^{(1)}$ and $\mathbf{n}^{(2)}$ on $\mathcal{S}^{(2)}$

[b] again, the integration over $\partial\Omega$ has been replaced by the integration over $\mathcal{S}^{(1)} \cup \mathcal{S}^{(2)}$ since the line integral over Γ is zero.

that the density $\tau^{(\alpha)}, (\alpha), = 1, 2$ depends, also in the case of body environment pairs of grade $(1, 2)$, explicitly also on $\mathbf{n}$. Thus, the surface potential density is assumed to be of the form

$$\tau^{(\alpha)} = \tau^{(\alpha)}(x, \mathbf{n}^{(\alpha)}, f, \nabla f, \nabla^2 f) \quad , \quad on\ \mathcal{S}^{(\alpha)} \quad , \quad \alpha = 1, 2; \tag{12}$$

which, on Γ, where the two different outer normal unit vectors $\mathbf{n}^{(1)}$ and $\mathbf{n}^{(2)}$ are both defined, hold both at the same time, where, $\tau^{(\alpha)}$ denotes the prolongation of $\tau^{(\alpha)}$ defined over $\mathcal{S}^{(\alpha)} \cup \Gamma$.

Now, the variation of the Surface Potential is needed to write in explicit form the equilibrium condition (1). Thus, it follows, when it is recalled that $\partial\Omega = \mathcal{S}^{(1)} \cup \mathcal{S}^{(2)}$,

$$\delta \mathrm{T}\{f\}[\mathbf{v}] = \int_{\mathcal{S}^{(1)} \cup \mathcal{S}^{(2)}} \left(\tau_f^{(\alpha)} \cdot \mathbf{v} + \tau_F^{(\alpha)} \cdot \nabla \mathbf{v} + \tau_{\mathcal{F}}^{(\alpha)} \cdot \nabla^2 \mathbf{v} \right) dA \tag{13}$$

Repeated applications of the Surface Divergence Theorem allow to find the equilibrium condition on $\mathcal{S}^{(1)} \cup \mathcal{S}^{(2)}$:

$$(\partial_F \sigma)\mathbf{n} = {}^s\mathrm{Div}\, {}^s\mathrm{Div}\, \tau_{\mathcal{F}}^{(\alpha)} - {}^s\mathrm{Div} \left(\tau_{\mathcal{F}}^{(\alpha)}[\mathbf{n}]\, {}^s\nabla \mathbf{n} \right) - {}^s\mathrm{Div}\, \tau_F^{(\alpha)} - \tau_{\mathcal{F}}^{(\alpha)} [{}^s\nabla H \otimes \mathbf{n}] \,. \tag{14}$$

The tangency conditions, according to their definition [1], on $\mathcal{S}^{(1)} \cup \mathcal{S}^{(2)}$ are

$$\tau_{\mathcal{F}}^{(\alpha)} \left[\mathbf{n}^{(\alpha)} \otimes \mathbf{n}^{(\alpha)} \right] = 0 \,, \quad \text{on} \quad \mathcal{S}^{(\alpha)} \quad , \quad \alpha = 1, 2 \tag{15}$$

and

$$\tau_F^{(\alpha)} [\mathbf{n}] - 2\, {}^s\mathrm{Div} \left(\tau_{\mathcal{F}}^{(\alpha)} [\mathbf{n}] \right) + \tau_{\mathcal{F}}^{(\alpha)} [{}^s\nabla \mathbf{n}] = 0 \qquad \text{on } \mathcal{S}^{(1)} \cup \mathcal{S}^{(2)} \quad , \tag{16}$$

and, again, as referring to body-environment pair $(1, 1)$, also further conditions on the discontinuity line, *Extra Line Conditions*, need to be imposed:

$$\tau_{\mathcal{F}}^{(\alpha)} \left[\mathbf{n}^{(\alpha)} \otimes \mathbf{n}^{(\alpha)} \right] = 0 \,, \quad \alpha = 1, 2 \ \text{ on } \ \Gamma \,. \tag{17}$$

furthermore, recalling (15), on Γ, it holds:

$$\tau_{\mathcal{F}}^{(\alpha)} \left[\mathbf{n}^{(2)} \otimes \mathbf{n}^{(1)} \right] = \tau_{\mathcal{F}}^{(\alpha)} \left[\mathbf{n}^{(1)} \otimes \mathbf{n}^{(2)} \right] = 0 \,, \quad \alpha = 1, 2 \quad \text{on } \Gamma \ . \tag{18}$$

and

$$
\begin{aligned}
&-\left\{\tau_F{}^{(1)}\left[\mathbf{t}\times\mathbf{n}^{(1)}\right]-\tau_F{}^{(2)}\left[\mathbf{t}\times\mathbf{n}^{(2)}\right]\right\}+\\
&+\left\{\tau_{\mathcal{F}}^{(1)}\left[{}^s\nabla\mathbf{n}^{(1)}\left[\mathbf{t}\times\mathbf{n}^{(1)}\right]\otimes\mathbf{n}^{(1)}\right]-\tau_{\mathcal{F}}^{(2)}\left[{}^s\nabla\mathbf{n}^{(2)}\left[\mathbf{t}\times\mathbf{n}^{(2)}\right]\otimes\mathbf{n}^{(2)}\right]\right\}+\\
&-\frac{\partial}{\partial\mathbf{t}}\left\{\tau_{\mathcal{F}}^{(1)}\left[\mathbf{t}\otimes\mathbf{t}\times\mathbf{n}^{(1)}\right]-\tau_{\mathcal{F}}^{(2)}\left[\mathbf{t}\otimes\mathbf{t}\times\mathbf{n}^{(2)}\right]\right\}+\\
&-\left\{{}^s\mathrm{Div}^{(1)}\tau_{\mathcal{F}}^{(1)}\left[\mathbf{t}\times\mathbf{n}^{(1)}\right]-{}^s\mathrm{Div}^{(2)}\tau_{\mathcal{F}}^{(2)}\left[\mathbf{t}\times\mathbf{n}^{(2)}\right]\right\}=0\,.
\end{aligned}
\tag{19}
$$

A comparison between the extra line conditions obtained in the two cases of body-environment pairs of grade $(1,1)$ and $(1,2)$, given, respectively, by (11) and by (17) (18) and (19), shows that, when the Surface Interaction Potential is assumed to depend on the first deformation gradient, on the discontinuity line only one further condition needs to be imposed to obtain the equilibrium solution. Moreover, it is an algebraic condition. Conversely, when also second deformation gradients are involved, the extra conditions on the discontinuity line to be satisfied, in addition to those which are required when the body has a smooth boundary, are three and, notably, they are not algebraic. Indeed, the geometrical structure of the boundary of Ω turns out to be crucial [c].

Acknowledgment Research partially supported by G.N.F.M. - I.N.D.A.M. and Italian M.U.R.S.T. through the project "*Mathematical Models for Materials Science*".

1. S. Carillo, P. Podio-Guidugli, G. Vergara Caffarelli Second-Order Surface Potentials in Finite Elasticity. in *Rational Continua, classical and new*, M. Brocato e P. Podio-Guidugli Ed.s, Springer-Verlag, 2001;
2. P.G. Ciarlet, *Mathematical Elasticity, Vol.I: Three-Dimensional Elasticity*, North Holland, Amsterdam, 1988;
3. P. Podio-Guidugli, Elasticity with live loads, *Atti dei Convegni Lincei N.148, Giornata Lincea su "Interactions between Analysis and Mechanics. The Legacy of Gaetano Fichera"* (Roma, 22-23 April 1998), 127-142, 1999
4. P. Podio-Guidugli, G. Vergara Caffarelli Surface interaction potentials in elasticity. *Arch. Rational Mech. Anal.* **109**, 343-383, 1990;
5. C. Truesdell, W. Noll, *The Non-Linear Field Theories of Mechanics.* Handbuch der Physik III/3, S. Flügge ed., Springer-Verlag, Berlin·Heidelberg·New York, 1965

[c] Special examples, such as when Ω is supposed to be an emisphere and, consequently, $\mathcal{S}^{(1)}$ half on the spheric surface and $\mathcal{S}^{(2)}$ the plane circular domain of radius equal to that one of the sphere Ω, can be easily worked out.

WAVELET SOLUTIONS IN ELASTIC NONLINEAR OSCILLATIONS

CARLO CATTANI

Dipartimento di Matematica "G.Castelnuovo",
Univ. di Roma "La Sapienza", P.le A. Moro 5, I-00185 Roma, Italy
E-mail: carlo.cattani@uniroma1.it

This paper deals with the oscillations of a mechanical system in presence of relay forces (string with monotone unilateral friction). The fundamental equation is studied in the MRA (Multi Resolution Analysis) framework, representing functions and differential operators with Haar wavelet bases and with the spline derivative of Haar series[2]. The nonlinear equation of the mechanical system is transformed into a linear ordinary differential system, in the wavelet coefficients and a numerical approximate solution is eventually given.

1 Introduction

In this paper is considered a mechanical system, in presence of the restricted disturbance of constraints arising from a unilateral friction. The resulting partial differential equation is nonlinear and contains relay forces[1],

$$\frac{1}{a^2}\frac{\partial^2 u}{\partial t^2} - \frac{\partial^2 u}{\partial x^2} + \varepsilon u^+ = 0 \ , \quad (x \in \Re, t \in (0,T], T < \infty, a \neq 0, \varepsilon > 0) \tag{1}$$

with $u = u(x,t)$, $u^+ = u^+(x,t) = \max\{0, u(x,t)\}$. The initial conditions are

$$u|_{t=0} = u_0(x) \quad , \left.\frac{\partial u}{\partial t}\right|_{t=0} = \dot{u}_0(x) \ , \tag{2}$$

and the relay force u^+ is defined under the hypothesis

$$u^+\big|_{x=x_1} \geq u^+\big|_{x=x_2} \ , \ x_1 < x_2 \ , \tag{3}$$

so that the friction is a decreasing function from a maximum value in 0.

Using the Haar wavelets and the spline derivative [2] of the Haar series is proposed the wavelet solution of the nonlinear problem (1)-(2), showing that the differential equations for the wavelet coefficients are linear equations.

Haar wavelets[5] are C^0-functions (piecewise constant functions) depending on two parameters, for each couple of their values they are compactly supported on a finite interval. Since wavelets squeeze or shrink according to the scaling parameter, and translates along the x-axis, they seem the best basis for the representation of functions with singularities, and for localized functions. The Haar wavelets are the only symmetric wavelets, localized in space and frequency which are also analytically defined. The main drawback of the Haar wavelets is their lacking in smoothness, but this was solved[2] defining the spline derivative of piecewise constant functions. This method consists in smoothing the Haar wavelets using suitable order spline, deriving the spline and assuming the Haar representation of the derivative of the spline as the derivative of the original Haar wavelet.

Using the spline derivative of the Haar wavelets [2], it is given the linear system for the wavelet solution of the nonlinear string with friction.

2 Haar wavelet basis

The Haar function $\Phi_k^n(x) \stackrel{def}{=} 2^{n/2}\Phi(2^n x - k)$ is a function with a compact support on the dyadic interval $D_k^n \stackrel{def}{=} \left[\frac{k}{2^n}, \frac{k+1}{2^n}\right)$, where its constant value is 1. The Haar family of wavelets

$$
\begin{aligned}
&\Psi_k^n(x) \stackrel{def}{=} \{2^{n/2}\Psi(2^n x - k)\}_{k,n\in Z} \qquad , ||\Psi_k^n(x)||_{L^2} = 1 \ , \\
&\Psi_k^n(x) = \begin{cases} 2^{n/2} & , x \in \left[\frac{k}{2^n}, \frac{k+1/2}{2^n}\right) \\ -2^{n/2} & , x \in \left[\frac{k+1/2}{2^n}, \frac{k+1}{2^n}\right) \\ 0 & , \text{ elsewhere } , \end{cases}
\end{aligned} \tag{4}
$$

with

$$
\Psi_k^n(x) = 2^{n/2}\left(\Phi_k^{n+1}(x) - \Phi_{k+1}^{n+1}(x)\right) \ (k, n \in Z) \quad , \tag{5}
$$

is a complete orthonormal system for the (finite energy) space of $L^2(\Re)$-functions[5]. According to the multi resolution axioms[5] this space is decomposed as

$$
\begin{cases} L^2(\Re) = \bigoplus_{n\in Z} W_n = V_q \oplus \bigoplus_{j\geq q} W_j \ , & q \in Z \\ V_{n+1} = V_n \oplus W_n & . \end{cases} \tag{6}
$$

where V_n, $n \in Z$ is the subspace of the piecewise constant functions $f(x)$ compactly supported on D_k^n (n fixed) , W_n is the complementary subspace (of wavelet), and $\oplus$ is the direct sum of orthogonal subspaces.

As a consequence any function of the space $L^2(\Re)$ is "reconstructed" as

$$
f(x) = \sum_{n,k\in Z} \beta_k^n \Psi_k^n(x) \quad , \tag{7}
$$

while, fixing the resolution value $N < \infty$, in (6), the $L^2(\Re)$ space it is approximated, at the resolution N, by

$$
\begin{aligned}
L^2(\Re) &\cong V_0 \oplus \bigoplus_{n=0}^{N-1} W_n \\
f(x) &\cong \pi^N f(x) \stackrel{def}{=} \alpha_0^0 + \sum_{n=0}^{N-1} \sum_{k=0}^{2^n-1} \beta_k^n \Psi_k^n(x) \quad .
\end{aligned}
\tag{8}
$$

being $\pi^n : L^2(\Re) \to V_{n+1}$ a projection operator into V_{n+1}.

In a discrete approach, choosing a number of (dyadic) nodes $x_k \stackrel{def}{=} \dfrac{h+k}{2^n}$, $(k = 0, \ldots, 2^n - 1)$, in the dyadic interval D_h^n, the dyadic discretization is the operator $\nabla^n : L^2(\Re) \to L^2(Z(2^{-n}))$, being $L^2(Z(2^{-n})) \subset L^2(\Re))$ the set of $L^2(\Re)$-functions with jumps at x_k. The action of ∇^n on $f(x)$ is such that $\nabla^n f(x) = \mathbf{f}^n$ with $\mathbf{f}^n = \{f_0, f_1, \ldots, f_{2^n-1}\}$ and $\{f_k \stackrel{def}{=} f(x)\,|_{x=x_k} \ , 0 \leq k \leq 2^n - 1\}$. The fast Haar-wavelet transform $\mathcal{H}$ of $\mathbf{f}^n$ is the operator[2,3] giving α_0^0 , β_k^n by

$$
\begin{cases}
\mathcal{H} \quad : \ L^2(Z(2^{-n})) \to V_{n+1} \mid \mathbf{f}^n \mapsto \mathcal{H}\mathbf{f}^n = \left\{\alpha_0^0, \beta_k^r\right\}_{k=0,\ldots,2^r-1}^{r=0,\ldots,n-1} , \\
\mathcal{H}\mathbf{f}^n \stackrel{def}{=} \left[\prod_{k=1}^{n} \left((P_{2^k} \oplus I_{2^n-2^k})(H_{2^k} \oplus I_{2^n-2^k})\right)\right] \mathbf{f}^n \ .
\end{cases}
\tag{9}
$$

In particular, matrix $H_{\ldots}$ contains the coefficients of (5), $P_{\ldots}$ is the permutation (or shuffle) matrix and $I_{\ldots}$ is the identity matrix [2,3,5]. Indeed the Haar transformation gives the components $\left\{\alpha_0^0, \beta_k^r\right\}_{k=0,\ldots,2^r-1}^{r=0,\ldots,n-1}$ of piecewise constant function with respect to the wavelet basis in V_{n+1}, but since the basis is univocally defined we assume that $\mathcal{H}$ defines also the function in V_{n+1}.

3 Spline operator

A p-order natural spline on the interval $\Omega \supseteq D_h^n$, is the $C^{p-1}(\Omega)$ differentiable operator

$$
\mathcal{S}^p : L^2(Z(2^{-n})) \to C^{p-1}(\Omega) : \ \mathbf{f}^n \mapsto s(x) \stackrel{def}{=} \mathcal{S}^p \mathbf{f}^n \quad ,
$$

such that $s(x)$ is a p-order polynomial based on the dyadic nodes. Using the splines we have

Definition 1 *The spline derivative[2] operator L_V, associated to the differential operator L and to a p-order spline, is a discrete operator acting on the*

piecewise constant functions of V_{n+1} *as*

$$\begin{cases} L_V : V_{n+1} \to V_{n+1} : \ \mathcal{H}\mathbf{f}^n \mapsto L_V\mathcal{H}\mathbf{f}^n \quad , \\ L_V \stackrel{def}{=} \pi^n L \mathcal{S}^p \mathcal{H}^{-1} \ . \end{cases} \tag{10}$$

The operator L_V commutes with the projection operator:

$$\pi^n L = L_V \pi^n \tag{11}$$

in a such a way that, given the set $\mathbf{f}^n$ and computed the spline of sufficiently large order p, the spline-derivative L_V of $\mathcal{H}\mathbf{f}^n$ it belongs to the same space of $\mathcal{H}\mathbf{f}^n$. For a fixed dyadic set, the operator L_V is univocally defined when is given the order of the spline and, according to $(10)_2$-(11), it is assumed $\pi^{-n} \stackrel{def}{=} (\pi^n)^{-1} = \mathcal{S}^p\mathcal{H}^{-1}$, i.e. the interpolation is done by splines.

4 Wavelet solution

Theorem 1 *The wavelet solution of (1)-(2)-(3) is the wavelet series*

$$u(x,t) = \sum_{n,k\in Z} \beta^n_k(t)\Psi^n_k(x) \tag{12}$$

where $\beta^n_k(t)$ *is obtained by solving the following Cauchy problem*

$$\begin{cases} \dfrac{1}{a^2}\ddot{\beta}^n_k(t) = \left[\displaystyle\sum_{m,h\in Z} \gamma^{nh}_{km}\beta^m_h(t) \right] + \varepsilon 2^{n(n+1)/2}\left(\beta^{n+1}_k(t) - \beta^{n+1}_{k+1}(t)\right) \\ \beta^n_k(0) \quad = \displaystyle\int_{D^n_k} u_0(x)\Psi^n_k(x)\,dx \ , \\ \dot{\beta}^n_k(0) \quad = \displaystyle\int_{D^n_k} \dot{u}_0(x)\Psi^n_k(x)\,dx \ , \qquad (n,k\in Z) \ , \end{cases} \tag{13}$$

and the connection (constant) coefficients γ^{nh}_{km} *are*

$$\gamma^{nh}_{km} = \langle L_V\Psi^n_k(x), \Psi^h_m(x)\rangle \stackrel{def}{=} \int_{-\infty}^{\infty} \left(L_V\Psi^n_k(x)\right)\Psi^h_m(x)\,dx \ ,$$

with

$$L_V\Psi^n_k(x) = \pi^n \frac{\partial^2}{\partial x^2}\left(\mathcal{S}^p\mathcal{H}^{-1}\Psi^n_k(x)\right) .$$

Inserting (12) in (1), and using (11) it immediately follows the expression (in square brackets) corresponding to the discretization of $\partial^2/\partial x^2$ by the Galerkin method and the orthonormality of the $\Psi_k^n(x)$.

The wavelet reconstruction of the relay force is, according to (12),

$$u(x,t)^+ = \left[\sum_{n,k\in Z} \beta_k^n(t)\Psi_k^n(x)\right]^+ = \sum_{n,k\in Z} \beta_k^n(t)\left[\Psi_k^n(x)\right]^+ \tag{14}$$

and taking into account equations (5) we have $[\Psi_k^n(x)]^+ = \Phi_k^{n+1}(x)$ it follows

$$u(x,t)^+ = \sum_{n,k\in Z} 2^{n/2}\beta_k^n(t)\Phi_k^{n+1}(x) \quad . \tag{15}$$

The inner product with $\Psi_h^m(x)$ gives

$$\langle\Phi_k^{n+1}(x), \Psi_h^m(x)\rangle = 2^{m/2}\langle\Phi_k^{n+1}(x), \Phi_h^{m+1}(x)\rangle - \langle\Phi_k^{n+1}(x), \Phi_{h+1}^{m+1}(x)\rangle \ ,$$

and since the characteristic functions $\Phi_k^n(x)$ are compact function on D_k^n, having the constant value 1, the last product vanishes when the shift parameters (h,k) differ from each other, thus we obtain[a]

$$\langle\Phi_k^{n+1}(x), \Psi_h^m(x)\rangle = 2^{m/2}2^{\min(n+1,m+1)/2}\left(\delta_{kh} - \delta_{k\ (h+1)}\right)$$

δ_{kh} being the Kroenecker symbol. Therefore equation $(13)_1$ follows by the Galerkin method and by a suitable change of the indices.

Equation (14) is a consequence of the hypothesis (3) together with

Theorem 2 *For a monotone decreasing [increasing] function, the coefficients* β_k^n *are all positive [negative]*

The proof easily follows from the definition of the fast Haar transform.

5 Numerical solution using wavelets

A simple numerical approximation of $u(x,t)$ might be obtained as follows. Let us consider the following problem on $\Re \times \Re^+$,

$$\begin{cases} \dfrac{\partial^2}{\partial t^2}u &= \dfrac{\partial^2}{\partial x^2}u + \varepsilon u^+ \\ \left.\dfrac{\partial u}{\partial t}\right|_{t=0} &= 1 \\ u(x,0) &= \sin 2\pi x \end{cases} ,$$

[a] since it is

$$\langle\Phi_k^n, \Phi_h^m\rangle = 2^{\min(n,m)/2}\delta_{kh}$$

and let us use the stepwise forward method

$$\begin{cases} u(x,t+2\Delta t) &= 2u(x,t+\Delta t) - u(x,t) + \Delta t^2 \left[\dfrac{\partial^2}{\partial x^2} u(x,t) + \varepsilon[u(x,t)]^+ \right] \\ u(x,\Delta t) &= u(x,0) + \Delta t \\ u(x,0) &= \sin 2\pi x\ . \end{cases} \tag{16}$$

The approximate wavelet solution up to the resolution N, is the vector $\pi^N u(x,t)$. Using the (cubic) spline derivative (10) and (11) it is:

$$\pi^N \left(\frac{\partial^2}{\partial x^2} u(x,t) \right) = \pi^N \frac{\partial^2}{\partial x^2} \mathcal{S}^3 \mathbf{u}_t^N \qquad (\mathbf{u}_t^N \stackrel{def}{=} \nabla^N u(x,t))$$

and the projection of the relay force is $\pi^N u^+ = (\pi^N u)^+$. There follows the numerical wavelet approximation of system (16) into V_{N+1}

$$\begin{cases} \mathbf{u}_{t+2\Delta t}^N &= 2\mathbf{u}_{t+\Delta t}^N - \mathbf{u}_t^N + \Delta t^2 \left[\pi^N \dfrac{\partial^2}{\partial x^2} \mathcal{S}^3 \mathbf{u}_t^N + \varepsilon[\mathbf{u}_t^N]^+ \right] \\ \mathbf{u}_{\Delta t}^N &= \mathbf{u}_0^N + \Delta t \\ \mathbf{u}_0^N &= \pi^N \sin 2\pi x \end{cases} .$$

Assuming that the unknown function is expressed in the Haar wavelet basis, we start from the initial function $\sin 2\pi x$ represented as Haar series, then we derive it using the spline derivative and we compute the non negative part of the function at time $t = 0$. By the previous recursive (explicit) formula, we obtain the explicit value of $u(x,t)$ at time $t = 1$, and so on.

References

1. I.M. Babakov, The Theory of Oscillations,Moskow, Nauka,(1968).
2. C. Cattani, "Haar Wavelet Spline", Journal of Interdisciplinary Mathematics vol.4, No. 1, (2001), 35-47.
3. C. Cattani and M.Pecoraro, "Non Linear Differential Equations in Wavelet Basis", Acoustic Bulletin,vol. 3, (4), (2000), 4-10.
4. C. Cattani and L.Toscano, "Hyperbolic Equations in Wavelet Bases", Acoustic Bulletin, vol. 3, (2), (2000), 4-10.
5. I.Daubechies, Ten lectures on wavelets, CBMS-NSF Regional Conference Series in Applied Mathematics, SIAM,Philadelphia,(1992).

CONTINUUM EQUATIONS FOR RAREFIED GASES

XINZHONG CHEN AND EDWARD A. SPIEGEL
Astronomy Department, Columbia University, New York, NY 10027

We sketch the derivation of continuum equations from the relaxation model of kinetic theory. The equations found involve a pressure tensor and a heat flux whose approximate expressions contain the time derivatives of the macroscopic fields. We may eliminate those derivatives by using the fluid equations themselves. If we use the Euler equations for this, as in the Chapman-Enskog procedure, we obtain the Navier-Stokes equations. However, we advocate using the full fluid equations and so are led to a generalization of the Navier-Stokes equations that provides improved agreement with experiment in the case of long mean free paths of particles.

1 General Discussion

The Navier–Stokes equations have played a key role in all discussions of fluid dynamical phenomena. However, in certain contexts, they do not provide adequate descriptions of processes that occur in the motions of real fluids. One of the chief failures of these basic equations occurs in the regime where the mean free paths of the constituent particles are comparable to (or larger than) the characteristic macroscopic scales of the motion. This happens in shock waves because the thickness of a shock layer is typically of the order of the mean free paths of the particles making up the gas in which the wave occurs. Another instance of failure is in the description of the propagation and dissipation of ultrasound, sound waves whose periods are of the order of or less than the mean flight times of the constituent particles.

To deal with such problems, one may study them in terms of the basic equations of kinetic theory. The drawback to this approach is that, even in the simplest case, one has to deal with the evolution of a probability density in a six-dimensional phase space whose coordinates are the components of the velocities (or momenta) and the spatial coordinates of the constituent particles. This approach has been described, for example, in the lectures of Uhlenbeck [1] and it is made difficult by the high dimension of the phase space. As in many other mathematical problems, it is useful to introduce some method of dimensional reduction to make deeper inroads into the problem. That is why one turns to the macroscopic description in terms of fluid equations in this context. In this paper we describe a way to go beyond the Navier–Stokes equations to derive more general macroscopic equations that may be effective in situations where the mean free paths of the particles are of the order of the scales of the macroscopic motion.

The original derivation of the basic fluid equations by Navier and by Stokes was phenomenological. Even if one is satisfied with such an approach, which is indeed very convincing, it leaves open the determination of the basic transport coefficients in terms of the underlying physical quantities such as temperature and density. To get to the heart of the matter, one has recourse to derivations from the basic equations of kinetic theory. These equations are the Boltzmann equation and other models inspired by it, such as the relaxation (or BGK) model [2, 3].

Derivations from the kinetic equations normally proceed by way of expansions in terms of the nondimensional mean free paths of the constituent particles, which we shall call ϵ. Such an expansion was used by Hilbert [4] to find an approximate solution of the kinetic equation as a formal series in ϵ. There followed the now standard Chapman–Enskog procedure, in which expansions in ϵ are employed to develop equations for the macroscopic fields [5, 6].

In seeking to improve on these earlier treatments, we have first considered the simplest case: a gas of hard spheres with no internal degrees of freedom. The interactions among the constituent particles conserve the total mass, momentum and energy of the particles involved in collisions. Because of this, in the Boltzmann theory, the linearization of the term in the kinetic equations representing the effect of collisions among particles produces an operator for which those five basic quantities are null vectors. All the other eigenvectors of the linearized operator have negative eigenvalues. There is here a resemblance to certain problems in multiple bifurcation theory.

When we develop the solution of the kinetic equation in terms of eigenvectors of the linearized collision operator, the coefficients of the null vectors are the slow variables of the conventional asymptotic treatment of this problem. Suitable combinations of these five quantities form the standard fluid dynamical fields: density, velocity and temperature. This parallel of kinetic theory to conventional asymptotic theory can be seen in the work of Bogoliubov and van Kampen, and it provides a vision that motivates our reexamination of the asymptotic treatment of the equations of kinetic theory with H. Rao [7–9].

Though the formalism of the Boltzmann equation suggests this helpful vision, we shall nevertheless describe our first efforts in this problem in terms of the relaxation model. We have, in the initial phase, confined ourselves to working at the classical level including neither relativity or quantum mechanics. However, even at the present stage of the work, there have appeared modified fluid equations that we present below, together with brief mention of the first applications.

We recognize that in suggesting that there exists a generalization of the

Navier–Stokes equations that works better than those hallowed equations, we are likely to encounter skepticism. Nevertheless, we contend that the way the asymptotic theory of the kinetic equations is normally developed, though formally correct, is not optimal. This is brought out by the second-order extension of the Chapman–Enskog procedure that produces the Burnett equations. These are no better than the Navier–Stokes equations, and they are perhaps worse because, as Grad [10] has emphasized, the expansions involved are asymptotic rather than convergent.

On the other hand, in theoretical physics problems, good results have been achieved when an approximation of a given accuracy contains terms of all orders in the expansion parameter in the final formulae. This is the feature of the asymptotics that we aim to incorporate in our treatment. Of course, in such murky waters one needs a reliable criterion of validity and ours is in the comparison to experiment. What we have found so far is that, in the experimental comparisons we have made, our results are better than those found from the Navier–Stokes equations. Here we describe the first results of our approach, which is still under development. We shall describe further developments elsewhere and we hope also to use the methods outlined here in other contexts such as the derivation of pattern equations for bifurcating systems with continuous spectra [11].

2 The Relaxation Model

We work in the conventional six-dimensional phase space, or μ-space, whose coordinates are the three spatial coordinates and the three velocity components of individual particles making up a gas of many identical particles of mass m. The kinetic equation is then a continuity equation for the one-particle distribution, which is a probability density in the phase space. This one-particle distribution function is defined such that the expected number of particles, dN, in an element of volume, $d^3\mathbf{x}d^3\mathbf{v}$, is given by $dN = f(\mathbf{x}, \mathbf{v}, t)\, d^3\mathbf{x}d^3\mathbf{v}$. We assume that the particles satisfy Hamilton's equations so that the velocity in six dimensions is solenoidal. Collisions, which rapidly change the velocities of the particles, act as sources and sinks and so f satisfies a kinetic equation of the form

$$\mathcal{D}f = \mathcal{C}[f] \tag{1}$$

where $\mathcal{D} \equiv \partial_t + v^i\, \partial_{x^i} + a^i\, \partial_{v^i}$, $\mathbf{a}$ is an acceleration due to an external force, and $\mathcal{C}$ is a functional that describes the effects of collisions on the distribution function. Summation over $1, 2, 3$ of repeated indices is understood.

There are several models for the collision term and we shall be especially concerned here with the relaxation model [2, 3], the simplest one to work with. It is useful for testing methods and its form is intuitively suggested by radiative transfer with coherent scattering. In both cases, we suppose that particles entering a sufficiently small volume of physical space in a very short time interval experience interactions that may change their momenta and energies individually while the total momenta and energies are conserved over all the particles. The process may be visualized as if the particles entering the interaction are destroyed and newly emergent particles are created in the same number as a result of the interactions. In the style of radiative transfer theory, we can express this as

$$\mathcal{C}[f] = \alpha - \beta f \ , \tag{2}$$

where the rate of destruction is proportional to f itself.

As in the radiative case, we think of α and β as functions of the macroscopic fields, such as density and temperature. These are in turn functionals of f itself but, in the modeling process, we do not allow for that explicitly. (This in fact is done in constructing model stellar atmospheres in astrophysical studies.) Instead, we imagine a local equilibrium solution f_0 and this is given by $\mathcal{C}[f_0] = 0$, or

$$f_0 = \frac{\alpha}{\beta} \ , \tag{3}$$

which is analogous to the Kirchhoff-Planck law of the radiative case. In both cases (massive or massless particles) we need to prescribe how f_0 and β in the model collision term depend on the macroscopic fields.

3 Macroscopic Equations

Hilbert [4] developed solutions of the kinetic equation in ϵ; Chapman and Enskog [6] used such expansions to develop fluid equations directly. In the simplest cases, where the total masses, momenta and energies of the particles involved in collisions are conserved in the process, we have

$$\int \psi^\alpha \, \mathcal{C} \, d^3 v = 0 \tag{4}$$

where

$$\psi^\alpha = \left(m, m\mathbf{v}, \tfrac{1}{2}m\mathbf{v}^2\right) \tag{5}$$

with $\alpha = 0, 1, 2, 3, 4$. (There may be other independent conserved quantities, such as angular momentum in the quantum case, but we shall not deal with these additional quantities here.)

If we multiply the kinetic equation by ψ^α and integrate over $d^3\mathbf{v}$ we obtain the fluid equations

$$\partial_t \rho + \nabla \cdot (\rho \mathbf{u}) = 0 \tag{6}$$

$$\rho \left(\partial_t \mathbf{u} + \mathbf{u} \cdot \nabla \mathbf{u}\right) + \nabla \cdot \mathbb{P} = \rho \mathbf{a} \tag{7}$$

$$\tfrac{3}{2}\rho R \left(\partial_t T + \mathbf{u} \cdot \nabla T\right) + \mathbb{P} : \nabla \mathbf{u} + \nabla \cdot \mathbf{Q} = 0, \tag{8}$$

where we have introduced these macroscopic quantities:

$$\text{Mass density:} \qquad \rho \equiv \int m f \, \mathrm{d}^3 v, \tag{9}$$

$$\text{Mean velocity:} \qquad \mathbf{u} \equiv \frac{1}{\rho} \int m \mathbf{v} f \, \mathrm{d}^3 v, \tag{10}$$

$$\text{Temperature:} \qquad T \equiv \frac{m}{3R\,\rho} \int c^2 f \, \mathrm{d}^3 v, \tag{11}$$

and where the peculiar velocity is $\mathbf{c} = \mathbf{v} - \mathbf{u}$, $R = k/m$ is the gas constant, and k is the Boltzmann constant. We assume throughout this work that f goes rapidly to zero as $|\mathbf{v}| \to \infty$ so that f-weighted integrals over velocity space are finite and well defined. The additional quantities in the fluid equations are the pressure tensor and heat flux

$$\mathbb{P} \equiv \int m \mathbf{c}\mathbf{c} f \, \mathrm{d}^3 \mathbf{v}, \qquad \mathbf{Q} \equiv \int \tfrac{1}{2} m c^2 \mathbf{c} f \, \mathrm{d}^3 \mathbf{v}, \tag{12}$$

whose determination is a main goal of the theory.

For much of the last century, the two principal approaches to this problem were the methods of Chapman and Enskog [6] and the moment procedure following from the work of Grad [10]. Van Kampen [12] has described how the Chapman–Enskog approach may be formulated in terms of a standard two-time approach, in particular in the way in which he himself has formulated this method. He has expressed misgivings about extending Grad's work to the case where the kinetic equation is used to derive equations for higher moments such as $\mathbb{P}$ and $\mathbf{Q}$ with closure being made by giving approximations for yet higher moments. Van Kampen argues that, in the context of the Boltzmann theory in particular, the higher moments are not slow variables, as ρ, $\mathbf{u}$ and T are, so that little is gained by introducing higher moments as basic variables in a reduced system. There are of course other views and approaches to these matters that we have not space to review but we may cite a few [13–17].

In our approach, we keep to the description in terms of slow variables but allow the inclusion of their derivatives, which are also slow variables, so as to improve the macroscopic description. As we have already suggested, the use of the term slow variables is natural in the context of the Boltzmann theory for anyone steeped in the lore of dynamical systems. If we let $f = f_0 + \epsilon\varphi$ with $\mathcal{C}[f_0] = 0$, we can write $\mathcal{C}[f] = \mathcal{L}\varphi + \cdots$ where $\mathcal{L}$ is the linearized collision operator. In the case of the Boltzmann model, $\mathcal{L}$ has as null vectors ψ^α while the rest of its eigenvectors all have negative eigenvalues. The qualitative expectation is then that, left to itself, the system will head into its null space much as a simple dynamical system proceeds to its center manifold. In this loose analogy, we may think of Boltzmann's $H-$theorem as providing the same kind of moral support for what follows as the center manifold theorem does in the analogous case of a bifurcation.

Van Kampen's formulation of the Chapman–Enskog approach brings out how the proximity to equilibrium assumed for the system is enforced very strongly by the recycling of the zeroth order results in the development of the first order results. There are terms thrown away in this process that we strive to keep. The basis of the recycling procedure is in the application of solvability conditions at every order. That, in essence, is the feature of the older asymptotics that we would like to avoid. In our formulation, we regard the fluid equations themselves as global solvability conditions on the kinetic theory. Our point is that, by introducing further solvability conditions, one loses generality. This is strongly felt in the initial conditions on the fluid variables whose distribution among the different orders of approximation of the fluid fields should be left open and not forced, as it is in the Chapman–Enskog procedure. This is also a familiar problem in center manifold theory.

A similar sacrifice of generality occurs in the standard two-timing procedure when applied to simple multiple bifurcation problems [18]. The limitations of the two-time methods in the contexts of multiple bifurcation theory [19] lead to normal forms of amplitude equations with mixed orders. In the context of kinetic theory, a mixing of orders arises in the Navier–Stokes equations where the viscosity and conductivity are of order ϵ compared to the other terms in the equations. These features have not been seen so clearly in the context of kinetic theory because there is no suitable unfolding parameter in this problem.

4 First-Order Theory

In this section, by way of clarifying our general remarks, we describe our approach for the case of the relaxation model [7–9, 20, 21]. For this purpose, we use the nondimensionalized kinetic equation based on characteristic macroscopic time and length scales as the natural units. We assume moreover that the Mach numbers of the flows encountered are of order unity so that there is only one basic nondimensional quantity, the Knudsen number, ϵ. This is the ratio of the mean free path of the constituents to a characteristic macroscopic length, such as an acoustic wave length. It is also the ratio of the mean flight time to a macroscopic time, which is a nondimensional β^{-1}. Then, with (2) and (3), the kinetic equation is

$$\epsilon \mathcal{D} f = f_0 - f \tag{13}$$

where f_0 is the local equilibrium toward which the gas is driven by its self-interaction.

We adopt the Maxwell-Boltzmann distribution as the equilibrium, namely

$$f_0 = n(2\pi RT)^{-3/2} \exp\left(-\frac{c^2}{2RT}\right). \tag{14}$$

Though f_0 is annihilated by the collision operator, it depends on T and ρ, so f_0 is not a strict equilibrium but a local equilibrium that may depend on $\mathbf{x}$ and t.

The fluid fields are defined, as we saw, in terms of f itself, so there exists an implicit relation between f and f_0. The explicit form of this relation that is usually found in kinetic theory is called the matching condition and, if it is not a consequence of the model in use, it is usually imposed to force consistency of the results. In the present case, the conditions of conservation of mass, momentum and energy, (4), ensure the matching conditions.

If we approximate f by f_0, we obtain $\mathbb{P} = p\mathbb{I}$ and $\mathbf{Q} = \mathbf{0}$ where $p = R\rho T$ and $\mathbb{I}$ is the unit dyad. This approximation gives the Euler equations of fluid dynamics. More generally, we seek higher approximations based on the development in ϵ: $f = f_0 + \epsilon f_1 + \cdots$.

In first order in ϵ, we have $f_1 = -\mathcal{D} f_0$, hence

$$f_1 = -f_0 \left[\mathcal{D} \ln \rho + \left(\frac{\mathbf{c}^2}{2RT} - \tfrac{3}{2}\right) \mathcal{D} \ln T + \frac{1}{RT} \mathbf{c} \cdot \mathcal{D}\mathbf{u}\right]. \tag{15}$$

As the macroscopic fields do not depend on $\mathbf{v}$, the streaming operator is here just $\mathcal{D} = \partial_t + \mathbf{v} \cdot \nabla$ and it is convenient to make the substitution $\mathcal{D} = D/Dt + \mathbf{c} \cdot \nabla$ where $D/Dt = \partial_t + \mathbf{u} \cdot \nabla$. Then, when we put the approximation $f = f_0 + \epsilon f_1 + \mathcal{O}(\epsilon^2)$ for f into (12), the integrand depends entirely on $\mathbf{c}$ so that we can replace $d\mathbf{v}$ in the integrals by $d\mathbf{c}$.

From the expression for f_1, the evaluation of the pressure tensor can be carried out quite easily and we find

$$\mathbb{P} = p\mathbb{I} - \left[\mu\left(\frac{D\ln T}{Dt} + \tfrac{2}{3}\nabla\cdot\mathbf{u}\right)\right]\mathbb{I} - \mu\,\mathbb{E} + \mathcal{O}(\epsilon^2) \tag{16}$$

where $p - R\rho T$, $\mu = \epsilon p$, and

$$E_{ij} = \frac{\partial u_i}{\partial x_j} + \frac{\partial u_j}{\partial x_i} - \tfrac{2}{3}\nabla\cdot\mathbf{u}\,\delta_{ij} \tag{17}$$

is the rate-of-strain tensor in traceless form. In a similar manner, the heat flux is found to be

$$\mathbf{Q} = -\eta\nabla T - \left[\tfrac{5}{2}\mu\left(\frac{D\mathbf{u}}{Dt} - \mathbf{a}\right) + \eta T\nabla\ln p\right] + \mathcal{O}(\epsilon^2) \tag{18}$$

where $\eta = \frac{5}{2}\mu R$.

5 The New Terms

The terms in square brackets in expressions (16) and (18) for $\mathbb{P}$ and $\mathbf{Q}$ are not normally seen in the fluid equations. Those extra terms are usually simplified, somewhere in the course of the derivation, by introducing the fluid equations to eliminate the time derivatives in the brackets. However, in the Chapman-Enskog method, the "fluid equations" used for this purpose are the Euler equations. When the Euler equations are used to eliminate the time derivatives from (16) and (18), the terms in square brackets turn out to be $\mathcal{O}(\epsilon^2)$ and are usually neglected. This procedure gives rise to the Navier-Stokes equations. But if it is desired to eliminate the time derivatives (though there is no real need to do this) we would prefer instead to use the full fluid equations for the purpose. To do this, we may introduce $\mathbb{D} = \mathbb{P} - p\mathbb{I}$. Then (7) and (8) become

$$\rho\left(\partial_t\mathbf{u} + \mathbf{u}\cdot\nabla\mathbf{u}\right) + \nabla p - \rho\mathbf{a} = -\nabla\cdot\mathbb{D} \tag{19}$$

$$\tfrac{3}{2}\rho R\left(\partial_t T + \mathbf{u}\cdot\nabla T\right) + p\nabla\cdot\mathbf{u} = \mathbb{D} : \nabla\mathbf{u} - \nabla\cdot\mathbf{Q}\ , \tag{20}$$

while (16) and (18), are written as

$$\mathbb{D} - \mu\mathbb{E} = -\tfrac{2}{3}\epsilon\mathbb{D} : \nabla\mathbf{u} - \tfrac{2}{3}\epsilon\nabla\cdot\mathbf{Q}\mathbb{I} + \mathcal{O}(\epsilon^2) \tag{21}$$

$$\mathbf{Q} + \eta\nabla T = -\tfrac{5}{2}\epsilon RT\nabla\cdot\mathbb{D} + \mathcal{O}(\epsilon^2). \tag{22}$$

If we were to omit the terms on the right sides of (21) and (22) we would have the usual Navier-Stokes expressions. Those terms are $\mathcal{O}(\epsilon^2)$ since $\mathbb{D}$ and

$\mathbf{Q}$ are both $\mathcal{O}(\epsilon)$. Thus, from the linear approximation to f, we obtain two theories that differ only by terms that are formally of order ϵ^2.

Equations (16) and (18) are not closure relations in the usual sense since they do not provide explicit expressions in terms of the fluid fields. These expressions are process-dependent since they involve derivatives of the fluid fields. Hence, they have a more important dynamical content than the usual closure approximations. When we eliminate the time derivatives using the equations of motion to get (21) and (22), we still do not have explicit formulae for $\mathbb{D}$ and $\mathbf{Q}$. However, we may combine (21) and (22) to obtain what meteorologists would call diagnostic equations, which may be written as differential equations for the unknown moments. For example, if we use (22) to replace $\mathbf{Q}$ in (21) we find an equation of the form

$$\nabla \cdot (\nu \nabla \cdot \mathbb{D}) + \Re : \mathbb{D} = \text{R.H.S.} \tag{23}$$

where $\Re$ is a certain fourth rank tensor, $\nu = \mu/\rho$ and the the right side of the equation (R.H.S.) involves derivatives of T and $\mathbf{u}$.

Inversion of the linear operators that appear in the diagnostic equations may be carried out by developing (21) and (22) in ϵ. To get explicit expressions, we include all orders in ϵ; to recover the Navier-Stokes expression, we would drop terms of second order. We are not advocating this favoring of space over time but merely wish to show where the difference in the two theories originates and to suggest that, when an asymptotic approximation gives an expression that has all orders of the parameter in it, this is sometimes a sign that it may be a good approximation. Our experience with the present form has so far shown it to be more accurate than the Navier-Stokes equations when ϵ is not infinitesimal.

6 Conclusion

We have described two versions of the fluid dynamical equations. In the Navier-Stokes case, as derived by the Chapman-Enskog scheme, results from lower order are used to simplify the higher moments. In our procedure, this recycling of lower order results is avoided. The two versions of fluid dynamics, with and without the insertion of results from lower orders, formally differ only by terms of order ϵ^2, which is permissible in first order theories. Yet they are markedly different in their predictions when ϵ is not infinitesimal.

A sample of the difference is seen in the linear theory of sound propagation. In that case, ϵ is the product of the mean flight time of particles and the frequency of the sound wave. Our version of the fluid equations gives a finite phase velocity for large ϵ that is in good agreement with the experiments [24]

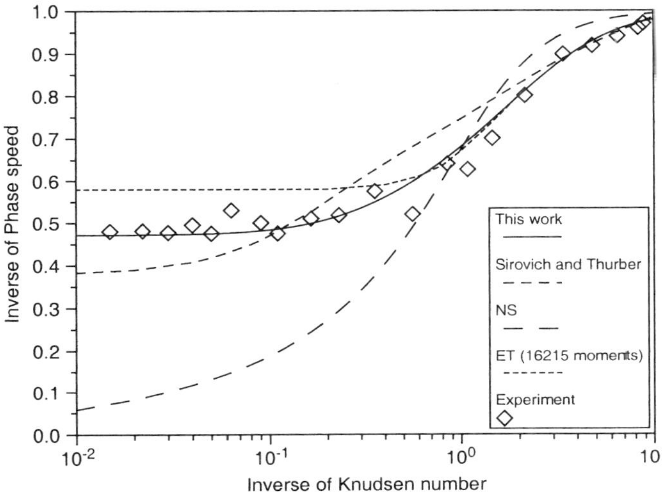

Figure 1. Comparison of the present theory with those from the Navier-Stokes equations for the linear theory of sound propagation. Also shown are results from [22, 23].

whereas the phase velocity predicted by the Navier-Stokes diverges at large ϵ [8]. This is not just a feature of linear theory since we also get a much better representation of the structure of shock waves from our theory than from the Navier-Stokes equations [9].

We may add that the agreement with the experimental acoustic phase velocity is almost perfect because we adjusted the Prandtl number in the equations to be that of the gas in question. If we do not do that, the results are very close to those of Sirovich and Thurber [22] in the figure. The slight difference from experiment is of interest because it makes it possible to see what the second order theory can do. Recently, J.-L. Thiffeault has examined the second-order terms in the expansion of f and finds that, if the Prandtl number is left at the value given by the relaxation model, the second order theory brings the theoretical phase velocity close to the empirical one. On the other hand, the prediction of the dissipation in the wave is good only out to values of ϵ of order unity so this is an interesting feature of the theory to think about next. However that may be, what we can say already is that the relaxation model with our procedure provides some notable improvements in the predictions of fluid mechanics in the regime of long mean free paths.

We are grateful to Hongling Rao and Jean-Luc Thiffeault for their help.

References

1. G. E. Uhlenbeck and G. W. Ford, *Lectures in Statistical Mechanics* (American Mathematical Society, Providence, Rhode Island, 1963).
2. P. L. Bhatnager, E. P. Gross, and M. Krook, Phys. Rev. **94**, 511 (1954).
3. P. Welander, Appl. Phys. (N.Y.) **7**, 507 (1954).
4. D. Hilbert, *Grundzüge einer allgemeinen Theorie der linearen Integralgleichungen* (Chelsea Pub. Co., New York, 1912).
5. C. Cercignani, *The Boltzmann equation and its applications* (Springer-Verlag, New York, 1988).
6. S. Chapman and T. G. Cowling, *The Mathematical Theory of Non-uniform Gases*, 2nd ed. (Cambridge University Press, Cambridge, U.K., 1960).
7. X. Chen, H. Rao, and E. A. Spiegel, Phys. Rev. E **64**, 046308 (2001).
8. X. Chen, H. Rao, and E. A. Spiegel, Phys. Rev. E **64**, 046309 (2001).
9. X. Chen, H. Rao, and E. A. Spiegel, Phys. Rev. E (2002), in submission.
10. H. Grad, Phys. Fluids **6**, 147 (1963).
11. P. H. Coullet and E. A. Spiegel, in *Energy, Stability, and Convection*, Vol. 168 of *Pittman Research Notes in Mathematics*, edited by S. P. Galdi and B. Straughen (Longman Science & Technical, New York, 1988), p. 22.
12. N. G. van Kampen, J. Stat. Phys. **46**, 709 (1987).
13. C. D. Levermore, J. Stat. Phys. **83**, 1021 (1996).
14. D. Jou, J. Casas-Vàzquez, and G. Lebon, *Extended Irreversible Thermodynamics*, 2nd ed. (Springer-Verlag, Berlin, 1996).
15. I. Müller and T. Ruggieri, *Rational Extended Thermodynamics*, Vol. 37 of *Springer Tracts in Natural Philosophy*, 2nd ed. (Springer-Verlag, New York, 1998).
16. P. Rosenau, Phys. Rev. A **40**, 7193 (1989).
17. M. Slemrod, Physica D **109**, 257 (1997).
18. P. H. Coullet and E. A. Spiegel, SIAM J. Appl. Math. **43**, 776 (1983).
19. P. H. Coullet and E. A. Spiegel, in *Proceedings of the 1981 Summer Program in Geophysical Fluid Dynamics* (Woods Hole Oceanographic Institute, Woods Hole, MA, 1981), p. 276.
20. X. Chen, Ph.D. thesis, Columbia University Astronomy Department, 2000.
21. X. Chen, H. Rao, and E. A. Spiegel, Phys. Lett. A **271**, 87 (2000).
22. L. Sirovich and J. Thurber, J. Acoust. Soc. Am. **38**, 478 (1966).
23. H. Alsmeyer, J. Fluid Mech. **74**, 497 (1976).
24. E. Meyer and G. Sessler, Z. Phys. **149**, 15 (1957).

DETONATION WAVE STRUCTURE ARISING FROM THE KINETIC THEORY OF REACTING GASES

F. Conforto
Dipartimento di Matematica, Università di Messina, Italy

R. Monaco
Dipartimento di Matematica, Politecnico di Torino, Italy

F. Schürrer and I. Ziegler
Institut für Theoretische Physik, Technische Universität Graz, Austria

Starting from the Boltzmann equation for a reacting gas mixture, we formulate the problem of the propagation of a steady detonation wave. Numerical solutions of the derived macroscopic equations are provided in order to show detonation profiles and to recover the relevant physical quantities of the chemical process.

1 Introduction

Detonation is a rapid and violent process of combustion generating a strong shock wave which is sustained by chemical reactions. The most easily measurable characteristic quantity of a detonation is the velocity at which the shock propagates into the explosive. The front of a detonation wave approximately has a plane shape and moves with constant supersonic velocity. According to such a scheme it is reasonable[5], on a mathematical ground, to treat detonation as a one–dimensional steady propagation problem of a shock front moving with constant velocity $\mathcal{D}$.

More in details, the steady detonation process can be represented[2] by the following overall picture:

- In a region $D_1 = [x_0, +\infty)$, ahead of the shock, an explosive mixture is in thermodynamical equilibrium at rest, with negligible reaction rate.

- At the space-point x_0 there is a jump discontinuity. On the edge of such a discontinuity (*Von Neumann point*) no reaction takes place.

Starting from this point the chemical reaction (in general an exothermic reaction) is initiated by the pressure jump, so that the reaction itself proceeds with a finite reaction rate and is completed in the *final state point* $x_F < x_0$, where the gas mixture reaches chemical equilibrium. Hereinafter, we will indicate with $D_2 = (x_F, x_0)$ such a region, the so-called *reaction zone*, characterized by strong chemical disequilibrium. The flow in the reaction zone is steady and the shock front propagates with constant detonation velocity $\mathcal{D}$ in the positive direction of the x-axis.

According to this phenomenology it is possible to state the following mathematical problem: *Solve the shock wave problem between the point $x_0 \in D_1$ and any point $x_R \in D_2$, namely find all possible steady solutions for the reaction zone, treating $\mathcal{D}$ as a parameter, once the physical state of the explosive mixture in D_1 has been stated.*

The solution of such a problem provides then the thickness $|x_0 - x_F|$ of the reaction zone. This problem is known in literature as the *$\mathcal{D}$–discussion* [2].

In the present paper, the procedure to study the detonation problem consists in starting from the Boltzmann equation for a reacting mixture, undergoing a reversible bimolecular reaction for four gas components, providing the explicit expression of the reaction rate. We then derive a set of macroscopic equations at Euler level and provide the mathematical formulation of the above problem, proposing its solution procedure and determining the detonation wave structure via numerical computations.

2 The kinetic model

Let us consider a mixture of four gases, say 1, 2, 3 and 4, undergoing the reversible bi–molecular chemical reaction

$$1 + 2 \rightleftharpoons 3 + 4 . \tag{1}$$

Each gas particle is characterized by its mass m_i and by its bond energy E_i. Mass conservation implies $m_1 + m_2 = m_3 + m_4$, and the subscript i can be ordered in such a manner that $E = E_3 + E_4 - E_1 - E_2 \geq 0$. The microscopic state of the gas is defined by the one–particle distribution functions f_i, for each gas species $i = 1, \ldots, 4$, where $f_i = f_i(t, \mathbf{x}, \mathbf{v})$, $t \in \mathbb{R}^+$, $\mathbf{x} \in \mathbb{R}^3$, $\mathbf{v} \in \mathbb{R}^3$. The kinetic equations (Boltzmann equation) for the reactive gas mixture, governing the space-time evolution of the system, are given by

$$\frac{\partial f_i}{\partial t} + \mathbf{v} \cdot \nabla f_i = Q_i[\mathbf{f}] + \mathcal{R}_i[\mathbf{f}], \quad i = 1, \ldots, 4, \tag{2}$$

where $\mathbf{f} = \{f_1, f_2, f_3, f_4\}(t, \mathbf{x}, \mathbf{v})$. In (2) the quantities Q_i and $\mathcal{R}_i$ are the collisional terms due, respectively, to elastic and inelastic collisions. The elastic collisional terms are well-known in literature when one consider the full Boltzmann equation extended to a mixture of four inert gases [4]. The inelastic collisional terms have been derived in paper [6].

Kinetic equations (2) admit a stable equilibrium solution if and only if

$$Q_i[\mathbf{f}] = \mathcal{R}_i[\mathbf{f}] = 0 , \qquad \forall t, \mathbf{x}, \mathbf{v}, \ i = 1, \ldots, 4.$$

In particular,

$$Q_i[\mathbf{f}] = 0 \Longrightarrow f_i(\mathbf{v}) = \widetilde{f}_i(\mathbf{v}) = n_i \left(\frac{m_i}{2\pi\kappa T}\right)^{3/2} \exp\left(-\frac{m_i(\mathbf{v}-\mathbf{u})^2}{2\kappa T}\right), \quad (3)$$

$$\mathcal{R}_i[\mathbf{f}] = 0 \Longrightarrow \frac{n_1 n_2}{n_3 n_4} = \left(\frac{m_1 m_2}{m_3 m_4}\right)^{3/2} \exp\left(\frac{E}{\kappa T}\right), \quad (4)$$

κ being the Boltzmann constant, n_i the number densities of the four species, $\mathbf{u}$ the mean velocity and T the temperature. Eqs.(3) provide the well-known expressions of Maxwellian distribution functions, governing mechanical equilibrium only, while Eq.(4) expresses the mass-action-law ruling chemical equilibrium.

In the present paper, the distribution functions are always assumed as Maxwellians and chemical non-equilibrium is simply characterized by the fact that the number densities n_i do not fulfill Eq.(4). Under this assumption (Euler level), it is standard [6] to derive a closed set of macroscopic equations. These, in one dimension, assume the form

$$\frac{\partial n_i}{\partial t} + \frac{\partial}{\partial x}(n_i u) = S_i, \qquad i = 1, \ldots, 4, \quad (5)$$

$$\frac{\partial}{\partial t}\left[u \sum_i m_i n_i\right] + \frac{\partial}{\partial x}\left[u^2 \sum_i m_i n_i + \kappa T \sum_i n_i\right] = 0, \quad (6)$$

$$\frac{\partial}{\partial t}\left[u^2 \sum_i m_i n_i + 3\kappa T \sum_i n_i + 2\sum_i E_i n_i\right] + \\ + \frac{\partial}{\partial x}\left[u\left(u^2 \sum_i m_i n_i + 5\kappa T \sum_i n_i + 2\sum_i E_i n_i\right)\right] = 0. \quad (7)$$

The quantities S_i, appearing in Eqs.(5) are expressed by $S_i = \lambda_i S$ ($\lambda_i = 1$ for $i = 1, 2$ and $\lambda_i = -1$ for $i = 3, 4$), where

$$S(n_1, \ldots, n_4, T) = 4\pi\beta \frac{\mu_{34}}{\kappa T}\left[n_1 n_2 \left(\frac{m_3 m_4}{m_1 m_2}\right)^{3/2} \exp\left[-\frac{E}{\kappa T}\right] - n_3 n_4\right] \times \\ \times \left\{\frac{\chi}{\pi}\sqrt{\frac{2\pi\kappa T}{\mu_{34}}} \exp\left[-\frac{\mu_{34}\chi^2}{2\kappa T}\right] + \left(\frac{\kappa T}{\mu_{34}} - \chi^2\right)\left[1 - \mathrm{erf}\left(\sqrt{\frac{\mu_{34}}{2\kappa T}}\chi\right)\right]\right\}. \quad (8)$$

S is deduced assuming that the exothermic reaction is ruled by a reactive cross-section model proposed in paper [3], where χ denotes a threshold-velocity, β a scale factor and $\mu_{34} = m_3 m_4/(m_3 + m_4)$.

Eqs.(5–7), with S expressed by (8), will be the set of equations describing the physical state of the detonation process in the two regions D_1 and D_2.

3 Analysis and solution of the detonation problem

The analysis of the detonation problem will be carried out assuming that: *in the reaction zone the gas-mixture is in chemical non-equilibrium, but in mechanical equilibrium. The macroscopic equations (5–7) actually represent such a physical situation.*

This assumption seems to be physically consistent since the typical time of elastic relaxation is of several order of magnitude shorter than the one of chemical relaxation.

Since we are searching for steady solutions to the detonation problem, let us change, as usual [2], our frame from laboratory coordinates to steady ones, attached to the shock wave moving with constant velocity $\mathcal{D}$, through the transformation

$$z = x - \mathcal{D}t \ . \tag{9}$$

Let us then indicate, respectively, with $z_0 \equiv x_0^+ - \mathcal{D}t$ and $z_N \equiv x_0^- - \mathcal{D}t$, the downstream and upstream (Von Neumann point) coordinates of the jump. In the same fashion, we will call with z_F the coordinate of the final point of the reaction zone. Moreover, we can now characterize the physical states of the two regions D_1 and D_2.

- In region $D_1 = [z_0, +\infty)$, where the mixture is at rest in absolute equilibrium and with negligible reaction rate, the physical state is given by

$$\begin{cases} n_i(z) = n_{i0} \\ u(z) = u_0 \\ T(z) = T_0 \\ S(z) = 0 \end{cases} \qquad \forall z \in D_1 .$$

- In region $D_2 = (z_F, z_N)$, the state variables $n_1(z), \dots, n_4(z)$, $u(z)$ and $T(z)$ change their values under the influence of chemical disequilibrium ($S(z) \neq 0$). For $z = z_F$, the physical state is characterized by $S = 0$, since there the chemical reaction is completed and the reaction rate vanishes. To determine the physical state of the system in $z = z_N$, we apply the Rankine-Hugoniot conditions to Eqs.(5-7), obtaining

$$(\mathcal{D} - u_N)n_{iN} = (\mathcal{D} - u_0)n_{i0}, \qquad i = 1, \dots, 4, \tag{10}$$

$$(\mathcal{D} - u_N)\rho_N u_N - \kappa n_N T_N = (\mathcal{D} - u_0)\rho_0 u_0 - \kappa n_0 T_0, \tag{11}$$

$$(\mathcal{D} - u_N)(\rho_N u_N^2 + 3\kappa n_N T_N + 2\sum_i E_i n_{iN}) - 2\kappa n_N T_N u_N =$$

$$= (\mathcal{D} - u_0)(\rho_0 u_0^2 + 3\kappa n_0 T_0 + 2\sum_i E_i n_{i0}) - 2\kappa n_0 T_0 u_0, \tag{12}$$

where $n = \sum_i n_i$, $\rho = \sum_i m_i n_i$. Eqs.(10-12) provide the macrcscopic variables computed in z_N in terms of those computed in z_0. After some algebra on (10-12), we get

$$n_{iN} = \frac{4\rho_0(\mathcal{D} - u_0)^2}{\rho_0(\mathcal{D}^2 - u_0^2) + 5\kappa n_0 T_0}\, n_{i0}, \qquad i = 1, \ldots, 4, \tag{13}$$

$$u_N = \frac{1}{4}\left[3\mathcal{D} - u_0 - \frac{5\kappa n_0 T_0}{\rho_0(\mathcal{D} - u_0)}\right], \tag{14}$$

$$T_N = \frac{[\rho_0(\mathcal{D}^2 - u_0^2) + 5\kappa n_0 T_0][\rho_0(\mathcal{D} - u_0)(3\mathcal{D} - 5u_0) - \kappa n_0 T_0]}{16\kappa n_0 \rho_0(\mathcal{D} - u_0)^2}. \tag{15}$$

Regarding the physical consistence of the above solution, we can observe that positivity of n_{iN} is assured by the detonation compatibility condition [2] $0 < u < \mathcal{D}$. For the positivity of u_N and T_N, it is sufficient that

$$\mathcal{D} > \max\left[\frac{4\rho_0 u_0 + \sqrt{\rho_0^2 u_0^2 + 3\kappa n_0 \rho_0 T_0}}{3\rho_0}, \frac{2\rho_0 u_0 + \sqrt{\rho_0^2 u_0^2 + 15\kappa n_0 \rho_0 T_0}}{3\rho_0}\right].$$

In order to determine the physical state of the system in the reaction zone D_2, we use Eqs.(5-7) re-written, after application of transformation (9), in normal form, i.e.

$$\frac{dn_i}{dz} = \frac{S}{\mathcal{D} - u}\left[\frac{2En_i}{5\kappa nT - 3\rho(\mathcal{D} - u)^2} - \lambda_i\right], \tag{16}$$

$$\frac{du}{dz} = \frac{2ES}{5\kappa nT - 3\rho(\mathcal{D} - u)^2}, \tag{17}$$

$$\frac{dT}{dz} = \frac{2ES[\rho(\mathcal{D} - u)^2 - \kappa nT]}{\kappa n(\mathcal{D} - u)[5\kappa nT - 3\rho(\mathcal{D} - u)^2]}. \tag{18}$$

Numerical integration of system (16-18) provides the state of the gas mixture at every point $z \in (z_N, z_F)$ of the reaction zone. Such an integration must be started at z_N using the initial data provided by (13-15) and stopped when the right-hand-sides of (16-18) vanish, namely when the state variables become constant, since the reaction is completed and the gas reaches chemical equilibrium. When such a condition occurs, the corresponding coordinate z_F is determined and the distance $|z_F - z_N|$ provides the thickness of the reaction zone. To complete the $\mathcal{D}$-discussion such computations must be repeated for several $\mathcal{D}$, once the physical and chemical state of the unreacted explosive mixture has been fixed, in order to find the minimum value $\mathcal{D}$ giving rise to detonation. This minimum value, indicated by $\mathcal{D}_J$, is known in literature[5] as the Chapman-Jouguet velocity.

In a very recent paper[1], the theoretical value of $\mathcal{D}_J$ has been determined through

$$\mathcal{D}_J = \sqrt{\frac{15p_0 + 16\rho_0 q_F + 4\sqrt{2\rho_0 q_F(8\rho_0 q_F + 15p_0)}}{9\rho_0}}, \tag{19}$$

where p_0 is the pressure in region D_1 and q_F is a quantity depending on the state of the gas at point z_F.

Let us comment that the solution procedure here developped is similar to that used in the above paper[1]. Although the present approach, conversely to the other, does not allow to get solution in an explicit form, both procedures provide quantitative results very close to each other for values of $\mathcal{D}$ not so near to $\mathcal{D}_J$. On the other hand, the present method has the advantage to permit a numerical computation of D_J which, as it will be shown in next section, results to be very closed to the value prescribed by Eq.(19).

4 Numerical results

We propose here some numerical experiments to obtain detonation profiles. At this end, we have chosen the following two chemical reactions:

(a) $\quad H_2O \ + \ H \ \rightleftharpoons \ OH \ + \ H_2$

(b) $\quad O \ + \ N_2 \ \rightleftharpoons \ NO \ + \ N$

which are characterized[5] by E =63311 J/mole, χ =3927.28 m/sec for (a) and E =313780 J/mole, χ =647.34 m/sec for (b). Moreover, according to[3], we set $\beta = 2 \times 10^9$ l/(mole $\cdot$ sec). The physical state of the gas in region D_1 is fixed as follows:

- $n_{10} = 0.03$ mole/l, $n_{20} = 0.02$ mole/l, $n_{30} = 0.1$ mole/l, $n_{40} = 0.2$ mole/l, $u_0 = 0$ m/sec, $T_0 = 298.15$ K.

In Fig.1 and Fig.2, we show the profiles of total density n and temperature T versus z for reactions (a) and (b), respectively. We have set $\mathcal{D} = 3500$ m/sec for reaction (a) and $\mathcal{D} = 4100$ m/sec for reaction (b), since the last one requires larger values of $\mathcal{D}$ to give rise to a detonation which is also characterized by a higher jump in the Von Neumann point and presents a small overshoot on the temperture profile (see Fig.2b).

The minimum numerical values of $\mathcal{D}$, capable to provide detonation, are $\mathcal{D} = 2974$ m/sec, for reaction (a), and $\mathcal{D} = 4045$ m/sec, for reaction (b). The

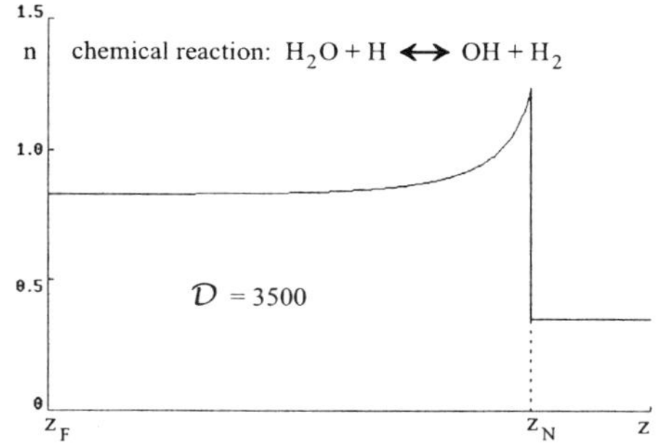

Figure 1a

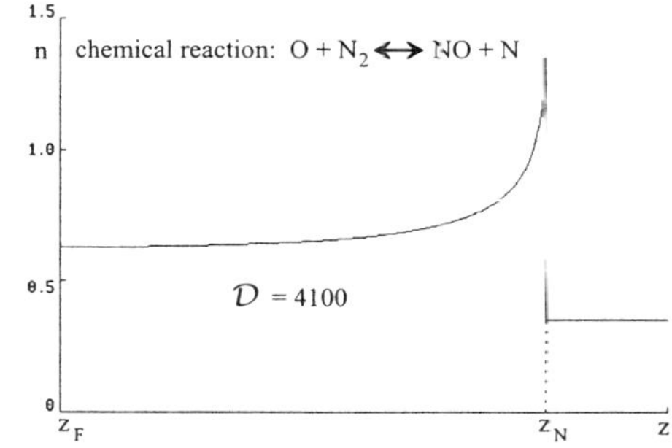

Figure 1b

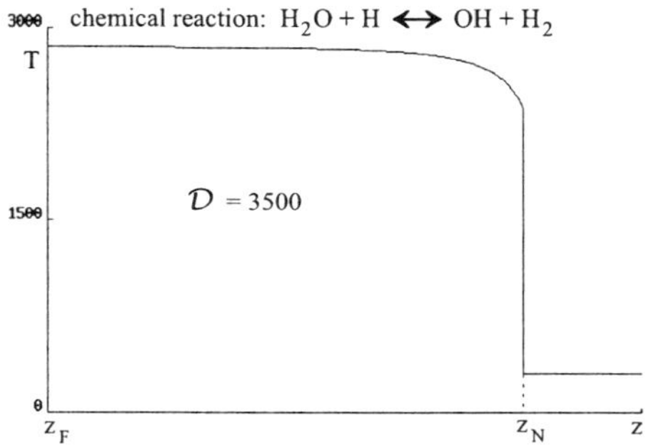

Figure 2a

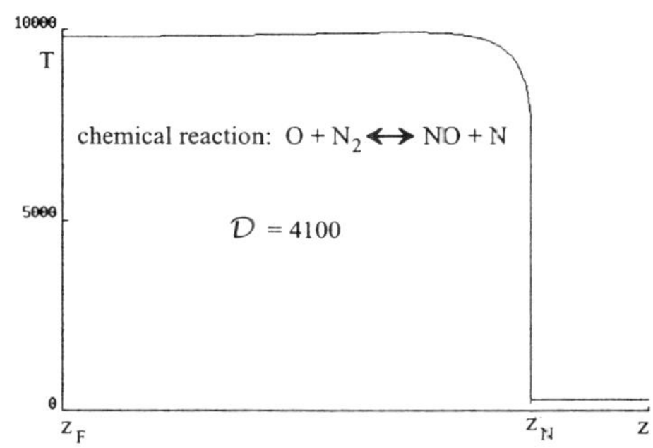

Figure 2b

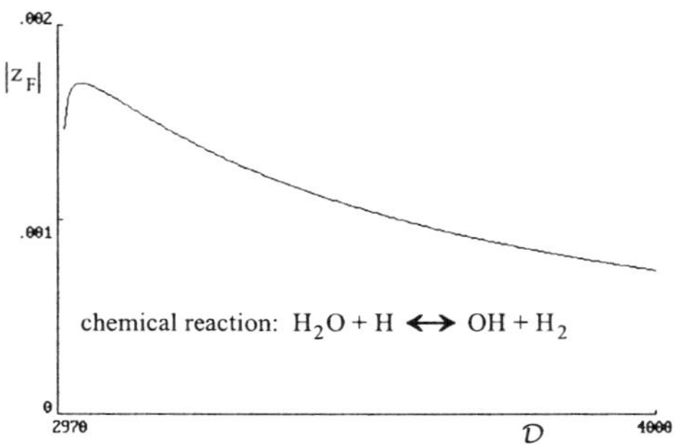

Figure 3a

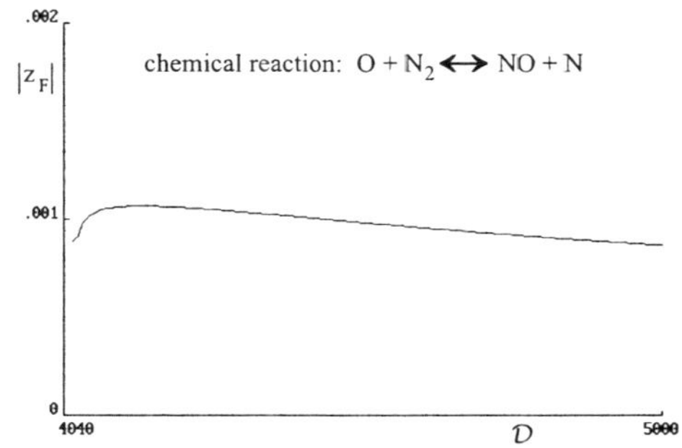

Figure 3b

prescribed values of $\mathcal{D}_J$, given by (19), are, respectively, 2967 m/sec and 4044 m/sec.

In Fig.3a and Fig.3b, putting $z_N = 0$, we plot $|z_F|$ versus $\mathcal{D}$. Although the shapes of the graphics, for the two reactions, are rather different, we observe that for values close to $\mathcal{D}_J$ the thickness grows increasing $\mathcal{D}$; afterwards, $|z_F|$ decreases.

Acknowledgements: The present paper has been supported by PRIN 2000-2001 "Problemi Matematici Non Lineari di Propagazione e Stabilità nei Modelli del Continuo" (coord. Prof. T. Ruggeri).

References

1. F. Conforto, R. Monaco, F. Schürrer and I. Ziegler, *Steady detonation waves via the Boltzmann equation for a reacting mixture*, Report N.39-2001 Dip. Mat. Politecn. Torino, to be published.
2. W. Fickett and W.C. Davis, *Detonation*, University of California Press, Berkeley, 1979.
3. P. Griehsnig, F. Schürrer, G. Küger, in *Rarefied Gas Dynamics: Theory and Simulations*, Eds. B.D. Shizgal & D.P. Weaver, vol.159, AIAA, Washington, 1992, 581-589.
4. C.R. Garibotti and G. Spiga, *J. Phys. A: Math. Gen.*, vol.27 (1994), 2709-2721.
5. K.K. Kuo, *Principles of Combustion*, John Wiley, New York, 1986.
6. A. Rossani, G. Spiga, *Physica A*, vol.272 (1999), 563-573.

LARGE SIZE PLANAR DVMs AND TWO CONTINUOUS RELATIONS

H. CORNILLE
Service de Physique théorique, CE Saclay,
F-91191 Gif-sur-Yvette, France
E-mail:cornille@spht.saclay.cea.fr

For two ClassesI-II of p-th squares Physical DVMs, planar $D = 2$ Discrete Velocity Models (generalizations of previous models[1-3]), we try to see whether they can satisfy two continuous relations deduced from Maxwellian equilibrium states.

1 Introduction and Results for Physical ClassesI-II models

For ClassI, fig.1a, when p increases, all the integer coordinates of the two axes and of the bisectors (like a 1-dimensional lattice) of the plane are filled while for ClassII, fig.1b, all integer coordinates of the plane (like a 2-dimensional lattice). We study one-dimensional travelling waves along the $\vec{x}$-axis with isotropic downstream (i) state and, at the upstream (ii) state, nonzero densities fixed along one $\vec{x}$ semi-axis. For the first Maxwellian continuous relation with mass ratio $\delta_M = M^{(i)}/M^{(ii)}$ ($D = 2$ dimension) across the shock, we want that it is satisfied for these DVMs. The second continuous Maxwellian relation in d-dimensions, (built-up with gaussians $\simeq e^{-\vec{v}^2 b}$), giving the energy to the mass ratio at the (i) state $2E^{(i)}/M^{(i)} = \int \vec{v}^2 e^{-b\vec{v}^2} d\vec{v} / \int e^{-b\vec{v}^2} d\vec{v} = d/2b$ (b continuous), was discussed, as a drawback for DVMs, by Cercignani[2]. The problem is whether the $D = 2$, DVMs solutions satisfy $d \simeq 2$ or not.

The ClassesI-II models are physical[4] (only mass, momentum and energy invariants). The tools[4] are that, for a single-gas (a species for a mixture), starting from a collision with velocities along rectangles (or squares) and 3 belonging to a previous physical model, we can add the last one. *First* we consider only the second Maxwellian relation, for the DVMs (i) state, deduce curves $d_p(b)$ for p-th squares fixed and choose a sufficient number of squares such that the d_p, b values remain stable when p increases but $p \to \infty$ when $b \to 0$. *The three DVMs conservation laws* give restrictions on these $d_p(b)$ curves: **For ClassII, we find $d \simeq 2$ while for ClassI $d \neq 2$ and $d \simeq 1$ only for b small.** We determine infinite and semi-infinite shock solutions (one or two densities at the upstream). *Second*, requiring the first $D = 2$ Maxwellian relation, we get new constraints but still $d \simeq 2$ for ClassII.

ClassI p-squares fig.1a and $\vec{v}_i(x, y)$ velocities with x, y integers or 0:

$$ClassI : \vec{v}_i : (0,0),\ (\pm q, 0),\ (0, \pm q),\ (\pm q, \pm q),\ (\pm q, \mp q),\ q = 1, 2, ..p \qquad (1)$$

Lemma1: **Physical ClassI models**. We assume that the $p = 2q-1$-square is physical. From the collision with $\vec{v}_{x,y}$: $(2q-1, 2q-1) + (1,-1) = (0, 2q-2) + (2q, 0)$ we add $(2q, 0)$ ($(0, 2q)$ with x, y symmetry). From $(2q, 0) + (0, 2q) = (0, 0) + (2q, 2q)$, we add $(2q, 2q)$. We add $(2q+1, 0)$ from $(2q, 2q) + (1, -1) = (0, 2q-1) + (2q+1, 0)$, $(0, 2q+1)$, $(2q+1, 2q+1)$ with the x, y symmetries and finally $x \to -x, y \to -y$. The $p = 2q, = 2q+1$ models are physical. Starting from $q = 1, p = 1, 9\vec{v}_i$ physical, all p-th models are physical.

To the $\vec{v}(x = q, y = s)$, we associate the densities $n_{q,s}^{(i)}$, $n_{q,s}^{(ii)}$ at the (i),(ii) states. At the isotropic (i) state, all $n_{x,y}^{(i)}$ with the same $x^2 + y^2$ are equal or multiplicity 4 for $n_{q,0}^{(i)}$, $n_{q,q}^{(i)}$. With the momentum $J^{(i)} = 0$, we write the mass $M^{(i)}$ and energy $E^{(i)}$ with independent $n_{x,y}^{(i)}$ and $r^{(i)}$ (rest-particle):

$$M^{(i)} = r^{(i)} + \sum_{q=1}^{p} 4[n_{q,0}^{(i)} + n_{q,q}^{(i)}], \quad E^{(i)} = \sum_q 2q^2[n_{q,0}^{(i)} + 4n_{q,q}^{(i)}] \qquad (2)$$

Assuming one density normalized to 1, another to $m > 1$, we define $b := log\ m$ and deduce[1−3], with collisions, all other densities functions of b. For p-fixed, the DVMs ratio $2E^{(i)}/M^{(i)} = d_p/2b$ depends on b, p, **giving a fictitious DVMs dimension d_p for the second Maxwellian relation.**

$$n_{1,0}^{(i)} = m = e^b > 1,\ n_{1,1}^{(i)} = 1,\ n_{q,0}^{(i)} = m^2/m^{q^2},\ n_{q,q}^{(i)} = m^2/m^{2q^2},\ r^{(i)} = m^2$$
$$\Delta_p =: 1/4 + \sum_{q=1}^{p} e^{-bq^2}(1 + e^{-bq^2}),\ 2E^{(i)}/M^{(i)} = -\partial_b\Delta_p/\Delta_p = d_p/2b > 0 \ (3)$$

Lemma2: For b fixed, $d_p \nearrow$ with p or $d_p - d_{p-1} > 0$. We write $-\partial_b\Delta_p = \sum_{q=1}^{p} q^2 e^{-bq^2}(1 + 2e^{-bq^2})$ and for $d_p > d_{p-1}$, it is sufficient to prove $Z := \sum_{q=1}^{q=p-1} e^{-bq^2}[e^{-bq^2}(p^2 - 2q^2) + e^{-bp^2}(2p^2 - q^2) + (1 + 2e^{-b(p^2+q^2)})(p^2 - q^2)] > 0$

Only the first term can be negative, written with the change $q \to p - q$.
$Z > e^{-bp^2}\sum_{q=1}^{q=p-1}[(p^2 - q^2)e^{b(p^2-q^2)} - Xe^{-bX}]$ with $X = p^2 + 2q^2 - 4pq$. (i) If $q/p > 1 - \sqrt{2}/2 \to X < 0 \to Z > 0$. (ii) It remains $q/p < 1 - \sqrt{2}/2 \to X > 0$. Due to $p^2 - q^2 - X = q(4p - 3q) > 0$ we get $Z > 0 \to d_p > d_{p-1}$.

ClassII p-squares fig.1b and $\vec{v}_i(x, y)$ velocities with x, y integers or 0:
$$ClassII : (0,0), (\pm q, \pm s), (\pm s, \pm q), (\mp q, \pm s), (\mp s, \pm q), s = 0, 1..q, q = 1..p \quad (1')$$

Lemma1': **Physical ClassII models**. We start with a physical p-th square, add: $(p+1, 0)$, $(0, p+1)$, $(p+1, p+1)$ and x, y symmetric (ClassI collisions). With new collisions $(p+1, 0) + (p, j) = (p, 0) + (p+1, j), j = 1, 2, ..p$, we add: $(p+1, j)$. Similarly with the symmetric x, y and $x \to -x$, $y \to -y$, the $p+1$-square is physical. Starting with $p = 1$, $\vec{9v}_i$, all p-th squares are physical.

From $q^2 + s^2$, we have a multiplicity 8 for $n_{q,s}^{(i)}$, $q \neq 0$, $s = 1, ..q-1$.

$$M^{(i)} = r^{(i)} + \sum_{q=1}^{p} 4[n_{q,0}^{(i)} + n_{q,q}^{(i)} + 2\sum_{s=1}^{q-1} n_{q,s}^{(i)}], \quad J^{(i)} = 0,$$

$$E^{(i)} = \sum_q [2q^2(n^{(i)}_{q,0} + 4n^{(i)}_{q,q}) + 4\sum_{s=1}^{q-1}(q^2+s^2)n^{(i)}_{q,s}] \qquad (2')$$

With $n^{(i)}_{1,s}$, $s = 0, 1$, $b = log\ m$ defined in ClassI, we find $n^{(i)}_{q,s} = m^2 e^{-b(q^2+s^2)}$. We write the second Maxwellian relation $2E^{(i)}/M^{(i)} = -\partial_b \Delta_p/\Delta_p$ with:

$$\Delta_p =: 1/4 + \sum_{q=1}^{p} e^{-bq^2}[1 + e^{-bq^2} + 2\sum_{s=1}^{q-1} e^{-bs^2}],\ 2E^{(i)}/M^{(i)} = d_p/2b \quad (3')$$

For *Lemma1*, we have verified numerically that, for b fixed, $d_p \nearrow$ when $p \nearrow$.

2 DVMs for the second continuous relation alone

Only isotropic (i) state, fig.2a. From p and b fixed we calculate, in (3-3'), $d_p(b)$ and consider the minimal p-square such that $d_p(b) \simeq d(b)$ remains stable when $p \nearrow$. We present the ClassesI-II curves and add the smallest common $p = 1$, $9\vec{v}_i$, $d_1(b)$. **If we except the interval $b \in [0, 1]$ for ClassII with** $d \simeq 2$, the $d(b)$ curves are $\neq 2$ and not constant. The limits for the 3 curves decrease and are the same for b large, **while for $b \to 0$ they are** 0, 1, 2 **like the associated lattices**. For all p-squares smaller than the minimal p associated to the presented ClassesI-II curves, the $d_p(b)$ curves are in the intermediate domains between the $9\vec{v}_i$ curve and either the ClassI or ClassII curve. For stable $d(b)$ and b varying, we cannot consider a fixed number of p-squares (At the limit $b \to 0$ then $p \to \infty$, while for b large $d_1(b)$ is sufficient). When $b \searrow$ then $p \nearrow$ for the stability. For ClassI we have $(8p+1)\vec{v}_i$ and $(4p(p+1)+1)\vec{v}_i$ for ClassII. In both classes for b fixed, we find a similar p minimal value for stable $d_p(b)$. We give some b, p values in particular for $b = 10^{-4}$ fixed, and $p \nearrow$, the $d_p(b) \nearrow$ values go to $1, 2$ for ClassesI-II.

b	> 3	1 0.9 .8 .7	.5. 3. 2	$10^{-1}10^{-2}$	$10^{-3}10^{-6}$	10^{-8}
p	1	2 3 3 4	5 7 8	11 38	120 3800	60000
$b = 10^{-4}$	$p = 1$	$= 40$	$= 120$	$= 240$	$= 380$	
$Cl - I\ d_p$	2.610^{-4}	< 0.15	< 0.76	< 1.005	< 1.0049	
$Cl - II\ d_p$	2.610^{-4}	< 0.209	< 1.3	< 1.983	$\simeq 2.$	

(ii) upstream and Conservation Laws for densities $n_{x,y} = n^{(ii)}_{x,y} - n^{(i)}_{x,y}$ leading to: $d \geq 2b$ **and** $d \simeq 2 = D$ **only for ClassII**. For *DVMs travelling waves* $\eta = x - \zeta t$, *(speed* ζ*)*, we recall[1-3] the mass, momentum (only macroscopic quantities) and energy (also microscopic terms) conservation laws. We retain a fixed number of $n^{(ii)}_{-q,0} \neq 0$: $0 < q < p$, $1 \leq q_{min} \leq q \leq q_{max}$.

$$J^{(ii)} = \zeta(M^{(ii)} - M^{(i)}),\ 2E^{(ii)} = \zeta J^{(ii)} + E^{(i)}, 2\zeta(E^{(i)} - E^{(ii)}) = \sum_q n^{(ii)}_{-q,0} q^3 \, (4)$$

$$M^{(ii)} = \sum n^{(ii)}_{-q,0},\ \ J^{(ii)} = -\sum q n^{(ii)}_{-q,0},\ \ 2E^{(ii)} = \sum q^2 n^{(ii)}_{-q,0}$$

With z, $\delta_E = E^{(i)}/E^{(ii)}$, we eliminate $J^{(ii)}$ from the 2 first relations:

$$z := M^{(i)}/E^{(i)} \to 2\delta_M/\delta_E = \delta_M + z\zeta^2(1-\delta_M) \text{ and } d/2b = 2/z \qquad (5)$$

In (4) we define $\overline{n}_i := n^{(ii)}_{-q_i,0}/E^{(i)}$, $1 \le q_1 < ..q_i.. < q_{max}$, $A_i := \overline{n}_i(\zeta + q_i)$,

$$1 = \sum_i A_i/\zeta z = \sum_i q_i A_i = \sum_i q_i^2 A_i/2\zeta \quad \text{(a): } \sum_{q_1}^{q_i} A_j + \sum_{q_{i+1}}^{q_{max}} A_j = \zeta z,$$
$$\text{(b): } \sum_{q_1}^{q_i} A_j q_j + \sum_{q_{i+1}}^{q_{max}} A_j q_j = 1, \quad \text{(c): } \sum_{q_1}^{q_i} A_j q_j^2 + \sum_{q_{i+1}}^{q_{max}} A_j q_j^2 = 2\zeta. \ (6)$$

Lemma3: From positivity $z > 0, \overline{n}_i > 0$, we get $\zeta > 0, A_j > 0$, $z \le 2$, $d_p \ge 2b$: First, if $q_1 + \zeta > 0$, we get $A_j > 0$, $\zeta > 0$ from (a). Second, if $q_{max} + \zeta < 0$, we get all $A_j < 0$, not possible with the rhs of (b). Third, if for one q_i: $A_i < 0, A_{i+1} > 0 \to \zeta < 0$, $A_{q_1}, ..A_i < 0$, $A_{i+1} > 0, ..A_{q_{max}} > 0$ and from (b-c): $\sum_{q_1}^{q_i} q_j(q_{i+1} - q_j)A_j + \sum_{q_{i+1}}^{q_{max}} q_j(q_{i+1} - q_j)A_j = q_{i+1} - 2\zeta$, not possible because the rhs is > 0 and the lhs < 0 and so all $A_j > 0$. From (c), we deduce $\zeta > 0$ and from (a-c): $q_{max}^{-2} < \sum A_j / \sum A_j q_j^2 = z/2 < q_1^{-2}$.

$$\to \quad 2/q_{max}^2 \le z \le 2/q_1^2 \le 2, \quad \to \mathbf{d_p = 4b/z \ge 2}b \qquad (7)$$

This connection between DVMs and the second Maxwellian relation with $z \le 2$ gives a **strong constraint** $d \ge 2b$. It remains (figs.2b-c) 2 domains: b large or $b < 0.1$ small. For ClassII we have $d \simeq 2$ (second Maxwellian relation) while for ClassI, $d \searrow$ when $b \searrow$ and $d \simeq 1$ for $b < 10^{-3}$. We give some stable b, d values for ClassesI-II ($d \simeq 2$ when $2 - d \le 10^{-6}$), but $p \nearrow$ when $b \searrow$:

b	0.95	0.9	0.7	0.5	0.3	10^{-1}	10^{-4}	$< 10^{-5}$
I, d	1.834	1.81	1.69	1.538	1.372	1.186	1.005	$\simeq 1$
II, d	1.997	1.998	1.99992	1.999994	$\simeq 2$	$\simeq 2$	$\simeq 2$	$\simeq 2$

For ClassII, $d \simeq 2 = D$**, we answer** (not ClassI with only $d \simeq 1$ for $b \ll 1$) **to the Cercignani criticism**, but b for DVMs is not continuous in general. We study solutions with 1 or 2 $n^{(ii)}_{-q,0} \ne 0$ (more general solutions exist[1–3]).

Infinite shock Solutions. In (6) with $\overline{n}_q = \overline{n}$, $A_q = A = \overline{n}(q + \zeta)$, $q \ge 1$, $1 = A/z\zeta = qA = q^2 A/2\zeta$ we deduce $\overline{n} = 2/3q^2$, $z\zeta^2 = 1/2$ and:

$$z = 2/q^2 \le 2, \quad \zeta = q/2 \ge 1/2, \quad \delta_E = \delta_M = 3, \quad z\zeta^2(1-\delta_M) + 1 = 0 \qquad (8)$$

ClassII: Except for $z = 2, q = 1$, $d = 1.9959, b = 0.9979$, for $q > 1$ we have $d \simeq 2$ or $b \simeq q^{-2}$ from (5-8). For $b > 0.1$ we have 2 other solutions $b, q = (1/4, 2), (1/9, 3)$.. 22 for $10^{-2} \le b < 10^{-1}$ with $q = 11, ..31..$ and an infinite number of $d \simeq 2 = D$ when $p \to \infty$. **ClassI**: some stable b, d_p values:

$q,\ z = 2q^{-2}$		1, 2	2, 1/2	3, 2/9	4, 1/8	6, 1/18	7, 2/49
I	b	0.907	0.1554	0.0634	0.0344	0.01478	0.01075
I	d_p	1.814	1.243	1.143	1.10	1.0641	1.0542

Here (fig.2b), $d(b) \ne 2$, but for $b < 10^{-3}$ we have $d \simeq 1$, $b \simeq 1/2q^2$ and an infinite number of solutions (discrete b) when $p \to \infty$.

Semi-Infinite Solutions. We assume in (6) two $\overline{n}_i$,, two q_i fixed: $1 \le q_1 < q_2$, two A_i and deduce one $z(\zeta)$ curve and inequalities:

$$1 = \sum_1^2 A_i/z\zeta = \sum_1^2 q_i A_i = \sum_1^2 A_i q_i^2/2\zeta \to z(\zeta) = (q_1 + q_2 - 2\zeta)/q_1 q_2 \zeta,$$
$$A_i = (q_j - 2\zeta)/q_i(q_j - q_i) = (q_j\zeta z - 1)/(q_j - q_i) = \overline{n}_{q_i}(\zeta + q_i) > 0$$
$$\to 1 \le q_1 \le 2\zeta \le q_2, \quad 2/q_2^2 \le z \le 2/q_1^2 \tag{9}$$

To the Lemma3 A_i, z inequalities, we add $q_j \gtrless 2\zeta$ if $q_j \gtrless q_i$. For $q_i = 1,\ 2$, we have $\zeta \in (1/2,\ 1)$, $z \in (1/2,\ 2)$ and a continuum of $(b,\ d)$ solutions linked by the infinite ones for $z = 2,\ 1/2$. For ClassII we get $b \in (0.9979,\ 0.25)$, $d \in (1.9959,\ 2)$ while for ClassI $b \in (0.907,\ 0.1554)$, $d \in (1.814,\ 1.243)$, similarly for any $q_1 < q_2$ integers. **For ClassII we have a continuum of solutions with $D = d \simeq 2$ and this answers to the Cercignani criticism.**

ClassII with a fixed number of p-squares (fig.2d). For stable d_p values, $p \nearrow$ when $b \searrow$, so that we expect more and more changes with figs.2b-c when $b \searrow$. We present curves for $p = 3,\ 4,\ 5,\ 6$ and for $d_p(b) \simeq 2$, an interval $b \in (b_{Min},\ b_{Max})$ (b_{Max} close to 0.95 for the Broadwell solution) and these intervals $\nearrow$ when $p \nearrow$. **For $b \in (0,\ b_{Min})$, the $d_p(b) \to 0$ when $b \to 0$ and $d \simeq 2$ solutions are only in fixed b intervals with b not too small.**

Extended ClassI Physical models (fig.2d), intermediate between ClassesI-II. For ClassI with $d < 2$, for p fixed, $d \to 0$ when $b \to 0$. To ClassI, we add the integers along $x = \pm r$, $y = \pm r$ and for $r = 1,\ 2,.,$ the $p = 2,\ 3,..$ squares are common to ClassII. Then, for b not too small, we will find b intervals with $d \simeq 2$ while $d \to 1$ when $b \to 0$. We present, in fig.2d, an example with $r = 1, 2$ and for $b \in [0.3,\ 0.95]$ we have $d \simeq 2$. (larger intervals when $r \nearrow$).

3 DVMs for the first and second Maxwellian relations

First D-dimensional continuous Maxwellian $M(2\pi T)^{-D/2}exp - [(v_1 - V)^2 + \sum_2^D v_i^2]/2T$ relations ((i) and (ii) states) between δ_M and the temperatures ratios[1-3]. Then $MV, M(V^2+T), MV(V^2+(D+2)T) =$ are constants, $(V^{(i)})^2\ (\delta_M - 1) = T^{(i)} - T^{(ii)}/\delta_M$, $\delta_M = D+1$ for $\delta_T = 0$ and with $J^{(i)} = 0$:

$$1 < \delta_M = (D+1+1/\delta_M\delta_T)/(1+(D+1)/\delta_M\delta_T) \le D+1, \tag{10}$$
$$[2\delta_M/\delta_E - (J^{(ii)}/M^{(ii)})^2 M^{(i)}/E^{(i)}][(D+1)\delta_M - 1] = 2\delta_M[D+1-\delta_M] \tag{11}$$

For planar $D = 2$ models, we get with (5-11):

$$[1 + z\zeta^2(1-\delta_M)][1 - 3\delta_M] + 2[3 - \delta_M] = 0 \tag{12}$$

For DVMs infinite shocks (see (8)), this relation is automatically satisfied.

Semi-infinite with $\mu = q_2/q_1$, (12-9-7), lead to fixed $\overline{\zeta} = \zeta/q_1$, z values:

$$P_2(\overline{\zeta}) = 4\overline{\zeta}^2 + \overline{\zeta}[\mu + 2(1+\mu^2)]/(1+\mu) - 2(1+\mu^2) + \mu = 0 \ \to \zeta, z, (d, b) \tag{13}$$

For q_i fixed, in the ζ, z intervals (9), remains only one value for $\zeta > 0$, z.

Lemma5[1-3]. i): The semi-infinite with $\mu = 2$ are infinite ones with q_2.
ii): For the really semi-infinite, positivity requires $\mu > 2$.

i): $q_2 = 2q_1$, $\mu = 2$: We get $\zeta = q_1, z = 2/q_2^2$ like the q_2 infinite.

ii): The $\overline{\zeta}, z$ solutions of (13), $\mu > 1$, satisfy (9): $1/2 \leq \overline{\zeta} \leq \mu/2$. First for $\overline{\zeta} > 0$, $P_2(\overline{\zeta})$ increases with $\overline{\zeta}$. Second, *sign* $P_2(1/2) = 1-4\mu^2 < 0$, *sign* $P_2(\mu/2) = \mu^2 - 4 \gtrless 0$ if $\mu \gtrless 2$. The $P_2 = 0$ root does not satisfy (9) if $\mu < 2$ and so $\mu \geq 2$. With $1 \leq q_i \leq 5$ we have 10 couples but only 4 semi-infinite: $\mu = 3, 4, 5, 5/2$.

ClassII with $d \simeq 2$, $b = z/2 \simeq q^{-2}$. For the semi-infinite, we get ζ from (13), z from (9) and b from (3'). **ClassI,** $d \neq 2$: The calculations are not so simple but we get the same numbers of solutions (smaller b values (4.2)). For both ClassesI-II, the number of semi-infinite $\nearrow$ in intervals smaller and smaller, when $b \searrow$ and $\to \infty$ when $b \to 0$. CLASSII, $10^{-r} \geq b > 10^{-(r+1)}$: numbers of semi-infinite and infinite solutions for $r = 0, 1, 2, 3$: (1,3),(19,6),(235,22),(2381,69). We give semi-infinite ClassesI-II examples:

q_1	q_2	μ	ζ	z	I	d	b	II	$b = z/2$	d
1	3	3	1.4645	0.24376	I	1.1511	0.0701	II	0.12188	$\simeq 2$
1	4	4	1.9487	0.14145	I	1.1088	0.0392	II	0.07072	$\simeq 2$
1	5	5	2.4402	0.09176	I	1.08484	0.02488	II	0.0458	$\simeq 2$
2	5	5/2	2.4564	0.08496	I	1.08123	0.02296	II	0.04248	$\simeq 2$

Here we find $d \leq 2$, but previously[3], with a *nonuniform distribution* $\vec{v}_i$: $\lambda_p 2^{q-1}(0, \eta_1)$, $\lambda_p 2^{q-1}(\eta_1, \eta_2)$, $\eta_i^2 = 1, \lambda_p = 1, 3/2, 5/4..$ and (x,y) symmetric, we found (binary collisions) solutions $d < 2$ for $p = 1, 2, .$ but for mixed models (multiple collisions) $p = 1 \cup 2$, $p = 1 \cup 2 \cup 3$ both $d < 2$ and > 2.
Finally we recall that in the continuous theory, b is inversely proportional to T (or E_I). For ClassII with finite DVMs, then $d \simeq 2$ only for b not too small.

4 References

(1) H. Cornille, TTSP **26**(3),359, 1997; VIII Int.Conf. Waves Cont. Media, rendiconti del Circolo matem. di Palermo,II45, 157,1996.
(2) C.Cercignani, TTSP **23**, 1, 1994; J. Stat.Phys.**87**, 1097, 1997.
(3) H. Cornille, IX Int.Conf.Waves Cont.Media 97, Bari, Rendiconti Del Circoulo Matem. di Palermo, SerieII, Suppl.**57**, 177 1998.
(4) H. Cornille, C. Cercignani, ECMI2000, ed. Muscato; J.Phys.A34,2985 2001 Wascom99, World Scientific,Singapore p.119; J.Stat.Phys.99,967 2000; C.Cercignani, A.V. Bobylev TTSP29,209, 1999;Rest-particle suggested in R. Monaco,L.Preziozi, Fluid Dynam. Appl. D.B.E, World Scien. Singapore 1991.

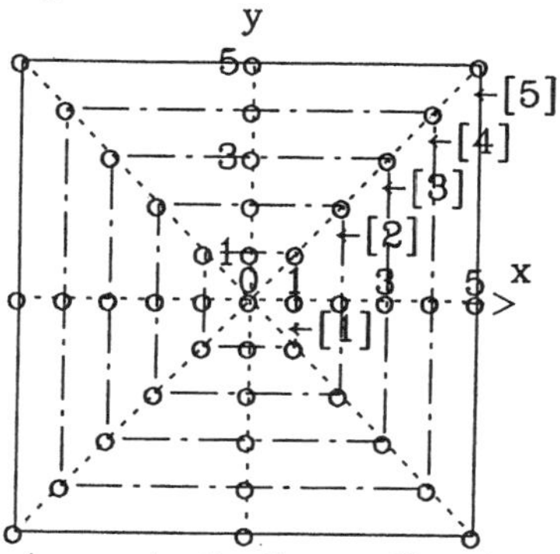

fig.1a p=1,..5−th sq ClassI

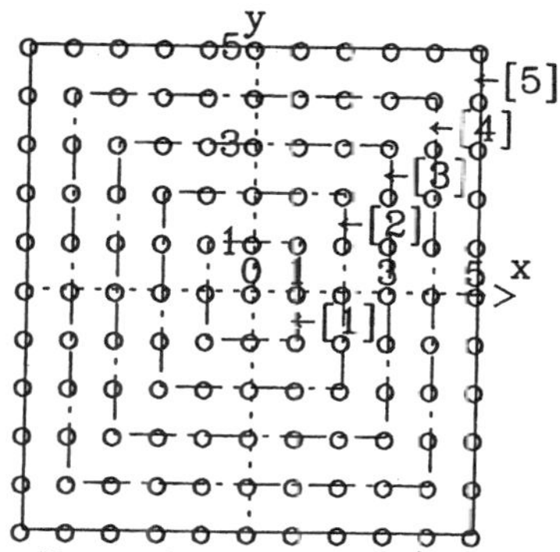

fig.1b p=1,..5−th sq ClassII

d(b): (i) Isotropic

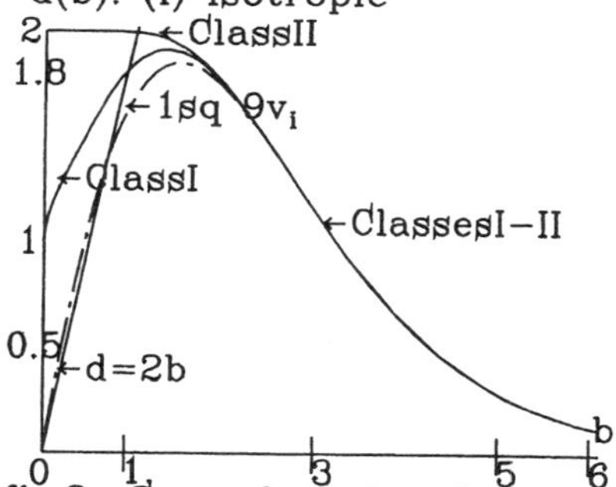

fig.2a Second cont. rel.

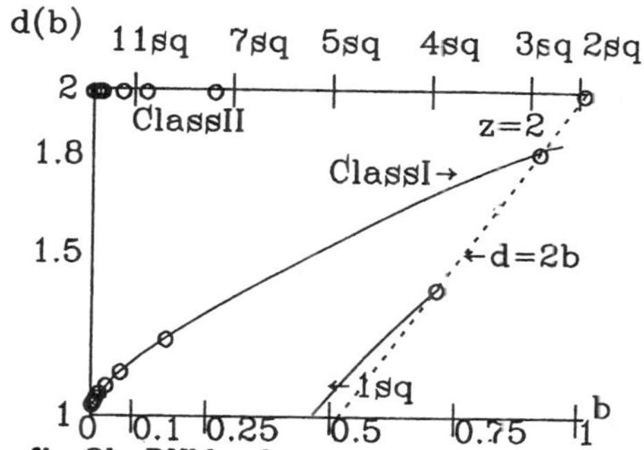

fig.2b DVMs: b not small o: infin.

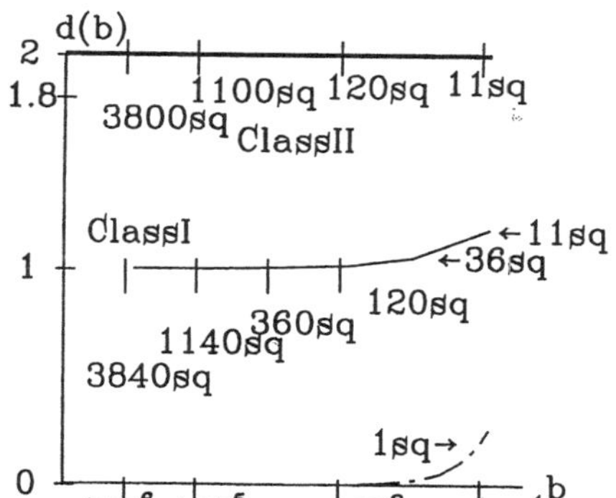

fig.2c DVMs: small $b=10^{-r}$,r=1,.6

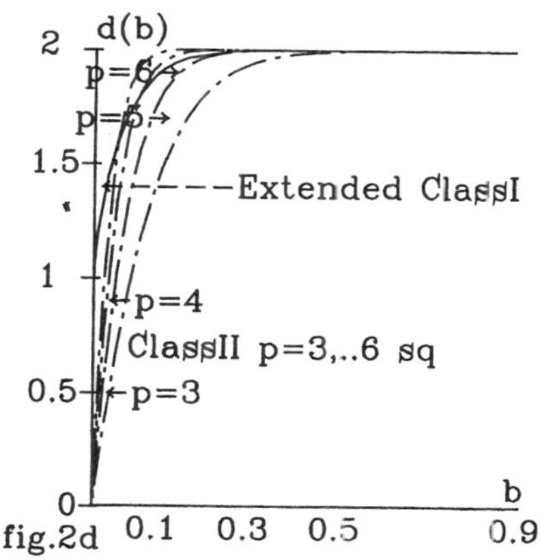

fig.2d

A Constitutive Equation for the Pressure Tensor Leading to the Hydrodynamic Form of a Nonlinear Schrödinger Equation

S. DE MARTINO, S. DE SIENA, F. ILLUMINATI

Dipartimento di Fisica dell'Università di Salerno;
INFM Unità di Salerno and INFN Sezione di Napoli, Gruppo collegato di Salerno;
Via S. Allende, I-84081 Baronissi (SA), Italy;
Email: demartino@sa.infn.it, desiena@sa.infn.it, illuminati@sa.infn.it

G. LAURO

Dipartimento di Matematica Applicata dell'Universitá di Firenze;
GNFM;
Via S.Marta 3, 50139 Firenze, Italy;
Email:lauro@dma.unifi.it

A system of two equations governing the irrotational flow of an isothermal fluid whose pressure tensor depends, in a suitable form, on the density and its first and second space derivatives, is shwon to reduce to a nonlinear Schrödinger equation.

1 Introduction

The hydrodynamic form of Schrödinger equation has its origin in a work of Madelung [1]in 1926, and has subsequently been the subject of investigation in connection with problems spacing in different branches of mathematical-physics. For instance, we quote paper [2] on the Aharonov-Bohm effect and paper [3] on a model of capillary fluid. It is well known [4] that, to make the hydrodynamical equations in a tractable form, one has to supplement them by constitutive equations. The crucial point, in the connection between hydrodynamics and Schrödinger-type equations, is the the choice of the constitutive equation for the pressure tensor. In the present paper we consider an isothermal fluid with the effect of the viscosity neglected and assume that the pressure tensor is given in terms of the density and the first and second space-derivatives of the density. We show, that the conservation equations of mass and momentum for such a fluid are equivalent to a nonlinear Schroedinger-type equation with logarithmic non linearity. It is worthwhile to stress that such an equation is integrable and furnishes soliton-like solutions, the so called gaussons, as proved in [5]. Finally, we observe that our model,at equilibrium,allows the hydrostatic pressure to be compatible with the law of ideal gases as it happens for a Boltzmann gas.

2 Model Derivation

The classical conservation equations of mass and momentum for an isothermal fluid,in absence of external forces, are:

$$\partial_t \rho = -\nabla \cdot \rho \boldsymbol{u} \tag{1}$$

$$\partial_t \rho \boldsymbol{u} = -\nabla \cdot (\rho \boldsymbol{u}\boldsymbol{u} + \boldsymbol{P}) \tag{2}$$

where $\rho(\boldsymbol{x}, t)$ is the density,$\boldsymbol{u}$ the velocity, $\boldsymbol{P}$ the pressure tensor, $\boldsymbol{x} \in R^3$ the position and $t \in R^+$ the time variables.

In order to close system (1),(2) we assigne the following constitutive relation to the pressure tensor $\boldsymbol{P}$:

$$P_{ij} = (\alpha\rho - \beta^2 \rho \partial_{x_i} \partial_{x_j} \log \rho)\delta_{ij} - (1 - \delta_{ij})\beta^2 \rho \partial_{x_i} \partial_{x_j} \log \rho \tag{3}$$

where α and β are positive dimensional parameters and i,j=1,2,3. Namely, we are assuming $P_{i,j}$ functions of the density ρ and of the immediate environment which is modelled by first and second space derivatives of ρ.

We remark that in kinetic theory of gases eqns.(1),(2) are a simple mathematical consequence of the Boltzmann equation [6]. In particular, as it is well known, the equilibrium state (local or absolute) of a Boltzmann gas is characterized by the constitutive equations of a perfect (Euler) fluid,i.e.,for the hydrostatic pressure we have:

$$\P_{eq} = \frac{1}{3} P_{i,i} = \rho_{eq} k_B T_{eq} \tag{4}$$

where k_B is the Boltzmann constant and T the temperature.

From (3) we obtain the following expression for the scalar pressure:

$$P = \frac{1}{3} P_{i,i} = \alpha\rho - \frac{\beta^2}{3} \nabla \log \rho = \alpha\rho - \frac{\beta^2}{3} \rho \left(\frac{\Delta\rho}{\rho} - \frac{(\nabla\rho)^2}{\rho^2}\right) \tag{5}$$

Hence, the second term of the rhs. of (5) can be interpreted as the deviation from the ideal gas pressure due to the fluctuations in the density of the environment.

3 Connecting Hydrodynamic and Schrödinger Equations

By taking in account (1)and (3)we can write (2)as follows:

$$\partial_t \boldsymbol{u} + [\boldsymbol{u} \cdot \nabla]\boldsymbol{u} = -\alpha \nabla \log \rho + \frac{\beta^2}{2} \nabla \left[\frac{\Delta\sqrt{\rho}}{\sqrt{\rho}}\right] \tag{6}$$

Now if we assume the fluid be irrotational, $\boldsymbol{u} = \nabla S$, S real function, then (6) becomes:

$$\partial_t S + \frac{1}{2}(\nabla S)^2 + \alpha \log \rho - \frac{\beta^2}{2}\frac{\Delta\sqrt{\rho}}{\rho} = 0 \tag{7}$$

It is easy to check that (1),(7)are the hydrodynamic form of a nonlinear Schroedinger-like equation with logarithmic non-linearity, where the fundamental unit of action $\hbar$ has been replaced by a suitable dimensional constant β.

Namely:

$$\imath\beta\partial_t\Psi + \frac{\beta^2}{2}\Delta\Psi + \alpha\log(|\ \Psi\ |)^2\Psi = 0 \tag{8}$$

with Ψ given by

$$\Psi(\boldsymbol{x},t) = \sqrt{\rho}\ exp[\frac{\imath}{\beta}S(\boldsymbol{x},t)] \tag{9}$$

We finally note that equation (8) is integrable and furnishes solitary waves as proven in [5].

4 Conclusions

The central assumption on the pressure tensor, (3), states that the scalar pressure depends not only on the state of the fluid, which can be described solely by density (being the fluid isothermal), but also on the state of its local enviroment which is here modelled by first and second derivatives of the density. Such a choice is in line with capillarity models as proposed in [3] and fits the the equilibrium state of a Boltzmann gas.

We are able to integrate the hydrodynamical equations associated to this model of fluid by showing that they are the hydrodynamic form of a Schrödinger-like equation with logarithmic nonlinearity whose solutions are expressed by solitary waves, the so called gaussons, as proven in [5].

1. E. Madelung, Z. Physik **40**, 332 (1926)
2. T.Takabayasi, Prog. Theor. Phys. **69**,1323 (1983)
3. L.K.Antanovskii, C.Rogers, W.K.Schief, J. Phys. A Math. Gen. **30**, L555 (1997)
4. P. Resibois, M. De Leener *Classical Kinetic Theory of Fluids*, (John Wiley and Sons,N.Y. 1974)
5. I. Bialynicki-Birula, J. Mycielski, Ann.of Phys. **100** 62, (1976)
6. R.Balescu, *Equilibrium and Nonequilibrium Statistical Mechanics*,(John Wiley and Sons, N.Y.1975)

SOME APPLICATIONS OF STABILITY SWITCH CRITERIA IN DELAY DIFFERENTIAL SYSTEMS

P.FERGOLA, M.CERASUOLO
Dipartimento di Matematica e Applicazioni "R. Caccioppoli",
Università degli Studi di Napoli Federico II,
Via Cintia, 80126 Napoli (Italy)
E-mail: paolo.fergola@dma.unina.it; marianna.cerasuolo@dma.unina.it

In this paper we present two new models, consisting of Delay Differential Equations with parameters independent or dependent on delay. The first system, in the Population Dynamics context, is actually a new type of polluted chemostat model, describing the allelopathic competition of two algal phytoplanktonic species. The second one, belongs to the context of Innovation Diffusion Theory and models the problem of the reaction of two different social systems to the offer of a new product. Stability properties of steady state solutions have been investigated and, by means of geometric criteria recently proposed in the literature, the problem of their stability switches has been solved. The results, based both on analytical and numerical computation, confirm that large size delays are ultimately stabilizing for systems with delay dependent parameters.

1 Introduction

Delay Differential Equation Systems play an important role in several evolutive phenomena studied in Mathematical Physics. In particular, in Population Dynamics or Innovation Diffusion theory, systems with delay terms are introduced in order to model, for instance, processes such as growth of populations, nutrient recycling, testing period of an innovation in its diffusion process and so on. These terms are represented by means of discrete or distributed delays and their coefficients are usually assumed to be constant, or time varying, or depending on delays. The interest for such these systems comes from applied and theoretical reasons because their analysis, very often, allows to get not only useful information on the dynamics of the considered processes but also gives new contributions in the context of mathematical modelling, proposing new mathematical approaches and procedures.

An interesting question concerning the stability properties of the solutions of Finite Delay Systems is related to the influence of the size of delay on the stability switches. Recently, geometric stability criteria have been proposed [1] to study this problem for systems involving delay dependent or independent parameters. In this paper we apply this geometric approach to investigate stability switches of the solutions of two new delay differential systems. The first model, in Population Dynamics context, can be considered as a new polluted chemostat model and describes the allelopathic competition of two algal species in a chemostat-like aquatic environment. The second model, in the context of Innovation Diffusion Theory, deals with the problem of the reaction of two different social systems to the proposal of a new product.

2 Algal allelopathic competition in a chemostat-like environment

In this Section we consider the problem of the density fluctuations of some phytoplanktonic communities due to the increase of competing populations. By supposing that the competition takes place in a chemostat-like environment [2-7] and that the unique source of pollution is the production of toxins from the competing populations, we mathematically represent the process by means of a new type of polluted chemostat model [8-15]. Such a model can, for example, describe the interaction between the unicellular green alga Chlorella vulgaris producing a toxin that inhibits the growth of the planktonic algae Asterionella formosa [16-17]. Mathematical contributions on this subject until now present in literature [18-21] never concern the case, here considered, that the allelopathic competition happens in a chemostat-type aquatic system.

By assuming that both species instantaneously produce allelopathic toxins the model is as follows

$$\begin{cases} \dot{S} = (S^0 - S)D - \sum_{i=1}^{2} m_i R(S) N_i \\ \dot{N}_1 = N_1(m_{01}R(S) - \gamma_1 N_1 N_2) - DN_1 \\ \dot{N}_2 = N_2(m_{02}R(S) - \gamma_2 N_1 N_2) - DN_2 \end{cases} \tag{1}$$

where

- $S(t)$ is the concentration of the nutrient;
- $N_i(t)$ is the measure of the species i $(i = 1, 2)$ in competition;
- $D\,(> 0)$ is the washout or dilution rate;
- $m_i\,(> 0)\,(i = 1, 2)$ is the maximum specific yield rate of nutrient;
- $S^0\,(> 0)$ is the constant input concentration of the nutrient;
- $m_{0i}\,(> 0)\,(i = 1, 2)$ is the maximum specific growth rate of the species i;
- $\gamma_i\,(> 0)$ $(i = 1, 2)$ is the rate of toxic inhibition of a species by the other one and vice versa.

Furthermore, we assume that the uptake function $R(S)$, describing the absorption of the nutrient S by the microorganisms of each species, is given by the Michaelis-Menten function $R(S) = \frac{S}{a+S}$ where a is a positive constant. It is immediately checked that $E_1 = E_{+00} = \left(S^0, 0, 0\right)$ is an equilibrium of system (1) whatever be the value of S^0. It is easy to prove that system (1) admits the further steady state solutions

(i) $E_2 = E_{+0+} = \left(\frac{Da}{m_{02}-D}, 0, \frac{m_{02}}{m_2}\left[S^0 - \frac{Da}{m_{02}-D}\right]\right)$ if $m_{02} > D$, $S^0 > \frac{Da}{m_{02}-D}$

(ii) $E_3 = E_{++0} = \left(\frac{Da}{m_{01}-D}, \frac{m_{01}}{m_1}\left[S^0 - \frac{Da}{m_{01}-D}\right], 0\right)$ if $m_{01} > D$, $S^0 > \frac{Da}{m_{01}-D}$

(iii) $E_4 = E^{(1)}_{+++} = (\bar{S},\ \bar{N}_1^-,\ \bar{N}_2^-)$ and $E_5 = E^{(2)}_{+++} = (\bar{S},\ \bar{N}_1^+,\ \bar{N}_2^+)$ with

$$\bar{S} = \frac{(\gamma_2-\gamma_1)aD}{[\gamma_2 m_{01}-\gamma_1 m_{02}]-(\gamma_2-\gamma_1)D} \qquad \bar{N}_1^{\mp} = \frac{m_{01}R(\bar{S})-D}{\gamma_1 \bar{N}_2^{\mp}}$$

$$\bar{N}_2^{\mp} = \frac{\gamma_1(S^0-\bar{S})D \mp \sqrt{\gamma_1^2(S^0-\bar{S})^2D^2-4\gamma_1 m_1 m_2 R^2(\bar{S})[m_{01}R(\bar{S})-D]}}{2\gamma_1 m_2 R(\bar{S})}$$

if
$\gamma_1(S^0 - \bar{S})^2D^2 > 4m_1m_2R^2(\bar{S})[m_{01}R(\bar{S}) - D]$
$\gamma_2/\gamma_1 > \frac{m_{02}-D}{m_{01}-D}, \quad m_{02} > m_{01}.$

The stability analysis of the generic equilibrium $\bar{E} = (\bar{S}, \bar{N}_1, \bar{N}_2)$ can be performed by means of the characteristic equation associated to the linearized system of (1) in $\bar{E}$

$$\begin{array}{l} \lambda^3 - (h + k_1 + k_2)\lambda^2 + (hk_1 + hk_2 + d_1 + d_2)\lambda + \\ -(d_1k_2 + d_2k_1 + \gamma_2\frac{m_2}{m_1}d_1\bar{N}_2^2 + \gamma_1\frac{m_1}{m_2}d_2\bar{N}_1^2) = 0 \end{array} \tag{2}$$

where

$h = -D - \frac{dR(\bar{S})}{d\bar{S}}(m_1\bar{N}_1 + m_2\bar{N}_2)$, $k_1 = m_{01}R(\bar{S}) - 2\gamma_1\bar{N}_1\bar{N}_2 - D$,
$k_2 = m_{02}R(\bar{S}) - 2\gamma_2\bar{N}_1\bar{N}_2 - D, \quad d_1 = R(\bar{S})\frac{dR(\bar{S})}{d\bar{S}}m_1m_{01}\bar{N}_1$,
$d_2 = R(\bar{S})\frac{dR(\bar{S})}{d\bar{S}}m_2m_{02}\bar{N}_2$.

For sake of brevity we omit the proof of the following theorem based on the analysis of eigenvalues and Routh-Hurwitz criterium.

Theorem 2.1 *The following statements hold true*

(i) If $m_{01} < m_{02}$ *then the equilibrium* E_{+0+} *is asymptotically stable;*

(ii) If $m_{02} < m_{01}$ *then the equilibrium* E_{++0} *is asymptotically stable;*

(iii) If $(\bar{N}_2 - \frac{m_1}{m_2}\bar{N}_1)(d_2\gamma_1\bar{N}_1 - d_1\gamma_2\bar{N}_2\frac{m_2}{m_1}) > 0$ *then the equilibrium* $E^{(i)}_{+++}$ $(i = 1, 2)$ *is asymptotically stable.*

Now, if we assume that the production of the toxic substances for the competing species is not instantaneous, but mediated by some finite time period required for maturity of the species, we are pushed to consider a model with delay. Precisely, here, we suppose that all the parameters of the system are constants and that only one species (let us assume, for instance, the species 1) has a delayed production of toxicant. Denoting by τ the time lag, the model can be written in the form

$$\begin{cases} \dot{S} = (S^0 - S)D - \sum_{i=1}^{2} m_i R_i(S) N_i \\ \dot{N}_1 = N_1(m_{01}R(S) - \gamma_1 N_1 N_2) - DN_1 \\ \dot{N}_2 = N_2(m_{02}R(S) - \gamma_2 N_1(t-\tau)N_2) - DN_2 \end{cases} \tag{3}$$

We observe that system (3) admits the same steady states E_j $(j = 1, \dots, 5)$ already computed for system (1). For the generic equilibrium $\bar{E}$ of system (3) the characteristic equation is

$$\begin{aligned} &\lambda^3 - (h + k_1 + k_2)\lambda^2 + (k_1k_2 + hk_2 + hk_1 + d_1 + d_2)\lambda - (hk_1k_2 + \\ &+ d_2k_1 + d_1k_2 + \gamma_1 d_2 \tfrac{m_1}{m_2}\bar{N}_1^2) - (k_1k_2\lambda + \gamma_2\tfrac{m_2}{m_1}d_1\bar{N}_2^2 - hk_1k_2)e^{-\lambda\tau} = 0. \end{aligned} \tag{4}$$

This characteristic equation computed in correspondence of the equilibria E_1, E_2, E_3 reduces to (2), therefore the stability properties of these equilibria do not change. In other words the delay τ has no influence on the stability properties. In order to study the stability of E_4 and E_5 we will use some new criteria [1], concerning the problem of stability switches for delay differential systems with delay dependent or independent parameters. These geometric criteria require both analytical computation and graphical information to study local stability. More precisely, they allow to study stability switches of a given steady state, simply determining the graphs of some functions of delay. We start by observing that equation (4) is of type

$$D(\lambda) = P(\lambda) + Q(\lambda)e^{-\lambda\tau} = 0 \tag{5}$$

where

$$\begin{aligned} P(\lambda) &= \lambda^3 - (h + k_1 + k_2)\lambda^2 + (k_1k_2 + hk_2 + hk_1 + d_1 + d_2)\lambda + \\ &\quad -(hk_1k_2 + d_2k_1 + d_1k_2 + \gamma_1 d_2\tfrac{m_1}{m_2}\bar{N}_1^2) \\ Q(\lambda) &= -(k_1k_2\lambda + \gamma_2\tfrac{m_2}{m_1}d_1\bar{N}_2^2 - hk_1k_2) \end{aligned}$$

We note that in (4) the coefficients are independent of delay τ. Setting $\lambda = i\omega$, according to the Beretta-Kuang procedure [1], we obtain

$$\begin{aligned} F(\omega) :&= |P(i\omega)|^2 - |Q(i\omega)|^2 = \\ &= \omega^6 + [(-h-k_1-k_2)^2 - 2(k_1k_2 - hk_1 - hk_2 - d_1 - d_2)]\omega^4 + \\ &-[2(-h-k_1-k_2)(-hk_1k_2 - k_1d_2 - d_1k_2 - \gamma_1 d_2 \bar{N}_1^2 \tfrac{m_1}{m_2}) + \\ &-(k_1k_2 + hk_1 + hk_2 + d_1 + d_2)^2 + k_1^2k_2^2]\omega^2 + (-hk_1k_2 - k_1d_2 + \\ &-k_2d_1 - \gamma_1 d_2 N_1^2 \tfrac{m_1}{m_2})^2 - (-\gamma_2 d_1 N_2^2 \tfrac{m_1}{m_2} + hk_1k_2)^2 \end{aligned} \tag{6}$$

If ω is a positive root of $F(\omega) = 0$, we define $\vartheta \in [0, 2\pi]$ such that

$$\begin{aligned} \sin\vartheta &= \frac{-P_R(i\omega)Q_I(i\omega) + P_I(i\omega)Q_R(i\omega)}{|Q(i\omega)|} \\ \cos\vartheta &= -\frac{P_R(i\omega)Q_R(i\omega) + P_I(i\omega)Q_I(i\omega)}{|Q(i\omega)|} \end{aligned} \tag{7}$$

then we look for the zeros τ^* of the functions $S_n(\tau) := \tau - \frac{\vartheta + 2n\pi}{\omega}$ where stability switches can happen. We observe that S_n are linear functions of ϑ and that the $sign \left\{ \frac{dS_n(\tau)}{d\tau}\Big|_{\tau=\tau^*} \right\} = 1$ for every $\tau^* \in [0, +\infty[$. The computation of the function $F(\omega)$ with respect to the equilibrium E_5 has been performed by means of "Maple" giving the following numerical values to the parameters of system (3)

$$D = 1,\ m_1 = 3,\ m_2 = 1.6,\ m_{01} = 3,\ m_{02} = 3.2,\ \gamma_1 = 0.08, \gamma_2 = 0.3,\ a = 0.8.$$

We have obtained for the function $F(\omega)$ the following expression

$$F(\omega) := \omega^6 - 2.791729518\omega^4 + 7.135539643\omega^2 + 0.02362746778$$

and all roots of equation $F(\omega) = 0$ are complex conjugates. So, we can conclude that there are no stability switches and E_5 does not change its asymptotic stability property. With respect to the equilibrium E_4 we find that

$$F(\omega) = \omega^6 - 3.083805834\omega^4 + 8.259002769\omega^2 - .6087046904.$$

In this case equation $F(\omega) = 0$ has two real roots and only one, $\omega = 0.2753075784$, is positive. In correspondence of this ω (7) gives

$$\sin(\vartheta) = -.9983095175 \quad \cos(\vartheta) = 0.5974382262$$

Since $\cos(\vartheta) > 0$ and $\sin(\vartheta) < 0$, ϑ is given by

$$\vartheta = 2\pi + \arctan\left(\frac{\sin(\vartheta)}{\cos(\vartheta)}\right) = 2\pi - 1.511022628$$

and the graphs of functions S_n, $n = 0, 1, 2$, are the straight lines in Fig.1.

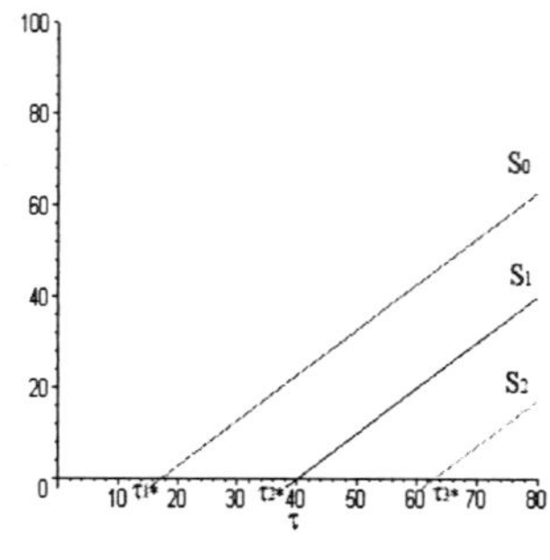

Figure 1: Graphs of functions S_0, S_1, S_2

According to the mentioned procedure, since

$$sign\left\{ \frac{d\mathbb{R}e\lambda}{d\tau}\Big|_{\lambda=i\omega} \right\} = sign\{F'(\omega)\}\, sign\{S_n'(\tau)\} = sign\{F'(\omega)\} = 1 \qquad (8)$$

the equilibrium E_4 keeps its instability property whatever be the intersection τ^* with the τ axis. In conclusion, the delay τ has no influence even on the stability properties of steady states E_4 and E_5.

Finally, if we suppose that the process of production of the toxic substances is delayed for both species, but with the same delay, the mathematical model can be written in the form

$$\begin{cases} \dot{S} = (S^0 - S)D - \sum_{i=1}^{2} m_i R_i(S) N_i \\ \dot{N}_1 = N_1(m_{01}R(S) - \gamma_1 N_1 N_2(t-\tau)) - DN_1 \\ \dot{N}_2 = N_2(m_{02}R(S) - \gamma_2 N_1(t-\tau) N_2) - DN_2 \end{cases} \qquad (9)$$

If we denote with $\bar{E} = (\bar{S}, \bar{N}_1, \bar{N}_2)$ a non-negative equilibrium of system (9) and compute the corresponding characteristic equation we obtain

$$\lambda^3 - (h + k_1 + k_2)\lambda^2 + (hk_1 + hk_2 + k_1k_2 + d_1 + d_2)\lambda - (hk_1k_2 + d_1k_2 + \\ +d_2k_1) - (\gamma_2 \tfrac{m_2}{m_1} d_1 \bar{N}_2^2 + \gamma_1 \tfrac{m_1}{m_2} d_2 \bar{N}_1^2)e^{-\lambda\tau} - k_1k_2(\lambda - h)e^{-2\lambda\tau} = 0 \qquad (10)$$

where h, k_1, k_2, d_1, d_2 have the same expression as in (2). This equation can be written in the following form

$$D(\lambda) = P(\lambda) + Q_1(\lambda)e^{-\lambda\tau} + Q_2(\lambda)e^{-2\lambda\tau} = 0 \qquad (11)$$

where

$$
\begin{aligned}
P(\lambda) = \; & \lambda^3 - (h + k_1 + k_2)\lambda^2 + (hk_1 + hk_2 + k_1k_2 + d_1 + d_2)\lambda + \\
& -(hk_1k_2 + d_1k_2 + d_2k_1) \\
Q_1(\lambda) = & -(\gamma_2 \tfrac{m_2}{m_1} d_1 \bar{N}_2^2 + \gamma_1 \tfrac{m_1}{m_2} d_2 \bar{N}_1^2) \\
Q_2(\lambda) = & -k_1k_2(\lambda - h)
\end{aligned}
$$

The procedure used for the (5) does not apply to (11). The problem of the extension of such a procedure to a characteristic equation of type (11) has not yet been solved.

3 An innovation diffusion model for a two patches environment

A new mathematical model has been proposed[22] to describe the dynamics of a diffusion process of one product in two different patches. This problem can be formulated in terms of innovation diffusion theory according to the modelling ideas developed in many recent papers[23-27]. For every patch we suppose that the channels of communication are represented by external sources (mass media like radio, TV, newspaper, poster, internet, etc.) and interpersonal contacts (word-of-mouth). Furthermore, we assume that the overall population at time t can be divided into two different classes, non-users of innovation and users of innovation. An important feature related to this problem is that of duration of product. Indeed, for products with long life-span, the problem of passage from users class to the unusers class may become relevant and the returning rates can be mathematically modelled by means of delayed or instantaneous terms[26-27]. For products with a short life-span, as we assume in this case, the previous feature can be neglected and the duration of products becomes weighty. Therefore our model can be written in the form of the following delay dependent parameters system

$$
\left\{
\begin{aligned}
\frac{dN_1}{dt} = & -\gamma_1 N_1(t) - \lambda_1 A_1(t) N_1(t) - \delta_1 N_1(t) + e^{-\delta_1\tau_1}(\gamma_1 N_1(t-\tau_1) + \\
& + \lambda_1 A_1(t-\tau_1) N_1(t-\tau_1)) + \beta_1 + \theta_2(N_2(t) + (1-k_2)A_2(t)) + \\
& -\theta_1 N_1(t) \\
\frac{dA_1}{dt} = & (\gamma_1 + \lambda_1 A_1) N_1(t) - \delta_1 A_1(t) - e^{-\delta_1\tau_1}(\gamma_1 N_1(t-\tau_1) + \lambda_1 \cdot \\
& \cdot A_1(t-\tau_1) N_1(t-\tau_1)) - \theta_1 A_1(t) + k_2\theta_2 A_2(t) \\
\frac{dN_2}{dt} = & -\gamma_2 N_2(t) - \lambda_2 A_2(t) N_2(t) - \delta_2 N_2(t) + e^{-\delta_2\tau_2}(\gamma_2 N_2(t-\tau_2) + \\
& + \lambda_2 A_2(t-\tau_2) N_2(t-\tau_2)) + \beta_2 + \theta_1(N_1(t) + (1-k_1)A_1(t)) + \\
& -\theta_2 N_2(t) \\
\frac{dA_2}{dt} = & (\gamma_2 + \lambda_2 A_2) N_2(t) - \delta_2 A_2(t) - e^{-\delta_2\tau_2}(\gamma_2 N_2(t-\tau_2) + \lambda_2 \cdot \\
& \cdot A_2(t-\tau_2) N_2(t-\tau_2)) - \theta_2 A_2(t) + k_1\theta_1 A_1(t)
\end{aligned}
\right. \tag{12}
$$

where for $i = 1, 2$

- $\gamma_i\,(> 0)$ represents the intensity of advertisement in the two patches;

- $\lambda_i\,(>0)$ represents the contact rate of users of the product with not-users in the two patches;
- $\delta_i\,(>0)$ represents the migration or death rate of the populations in the two patches;
- $\beta_i\,(>0)$ represents the immigration or birth rate of the populations in the patches;
- $\theta_i\,(>0)$ represents the probability that an individual of the population migrates from the patch i to the other one;
- $k_i\,(\geq 0)$ represents the fraction of the immigrants from the patch i who remain in the users class.

Here τ_1 represents the life-span of the product in the first patch and τ_2 represents the life-span of the product in the second patch. We will suppose that all the parameters are constants. Equation (12), setting $T_i = N_i + A_i$ ($i = 1,2$), implies that

$$\begin{aligned}\dot{T}_1 &= \beta_1 - (\delta_1 + \theta_1)T_1 + \theta_2 T_2 \\ \dot{T}_2 &= \beta_2 - (\delta_2 + \theta_2)T_2 + \theta_1 T_1\end{aligned} \tag{13}$$

Let us define

$$C_1 = \frac{\delta_2\beta_1 + \theta_2\beta_1 + \theta_2\beta_2}{\delta_1\delta_2 + \delta_1\theta_2 + \theta_1\delta_2} \qquad C_2 = \frac{\delta_1\beta_2 + \theta_1\beta_2 + \theta_1\beta_1}{\delta_1\delta_2 + \delta_1\theta_2 + \theta_1\delta_2}$$

It is easy to see that any positive solution $(T_1(t), T_2(t))$ of (13) satisfies

$$T_i(t) \to C_i \quad \text{as} \quad t \to \infty,\ i = 1,2)$$

Since we are interested in the asymptotic behavior of (12), we may regard $N_i + A_i$ as C_i for $i = 1,2$. Thus, we have

$$\left\{\begin{aligned}\tfrac{dA_1}{dt} &= (\gamma_1 + \lambda_1 A_1)(C_1 - A_1(t)) - (\delta_1 + \theta_1)A_1(t) - e^{-\delta_1\tau_1}(\gamma_1 + \\ &\quad +\lambda_1 A_1(t-\tau_1)(C_1 - A_1(t-\tau_1))) + k_2\theta_2 A_2(t) \\ \tfrac{dA_2}{dt} &= (\gamma_2 + \lambda_2 A_2)(C_2 - A_2(t)) - (\delta_2 + \theta_2)A_2(t) - e^{-\delta_2\tau_2}(\gamma_2 + \\ &\quad +\lambda_2 A_2(t-\tau_2)(C_2 - A_2(t-\tau_2))) + k_1\theta_1 A_1(t)\end{aligned}\right. \tag{14}$$

Here we will limit our attention to the case $k_1 = k_2 = 0$, which means that all the new immigrants are the non-users of the product. In this case, (14) reduces to the following independent equations

$$\begin{aligned}\tfrac{dA_i}{dt} &= (\gamma_i + \lambda_i A_i)(C_i - A_i(t)) - p_i A_i(t) - \eta_i[\gamma_i + \\ &\quad +\lambda_i A_i(t-\tau_i)(C_i - A_i(t-\tau_i))] \qquad i = 1,2\end{aligned} \tag{15}$$

where $p_i = \delta_i + \theta_i$ and $\eta_i = e^{-\delta_i \tau_i}$, $i = 1, 2$. It is easy to see that (15) for $\tau_i > 0\,(i = 1, 2)$ has one and only one positive equilibrium (A_1^*, A_2^*), where

$$
\begin{aligned}
A_i^* = & \frac{1}{2(1-\eta_i)\lambda_i}\{-(1-\eta_i)\gamma_i + (1-\eta_i)\lambda_i C_i - p_i + [(1-\eta_i)^2 \cdot \\
& \cdot \gamma_i^2 + 2(1-\eta_i)^2 \lambda_i \gamma_i C_i + 2(1-\eta_i)\gamma_i p_i + (1-\eta_i)^2 \lambda_i^2 C_i^2 + \\
& -2(1-\eta_i)\lambda_i C_i p_i + p_i^2]^{\frac{1}{2}}\} \qquad i = 1, 2
\end{aligned} \tag{16}
$$

Linearizing (15) at (A_1^*, A_2^*), we obtain

$$\frac{dB_i}{dt} = q_i B_i(t) - \eta_i(q_i + \gamma_i + p_i)B_i(t - \tau_i) \qquad i = 1, 2 \tag{17}$$

where

$$q_i = \lambda_i C_i - \gamma_i - p_i - 2\lambda_i A_i^* \qquad i = 1, 2.$$

Note that the two equations of (17) are independent and have the same form. Therefore, we will concentrate only on the first one. The characteristic equation is

$$\lambda = q_1 - \eta_1(q_1 + \gamma_1 + p_1)e^{-\lambda\tau_1} \tag{18}$$

i.e. of type

$$D(\lambda, \tau) = P(\lambda, \tau) + Q(\lambda, \tau)e^{-\lambda\tau} = 0 \tag{19}$$

with $\tau = \tau_1$, $P(\lambda, \tau) := q_1 - \lambda$, $Q(\lambda, \tau) := -\eta_1(q_1 + \gamma_1 + p_1)$. Even if the coefficients of equation (19) explicitly depend on delay τ_1, we can still apply the procedure used in Sec. 2 for equation (5). Therefore we set $\lambda = i\omega$ and obtain

$$F(\omega(\tau), \tau) := \omega^2 + q_1^2 - \eta_1^2(q_1 + \gamma_1 + p_1)^2 \tag{20}$$

If ω is a positive root of $F(\omega(\tau), \tau) = 0$, we define $\vartheta \in [0, 2\pi]$ such that

$$\sin(\vartheta(\tau)) = \frac{\omega(\tau)}{\eta_1(q_1 + \gamma_1 + p_1)} \qquad \cos(\vartheta(\tau)) = \frac{p_1}{\eta_1(q_1 + \gamma_1 + p_1)} \tag{21}$$

The computation of the function $F(\omega(\tau_1), \tau_1)$ with respect to the equilibrium A_1^* given in (16) has been performed by means of "Maple" giving the following numerical values to the parameters of system (12)

$$\delta_1 = 0.5,\ \theta_1 = 0.8,\ \beta_1 = 3,\ \lambda_1 = 3,\ \gamma_1 = 0.5,\ C_1 = 4.153846154.$$

We have obtained

$$
\begin{aligned}
F(\omega(\tau_1), \tau_1) := \omega^2 - & \left(10.66153846 - \frac{10.66153846 - 11.96153846e^{(-.5\tau_1)} + R}{(1 - e^{(-.5\tau_1)})}\right)^2 \\
-e^{(-.5\tau_1)} & \left(12.46153846 - \frac{10.66153846 - 11.96153846e^{(-.5\tau_1)} + R}{(1 - e^{(-.5\tau_1)})}\right)^2
\end{aligned} \tag{22}
$$

where

$$R = \sqrt{(168.0014793(1-e^{(-.5\tau_1)})^2 - 29.41000000 + 31.10000000e^{(-.5\tau_1)})}.$$

The equation $F(\omega(\tau_1),\tau_1) = 0$ admits two real roots for every $\tau_1 \in]0,+\infty[$. Only one of these roots is positive. Since in the interval $]0, 0.364663607[$ both $\sin(\vartheta(\tau_1))$, and $\cos(\vartheta(\tau_1))$ are positive, ϑ is given by $\vartheta = \arctan \frac{\sin(\vartheta(\tau_1))}{\cos(\vartheta(\tau_1))}$. Being $S_n(\tau_1) := \tau_1 - \dfrac{\vartheta(\tau_1)+2n\pi}{\omega(\tau_1)}$ we obtain the following graphs for $n = 0, 1$.

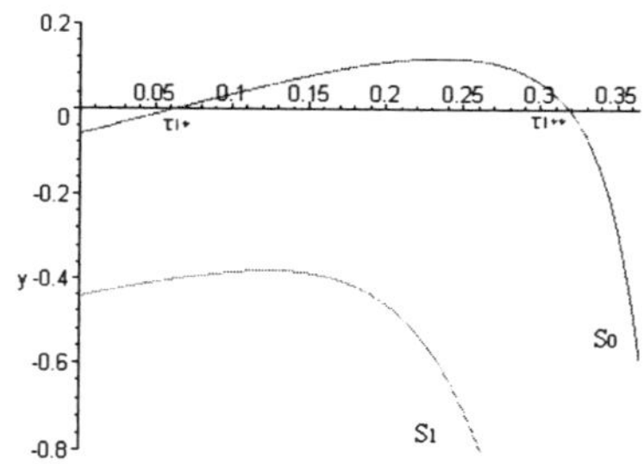

Figure 2: Graphs of functions S_0, S_1

We observe that the line S_0 intersects the τ_1-axis in the two points $\tau_1^* = 0.06328654373$, $\tau_1^{**} = 0.3183000096$. If we set

$$\delta(\tau^*) = sign\left\{ \frac{d\mathbb{R}e\lambda}{d\tau}\bigg|_{\lambda=i\omega(\tau^*)} \right\} = sign\left\{ [F'_\omega(\omega(\tau),\tau)]|_{\tau=\tau^*}\right\} sign\left\{S'_n(\tau^*)\right\},$$

we obtain $\delta(\tau_1^*) = 1$ and $\delta(\tau_1^{**}) = -1$. Therefore, according to [22], we conclude that the equilibrium A_1^* is asymptotically stable for $\tau_1 \in]0, 0.06328654373[$, unstable for $\tau_1 \in]0.06328654373, 0.3183000096[$ and again asymptotically stable for $\tau_1 > 0.3183000096$.

4 Conclusions

We observe that the obtained results of the performed simulations on systems (3) and (12), completely agree with the general statement contained in [1] and concerning the influence of the size of delay on the stability properties of solutions of these systems: for Delay Differential Systems with delay dependent parameters "the common scenario is that as time delay increases, stability changes from stable to unstable to stable, implying that large delay can be stabilizing. This scenario often contradicts the one provided by similar models with only delay independent parameters".

References

1. E. Beretta, Y. Kuang, *Geometric Stability Switch Criteria in Delay Differential Systems with Delay Dependent Parameters*, to appear on SIAM Journal Mathematical Analysis.
2. H.I.Freedman, J.W.H. So, P.Waltman, *Chemostat competition with time delays.* In IMACS 1988, 12th World Congress on Scientific Computer, Proceedings, Vol.4, Ed. By R.Vinchnevetski, P.borne, J.Vignes, 1988.
3. L.A.Segel, Modelling dynamic phenomena in moleculas and cellular biology, Cambridge University Press, 1989.
4. P.Fergola, S.Rionero, C.Tenneriello, *A qualitative analysis of an additive chemostat model of microorganisms competing for two complementary nutrients with delayed recycling*, Proc. of VII Int. Conf. on Waves and Stability in continuous media,World Scientific, 1993.
5. H.I.Freedman, Y.Xu, *Models of competition in the chemostat with instantaneous and delayed nutrient recycling*, J.Math.Biol., Vol.31, n.5, 1993.
6. E.Beretta, Y.Takeuchi, *Qualitative properties of chemostat equations with time delays: Boundedness, local and global asymptotic stability*. Differential Equations Dynam. Systems, 2, 1994.
7. H. Smith, P.Watmann, The theory of the chemostat, Cambridge Studies in Mathematical Biology, 1995.
8. P.Fergola, C.Tenneriello, Z.Ma, X.Wen, *Effects of toxicants on chemostat model*, Cybernetics and Systems '94, Vienna, Aprile 5-8, 1994.
9. P.Fergola, C.Tenneriello, Z.Ma, X.Wen, *Polluted chemostat models with periodic nutrient and fluctuatig toxicants input*, Proc. of VIII Int. Conf. on Waves and Stability in continuous media, Palermo, 1995.
10. P. Fergola, B. Buonomo, C. Ruggieri, *Chemostat type equations modelling a polluted environment*, Math. Comput. Modelling Vol. 24, 1996.
11. P.Fergola, C.Tenneriello, B.Buonomo, *Survival of populations in polluted chemostat environment and partial ultimate boundedness,* Ecology of Industrial Regions, Vol.2 (1-2), 1996.
12. P. Fergola, C. Tenneriello, B. Buonuomo, *On the survival ofpopulations in a polluted chemostat environment*, Atti del Convegno "La Matematica nei problemi dell'Ambiente, della Biologia e della Medicina: aspetti modellistici, analitici e computazionali, Studi Urbinati, Anno I,Serie A n.1, 1997.

13. Z. Ma, S. Guo, P. Fergola, C.Tenneriello, *Effects of Toxicants on Chemostat Model with Time Variable Nutrient Input and Washout*, Systems Science and Math. Sciences, Vol.11, n.4, 1998.
14. Z. Ma, S. Guo, P. Fergola, C.Tenneriello, *The Threshold of Population Survival in a Polluted Chemostat Model*, Math. Comp.Modelling, Vol.28, n.9, 1998.
15. P.Fergola, C.Tenneriello, M.Cerasuolo, *A survey of polluted chemostat models*, to appear.
16. R.Pratt, *Influence of the size of the inoculum on the growth of Chlorella vulgaris in freshly prepared culture medium*, Am. J. Bot. 27 (1940) 52.
17. J.Maynard-Smith, *Models in Ecology*, Cambridge University, Cambridge, 1974, p. 146.
18. J.Chattopadhyay, *Effects of toxic substances on a two-species competitive system*, Eco.Mod. 84, 1996.
19. A.Mukhopadhyay, J.Chattopadhyay, P.K.Tapaswi *A delay differential equations model of plankton allelopathy*, Math. Biosc. 149 (1998) 167.
20. P.K.Tapaswi, A.Mukhopadhyay, *Effects of environmental fluctuation on plankton allelopathy*, J.Math.Biol. 39, 1999.
21. J.Chattopadhyay,, E.Beretta, F.Solimano, *Role of Time-delay in an ecotoxicological problem*, Canadian Applied Mathematics Quarterly, vol.5, n.1, 1997.
22. C. Tenneriello, P. Fergola, Z.Ma, W.Wang, *Innovation Diffusion Model in patch environment*, to appear.
23. V.Mahajan, E.Muller, R.A.Kerin, *Introduction strategy for a new products with positive and negative word-of-mouth*, Management Science. 30 (1984), 1389-1404.
24. V.Mahajan, R.Peterson, *Models for innovation diffusion*, Sage Publication, Beverly Hills, 1985.
25. V.Mahajan, Y.Wind, *Innovation diffusion models of a new product acceptance*, Bellinger Pub. Co., Cambridge Press, 1986.
26. P.Fergola, C.Tenneriello, Z.Ma, F.Petrillo *Delayed innovation diffusion processes with positive and negative word-of-mouth*, Int. J. Diff. Equa. Appl. 1 (2000), 131-147.
27. P.Fergola, C.Tenneriello, Z.Ma, F.Petrillo *An innovation diffusion model with time delay: positive and negative word-of-mouth*, to appear.

DECAY AND OTHER PROPERTIES OF CROSS-SECTIONAL MEASURES IN ELASTICITY

J.N. FLAVIN
Department of Mathematical Physics
National University of Ireland, Galway, Ireland
E-mail: james.flavin@nuigalway.ie

The paper commences with a comparison between the Liapunov method for stability in connection with parabolic systems and the cross-sectional method for elliptic systems. The two approaches are illustrated using essentially the same Liapunov functional in two different, yet related, contexts: unsteady and steady nonlinear diffusion. The paper proceeds to use the cross-sectional method in two different contexts in elasticity: A traction problem for a semi-infinite, smoothly varying inhomogeneous, isotropic strip is considered, and a cross-sectional estimate is obtained reflecting Saint-Venant's Principle; cross-sectional estimates are also obtained for a displacement type problem for an isotropic right cylinder, both homogeneous, and smoothly varying inhomogeneous incompressible, material.

1 Introduction

The paper commences with some remarks outlining some formal similarities between the Liapunov functional approach to stability – typically for first order in time parabolic differential equations – and the "cross-sectional method" – typically for second order, in the axial coordinate x_3, elliptic partial differential equations. This is done in order to relate the principal theme of the paper to one of the principal themes of the conference (stability). See also [1].

The paper then illustrates the two issues by applying essentially the same Liapunov functional

(i) to a stability issue for a nonlinear diffusion equation in a space-time cylinder with time-dependent boundary conditions;

(ii) to a steady-state boundary value problem for a nonlinear diffusion equation in a (spatial) right cylinder with boundary conditions on its lateral boundary independent of the axial coordinate.

The role of the time in the former case corresponds to that of the axial coordinate in the latter.

The remainder – and heart of – the paper concerns the application of the "cross-sectional method" to two issues in elasticity. The first considers a (semi-infinite) rectangular region consisting of isotropic, smoothly varying laterally inhomogeneous elastic material: a self-equilibrated load is applied to one edge, the other boundaries being traction-free, and a decay estimate is derived which reflects Saint-Venant's principle; furthermore, some attention is devoted to the dependence of the estimated decay rate on the constitutive profile. The second issue concerns a right cylinder, of essentially arbitrary cross-section, consisting of homogeneous isotropic elastic material whose lateral boundary is subject to a non-null displacement boundary condition: an inequality estimate is derived for a cross-sectional measure, which proves to be sharp. A similar issue is considered when the material is incompressible and cross-sectionally inhomogeneous.

2 Liapunov Stability and the Cross-Sectional Method: A Comparison

The Liapunov stability approach essentially considers a "perturbation" $u(\mathbf{x}, t)$ satisfying a parabolic equation, or system of equations, first order in the time variable t,

$$L[u] = 0 \tag{1}$$

in a space-time (right) cylinder

$$D(t) \times 0 < t < \infty,$$

typically satisfying a boundary condition

$$u = 0 \text{ on } \partial D \times 0 < t < \infty, \tag{2}$$

and an initial condition

$$u(\mathbf{x}, 0) = \text{ specified on } D(0). \tag{3}$$

One defines a Liapunov functional

$$F(t) = \int_{\mathrm{D}(t)} P(u, \cdot)\, d\mathbf{x} \tag{4}$$

where P depends on u and/or its spatial derivatives, and is positive-definite therein. Typically, one shows that

$$\frac{dF}{dt} \leqslant 0 \Rightarrow F(t) \leqslant F(0) \; (\textit{stability}) \tag{5}$$

or, more generally,

$$\frac{dF}{dt} + mF(t) \leqslant 0 \Rightarrow F(t) \leqslant F(0)\, e^{-mt} \; (\textit{asymptotic stability}) \tag{6}$$

where m is a positive constant, $F(0)$ being available from the initial condition (3).

The "cross-sectional method" typically deals with a vector dependent variable $u(\mathbf{x}, x_3)$ satisfying an elliptic system of p.d.e.S (second order in x_3)

$$L[u] = 0 \tag{1'}$$

in a spatial right cylinder

$$D(x_3) \times 0 < x_3 < l,$$

$\mathbf{x} = (x_1, x_2), x_3$, corresponding to rectangular cartesian coordinates and l being a constant, typically (though not necessarily) subject to a null boundary condition on the lateral boundary

$$u = 0 \text{ on } \partial D \times 0 < x_3 < l, \tag{2'}$$

and boundary conditions on the ends $x_3 = 0, l$

$$f_0(u, \cdot),\ f_l(u, \cdot) \text{ specified respectively on } x_3 = 0, l, \tag{3'}$$

where the boundary conditions involve the specification of u and/or the gradients thereof. One defines – analogous to the Liapunov functional in the previous context –

$$F(x_3) = \int_{\mathrm{D}(x_3)} P(u, \cdot)\, d\mathbf{x} \tag{4'}$$

where P depends on u and/or its spatial gradients, and is (normally) positive-definite therein. Typically, one has

$$\frac{d^2 F}{dx_3^2} \geqslant 0 \Rightarrow F(x_3) \leqslant F(0) + [F(L) - F(0)](x_3/l)\ (\textit{convexity}) \tag{5'}$$

or, more generally, m being a positive constant,

$$\frac{d^2 F}{dx_3^2} - m^2 F \geqslant 0 \Rightarrow F(x_3) \leqslant G(x_3)\ (\textit{generalised convexity}) \tag*{$(6')_1$}$$

where $G(x_3)$ satisfies

$$\frac{d^2 G}{dx_3^2} - m^2 G = 0,\ G(0) = F(0),\ G(l) = F(l) \tag*{$(6')_2$}$$

$F(0), F(l)$ being available from the boundary conditions $(3')$ on the plane ends.

The points of similarity between the two contexts are suggestively conveyed by the numbering of the various equations etc.: (n') in the context of the "cross-sectional method" is the counterpart of (n) in the Liapunov stability context.

Moreover, a further similarity arises if in the second context one is considering a semi-infinite cylinder for which

$$l \to \infty \text{ and } F(l) \to 0.$$

In this case, $(5'), (6')$ give rise to

$$F(x_3) \leqslant F(0), \tag{5''}$$

$$F(x_3) \leqslant F(0)\, e^{-mx_3}, \tag{6''}$$

respectively. These provide a closer correspondence to $(5), (6)$ than $(5'), (6')$.

We proceed to illustrate the aforementioned two contexts using essentially the same Liapunov functional in the context of related non-linear diffusion equations.

We consider an initial boundary value problem, $\mathbf{x} = (x_1, x_2, x_3)$ being rectangular cartesian coordinates, t being the time, in a space-time cylinder

$$D(t) \times 0 < t < \infty :$$

$T(\mathbf{x}, t)$ is a smooth solution of

$$\frac{\partial T}{\partial t} = \boldsymbol{\nabla} \cdot \{k(T)\, \boldsymbol{\nabla} T\} \tag{7}$$

subject to

$$T = T_0(\mathbf{x}) \text{ on } \partial D(t) \times 0 < t < \infty \tag{8}$$

and

$$T(\mathbf{x}, 0) = \text{ specified.} \tag{9}$$

The conductivity $k(\cdot)$ is assumed to satisfy

$$k(\cdot) \geqslant k_0, \tag{10}$$

k_0 being a given positive constant.

Writing

$$T(\mathbf{x}, t) = U(\mathbf{x}) + u(\mathbf{x}, t) \tag{11}$$

where the steady state temperature $U(\mathbf{x})$ satisfies

$$\nabla \cdot \{k(U)\nabla U\} = 0 \text{ in D}, \tag{12}$$

subject to

$$U = T_0(\mathbf{x}) \text{ on } \partial D. \tag{13}$$

Defining

$$\Phi(u,U) = \int_0^u d\bar{u} \int_0^{\bar{u}} k(\zeta + U)\, d\zeta \tag{14}$$

for which (see [2])

$$\left.\begin{array}{r} (i)\ \Phi(0,\cdot) = \Phi_u(0,\cdot) = 0; \\ (ii)\ \Phi_{uu} = k(u+U)\,; \\ (iii)\ \Phi_u^2 \geqslant \frac{1}{2}k_0 u^2; \\ (iv)\ \Phi_u^2 \geqslant 2k_0\Phi; \\ (v)\ \Phi \text{ is positive-definite in } u, \end{array}\right\} \tag{15}$$

one finds

$$u_t = \nabla^2 [\Phi_u] \tag{16}$$

subject to

$$u = 0 \text{ on } \partial D(t) \times 0 < t < \infty \tag{17}$$

and

$$u(\mathbf{x},0) = \text{ specified.} \tag{18}$$

One defines the Liapunov functional (positive-definite in u)

$$F(t) = \int_{\mathrm{D}(t)} \Phi(u;U)\, d\mathbf{x} \tag{19}$$

and one readily proves using (15) *inter alia* (for details, see [2])

$$\frac{dF}{dt}(t) + 2\lambda_1 k_0 F(t) \leqslant 0 \tag{20}$$

implying

$$F(t) \leqslant F(0)\exp[-2\lambda_1 k_0 t] \tag{21}$$

where λ_1 is the lowest (fixed-membrane) eigenvalue of

$$\nabla^2\chi + \lambda\chi = 0 \text{ in } D,\ \chi = 0 \text{ on } \partial D. \tag{22}$$

The inequality (21) essentially conveys the exponential asymptotic stability of the steady state solution.

We now turn to a steady state problem for an analogous nonlinear diffusion equation. Let $\mathbf{x} = (x_1, x_2), x_3$ denote rectangular cartesian coordinates and we consider the spatial cylinder

$$D(x_3) \times 0 < x_3 < l,$$

$\mathrm{D}(x_3)$ being the cylinder cross-section (at x_3), l being a constant. In this context x_3 plays the role of time in the previous one. We consider the steady state boundary value problem: $T(\mathbf{x}, x_3)$ is a smooth solution of

$$\boldsymbol{\nabla}_1 \cdot [k(T)\,\boldsymbol{\nabla}_1 T] + T_{,33} = 0 \tag{23}$$

subject to

$$T = T_0(\mathbf{x}) \text{ on } \partial D(x_3) \times 0 < x_3 < l \tag{24}$$

and

$$T(\mathbf{x}, 0), T(\mathbf{x}, l) = \text{ specified,} \tag{25}$$

$\boldsymbol{\nabla}_1$ denoting the gradient operator in the (x_1, x_2) plane. The conductivity $k(\cdot)$ again satisfies (10) in the new circumstances.

Writing

$$T(\underline{x}, x_3) = U(\underline{x}) + u(\underline{x}, x_3) \tag{26}$$

one finds that

$$\nabla_1^2 \Phi_u + u_{,33} = 0 \tag{27}$$

subject to

$$u = 0 \text{ on } \partial D(x_3) \times 0 < x_3 < l \tag{28}$$

and

$$u(\mathbf{x}, 0), u(\mathbf{x}, l) \text{ specified.} \tag{29}$$

Defining (by analogy with (19)) the cross-sectional measure of u :

$$F(x_3) = \int_{\mathrm{D}(x_3)} \Phi(u, U)\, d\mathbf{x}$$

where Φ has the properties previously listed, *mutatis mutandis.* One may prove (see [3]) that $F(x_3)$ satisfies the generalised convexity property

$$F''(x_3) - m^2 F(x_3) \geqslant 0 \tag*{$(31)_1$}$$

where

$$m^2 = 2\lambda_1 k_0 \tag*{$(31)_2$}$$

λ_1 being the lowest fixed-membrane eigenvalue of the cross-section [cf. (22)]. Using the fact (see [4]) that $F(x_3)$ is bounded above by the solution $G(x_3)$ of the differential equation corresponding to the differential inequality $(31)_1$, such that $G(0), G(l)$ have the values $F(0), F(l)$ respectively, we have the result

$$F(x_3) \leqslant [F(0) \sinh\{m(l - x_3)\} + F(l) \sinh(mx_3)] / \sinh(ml). \qquad (32)$$

By deduction from this by elementary means, one obtains the cruder, but more transparent, result

$$F(x_3) \leqslant F(0) \exp(-mx_3) + F(l) \exp\{-m(l - x_3)\}, \qquad (33)$$

$F(0), F(l)$ being computable from data. Essentially, (33) shows that end effects fall off exponentially away from the ends. Improvements of (31) – (33) may be obtained: see [3].

3 Examples of the Cross-Sectional Method in Elasticity

In this section a rectangular region is considered

$$0 < x_1 < l, 0 < x_2 < 1,$$

where $l \to \infty$ ultimately, and where (x_1, x_2) denote rectangular cartesian coordinates. We suppose that the region is occupied by smoothly varying transversely inhomogeneous isotropic elastic material in a state of plane stress, due to an (in-plane) self-equilibrated load on the edge $x_1 = 0$, the remaining edges being free. The material in question may model a functionally graded elastic material (FGM). The general objective is to define a positive-definite cross-sectional measure of stress and to prove that it decays at least exponentially fast away from the loaded end, reflecting Saint-Venant's principle; and to examine how the estimated decay rate depends on the constitutive profile.

The elastic moduli E (Young's Modulus), σ (Poisson's ratio) are smooth functions of y, but instead of these it proves convenient to work with alternative moduli $\varepsilon(y), \underline{\varepsilon}(y)$ related to the foregoing by means of

$$\varepsilon = \left(1 - \sigma^2\right) E^{-1}, \underline{\varepsilon} = \sigma(1 - \sigma)^{-1} \varepsilon. \qquad (34)$$

It is assumed throughout that

$$E > 0, 0 \leqslant \sigma \leqslant \tfrac{1}{2} \qquad (35)$$

and that thereby

$$\varepsilon > 0, \underline{\varepsilon} \geqslant 0. \qquad (36)$$

It should be noted that in the case of incompressibility $(\sigma \to 1/2)$ $\underline{\varepsilon} \to \varepsilon$.

The Airy stress function $\phi(x, y)$ is introduced to simplify the analysis, and is such that the (relevant) stress components $\tau_{xx}, \tau_{xy}, \tau_{yy}$ are given by

$$\tau_{xx} = \phi_{yy},\ \tau_{yy} = \phi_{xx},\ \tau_{xy} = -\phi_{xy} \tag{37}$$

where subscripts attached to ϕ mean partial differentiation with respect to these variables, both here and subsequently. It is readily verified that ϕ satisfies

$$(\varepsilon\phi_{xx})_{xx} + 2(\varepsilon\phi_{xy})_{xy} + (\varepsilon\phi_{yy})_{yy} - \ddot{\underline{\varepsilon}}\phi_{xx} = 0 \tag{38}$$

where superposed dots mean differentiation with respect to y. It is easily shown (exploiting the arbitrariness in ϕ) that the boundary conditions on the free edges may be written

$$\left.\begin{array}{l} \phi = \phi_y = 0 \text{ on the edges } y = 0, 1, \\ \phi = \phi_x = 0 \text{ on the edge } x = l. \end{array}\right\} \tag{39}$$

It shall be assumed that $\phi \in C^4$ in the closed region.

We defined the cross-sectional measure of stress as follows

$$F(x) = \int_0^1 \varepsilon\left(\phi_{xx}^2 + \phi_{yy}^2\right) dy. \tag{40}$$

Straightforward analysis gives

$$F''(x) = 2\int_0^1 \varepsilon\left(\phi_{xxx}^2 + \phi_{xyy}^2 + 2\phi_{xxy}^2\right) dy + 2\int_0^1 \ddot{\underline{\varepsilon}}\phi_{xx}^2 dy \tag{41}$$

primes denoting (ordinary) differentiation with respect to x, superposed dots differentiation with respect to y. Let us assume henceforward that

$$\ddot{\underline{\varepsilon}} \geqslant 0 \tag{42}$$

i.e. that $\underline{\varepsilon}$ is convex. In these circumstances, F is convex in x. One may deduce that

$$F''(x) \geqslant 2\int_0^1 \varepsilon\left(\phi_{xxx}^2 + \phi_{xyy}^2 + 2\lambda_1\phi_{xx}^2\right) dy \tag{43}$$

where λ_1 is the lowest positive eigenvalue of

$$\ddot{u} + \left[\lambda - \left(\varepsilon^{1/2}\right)^{\cdot\cdot}\varepsilon^{-1/2}\right] u = 0,\ y \in (0,1),\ u(0) = u(1) = 0; \tag{44}$$

this is done using the results of the Appendix.

To establish "generalised convexity" of $F(x)$ we need the following conservation law:

$$\int_0^1 \left[\varepsilon\left\{2\phi_x\phi_{xxx} - \phi_{xx}^2 + \phi_{yy}^2 - 2\phi_{xy}^2\right\} - \ddot{\underline{\varepsilon}}\phi_x^2\right] dy = E \tag{45}$$

where E is a constant (see [5]).

One may deduce from (40), (43), (45) the following generalised convexity inequality for $F(x)$:

$$F''(x) - r\lambda_1 F(x) \geqslant 0 \tag{46}$$

where

$$r = 2\left\{\sqrt{(1+2\Sigma)^2+1} - (1+2\Sigma)\right\} \qquad (47)_1$$

in which

$$\Sigma = \sigma_M (1-\sigma_M)^{-1}, \qquad (47)_2$$

σ_M denoting the maximum value of σ (Poisson's ratio) arising. One may deduce from (46) by means of a comparison principle (cf. (33))

$$F(x) \leqslant F(0)\exp\left(-\sqrt{r\lambda_1}x\right) + F(l)\exp\left(-\sqrt{r\lambda_1}\{l-x\}\right). \tag{48}$$

On letting $l \to \infty$ we obtain a decay result (reminiscent of a Phrágmen-Lindelöf theorem):

Theorem 1. *In the context of a rectangular strip* $0 < x < l, 0 < y < 1$, *for which* $l \to \infty$, *consisting of isotropic elastic material, smoothly varying inhomogeneous with respect to the* y *direction for which* (42) *holds, in a state of plane stress, due to a self-equilibrated load on the edge* $x = 0$, *the remaining edges being free, the cross-sectional stress measure* (40) *satisfies the decay law*

$$F(x) \leqslant F(0)\exp\left(-\sqrt{r\lambda_1}x\right) \tag{49}$$

where r *is defined by* (47) *and* λ_1 *is the lowest positive eigenvalue of* (44), *provided that*

$$\lim_{l\to\infty} e^{-\sqrt{r\lambda_1}l} \int_0^1 \tau_{yy}^2(l,y)\,dy = 0. \tag{50}$$

Remark 1. It will be noted that (50) is satisfied in particular if

$$\phi_{xx} \to 0 \text{ as } x \to \infty. \tag{51}$$

Remark 2. Note that the decay rate given in (49) coincides with that of Knowles [6] in the case of homogeneous materials with vanishing Poisson's ratio. Whereas it is possible to do better than this in particular contexts our aim is to get a relatively simple result which will give an overall view of how the estimated decay rate depends on varying constitutive profiles e.g. [1].

Remark 3. The estimated decay rate is essentially the eigenvalue λ_1 of (44). The dependence of λ_1 on the constitutive profile is discussed in [7]. In the present context, for example, λ_1 is, in a sense, a monotonically increasing function of $\left(\varepsilon^{1/2}\right)^{\cdot\cdot}\varepsilon^{-1/2}$.

Remark 4. The estimate (49) can be made explicit in terms of conventional data in either of the following two circumstances (cf. [5]):

(a) There is normal loading (i.e. $\tau_{xy}=0$) on the edge $x=0$ and the (additional) asymptotic condition (51) obtains. One then has (via the Conservation Law (45))

$$F(0)=\int_0^1 \tau_{xx}^2(0,y)\,dy. \tag{52}$$

(b) The normal stress component and the complementary, tangential displacement component are specified on the edge $x=0$.

In the remainder of the paper we present an example (examples) of the cross-sectional method in elasticity where the boundary conditions on the lateral boundary are non-null. We consider a right cylinder of isotropic material subject to a class of non-zero displacements on its lateral boundary, and to suitable actions on its plane ends. Two cases are considered in succession, that of a (compressible) homogeneous material, followed by that of an incompressible material which is smoothly varying inhomogeneous in the cross-section.

Consider solutions $u_i\in C^3\left(D\times 0\leqslant x_3\leqslant l\right)$ of the displacement equations

$$u_{i,jj}+\alpha u_{j,ji}=0 \text{ in } D\times 0\leqslant x_3\leqslant l \tag{53}$$

where D is the cylinder cross-section, α a positive material constant, and where the usual indicial notation and summation convention are used throughout; the solutions are subject to the boundary conditions on the lateral surface

$$u_\beta=f_\beta(x_1,x_2)\,,u_3=g(x_1,x_2)+x_3h(x_1,x_2) \text{ on } \partial D\times 0\leqslant x_3\leqslant l, \tag{54}$$

Greek indices taking the values 1, 2, Latin indices taking the values 1, 2, 3 throughout.

We consider the cross-sectional measure of deformation

$$F(x_3)=\int_{\mathrm{D}(x_3)}\left[u_{\beta,3}u_{\beta,3}+u_{3,\beta}u_{3,\beta}\right]dA. \tag{55}$$

This measure is non-negative, and positive-definite if $f_\beta\equiv g\equiv h\equiv 0$. Elementary analysis shows

$$\begin{aligned}F''(x_3)=2\int_{\mathrm{D}(x_3)}&\left[u_{\beta,33}u_{\beta,33}+u_{3,\beta3}u_{3,\beta3}+u_{\beta,\gamma3}u_{\beta,\gamma3}\right.\\&\left.+\alpha(\alpha+1)^{-1}u_{\beta,\gamma3}u_{\beta,\gamma3}+(\alpha+1)^{-1}u_{3,\beta\beta}u_{3,\gamma\gamma}\right]dA.\end{aligned} \tag{56}$$

[For the case of homogeneous lateral boundary conditions, see [1]]. While it is evident that F is convex, a stronger result is possible (using Schwarz's inequality etc.):

$$\left[F^{1/2}(x)\right]'' \geqslant 0 \qquad (F \neq 0) \tag{57}$$

whence

$$F^{1/2}(x_3) \leqslant F^{1/2}(0) + \left[F^{1/2}(l) - F^{1/2}(0)\right] x_3/l. \tag{58}$$

Remark 5. $F(0)$, $F(l)$ are available in terms of conventional data on $x_3 = 0, l$ if u_3 and the complementary components of shear stress $\tau_{3\beta}$ are prescribed (as smooth functions).

Remark 6. Both the inequality (57) and its concomitant (58) are sharp in the sense that the equality sign occurs therein for the following (non-trivial) field:

$$u_3 = U(x_1, x_2) + x_3 W(x_1, x_2), u_\beta \equiv 0 \tag*{$(59)_1$}$$

where

$$\left.\begin{array}{l} U_{,\beta\beta} = 0, \\ W_{,\beta\beta} = 0, \end{array}\right\} \text{ in } D(x_3), \quad \left.\begin{array}{l} U = g(x_1, x_2) \\ W = h(x_1, x_2) \end{array}\right\} \text{ on } \partial D. \tag*{$(59)_2$}$$

One may augment this field by a suitable (non-zero) plane strain field u_β.

One may similarly deal with analogous issues for a class of incompressible, isotropic elastic right cylinders with smoothly varying transverse inhomogeneity. We consider the solutions $u_i \in C^3, p \in C^1$ (p denoting the 'pressure') of

$$-p_{,i} + \mu u_{i,jj} + \mu_{,\gamma}(u_{i,\gamma} + u_{\gamma,i}) = 0, \tag{60}$$

$$u_{i,i} = 0, \tag{61}$$

in $D \times 0 < x_3 < l$, where the (positive) rigidity modulus $\mu(x_1, x_2)$ is supposed to be a C^2 function. These solutions are subject to the boundary conditions on the lateral boundary

$$u_\beta = f_\beta(x_1, x_2), u_3 = g(x_1, x_2) + h(x_1, x_2) x_3 \quad \text{on } \partial D \times 0 < x_3 < l. \tag{62}$$

We define the cross-sectional measure (cf. (55)) as follows:

$$F(x_3) = \int_{D(x_3)} \mu\left[u_{\beta,3} u_{\beta,3} + u_{3,\beta} u_{3,\beta}\right] dA. \tag{63}$$

Again this measure is non-negative, being positive-definite when

$$f_\beta \equiv g \equiv h \equiv 0.$$

One may prove (see [8]) that

$$F''(x_3) = 2\int_{D(x_3)} \left[\mu\left(u_{\beta,33}u_{\beta,33} + u_{3,\beta 3}u_{3,\beta 3} + u_{\beta,\gamma 3}u_{\beta,\gamma 3} + u_{3,33}^2\right) + \mu_{,\beta\gamma}u_{\beta,3}u_{\gamma,3}\right] dA. \tag{64}$$

In view of the non-negativity of the first four terms in the integrand we have the following result:

Theorem 2. *The cross-sectional measure* (63)*, in the context of smooth solutions of* (60), (61) *subject to* (62)*, is a convex function provided that* μ *is a convex function of the cross-sectional coordinates* x_1, x_2*, i.e.*

$$\mu_{,11} \geqslant 0,\ \mu_{,22} \geqslant 0,\ \mu_{,11}\mu_{,22} - \mu_{,12}^2 \geqslant 0.$$

One may obtain a stronger result using Schwarz's inequality:

Theorem 2a. *The cross-sectional measure* $F(x_3)$*, under the same conditions as in Theorem 2, is such that* $F^{1/2}(x_3)\left(= \sqrt{F(x_3)}\right)$ *is convex in* x_3 $(F \neq 0)$.

One may in this case draw conclusions similar to those drawn in the case of a *compressible* elastic cylinder.

Appendix

Any smooth function $\Phi(y)$, $y\,(0,1)$, such that $\Phi(0) = \Phi(1) = 0$, satisfies

$$\int_0^1 \varepsilon(y)\,\dot{\Phi}^2 dy \geqslant \lambda_1 \int_0^1 \varepsilon(y)\,\Phi^2 dy \tag{A1}$$

where λ_1 is the lowest eigenvalue of

$$\left(\varepsilon\dot{\Phi}\right)^{\cdot} + \lambda\varepsilon\Phi = 0,\ \Phi(0) = \Phi(1) = 0,$$

or (via the transformation $u = \varepsilon^{1/2}\Phi$) of

$$\ddot{u} + \left[\lambda - \left(\varepsilon^{1/2}\right)^{\cdot\cdot}\varepsilon^{-1/2}\right]u = 0,\ u(0) = u(1) = 0.$$

References

1. J.N. Flavin, *Proceedings of the VIII International conference on Waves and Stability in Continuous Media*, Supplemento ai Rendiconti del Circolo Matematico di Palermo, Serie II, Numero 45 (1996).

2. J.N. Flavin and S. Rionero, *J. Math. Anal. Appl.*, **228**, 119 (1998).

3. J.N. Flavin and S. Rionero, *Math. Meth. Appl. Sc.*, (forthcoming).

4. J.N. Flavin and S. Rionero in *Qualitative Estimates for Partial Differential Equations: An Introduction* (CRC Press, Boca Raton, 1995).

5. J.N. Flavin, Qualitative estimates for a generalised biharmonic equation for inhomogeneous elastic materials (forthcoming).

6. J.K. Knowles, *Arch. Ration. Mech. Anal.*, **21**, 1 (1966).

7. J.N. Flavin, Symposium on Trends in the Application of Mathematics to Mechanics 2000, Elsevier, 72 (2000).

8. B. Gleeson and J.N. Flavin, A Convexity property for inhomogeneous incompressible elastic cylinders, *P. Roy. Irish Acad.* (2001) (forthcoming).

A REDUCTION METHOD FOR MULTIPLE WAVE SOLUTIONS TO QUASILINEAR DISSIPATIVE SYSTEMS

D. FUSCO AND N. MANGANARO

Department of Mathematics, University of Messina
Contrada Papardo, Salita Sperone 31,
98166 Vill. Sant'Agata, Messina, Italy.
E-mail: fusco@mat520.unime.it
nat@mat520.unime.it

A reduction method is worked out for determining a class of exact solutions to nonlinear second order PDEs involving two independent variables. The approach in point is based upon appending suitable constraint equations to the governing equations to be satisfied by the dependent field variables. That permits to introduce a subsystem of equations which rules the evolution of a "reduced field". The integration of this auxiliary subsystem provides the searched solution to the full governing model. In the process appropriate classes of material response functions can be determined in order to solve in a closed form initial and/or boundary value problems associated to nonlinear equations of interest in wave propagation. The present method can be also used for determining exact wave-like solutions to mathematical models exhibiting a "wave hierarchy" structure, typically accounting for relaxation effects in the field evolution, as those arising from the theory of irreversible processes.

1 Introduction

Within the theoretical framework of nonlinear wave propagation, much attention has been paid to systems of higher order PDEs. Actually such mathematical models endowed by initial or boundary data rule the evolution of the concerned field variables in several areas of science and engineering interest. In most cases their structure incorporate nonlinearity in the wave dynamics and dissipation or dispersion. Wave behaviors for the models in point were investigated by searching for asymptotic solutions (Fusco)[1], (Whitham)[2], (Jeffrey & Kawahara)[3], while results based upon group or symmetry analysis can be found in (Ibragimov) [4],[5],[6].

In this paper we work out a reduction method for determining exact wave–like solutions to the following nonlinear second order model

$$F(x,t,U,U_t,U_x,U_{tt},U_{xx},U_{xt}) = 0 \tag{1}$$

where x, and t are space and time coordinates, respectively, while $F, U \in \Re^N$. Hereafter subscript means for derivative with respect to the indicated variable.

By setting $U_t = R$ and $U_x = S$ the set of equations (1) can be recast into the system of first order PDEs

$$U_t = R \tag{2}$$

$$S_t - R_x = 0 \tag{3}$$

$$F(x,t,U,S,R,R_t,S_x,R_x) = 0 \tag{4}$$

provided that $S(x,0) = U_x(x,0)$ in order that $S(x,t) = U_x(x,t)\ \forall t$. The present approach is based upon searching for solutions to (2)–(4) obeying the finite relantioship

$$R = \hat{R}(x,t,U,S) \tag{5}$$

In terms of the field variables R, S, U the ansatz (5) includes that of multiple wave (Ovsiannikov)[7], (Meleshko)[8], [9].

By virtue of (5) the set of equations (2)–(4) writes

$$U_t = \hat{R}(x,t,U,S) \tag{6}$$

$$S_t - \hat{R}_S S_x = \hat{R}_x + \hat{R}_U S \tag{7}$$

$$\hat{F}(x,t,U,S,R,S_x) = 0 \tag{8}$$

Provided that the matrix $\|\frac{\partial \hat{R}}{\partial S}\|$ admits N real eigenvalues to which there correspond N left eigenvectors spanning the Euclidean space $\Re^N$, the pair of relations (6) and (7) represents a quasilinear hyperbolic system of first order PDEs governing the "reduced" field variables U and S which, in turn, are also required to satisfy the set of N differential constraints determined by the equation (8).

Therefore, despite of the original governing model (1) being hyperbolic or parabolic, the method developed herein falls into the subject of quasilinear first order hyperbolic systems of PDEs endowed by differential constraint equations (Janenko)[10], (Sidorov, Shapeev & Janenko) [11],(Rozhdestvenskii & Janenko)[12]. If the equation (8) is linear with respect to S_x then for the full set of differential conditions arising from requiring the consistency of (6), (7) and (8) the closure algorithm developed in (Fusco & Manganaro)[13] holds. In a given governing model belonging to the class (1) these closure processe also provides a vehicle for selecting forms of the material response functions involved therein as well as for determining the vector function $\hat{R}$ characterizing the ansatz (5).

Bearing in mind (6), we notice that it is possible to look for exact solutions to (1) exhibiting expected wave–like features by prescribing a special structure to the function $\hat{R}$.

This aspect will be illustrated later in sections 2 and 3, respectively, by means of two main models: the hyperbolic Monge–Ampère equation and a parabolic equation accounting for diffusion with absorption.

Finally, in the last section we will show how the leading ideas of the present approach also permit to investigate quasilinear hyperbolic dissipative systems of first order PDEs allowing for wave hierarchies (Whitham)[2].

2 Monge–Ampère equation

Let us consider the nonlinear equation

$$u_{tt}u_{xx} - u_{tx}^2 + \alpha u_{tt} + \beta u_{tx} + \gamma u_{xx} + \delta = 0 \tag{9}$$

where the coefficients α, β, γ, δ are functions of t, x, u, u_t, u_x.

As well known (Boillat)[14] the equation (9) represents the more general equation of the form

$$u_{tt} + f(x, t, u, u_x, u_{tx}, u_{xx}) = 0 \tag{10}$$

which turns out to be completely exceptional (Boillat)[15].

Completely exceptional Monge–Ampère equations involving more space variables were considered in (Ruggeri)[16], (Boillat)[17],[18], while in (Donato, Ramgulam & Rogers)[19], (Oliveri)[20] linearization results on this subject were obtained.

In line with the assumption (5) here the system (6)–(8) specializes to

$$u_t = \hat{r}(s, u, x, t) \tag{11}$$

$$s_t - \hat{r}_s s_x = \hat{r}_u s + \hat{r}_x \tag{12}$$

$$\begin{aligned} &\{\hat{r}_u\hat{r} + \hat{r}_t - \hat{r}_s(\hat{r}_u s + \hat{r}_x) + \alpha\hat{r}_s^2 + \beta\hat{r}_s + \\ &+\gamma\}s_x - (\hat{r}_u s + \hat{r}_x)^2 + \alpha\{\hat{r}_s(\hat{r}_u s + \hat{r}_x) \\ &+\hat{r}\hat{r}_u + \hat{r}_t\} + \beta(\hat{r}_u s + \hat{r}_x) + \delta = 0 \end{aligned} \tag{13}$$

Requiring the compatibility of (11), (12) and (13) leads to specializing the functions α, β, γ, δ or, in other words, the structure of the evolution equation (9) as well as the function $\hat{r}(u, s, x, t)$ which characterizes the ansatz (5). Hence, the searched solution $u(x, t)$ to the equation (9) is obtained by integrating the PDEs (11), (12), (13).

By way of illustration, we consider the following Monge–Ampère equation

$$u_{tt}u_{xx} - u_{tx}^2 + \hat{\delta} = 0; \qquad \hat{\delta} = const. > 0 \tag{14}$$

which was investigated in (Mokhov & Nutku)[21].

Here we limit ourselves to assuming $\hat{r}_u = 0$ whence by requiring the equation (13) to hold $\forall s_x$, we obtain:

$$\hat{r}(s, x, t) = ax + \pi(\eta) \tag{15}$$

where $a = \pm\sqrt{\hat{\delta}}$, $\eta = s + at$ and π is an arbitrary function.

Owing to (15) the pair of equations (11), (12) writes

$$u_t = ax + \pi(\eta) \tag{16}$$

$$\eta_t - \pi'(\eta)\eta_x = 2a \tag{17}$$

where hereafter the prime means for ordinary differentiation.

Integration of equation (17) by method of characteristics yields:

$$\eta = 2at + \Lambda(\xi) \tag{18}$$

$$\xi = 2ax + \pi(\eta) \tag{19}$$

where Λ is an arbitrary function, so that the equation (16) provides the exact solution to the governing model (14):

$$u(x,t) = axt + \int \pi(\eta)dt \tag{20}$$

which also satisfies $u_x(x,t) = s(x,t)\ \forall t \geq 0$.

We notice that the free functions Λ and π which are involved in the solution (20) through the relations (18) and (19) make the exact solution in point flexible enough to fit a given class of initial or boundary value problems.

Assume we are given the following initial data:

$$u(x,0) = \Phi(x) \tag{21}$$

$$u_t(x,0) = \psi(x) \tag{22}$$

Because of (21), by inserting $\eta(x,0) = s(x,0) = \Phi'(x)$ into (18) and (19) we have:

$$\eta = 2at + \Phi'(\xi) \tag{23}$$

$$\xi = x + \frac{1}{2a}\left[\pi\left(2at + \Phi'(\xi)\right) - \pi\left(\Phi'(\xi)\right)\right] \tag{24}$$

whereupon equation (20) evaluated at $t = 0$ provides

$$\psi(x) = ax + \pi\left(\Phi'(x)\right) \tag{25}$$

Once the function π has been specified, the relation (25) characterizes initial data (21), (22) which can be fitted by the exact solution (20).

Furthermore we remark that in (Mokhov)[21] there was shown that by means of a Bianchi–like variable transformation the Monge–Ampère equation (14) can be transformed into the Born–Infeld equation

$$(1 + \phi_x^2)\phi_{tt} - 2\phi_t\phi_x\phi_{tx} - (1 - \phi_t^2)\phi_{xx} = 0 \tag{26}$$

for which the results obtained herein also hold.

3 Nonlinear diffusion with absorption

In many areas of engineering interest the parabolic equation

$$u_t = \left(D\left(u\right)u_x\right)_x + A(u) \tag{27}$$

is assumed to model nonlinear diffusion with absorption for appropriate form of the material response functions $D(u)$ and $A(u)$.

Whithin the theoretical framework outlined in section 1, by requiring the double wave–like ansatz $r = R(u, s)$ the system (6)–(8) specializes to

$$s_t - R_s s_x = R_u s \tag{28}$$
$$u_t = R(u, s) \tag{29}$$
$$D(u)s_x = R(u, s) - D'(u)s^2 - A(u) \tag{30}$$

Next we assume

$$R(u, s) = a(u)s + b(u) \tag{31}$$

that is tantamount to searching for solutions to (27) which also satisfy the first order PDE

$$u_t = a(u)u_x + b(u) \tag{32}$$

As well known, the equation (32) is the simplest hyperbolic model which incorporates both nonlinearity in the wave dynamics and dissipation. In passing we notice that appending the equation (32) to (27) is in fact equivalent to search for solutions by non classical Lie group approach (Bluman & Cole)[22]. The latter investigation for equation (27) was carried on by Arrigo & Hill[23].

Owing to relation (31), after cross differentiation the consistency of the set of equations (28), (29), (30) requires the following conditions to hold

$$D = k_0 a' \tag{33}$$
$$\frac{2a}{k_0} + b'' + \frac{d}{du}\left(b\frac{D'}{D}\right) = 0 \tag{34}$$
$$\frac{A}{k_0} = 2qa'b + \frac{baD'}{D^2} \tag{35}$$
$$b = \frac{A}{1 - qD} \tag{36}$$

where $k_0 \neq 0$ and q are constants and the prime means differentiation.

The system of conditions (33) to (36) determines the material response funtions $D(u)$ and $A(u)$ as well as the coefficients $a(u)$ and $b(u)$ which characterize the first order wave equation (32).

By way of simplicity we take $q = 0$ so that the set of relations (33)–(36) yields:

$$a(u) = k_1 e^{k_2 u} \tag{37}$$

$$D(u) = k_0 a'(u) \tag{38}$$

$$A(u) = b(u) = k_3 e^{-k_2 u} - \frac{k_1}{k_0 k_2^2} e^{k_2 u} + k_4 \tag{39}$$

where k_1, k_2, k_3, k_4 are constants and $k_2 \neq 0$.

Owing to the relations (37)–(39), the integration of the resulting consistent system of equations (28)–(30) along with (31) shows that the parabolic equation

$$u_t = \left(k_0 k_1 k_2 e^{k_2 u} u_x\right)_x + k_3 e^{-k_2 u} - \frac{k_1}{k_0 k_2^2} e^{k_2 u} + k_4 \tag{40}$$

shares with the quasilinear wave equation

$$u_t - k_1 e^{k_2 u} u_x = k_3 e^{-k_2 u} - \frac{k_1}{k_0 k_2^2} e^{k_2 u} + k_4 \tag{41}$$

the class of exact solutions defined by

$$e^{k_2 u} = \phi(t) + \psi(t) e^{\bar{k}_0 x} \tag{42}$$

where the functions $\phi(t)$ and $\psi(t)$ must satisfy the system of ODEs:

$$\psi' = (\bar{k}_4 - \bar{k}_0 \phi)\psi \tag{43}$$

$$\phi' = -\bar{k}_0 \phi^2 + \bar{k}_3 + \bar{k}_4 \phi \tag{44}$$

with $\bar{k}_0 = \frac{1}{k_0 k_1 k_2}$, $\bar{k}_3 = k_2 k_3$, $\bar{k}_4 = \frac{k_2 k_4}{k_1}$.

According to different choices of the material parameters involved in the model constitutive laws (37)–(39), several forms of solutions to (40) arise from the integration of (43) and (44).

i) If $\bar{k}_3 + \frac{\bar{k}_4^2}{4\bar{k}_0} = 0$, then

$$\phi = \frac{\bar{k}_4}{2\bar{k}_0} + \frac{1}{\bar{k}_0 (t - t_0)} \tag{45}$$

$$\psi = \frac{k_5}{t - t_0} e^{\frac{\bar{k}_4}{2} t} \tag{46}$$

ii) If $\frac{1}{\bar{k}_0 \left(\bar{k}_3 + \frac{\bar{k}_4^2}{4\bar{k}_0}\right)} > 0$, then

$$\phi = \frac{\bar{k}_4}{2\bar{k}_0} + \bar{h} \tanh(\bar{h} \bar{k}_0 (t - t_0)) \tag{47}$$

$$\psi = \frac{k_5 e^{\frac{\bar{k}_4}{2}t}}{\cosh(\bar{h}\bar{k}_0(t-t_0))} \tag{48}$$

where $\bar{h} = \sqrt{\frac{1}{\bar{k}_0\left(\bar{k}_3+\frac{\bar{k}_4^2}{4\bar{k}_0}\right)}}$.

iii) If $\frac{1}{\bar{k}_0\left(\bar{k}_3+\frac{\bar{k}_4^2}{4\bar{k}_0}\right)} < 0$, then

$$\phi = \frac{\bar{k}_4}{2\bar{k}_0} + \bar{h}\tan(-\bar{h}\bar{k}_0(t-t_0)) \tag{49}$$

$$\psi = \frac{k_5 e^{\frac{\bar{k}_4}{2}t}}{\cos(\bar{h}\bar{k}_0(t-t_0))} \tag{50}$$

where $\bar{h} = \sqrt{-\frac{1}{\bar{k}_0\left(\bar{k}_3+\frac{\bar{k}_4^2}{4\bar{k}_0}\right)}}$.

In the relations (45)–(50) t_0 represents an arbitrary constant.

4 Exact solutions to wave–hierarchies

The leading idea of the analysis worked out in section 1 for the higher order model (1) was to assume a generalized multiple wave–like ansatz for the field variables which were associated to the corresponding first order system (2)–(4). That assumption permitted to select a reduced set of field equations and led to considering a hyperbolic subsystem of field equations which is endowed by a set of differential constraints.

Along the guide lines of the approach in point, here we aim at outlining a reduction method for determining exact wave–like solutions to quasilinear systems of first order PDEs of the form

$$\frac{\partial U}{\partial t} + A(U,V)\frac{\partial U}{\partial x} + B(U,V)\frac{\partial V}{\partial x} = C(U,V) \tag{51}$$

$$\frac{\partial V}{\partial t} + G(U,V)\frac{\partial U}{\partial x} + L(U,V)\frac{\partial V}{\partial x} = P(U,V) \tag{52}$$

with the vector functions $U \in \Re^N$, $V \in \Re^M$, while $A(U,V)$, $B(U,V)$, $C(U,V)$, $G(U,V)$, $L(U,V)$, $P(U,V)$, are matrix coefficients of suitable dimensions.

Actually these models involving source–like terms play a dominant role in dissipative theories allowing for finite speeds to propagating disturbances. Usually they arise from hyperbolic dissipative theories as those concerning extended thermodynamics (Müller & Ruggeri)[24] or active media

(Engelbrecht)[25]. In most cases their structure is consistent with different levels of description of the wave processes so that "wave hierarchy" problems are to be considered (Whitham)[2], (Fusco)[1].

We search for exact solutions to (51) (52) obeying the finite relationship

$$F(U, V, x, t) = 0 \tag{53}$$

with the vector function $F(U, V, x, t) \in \Re^M$ satisfies

$$\left| \frac{\partial F}{\partial V} \right| \neq 0 \tag{54}$$

and it is to be determined in the subsequent process.

Before proceeding further two remarks are in order about the assumption (53).

i) The ansatz (53) generalizes that one of the classical multiple wave solution given by $V = V(U)$.

ii) If the system (51) and (52) is hyperbolic and is also consistent with wave hierarchies then the so called lower order approximation (equilibrium situation) is given by the equation (51) endowed by

$$P(U, V) = 0 \tag{55}$$

which is tantamount to assuming $V = V(U)$ (i.e. a multiple wave solution) whereas the full system (51), (52) represents the higher order description of the wave process. It is of relevant interest to highlight the influence of the higher order terms involved in the left hand side of the equation (52) on the wave motion ruled by the lower order system. In the nonlinear case such an analysis is usually carried on by means of asymptotic or iterative methods (Whitham)[2], (Fusco)[1].

Therefore, within the framework of the wave hierarchies, the exact solutions to the full system (51) and (52) defined by (53) would be able to describe a wave process when higher order effects included in the left hand side of the equation (52) are not negligible.

By virtue of the ansatz (53), and in line with the assumption (54), the system of equations (51) and (52) can be recast into the form

$$\frac{\partial U}{\partial t} + \bar{A}(U, V, x, t)\frac{\partial U}{\partial x} = \bar{C}(U, V, x, t) \tag{56}$$

$$H(U, V, x, t)\frac{\partial U}{\partial x} = \bar{P}(U, V, x, t) \tag{57}$$

where $V = V(U, x, t)$ is defined by (53) and

$$\bar{A}(U, V, x, t) = A(U, V) - B(U, V)Q(U, V, x, t)\frac{\partial F}{\partial U} \tag{58}$$

$$\bar{C} = C(U,V) + B(U,V)Q(U,V,x,t)\frac{\partial F}{\partial x} \tag{59}$$

$$H(U,V,x,t) = \frac{\partial F}{\partial V}G - \frac{\partial F}{\partial U} - \left(\frac{\partial F}{\partial V}L\right)\left(Q\frac{\partial F}{\partial U}\right) \tag{60}$$

$$\bar{P}(U,V,x,t) = \frac{\partial F}{\partial V}P - \frac{\partial F}{\partial t} - \left(\frac{\partial F}{\partial V}L\right)\left(Q\frac{\partial F}{\partial x}\right) \tag{61}$$

Moreover in (58)–(61) $Q(U,V,x,t)$ denotes the inverse of the matrix $||\partial F/\partial V||$.

The "subsystem" of N equations (56) represents the governing model for the "reduced field" $U(x,t)$ whereas the M equations (57) in fact plays the role of a system of differential constraints appended to (56).

In order to characterize exact solutions to (51) and (52) of interest in wave problems as those arising from wave hierarchies connected to hyperbolic dissipative theories we assume the reduced system (56) to be hyperbolic so that the consistency of the overdetermined system (56), (57) can be studied along the lines of the approach developed by Fusco & Manganaro[13]. Of course the latter assumption gives rise to a restriction on the function $F(U,V,x,t)$ which is involved in the relation (53).

Acknowledgments

This work was partially supported by Fondi MURST Progetto di Cofinanziamento 2000 "Problemi Matematici Non Lineari di Propagazione e Stabilità nei Modelli del Continuo" and by Programma di Ricerca di Ateneo "Tecniche di Riduzione e Metodi Qualitativi Non Lineari di Evoluzione della Fisica Matematica" 2000.

References

1. D. Fusco, *Meccanica* 17, 128 (1982).
2. G. B. Whitham, *Linear and Nonlinear Waves*, Wiley–Interscience, New–York (1974).
3. A. Jefrrey & T. Kawahara, *Asymptotic Methods in Nonlinear Wave Theory*, Pitman, London (1982).
4. N. Ibragimov ed., *CRC Handbook of Lie Group Analysis of Differential Equations*, vol. 1, CRC press, Boca Raton (1996).
5. N. Ibragimov ed., *CRC Handbook of Lie Group Analysis of Differential Equations*, vol. 2, CRC press, Boca Raton (1996).
6. N. Ibragimov ed., *CRC Handbook of Lie Group Analysis of Differential Equations*, vol. 3, CRC press, Boca Raton (1996).

7. L. V. Ovsiannikov, *Group Analysis of Differential Equations*, Nauka, Moscow (1978).
8. S. V. Meleshko, *Prokl. Matem. and Mech.* 49, 3, 406 (1990).
9. S. V. Meleshko, *Differential Equations* 30, 10, 1690 (1994).
10. N. N. Janenko, in *Proceeding of the Fourth All–Union Mathematics Congress*, Nauka, Leningrad (1964).
11. A. F. Sidorov, V. P. Shapeev & N. N. Janenko, *The Method of Differential Constraints and Its Applications in Gas Dynamics*, Nauka, Novosibirsk (1984).
12. B. L. Rozhdestvenskii & N. N. Janenko, *Systems of Quasilinear Equations and Their Applications to Gas Dynamics*, Nauka, Moscow (1978).
13. D. Fusco & N. Manganaro, *IMA J. Appl. Math.* 57, 223 (1996).
14. G. Boillat, *Det. Kgl. Norske Vid. Selsk. Forth.* 41, 78 (1968).
15. G. Boillat, *La Propagation des Ondes*, Gauthier–Villars, Paris (1965).
16. T. Ruggeri, *Rend. Acc. Naz. Lincei* 55, 445 (1973).
17. G. Boillat, *C. R. Acad. Sci. Paris Ser. I Math.* 313, 805 (1991).
18. G. Boillat, *C. R. Acad. Sci. Paris Ser. I Math.* 315, 1211 (1992).
19. A. Donato, U. Ramgulam & C. Rogers, *Meccanica* 27, 257 (1992).
20. F. Oliveri, *J. Math. Anal. Appl.* 218, 329 (1998).
21. O. I. Mokhov & Y Nutku, *Letters in Math. Phis.* 32, 121 (1994).
22. G. W. Bluman & J. D. Cole, *Similarity Methods for Differential Equations*, Springer, Berlin (1974).
23. D. G. Arrigo & J. M. Hill, *Studies Appl. Math.* 94, 21 (1995)
24. I. Müller & T. Ruggeri, *Rational Extended Thermodynamics*, Springer Tracts in Natural Philosophy, 37, Spinger–Verlag, New–York (1998).
25. J. Engelbrecht ed., *Nonlinear Waves in Active Media*, Springer-Verlag (1989).

STABILITY RESULTS FOR PENETRATIVE CONVECTION IN POROUS MEDIA FOR FLUIDS WITH CUBIC DENSITY

M. GENTILE, S. RIONERO

Dipartimento di Matematica e Applicazioni, Università di Napoli "Federico II", Complesso Universitario Monte S. Angelo - Via Cintia, 80126 Napoli - ITALY

E-mail: m.gentile@unina.it; rionero@unina.it

The penetrative convection in a fluid-saturated porous medium with a cubic equation of state for the density is considered, according to the Darcy model. Here, the stability analysis of the conduction solution by using normal modes analysis and the Lyapunov direct method is performed.

1 Introduction

Let Σ be a horizontal fluid saturated porous layer, bounded from below by a plane Π_1 and from above by a plane Π_2. Π_1 is kept at constant temperature T_1, while Π_2 at temperature $T_2 > T_1$. Suppose, moreover, that the fluid has a density depending nonlinearly on the temperature (non-Boussinesq approximation), with a maximum in the open interval (T_1, T_2). In such a way the maximum is attained over an intermediate plane Π_3. Therefore, owing to the thermal expansion, the fluid near Π_1 will be lighter than that near Π_3. Because of this, there will be a tendency on the part of the fluid to redistribute itself and remedy the weakness in its arrangement. So, small perturbations give rise to convective motions, where the fluid particles near Π_3 go downward and the ones near Π_1 go upward, beyond Π_3, penetrating into the stable upper layer. This phenomenon is known as penetrative convection [6]. The aim of this paper is to analyze the above problem in the setting of porous media [4]. In such a case the equations of motion in the layer differ from those of Oberbeck-Boussinesq-Darcy only in the form of dependence of density on temperature in the body force term. In the following we will consider the equation of state:

$$\rho = \rho_0 \left[1 + \varepsilon_1\alpha_1(T - T_0) + \varepsilon_2\alpha_2(T - T_0)^2 + \varepsilon_3\alpha_3(T - T_0)^3\right], \tag{1}$$

$$\rho_0 = const\,, \quad \varepsilon_i = \pm 1\,, \quad \alpha_i > 0 \quad (i = 1, 2, 3). \tag{2}$$

The scheme of the paper is as follows. In Section 2, we introduce the model and the perturbation equations to the nonconvective stationary solution. Then, Section 3, we perform a linear stability analysis showing that the strong version of the principle of exchange of stabilities holds true. In Section 4, we study the nonlinear stability of motionless state by using a weighted energy technique.

By means of a variational method, we find conditions assuring global nonlinear stability.

2 Statement of the problem

Let us consider a homogeneous fluid contained in a horizontal infinite layer $\{(x,y,z) \in \mathbf{R}^3 : 0 \le z \le 1\}$ of a porous medium, under the action of a vertical gravity field $\mathbf{g} = -g\mathbf{k}$, with assigned temperatures $T(0) = T_1$ and $T(d) = T_2$, where $T_1 < T_2$. Furthermore we suppose that the density of the fluid, given by (1), has a maximum in (T_1, T_2). The Darcy equations are:

$$\begin{aligned} &\nabla p = -\frac{\mu}{k}\mathbf{v} - g\rho\mathbf{k} \\ &\nabla \cdot \mathbf{v} = 0 \\ &(\rho_0 c_0)_m T_t + (\rho_0 c_0)_f \mathbf{v} \cdot \nabla T = \kappa_m \Delta T, \end{aligned} \tag{3}$$

with the constitutive law for the density of the fluid given by (1) and where: z is the upward vertical, $\mathbf{v}$ is the velocity field, p is the pressure, μ is the constant dynamic viscosity, g is the gravity, k is the permeability, κ_m is the overall thermal conductivity. Here we study the stability of nonconvecting stationary solution $m_0(\bar{\mathbf{v}} = \mathbf{0}, \bar{T} = \beta z + T_1, p)$, with $\beta = (T_2 - T_1)/d$. Let us denote by $\mathbf{u} = (u, v, w)$, θ, π respectively the perturbation to the velocity, the temperature and the pressure fields, and introduce the dimensionless quantities:

$$\begin{aligned} &t = t^*\frac{d^2}{\kappa}, \ \mathbf{x} = \mathbf{x}^* d, \ \mathbf{u} = \mathbf{u}^*\frac{\kappa}{d}, \ \pi = \pi^*\frac{\mu\kappa}{k}, \ \theta = \theta^* T^\sharp, \ \kappa = \frac{\kappa_m}{(\rho_0 c_0)_f}, \\ &A = \frac{(\rho_0 c_0)_m}{(\rho_0 c_0)_f}, \ \xi = \frac{T_1 - T_0}{T_2 - T_1}, \ T^\sharp = \frac{\kappa}{d}\sqrt{\frac{\beta\mu d^2}{kg\alpha\kappa\rho_0}}, \ R = \sqrt{\frac{\alpha_1 g\beta d^2 k\rho_0}{\kappa\mu}}, \end{aligned} \tag{4}$$

and where R^2 is the Rayleigh number. Dropping asterisks, the equations governing the perturbation in the strip $\mathbf{R}^2 \times [0,1]$ become:

$$\begin{aligned} &\nabla \pi = -\mathbf{u} - R\,h_2(z)\theta\mathbf{k} - h_1(z)\theta^2\mathbf{k} - \frac{\varepsilon_3 a_2}{R}\theta^3\mathbf{k} \\ &\nabla \cdot \mathbf{u} = 0 \\ &A\,\theta_t = \Delta\theta - Rw - \mathbf{u} \cdot \nabla\theta, \end{aligned} \tag{5}$$

where, setting $a_1 = (\alpha_2 \Delta T)/\alpha_1,\ a_2 = (\alpha_3(\Delta T)^2)/\alpha_1$:

$$
\begin{aligned}
h_1(z) &= \varepsilon_2 a_1 + 3\varepsilon_3 a_2(\xi + z), \\
h_2(z) &= \varepsilon_1 + 2\varepsilon_2 a_1(\xi + z) + 3\varepsilon_3 a_2(\xi + z)^2,
\end{aligned}
\tag{6}
$$

To the previous system we append the following boundary conditions:

$$
w = \theta = 0 \quad \text{on } z = 0,\, 1. \tag{7}
$$

In the sequel we assume, as usual, that the perturbation fields $\mathbf{u}$, θ, π are sufficiently smooth, that they are periodic functions of x and y, of periods $2\pi/a_1$, $2\pi/a_2$. We shall denote by $\Omega = [0,\, 2\pi/a_1] \times [0,\, 2\pi/a_2] \times [0,\, 1]$ the periodicity cell, and $a = (a_1^2 + a_2^2)^{\frac{1}{2}}$ the wave number, by $< \cdot >$ and $\|\cdot\|$ respectively the integral and the L^2-norm on Ω. Finally, taking into account the fact that the stability of m_0 makes sense only in a class of solutions of (5) in which m_0 is unique, we exclude any other rigid solution requiring $< u > = < v > = 0$.

3 Linear stability analysis: Principle of Exchange of Stabilities

First of all, we point out that the linear part $\mathcal{L}$ of the operator governing the problem (5) is not symmetric with respect to the usual scalar product in the Lebesgue space L^2. Nevertheless we will prove that the all the eigenvalues of $\mathcal{L}$ are real.

Theorem 1 Let us consider the linear system

$$
\begin{aligned}
&\nabla \pi = -\mathbf{u} - R\,H(z)\theta \mathbf{k} \\
&\nabla \cdot \mathbf{u} = 0 \\
&A\,\theta_t = \Delta\theta - Rw\,,
\end{aligned}
\tag{8}
$$

where $H(z)$ is a real function. Then *the Strong Principle of Exchange of Stabilities* holds.

Proof. By taking the double curl of $(8)_1$ and projecting it on the z axis, we obtain the following system:

$$
\begin{aligned}
&\Delta w + R\,H(z)\Delta_1 \theta = 0 \\
&A\theta_t = \Delta\theta - Rw\,,
\end{aligned}
\tag{9}
$$

where $\Delta_1 = \partial_{xx} + \partial_{yy}$. Next, consider perturbations of the type:

$$\mathbf{u}(\mathbf{x},t) = \mathbf{u}(\mathbf{x})e^{\sigma t} \quad \theta(\mathbf{x},t) = \theta(\mathbf{x})e^{\sigma t}, \tag{10}$$

with σ a priori complex number. Substituting in (9), multiplying by θ^* (the complex conjugate of θ) and integrating by parts over Ω, we find out:

$$\sigma A \parallel \nabla\theta \parallel^2 = R^2 < H(z)|\nabla_1\theta|^2 > - \parallel \Delta\theta \parallel^2, \tag{11}$$

where $\nabla_1 = \partial_x \mathbf{i} + \partial_y \mathbf{j}$. From (11) it follows that $Im(\sigma) = 0$, and hence all the eigenvalues are real numbers.

In order to study the linear stability [1] of m_0, we observe that by virtue of Theorem 1, when $H = h_2$, the critical perturbations verify the linear stationary equations:

$$\begin{aligned} &\Delta w + R\,h_2(z)\Delta_1\theta = 0 \\ &\Delta\theta - Rw = 0 \end{aligned} \tag{12}$$

In the case of the water a good approximation for the density, in the interval $[0°C, 40°C]$ is given by (1) [5], where $T_0 = 0$, ρ_0 is the density at $0°C$ and:

$$\begin{aligned} &\alpha_1 = 6.8565 \times 10^{-5}(°C^{-1}), \quad \alpha_2 = 8.8206 \times 10^{-6}(°C^{-2}), \\ &\alpha_3 = 4.1666 \times 10^{-8}(°C^{-3}), \quad \varepsilon_1 = 1, \quad \varepsilon_2 = -1, \quad \varepsilon_3 = 1. \end{aligned} \tag{13}$$

Therefore, substituting normal modes perturbations

$$w = W(z)\exp[i(a_1x + a_2y)], \quad \theta = \Theta(z)\exp[i(a_1x + a_2y)], \tag{14}$$

in (12), by using compound matrix method and the golden section search [5], we found the numerical values listed in Table 1.

4 Global nonlinear stability

The aim of this section, is to obtain a nonlinear global stability result, about the motionless state m_0, by using the Lyapunov direct method. To this end, let us introduce the weighted energy functional:

$$V = A\left[\frac{\lambda}{2} < \theta^2 > + \frac{1}{3\,R} < g(z)\,\theta^3 > + \frac{\mu}{4\,R^2} < \theta^4 >\right], \tag{15}$$

where λ and μ are positive coupling parameters to be selected, while $g(z)$ is a linear function in z to be chosen suitably later. Of course, λ, μ and $g(z)$ have to be chosen in such a way that $V(t)$ is positive definite. From (5), taking into account the boundary conditions (7), on choosing $g(z) = 3\,(\mu + \varepsilon_3 a_2)\; z\,,$ it follows that:

$$\dot{V} = R\mathcal{I} - \mathcal{D}\,, \tag{16}$$

where

$$\begin{aligned} \mathcal{I} &= - < f_2(z)\theta w > - < f_1(z)\psi w >, \\ \mathcal{D} &= \| \mathbf{u} \|^2 + \lambda \| \nabla\theta \|^2 + < g(z)\nabla\theta \cdot \nabla\psi > + \tfrac{3}{4}\mu \| \nabla\psi \|^2 > . \end{aligned} \tag{17}$$

with $\psi = \theta^2/R\,,\ f_1(z) = g(z) + h_1(z)\,,\ f_2(z) = \lambda + h_2(z)\,.$
Now for any λ such that:

$$\lambda > \lambda^* = \begin{cases} 0 & \text{if } \varepsilon_3 = -1 \\ 12\,a_2 & \text{if } \varepsilon_3 = 1, \end{cases} \tag{18}$$

and choosing $\mu = (\lambda - 6\varepsilon_3 a_2)/6 (> 0)\,,$ it is easily seen that: $V \geq 0\,,\ \mathcal{D} \geq 0.$ By following the standard energy method [2 5], we define:

$$\frac{1}{R_w} = \max_{\mathcal{H}} \frac{\mathcal{I}}{\mathcal{D}}\,, \tag{19}$$

where $\mathcal{H}$ is the space of admissible perturbations $(\mathbf{u}, \theta, \psi):$ i) regular in Ω, ii) periodic in x and y, iii) $\nabla \cdot \mathbf{u} = 0$, iv) $\mathcal{D} < \infty\,,$ v) $\theta^2 = R\psi$. By means of Poincaré inequality, there exists a positive constant h such that $h\,V \leq \mathcal{D}$, then from (16) and (19) it follows:

Theorem 2 If $R < R_w$ then the basic motion is asymptotically exponentially stable with respect to the weighted norm (15), and for any time t the following inequality holds:

$$V(t) \leq V(0) \exp\left[-h \frac{R_w - R}{R_w} t \right]\,. \tag{20}$$

Next, consider the following variational problem:

$$\frac{1}{R^*} = \max_{\mathcal{H}^*} \frac{\mathcal{I}}{\mathcal{D}}\,, \tag{21}$$

where $\mathcal{H}^*$ is defined by i),..., iv) of the previous section, and hence $\mathcal{H} \subset \mathcal{H}^*$. Comparing (19) and (21) we can conclude that $R^* \leq R_w$. So, if we assume $R < R^*$ then the Theorem 2 holds. On considering normal modes perturbations and by using the compound matrix method and the golden section search [5], in the case of the water we found the numerical values listed in table 1.

Finally, we point out that, by using a nonweighted L^2-energy, critical Rayleigh numbers R_E^2 of nonlinear stability closer to R_L^2 can be obtained (table1). But, they only guarantee the nonlinear stability under restrictions on the initial data which go to zero when R^2 tends to R_E^2. Further details can be found in [3].

Table 1: Critical Rayleigh numbers in the case of water

ΔT	R_L^2	R^{*2}	R_E^2
1	45.26	43.41	45.26
2	52.85	45.40	52.73
3	63.23	45.48	62.81
4	78.02	44.36	76.71
5	100.11	42.63	96.34

Aknowledgements

This work has been performed under the auspices of the G.N.F.M. of I.N.D.A.M. and M.U.R.S.T. (P.R.I.N.): "Nonlinear mathematical problems of wave propagation and stability in continuous media".

References

1. Chandrasekhar, S. *Hydrodynamic and hydromagnetic stability*. S. New York: Dover 1961.
2. Flavin, J., Rionero,S. *Qualitative estimates for partial differential equations*. Boca Raton, Florida: CRC Press 1996.
3. Gentile, M., Rionero, S. *Rend. Acc. Sc. fi. mat.*, vol. LXVII, 129-142 (2000).
4. Nield, D.A., Bejan, A. *Convection in porous media*. Berlin Heidelberg New York: Springer-Verlag, 1992.
5. Straughan, B.*The energy method, stability, and nonlinear convection*. Berlin Heidelberg New York Tokyo: Springer 1992.
6. Veronis, G. *Astrophys. J.* **137**, 641-663(1963).

DYNAMICS OF LINES IN THE SPREADING OF LIQUIDS ON SOLID SURFACES

H. GOUIN

Laboratoire de Modélisation en Mécanique et Thermodynamique, EA 2596
Université d'Aix-Marseille, 13397 Marseille Cedex 20, France
E-mail: henri.gouin@univ.u-3mrs.fr

The spreading of fluids on solid surfaces constitutes a significant field of research into the processes met in nature, biology and modern industry. It is noteworthy that since the publication of Young on capillarity, the understanding of this phenomenom has remained incomplete. The equation of the contact-line motion is the key of our study. An example of non-Newtonian fluid motions exhibits a flat liquid interface moving steadily over a flat solid in creeping flow approximation and theory of capillarity. The kinematics of liquids at the contact-line and equations of motion are revisited. Adherence conditions are required except at the contact-line. The velocity field appears to be realistic but at some instant a liquid material point may leave the surface. The velocity field discontinuous at the contact-line generates a new concept of line viscosity but stresses and viscous dissipation remain bounded. A Young-Dupré equation for the dynamic contact angle between the interface and the solid is proposed.

1 A creeping flow example of non-Newtonian liquid

The notations are those of ordinary Cartesian vector analysis. In a fixed coordinate system, the components of vector (covector) **a** are denoted by a^i, (a_i), where $i = 1, 2, 3$. In order to describe the fluid motion analytically, we refer to the coordinates $\mathbf{x} \equiv (x^1, x^2, x^3)$ as the particle position (Eulerian variables). Reference positions are the coordinates $\mathbf{X} \equiv (X^1, X^2, X^3)$ (Lagrangian variables).
The motion of a fluid is classicaly represented by the transformation $\mathbf{x} = \varphi(t, \mathbf{X})$ or $x^i = \varphi^i(t, \mathbf{X})$. It is assumed that φ possesses an inverse $\mathbf{X} = \Phi(t, \mathbf{x})$ and continuous derivatives up to the second order except for space variables at certain surfaces and curves[1].
The whole domain occupied by the fluid in Lagrangian variables is D_o. The boundary of D_o is a surface Σ_o. The fluid occupies a volume D_t in Eulerian variables with a boundary Σ_t corresponding to the fixed regions D_o, Σ_o in the **X**-variables. A moving curve Γ_t on Σ_t corresponds to a moving curve Γ_{ot} on Σ_o. The domains $D_o, D_t, \Sigma_o, \Sigma_t, \Gamma_{ot}, \Gamma_t$ are differentiable oriented manifolds. To each point of Σ_t corresponds an external unit normal vector $\mathbf{n}$ (n^i) and a mean radius of curvature R_m. We note **Id** the identity tensor with components δ_i^j. Then $\mathbf{Id} - \mathbf{n} \otimes \mathbf{n}$ (components $\delta_i^j - n^j n_i$) is the projection operator on

the tangent plane to the surface Σ_t; $\mathbf{t}$ denotes the unit tangent vector to Γ_t oriented following the external normal vector $\mathbf{n}$ to Σ_t; $\mathbf{n}' = \mathbf{n} \times \mathbf{t}$ is the binormal vector to Γ_t with respect to Σ_t.

To understand the behaviour of a liquid near a moving solid/liquid/fluid contact line, we reconsider the simple situation proposed by Huh & Scriven[2]. Let us recall the main results of their paper:
A flat solid surface translating with a steady velocity $\mathbf{U} = U\ \mathbf{i}$ is inclined at the angle θ from a flat interface between a liquid A and a fluid B (the contact-line celerity with respect to the solid is $-U\ \mathbf{i}$). In a two-dimensional situation (plane $0\ \mathbf{i}\ \mathbf{j}$), it is convenient to take the contact-line as the origin of a polar coordinate system (r, φ). The two bulks are incompressible Newtonian fluids. In term of the stream function $\psi(r, \varphi)$ for two-dimensional steady flows the velocity components are $v_r = -r^{-1}\partial\psi/\partial\varphi$ and $v_\theta = \partial\psi/\partial r$. In the creeping flow approximation[3] the linearized Navier-Stokes equation leads to the biharmonic equation $\nabla^4\psi = 0$. A solution is

$$\psi(r, \varphi) = r(a \sin \varphi + b \cos \varphi + c\ \varphi \sin \varphi + d\ \varphi \cos \varphi) \tag{1}$$

The boundary conditions at the solid wall and the fluid interface are:
(i) vanishing normal component of the velocity at the solid surface and interface (ii) continuity of velocity at the interface (iii) continuity of tangential stress at the interface (iv) non tangential relative motion of fluid at the solid surface[2].
These eight linear conditions lead to the values of coefficients a, b, c, d for the two fluids A and B [4]: $a_A = UD(\theta)[\ (\pi - \theta)(\sin\theta \cos\theta - \theta) + \pi\ \theta \sin^2\theta\ ]$, $a_B = -U\theta D(\theta)[\ \pi - \theta + \sin\theta \cos\theta\ ]$, $b_A = UD(\theta)[\ \pi\ \theta \sin\theta \cos\theta - \pi \sin^2\theta\ ]$, $b_B = 0$, $c_A = -UD(\theta)\ \theta \sin^2\theta$, $c_B = UD(\theta)\ (\pi - \theta) \sin^2\theta$, $d_A = UD(\theta)[\ -\theta \sin\theta \cos\theta + \sin^2\theta\ ]$, $d_B = UD(\theta)[\ (\pi - \theta) \sin\theta \cos\theta + \sin^2\theta\ ]$, with $D(\theta) = [\ \theta\ (\pi - \theta) - (\pi - 2\theta) \sin\theta \cos\theta - \sin^2\theta\ ]^{-1}$.
The form of the streamlines issued from Huh & Scriven paper is schematized on figure 1. If $\mu_A = \mu_B = \mu$ (we assume that the viscosity coefficients are identical but no difficulty should arise if they are different), for all the values of the dynamic contact angle θ, the viscous stress components are $\tau_{r\theta} = 2\mu\ r^{-1}(c \cos\varphi - d \sin\varphi)$, $\tau_{rr} = \tau_{\theta\theta} = 0$ and the pressure field is $p = p_0 - 2\mu\ r^{-1}(c \sin\varphi + d \cos\varphi)$ where p_0 is the hydrostatic pressure (in both formulas phase subscripts have been omitted).
As proved by Huh and Scriven, the dissipative function and the total force exerted on the solid surface are logarithmically infinite. Moreover the normal stress across the fluid interface varies as r^{-1}; the stress jump should be balanced by the Laplace interfacial tension $2\sigma R_m^{-1}$ and the curvature R_m^{-1} should increase infinitely at the contact-line (σ is the surface tension).

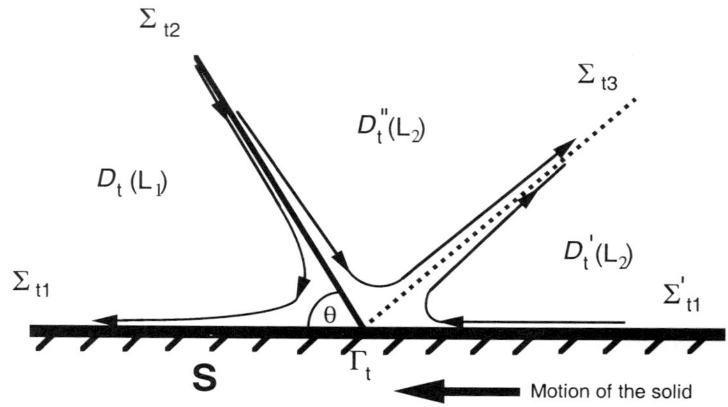

Figure 1. Typical motion of fluids in contact on solid surface with stationary contact-line.

To avoid these physical paradoxes, we consider non-Newtonian fluids with a convenient viscous stress tensor behaviour. Moreover, the adherence conditions on solid are required.

The viscous stress tensor Q_v is assumed to be a function of the velocity stress tensor Δ. For moderate values of Δ the fluid is Newtonian and the function is linear. The function deviates from this classical behaviour for large values of Δ. For an isotropic stress tensor of *two-dimensional* flow, the Rivlin-Ericksen representation theorem[5] leads to a viscous stress tensor $Q_v = \lambda\, \mathbf{I} + 2\mu\, \Delta$ but λ and μ are non-constant functions of invariants of Δ and $\mu\Delta$ is a functional of Δ equivalent to $\mu_0\Delta$ (μ_0 constant) when Δ tends to zero.

In wide variety of technological applications, liquids are subjected to large shear strain forces. A molecular dynamics investigation of liquids subjected to large shear strain rates is done in[6]. The shearing action has been found to change the liquid structure and reveals a tendency to reduce shear viscosity. In[7], a viscosity behaving closely like an exponential function for large shear strain rates is obtained. If we consider a *viscosity* μ in the form

$$\mu = (1 - e^{-\gamma})\, \mu_0 \quad \text{with} \quad \gamma = \frac{1}{\varepsilon}\left(\frac{-1}{\varepsilon\, \tau\, \det\Delta}\right)^{\frac{\varepsilon}{2}} \tag{2}$$

where τ is a characteristic time of the fluid and $0 < \varepsilon \ll 1$, then for convenient values of τ and ε the μ-curve as a function of $\det\Delta$ fits with the curve of fig. 2 in[7]; μ is equal to μ_0 except for large values of Δ (as near the contact-line).

Consequently, Q_v is a function of Δ which tends to infinity with Δ but weaker than a linear function.
In the incompressible case, the total stress tensor is $Q = -P\,\mathbf{I} + 2\mu\,\Delta$ with $P = p - \lambda$, (p is the hydrostatic pressure). The steady flow equation is

$$\rho_0 \text{ grad } (\frac{1}{2}\mathbf{V}^2 + \frac{p}{\rho_0}) + \rho_0 \text{ rot } \mathbf{V} \times \mathbf{V} = \mu\nabla^2\mathbf{V} + 2\,\Delta \text{ grad } \mu$$

Consequently,

$$\text{rot } (\mu \text{ rot } (\text{rot } \mathbf{V})) - 2 \text{ rot } (\Delta \text{ grad}\,\mu) + \rho_0 \text{ rot } (\text{rot } \mathbf{V} \times \mathbf{V}) = 0 \qquad (3)$$

For a stream function $\psi(r,\varphi) \equiv r f(\varphi)$ given by expression (1), we get

$$\text{rot } (\mu \text{ rot rot } \mathbf{V}) - 2 \text{ rot } (\Delta \text{ grad}\,\mu) + \rho_0 \text{ rot } (\text{rot } \mathbf{V} \times \mathbf{V}) =$$

$$-\varepsilon\,\mu_0\,r^{-3}\,\gamma\,e^{-\gamma}\left(f + f" - \frac{(f' + f^{(3)})^2}{(f + f")}\right)(1 - \varepsilon\,(1-\gamma))\mathbf{k}$$

$$+\left(\rho_0\,r^{-2}\frac{d}{d\varphi}(f^2 + f f") - \mu\,r^{-3}\,(f + 2f" + f^{(4)})\right)\mathbf{k}$$

where $\mathbf{k}$ is the normal vector to the plane of flows.
It is easy to prove that the main term of Eq. (3) is $-\,\mu\,r^{-3}\,(f + 2f" + f^{(4)})\,\mathbf{k}$ and in creeping flow approximation the trajectories and the velocities are those of a fluid with constant viscosity. By using expression (2), straightforward calculations prove that the total force exerted on the solid surface and the dissipative function near the contact-line are finite. Through the interface $p_A - p_B \equiv -2\mu\,r^{-1}\,UD(\theta)\pi \sin\theta = 2\sigma R_m^{-1}$ and the curvature increase when the contact line is approached. However, following the arguments in Huh & Scriven, R_m would be about 10^5 times greater than the distance to the contact-line when water at a moderate dynamic contact angle wets a surface at 6 mm/min and in our case, variations of θ are bounded and small enough; consequently the flat interface near the solid wall is a good approximation.
The velocity of a fluid is discontinuous on the contact-line. Let us note $[\mathbf{V}] = \mathbf{V}_2 - \mathbf{V}_1$ where $\mathbf{V}_1$ and $\mathbf{V}_2$ are respectively velocities on the fluid interface and the solid,

$$[\mathbf{V}] = -U\,(1 + D(\theta)\,(\pi - 2\theta)\,\sin\theta\,\cos\theta)\,\sin^2\theta\,\mathbf{i}$$

$$+\,U\left(\cos\theta - D(\theta)\,(\pi - 2\theta)\,\sin^3\theta\right)\sin\theta\,\mathbf{j}$$

and the variations of $g(\theta) = [\mathbf{V}]\,\mathbf{i}$ belong to the interval $[-1.5\,U, -U]$.

2 Kinematics of a liquid at a contact-line

Following the example of paragraph 1 and experiments by Dussan V and Davis[8,9], for contact-line motion, the usual stick-adhesive point of view of fluid adherence at a solid wall is disqualified. A liquid which does not slip on a solid surface does not preclude the possibility that at some instant a liquid material point may leave the surface. *The no-slip condition is expressed as follows: the velocity of the liquid must equal the solid velocity at the surface.*

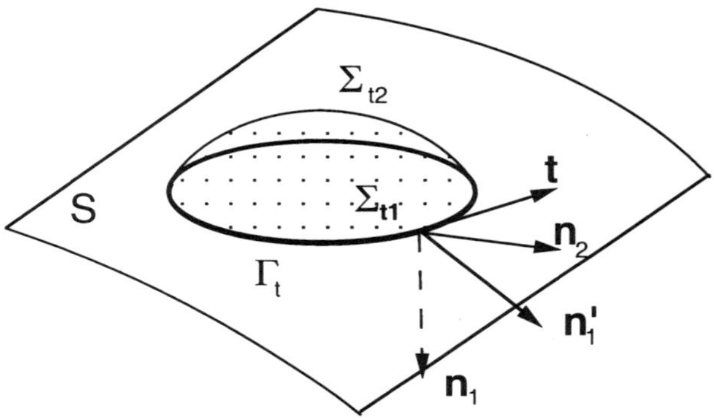

Figure 2. A liquid (in drop form) lies on a solid surface S. Σ_{t1} and Σ_{t2} are the fluid boundaries. The edge Γ_t (or contact-line) is common to Σ_{t1} and Σ_{t2}.

A liquid L_1 is in contact with a solid surface S on an imprint Σ_{t1} and a liquid L_2 along an interface Σ_{t2} (figure 2). The mobile surfaces $\Sigma_{t\alpha}$ $(\alpha = 1, 2)$ having for Cartesian equations $f_\alpha(t, \mathbf{x}) = 0$ $(\alpha = 1, 2)$ are considered in the same reference space. Equations of the common curve $\Gamma_t = \Sigma_{t1} \cap \Sigma_{t2}$ are

$$f_1(t, \mathbf{x}) = f_2(t, \mathbf{x}) = 0$$

For a geometric point M of $\Sigma_{t\alpha}$ with velocity $\mathbf{W}_\alpha$ we obtain the relation

$$\frac{\partial f_\alpha}{\partial x^i} W_\alpha^i + \frac{\partial f_\alpha}{\partial t} = 0$$

If we observe that $n_{\alpha i} = \mu_\alpha \dfrac{\partial f_\alpha}{\partial x^i}$ where μ_α is a suitable scalar, the celerity of surface $\Sigma_{t\alpha}$ has the value $c_\alpha = n_{\alpha i} W_{\alpha i} = -\mu_\alpha \dfrac{\partial f_\alpha}{\partial t}$. This celerity depends only on coordinates $(t, \mathbf{x})$ of M.

The velocity of a point of Γ_t is noted $\mathbf{W}$. The unit tangent vector to Γ_t oriented by $\mathbf{n}_2$ is $\mathbf{t}_2$ simply noted $\mathbf{t}$ ($\mathbf{t} = \mathbf{t}_2 = -\mathbf{t}_1$). The celerity of the common line is

$$\mathbf{u} = (\mathbf{Id} - \mathbf{t} \otimes \mathbf{t})\mathbf{W}$$

It depends only on coordinates $(t, \mathbf{x})$ of the point on Γ_t and is orthogonal to Γ_t. Along Γ_t, $n_{\alpha i} W^i = c_\alpha$ and

$$\mathbf{u} = \frac{(c_2\, \mathbf{n}'_1 + c_1\, \mathbf{n}'_2)}{(\mathbf{n}_1, \mathbf{n}_2, \mathbf{t})}$$

Following the example proposed in paragraph 1, the kinematics of liquids at the vicinity of the contact-line is axiomatised as follows:
Σ_{t1} is a part of the surface of solid S. Liquid (L_1) adheres to S in the sense of the no-slip condition previously proposed. Σ_{t2} is a material surface of liquid (L_1).

In a reference frame related to S,

$$\mathbf{u} = \frac{c_2\, \mathbf{n}'_1}{(\mathbf{n}_1, \mathbf{n}_2, \mathbf{t})} \equiv \frac{(n_{2i} V_2^i)\, \mathbf{n}'_1}{(\mathbf{n}_1, \mathbf{n}_2, \mathbf{t})} \equiv u\, \mathbf{n}'_1$$

where u notes the value of the contact-line celerity on $\mathbf{n}'_1$ and $\mathbf{V}_2$ the common velocity of liquids on Σ_{t2}. The contact-line Γ_t is not a material line of (L_1); its celerity is different from the velocities of the liquid on S and on Σ_{t2}.

The motion of the particles of (L_1) on S and Σ_{t2} is comparable with that of an adhesive tape stuck on a wall, the other edge of the adhesive tape being mobile (figure 1): for $u > 0$ (either $c_2 > 0$) the particles of (L_1) belonging to Σ_{t2} are driven towards Γ_t and necessarily adhere at S along Σ_{t1}; consequently on the contact-line $n_{1i} V_2^i > 0$. For $u < 0$ (either $c_2 < 0$) the result is reversed: the particles of (L_1) belonging to Σ_{t1} reach Γ_t and are driven towards Σ_{t2}; consequently on the contact-line $n_{1i} V_2^i < 0$. On figure 3, the two manifolds Σ_{t1} and Σ_{t2} constitute two sheets of the same material surface. The motion of the liquid (L_1) is represented by using a continuous mapping φ from a half reference space $D_0(L_1)$ of boundary S_0 on the actual domain $D_t(L_1)$ occupied by (L_1). The domain $D_t(L_1)$ is included in the dihedral angle formed by Σ_{t1} and Σ_{t2}. The contact-line Γ_t is the image of the mobile curve Γ_{0t} in S_0. Outside Γ_{0t}, the application φ is C^2-differentiable.

A second liquid (L_2) occupies the supplementary dihedral angle $(\Sigma'_{t1}, \Sigma_{t2})$. The conditions of motion are the opposite of those of (L_1). Material surface Σ_{t2} is the interface common to (L_1) and (L_2). The velocities of the liquids (L_1) and (L_2) are equal along Σ_{t2}. Moreover, for liquid (L_1), if $u > 0$, we

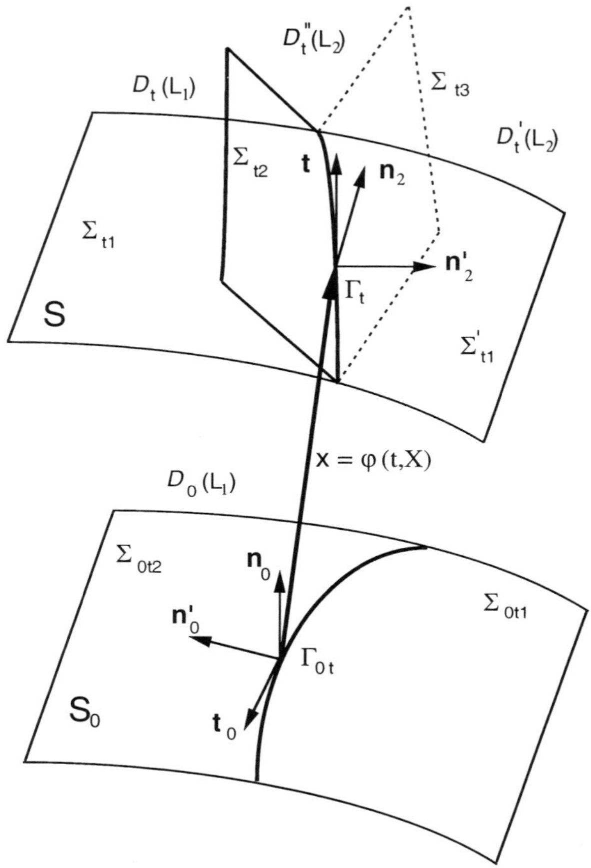

Figure 3. In Lagrangian variables, the two sheets of the same material surface of fluid (L_1) are represented by a differentiable manifold S_0. Its image in the physical space D_t is divided into two parts of a dihedral angle $(\Sigma_{t1}, \Sigma_{t3})$. The common edge Γ_t is the image of a moving curve Γ_{0t} in S_0.

obtain $n_{1i}\ V_2^i > 0$ but u is the opposite for liquid (L_2) and $n_{1i}V_2^i < 0$. To remove this contradiction, the example of paragraph 1 exhibits a material surface Σ_{t3} associated with the liquid (L_2) (see figures 1 and 3). Σ'_{t1} and Σ_{t3} are the two sheets of the same material surface for a domain $D'_t(L_2)$ of the liquid (L_2) included in the dihedral angle $(\Sigma'_{t1}, \Sigma_{t3})$. The sheets Σ_{t2} and Σ_{t3} constitute two parts of the same material surface for a domain $D_t"(L_2)$ of the liquid (L_2) included in the dihedral angle $(\Sigma_{t2}, \Sigma_{t3})$. The two domains

$D'_t(L_2)$ and $D_t"(L_2)$ with common material surface Σ_{t3} -on which velocity is continuous- constitute two independent liquid domains which do not mix. For the domain $D'_t(L_2)$, the conditions in the vicinity of the contact system are similar to those studied for liquid (L_1). Velocities which are discontinuous and multi-valued on Γ_t are compatible with the motions of liquids (L_1) and (L_2) associated to the domains $D_t(L_1)$, $D'_t(L_2)$ and $D_t"(L_2)$.

3 Equations of motion and boundary conditions revisited

The fundamental law of dynamics is expressed in the form of the Lagrange-d'Alembert principle of virtual works.
Let us recall the main sketch of virtual displacements. A varied motion is given by $\mathbf{x} = \varphi(t, \mathbf{X}, \beta)$ where β belongs to a neighbourhood of zero in the real number set and $\beta = 0$ corresponds to the medium motion. The virtual displacement corresponding to a varied motion is defined as $\boldsymbol{\zeta} \equiv \dfrac{\partial \varphi}{\partial \beta}(t, \mathbf{X}, 0)$.

All the motions are isothermal (consequently, we use free energies). In Galilean frame, the virtual work of forces applied to L_1 (including inertial forces but without forces due to capillarity) is in the form

$$\delta T_1 = \int_{D_t} (\phi_i - \rho\ a_i)\ \zeta^i + (p\ \delta_i^j - Q_i^j)\zeta^i_{,j} + \sum_{\alpha=1}^{2} \int_{\Sigma_{t\alpha}} P_{\alpha i}\ \zeta^i$$

where we denote by $\boldsymbol{\phi}$ the volumic forces, ρ the density, $\mathbf{a}$ the acceleration, Q the viscous stress tensor, p the pressure and $\mathbf{P}_\alpha$ $(\alpha = 1, 2)$ describe the actions of external stress vectors on $\Sigma_{t\alpha}$. For the sake of simplicity, in unambiguous case, the measure of volume, surfaces or line will not be indicated. The virtual work is written in a *separated form*

$$\delta T_1 = \int_{D_t} (\phi_i - \rho\ a_i - p_{,i} + Q^j_{i,j})\zeta^i + \sum_{\alpha=1}^{2} \int_{\Sigma_{t\alpha}} (P_{\alpha i} + p\ n_{\alpha i} - n_{\alpha j}\, Q_i^j)\zeta^i \quad (4)$$

where $\mathbf{n}_\alpha$ is the external unit normal vector to $\Sigma_{t\alpha}$ (figure 2).

In continuum mechanics, it is usual to define a measure of energy (denoted by $\sigma\ ds$) on the surfaces. The energy

$$E = \int_{\Sigma_t} \sigma\ ds$$

is the total surface energy on Σ_t. Fluid-fluid and fluid-solid interfaces are endowed with surface free-energies (in fact only the difference between the

surface energy of liquid-surface S and the surface energy of gas-surface S is taken into account).
For any virtual displacement field, the variation of E is[10]

$$\delta E = \int_{\Sigma_t} \delta\sigma - \left(\frac{2\sigma}{R_m}\, n_i + \left(\delta_i^j - n^j n_i\right) \sigma_{,j} \right) \zeta^i + \int_{\Gamma_t} \sigma\, n'_i\, \zeta^i$$

The scalar $\delta\sigma$ notes the variation of the surface energy σ associated with the displacement $\boldsymbol{\zeta}$.
Generally, the surface energy depends on the geometrical and physico-chemical properties of surface, the microscopic asperities or the presence of surfactant[9]. The simplest case is that for which the surface energy is defined as a function of the position on the surface ($\mathbf{x} \in \Sigma_{t1} \to \sigma_1(\mathbf{x})$), value σ_2 being constant positive along the fluid-fluid interface. It is the case we shall consider hereafter. Energies of the interfaces have the values

$$E_1 = \int_{\Sigma_{t1}} \sigma_1 \text{ and } E_2 = \int_{\Sigma_{t2}} \sigma_2$$

To the potential energies E_1 and E_2 are associated two variations $\delta T_{\alpha c} = -\delta E_\alpha$, $(\alpha = 1, 2)$ such that

$$\delta T_{\alpha c} = \int_{\Sigma_{t\alpha}} \frac{2\sigma_\alpha}{R_{m\alpha}}\, n_{\alpha i}\, \zeta^i - \int_{\Gamma_t} \sigma_\alpha\, n'_{\alpha i}\, \zeta^i \tag{5}$$

For dissipative fluids, a viscous stress tensor Q_i^j is added to the pressure term. In the expression of virtual work, the dissipative terms are distributed in volume as $Q_{i,j}^j$ and on the surfaces as $-n_{\alpha j} Q_i^j$. Expression (5) introduces a new term associated with the contact-line Γ_t which does not appear in flows without capillarity. In dissipative motions, it is necessary to add to the line tension a *dissipative vector* $\boldsymbol{\tau_v}$ having the dimension of a surface tension. The velocity of a liquid is discontinuous along the contact line; in the simplest dissipative model we assume the $\boldsymbol{\tau_v}$ behaviour as

$$\boldsymbol{\tau_v} = -\lambda\, [\mathbf{V}] \tag{6}$$

where λ is constant. Relations (4-6) lead to the expression of virtual work of forces applied to (L_1).

$$\delta T = \int_{D_t} (\phi_i - \rho a_i - p_{,i} + Q_{i,j}^j)\zeta^i + \tag{7}$$

$$\sum_{\alpha=1}^{2} \int_{\Sigma_{t\alpha}} \left(\left(\frac{2\sigma_\alpha}{R_{m_\alpha}} + p\right) n_{\alpha i} + P_{\alpha i} - n_{\alpha j}\, Q_i^j \right) \zeta^i + \int_{\Gamma_t} (\tau_{vi} - \sigma_1 n'_{1i} - \sigma_2 n'_{2i})\, \zeta^i$$

Virtual displacements on Σ_{t1} are tangent to the surface S (on Σ_{t1}, $n_{1i}\zeta^i = 0$). We emphasize it is not necessary that D_t, $\Sigma_{t\alpha}$ and Γ_t are material domain, surfaces and curve. For any ζ, $\delta T = 0$; the equations of motion and the boundary conditions are the following:

3.1 Equations of motion

$$\rho a_i + p_{,i} = \phi_i + Q^j_{i,j} \tag{8}$$

3.2 Conditions on the free surface Σ_{t2}

$$P_{2i} = n_{2j}\, Q^j_i - n_{2i}\, (p + \frac{2\sigma_2}{R_{m2}}) \tag{9}$$

When Σ_{t2} is an interface separating liquid and gas, $\mathbf{P}_2 = -P_0\, \mathbf{n}_2$ where P_0 denotes the gas pressure. The projection of the relation (9) on normal $\mathbf{n}_2$ implies the dynamic form of the Laplace equation

$$P_0 = p + \frac{2\sigma_2}{R_{m2}} - n_{2j} Q^j_i n^i_2$$

The projection of the relation (9) on the tangent plane to Σ_{t2} expresses that the tangential part of the viscous stress vector is null

$$n_{2j}(Q^j_i - Q^j_k n^k_2 n_{2i}) = 0$$

3.3 Conditions on the surface Σ_{t1}

By taking into account that on Σ_{t1}, $n_{1i}\zeta^i = 0$, there exists a scalar field μ of Lagrange multipliers defined on Σ_{t1} such that

$$P_{1i} + \mu\, n_{1i} = n_{1j}\, Q^j_i - n_{1i}\, (p + \frac{2\sigma_1}{R_{m1}}) \tag{10}$$

The projections of the relation (10) imply

$$P_{1i}\, n^i_1 + \mu \;=\; n_{1j}\, Q^j_i\, n^i_1 - (p + \frac{2\sigma_1}{R_{m1}})$$

$$P_{1j}\, (\delta^j_i - n^j_1 n_{1i}) \;=\; n_{1j}(Q^j_i - Q^j_k n^k_1 n_{1i})$$

($\mathbf{P}_1$ is non colinear to $\mathbf{n}_1$ and μ is an additional unknown scalar).

3.4 Conditions on the contact-line Γ_t

At any point of the contact-line Γ_t, a virtual displacement is in the form

$$\boldsymbol{\zeta} = \alpha\, \mathbf{t} + \beta\, \mathbf{n}'_1 \tag{11}$$

The two scalar fields α and β are defined on Γ_t. For any field $\boldsymbol{\zeta}$ in the form (11), relation (7) yields $\int_{\Gamma_t} (\sigma_1 n'_{1i} + \sigma_2 n'_{2i} - \tau_{vi})\, \zeta^i = 0$. In general case, from $\mathbf{n}'_1 = \mathbf{t} \times \mathbf{n}_1$, $\mathbf{n}'_2 = -\mathbf{t} \times \mathbf{n}_2$, expression (11) implies that $\boldsymbol{\tau_v}$ is normal to Γ_t and

$$\sigma_2 \cos\theta + \sigma_1 + \lambda\, n'_{1i}[V^i] = 0$$

where θ is the dynamic Young angle associated with interfaces Σ_{t1} and Σ_{t2}. For the flow studied in paragraph 1 the contact-line celerity is $u = -U$ and $n'_{1i}[V^i] = u\, g(\theta)$. We note $\nu = \lambda\, g(\theta)$; the scalar ν is bounded and

$$\sigma_2 \cos\theta + \sigma_1 + \nu\, u = 0 \tag{12}$$

Relation (12) is the linear form of a *Young-Dupré dynamic condition.* The scalar ν is positive which is in accordance with the second law of thermodynamics. We call ν the *line viscosity.* More complex expressions of the line viscosity may be proposed. Some other attempt is done in[12].

For any value of the contact-line celerity, relation (12) yields the contact angle θ. Then, the system made up of equation (8), boundary conditions (9) and (10), adherence assumption on solid surfaces and contact-line condition (12), well pose the fluid motion problem of the non-Newtonian fluid presented in paragraph 1.

In case of equilibrium, relation (12) yields the static Young-Dupré relation[13]

$$\sigma_2 \cos\theta_e + \sigma_1 = 0$$

where θ_e is the balance Young angle.

4 Concluding remarks

A creeping flow of Newtonian fluid is unrealistic at the corner of a moving contact-line. The following questions merge in this problem:

What is the kinematics of the contact-line? Can the fluid velocity fields be multivalued on this line? What is the work of dissipative forces in its vicinity? Is there slip of the contact-line on the solid wall?

We propose a dynamic model of the contact system between two convenient non-Newtonian fluids and a solid surface. The equations of the motion are in the classical form. Adherence and boundary conditions on surfaces and

interfaces are conserved but a dynamic Young-Dupré relation issuing from the virtual work principle yields the contact angle as a function of the contact-line celerity where the contact-line acts in a similar way as a *shock line* and the velocity fields are multivalued on the line. The paradox of infinite viscous dissipation is removed.
Relation (12) yields a simple explanation of a well-known experimental result: with the advance of the contact-line, u is positive and the dynamic angle θ is higher than the equilibrium angle θ_e. This result is reversed when u is negative.
The contact angle hysteresis phenomenon and the modelling of experimentally well-known results expressing the dependence of the dynamic contact angle on the celerity of the contact-line are important phenomena. Eq. (12) allows to obtain an explanation of the contact-angle hysteresis in the advance and retreat of a fluid on a solid surface[14].

Acknowledgments

I am grateful to Professor Ruggeri for his invitation to the "Wascom 2001", to Professor Teshukov for his criticisms and Dr Garajeu for his help.

References

1. J. Serrin in *Mathematical principles of classical fluid mechanics*, Encyclopedia of Physics, VIII/1 (Springer, Berlin, 1959).
2. C. Huh and L.E. Scriven, *J. Colloid Interface Sci.* **35**, 85 (1971).
3. K. Moffat, *J. fluid Mech.* **18**, 1 (1964).
4. P. Seppecher, *Int. J. Engng Science* **34**, 977 (1996).
5. C. Truesdell and W. Noll in *The non-linear field theories of mechanics*, Encyclopedia of Physics, III/3 (Springer, Berlin, 1965).
6. D.M. Heyes *et al*, *J. Chem. Phys.* **73**, 3987 (1980)
7. J.P. Ryckaert *et al*, *Phys. Rev. Lett.* **60**, 128 (1988)
8. E.B. Dussan V and S.H. Davis, *J. fluid Mech.* **65**, 71 (1974).
9. E.B. Dussan V, *Annual Rev. Fluid Mech.* **11**, 371 (1979).
10. H. Gouin and W. Kosinski, *Archives of Mechanics* **50**, 907 (1998).
11. E.B. Dussan V, E. Ramé and S. Garoff, *J. fluid Mech.* **230**, 97 (1991).
12. H. Fan, Y. X. Gao and X.Y. Huang, *Physics of fluids* **13**, 1615 (2001).
13. V. Levitch, *Physicochemical hydrodynamics* (Prentice Hall, Englewood Cliffs, 1962).
14. H. Gouin, *J. de Physique IV* **11, Pr 6**, 261 (2001).

ON THE BOUSSINESQ HIERARCHY

ANTONIO M. GRECO *
Dept. of Mathematics, University of Palermo,
Via Archirafi 34, 90123 Palermo (Italy)
E-mail: greco@math.unipa.it

GAETANA GAMBINO
Dept. of Mathematics, University of Palermo,
Via Archirafi 34, 90123 Palermo (Italy)
E-mail: gaetana@math.unipa.it

A new sequence of nonlinear evolution systems satisfying the zero curvature property is constructed, by using the invariant singularity analysis. All these systems are completely integrable and a *pseudo-potential* (linearization) is explicitly determined for each of them. The second system of the sequence is the *Broer-Kaup* system, which, as is well known, corresponds to the higher order Boussinesq approximation in describing shallow water waves.

1 Introduction

Recently many efforts were given to construct hierarchies of nonlinear evolution completely integrable equations.

Usually, starting from a well known given nonlinear evolution equation owing the complete integrability property, one looks for a sequence of equations to which the given one belongs and all the other being obtained in a *recursive* way. More precisely each new flow comes out from the previous one in a well defined manner, giving rise to a new nonlinear evolution completely integrable equation.

The first seminal work by Lax [1] concerns the KdV hierarchy algebraically obtained by requiring the operator defining the Lax pair to be consistent at any order. *Consistent* here means that the operator evolution equation

$$L_t = [B_n, L]$$

reduces to an evolution equation for the unknown field u_n free from the spectral parameter at any order n. In the above equation $L(t)$ denotes a family of

* Supported by MURST under grant *Nonlinear Mathematical Problems of Wave Propagation and Stability in Models of Continuous Media*

unitarily equivalent operators depending differentially on t, B_n the n-th order skew-symmetric differential operator, the first one being ∂_x.

After Lax, many tools were developed to construct the hierarchy to which a nonlinear evolution completely integrable equation belongs. Among them we recall the trace-less operator method strictly connected with the role that the $SU(n)$ group plays for the existence of any kind of hierarchy, the recursion pseudo-operator method and so on.

As far as we know, there are very few examples of hierarchies, explicitly worked out, concerning nonlinear systems.

In this paper we consider two nonlinearly coupled evolving scalar fields and we construct a hierarchy of nonlinear evolution systems, which rules the behaviour of the considered fields at any order, by using the invariant singularity analysis, as given by Conte, Musette and Pickering [2].

The expansion function is chosen so that its dependence on x and t is ruled by two Riccati equations. We construct a sequence of pair of spectral matrices $\mathcal{A}_n, \mathcal{M}_n$ connecting them to the coefficients of the above Riccati equations and to the unknown evolving fields. We prove that

$$\mathcal{A}_{n_t} - \mathcal{M}_{n_x} - [\mathcal{A}_n, \mathcal{M}_n] = 0$$

turns out to be free of the spectral parameter at any order n, giving rise to the sequence of nonlinear integrable systems.

The method we follow uses the procedure given by Pickering [3] to show the integrability of the Broer-Kaup system. This approach, as we will show, can be used also to construct the hierarchy and, in our case, the Broer-Kaup system turns out to be the second system of the hierarchy.

2 Invariant singularity analysis and Riccati system

Invariant singularity analysis is a variant of the classical Painlevé analysis, easily achieved by choosing the expansion variable in a suitable way. Usually in the Painlevé analysis [4] the expansion variable is the function φ giving the singularity manifold via the equation $\varphi = 0$.

As matter of fact, we can choose any function χ satisfying $\lim_{\varphi\to 0}\chi = 0$, provided that the type of singularities will not change.

Supposing, without loss of generality, $\varphi_x \neq 0$, let us choose

$$\chi = (\frac{\varphi_x}{\varphi} - \frac{\varphi_{xx}}{2\varphi_x})^{-1}$$

and, in this way, the coefficients of the resulting expression $u = \sum_{j=0}^{\infty} u_j \chi^{j+p}$ (here p is a negative integer) [5] are invariant under the six-parameters homographic group:

$$\varphi \longrightarrow \frac{a\varphi + b}{c\varphi + d},$$

$ad - bc = 1,\ a, b, c, d \in \mathbb{C}$.

A straightforward calculation leads to the Riccati system:

$$\begin{aligned} \chi_x &= 1 + \frac{S}{2}\chi^2, \\ \chi_t &= C + C_x \chi - \frac{1}{2}(C_{xx} + CS)\chi^2, \end{aligned} \tag{2.1}$$

where

$$\begin{aligned} C &= -\frac{\varphi_t}{\varphi_x} & &\textit{celerity}, \\ S &= \frac{\varphi_{xxx}}{\varphi_x} - \frac{3}{2}\left(\frac{\varphi_{xx}}{\varphi_x}\right)^2 & &\textit{schwarzian}. \end{aligned}$$

The compatibility condition on the Riccati system gives us the singularity manifold equation:

$$S_t + C_{xxx} + 2SC_x + S_x C = 0.$$

Now, no matter what is φ, apart the assumed regularity (analiticity), S and C are classical homographic invariants. Then the Riccati system turns out to be invariant, too.

This choice of the expansion variable provides advantages: the expression for the determining equations are greatly simplified and it is possible to obtain exact solutions from a higher order (at a positive power) truncation for the expansion in the new variable χ [6].

3 Zero-curvature method

It is a generalization of the Lax representation.

Consider the two linear equations for the field v [7]:

$$
\begin{aligned}
v_x &= \mathcal{A}v \qquad\qquad & \mathcal{A} &\equiv \mathcal{A}(v, Dv, \lambda),\\
v_t &= \mathcal{M}v \qquad\qquad & \mathcal{M} &\equiv \mathcal{M}(v, Dv, \lambda),
\end{aligned}
$$

where λ is the spectral parameter and D denotes any spatial differentiation.

Compatibility condition gives:

$$
\mathbf{M} \equiv \mathcal{A}_t - \mathcal{M}_x + [\mathcal{A}, \mathcal{M}] = 0.
$$

If $\mathbf{M}$ is a trace-less matrix, free of the spectral parameter, by putting its elements equal to zero one obtains the evolution equations for v, that result completely integrable, i.e. exactly linearizable. We mean that, by inverting the procedure, it transforms in the above linear system.

Given a nonlinear evolution equation (or system), to find the matrices that realize the zero-curvature condition one has to linearize the Riccati system (2.1) [3].

If we want to utilize the invariant Painlevé analysis and to truncate the expansion at positive integers it is useful to introduce a new expansion variable Z, related to the older one via the gauge $Z^{-1} = \chi^{-1} + \frac{1}{2}A$, satisfying the generalized Riccati system :

$$
\begin{aligned}
Z_x &= 1 - AZ - BZ^2, \\
Z_t &= -C + (AC + C_x)Z - (D - BC)Z^2
\end{aligned}
\tag{3.1}
$$

and linearize it via the gauge $Z^{-1} = \frac{\psi_1}{\psi_2}$:

$$
\begin{aligned}
\begin{pmatrix} \psi_1 \\ \psi_2 \end{pmatrix}_x &= \begin{pmatrix} -\frac{1}{2}A & 1 \\ B & \frac{1}{2}A \end{pmatrix} \begin{pmatrix} \psi_1 \\ \psi_2 \end{pmatrix}, \\
\begin{pmatrix} \psi_1 \\ \psi_2 \end{pmatrix}_t &= \begin{pmatrix} -\frac{1}{2}(AC + C_x) & -C \\ (D - BC) & -\frac{1}{2}(AC + C_x) \end{pmatrix} \begin{pmatrix} \psi_1 \\ \psi_2 \end{pmatrix}.
\end{aligned}
\tag{3.2}
$$

The compatibility condition for (3.2) enjoys the zero-curvature property, so they constitute a pseudo-potential for the given evolution equation [3].

Summarize the method: consider a nonlinear evolution system and apply the invariant singularity analysis, choosing the coefficients in such a way that the truncation of Laurent series at appropriate positive terms leads to a Darboux transformation or, better, to an auto-Backlünd transformation. If this happens, the same coefficients permit us to construct matrices depending on a parameter (spectral parameter) which:

- give the linear spectral problem
- satisfy the zero-curvature condition.

Complete integrability is achieved.

4 Application of the method to verify complete integrability

Let us consider the system

$$\begin{aligned} u_{xt} &= -u_{xxx} - u_x u_{xx} + u_{xx} - v_x, \\ v_t &= -2u_{xxx} + u_x u_{xx} - v u_{xx} + u_{xx} - u_x v_x + v_{xx} - v_x. \end{aligned} \tag{4.1}$$

In our knowledge it does not appear in literature until now. We shall show that it belongs to a sequence of systems all enjoying complete integrability property.

Let us start with the invariant Painlevé analysis where the expansion variable Z satisfies the Riccati system (3.1).

Suppose the solution admits the expansions $u = \sum_{j=0}^{+\infty} u_j Z^{j+p}$ and $v = \sum_{j=0}^{+\infty} v_j Z^{j+q}$. First of all verify that p and q can be determined in such a way the leading order terms "balance".

We find the values $p = 0$ and $q = -1$, therefore the above proposed expansions take the form $u = u_0 \log Z + U$ and $v = \sum_{j=0}^{3} v_j Z^{j-1}$.

At this point it is possible to go ahead by hands, but it is better to use a symbolic package as REDUCE or MACSYMA. We find:

$$\begin{aligned} A &= \frac{\lambda + u_x}{2}, \\ B &= -\frac{v - u_x}{4}, \\ C &= \frac{-\lambda + u_x}{2}, \\ D &= B_x. \end{aligned} \tag{4.2}$$

Linearize the Riccati system via the gauge $Z^{-1} = \frac{\psi_1}{\psi_2}$ obtaining:

$$\begin{pmatrix}\psi_1\\ \psi_2\end{pmatrix}_x = \begin{pmatrix}-\frac{\lambda+u_x}{4} & 1\\ -\frac{v-u_x}{4} & \frac{\lambda+u_x}{4}\end{pmatrix}\begin{pmatrix}\psi_1\\ \psi_2\end{pmatrix},$$

$$\begin{pmatrix}\psi_1\\ \psi_2\end{pmatrix}_t = \begin{pmatrix}\frac{2u_{xx}+u_x^2-\lambda^2}{8} & \frac{\lambda-u_x}{2}\\ -\frac{-2v_x+2u_{xx}+u_xv-\lambda v-u_x^2+\lambda u_x}{8} & -\frac{2u_{xx}+u_x^2-\lambda^2}{8}\end{pmatrix}\begin{pmatrix}\psi_1\\ \psi_2\end{pmatrix}.$$

Substitute the above values of A, B, C, D and write compatibility condition and find just the starting system (4.1).

5 From a system to a hierarchy

Make the substitution $u \to u_{nx}$ in the expression of the coefficient B in (4.2) and write exactly the same above linear spectral system:

$$\begin{pmatrix}\psi_1\\ \psi_2\end{pmatrix}_x = \mathcal{A}_n\begin{pmatrix}\psi_1,\\ \psi_2\end{pmatrix},$$
$$\begin{pmatrix}\psi_1\\ \psi_2\end{pmatrix}_t = \mathcal{M}_n\begin{pmatrix}\psi_1\\ \psi_2\end{pmatrix}, \qquad (5.1)$$

where

$$\mathcal{A}_n = \begin{pmatrix}-\frac{\lambda+u_x}{4} & 1\\ -\frac{v-u_{nx}}{4} & \frac{\lambda+u_x}{4}\end{pmatrix},$$

$$\mathcal{M}_n = \begin{pmatrix}\frac{2u_{xx}+u_x^2-\lambda^2}{8} & \frac{\lambda-u_x}{2}\\ \frac{2u(n+1)x-u_xu_{nx}+u_xv-2v_x+\lambda u_{nx}-\lambda v}{8} & -\frac{2u_{xx}+u_x^2-\lambda^2}{8}\end{pmatrix}.$$

Write again compatibility condition and find:

$$(\begin{matrix}M_{11} & M_{12}\\ M_{21} & M_{22}\end{matrix}) = 0,$$

where

$$M_{11} \equiv u_{xt_n} + u_{xxx} + u_xu_{xx} + v_x - u_{(n+1)x},$$
$$M_{12} \equiv 0,$$
$$M_{21} \equiv v_{t_n} + (u_xv)_x - v_{xx} - u_{nxt} + u_{(n+2)x} - u_xu_{(n+1)x} - u_{xx}u_{nx},$$
$$M_{22} \equiv -M_{11}.$$

Substitute the value of u_{xt} from $M_{11}=0$ in $M_{21}=0$ and obtain the $n-$th system of the sequence:

$$\begin{aligned} u_{xt_n} &= -u_{xxx} - u_x u_{xx} - v_x + u_{(n+1)x} \\ v_{t_n} &= v_{xx} - (u_x v)_x - 2u_{(n+2)x} + u_{2nx} - v_{nx} \\ &\quad +(u_x u_{nx})_x - (u_x u_{xx})_{(n-1)x}. \end{aligned} \tag{5.2}$$

These systems are completely integrable for all n because they all enjoy the zero-curvature condition.

Besides, for $n=1$ we obtain the system of the previous paragraph; for $n=2$ the Broer-Kaup system:

$$\begin{aligned} u_t &= -uu_x - v_x, \\ v_t &= -u_{xxx} - (uv)_x, \end{aligned} \tag{5.3}$$

which is equivalent to the classical *higher order* Boussinesq equation:

$$\begin{aligned} &\frac{a^2}{3}U_{4x} - 2U_x^2 U_{xx} + 2b_0(U_t U_{xx} + 2U_x U_{xt}) - \\ &4b_2 U_x U_{xx} + 2b_3 U_{xx} + 4b_0 b_2 U_{xt} - 3b_0^2 U_{yy} = 0. \end{aligned} \tag{5.4}$$

Remember that the Boussinesq higher order equation represents how an inviscid, incompressible, irrotational fluid evolves in a narrow infinite (long enough) channel with horizontal bottom, of mean density ρ, free surface τ.

Under the above conditions, in the framework of the nonlinear shallow water theory for the one-dimensional waves of

- small enough amplitude
- large enough wavelength

two parameters must be considered fundamental:

$$\frac{wave-amplitude}{depth} = \frac{a}{h} = \epsilon,$$

$$(\frac{depth}{wavelength})^2 = \frac{h^2}{\lambda^2} = \delta^2.$$

Then starting from the basic equations for the water waves, introducing the velocity potential U, expanding the equation and retaining the terms up to the order δ^4 and $\epsilon\delta^2$, one obtains the higher order Boussinesq equation (5.4), [8].

Without loss of generality we choose, as usually done, the coefficients $b_2 = b_3 = 0, a^2 = 1$ and $b_0 = 1$. The appropriate gauge

$$u = -\frac{2}{3}b_0(U_x + b_2),$$

$$v = \frac{2}{3}b_0(U_t + \frac{b_3 + b_2^2}{b_0}) - \frac{u^2}{2}$$

leads, via some straightforward algebra, to the Broer Kaup system (5.3)

6 From the zero-curvature to the higher order Lax pair

Make in the $n-$th system of the sequence the substitution $u \to u_x$

$$\begin{aligned} u_{t_n} &= -u_{xx} - uu_x - v_x + u_{nx}, \\ v_{t_n} &= v_{xx} - (uv)_x - 2u_{(n+1)x} + u_{(2n-1)x} - v_{nx} \\ &\quad +(uu_{(n-1)x})_x - (uu_x)_{(n-1)x}, \end{aligned} \tag{6.1}$$

(potential form) where the flows of u and v at the $n-$th order explicitly appear.

To determine the Lax pair we have to find two differential operators L_n and B_n, adjoint and skew-adjoint respectively, such that $L_{n_t} = [B_n, L_n]$ if the $n-$th system of the hierarchy is identically verified and vice versa.

Consider the $n-$th linearized Riccati system (5.1).

The operator L_n has to satisfy the eigenvalue equation $L_n\psi = \lambda\psi$. If we start from the (5.1), we obtain:

$$\begin{aligned} \lambda\psi_1 &= -u\psi_1 - 4\psi_{1x} + 4\psi_2, \\ \lambda\psi_2 &= v\psi_1 - u_{(n-1)x}\psi_1 + 4\psi_{2x} - u. \end{aligned} \tag{6.2}$$

Therefore:

$$L_n\psi = \lambda\psi \Rightarrow L_n = \begin{pmatrix} -4\partial_x - u & 4 \\ v - u_{(n-1)x} & 4\partial_x - u \end{pmatrix}.$$

To find B_n differentiate the relation $L_n\psi = \lambda\psi$ with respect to t and utilize the (5.1):

$$Ln_t\psi + L_n\psi_t = \lambda\psi_t \Rightarrow Ln_t\psi + L_n\mathcal{M}_n\psi = \lambda\mathcal{M}_n\psi$$
$$\Rightarrow Ln_t\psi = \mathcal{M}_nL_n\psi - L_n\mathcal{M}_n\psi \Rightarrow Ln_t\psi = [\mathcal{M}_n, L_n].$$

We can note that the operator B_n of the Lax pair is the same matrix $\mathcal{M}_n$ which realize the zero-curvature condition.

Now we have to free it of the spectral parameter by substituting the values of $\lambda\psi_1$ and $\lambda\psi_2$ given in (6.2). We obtain:

$$B_n = \begin{pmatrix} -2\partial_x^2 - u\partial_x - \frac{1}{4}u_x & 2\partial_x \\ -\frac{1}{2}u_x\partial_x + \frac{1}{2}v\partial_x - \frac{1}{4}u_{xx} + \frac{1}{4}v_x & 2\partial_x^2 - u\partial_x - \frac{3}{4}u_x \end{pmatrix}.$$

It remains to be verified the adjointness and skew-adjointness properties for L_n and B_n respectively. What we are able to prove with a straightforward calculation is not exactly the requested properties, but some very closed occurrence, that is

$$L_n = L_n^* \begin{pmatrix} 0 & 1 \\ 1 & 0 \end{pmatrix} \quad \text{and} \quad B_n = B_n^* \begin{pmatrix} 0 & 1 \\ 1 & 0 \end{pmatrix}.$$

7 Reduction to a single equation of the generic system

Consider the n-th system (non potential)(5.1) and calculate v from the first equation:

$$v = u_{nx} - u_t - u_{xx} - \frac{1}{2}u_x^2 + F(t), \tag{7.1}$$

in which $F(t)$ is an arbitrary integration function.

Substitute this value in the second equation of the system:

$$F'(t) - u_{tt} - 2u_xu_{xt} = -u_{4x} + \frac{3}{2}u_x^2u_{xx} - u_{xx}F(t) + u_{xx}u_t,$$

and obtain an expression free of n. In particular by putting equal to zero the integration function, we obtain:

$$-u_{tt} - 2u_x u_{xt} = -u_{4x} + \frac{3}{2}u_x^2 u_{xx} + u_{xx}u_t,$$

that is the classical Boussinesq equation.

Therefore the explicit solution of the system (5.1) is (u, v_n), where u is the solution of the Boussinesq equation and v_n the corrispondent value of v from the (7.1).

8 Unsolved question and work in progress

In this work we constructed a sequence of nonlinear evolution completely integrable systems. The "recursion operator" we proposed requires to be studied in deeper to obtain the possibility of generating the flow at any order from the previous one.

Moreover the Lax pair is "anomalous", L and B are not adjoint and skew-adjoint unless one carries out some variations. This, probably would be achieved by using an appropriate norm in the involved Banach space, giving rise to a suitable scalar product.

These aspects are object of current research.

At the state of survey it seems that any system of the sequence could be origin of a proper hierarchy. This aspect will be investigated, too.

References

1. P. D. Lax, Integrals of nonlinear equations of evolution and solitary waves, Comm. on pure and applied mathematics, **21** (1968), pp.467-490.
2. A. Pickering, R. Conte, M. Musette, The two singular manifold method: classical Boussinesq system, J. Phys., **28** (1995), pp.179-187.
3. A. Pickering, The singular manifold method revisited, J. Math. Phys., **37** (1996), pp. 1894-1926.
4. J. Weiss, M. Tabor, G. Carnevale, The Pailevé property for partial differential equation, J. Math. Phys., **24** (1983), pp.522-526.
5. R. Conte, Invariant Painlevé analysis of partial differential equations , Physics Letters A , **140** (1989), pp.383-390.
6. A. Pickering, A new truncation in Painlevé analysis, J. Phys. A, **26** (1993), pp. 4395-4405.

7. M. J. ABLOWITZ, P. A. CLARKSON, Solitons, nonlinear evolution equations and inverse scattering, London Mathematical Society, Cambridge University press (1991).
8. G. B. WHITHAM, Linear and nonlinear waves, Pure and Applied Mathematics, A. Wiley, Interscience series of texts monographs and tracts (1974).

BASIC PARAMETERS IN CONTINUUM MECHANICS

G. GRIOLI

University of Padua.
Address for correspondence:
v. XXVII Luglio is.192 n.54
98100 Messina, Italy.

It is pointed out the special meaning of some parameters which are present in the constitutive relations in Continuum Mechanics. Questions related to the stress in the rigid bodies, the incompressibility, the heat propagation and to the microstructures theory are considered (see also Grioli[1]).

1 Introduction

In my opinion, in The problems of Continuum Mechanics often it is convenient to pay attention to a special parameter introduced by us or present in the constitutive relations.

The quests of Antonio Signorini are well known, who studying the problem of the elastic equilibrium in finite Elasticity assumed that the external dead loads are proportional to a parameter h. Among other things, A. S. , supposing that the linear Elasticity be the first term of a Taylor's serie in h, showed that, in general non indetermination of an infinitesimal rigid displacement exists. Further, Signorini[2] showed that in special cases the meaning of linear Elasticity is doubtful.

The introduction of the parameter h by Signorini has a mere analytical meaning with the aim to change a non–linear problem in a set of linear ones. Others Authors continued his researchs (Grioli[3,7], Capriz et al. [4,6],Baratha et al.[5]).

In the following, briefly, I will consider some questions in which the presence of a particular parameter gives remarkable advantages. Precisely:

i) <u>The stress in rigid bodies.</u> If one considers a rigid body as a limit of a deformable elastic one, it is possible to show that operating on a particular parameter the stress ceases to be indeterminate (following a general opinion) and it is possible to write constitutive relations (Grioli[8]).

ii) <u>Incompressibility.</u> Some considerations may be made about the incompressibility. Every solutions tends to an incompressible solution when a special parameter in constitutive relations, tends to a certain limit with finite values for stress and strain.

iii) <u>Heat propagation.</u> Some critical observations to the Fourier law and to

the consequent parabolic heat propagation equation are well known. To overcome some difficulties regarding propagation problems, Carlo Cattaneo [9,10], in 1948 proposed a different connection between temperature and heat flux. Successively, the Extended Thermodynamics theory has been formulated. In my opinion, it is worthwile to observe that assuming that the Helmotz Free Energy depends on the heat flux, q, many difficulties are overcomed and a general connection between temperature and heat flux derive. The theory includes the Cattaneo law as a particular case. All thermodynamics general laws and Continuum Mechanics are deeply influenced by the new law (Grioli[11]).
iv) Microstructures. Recently, my attention has been attracted by the theory of Microstructures. In the theory a very subtle question exists: the problem of boundary and initial conditions necessary to make determinate the analytical problem. In general, it is very difficult (or impossible), from the physical point of view, to satisfy that conditions. Therefore, the Continuum model is a very mathematically interesting problem but has very little physical concreteness. It is possible to overcome the difficulty introducing in the continuum mathematical model a special parameter, h. To be clear, it is fundamental the remark that, considering the "elementary particles" by which the material is constituted, according to the corpuscular hypotesis of modern Physics, as a basis for constructing the continuum mathematical model, one sees that the deformation depends on the greatness of the "elementary particles". Therefore, a trace of that magnitude in that model must be present, in some manner. I saw that it will be sufficient to introduce as a basic parameter, h, the greatest dimension of the "elementary particles". Otherwise, in my opinion, to not take his into consideration lessens the theory (Grioli[12,13]).

Since I studied from new points of view the problem of Microstructures recently, speaking here and there about it, I will consider the problem at the end, for a question of time. I hope to be able to express at least some basic ideas on the subject.

Naturally, the above proposed procedure presupposes opportune analytical researches.

2 Rigidity property

Now and in the following, I will denote by C, C', C'' a reference configuration, the actual one and a very close position to C' of a threedimensional body, and denote by σ, σ', σ'' their boundaries. Let be X_r, x_r $(r = 1,2,3)$ the coordinates of two correspondent points P, P' in C, C' with respect to a rectangular Cartesian coordinate system, $u_r = x_r - X_r$ the components of the displacement $\underline{u} = PP'$. Assuming that the body is hysotropic and C a natural equilibrium position, let $(\epsilon = \epsilon_{rs})$ be the strain and $I_1(\epsilon)$, $I_2(\epsilon)$, $I_3(\epsilon)$ the three principal invariants of ϵ.

The Free Energy W, depends on $I_i(\epsilon)$, $(i = 1, 2, 3)$ and, in general, on the other variables such as Temperature, but now it is not necessary to point out this fact. Therefore, let us assume

$$W = W(I_1, I_2, I_3). \tag{1}$$

Therefore, the symmetrical lagrangian Piola Kirchoff stress matrix is

$$\begin{aligned} T_{rs} = -\frac{\partial W}{\partial \epsilon_{rs}} &= -\frac{\partial W}{\partial I_1}\delta_{rs} + \\ &-\frac{\partial W}{\partial I_2}(I_1\delta_{rs} - \epsilon_{rs}) - \frac{\partial W}{\partial I_3}\frac{\partial I_3}{\partial \epsilon_{rs}} \quad (\delta_{rs} = \text{Kronecker symbols}) \end{aligned} \tag{2}$$

where

$$\left(\frac{\partial W}{\partial I_1}\right)_{\epsilon=0} = 0. \tag{3}$$

Let us assume

$$2\mu = -\left(\frac{\partial W}{\partial I_2}\right)_{\epsilon=0}, \quad \beta\mu = \left(\frac{\partial^2 W}{\partial I_1^2}\right)_{\epsilon=0}, \tag{4}$$

$$\mu > 0, \qquad 3\beta - 4 > 0. \tag{5}$$

In general, according to the traditional procedure, one can write

$$W = \mu\left(\beta\frac{I_1^2}{2} - 2I_2\right) + W'(I_1, I_2, I_3), \tag{6}$$

where the function $W'(...)$ satisfies the conditions

$$\left(\frac{\partial W'}{\partial \epsilon_{rs}}\right)_{\epsilon=0} = \left(\frac{\partial W'}{\partial I_1}\right)_{\epsilon=0} = \left(\frac{\partial W'}{\partial I_2}\right)_{\epsilon=0} = \left(\frac{\partial^2 W'}{\partial I_1^2}\right)_{\epsilon=0} = 0 \tag{7}$$

According to (2), (6), it follows

$$T_{rs} = -2\mu\epsilon_{rs} + \frac{\beta - 2}{3\beta - 4}\left(I_1(T) - I_1(T')\right)\delta_{rs} - \frac{\partial W'}{\partial \epsilon_{rs}} \tag{8}$$

From (8) one deduces: every solution of the field and boundary equations whose stress and strain have finite limits when the parameter μ goes to infinity satisfies the property:

$$\lim_{\mu\to\infty} \epsilon_{rs} = 0, \quad \text{rigidity property.} \tag{9}$$

Let us denote by

$$\underline{\overline{u}}(X,t) = \lim_{h\to 0} \underline{u}(X,t;h), \quad h = \mu^{-1}, \tag{10}$$

the rigid displacement that is the limit of $\underline{u}(X,t,h)$ when h tends to zero and

$$u'_{r,s} = \lim_{h\to 0}\left(\frac{\partial u_{r,s}}{\partial h}\right). \tag{11}$$

Denoting by $R = (R_{rs})$ the rotation contained in the displacement $\underline{\overline{u}}$, one has

$$\lim_{h\to 0}\left(\frac{\epsilon_{rs}}{h}\right) = \frac{1}{2}\left(R_{ir}u'_{i,s} + R_{is}u'_{i,r}\right) \tag{12}$$

where the comma denotes partial derivatives with respect to X_r.

In conclusion, from (8) it follows:

$$\overline{T}_{rs} = \lim_{h\to 0} T_{rs} = (2-\beta)R_{ip}u'_{i,p}\delta_{rs} - \left(R_{ir}u'_{i,s} + R_{is}u'_{i,r}\right) \tag{13}$$

that may be interpreted as the constitutive relations of a rigid body, expressed in the unknown vector $\underline{u}'$. The lagrangian field and boundary equations are

$$\left(\overline{x}_{r,l}\overline{T}_{ls}\right)_{,s} = F_r - \gamma\ddot{\underline{\overline{u}}}_r, \quad (\text{in } C), \tag{14}$$

$$\overline{x}_{r,l}\overline{T}_{ls}N_s = f_r, \quad (\text{on } \sigma), \tag{15}$$

where N_r denotes the internal unit perpendicular to σ.

It is interesting to remark that the integrability condition for equations (14), (15) concide with the dynamical equations of a rigid body and determine the vector $\underline{\overline{u}}(X,t)$, depending on the initial conditions. Therefore, the analytical problem (14), (15) is of a static type. The variable t (time) is present as a parameter contained in the known displacement $\underline{\overline{u}}(X,t)$.

From the above considerations, one deduces the following information on the Free Energy: in general, the analytical structure of the potential energy must contain a parameter, h, such that when h tends to zero the strain goes to zero for every finite stress.

3 Incompressibility

Denoting by $D > 0$ the jacobian of the fransformation CC', one may express the potential elastic energy by the variables D, I_2, I_3, rather than I_1, I_2, I_3:

$$W(I_1, I_2, I_3) = V(D, I_2, I_3), \tag{16}$$

where

$$-2\mu = \left(\frac{\partial V}{\partial I_2}\right)_{\epsilon=0} = \left(\frac{\partial W}{\partial I_2}\right)_{\epsilon=0}, \tag{17}$$

$$\left(\frac{\partial V}{\partial D}\right)_{\epsilon=0} = 0, \quad \left(\frac{\partial^2 V}{\partial D^2}\right)_{\epsilon=0} = \left(\frac{\partial^2 W}{\partial I_1^2}\right)_{\epsilon=0} = \beta\mu. \tag{18}$$

In general, one may assume

$$V = \left(\beta\frac{(D-1)^2}{2} - 2I_2\right)\mu + V', \tag{19}$$

where

$$\left(\frac{\partial V'}{\partial \epsilon_{rs}}\right)_{\epsilon=0} = \left(\frac{\partial V'}{\partial D}\right)_{\epsilon=0} = \left(\frac{\partial V'}{\partial I_2}\right)_{\epsilon=0} = \left(\frac{\partial^2 V'}{\partial D^2}\right)_{\epsilon=0} = 0. \tag{20}$$

The correspondent expression for lagrangian stress is

$$T_{rs} = -\mu\left(\beta(D-1)\frac{\partial D}{\partial \epsilon_{rs}} - (I_1\delta_{rs} - \epsilon_{rs})\right) - \frac{\partial V'}{\partial \epsilon_{rs}}. \tag{21}$$

If the stress and the strain of a solution of a concrete problem are finite when β satisfies the condition (4), (5), from (21) follows:

$$\lim_{\beta\to\infty}(D-1)\frac{\partial D}{\partial \epsilon_{rs}} = 0. \tag{22}$$

It is easy to show that the only solution of (22) is a strain that renders $D = 1$. Therefore, the condition $\beta \to \infty$ characterize the incompressibility.

4 Heat Flux

Let us assume that the Free Energy, Z, depends on the variables e_{rs}, $(r, s = 1, 2...)$ that characterize the deformation, the Temperature, T, and on the heat flux, $\underline{q}$ (in objective manner).

Denoting by T'_{rs} and T''_{rs}, the lagrangean components of the reversible and irreversible parts of the stress, and assuming that there is no internal constraint, one has

$$T'_{rs} = -\frac{\partial Z}{\partial e_{rs}}, \quad H = -\frac{\partial Z}{\partial T}, \quad T''_{rs}\dot{e}_{rs} \leq 0 \tag{23}$$

and

$$\frac{\partial Z}{\partial q_i}\dot{q}_i + \frac{\underline{q}\cdot\underline{g}}{T} \leq 0, \tag{24}$$

where H denotes lagrangean density of the Entropy and $\underline{g}$ the gradient of the Temperature.

In each instant when $\underline{q}$ is different to zero, one may put

$$\frac{\partial Z}{\partial q_i} = Lq_i + e_{ilt}S_l q_t, \tag{25}$$

where L and S_l are parameters and e_{imt} denotes Ricci's tensor in the euclidean threedimensional space. Therefore, (24) becomes

$$\left(L\dot{q}_l + e_{tml}S_m\dot{q}_t + \frac{g_l}{T}\right) q_l \leq 0. \tag{26}$$

The relation (26) is equivalent to the equality

$$L\dot{q}_l + e_{tml}S_m\dot{q}_t + \frac{g_l}{T} = -mq_l + e_{lrs}M_r q_s, \tag{27}$$

where M_r are arbitrary parameters and $m > 0$.

It is evident that (27) is a more general relation than that Maxwell Cattaneo which is included in (27) for $S_i = M_i = 0$. In particular, if also $L = 0$ (that is Z is independent of $\underline{q}$) one has the Fourier law.

The general constitutive relations for the stress and the Entropy are influenced by the dependence of Z on $\underline{q}$ and one has a general complete interaction between mechanical and thermical phenomena with important consequences on the propagation problems.

The dependence of the Free Energy on the heat flux must satisfy the objectivity principle. In my opinion, it is sufficient that Z be a function of the magnitude of the vector $\underline{q}$. More generally, Z may depend on the vector $F\underline{q}$, where F is the displacement gradient in the transformation CC'.

5 Microstructure

As is well known, the theory of Cauchy for describing the behaviour of a material body is the basis of modern Continuum Mechanics. Nevertheless, that theory often is not sufficient for the study of many questions. The use of more complex mathematical models, called Microstructures, are necessary. The simplest but interesting model is Cosserat's in which the fundamental fields are two: a field of displacement, as in the Cauchy theory, and one of rotations, $R = (R_{rs})$. The corresponding mathematical model is very interesting but, in my opinion, in general has little physical concreteness. Whatsmore, that happens for more evolved microstructures as, for example, the affine structures.

A subtle and fundamental question regards the boundary and the initial conditions required by the mathematical model. For example, in the simplest case of microstructures, Cosserat's, it is necessary to know on the boundary certain functions, m_r, traditionally called surface couples, or, as substitute, some relations constructed with rotations. But, from the physical point of view, it is very difficult (or impossible) to give the m_r as known functions of the points of the boundary. Analogous considerations are valid for the initial values. In my opinion, that difficulty may be overcome recalling the corpuscular hypotesis of matter of modern Physics, before constructing the mathematical model.

With references to the simplest Cosserat case, let us denote by c, c_1 and c', c'_1 the position of two "elementary particles" in the reference and actual position and P, P' and P_1, P'_1 correspondent material points of c, c' and c_1, c'_1 respectively. Surely, in general, the lenght of $P'P'_1$ depends not only on the lenght of PP_1 but also on the size of the particles c, c_1. Therefore, in general, the deformation is dependent on that size. That is very important observation and I believe that in continuum model must be present a special parameter that recalls, in some way, the size of the elementary particles. The question is complicated but I saw that it is sufficient to introduce as basic parameter the greatest dimension, h, of the elementary particles, for constructing a mathematical model of physical concrete Microstructure.

In the hypotesis that the elementary particles are small rigid bodies (Cosserat bodies) the deformation is characterized by two field, the displacement, $\underline{u}$ and the rotation, R. In my opinion, three options are interesting:

1. the fields $\underline{u}$ and R are independent;
2. R coincides with the local rotation present in the polar decomposition of the displacement gradient of the field $\underline{u}$;
3. R satisfies the condition that in the linear transformation from the actual position C' to a very close one, C'', the rotation of an elementary particle, c' coincides with the local rotation present in the polar decomposition of the gradient of the field $\delta\underline{u}$ which characterizes the displacement $C'C''$.

The option 2) is very complex in the case of finite deformation. It coincides with 3) in the case of linear small deformations. Obviously, different constitutive relations correspond to the three options 1), 2), 3).

In my opinion, the case 3) gives more physical concreteness to the corresponding mathematical model.

Denoting by F the gradient of displacement, let us consider the matrices

$$\nu = F^T R, \quad \nu^i = R^T R_{,i}, \quad (i = 1, 2, 3). \tag{28}$$

It is possible to show that the linear dilatation coefficient, δ, in the transformation CC' is characterized by the equality

$$(1+\delta)^2 = \frac{\alpha(\epsilon) + h\beta(\nu,\nu^i) + h^2\gamma(\nu,\nu^i)}{\overline{\alpha} + h\overline{\beta} + h^2\overline{\gamma}}, \tag{29}$$

where β and γ are known functions of ν and ν^i, while $\overline{\alpha}$, $\overline{\beta}$ and $\overline{\gamma}$ are some constants.

The coefficient δ becomes a function of the only ϵ and the partial derivatives of any order with respect to h exist when h tends to zero. That is a very basic circumstance. Further, it is

$$\nu\nu^T = \underline{1} + 2\epsilon. \tag{30}$$

The eulerian density of the work of the internal forces when the body goes from C' to a very close position C'', as is well known, is a linear polynomial in certain variables deduced from the variables that characterize the deformation, that is, in our case, from ν and ν^i. The coefficients have, for us, no physical meaning. They are to be considered as simple multiplier parameters.

In the option 3), that us now consider, it is

$$dl^{(i)} = t^{(s)}_{rs}\,(\delta u_r)_{/s} + p_{rs}\omega_{r/s}, \tag{31}$$

where

$$\omega_r = \frac{1}{2} e_{rlm}\,(\delta u_m)_{/l} \tag{32}$$

characterizes the rotation in the transformation $C'C''$ characterized by the displacement $\underline{u}$. The symmetrical matrix $t^{(s)}_{rs}$ corresponds to the classical Cauchy stress. It is to be remarked that $dl^{(i)}$ does not depend on the asymmetrical part, $t^{(a)}_{rs}$, of the general stress matrix, t_{rs}. In fact, $t^{(a)}_{rs}$ has the meaning of without friction internal constraint reactions.

Assuming

$$t_{rs} = \frac{x_{r,l}x_{s,m}}{D} T_{lm}, \quad p_{rs} = \frac{x_{r,l}x_{s,m}}{D} P_{lm}, \tag{33}$$

the lagrangean parameters T_{rs}, P_{rs} in a equilibrium problem, satisfy the field and boudary equations

$$(x_{r,l}T_{ls})_{,s} = B_r, \quad (\text{in } C), \tag{34}$$

$$x_{r,l}T_{ls}N_s = b_r, \quad (\text{on } \sigma), \tag{35}$$

$$(x_{r,l}P_{ls})_{,s} + e_{rps}x_{p,l}x_{s,m}T^{a}_{ms} = M_r, \quad (\text{in } C\ ), \tag{36}$$

$$x_{r,l}P_{ls}N_s = m_r, \quad (\text{on } \sigma). \tag{37}$$

The following remark is fundamental: because of the property that the Free Energy, Z, depends on the deformation and keeping in mind that the deformation depends on the parameter h, therefore Z depends also on h:

$$Z = Z(\nu, \nu^i; h). \tag{38}$$

Further, as for the deformation, it is natural to attribute to Z the following property:

$$\lim_{h\to 0} Z = \Sigma(\epsilon) \tag{39}$$

and to admit that for $h = 0$ the partial derivatives of any order with respect to h exist.

With a lot of patience, one may give a lagrangean expression to the work of the internal forces and show that Z may be expressed as a function of ϵ_{rs}, $\epsilon_{rs,i}$ and h:

$$Z = \overline{\Sigma}(\epsilon_{rs}, \epsilon_{rs,i}; h) \tag{40}$$

Further, one obtains the constitutive relations for the parameters $T^{(s)}_{re}$, P_{rs}. One deduces

$$\lim_{h\to 0} T^{(s)}_{rs} = T^{(s)^{(0)}}_{rs}(\epsilon), \quad (\text{Cauchy model}), \tag{41}$$

$$\lim_{h\to 0} P_{rs} = 0. \tag{42}$$

The ralations (41), (42) are basic in order to show that the field and boundary problem in the vector $\underline{u}$ depends on a differential operator which is similar to that of Cauchy's problem and that the vector $\underline{m}$ is determined by the solution of the Cauchy's problem, as happens for the P_{rs}.

For example, when Taylor's series are valid:

$$\underline{u} = \underline{u}^{(0)}_r + h\underline{u}^{(1)}_r + ... \tag{43}$$

$$T_{rs} = T^{(0)}_{rs} + hT^{(1)}_{rs} + ... \tag{44}$$

$$P_{rs} = P^{(0)}_{rs} + hP^{(1)}_{rs} + ... \tag{45}$$

it follows:

$$P^{(0)}_{rs} = 0, \quad P^{(1)}_{rs} = P^{(1)}_{rs}(\underline{u}^0), ... \tag{46}$$

and, therefore,

$$\underline{m} = \underline{m}^{(0)} + h\underline{m}^{(1)} + ... \tag{47}$$

where

$$\underline{m}^{(0)} = 0, \quad \underline{m}_r^{(1)} = x_{r,l}^{(0)} P_{ls}^{(1)} (\underline{u}^{(0)} N_s, \ldots \tag{48}$$

That is, the vector $m^{(i)}$, $(i = 1.2....)$ are determined by the vectors $\underline{u}^{(0)}$, $\underline{u}^{(1)}$, $\underline{u}^{(2)}$,...,$\underline{u}^{(i-1)}$, which are solutions of Cauchy's type problem.

Therefore, it is not necessary to assign the vector $\underline{m}$ on the boundary: the mathematical model has physical concreteness.

References

1. G. Grioli, *Rend. di Matematica. S.* VII, vol.120, 281 (2000).
2. A. Signorini, *Ann. Mat. Pura e Appl. (memoria II)*, (4), 30, 1 (1947).
3. G. Grioli, *Mathematical Theory of Elastic Equilibrium (Recent Results)*, Springer–Verlag (1962).
4. G. Capriz & P. P. Podio Guidugli, *Arch. Rational Mech. An.*, 57, 1 (1974).
5. S. Baratha & Levinson *Arch. Rational Mech. An.*, 1, 365 (1977).
6. G. Capriz & P. P. Podio Guidugli, *Arch. Rational Mech. An.*, 261 (1979).
7. G. Grioli, *Applicable Analysis* (1983).
8. G. Grioli, *Mechanics*, 18, 3 (1983).
9. C. Cattaneo, *Atti del Seminario Matematico e Fisico dell'Università di Modena*, 3, 83 (1948).
10. C. Cattaneo, *Compt. Rend. Acad. Sc.*, 247 (1958).
11. G. Grioli, *Atti dell'Accademia Gioenia di Catania*, (1987).
12. G. Grioli, *Mathematical Reality and Physical Reality in Continuum Mechanics*, X Conferences on Waves and Stability in Continuous Media.
13. G. Grioli, *Cauchy theory and the continua of Cosserat: new points of view*, Atti del II simposio internazionale "Problemi attuali dell'Analisi e della Fisica Matematica", Taormina 15–17 ottobre 1995. Ed. Aracne Roma (2000).

PERTURBATIVE METHOD IN THE STUDY OF NONLINEAR EVOLUTION PROBLEMS IN THE DIFFUSION OF THE PARTICLES OF A MIXTURE

GABRIELE GUERRIERO

Dipartimento di Matematica e Applicazioni "R.Caccioppoli" dell'Università degli Studi di Napoli Federico II 80126 Napoli via Cinthia, Italy
E-mail:gabriele.guerriero@dma.unina.it

A perturbative method, in order to generate explicit solutions to a nonlinear equation of the particle transport theory in the stochastic models, is developed.

1 Introduction

Nonlinear evolution problems for the diffusion of the particles of a mixture are of great interest from both the basic and the applicative point of view, for example the diffusion of thermal neutrons in matter. We consider the following physical situation. A spatial uniform pulsed source,$Q^{\star}(\mathbf{v},t) = QS(\mathbf{v})\delta(t)$, emits certain $\mathcal{A}$ particles, of mass m, with a given distribution in velocity $S(\mathbf{v})$, nonnegative and normalized to unity, in the interior of an infinite host medium, consisting of certain other $\mathcal{B}$ particles having mass M and whose total density N is a constant fixed once for all. The $\mathcal{A}$ particles so emitted diffuse then in the host medium by binary collisions against the $\mathcal{B}$ particles as well as between themselves, the general collision resulting in either scattering or removal events. For describing this phenomenon the natural model appears to be the Boltzmann equation and recently very attention has been given to this equation in the frame of the so called *scattering kernel* for the collisional term. The physical problem described has been object of several investigation aimed at focusing the mathematical problems connected with existence, uniqueness, structure and form of the solutions [6–13]. We remark that the possibility of constructing explicit exact solutions to the nonlinear Boltzmann equation depends on both the cross sections and the scattering probability. In particular we recall the works of Krook and Wu and Bobylev who first obtained indipendently the exact solution for the isotropic distribution function in the special case of $1/v$ cross sections [10].

In order to describe this physical situation and to evaluate the distribution function $f(\mathbf{v},t)$ of the $\mathcal{A}$ particles considered we resort to the scattering kernel formulation of the Boltzmann equation, that in the limit of constant microscopic collision frequencies read as [3,8]:

$$\left(\partial_t + \frac{\mathbf{F}(t)}{m} \cdot \nabla_{\mathbf{v}}\right) f(\mathbf{v},t) = -[N\hat{C} + Cn(t)]f(\mathbf{v},t) +$$

$$N\hat{C}_s \int_{\Re^3} \hat{\pi}_s(\mathbf{v}',\mathbf{v}) f(\mathbf{v}',t) d\mathbf{v}' + \tag{1}$$

$$C_s \int_{\Re^3}\int_{\Re^3} \pi_s(\mathbf{v}',\mathbf{v}'',\mathbf{v}) f(\mathbf{v}',t) f(\mathbf{v}'',t) d\mathbf{v}' d\mathbf{v}''$$

and is to be integrated upon the initial condition

$$f(\mathbf{v},0) = QS(\mathbf{v}). \tag{2}$$

in the equation (1)

$$n(t) = \int_{\Re^3} f(\mathbf{v},t) d\mathbf{v} \tag{3}$$

denotes the unknown total density of the $\mathcal{A}$ particles considered, and

$$\hat{C} = \hat{C}_s + \hat{C}_r\,, \qquad C = C_s + C_r \tag{4}$$

denote the total microscopic frequencies (the subscripts s, r referring to scattering and removal events respectively), for the relevant cross sections, of the $\mathcal{A}$–$\mathcal{A}$ and $\mathcal{A}$–$\mathcal{B}$ collisions respectively. All the constants $\hat{C}_i$, C_i $(i = s, r)$ are real and nonnegative. The scattering probability distributions $\hat{\pi}_s(\mathbf{v}',\mathbf{v})$ and $\pi_s(\mathbf{v}',\mathbf{v}'',\mathbf{v})$ are nonnegative functions obeying the normalization conditions:

$$\int_{\Re^3} \hat{\pi}_s(\mathbf{v}',\mathbf{v}) = 1\,, \quad \int_{\Re^3} \pi_s(\mathbf{v}',\mathbf{v}'',\mathbf{v}) = 1\,. \tag{5}$$

Further π_s is symmetric with respect to the velocities $\mathbf{v}'$ and $\mathbf{v}''$ before the collisions, that is

$$\pi_s(\mathbf{v}',\mathbf{v}'',\mathbf{v}) = \pi_s(\mathbf{v}'',\mathbf{v}',\mathbf{v}). \tag{6}$$

Equation (1) is different, formally, from the usual kinetic formulation of the Boltzmann equation which is generally adopted [4].But, in the case of Maxwellian particles with a cut-off and for deterministic model based on momentum and energy conservation laws, the homogeneous Boltzmann equation is equivalent to equation (1) for a suitable specification of the scattering probability distributions [2,5]. In a previous paper we have proved, for the problem considered above, the global existence and uniqueness of the solutions to eq. (1).Then the problem of determining the distribution function $f(\mathbf{v},t)$ for $t \in [0,\infty)$, under the initial condition $f(\mathbf{v},0) = QS(\mathbf{v})$, arises. We recall that, in the limit of no background of $\mathcal{B}$ particles, i.e. when the $\mathcal{A}$ particles interact only through scattering collisions, the afore said problem has been studied in ref.12, where the global existence and uniqueness of $f(\mathbf{v},t)$ has been studied in the natural space $L_1(\Re^3) \times [0,\infty)$, while in Ref.13 we succeed in constructing explicit solutions for the considered problem. The aim of present paper is to extend the results obtained in Ref.7 for a mixture of two type of particles. Precisely, after some preliminaries (Sect.1 and Sect2), in Sect.3, in the framework of stochastic models for which the scattering probability distributions are expanding as power series of a parameter, we define an iterative constructive scheme leading to explicit solution of the form of power series.

2 Basic theory

If $f \in \mathbf{L}^1(\mathbf{R}^3)$ by integration of equation (1) over $\mathbf{v} \in \mathbf{R}^3$, we obtain the continuity equation:

$$\frac{\mathrm{d}\,n(t)}{\mathrm{d}\,t} + N\hat{C}_r n(t) + C_r n^2(t) = 0 \tag{7}$$

whose explicit solution, satisfying the initial conditions $n(0) = Q$ is

$$n(t) = QN\hat{C}_r[(N\hat{C}_r + QC_r)\exp(N\hat{C}_r t) - QC_r]^{-1} \; . \tag{8}$$

A meaningful case in which equation (1) can be solved analytically is when $\hat{\pi}_s$ and π_s are represented by model

$$\hat{\pi}_s(\mathbf{v}',\mathbf{v}) = \hat{\psi}_s(\mathbf{v}) \; , \quad \pi_s(\mathbf{v}',\mathbf{v}'',\mathbf{v}) = \psi_s(\mathbf{v}) \tag{9}$$

Equation (1) now becomes

$$\begin{aligned} &\frac{\partial f}{\partial t} + \frac{\mathbf{F}}{m}\nabla_{\mathbf{v}} f(\mathbf{v},t) + N[\hat{C} f(\mathbf{v},t) - \hat{C}_s\hat{\psi}_s(\mathbf{v})n(t)]+ \\ &+[Cf(\mathbf{v},t) - C_s\psi_s(\mathbf{v})n(t)]n(t) = 0. \end{aligned} \tag{10}$$

This equation can be easily integrated upon the initial condition $f(\mathbf{v},0) = QS(\mathbf{v})$, using standard Fourier transform, and yields [1]:

$$f_0(\mathbf{v},t) =$$
$$n(t)^{C/C_r} \exp\left(-N\frac{\hat{C}C_r - C\hat{C}_r}{C_r}t\right) \left[Q \exp(-C_s/C_r) S\left(\mathbf{v} - \int_0^t \frac{\mathbf{F}(t')}{m}\,\mathrm{d}\,t'\right) + \right.$$
$$N\hat{C}_s \int_0^t \hat{\psi}_s \left(\mathbf{v} - \int_{t'}^t \frac{\mathbf{F}(t'')}{m}\mathrm{d}\,t''\right) \exp\left(-N\frac{\hat{C}C_r - C\hat{C}_r}{C_R}t'\right) n(t')^{-C/C_r}\mathrm{d}\,t' +$$
$$\left. C_s \int_0^t \psi_s \left(\mathbf{v} - \int_{t'}^t \frac{\mathbf{F}(t'')}{m}\mathrm{d}\,t''\right) \exp\left(N\frac{\hat{C}C_r - C\hat{C}_r}{C_r}t'\right) n(t')^{(1-C_s/C_r)}\mathrm{d}\,t'\right]. \tag{11}$$

3 Solutions

In the sequel we shall assume:

Hypothesis. *The scattering probabilities* π_s $\hat{\pi}_s$ *are expanding about* $\hat{\Psi}(\mathbf{v})$, $\Psi(\mathbf{v})$ *as power series of a parameter* ε *according to*

$$\begin{cases} \pi_s(\mathbf{v}',\mathbf{v}'',\mathbf{v}) = \sum_{i=1}^{\infty} \varepsilon^i \pi_s(\mathbf{v}',\mathbf{v}'',\mathbf{v}) + \Psi_s(\mathbf{v}) \\ \hat{\pi}_s(\mathbf{v}',\mathbf{v}) = \sum_{i=1}^{\infty} \varepsilon^i \hat{\pi}_i(\mathbf{v}',\mathbf{v}) + \hat{\Psi}_s(\mathbf{v}) \end{cases} \tag{12}$$

In order to satisfy the normalization and symmetry conditions we require:

$$\int_{\Re^3} \pi_i(\mathbf{v}',\mathbf{v}'',\mathbf{v})\, d\mathbf{v} = 0 \qquad \int_{\Re^3} \hat{\pi}_i(\mathbf{v}',\mathbf{v})\, d\mathbf{v} = 0 \tag{13}$$

$$\pi_i(\mathbf{v}',\mathbf{v}'',\mathbf{v}) = \pi_i(\mathbf{v}'',\mathbf{v}',\mathbf{v}). \tag{14}$$

Because, when the scattering probabilities π_s, $\hat{\pi}_s$ are independent of the velocities before the collision

$$\hat{\pi}_s(\mathbf{v}',\mathbf{v}) = \hat{\Psi}_s(\mathbf{v}), \quad \pi_s(\mathbf{v}',\mathbf{v}'',\mathbf{v}) = \Psi_s(\mathbf{v}), \tag{15}$$

the solution to equation (1) is $f_0(\mathbf{v},t)$ of eq.(11), it is then quite natural to seek a solution to equation (1) in the form

$$f(\mathbf{v},t) = f_0(\mathbf{v},t) + \sum_{i=1}^{\infty} \varepsilon^i f_i(\mathbf{v},t) \tag{16}$$

with

$$\int_{\Re^3} f_i(\mathbf{v},t)\, d\mathbf{v} = 0 \qquad f_i(\mathbf{v},0) = 0 \tag{17}$$

It is easy to verify that evolution equations for f_i, $(i = 0, 1, 2, ...)$ are

$$\begin{cases} \dfrac{\partial f_0}{\partial t} + \dfrac{\mathbf{F}}{m} \nabla_{\mathbf{v}} f_0 + N[\hat{C} f_0 - \hat{C}_s \hat{\psi}_s(\mathbf{v}) n(t)] + \\ [C f_0 - C_s \Psi_s(\mathbf{v}) n(t)]\, n(t) = 0 \end{cases} \tag{18}$$

$$\begin{cases} \dfrac{\partial f_1}{\partial t} + \dfrac{\mathbf{F}}{m} \nabla_{\mathbf{v}} f_1 + N\hat{C} f_1 + Cn(t) f_1 = N\hat{C}_s \displaystyle\int_{\Re^3} \hat{\pi}_1(\mathbf{v}',\mathbf{v})\, f_0(\mathbf{v}',t)\, d\mathbf{v}' + \\ C_s \displaystyle\int_{\Re^3} \int_{\Re^3} \pi_1(\mathbf{v}',\mathbf{v}'',\mathbf{v})\, f_0(\mathbf{v}',t) f_0(\mathbf{v}'',t)\, d\mathbf{v}'\, d\mathbf{v}'' \end{cases} \tag{19}$$

$$\begin{cases} \dfrac{\partial f_2}{\partial t} + \dfrac{\mathbf{F}}{m} \nabla_{\mathbf{v}} f_2 + N\hat{C} f_2 + Cn(t) f_2 = \\ N\hat{C}_s \displaystyle\int_{\Re^3} \hat{\pi}_1(\mathbf{v}',\mathbf{v}) f_1(\mathbf{v}',t)\, d\mathbf{v}' + N\hat{C}_s \displaystyle\int_{\Re^3} \hat{\pi}_2(\mathbf{v}',\mathbf{v}) f_0(\mathbf{v}',t)\, d\mathbf{v}' + \\ C_s \displaystyle\int_{\Re^3} \int_{\Re^3} \pi_2(\mathbf{v}',\mathbf{v}'',\mathbf{v})\, f_0(\mathbf{v}',t) f_0(\mathbf{v}'',t)\, d\mathbf{v}'\, d\mathbf{v}'' + \\ 2C_s \displaystyle\int_{\Re^3} \int_{\Re^3} \pi_1(\mathbf{v}',\mathbf{v}'',\mathbf{v})\, f_0(\mathbf{v}',t) f_0(\mathbf{v}'',t)\, d\mathbf{v}'\, d\mathbf{v}'' \end{cases} \tag{20}$$

....

Recalling that π_i, $\hat{\pi}_i\,(i = 1, 2, ...)$ are known from equation (19) one obtains $f_1(\mathbf{v},t)$. Inserting $f_1(\mathbf{v},t)$ in (20) one obtain $f_2(\mathbf{v},t)$ and so on. The following theorem holds:

Theorem *If* $\exists M > 0$ *such that:*

$$\left| \int_{\Re^3} \pi_n(\mathbf{v}', \mathbf{v}'', \mathbf{v}) \, \Psi_1(\mathbf{v}')\Psi_2(\mathbf{v}'') \, d\mathbf{v}' \, d\mathbf{v}'' \right| \leq M^n \tag{21}$$

$$\left| \int_{\Re^3} \hat{\pi}_n(\mathbf{v}', \mathbf{v}) \, \Psi_0(\mathbf{v}')) \, d\mathbf{v}' \right| \leq M^n \tag{22}$$

$$\forall \Psi_i(\mathbf{v}), \quad \int_{\Re^3} \Psi_i(\mathbf{v}) \, d\mathbf{v} \leq Q \quad i = 0, 1, 2 \tag{23}$$

then the series

$$f(\mathbf{v}, t) = f_0(\mathbf{v}, t) + \sum_{i=1}^{\infty} \varepsilon^i \, f_i(\mathbf{v}, t), \qquad \varepsilon < \frac{1}{M}, \tag{24}$$

where f_i *are* L^1*-solutions to the initial value problem (19),(20),... and* $f_i(\mathbf{v}, 0) = 0$*), is a.e.uniformly convergent in* $\Re^3 \times [0, \infty)$*. Moreover (16) can be differentiated (with respect to t) term by term, is termwise* L^1*-summable (with respect to* $\mathbf{v}$*) and is solution to equation (1).* [a]

References

1. Boffi V.C., Spiga G., *Physical Review A*, **29** n.2, (1984).
2. Boffi V.C., Protopopescu V. e Spiga G., *Physica A* **164** , (1990).
3. Boffi V.C. e Spiga G., *J. Math. Phys.*, **23**, pp.2299-2303, (1982).
4. Cercignani C.,The Boltzmann Equation and Its Applications, Springer-Verlag, New York (1988).
5. Cercignani C., Lampis M., Sgarra C., *Il Nuovo Cimento*, **101** B, (1989).
6. Guerriero G., *Rend. Acc. Sc. Fis. Mat.*, Napoli Serie IV **LVI** , (1989).
7. Guerriero G., Advances in Mathematics for Applied Sciences Vol.4, World Scientific pp.162-167, (1991).
8. Guerriero G.,*Int.J.Engng.Sci.* **29**, No.9, pp.1077-1083, (1991).
9. Guerriero G. *Rend. Acc. Sc. Fis. Mat.* Napoli **LXII** , (1995).
10. Krook M.,Wu T.T.,*Phys. Rev. Lett.***36**,1107 (1976).
11. Rionero S., *Atti Acc. Sc. Torino*, Suppl. **120** p.199-209, (1986).
12. Rionero S.,Guerriero G., *Meccanica* **22**, pp.59-64, (1987).
13. Rionero S.,Guerriero G.,*Meccanica* **24**, pp.191-199, (1989).

[a] **Acknowledgments.** This research has been performed under the auspices of the G.N.F.M.of I.N.D.A.M. and has been partially supported by M.I.U.R., PRIN : Problemi matematici non lineari di propagazione e stabilità nei modelli del continuo.

ON THE GENERALIZED ENERGY METHOD FOR CHANNEL FLOWS

RALF KAISER
Department of Mathematics, Universität Bayreuth, D-95440 Bayreuth, Germany

ANDREAS TILGNER
Institute of Geophysics, Universität Göttingen, D-37075 Göttingen, Germany

The method of generalized energy functionals is discussed with respect to its applicability to channel flows. We present functionals which provide better stability boundaries than so far known. However, the rigid boundary conditions turn out to be incompatible with the standard procedure to arrive at a nonlinear stability result. Thus, the method does not seem to be appropriate for channel flows.

1 Introduction

Although the flow of a viscous incompressible fluid in plane parallel channels belongs to the simplest hydrodynamical systems the stability of the basic flow (Couette flow in the boundary driven case and Poiseuille flow in the pressure driven case) is still only unsatisfactorily understood. Typically, in these systems a quite direct transition from the laminar basic flow to a turbulent one is observed. Thus, the stability boundary of the basic flow represents likewise the onset of turbulence in these systems [5].

Linear stability analysis can only be performed with asymptotic or numerical methods and provide critical Reynolds numbers Re_c far above the experimentally observed onset of instability (see e.g. ref. [4]). For Couette flow Re_c is even infinite [11]. The energy method on the other side provides unconditional and monotonic stability with respect to the energy norm for Reynolds numbers below a number Re_E which, unfortunately, is far below the onset of instability [8,1]. So, obviously the former method overestimates and the latter one underestimates the stability of the basic flow and another method describing more accurately the stability region would be desirable.

The method of generalized energy functionals is a candidate for such a more appropriate method. In more complicated hydrodynamical systems the use of generalized energy functionals turned out to be quite successful (see e.g. ref. [6] or ref. [12]). In channel flows, however, no such functionals providing better stability results than the energy functional have been found so far. In this note we want to discuss the method for Couette flow and will argue that, unfortunately, no better stability results can be expected with this method.

A generalized energy functional $\mathcal{E}$ is a bilinear form of the dynamic variables of the problem. In comparison with the ordinary energy these variables are, however, differently weighted by additional coupling parameters and appear possibly in the form of higher derivatives. A first part $\mathcal{E}_1$ of the functional determines (analogously to the energy method) via a variational problem the stability boundary $\mathrm{Re}_{\mathcal{E}}$. The coupling parameters are chosen such that $\mathrm{Re}_{\mathcal{E}}$ becomes as large as possible. Contrary to the energy balance the nonlinear terms in general do not drop from the generalized energy balance. Therefore, a second part $\mathcal{E}_2$ involving higher derivatives of the dynamic variables is needed in $\mathcal{E}$ in order to dominate these terms. Note that only a conditional stability result can be expected if the nonlinear terms do not drop.

The stability of Couette – as well as of Poiseuille flow has also been investigated under the assumption of stress–free boundary conditions for the perturbations [10]. In that case the method of generalized energy functionals worked very well and furnished nonlinear conditional stability for all Reynolds numbers. For channel flows, however, stress–free boundary conditions are clearly not suitable.

In section 2 we set the mathematical framework for the subsequent discussion of the generalized energy method. In particular we introduce the socalled poloidal–toroidal decomposition of solenoidal vector fields which allows us to eliminate the divergence–free constraint. In section 3 possible candidates for $\mathcal{E}_1$ are discussed and the associated variational problem is solved. Section 4, finally, deals with $\mathcal{E}_2$ and the problems arising from the rigid boundary conditions.

2 Mathematical setting

Let us consider the following situation: In an infinite layer $\mathbb{R} \times (-\frac{1}{2}, \frac{1}{2})$ with horizontal coordinates x, y and vertical one z there is given a steady solution $\mathbf{v}_0$ of the Navier-Stokes system of the form

$$\mathbf{v}_0 = \mathbf{v}_0(z) = \mathrm{Re}(f(z),\, 0,\, 0)^{\mathrm{T}}, \tag{1}$$

i.e. $\mathbf{v}_0$ is a plane parallel shear flow. The function $f : [-\frac{1}{2}, \frac{1}{2}] \to \mathbb{R}$ is assumed to be sufficiently smooth and is called the shear profile. For Couette flow $f(z) = -z$ and for Poiseuille flow $f(z) = 1 - 4z^2$. $\mathrm{Re} > 0$ is the Reynolds number. In order to investigate the stability of $\mathbf{v}_0$ we impose x, y-periodic perturbations $\mathbf{u} = \mathbf{u}(x, y, z, t)$. These are governed by the system

$$\left.\begin{aligned} \partial_t \mathbf{u} - \Delta \mathbf{u} + \mathbf{u} \cdot \nabla \mathbf{u} + \mathrm{Re} f \partial_x \mathbf{u} + \mathrm{Re} f' u_z \mathbf{e}_x + \nabla p = 0 \\ \nabla \cdot \mathbf{u} = 0 \end{aligned}\right\} \tag{2}$$

in $\mathbb{R}^2 \times (-\frac{1}{2}, \frac{1}{2}) \times (0,T)$ and satisfy the boundary conditions

$$\mathbf{u}(x,y,z,t) = 0 \qquad \text{for } (x,y,z) \in \mathbb{R}^2 \times \{-\frac{1}{2}, \frac{1}{2}\},\ t > 0. \tag{3}$$

Here $\mathbf{e}_x = (1,0,0)^{\mathrm{T}}$. The initial value $\mathbf{u}(\cdot,0) = \mathbf{u}_0$ at time $t = 0$ is assumed to be given (and of course solenoidal). $\mathbf{u}$ corresponds to the velocity field of the perturbation and $p = p(x,y,z,t)$ denotes the pressure. Both $\mathbf{u}$ and p are x,y-periodic with respect to a rectangle $\mathcal{P} = (-\frac{\pi}{\alpha}, \frac{\pi}{\alpha}) \times (-\frac{\pi}{\beta}, \frac{\pi}{\beta})$ with wave numbers $\alpha, \beta \in (0,\infty)$. In the following it suffices therefore to consider functions over the box

$$\Omega = \mathcal{P} \times (-\frac{1}{2}, \frac{1}{2}) = (-\frac{\pi}{\alpha}, \frac{\pi}{\alpha}) \times (-\frac{\pi}{\beta}, \frac{\pi}{\beta}) \times (-\frac{1}{2}, \frac{1}{2}).$$

As basic function space we take $L^2(\Omega)$. In the sequel $\|\cdot\|$ always is the norm in $L^2(\Omega)$ except for the case when applied to a function defined on $(-\frac{1}{2}, \frac{1}{2})$. Then $\|\cdot\|$ means the norm in $L^2(-\frac{1}{2}, \frac{1}{2})$; the correct notion should be clear from the context. $(\cdot,\cdot)$ always denotes the scalar product associated with $\|\cdot\|$.

In order to cope with the solenoidality condition $(2)_2$ we make use of the poloidal-toroidal decomposition [13]:

$$\begin{aligned} \mathbf{u} &= \nabla \times (\nabla \times (\varphi\, \mathbf{e}_z)) + \nabla \times (\psi\, \mathbf{e}_z) + \mathbf{F} \\ &=: \boldsymbol{\delta}\varphi + \boldsymbol{\varepsilon}\psi + \mathbf{F}. \end{aligned} \tag{4}$$

Here $\mathbf{e}_z = (0,0,1)^{\mathrm{T}}$. The functions φ and ψ are determined uniquely if one requires them to be periodic with respect to $\mathcal{P}$ and to fulfill $\int_{\mathcal{P}} \varphi(x,y,z)\,dx\,dy = 0 = \int_{\mathcal{P}} \psi(x,y,z)\,dx\,dy$ for every $z \in (-\frac{1}{2}, \frac{1}{2})$. The first part in (4) is called the poloidal part of $\mathbf{u}$ and the second one the toroidal one. The third part, the mean flow, depends only on z and has constant third component. These three parts are mutually orthogonal in $L^2(\Omega)^3$. The vector operators $\boldsymbol{\delta}$ and $\boldsymbol{\varepsilon}$ have the form

$$\boldsymbol{\delta}\varphi = \begin{pmatrix} \partial_x\partial_z\varphi \\ \partial_y\partial_z\varphi \\ (-\Delta_2)\varphi \end{pmatrix}, \qquad \boldsymbol{\varepsilon}\psi = \begin{pmatrix} \partial_y\psi \\ -\partial_x\psi \\ 0 \end{pmatrix},$$

where $\Delta_2 = \partial_x^2 + \partial_y^2$ is the horizontal Laplacian. The boundary conditions (3) for $\mathbf{u}$ transform into

$$\varphi = \partial_z\varphi = 0, \quad \psi = 0, \quad F_x = F_y = 0 \qquad \text{for } z = \pm\frac{1}{2}, \tag{5}$$

and $F_z(z) \equiv 0$. Applying the operators $\boldsymbol{\delta}$ and $\boldsymbol{\varepsilon}$ to eq. $(2)_1$ as well as taking the mean with respect to $\mathcal{P}$ the system (2) can equivalently be formulated in terms of the new variables $(\varphi,\ \psi,\ F_x,\ F_y)$:

$$\begin{aligned}
&(-\Delta)(-\Delta_2)\partial_z\varphi + \Delta^2(-\Delta_2)\varphi \\
&\qquad + \mathrm{Re} f\,(-\Delta)(-\Delta_2)\partial_x\varphi + \mathrm{Re} f''(-\Delta_2)\partial_x\varphi + \boldsymbol{\delta}\cdot(\mathbf{u}\cdot\nabla\mathbf{u}) = 0, \\
&(-\Delta_2)\partial_t\psi + (-\Delta)(-\Delta_2)\psi \\
&\qquad -\mathrm{Re} f'(-\Delta_2)\partial_y\varphi + \mathrm{Re} f\,(-\Delta_2)\partial_x\psi - \boldsymbol{\varepsilon}\cdot(\mathbf{u}\cdot\nabla\mathbf{u}) = 0, \\
&\qquad \partial_t F_x + (-\partial_z^2)F_x + \tfrac{1}{|\mathcal{P}|}\textstyle\int_{\mathcal{P}} \tilde{\mathbf{u}}\cdot\nabla\tilde{u}_x\,dxdy = 0, \\
&\qquad \partial_t F_y + (-\partial_z^2)F_y + \tfrac{1}{|\mathcal{P}|}\textstyle\int_{\mathcal{P}} \tilde{\mathbf{u}}\cdot\nabla\tilde{u}_y\,dxdy = 0.
\end{aligned} \tag{6}$$

$\tilde{\mathbf{u}} := \boldsymbol{\delta}\varphi + \boldsymbol{\varepsilon}\psi$ is that part of $\mathbf{u}$ which has vanishing mean value over $\mathcal{P}$ and $|\mathcal{P}| := \frac{4\pi^2}{\alpha\beta}$ denotes the volume of $\mathcal{P}$.

The energy of the system (in the volume Ω) takes in these variables the form

$$E = \frac{1}{2}\,||\mathbf{u}||^2 = \frac{1}{2}\Big\{||\boldsymbol{\delta}\varphi||^2 + ||\boldsymbol{\varepsilon}\psi||^2 + |\mathcal{P}|||\mathbf{F}||^2\Big\}, \tag{7}$$

and the variational expression determining Re_E takes the form

$$\frac{|\Re(u_x\,,f'u_z)|}{||\nabla\mathbf{u}||^2} = \frac{|\Re((-\Delta_2)\varphi\,,f'(\partial_x\partial_z\varphi + \partial_y\psi + F_x))|}{||(-\Delta)\boldsymbol{\varepsilon}\varphi||^2 + ||\boldsymbol{\delta}\psi||^2 + |\mathcal{P}|||\partial_z\mathbf{F}||^2}\,. \tag{8}$$

For later convenience we admit here complex valued velocity fields. Thus, the real part (denoted by $\Re$) of the interaction term appears in the numerator of (8). Re_E is then given by

$$\mathrm{Re}_E^{-1} = \sup_{(\alpha,\beta)\in\mathbb{R}^2}\ \sup_{(\varphi,\psi)\in\mathcal{C}_{\alpha\beta}} \frac{|\Re((-\Delta_2)\varphi\,,f'(\partial_x\partial_z\varphi + \partial_y\psi))|}{||(-\Delta)\boldsymbol{\varepsilon}\varphi||^2 + ||\boldsymbol{\delta}\psi||^2} \tag{9}$$

with

$$\begin{aligned}
\mathcal{C}_{\alpha\beta} := \Big\{&(0,0)\not\equiv(\varphi,\psi)\,:\ \varphi,\psi \text{ periodic in } x \text{ and } y \text{ with periodicity lenghts} \\
&2\pi/\alpha \text{ and } 2\pi/\beta, \text{ sufficiently regular (cf. ref. }^{9}) \text{ and satisfying} \\
&\varphi = \partial_z\varphi = \psi = 0 \text{ at } z = \pm\tfrac{1}{2} \text{ as well as} \\
&\textstyle\int_{\mathcal{P}}\varphi\,dxdy = \int_{\mathcal{P}}\psi\,dxdy = 0 \text{ for } z\in(-\frac{1}{2},\,\frac{1}{2})\Big\}.
\end{aligned}$$

Note that $\mathbf{F}$ does not depend on x or y and, therefore, drops from the numerator of (8). Thus, $\mathbf{F}$ does not contribute to the supremum of (8) and can be omitted altogether.

If the class $\mathcal{C}_{\alpha\beta}$ of admissible functions is restricted to the class $\mathcal{C}_\alpha$ of functions depending only on x and z, or to the class $\mathcal{C}_\beta$ of functions depending only on y and z the corresponding 2–dimensional limits Re_E^x and Re_E^y are determined by the following simplified variational expressions:

$$1/\mathrm{Re}_E^y = \sup_{\beta \in \mathbb{R}} \sup_{(\varphi,\psi)\in \mathcal{C}_\beta} \frac{|\Re((-\Delta_2)\varphi\,,\, f'\partial_y\psi)|}{\|(-\Delta)\partial_y\varphi\|^2 + \|\boldsymbol{\delta}\psi\|^2}, \tag{10}$$

$$1/\mathrm{Re}_E^x = \sup_{\alpha \in \mathbb{R}} \sup_{(\varphi,0)\in \mathcal{C}_\alpha} \frac{|\Re((-\Delta_2)\varphi\,,\, f'\partial_x\partial_z\varphi)|}{\|(-\Delta)\partial_x\varphi\|^2}. \tag{11}$$

If the profile function f is an odd function of z the variational expression (11) expressed in terms of φ_e and φ_o, the even and odd parts of φ with respect to z, takes the form:

$$1/\mathrm{Re}_E^x = \sup_{\alpha \in \mathbb{R}} \sup_{\substack{(\varphi_e,0)\\(\varphi_o,0)}\in \mathcal{C}_\alpha} \frac{|\Re((-\Delta_2)\varphi_e\,,\, f'\partial_x\partial_z\varphi_o) + \Re((-\Delta_2)\varphi_o\,,\, f'\partial_x\partial_z\varphi_e)|}{\|(-\Delta)\partial_x\varphi_e\|^2 + \|(-\Delta)\partial_x\varphi_o\|^2}. \tag{12}$$

In the case of Couette flow it is well known that $\mathrm{Re}_E = \mathrm{Re}_E^y \approx 82.6$ and $\mathrm{Re}_E^x \approx 177.2$ (cf. ref. [2]).

3 Generalized functionals and variational problems for Couette flow

The usual method to proceed from the energy functional to a generalized one is to introduce additional coupling parameters and possibly additional derivatives in order to weigh the dynamic variables in an optimal way. For this purpose the generalized energy balance is considered and (analogously to the energy method) the ratio of interaction term over dissipative term is maximized with respect to the admissible functions. This maximum still depends on the coupling parameters and possibly discrete parameters counting the additional derivatives. Minimizing with respect to these parameters furnishes then optimal (generalized) energy limits. So, the first problem is finding functionals which furnish larger stability limits than those provided by the energy functional. We will explore this question in the following for the Couette profile $f(z) = -z$. Considering the functional (7) with $\mathbf{F} \equiv 0$ (the mean flow does not contribute to the maximum in the variational problem) there is however, not much freedom to introduce additional parameters. An obvious choice is the functional

$$E_\lambda(\varphi,\psi) := \frac{1}{2}\Big\{\|\boldsymbol{\delta}\varphi\|^2 + \lambda\,\|\boldsymbol{\varepsilon}\psi\|^2\Big\} \tag{13}$$

with $0<\lambda<\infty$, which is treated first. The generalized energy limit Re_{E_λ} is then determined by

$$\mathrm{Re}_{E_\lambda}^{-1}=\sup_{(\alpha,\beta)\in\mathbb{R}^2}\ \sup_{(\varphi,\psi)\in\mathcal{C}_{\alpha\beta}}\frac{|\Re((-\Delta_2)\varphi\,,\partial_x\partial_z\varphi)+\lambda\,\Re((-\Delta_2)\varphi\,,\partial_y\psi)|}{\|(-\Delta)\varepsilon\varphi\|^2+\lambda\,\|\delta\psi\|^2}\,. \tag{14}$$

Observe that a comparison of (14) with the 2–dimensional variational expressions (10) and (11) furnishes already some bounds on Re_{E_λ}: Setting $\varphi=\varphi(x,z)$, $\psi=0$ in (14) reduces the variational expression to that in (11) which implies the bound $\mathrm{Re}_{E_\lambda}\leq\mathrm{Re}_E^x\approx 177.2$ for all $0<\lambda<\infty$. For $\lambda>1$ the transformation $\tilde{\psi}:=\lambda\psi$ allows the estimate

$$\begin{aligned}&\frac{|\Re((-\Delta_2)\varphi\,,\partial_x\partial_z\varphi)+\lambda\,\Re((-\Delta_2)\varphi\,,\partial_y\psi)|}{\|(-\Delta)\varepsilon\varphi\|^2+\lambda\,\|\delta\psi\|^2}\\&=\frac{|\Re((-\Delta_2)\varphi\,,\partial_x\partial_z\varphi)+\Re((-\Delta_2)\varphi\,,\partial_y\tilde{\psi})|}{\|(-\Delta)\varepsilon\varphi\|^2+\frac{1}{\lambda}\|\delta\tilde{\psi}\|^2}\\&\geq\frac{|\Re((-\Delta_2)\varphi\,,\partial_x\partial_z\varphi)+\Re((-\Delta_2)\varphi\,,\partial_y\tilde{\psi})|}{\|(-\Delta)\varepsilon\varphi\|^2+\|\delta\tilde{\psi}\|^2}\end{aligned}$$

and restricting φ and $\tilde{\psi}$ to functions independent of x furnishes the bound $\mathrm{Re}_{E_\lambda}\leq\mathrm{Re}_E^y=\mathrm{Re}_E\approx 82.6$ for $\lambda>1$. Thus the question remains whether Re_{E_λ} does exceed Re_E for some $0<\lambda<1$. To answer this question we solve first the eigenvalue problem associated to the variational problem (14) with fixed periodicity cell $\mathcal{P}$ and perform subsequently the variation with respect to $\mathcal{P}$.

The Euler–Lagrange equations with Lagrange parameter μ read:

$$\begin{aligned}\Delta^2(-\Delta_2)\varphi-\frac{\mu}{2}\left(2\,(-\Delta_2)\partial_x\partial_z\varphi+\lambda\,(-\Delta_2)\partial_y\psi\right)&=0,\\ \lambda\,(-\Delta)(-\Delta_2)\psi+\frac{\mu}{2}\,\lambda\,(-\Delta_2)\partial_y\varphi&=0.\end{aligned} \tag{15}$$

Inserting the mode expansion $f(x,y,z)=\frac{1}{\sqrt{|\mathcal{P}|}}\sum_{\kappa\in\mathbb{Z}^2\setminus\{0\}}f_\kappa(z)e^{i(\alpha\kappa_1x+\beta\kappa_2y)}$ for the variables φ and ψ the system (15) becomes equivalent to:

$$\begin{aligned}D^2_{\kappa_1\alpha,\kappa_2\beta}\,\varphi_\kappa(z)-i\frac{\mu}{2}\left(2\alpha\kappa_1\partial_z\varphi_\kappa(z)+\lambda\,\beta\kappa_2\psi_\kappa(z)\right)&=0,\\ D_{\kappa_1\alpha,\kappa_2\beta}\,\psi_\kappa(z)+i\frac{\mu}{2}\beta\kappa_2\varphi_\kappa(z)&=0,\end{aligned}\qquad \kappa\in\mathbb{Z}^2\setminus\{0\} \tag{16}$$

with $D_{\tilde{\alpha},\tilde{\beta}}:=\tilde{\alpha}^2+\tilde{\beta}^2-\partial_z^2$. The system (16) has to be complemented with the boundary conditions

$$\varphi_\kappa=\partial_z\varphi_\kappa=\psi_\kappa=0\quad\text{at}\ \ z=\pm\frac{1}{2},\quad \kappa\in\mathbb{Z}^2\setminus\{0\}$$

in order to have a well–posed eigenvalue problem. Observe now that for a fixed periodicity cell $\mathcal{P}$ the supremum over $(\varphi,\psi) \in \mathcal{C}_{\alpha,\beta}$ of the variational expression in (14) is obtained by a single mode in the system (16) (cf. ref. [9]). Since we are ultimately interested in the supremum with respect to all perodicity cells it is, therefore, sufficient to consider the finite dimensional system

$$D^2_{\tilde{\alpha},\tilde{\beta}}\,\tilde{\varphi}(z) - i\frac{\mu}{2}\Big(2\tilde{\alpha}\partial_z\tilde{\varphi}(z) + \lambda\,\tilde{\beta}\tilde{\psi}(z)\Big) = 0, \qquad D_{\tilde{\alpha},\tilde{\beta}}\,\tilde{\psi}(z) + i\frac{\mu}{2}\tilde{\beta}\tilde{\varphi}(z) = 0 \tag{17}$$

together with

$$\tilde{\varphi} = \partial_z\tilde{\varphi} = \tilde{\psi} = 0 \quad \text{at} \quad z = \pm\frac{1}{2}\,. \tag{18}$$

Re_{E_λ} is then given by

$$\mathrm{Re}_{E_\lambda} = \min_{(\tilde{\alpha},\tilde{\beta})\in\mathbb{R}^2} \mu_0(\tilde{\alpha},\tilde{\beta},\lambda)$$

with μ_0 being the smallest eigenvalue in (17), (18). Due to the boundary conditions (18) the eigenfunctions of (17) cannot be expressed in a simple analytic way. Instead, we solve the (real 12–dimensional first order) system (17), (18) by a numerical procedure. Applying a standard shooting method based on a fourth order Runge–Kutta integration the lowest eigenvalue μ_0 is determined as a function of $\tilde{\alpha}$, $\tilde{\beta}$ and λ. Subsequent minimization with respect to $\tilde{\alpha}$ and $\tilde{\beta}$ furnishes Re_{E_λ} as a function of λ. The result is displayed in figure 1: With λ decreasing Re_{E_λ} increases from the ordinary energy limit $\mathrm{Re}_E \approx 82.6$ ($\lambda = 1$) up to the value $\mathrm{Re}_E^x \approx 177.2$ (fig. 1 left) and this value is in fact attained for finite λ ($\lambda \approx 0.042$, see fig. 1 right).

The question arises whether larger stability limits can be obtained with different functionals. Guided by the 2–dimensional expression (12) leading to $\mathrm{Re}_E^x \approx 177.2$ we consider the functional

$$E_{\tilde{\lambda},n}(\varphi_e,\varphi_o) := \frac{1}{2}\Big\{\|\nabla\partial_x^n\varphi_e\|^2 + \tilde{\lambda}\,\|\nabla\varphi_o\|^2\Big\} \tag{19}$$

with $\varphi_{e/o}$ being even/odd with respect to z and being independent of y. In comparison with (the poloidal part of) ordinary energy the even and odd part have been weighted differently by a coupling parameter $\tilde{\lambda} > 0$ and additionally by an x–derivative of order n ($n \in \mathbb{N}$). Note that common derivatives of φ_e and φ_o can be eliminated by redefinition. Higher order z–derivatives lead to problems which are discussed in the next section.

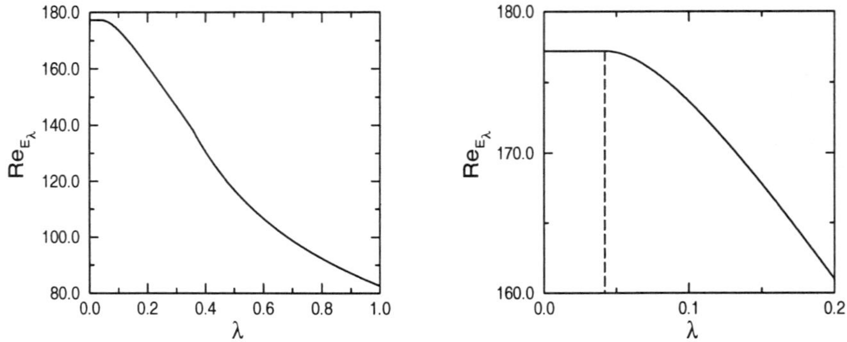

Figure 1. The generalized energy limit Re_{E_λ} versus coupling parameter λ with E_λ given in eq. (13). In the left figure λ covers the range between zero and 1 (corresponding to ordinary energy), the right figure magnifies the region close to $\lambda = 0$.

The associated (2–dimensional) stability limit is now given by

$$1/\mathrm{Re}^x_{E_{\tilde\lambda,n}} = \sup_{\alpha\in\mathbb{R}} \sup_{\substack{(\varphi_e,0)\\(\varphi_o,0)}\in\mathcal{C}_\alpha} \frac{|\Re((-1)^n\partial_x^{2n}\varphi_e\,,z(-\Delta)\partial_x\varphi_o) + \tilde\lambda\,\Re(\varphi_o\,,z(-\Delta)\partial_x\varphi_e)|}{\|(-\Delta)\partial_x^n\varphi_e\|^2 + \tilde\lambda\,\|(-\Delta)\varphi_o\|^2}. \tag{20}$$

The evaluation of (20) proceeds as above. The Euler–Lagrange equations with Lagrange parameter μ read:

$$\Delta^2(-1)^n\partial_x^{2n}\varphi_e - \frac{\mu}{2}\Big(z((-1)^n\partial_x^{2n} - \tilde\lambda)(-\Delta) + 2\,\lambda\partial_z\Big)\partial_x\varphi_o = 0,$$

$$\tilde\lambda\,\Delta^2\varphi_o - \frac{\mu}{2}\Big(z((-1)^{n+1}\partial_x^{2n} + \tilde\lambda)(-\Delta) + 2(-1)^n\partial_x^{2n}\partial_z\Big)\partial_x\varphi_e = 0.$$

For fixed α the supremum is again obtained by a single mode in the expansion $g(x,z) = \sqrt{\frac{\alpha}{2\pi}}\sum_{\kappa\in\mathbb{Z}\setminus\{0\}} g_\kappa(z)e^{i\alpha\kappa x}$ and the system to be solved reads now:

$$\begin{aligned} D^2_{\tilde\alpha}\,\tilde\varphi_e(z) - i\tilde\alpha\frac{\mu}{2}\Big(z(1-\lambda)D_{\tilde\alpha} + 2\,\lambda\,\partial_z\Big)\tilde\varphi_o = 0, \\ D^2_{\tilde\alpha}\,\tilde\varphi_o(z) - i\tilde\alpha\frac{\mu}{2}\Big(z(1-1/\lambda)D_{\tilde\alpha} + (2/\lambda)\partial_z\Big)\tilde\varphi_e = 0 \end{aligned} \tag{21}$$

with $D_{\tilde\alpha} := \tilde\alpha^2 - \partial_z^2$ and $\tilde\alpha = \kappa\alpha$. Note that $\tilde\lambda$ and n have been combined to the single parameter $\lambda := \tilde\lambda/\tilde\alpha^{2n}$. The system (21) has to be complemented

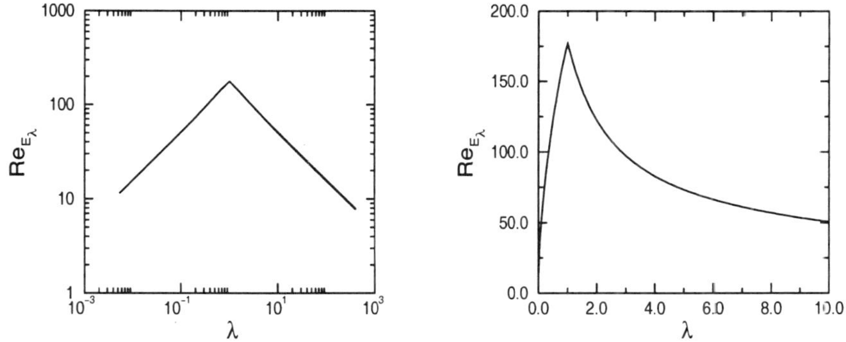

Figure 2. The same as fig. 1 for $E_{\tilde{\lambda},n}$ given in eq. (19) and with parameter $\lambda = \tilde{\lambda}/\tilde{\alpha}^{2n}$. Left and right figure differ only in the scale for λ.

with the boundary conditions

$$\tilde{\varphi}_{e/o} = \partial_z \tilde{\varphi}_{e/o} = 0 \quad \text{at} \quad z = \pm\frac{1}{2} \,. \tag{22}$$

The smallest eigenvalue $\mu_0(\tilde{\alpha}, \lambda)$ is again determined numerically and the stability limit

$$\mathrm{Re}^x_{E_\lambda} = \min_{\tilde{\alpha} \in \mathbb{R}} \mu_0(\tilde{\alpha}, \lambda)$$

is displayed in figure 2: Obviously, the ordinary energy limit $\mathrm{Re}^x_E \approx 177.2$ attained for $\lambda = 1$ cannot be improved. Whether still other functionals allow for better stability boundaries is an open problem. However, we will not pursue this any further. Instead, we will discuss in the next section problems associated with the second step of the procedure necessary to arrive at a nonlinear stability result.

4 Conditional nonlinear stability and rigid boundary conditions

For generalized energy functionals such as those discussed in the last section the nonlinear terms in general do not drop from the (generalized) energy balance. In that case estimates of the type

$$\mathcal{E} \leq c_1 D, \qquad I \leq c_2 D, \qquad N \leq c_3 D\, \mathcal{E}^{1/2} \tag{23}$$

with $c_i > 0$, $i = 1, 2, 3$ allow to handle the energy balance (cf. refs. [6,12]):

$$\partial_t \mathcal{E}(t) = -D(t) + I(t) + N(t) \leq -D(t)\Big(1 - c_2 - c_3\, \mathcal{E}^{1/2}(t)\Big)$$
$$\leq c_1^{-1} \mathcal{E}(t)\Big(1 - c_2 - c_3\, \mathcal{E}^{1/2}(0)\Big)$$

with the stability result

$$\mathcal{E}(t) \leq \mathcal{E}(0)\, e^{-c_1^{-1}(1-c_2-c_3\mathcal{E}^{1/2}(0))t}.$$

D denotes here the positive definite dissipation term, I the interaction term and N the nonlinear (in fact trilinear) term. The estimate $(23)_1$ is due to a Poincaré–type inequality, $(23)_2$ involves the solution of a variational problem as in sec. 3 and $(23)_3$ is based on the calculus inequality [7]

$$ess\text{–}sup\ f \leq c_4\, \|(-\Delta) f\|\,. \tag{24}$$

Inequality (24) is not the only possible in bounding the supremum of a function in terms of the L_2–norm; second–order derivatives on the right–hand side, in particular with respect to z, can however not be avoided.

Let us consider now the functional $\mathcal{E}_1 := E_\lambda$ of sec. 3. The nonlinear term takes here the form $N = -\big((\mathbf{u}\cdot\nabla)\mathbf{u}\,,\, \boldsymbol{\delta}\varphi\big) - \lambda\big((\mathbf{u}\cdot\nabla)\mathbf{u}\,,\, \boldsymbol{\varepsilon}\psi\big)$. Simply by counting derivatives it can be verified that the part $N_\varphi := (1-\lambda)\big((\boldsymbol{\delta}\varphi\cdot\nabla)\,\boldsymbol{\delta}\varphi\,,\, \boldsymbol{\varepsilon}\psi\big)$ does not allow an estimate of the type $(23)_3$ (note that the estimate (24) increases the number of necessary derivatives on the right–hand side of $(23)_3$ by two). Therefore, another part $\mathcal{E}_2$ bilinear in φ and involving higher derivatives is necessary in order to dominate N_φ. However, problems arise if z–derivatives are involved. Let us consider the simplest possibility, viz. $\mathcal{E}_2 = \frac{1}{2}\, \|\partial_z \boldsymbol{\delta}\varphi\|^2$. Performing the energy balance furnishes:

$$\partial_t \mathcal{E}_2 = -\|\nabla \boldsymbol{\delta}\, \partial_z \varphi\|^2 + \iint_{\mathcal{P}} \big[\partial_z^3 \partial_x \varphi\, \partial_z^2 \partial_x \varphi\big]_{z=-1/2}^{z=1/2} dx dy + I_2 + N_2\,. \tag{25}$$

Since there is no information about second or higher order z–derivatives of φ at the boundary the boundary term on the right–hand side of (25) cannot be eliminated. Moreover, the following argument shows that the term does not have a stabilizing sign. Consider the 2d–Couette problem for y–independent perturbations: The functional $\|\partial_x\psi\|^2 + |\mathcal{P}|\|F_y\|^2$ can easily be shown to decay unconditionally to zero. Once the variables ψ and F_y have died out one obtains for the rest:

$$\partial_t\, \frac{1}{2}\Big[\|(-\Delta)\partial_x\varphi\|^2 + |\mathcal{P}|\|F_x\|^2\Big] = -\Big[\|\nabla(-\Delta)\partial_x\varphi\|^2 + |\mathcal{P}|\|F_x'\|^2\Big]$$
$$+ \iint_{\mathcal{P}} \big[\partial_z^3 \partial_x \varphi\, \partial_z^2 \partial_x \varphi\big]_{z=-1/2}^{z=1/2} dx\, dy\,.$$

A stabilizing sign of the boundary term would thus imply global stability of the 2d–Couette problem. This, however, is known to be false at least on a numerical basis [3].

Consequently, only horizontal derivatives are allowed for constructing $\mathcal{E}_2$. Since horizontal derivatives occur in N_φ only in the operators $\boldsymbol{\delta}$, $\boldsymbol{\varepsilon}$ or ∇ it is sufficient to consider functionals of the form $\mathcal{E}_2^{(n)} := \frac{1}{2}||\nabla(-\Delta_2)^n\varphi||^2$ or $\tilde{\mathcal{E}}_2^{(n)} := \frac{1}{2}||\boldsymbol{\delta}(-\Delta_2)^n\varphi||^2$. Let us consider the nonlinear term in the energy balance of $\mathcal{E}_2^{(n)}$:

$$\partial_t\mathcal{E}_2^{(n)} = -||(-\Delta)(-\Delta_2)^n\varphi||^2 + I_2^{(n)} - (\mathbf{u}\cdot\nabla\mathbf{u}\,,\,\boldsymbol{\delta}(-\Delta_2)^{2n-1}\varphi)$$

The part $N_\varphi^{(n)} := ((\boldsymbol{\delta}\varphi\cdot\nabla)\,\boldsymbol{\delta}\varphi\,,\,\boldsymbol{\delta}(-\Delta_2)^{2n-1}\varphi)$ contains at most three z–derivatives. Thus, a superficial counting of z–derivatives does not prevent an estimate of the type $(23)_3$. Looking more closely into the structure of the term reveals, however, other reasons which prevent in fact such an estimate: Let us consider the term

$$((-\Delta_2)\varphi\;\boldsymbol{\varepsilon}\partial_z^2\varphi\,,\,\boldsymbol{\varepsilon}(-\Delta_2)^{2n-1}\partial_z\varphi), \tag{26}$$

which is part of $N_\varphi^{(n)}$. Considering the number of z–derivatives of the ϕ–factors in (26) as well as in $\mathcal{E}_2^{(n)}$ and in $D_2^{(n)} = ||(-\Delta)(-\Delta_2)^n\phi||^2$ the first factor in (26) must be estimated with (24) and then by $\sqrt{D_2^{(n)}}$, the second as well by $\sqrt{D_2^{(n)}}$ and the third by $\sqrt{\mathcal{E}_2^{(n)}}$. Counting the operator $\boldsymbol{\varepsilon}$ as $(-\Delta_2)^{1/2}$ the first and third factor carry a total of $2n{+}1/2$ of $(-\Delta_2)$–operators whereas $\sqrt{\mathcal{E}_2^{(n)}}$ and $\sqrt{D_2^{(n)}}$ provide a total of only $2n$. Rearranging the z–derivatives in (26) by partial integration does not save the story since then (after application of (24)) always terms appear with more than two z–derivatives and these cannot be estimated at all. The same arguments apply also to $\tilde{\mathcal{E}}_2^{(n)}$ which must, therefore, be discarded too.

In conclusion: Although there are functionals which provide better stability limits than the energy functional the rigid boundary conditions appear to be incompatible with the standard procedure to arrive at a nonlinear stability result.

References

1. Busse, F.H.: Bounds on the transport of mass and momentum by turbulent flow between parallel plates. *Z. Angew. Math. Phys.* **20**, 1–14 (1969).

2. Busse, F.H.: A property of the energy stability limit for plane parallel shear flow. *Arch. Rational Mech. Anal.* **47**, 28–35 (1972).
3. Cherhabili, A., Ehrenstein, U.: Spatially localized two–dimensional finite–amplitude states in plane Couette flow. *J. Mech. B/Fluids* **14**, 677–696 (1995).
4. Drazin, P.G., Reid, W.H.: *Hydrodynamic Stability.* Cambridge University Press: Cambridge 1981.
5. Grossmann, S.: The onset of shear flow turbulence. *Reviews of Modern Physics* **72**, 603–618 (2000).
6. Galdi, G.P., Padula, M.: A new approach to the energy theory in the stability of fluid motion. *Arch. Rational Mech. Anal.* **110**, 187–286 (1990).
7. Galdi, G.P., Straughan, B.: A nonlinear analysis of the stabilizing effect of rotation in the Bénard problem. *Proc. Roy. Soc. London Ser. A* **402**, 257–283 (1985).
8. Joseph, D.D.: Nonlinear stability of the Boussinesq equations by the method of energy. *Arch. Rational Mech. Anal.* **22**, 163–184 (1966).
9. Kaiser, R., Schmitt, B.: Bounds on the energy stability limit of plane parallel shear flows. *Z. Angew. Math. Phys.* (ZAMP) **52**, 573–596 (2001).
10. Rionero, S., Mulone, G.: On the nonlinear stability of parallel shear flows. *Continuum Mech. Thermodyn.* **3**, 1–11 (1991).
11. Romanov, V. A.: Stability of plane-parallel Couette flow. *Funct. Anal. Appl.* **7**, 137–146 (1973).
12. Straughan, B.: *The Energy Method, Stability and Nonlinear Convection.* Springer: New York 1992.
13. Schmitt, B.J., Wahl, W. von: Decomposition of solenoidal fields into poloidal fields, toroidal fields and the mean flow. Applications to the Boussinesq-equations. *The Navier-Stokes Equations II — Theory and Numerical Methods.* Proceedings, Oberwolfach 1991. Lecture Notes in Math. **1530** (eds. J.G. Heywood, K. Masuda, R. Rautmann, S.A. Solonnikov), 291–305. Springer: Berlin, Heidelberg, New York 1992.

QUASI-STATIC APPROXIMATIONS OF PHOTON TRANSPORT IN AN INTERSTELLAR CLOUD

MERI LISI AND SILVIA TOTARO
Dipartimento di Matematica "Roberto Magari",
Via del Capitano 15, 53100 Siena, ITALY
E-mail: totaro@unisi.it

We study a mathematical one-dimensional model of photon transport in a homogeneous interstellar cloud (nebula) with a source of photons inside (e.g. a star). Existence, uniqueness and positivity of the solution are proved by using the theory of semigroups of operators. Moreover, we define the quasi-static solution and we show that it is a good approximation to the exact solution. A numerical estimate of the relative error is also given.

1 Introduction

The stars are not the only "things" in the space: between a star and another one, there are lot of atoms, molecules and grains, which move in the vacuum, and form the *interstellar matter.* This represents an important part of the universe, because of its fundamental role in many space phoenomenons.

We can find interstellar matter in different shapes, but it usually creates an interstellar cloud, also called nebula.

A nebula is a mass of dust grains, hydrogen molecules (90%), and a small percentage of other molecules.

The dimension of a nebula is beetween 10^{-1} and 10 parsec, (1 parsec $= 3 \cdot 10^{13}$ km, the diameter of the solar system is about 10^{-4} parsec).

The numerical density of the particles inside a nebula is about 10^4 particles/cm^3, [1], the density of earth atmosphere, at sea level, is approximately 10^{19}particles/cm^3, whereas in the intergalactic vacuum one can find 10^0particles/cm^3, [5]. These facts imply that the matter which is inside a nebula is enough to make one or more stars.

There are many kinds of nebulae:

i) *dark nebulae*, which absorb the light of one or more stars that are behind or inside the clouds themselves: an example is given by the Horsehead nebula;

ii) nebulae which reflect light, for example, the Pleiades;

iii) nebulae which emit light, because they have some stars inside, for example the Orion nebula.

In this paper, we want to study a photon transport in a homogeneous dark nebula, which occupies a bounded convex region of the space. We assume that the transport phoenomenon is one-dimensional, i.e., we consider that the photon number density U depends on the space variable x, $x \in (-\infty, +\infty)$, on the "angle" variable μ, $\mu \in (-1, +1)$ and on time t, $t \geq 0$, [6].

We assume that the nebula is bounded by two surfaces $x = a(t)$ and $x = b(t)$. Setting for simplicity $a(t) = 0$, the boundary plane $x = b(t)$ moves with speed $\dot{b}(t)$; we assume that $b(t)$ is a continuously real function of $t \in [0, +\infty)$ and a positive constant γ exists such that:

$$|\dot{b}(t)| \leq \sup_{t \geq 0} |\dot{b}(t)| = \gamma < \infty. \tag{1}$$

According to our model, photons are monocromatic and can be captured or scattered by the particles of the nebula; we also assume that there is a source of photons inside the nebula.

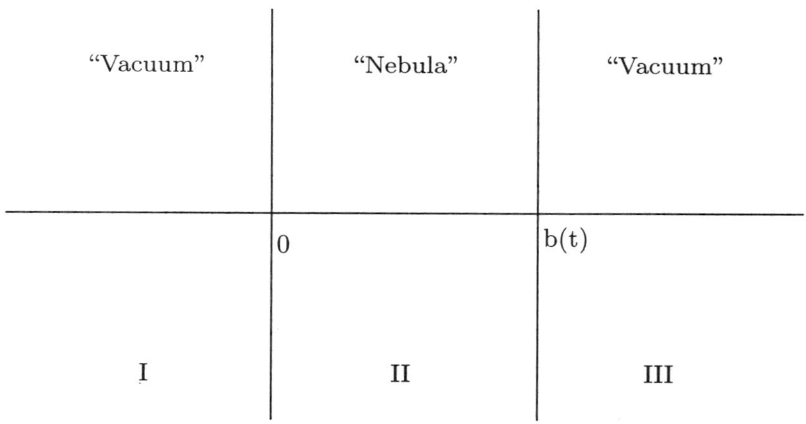

Figure 1

Figure 1 shows a sketch-plan of the situation. In part I and part III, the total cross section and the scattering cross section are zero, because the particle density is very low. In part II, the particle density is higher than that of the other parts. We assume that the cross sections of the transport process occurring in part II are costant.

If we indicate with σ, σ_s the total cross section and the scattering cross section respectively, we assume that $\sigma \geq \sigma_s \geq 0$.

Note that in part II, we also consider a source of photons, (e.g. a star), which can be modelled by a bounded positive function q, that depends on the space variable x and on time t.

By using the above considerations and assuming that the scattering is isotropic, the photon transport equation in the interstellar space reads as follows:

$$\frac{\partial}{\partial t} U(x,\mu,t) = -c\mu \frac{\partial}{\partial x} U(x,\mu,t) - c\sigma\, \chi(x,t)\, U(x,\mu,t) +$$
$$+\frac{1}{2} c\sigma_s\, \chi(x,t) \int_{-1}^{+1} U(x,\mu',t)\, d\mu' +$$
$$+q(x,t)\, \chi(x,t), \tag{2}$$

$$\forall x \in (-\infty,+\infty), \qquad \forall \mu \in (-1,1), \qquad \forall t \in (0,+\infty),$$

In (2), $\chi = \chi(x,t) = \chi_{[0,b(t)]}$ is the characteristic function of the interval $[0, b(t)]$. To deal with regular functions, we replace $\chi(x,t)$ with a mollified version $\hat{\chi}(x,t) = \hat{\chi}_{[0,b(t)]}$ of the characteristic function of $[0, b(t)]$, such that $\hat{\chi}(x,t)$ is a twice continuously differentiable function and there exists a positive constant β such that:

$$\hat{\chi}'(x,t) \le \beta, \qquad \forall x \in (-\infty,+\infty), \qquad \forall t \ge 0. \tag{3}$$

Equation (2) is supplemented with the initial condition:

$$U(x,\mu,0) = U_0(x,\mu), \qquad \forall x \in (-\infty,+\infty), \quad \forall \mu \in (-1,1), \tag{4}$$

where U_0 is a given positive function.

2 The Abstract Problem

In this section we study system (2)-(4) by using the theory of semigroups.

To do this, we consider the Banach space $X = L^1(\mathbb{R} \times (-1,1))$, with norm:

$$\|f\| = \int_{-\infty}^{+\infty} dx \int_{-1}^{+1} |f(x,\mu)|\, d\mu, \qquad \forall f \in X, \tag{5}$$

and the positive cone, [4]:

$$X^+ = \{ f \in X : f(x,\mu) \ge 0, \text{ a.e.}(x,\mu) \in (\mathbb{R} \times (-1,1)) \}. \tag{6}$$

We define the "free-streaming" operator:

$$S : D(S) \subset X \longrightarrow R(S) \subset X, \qquad D(S) = \{ f \in X : Sf \in X \},$$

$$Sf\,(x,\mu) = -\,c\,\mu\,\frac{\partial}{\partial x}\,f\,(x,\mu)\,, \qquad \forall f\in X,$$

and the operator:

$$J : D(J) = X \longrightarrow R(J) \subset X, \qquad Jf\,(x,\mu) = \frac{1}{2}\int_{-1}^{+1} f\,(x,\mu')\,d\mu', \quad \forall f \in X.$$

Moreover, we introduce the following notations:

$$\sigma(t) = \sigma\,\hat{\chi}\,(x,t\,)\,, \qquad \sigma_s(t) = \sigma_s\,\hat{\chi}\,(x,t\,)\,, \qquad Q(t) = q\,(x,t\,)\,\hat{\chi}\,(x,t\,)\,,$$

where $\sigma(t)$ and $\sigma_s(t)$ are now functions from $[0,+\infty)$ into $L^{\infty}\,(\mathbb{R})$, and $Q(t)$ is a function from $[0,+\infty)$ into X.

Then, system (2)-(4) becomes:

$$\frac{d}{dt}U(t) = [S - c\sigma(t)I + c\sigma_s(t)J]\,U(t) + Q(t), \quad \forall t > 0, \tag{7}$$

$$U(0) = U_0, \tag{8}$$

where $U\,(t)$ must be considered as a function from $[0,+\infty)$ into X.

Theorem 2.1 *If $U_0 \in D\,(S) \cap X^+$, problem (7)-(8) has a unique positive solution which is the continuous solution of the integral equation:*

$$U(t) = Z(t)U_0 + \int_0^t Z(t-s)\,\{[c\,(\sigma - \sigma(s))\,I + c\sigma_s(s)J]\,U(s) + Q(s)\}\,ds, \tag{9}$$

where $Z(t) = \exp\left[(S - c\sigma I)t\right].$

The solution $U\,(t)$ can be found by using the successive approximation method, [3].

By using the generalized Gronwall's lemma, [7], and the estimate of the norm of the semigroup $Z(t)$, we have also an estimate of the norm of the solution of (9):

$$\|U(t)\| \leq \|U_0\| + \frac{2\,\|q(t)\|}{c(\sigma - \sigma_s)}. \tag{10}$$

Inequality (10) has a precise physical meaning. In fact, if there is no source of photons (i.e $q\,(t) = 0$), the number of photons inside the cloud, at a time t, can be less than the initial number of photons.

3 The Quasi-Static Approximation

Since the computation of the exact solution of (9) is not simple, it is useful to introduce some approximate solution of problem (7).

For example, we can assume that the number of photons inside the cloud changes slowly in time: in this way, we shall study the so-called quasi-static approximation of system (7)-(8), [2].

We say that $u(t) = u(\cdot,\cdot,t)$ is the quasi-static approximation to the solution $U(t) = U(\cdot,\cdot,t)$ of problem (7)-(8) if it satisfies the following equation:

$$0 = [S - c\sigma(t)I + c\sigma_s(t)J]\,u(t) + Q(t), \quad \forall t > 0. \tag{11}$$

Note that equation (11) can be interpreted as an approximative form of (7) if $\dfrac{dU}{dt}$ is "small".

We remark that σ, σ_s and Q depend on t through $b(t)$; hence, $u(t)$ is a "good" approximation if $b(t)$ is a slowly varying function of time.

Theorem 3.1 *The unique solution of the quasi-static equation (11) reads:*

$$u(t) = -\left[S - c\sigma(t)I + c\sigma_s(t)J\right]^{-1} Q(t). \tag{12}$$

Moreover, for each $t \geq 0$, $u(t) \in D(S) \cap X^+$ *and it is such that:*

$$\|u(t)\| \leq \frac{2\|q(t)\|}{c(\sigma - \sigma_s)}, \quad t \geq 0. \tag{13}$$

To prove that $u(t)$ is a good approximation to $U(t)$, we can evaluate the "relative error" $\dfrac{\|\Delta(t)\|}{\|U(t)\|}$.

By using (1) and (3), we have after many computations,

$$\frac{\|\Delta(t)\|}{\|U(t)\|} \leq \frac{\alpha\gamma}{c\xi}, \tag{14}$$

where

$$\xi = 2\left[\|q(0)\| + \|q(t)\|\right] \qquad \text{and} \qquad \alpha = \frac{4\sigma}{(\sigma - \sigma_s)^2}\,\beta\,\|q(t)\|.$$

We can also evaluate numerically (14), in fact, we have the following data:

i) the speed of a "typical" cloud, in non critical conditions, is about 30 km/s, [5];

ii) the total cross section of a cloud is about $3 \cdot 10^{-13}$ cm^{-1}, [8];

iii) the light speed is $3 \cdot 10^{10}$ cm/s.

By using these values, (14) becomes:

$$\frac{\|\Delta(t)\|}{\|U(t)\|} \leq 10^{-7}. \tag{15}$$

Relation (15) shows that the quasi-static solution $u(t)$ can replace the exact solution $U(t)$.

The bounderies of an interstellar cloud move very slowly: an appreciable variation takes many years to be noticed. Moreover, the critical conditions, which bring a cloud to collapse and generate a star, are reached after about a million of years.

Acknowledgments

This work was partially supported by the Italian "Ministero dell'Università e della Ricerca Scientifica e Tecnologica" research funds, as well as by GNFM of Italian CNR.

References

1. A. Belleni Morante, A.Moro, S.Aiello and C. Cecchi Pestellini, "Photon transport in an interstellar cloud with stochastic clumps: the three - dimensional case", *Conferenze del seminario di matematica dell'Università di Bari*, 255, 1994.
2. A.Belleni Morante and S. Totaro, "Photon transport in a time - dependent region: a quasi - static approximation", *Conferenze del seminario di matematica dell'Università di Bari*, 276, 1999.
3. T.Kato, *Perturbation theory for linear operators*, Springer Verlag, New York, 1984.
4. M.Krasnosel'skii, *Positive solutions of operator equations*, P. NoordHoof, Groningen, 1964.
5. D.E.Osterbrock, *Astrophysics of gaseous nebulae and active galactic nuclei*, University Science Book, U.S.A., 1989.
6. G.C.Pomraning, *The equations of radiation hydrodinamics*, Pergamon Press, Oxford, 1973.
7. G.Sansone and R.Conti, *Equazioni differenziali non lineari*, Edizioni Cremonese, Roma, 1956.
8. L.Spitzer, *Physical processes in the interstellar medium*, John Wiley & Sons, Toronto, 1978.

DOUBLE-DIFFUSIVE CONVECTION IN POROUS MEDIA:THE DARCY AND BRINKMAN MODELS

S. LOMBARDO
Dipartimento di Matematica e Informatica
Viale A. Doria, 6, 95125 Catania, Italy
E-mail: lombardo@dmi.unict.it

G. MULONE
Dipartimento di Matematica e Informatica
Viale A. Doria, 6, 95125 Catania, Italy
E-mail: mulone@dmi.unict.it

The nonlinear stability of a horizontal layer of a binary fluid mixture in an isotropic and homogeneous porous medium heated and salted from below is studied, for the Oberbeck-Boussinesq - Darcy and Oberbeck-Boussinesq - Brinkman-Forchheimer models, through the Lyapunov direct method. This is an interesting geophysical case because the solute concentration gradient is stabilizing while heating from below provides a destabilizing effect. Unconditional nonlinear stability is found for any value of the porosity and the Lewis number. In particular, if the normalised porosity ϵ is equal to 1, a necessary and sufficient condition for nonlinear stability is proved: in this case the critical linear and nonlinear Rayleigh numbers coincide. For other values of ϵ a conditional stability theorem is shown and the coincidence of the critical parameters holds whenever the Principle of Exchange of Stabilities is valid.

1 Introduction

Double-diffusive flows in porous media are widely encountered in nature and technological processes. They have recently been the subject of intensive study in many areas, e.g. in geotechnical world, in environmental engineering, etc. (see Lombardo, Mulone and Straughan [8], the book by Nield and Bejan [11], the review article by Trevisan and Bejan [16], and the references therein).

In situations involving one or more solute fields the heat transfer aspect cannot completely describe the phenomenon. Both heat and mass transfer must be considered. The interesting effects arise from the diffusion rates of heat and solute which are usually different: heat diffuses more rapidly that a dissolved substance.

Due to the many applications it is important to have a good theoretical understanding of the processes occurring in double-diffusive porous media flow. In Lombardo, Mulone and Straughan [8] this has been addressed via a linear instability and nonlinear stability analysis in the case of the Darcy model.

When the velocity $\mathbf{v}$ and porosity ε are sufficiently large the Brinkman-Forchheimer law in the momentum equation is used (see [11] and the references therein).

Here we use the Oberbeck-Boussinesq approximation and both the Darcy and the Brinkman-Forchheimer laws to study double diffusive convection in porous media. ¿From a mathematical point of view the problem is interesting because the linear operator associated with the dynamical system of the perturbation equations to the basic conduction-diffusion solution is *not symmetric* with respect to the energy ($L_2(\Omega)$) norm.

If the OB approximation is not applicable, the compressible Navier-Stokes equations must be used and the problems discussed here are *still open*.

The plan of the work is the following: in Section 2 we give the mathematical models, in Section 3 we recall some linear instability results. Section 4 deals with unconditional nonlinear stability, in particular, when $\varepsilon = Le = 1$, we obtain also the coincidence of the linear and nonlinear stability boundaries. Finally, in Section 5, a conditional stability theorem is stated and the coincidence of the critical Rayleigh numbers is shown.

2 Mathematical models

2.1 Darcy's model

Henry Darcy's (1856) investigations into the hydrology of the water supply of Dijon and his experiments on steady-state unidirectional flow in a uniform medium revealed a proportionality between flow rate and the applied pressure differences. Darcy's law has been verified by the results of many experiments.

Let us consider a layer of a porous medium saturated by a binary fluid mixture heated and salted from below, bounded by two horizontal parallel planes. Let $d > 0$, $\Omega_d = \mathbb{R}^2 \times (-d/2, d/2)$ and $Oxyz$ be a cartesian frame of reference with unit vectors $\mathbf{i}, \mathbf{j}, \mathbf{k}$ respectively. We assume that the Oberbeck-Boussinesq approximation be valid and that the flow in the porous medium is governed by Darcy's law. Thus, if the plane $z = 0$ is parallel to the layer, the basic equations are:

$$\begin{cases} \nabla P = -\rho_f g \mathbf{k} - \dfrac{\mu}{K}\mathbf{v}, \quad \nabla \cdot \mathbf{v} = 0 \\ \dfrac{1}{M}\dfrac{\partial T}{\partial t} + \mathbf{v} \cdot \nabla T = k\Delta T, \quad \varepsilon \dfrac{\partial C}{\partial t} + \mathbf{v} \cdot \nabla C = k'\Delta C, \end{cases} \tag{1}$$

where $\rho_f = \rho_0[1 - \alpha(T - T_0) + \alpha'(C - C_0)]$. We have denoted with

$\mathbf{v} = (U, V, W), P, T, C$ the Darcy or seepage velocity, the pressure, the temperature and the concentration, respectively. M is the ratio of heat capacities. The quantities μ, ρ_f, denote the viscosity, density, and K is the permeability of the medium. Further, g is the gravitational acceleration, k' is the salt diffusivity, k is the effective thermal conductivity of the saturated porous medium defined by the ratio of the thermal diffusivity of the porous medium $(1-\varepsilon)k_s + \varepsilon k_f$ with $(\rho c_p)_f$. k_s and k_f are the thermal diffusivities of the solid and fluid, respectively. α and α' are the coefficients for thermal and solutal expansion, and ρ_0, T_0, C_0 are reference density, temperature and concentration. The parameter ϵ is the normalised porosity defined by $\epsilon = \varepsilon M$ where ε is the porosity and M is the ratio of heat capacities $M = \frac{(\rho c_p)_f}{(\rho c)_m}$.
c_p is the specific heat of the fluid at constant pressure, c is the specific heat of the solid, and $(\rho c)_m = (1-\varepsilon)(\rho c)_s + \varepsilon(\rho c_p)_f$ denotes the specific heat of the porous medium. The subscripts f, m denote fluid and porous medium values, respectively. Moreover $T_0 = (T_1 + T_2)/2, C_0 = (C_1 + C_2)/2$ with $T_1 > T_2, C_1 > C_2$.

The boundary conditions (for impermeable surfaces) of the problem are $W = 0, \quad T = T_0 \pm (T_1 - T_2)/2, \quad C = C_0 \pm (C_1 - C_2)/2 \qquad \text{at} \quad z = \mp d/2$.

The governing equations for nondimensional quantities take the following form

$$\begin{cases} \nabla P = (\mathcal{R}T - \mathcal{C}C)\mathbf{k} - \mathbf{v}, \quad \nabla \cdot \mathbf{v} = 0 \\ \dfrac{\partial T}{\partial t} + \mathbf{v} \cdot \nabla T = \Delta T, \quad \epsilon \dfrac{\partial C}{\partial t} + \mathbf{v} \cdot \nabla C = \dfrac{1}{Le} \Delta C \end{cases} \tag{2}$$

where

$$\mathcal{R} = \frac{\alpha g (T_1 - T_2) d K}{\nu k}, \quad \mathcal{C} = \frac{\alpha' g (C_1 - C_2) d K}{\nu k}$$

are the thermal and solutal Rayleigh numbers, respectively. The parameter Le is the Lewis number defined by $Le = k/k'$.

The boundary conditions become

$$W = 0, T = \pm 1/2, C = \pm 1/2, \text{at } z = \mp 1/2. \tag{3}$$

We study the stability of the motionless double-diffusive solution $m_0 = (\mathbf{v}, T, C, P)$ given by

$$\mathbf{v} := 0, T = -z, C = -z, \nabla P = z(-\mathcal{R} + \mathcal{C})\mathbf{k}. \tag{4}$$

The nondimensional equations which govern the evolution of a disturbance $(\mathbf{u}, \vartheta, \gamma, p_2)$ to m_0 are

$$\begin{cases} \nabla p_2 = (\mathcal{R}\vartheta - Le\mathcal{C}\gamma)\mathbf{k} - \mathbf{u}, \quad \nabla \cdot \mathbf{u} = 0 \\ \vartheta_t + \mathbf{u} \cdot \nabla\vartheta = w + \Delta\vartheta, \quad \epsilon Le\gamma_t + Le\mathbf{u} \cdot \nabla\gamma = w + \Delta\gamma, \end{cases} \tag{5}$$

where the subscript t denotes partial derivative and $\mathbf{u} = (u, v, w)$. The boundary conditions are

$$w = 0, \quad \vartheta = \gamma = 0, \qquad \text{at} \quad z = \mp 1/2.$$

We assume that the perturbations $\mathbf{u}, \vartheta, \gamma, p_2$ are periodic functions of x and y of periods $2\pi/a_x$, $2\pi/a_y$, respectively, $(a_x > 0,\ a_y > 0)$ and denote by Ω the periodicity cell: $\Omega = [0, 2\pi/a_x] \times [0, 2\pi/a_y] \times [-1/2, 1/2]$ and by $a = (a_x^2 + a_y^2)^{1/2}$ the wave number. The periodicity cell is that spatial domain which defines the pattern of convective overturning, with wavenumber a.

2.2 Brinkman-Forchheimer's model

The Brinkman-Forchheimer model is believed accurate when the flow velocity is too large for Darcy's law to be valid and additionally the porosity is not too small. This model has been justified theoretically with different approaches by Giorgi [5] and Whitakher [17], whereas Gilver and Altobelli [4] pay particular attention to estimating the effective viscosity in real situations. Much recent researches on the solution decay in space and time and stability of thermal convection have been done in Qin and Kaloni [14], Guo and Kaloni [6]. In Kaloni and Guo [7] the existence, regularity and uniqueness of steady weak solutions are studied. In Payne and Straughan [12] the structural stability, i.e., the continuous dependence on the model itself has been investigated.

The governing equations are (OBBF system):

$$\begin{cases} \rho_0 \dfrac{1}{\varepsilon}\dfrac{\partial \mathbf{v}}{\partial t} + \nabla P = -\rho_f g\mathbf{k} - \dfrac{\mu_1}{K}\mathbf{v} + \mu_2\Delta\mathbf{v} - \dfrac{c_F\rho_0}{K^{1/2}}|\mathbf{v}|\mathbf{v}, \quad \nabla \cdot \mathbf{v} = 0 \\ \dfrac{1}{M}\dfrac{\partial T}{\partial t} + \mathbf{v} \cdot \nabla T = k\Delta T, \quad \varepsilon\dfrac{\partial C}{\partial t} + \mathbf{v} \cdot \nabla C = k'\Delta C, \end{cases} \tag{6}$$

where c_F is a positive dimensionless form-drag constant, the quantities μ_1, μ_2 are viscosities, μ_2 is an effective viscosity (see [11]). Some authors include also the term $\mathbf{v} \cdot \nabla\mathbf{v}$ but it is generally small in comparison with the quadratic drag term $|\mathbf{v}|\mathbf{v}$ and its inclusion is not a satisfactory way of expressing the nonlinear drag (see [11], pag. 8). Moreover also the term $\rho_0 \dfrac{1}{\varepsilon}\dfrac{\partial \mathbf{v}}{\partial t}$ is small, see

[11]. We have included it here for the sake of generality. The stability results we shall obtain remain still valid if this term is dropped.

For impermeable "rigid", isothermal and of isoconcentration surfaces, the boundary conditions are (see [11], [6])

$$U = U_1(x,y),\ V = V_1(x,y),\ W = 0,$$

$$T = T_0 \pm (T_1 - T_2)/2,\ C = C_0 \pm (C_1 - C_2)/2$$

at $z = \mp d/2$,

for impermeable "stress-free", isothermal and of isoconcentration surfaces, the boundary conditions are (see Poulikakos [13], [6])

$$U_z = U_2(x,y),\ V_z = V_2(x,y),\ W = 0,$$

$$T = T_0 \pm (T_1 - T_2)/2,\ C = C_0 \pm (C_1 - C_2)/2$$

at $z = \mp d/2$.

Defining suitable nondimensional quantities the governing equations take the following form

$$\begin{cases} \gamma_a \mathbf{v}_t + \nabla P = (\mathcal{R}T - \mathcal{C}C)\mathbf{k} - \mathbf{v} + \tilde{\mathrm{D}}\mathrm{a}\Delta\mathbf{v} - \tilde{c}_F|\mathbf{v}|\mathbf{v}, \quad \nabla\cdot\mathbf{v} = 0 \\ \dfrac{\partial T}{\partial t} + \mathbf{v}\cdot\nabla T = \Delta T, \quad \epsilon\dfrac{\partial C}{\partial t} + \mathbf{v}\cdot\nabla C = \dfrac{1}{Le}\Delta C \end{cases} \tag{7}$$

where $\tilde{\mathrm{D}}\mathrm{a} = \dfrac{\mu_2 K}{\mu_1 d^2}$ is the Darcy number, $\gamma_a = \dfrac{kKM^2}{\nu d^2 \epsilon}$, is a nondimensional acceleration coefficient, $\nu = \dfrac{\mu_1}{\rho_0}$ is the kinematic viscosity, and $\tilde{c}_F = c_F\dfrac{K^{1/2}k}{\nu d}$, is a nondimensional Forchheimer coefficient.

The nondimensional equations which govern the evolution of a disturbance $(\mathbf{u}, \vartheta, \gamma, p_2)$ to m_0 are

$$\begin{cases} \gamma_a \mathbf{u}_t + \nabla p_2 = (\mathcal{R}\vartheta - Le\mathcal{C}\gamma)\mathbf{k} - \mathbf{u} + \tilde{\mathrm{D}}\mathrm{a}\Delta\mathbf{u} - \tilde{c}_F|\mathbf{u}|\mathbf{u}, \quad \nabla\cdot\mathbf{u} = 0 \\ \vartheta_t + \mathbf{u}\cdot\nabla\vartheta = w + \Delta\vartheta, \quad \epsilon Le\gamma_t + Le\mathbf{u}\cdot\nabla\gamma = w + \Delta\gamma. \end{cases} \tag{8}$$

The boundary conditions are $\mathbf{u} = \vartheta = \gamma = 0$, at $z = \pm 1/2$ in the rigid case and $u_z = v_z = w = \vartheta = \gamma = 0$, at $z = \pm 1/2$ in the stress free-case. In the last case we must also add the average velocity conditions $\int_\Omega u d\Omega = \int_\Omega v d\Omega = 0$.

3 Linear instability

3.1 The Darcy's case

The linear instability of m_0 can be studied by linearization and by using the standard normal modes analysis (see Chandrasekhar [1]). One obtains: $\epsilon Le \leq 1 \Longrightarrow \sigma \in \mathbb{R}$, (the strong Principle of Exchange of Stabilities holds) and the instability occurs via a stationary convection.

The critical linear instability Rayleigh number for the onset of stationary convection is

$$\mathcal{R}_c = Le\mathcal{C} + 4\pi^2. \tag{9}$$

As concerns the oscillatory convection it can be proved that

$$\mathcal{R}_{over} = \frac{\mathcal{C}}{\epsilon} + 4\pi^2 \frac{1+\epsilon Le}{\epsilon Le}. \tag{10}$$

and overstability is not excluded whenever

$$\epsilon Le > 1 \quad \text{and} \quad \mathcal{C} > \frac{4\pi^2}{(\epsilon Le - 1)Le}. \tag{11}$$

3.2 The Brinkman-Forchheimer's case

In the case of stress-free boundary conditions it can be proved that *whenever* $\epsilon Le \leq 1$, *the Principle of exchange of stabilities holds* for any $\mathcal{C}$, and the instability occurs via a stationary convection. The critical linear instability Rayleigh number for the onset of stationary convection (see [13]) is given by

$$\mathcal{R}_c = Le\mathcal{C} + \frac{(a_c^2+\pi^2)^2}{a_c^2}[1 + \tilde{\mathrm{D}}\mathrm{a}(a_c^2+\pi^2)] \tag{12}$$

where

$$a_c^2 = \frac{[-b + \sqrt{b^2 + 8b\tilde{\mathrm{D}}\mathrm{a}\pi^2}]}{4\tilde{\mathrm{D}}\mathrm{a}} \quad \text{and} \quad b = (\tilde{\mathrm{D}}\mathrm{a}\pi^2 + 1). \tag{13}$$

In the case $\epsilon Le > 1$ if $\mathcal{C} \leq \mathcal{C}_0$ where

$$\mathcal{C}_0 := \frac{(a_1^2+\pi^2)^2[1 + (\tilde{\mathrm{D}}\mathrm{a} + \gamma_a)(a_1^2+\pi^2)]}{(\epsilon Le - 1)a_1^2 Le} \tag{14}$$

PES still holds. It is easy to see that

$$a_1^2 = \frac{-(h\pi^2+1) + \sqrt{(h\pi^2+1)^2 + 8h\pi^2(h\pi^2+1)}}{4h}$$

and $h = (\tilde{\mathrm{D}}\mathrm{a} + \gamma_a)$.
In the case where

$$\epsilon Le > 1 \quad \text{and} \quad \mathcal{C} > \mathcal{C}_0 \tag{15}$$

overstability cannot be excluded.

4 Unconditional nonlinear stability

By using the classical energy method, an unconditional nonlinear stability result can be obtained for any ϵ, Le, (see [8]). In [8], in the Darcy model, it is proved that, whenever

$$\mathcal{R} < 4\pi^2, \tag{16}$$

the solution m_0 given by (2.5) is unconditionally, nonlinearly stable regardless of what value $\mathcal{C}(> 0)$ has, and for any ϵ and Le.
For the OBBF system, the solution m_0 is unconditionally nonlinearly stable for any ϵ, Le, whenever

$$\mathcal{R} < \mathcal{R}_{BB} := \frac{(a_c^2 + \pi^2)^2}{a_c^2}[1 + \tilde{\mathrm{D}}\mathrm{a}(a_c^2 + \pi^2)], \tag{17}$$

where a_c^2 is given by (13); (note that $\mathcal{R}_{BB} \to 4\pi^2$ when $\tilde{\mathrm{D}}\mathrm{a} \to 0$).

In order to improve these results when $\mathcal{R}_{BB} \le \mathcal{R} \le Le\mathcal{C} + \mathcal{R}_{BB}$, we first investigate the situation where $Le = \epsilon = 1$. In this case we may introduce the natural variable ϕ by

$$\phi = \mathcal{R}\vartheta - \mathcal{C}\gamma\,. \tag{18}$$

With the aid of (18) we can transform (8) to the equivalent system

$$\begin{cases} \gamma_a \mathbf{u}_t + \nabla p_2 = \phi\mathbf{k} - \mathbf{u} + \tilde{\mathrm{D}}\mathrm{a}\Delta\mathbf{u} - \tilde{c}_F|\mathbf{u}|\mathbf{u}, & \nabla\cdot\mathbf{u} = 0 \\ \phi_t + \mathbf{u}\cdot\nabla\phi = (\mathcal{R} - \mathcal{C})w + \Delta\phi, & \gamma_t + \mathbf{u}\cdot\nabla\gamma = w + \Delta\gamma, \end{cases} \tag{19}$$

where now $\phi = 0$ at $z = \pm 1/2$.

Since the first three equations of (19) do not contain γ, one may study the system

$$\begin{cases} \gamma_a \mathbf{u}_t + \nabla p_2 = \phi\mathbf{k} - \mathbf{u} + \tilde{\mathrm{D}}\mathrm{a}\Delta\mathbf{u} - \tilde{c}_F|\mathbf{u}|\mathbf{u}, & \nabla\cdot\mathbf{u} = 0 \\ \phi_t + \mathbf{u}\cdot\nabla\phi = (\mathcal{R} - \mathcal{C})w + \Delta\phi \end{cases} \tag{20}$$

separately from the third equation of (19).

Now, we observe that if $\mathcal{R}-\mathcal{C}\leq 0$ the flow is always stable. To see this, supposing in a first moment that $\mathcal{R}-\mathcal{C}<0$, we multiply $(20)_1$ by $(\mathcal{C}-\mathcal{R})\mathbf{u}$ and integrate over Ω, and likewise multiply $(20)_2$ by ϕ and integrate over Ω. Then we add the equations so obtained. We have

$$\begin{aligned}&\tfrac{1}{2}\frac{d}{dt}[||\mathbf{u}||^2\gamma_a(\mathcal{C}-\mathcal{R})+||\phi||^2]=-[(\mathcal{C}-\mathcal{R})||\mathbf{u}||^2+(\mathcal{C}-\mathcal{R})\tilde{\mathrm{D}}\mathrm{a}||\nabla\mathbf{u}||^2\\&+(\mathcal{C}-\mathcal{R})\tilde{c}_F\int_\Omega|\mathbf{u}|^3\,d\Omega+||\nabla\phi||^2].\end{aligned}\tag{21}$$

¿From this equation, the exponential decay of $||\mathbf{u}||$ and $||\phi||$ follows. Proceeding as in [8], it is easily to prove that the flow is always stable. In the case $\mathcal{R}=\mathcal{C}$, the exponentially decay of $||\phi||$ follows if we multiply $(20)_2$ by ϕ and integrate over Ω. Then we proceed as in [8].
If instead $\mathcal{R}-\mathcal{C}>0$ we multiply $(20)_1$ by $(\mathcal{R}-\mathcal{C})$ to obtain the system

$$\begin{cases}(\mathcal{R}-\mathcal{C})(\gamma_a\mathbf{u}_t+\nabla p_2)=(\mathcal{R}-\mathcal{C})(\phi\mathbf{k}-\mathbf{u}+\tilde{\mathrm{D}}\mathrm{a}\Delta\mathbf{u}-\tilde{c}_F|\mathbf{u}|\mathbf{u})\\ \phi_t+\mathbf{u}\cdot\nabla\phi=(\mathcal{R}-\mathcal{C})w+\Delta\phi\\ \nabla\cdot\mathbf{u}=0.\end{cases}\tag{22}$$

We can now verify that the linear operator associated with (22), viz.

$$LU:=(\mathcal{R}-\mathcal{C})\begin{pmatrix}\Pi(\phi\mathbf{k}-\mathbf{u}+\tilde{\mathrm{D}}\mathrm{a}\Delta\mathbf{u})\\ \mathbf{u}\cdot\mathbf{k}+\dfrac{1}{\mathcal{R}-\mathcal{C}}\Delta\phi\end{pmatrix}\tag{23}$$

where $U=(\mathbf{u},\phi)^T$ and Π is the projector of $[L_2(\Omega)]^3$ onto the subspace of the divergence free vectors, is symmetric with respect to the scalar product associated to the norm $E(t)=\dfrac{1}{2}[\gamma_a(\mathcal{R}-\mathcal{C})||\mathbf{u}||^2+||\phi||^2]$. Hence, by the results of Galdi and Straughan [3], *the coincidence of the critical linear and nonlinear stability Rayleigh number* follows in both cases of rigid and stress-free boundaries.

Now we assume that $Le\neq 1$ and $\epsilon=1$. By denoting $\hat{\mathcal{C}}=Le\mathcal{C}$ and introducing the change of variables

$$\phi=\mathcal{R}\vartheta-\hat{\mathcal{C}}\gamma,\quad \psi=\mathcal{R}\vartheta-Le\delta\hat{\mathcal{C}}\gamma\tag{24}$$

with $Le\delta\neq 1$, system (8) becomes

$$
\begin{cases}
\gamma_a \mathbf{u}_t + \nabla p_2 = \phi \mathbf{k} - \mathbf{u} + \tilde{\mathrm{D}}\mathrm{a}\Delta \mathbf{u} - \tilde{c}_F |\mathbf{u}|\mathbf{u}, \quad \nabla \cdot \mathbf{u} = 0 \\
Le(\phi_t + \mathbf{u} \cdot \nabla \phi) = (Le\mathcal{R} - \hat{\mathcal{C}})w + \dfrac{Le^2\delta - 1}{Le\delta - 1}\Delta\phi - \dfrac{Le - 1}{Le\delta - 1}\Delta\psi \\
\psi_t + \mathbf{u} \cdot \nabla \psi = (\mathcal{R} - \delta\hat{\mathcal{C}})w + \dfrac{\delta(Le - 1)}{Le\delta - 1}\Delta\phi + \dfrac{\delta - 1}{Le\delta - 1}\Delta\psi
\end{cases}
\tag{25}
$$

By introducing the norm

$$E(t) = \frac{1}{2}[Le||\phi||^2 + \mu||\psi||^2 + \lambda\gamma_a||\mathbf{u}||^2],$$

we easily obtain the evolution equation

$$\dot{E}(t) = I - D - \lambda\tilde{c}_F \int_\Omega |\mathbf{u}|^3 \, d\Omega \leq I - D \tag{26}$$

where

$$
\begin{cases}
I = \tilde{\alpha}(\phi, w) + \beta(\psi, w) \\
D = \lambda||\mathbf{u}||^2 + \lambda\tilde{\mathrm{D}}\mathrm{a}||\nabla\mathbf{u}||^2 + \boldsymbol{\Gamma}||\nabla\phi||^2 + \zeta||\nabla\psi||^2 + \eta(\nabla\phi, \nabla\psi)
\end{cases}
\tag{27}
$$

and

$$
\begin{cases}
\tilde{\alpha} = Le\mathcal{R} - \hat{\mathcal{C}} + \lambda, \quad \beta = \mu(\mathcal{R} - \delta\hat{\mathcal{C}}) \\
\Gamma = \dfrac{Le^2\delta - 1}{Le\delta - 1}, \quad \zeta = \mu\dfrac{\delta - 1}{Le\delta - 1}, \; \eta = \dfrac{(Le - 1)(\mu\delta - 1)}{Le\delta - 1}.
\end{cases}
\tag{28}
$$

In order to assure that D is positive definite, we assume that μ and δ are chosen in such a way to verify the following inequalities:

$$\frac{Le^2\delta - 1}{Le\delta - 1} > 0, \quad \frac{\delta - 1}{Le\delta - 1} > 0,$$

$$(Le - 1)^2(\mu\delta - 1)^2 - 4\mu(Le^2\delta - 1)(\delta - 1) < 0.$$

Defining

$$m = \max_{\mathcal{H}} \frac{I}{D}, \tag{29}$$

from (26) we obtain the inequality

$$\dot{E}(t) \leq (m - 1)D. \tag{30}$$

By using the Poincaré's and the Wirtinger's inequalities, and integrating the differential inequality we deduce the following theorem.

Theorem 4.1 *If $m < 1$, the basic state is globally stable with decay at least like*

$$E(t) \leq E(0)\exp[(m-1)\nu_1 t], \tag{31}$$

where ν_1 is a suitable positive constant.

In order to obtain the nonlinear critical Rayleigh number $\mathcal{R}_E$ - *in the stress-free boundary and rigid cases* - we apply the optimum Lyapunov parameters method used in the fluid-dynamic case by Mulone and Rionero [10], see also the books by Flavin and Rionero [2] and Straughan [15].

We write the Euler-Lagrange equations for the maximum problem (29). Then we solve with respect to the eigenvalue of the maximum problem; we maximize with respect to n and the wave number a^2 and minimize with respect to the Lyapunov parameters λ and μ. We obtain: $m^2 = (\mathcal{R}-\hat{\mathcal{C}})/\mathcal{R}_{BB}$. Taking into account that $\hat{\mathcal{C}} = Le\mathcal{C}$, we find $\mathcal{R} < \mathcal{R}_E := Le\mathcal{C} + \mathcal{R}_{BB}$ whenever $m < 1$. Therefore *the critical nonlinear Rayleigh number coincides with the linear one.* The coincidence between the nonlinear critical Rayleigh number $\mathcal{R}_E$ with the linear one $\mathcal{R}_C$, still holds for $Le > 1$ and $\mathcal{C} < \dfrac{\mathcal{R}_{BB}}{Le(Le^2-1)}$. To prove this we follow the technique described by Lombardo, Mulone and Rionero [9].

5 The general case: conditional stability and coincidence of critical Rayleigh numbers

Now we consider again the system (5). We shall rewrite it as

$$\begin{cases} \nabla p_2 = (\mathcal{R}\vartheta - \hat{\mathcal{C}}\gamma)\mathbf{k} - \mathbf{u}, & \nabla\cdot\mathbf{u} = 0 \\ \vartheta_t + \mathbf{u}\cdot\nabla\vartheta = w + \Delta\vartheta, & \tau\gamma_t + p\mathbf{u}\cdot\nabla\gamma = w + \Delta\gamma, \end{cases} \tag{32}$$

where $p \equiv Le$, $\tau \equiv \epsilon Le$, $\hat{\mathcal{C}} \equiv Le\mathcal{C}$. In the sequel we shall omit the accent *hat*. By applying the transformation

$$\phi = \mathcal{R}\vartheta - \mathcal{C}\gamma, \quad \psi = \mathcal{R}\vartheta - \delta\tau\mathcal{C}\gamma \tag{33}$$

with $\delta\tau \neq 1$, (32) becomes

$$\begin{cases} \nabla p_2 = \phi\mathbf{k} - \mathbf{u}, \quad \nabla\cdot\mathbf{u} = 0 \\ \tau\phi_t + \frac{\delta\tau^2-p}{\delta\tau-1}\mathbf{u}\cdot\nabla\phi + \frac{p-\tau}{\delta\tau-1}\mathbf{u}\cdot\nabla\psi = (\tau\mathcal{R}-\mathcal{C})w + \frac{\delta\tau^2-1}{\delta\tau-1}\Delta\phi + \frac{1-\tau}{\delta\tau-1}\Delta\psi \\ \psi_t + \frac{\delta(\tau-p)}{\delta\tau-1}\mathbf{u}\cdot\nabla\phi + \frac{p\delta-1}{\delta\tau-1}\mathbf{u}\cdot\nabla\psi = (\mathcal{R}-\delta\mathcal{C})w + \frac{\delta(\tau-1)}{\delta\tau-1}\Delta\phi + \frac{\delta-1}{\delta\tau-1}\Delta\psi. \end{cases} \tag{34}$$

To establish nonlinear stability we consider the Lyapunov function

$$V(t) = \frac{1}{2}[\tau\|\phi\|^2 + \mu\|\psi\|^2] + \frac{b}{2}[\|\nabla\phi\|^2 + \|\nabla\psi\|^2] \tag{35}$$

where b is positive, and write its evolution equation:

$$\dot{V}(t) = I_0 - D_0 + N_0 + bI_1 - bD_1 + bN_1$$

where

$$\begin{cases} I_0 = (\tau\mathcal{R} - \mathcal{C} + \lambda)(\phi, w) + \mu(\mathcal{R} - \delta\mathcal{C})(\psi, w) \\ D_0 = \lambda\|\mathbf{u}\|^2 + \Gamma\|\nabla\phi\|^2 + \zeta\|\nabla\psi\|^2 + \eta(\nabla\phi, \nabla\psi) \\ N_0 = \frac{(p-\tau)(1+\mu\delta)}{\delta\tau-1}(\mathbf{u}\cdot\nabla\phi, \psi) \\ I_1 = -\frac{1}{\tau}(\tau\mathcal{R} - \mathcal{C})(\Delta\phi, w) - (\mathcal{R} - \delta\mathcal{C})(\Delta\psi, w) \\ D_1 = \bar{\Gamma}\|\Delta\phi\|^2 + \bar{\zeta}\|\Delta\psi\|^2 + \bar{\eta}(\Delta\phi, \Delta\psi) \\ N_1 = \frac{\delta\tau^2-p}{\tau(\delta\tau-1)}(\Delta\phi, \mathbf{u}\cdot\nabla\phi) + \frac{p-\tau}{\tau(\delta\tau-1)}(\Delta\phi, \mathbf{u}\cdot\nabla\psi) \\ +\frac{\delta(\tau-p)}{\delta\tau-1}(\Delta\psi, \mathbf{u}\cdot\nabla\phi) + \frac{p\delta-1}{\delta\tau-1}(\Delta\psi, \mathbf{u}\cdot\nabla\psi), \end{cases}$$

with $\Gamma = (\delta\tau^2 - 1)/(\delta\tau - 1)$, $\zeta = \mu(\delta - 1)/(\delta\tau - 1)$, $\eta = (\mu\delta - 1)(\tau - 1)/(\delta\tau - 1)$, $\bar{\Gamma} = \Gamma/\tau$, $\bar{\zeta} = \zeta/\mu$, $\bar{\eta} = (\tau - 1)/\tau$.

We observe that the maximum problem $m = \max\limits_{\mathcal{G}} I_0/D_0$, with $\mathcal{G}$ = the space of the admissible fields , is the same as that which arises in the fluid-dynamic case, see [10]. The parameter $\tau \equiv \epsilon Le$ replaces the ratio P_C/P_T (of the Prandtl numbers) which in [10] had been denoted by p.

We assume $m < 1$ (*stability condition*) and, in a first step, we consider only the linearized system associated to (34).

In the case $\tau \leq 1$, as in the previous section, it is possible to show that the condition

$$\mathcal{R} < \mathcal{R}_E = 4\pi^2 + Le\mathcal{C}, \tag{36}$$

implies linear stability with respect to the Lyapunov function

$$V_0(t) = \frac{1}{2}[\tau\|\phi\|^2 + \mu\|\psi\|^2].$$

In order to prove that condition (36) implies also nonlinear conditional stability with respect to the Lyapunov function $V(t)$, we estimate $\dot{V}$. ¿From $(34)_1$ we have $\|\mathbf{u}\| \leq \|\phi\|$, so if

$$k_1 = max(|\mathcal{C}/\tau - \mathcal{R}|, |\mathcal{R} - \delta\mathcal{C}|, 1)$$

then $I_1 \leq k_1[||\phi||||\Delta\phi|| + ||\phi||||\Delta\psi||]$. Defining

$$D_2 = \frac{1-m}{2}D_0 + \frac{b}{2}D_1$$

where $0 < m < 1$, by means of Schwarz and Poincaré-type inequalities, there follows $bI_1 \leq D_2$ with $b = \frac{1-m}{2} \cdot \frac{\pi^2\Gamma}{k_1^2} \cdot \frac{\bar{\Gamma}\bar{\zeta}}{\bar{\Gamma}+\bar{\zeta}}$.

Moreover, there exists a positive "computable" constant A :

$$\begin{cases} A = \frac{(2)^{3/2}}{\sqrt{b}}\left[\frac{k_2}{\pi^2(1-m)\Gamma} + \frac{k_3}{\sqrt{\tau(1-m)}}((\bar{\Gamma}\Gamma)^{-1/2}\right. \\ \left. +(\bar{\Gamma}\zeta)^{-1/2} + (\bar{\zeta}\Gamma)^{-1/2} + (\bar{\zeta}\zeta)^{-1/2})\right] \end{cases} \tag{37}$$

where

$$k_2 = |\frac{(p-\tau)(1+\mu\delta)}{\delta\tau - 1}|,$$

$$k_3 = max(|\frac{\delta\tau^2 - p}{\tau(\delta\tau - 1)}|, |\frac{p-\tau}{\tau(\delta\tau-1)}|, |\frac{\delta(\tau-p)}{\delta\tau-1}|, |\frac{\delta p - 1}{\delta\tau - 1}|), \tag{38}$$

such that

$$bN_1 + N_0 \leq AD_2V^{\frac{1}{2}}.$$

¿From the previous inequalities we obtain

$$\dot{V} \leq -D_2(1 - AV^{\frac{1}{2}})$$

which leads to the following theorem.

Theorem 5.1 *If $m < 1$, $V(0) < A^{-2}$, then the motionless solution m_0 is nonlinearly asymptotically stable, and there exists a positive constant k_0 depending on the parameters $\epsilon, Le, \mathcal{C}, \mathcal{R}$ such that*

$$V(t) \leq V(0)\exp[-k_0(1 - AV(0)^{\frac{1}{2}})t], \qquad t \geq 0.$$

The proof of the theorem follows by applying a recursive method.

Acknowledgments

This research has been partially supported by the University of Catania under the local contract "Analisi qualitativa in problemi di meccanica dei continui e di biomatematica", the Italian Ministry for University and Scientific Research (M.U.R.S.T.), PRIN " Problemi matematici non lineari di propagazione e stabilità nei modelli del continuo", Director Prof. Tommaso Ruggeri, and by "Gruppo Nazionale della Fisica Matematica" of the "Istituto Nazionale di Alta Matematica". *Last but not least*, we should like to thank the local organisers of Wascom01, Professors Roberto Monaco and Miriam Pandolfi Bianchi.

References

1. S. Chandrasekhar, *Hydrodynamic and hydromagnetic stability*. Oxford: Clarendon Press, 1961.
2. J. Flavin and S. Rionero, *Qualitative estimates for partial differential equations. An introduction*, Boca Raton, Florida: CRC Press, 1996.
3. G. P. Galdi and B. Straughan, *Arch. Rational Mech. Anal.* **89**, 211 (1985).
4. L.C. Gilver and S.A. Altobelli, *J. Fluid Mech.* **258**, 355 (1994).
5. T. Giorgi, *Transport in Porous Media* **29**, 191 (1997).
6. J. Guo and P.N. Kaloni, *J. Math. Anal. Appl.* **190**, 373 (1995).
7. P.N. Kaloni and J. Guo, *J. Mat. Anal. Appl.* **204**, 138 (1996).
8. S. Lombardo, G. Mulone and B. Straughan, Nonlinear stability in the Bénard problem for a double-diffusive mixture in a porous medium, *Math. Met. Appl. Sciences,* (2001, to appear).
9. S. Lombardo, G. Mulone and S. Rionero, Global nonlinear exponential stability of the conduction-diffusion solution for Schmidt numbers greater than Prandtl numbers, *J. Mat. Anal. Appl.* (to appear).
10. G. Mulone and S. Rionero, *Rend. Mat. Acc. Lincei. s. 9* **9**, 221 (1998).
11. D.A. Nield and A. Bejan, *Convection in porous media*, Second Ed., Springer-Verlag, New-York, 1992.
12. L.E. Payne and B. Straughan, *Stud. Appl. Math.* **102**, 419 (1999).
13. D. Poulikakos, *Int. Comm. Heat Mass Transfer* **13**, 587 (1986).
14. Y. Qin and P.N. Kaloni, *Quart. Appl. Math.* **96**, 189 (1994).
15. B. Straughan, *The Energy Method, Stability, and Nonlinear Convection*. Springer-Verlag: Ser. in Appl. Math. Sci., **91**, 1992.
16. O.V. Trevisan and A. Bejan, *Adv. in Heat Transfer* **20**, 315 (1990).
17. S. Whitaker, *Transport in Porous Media* **25**, 27 (1996).

EXISTENCE AND UNIQUENESS FOR PRANDTL EQUATIONS AND ZERO VISCOSITY LIMIT OF THE NAVIER-STOKES EQUATIONS

MARIA CARMELA LOMBARDO AND MARCO SAMMARTINO

Department of Mathematics and Applications, University of Palermo
Via Archirafi 34, 90123 Palermo, Italy
E-mail: lombardo@math.unipa.it, marco@math.unipa.it

The existence and uniqueness of the mild solution of the boundary layer (BL) equation is proved assuming analyticity of the data with respect to the tangential variable. Moreover we use the well-posedness of the BL equation to perform an asymptotic expansion of the Navier-Stokes equations on a bounded domain.

1 Introduction

Prandtl's boundary layer equations were first formulated in 1904 in order to solve the differences between the viscous and inviscid theory of fluids. In particular inviscid flow does not account for the total drag on a body. Moreover, in presence of a boundary, a perfect flow allows only vanishing normal velocity, while a viscous flow imposes all the components of the velocity to be zero on the surface of a stationary object.

Introducing the proper scaling to make the viscous forces to be of the same order of magnitude of the inertial forces, one derives Prandtl equations. They hold inside a narrow "boundary layer" region of thickness $O(\sqrt{\nu})$ where viscous drag and no-slip boundary conditions occur. The BL equations are:

$$(\partial_t - \partial_{YY})\, u^P + u^P \partial_x u^P + v^P \partial_Y u^P + \partial_x p^P = 0 \; , \tag{1}$$

$$\partial_Y p^P = 0 \; , \tag{2}$$

$$\partial_x u^P + \partial_Y v^P = 0 \; , \tag{3}$$

$$u^P(x, Y = 0, t) = 0 \; , \tag{4}$$

$$u^P(x, Y \to \infty, t) \longrightarrow U(x,t) \; , \tag{5}$$

$$p^P(x, Y \to \infty, t) \longrightarrow p^E(x, y = 0, t) \; , \tag{6}$$

$$u^P(x, Y, t = 0) = u^P_{in} \; . \tag{7}$$

In the above equations (u^P, v^P) and p^P represent the components of the fluid velocity and the pressure inside the BL, Y is the rescaled normal variable $Y = y/\sqrt{\nu}$. Equation (5) is the matching condition between the velocity of inside the BL and the outer Euler flow; $U(x,t)$ is the tangential component of the Euler flow at the boundary. Up to date the well-posedness of the above system of equations is an open question and an exhaustive theory of the Prandtl equations is far from being achieved (see [1]). Relevant results have been obtained by Oleinik and coworkers (see [5] for an update review) but they have to require quite restrictive monotonicity assumptions on the initial data.

Existence and uniqueness of the Prandtl system Eqs. (1)-(7) was proved by Caflisch and Sammartino in [6] without any monotonicity restriction but imposing analitycity on the initial data with respect to both the tangential and the normal component of the velocity. The main tool of their proof was the abstract formulation of the Cauchy Kowalewski theorem (ACK) in the Banach spaces of analytic functions.

In this paper we extend the results of [6], proving existence and uniqueness for the Prandtl equations requiring analyticity only with respect to the tangential variable. The well-posedness of the Boundary Layer equations is then used to address the study of the zero viscosity limit of the Navier-Stokes equations on a bounded domain. The results of this paper are valid in $3D$ as well as in $2D$. To simplify the notation we shall restrict to the $2D$ case.

2 Function spaces

In this section we introduce the function spaces used in the proof of existence and uniqueness of the Prandtl equations.

Definition 2.1 $K^{l,\rho}$ is the space of the analytic functions $f(x)$ defined in $\{x \in \mathbb{C} : \Im x \in (-\rho, \rho)\}$ such that:

- if $\Im x \in (-\rho,\rho)$ and $0 \le j \le l$, then $\partial_x^j f(\Re x + i\Im x) \in L^2(\Re x)$;
- $|f|_{l,\rho} \equiv \sum_{j=0}^{l} \sup_{\Im x \in (-\rho,\rho)} \|\partial_x^j f(\cdot + i\Im x)\|_{L^2(\Re x)} < \infty$.

Definition 2.2 $K^{l,\rho,\mu}$, with $\mu > 0$, is the space of the functions $f(Y,x)$ such that:

- $e^{\mu Y}\partial_x^i \partial_Y^j f \in L^\infty(\mathbb{R}^+, K^{0,\rho})$ when $i+j \le l$ and $j \le 2$;
- $|f|_{l,\rho,\mu} \equiv \sum_{j\le 2}\sum_{i\le l-j} \sup_{Y\in\mathbb{R}^+} e^{\mu Y}\ |\partial_Y^j \partial_x^i f(Y,\cdot)|_{0,\rho} < \infty$.

Definition 2.3 $K^{l,\rho}_{\beta,T}$,with $\beta > 0$, is the space of the functions $f(x,t)$ such that:

- $\partial_t^i \partial_x^j f(x,t) \in K^{l,\rho-\beta t} \quad \forall 0 \le t \le T$, where $0 \le i+j \le l$ and $0 \le i \le 1$;
- $|f|_{l,\rho,\beta,T} \equiv \sum_{0\le j\le 1}\sum_{i\le l-j} \sup_{0\le t\le T} |\partial_t^j \partial_x^i f(\cdot,t)|_{0,\rho-\beta t} < \infty$.

Definition 2.4 $K^{l,\rho,\mu}_{\beta,T}$ is the space of the functions $f(x,Y,t)$ such that:

- $f \in K^{l,\rho-\beta t,\mu-\beta t}$; $\partial_t \partial_x^i f \in K^{0,\rho-\beta t,\mu-\beta t} \quad \forall 0 \le t \le T$, for $0 \le i \le l-2$;
- $|f|_{l,\rho,\mu,\beta,T} \equiv \sum_{0\le j\le 2} \sum_{i\le l-j} \sup_{0\le t\le T} |\partial_Y^j \partial_x^i f(\cdot,\cdot,t)|_{0,\rho-\beta t,\mu-\beta t}$ $+ \sum_{i\le l-2} \sup_{0\le t\le T} |\partial_t \partial_x^i f(\cdot,\cdot,t)|_{0,\rho-\beta t,\mu-\beta t} < \infty$.

3 A parabolic equation

In this section we shall be concerned with the following equations:

$$(\partial_t - \partial_{YY})u + \alpha(x,t)\, Y\, \partial_Y u = f, \tag{1}$$

$$u(x, Y=0, t) = g, \tag{2}$$

$$u(x, Y, t=0) = u_0. \tag{3}$$

To solve the above system, we first introduce the following kernel:

$$F_\alpha(x,Y,t) = \frac{1}{\sqrt{4\pi}} \frac{1}{\left(\int_0^t d\tau\; e^{-2A(x,\tau)}\right)^{1/2}} \exp\left(\frac{Y^2 e^{-2A(x,t)}}{4\left(\int_0^t d\tau\; e^{-2A(x,\tau)}\right)}\right), \tag{4}$$

where $A(\tau)$ and E_α are defined as:

$$A(x,\tau) = \int_0^\tau d\theta\; \alpha(x,\theta)\; ;\; E_\alpha(x,Y,t) = \int_0^\infty dY' \left[F_\alpha(x, Y-Y', t) - F_\alpha(x, Y+Y', t)\right].$$

We can introduce the following operators:

$$M_0 u_0 = \int_0^\infty dY' \; [F_\alpha(Y-Y',t) - F_\alpha(Y+Y',t)]\; u_0(x,Y'), \tag{5}$$

$$M_2 f = \int_0^t ds \int_0^\infty dY' \; [F_\alpha(Y-Y',t-s) - F_\alpha(Y+Y',t-s)]\; f(x,Y',s), \tag{6}$$

$$M_1 g = 2\int_0^t ds \left(-2\frac{\partial F_\alpha}{\partial Y} + 2Y\alpha F_\alpha - \alpha E_\alpha\right)(x,Y,t-s)\; g(x,s), \tag{7}$$

$$M_3 h = \int_0^t ds \int_0^\infty dY' \partial_Y \left[F_\alpha(Y-Y',t-s) - F_\alpha(Y+Y',t-s)\right] f(x,Y',s). \tag{8}$$

Notice that, if $h(x, Y=0, t) = 0$, then, integrating by parts, one gets $M_3 h \equiv M_2 \partial_Y h$. Using the operators M_0, M_1 and M_2 given by (5), (6) and (7), one can write the explicit expression of the initial boundary value problem (1)-(3), and prove the following Proposition:

Proposition 3.1 *Suppose* $\alpha, g \in K^{l,\rho}_{\beta,T}$, $f \in K^{l,\rho,\mu}_{\beta,T}$ *and* $u_0 \in K^{l,\rho,\mu}$. *Moreover let the compatibility condition* $g(x,t=0)=0$ *be satisfied. Then the solution* u *of Eqs. (1)-(3) is in* $K^{l,\rho,\mu}_{\beta,T}$, *and the following estimate holds:*

$$|u|_{l,\rho,\mu,\beta,T} \le c\left(|\alpha|_{l,\rho,\beta,T} + |f|_{l,\rho,\mu,\beta,T} + |g|_{l,\rho,\beta,T}\right) .$$

We will also need the following estimates for the operators M_2 and M_3:

Lemma 3.1 *Suppose* $\alpha \in K^{l,\rho}_{\beta,T}$ *and* $f \in K^{l,\rho,\mu}_{\beta,T}$ *with* $f(x,Y=0,t)=0$. *If* $\rho' < \rho - \beta t$ *and* $\mu' < \mu - \beta t$ *then the following estimate holds:*

$$|M_2 f|_{l,\rho',\mu'} \le c\int_0^t ds |f(\cdot,\cdot,s)|_{l,\rho',\mu'} \le c|f|_{l,\rho,\mu,\beta,T} .$$

Lemma 3.2 *Suppose* $\alpha \in K^{l,\rho}_{\beta,T}$, $h \in K^{l,\rho,\mu}_{\beta,T}$ *with* $h|_{Y=0}=0$, $\partial_Y h|_{Y=0}=0$. *If* $0<\mu'<\mu(s)<\mu-\beta s$ *then* $M_3 h \in K^{l,\rho,\mu'}$ *for each* $0<t<T$ *and the following estimate holds:*

$$|M_3 h|_{l,\rho,\mu'} \le c\int_0^t ds\left(\frac{|h(\cdot,\cdot,s)|_{l,\rho,\mu'}}{\sqrt{t-s}} + \frac{|h(\cdot,\cdot,s)|_{l,\rho,\mu(s)}}{\mu(s)-\mu'}\right) .$$

For the details of the proofs of the above statements see [3].

4 The mild form of the Prandtl equations

Following [3], we introduce the new variable $u = u^P - U$, so that using the Euler equation at the boundary, Eqs. (1)-(7) can be written in the form:

$$u = F(u,t) , \tag{9}$$

where:

$$F(u,t) = M_2 K_1(u,t) + M_3 K(u,t) + \mathcal{C} , \tag{10}$$

$$K_1(u,t) = -\left(2u\,\partial_x u + U\,\partial_x u + u\,\partial_x U\right) , \tag{11}$$

$$K(u,t) = u\int_0^Y dY'\,\partial_x u, \tag{12}$$

where $\mathcal{C}$ takes into account the initial and boundary conditions and we have identified the α appearing in the kernel F_α with the Euler datum at the boundary U. We shall call the Eq. (9) the mild form of the Prandtl equations.

5 The main result

To prove the existence and the uniqueness of the mild solution of the Prandtl equations, we shall use a slightly modified version of the Abstract Cauchy-Kowalewski Theorem (ACK) as given in [6]. We refer to [2] and [3] for the formal statement and the proof of the Modified ACK theorem while here we shall only give an informal version.

Theorem 5.1 (Modified ACK) *Let $\{X_\rho : 0 < \rho \le \rho_0\}$ be a Banach scale with norms $|\ |_\rho$, such that $X_{\rho'} \subset X_{\rho''}$ and $|\ |_{\rho''} \le |\ |_{\rho'}$ when $\rho'' \le \rho' \le \rho_0$.*

For t in $[0,T]$, consider the equation

$$u + F(t,u) = 0\,. \tag{13}$$

Suppose that the function $F(t,u)\ :\ [0,\tau] \mapsto X_{\rho'}$ is continuous and $|F(t,0)|_{\rho_0-\beta_0 t} \le R_0 < R$. Moreover $\forall 0 < \rho' < \rho(s) \le \rho_0 - \beta_0 s$ and $\forall\ u^1$ and $u^2 \in \{u \in X_\rho : \sup\limits_{0\le t\le T} |u(t)|_{\rho-\beta_0 t} \le R\}$,

$$|F(t,u^1) - F(t,u^2)|_{\rho'} \le C \int_0^t ds \left(\frac{|u^1-u^2|_{\rho(s)}}{\rho(s)-\rho'} + \frac{|u^1-u^2|_{\rho'}}{\sqrt{t-s}} \right). \tag{14}$$

Then $\exists \beta > \beta_0$ and $0 < T_1 < T$ such that Eq. (13) has a unique solution $u(t) \in X_{\rho_0-\beta t}$ with $t \in [0,T_1]$; moreover $\sup\limits_{0\le t\le T_1} |u(t)|_{\rho_0-\beta t} \le R$.

To apply the ACK theorem to the Prandtl equation, we have to prove that the right hand side of Eq. (10) satisfies an estimate like (14). Using Proposition 3.1, Lemma 3.1, Lemma 3.2 and through the Cauchy estimate for the derivative of an analytic function (see also [6]), one can easily prove the following Proposition.

Proposition 5.1 *Suppose that u^1 and u^2 are in $K^{l,\rho_0,\mu_0}_{\beta_0,T}$. Suppose $0 < \rho' < \rho(s) < \rho_0 - \beta_0 s$ and $0 < \mu' < \mu(s) < \mu_0 - \beta_0 s$. Then the following estimate holds:*

$$\begin{aligned} &\left|\ F(u^1,t) - F(u^2,t)\ \right|_{l,\rho',\mu'} \\ &\le c \int_0^t ds \ \left(\frac{|u^1-u^2|_{l,\rho(s),\mu}}{\rho(s)-\rho'} + \frac{|u^1-u^2|_{l,\rho,\mu(s)}}{\mu(s)-\mu'} + \frac{|u^1-u^2|_{l,\rho',\mu'}}{\sqrt{t-s}} \right). \end{aligned} \tag{15}$$

Therefore using the above Proposition and the ACK Theorem, the main result of this paper can be proved.

Theorem 5.2 *Suppose $U \in K^{l,\rho_0}_{\beta_0,T}$ and $u^P_{in} - U \in K^{l,\rho_0,\mu_0}$. Moreover let the compatibility conditions $u^P_{in}(x,Y=0) = 0$; $u^P_{in}(x,Y\to\infty) - U \longrightarrow 0$ hold. Then there exists a unique mild solution u^P of Eqs. (1)-(7), which can be written as $u^P(x,Y,t) = u(x,Y,t) + U$ where $u \in K^{l,\rho_1,\mu_1}_{\beta_1,T_1}$ with $0 < \rho_1 < \rho_0$, $0 < \mu_1 < \mu_0$, $\beta_1 > \beta_0 > 0$ and $0 < T_1 < T$.*

6 Zero Viscosity Limit on Bounded Domains

In this section we shall show how one could address the problem of the zero-viscosity limit of the Navier-Stokes equations in domains with curved boundaries. We define adapted coordinates (s, n) and, as we can allow non-analytic initial data with respect to the normal variable, we introduce a C^∞ cut-off function $m(n/\varepsilon^\alpha)$, with $0 < \alpha < 1$, such that:

$$m(n/\varepsilon^\alpha) = \begin{cases} 1 & \text{for } 0 < n < \varepsilon^\alpha, \\ 0 & \text{for } n > 2\,\varepsilon^\alpha. \end{cases}$$

We thus construct an approximate solution to the Navier-Stokes equations of the form:

$$\boldsymbol{u}^\varepsilon = (1-m)\left(\boldsymbol{u}_0^E + \varepsilon\,\boldsymbol{u}_1^E + \varepsilon^2\,\boldsymbol{u}_2^E\right) + m\left(\boldsymbol{U}_0^P + \varepsilon\,\boldsymbol{U}_1^P + \varepsilon^2\,\boldsymbol{U}_2^P\right) ,$$

where the $\boldsymbol{u}_i^E$ terms satisfy the $i-th$ order Euler equations and the $\boldsymbol{U}_i^P$ terms satisfy the $i-th$ order Prandtl equations with proper boundary and matching conditions (see [4]). Then the solution to the Navier-Stokes equation can be written as:

$$\boldsymbol{u}^{NS} = \boldsymbol{u}^\varepsilon + \varepsilon^2 \boldsymbol{w} ,$$

where the correction term $\boldsymbol{w}$ satisfies a Navier-Stokes-type equation with bounded source term. The possibility of giving a rigorous estimate of the norm of $\boldsymbol{w}$ in the appropriate function space is under current investigation.

Acknowledgments

This paper was supported in part by the MURST, under grant "Problemi matematici Non Lineari di Propagazione e Stabilità nei Modelli del Continuo".

References

1. Caflisch R.E., Sammartino M., *Existence and singularities for the Prandtl boundary layer equations,* Z. Angew. Math. Mech., **80**, 733–744, (2000)
2. Cannone, M.; Lombardo, M.C.; Sammartino, M. *Existence and uniqueness for the Prandtl equations* C. R. Acad. Sci. Paris Sér. I Math., **332**, n 3, 277–282, (2001)
3. Cannone M., Lombardo M.C. and Sammartino M., *Well-posedness of the boundary layer equations with a non-analyticity hypothesis,* submitted.
4. Lombardo M.C. and Sammartino M., *Zero viscosity limit of the Navier-Stokes equations on bounded domains,* in preparation.
5. Oleinik O.A., Samokhin V.N., *Mathematical models in boundary layer theory,* ed. Chapman & Hall/CRC, (1999)
6. Sammartino M., Caflisch R.E., *Zero Viscosity Limit for Analytic Solutions of the Navier-Stokes Equation on a Half-Space I. Existence for Euler and Prandtl Equations,* Comm. Math. Phys., **192**, 433–461 (1998)

SPACE-TIME FRACTIONAL DIFFUSION: EXACT SOLUTIONS AND PROBABILITY INTERPRETATION

Francesco MAINARDI
Dipartimento di Fisica, Università di Bologna
Via Irnerio 46, I-40126 Bologna, Italy
Centro Interdipartimentale di Ricerca per le Applicazioni della Matematica,
Università di Bologna, Via Saragozza 8, 40123 Bologna, Italy
E-mail: mainardi@bo.infn.it URL: www.fracalmo.org

Gianni PAGNINI
Istituto per le Scienze dell'Atmosfera e del Clima del CNR,
Via Gobetti 101, I-40129 Bologna, Italy
E-mail: g.pagnini@isao.bo.cnr.it

The fundamental solutions (Green functions) for the Cauchy problems of the space-time fractional diffusion equation are investigated with respect to their scaling and similarity properties, starting from their composite Fourier-Laplace representation. By using the Mellin transform, a general representation of the Green functions in terms of Mellin-Barnes integrals in the complex plane is presented, that allows us to obtain their computational form in the space-time domain and to analyse their probability interpretation.

Mathematics Subject Classification 2000: 26A33, 33E12, 33C60, 44A10, 45K05, 60G18.

1 Introduction

By replacing in the standard diffusion equation

$$\frac{\partial}{\partial t}u(x,t) = \frac{\partial^2}{\partial x^2}\, u(x,t)\,, \quad -\infty < x < +\infty\,, \quad t \geq 0\,, \tag{1.1}$$

where $u = u(x,t)$ is the (real) field variable, the second-order space derivative and the first-order time derivative by suitable *integro-differential* operators, which can be interpreted as a space and time derivative of fractional order, we obtain a sort of "generalized diffusion" equation. Such equation may be referred to as the *space-time fractional diffusion* equation when its fundamental solution (see below) can be interpreted as a probability density. We write

$${}_tD_*^{\beta}\, u(x,t) \;=\; {}_xD_\theta^{\alpha}\, u(x,t)\,, \quad -\infty < x < +\infty\,, \quad t \geq 0\,, \tag{1.2}$$

where the α, θ, β are real parameters restricted as follows

$$0 < \alpha \leq 2\,, \quad |\theta| \leq \min\{\alpha, 2-\alpha\}\,, \quad 0 < \beta \leq 2\,. \tag{1.3}$$

In (1.2) ${}_xD_\theta^\alpha$ is the space fractional *Riesz-Feller derivative* of order α and skewness θ, and ${}_tD_*^\beta$ is the time fractional *Caputo derivative* of order β. The definitions of these fractional derivatives are more easily understood if given in terms of Fourier transform and Laplace transform, respectively.

For the space fractional *Riesz-Feller derivative* we have

$$\mathcal{F}\{{}_xD_\theta^\alpha f(x);\kappa\} = -\psi_\alpha^\theta(\kappa)\,\widehat{f}(\kappa), \quad \psi_\alpha^\theta(\kappa) = |\kappa|^\alpha\, \mathrm{e}^{i(\mathrm{sign}\,\kappa)\theta\pi/2}, \tag{1.4}$$

where $\kappa \in \mathbb{R}$ and $\widehat{f}(\kappa) = \mathcal{F}\{f(x);\kappa\} = \int_{-\infty}^{+\infty} \mathrm{e}^{+i\kappa x} f(x)\,dx$. In other words the symbol of the pseudo-differential operator[a] ${}_xD_\theta^\alpha$ is required to be the logarithm of the characteristic function of the generic *stable* (in the Lévy sense) probability density, according to the Feller parameterization [3]. For $\alpha = 2$ (hence $\theta = 0$) we have $\widehat{{}_xD_0^2}(\kappa) = -\kappa^2 = (-i\kappa)^2$, so we recover the standard second derivative. More generally for $\theta = 0$ we have $\widehat{{}_xD_0^\alpha}(\kappa) = -|\kappa|^\alpha = -(\kappa^2)^{\alpha/2}$ so

$${}_xD_0^\alpha = -\left(-\frac{d^2}{dx^2}\right)^{\alpha/2}. \tag{1.5}$$

In this case we call the LHS of (1.5) simply the *Riesz fractional derivative* operator of order α. For the explicit expressions in integral form of the general *Riesz-Feller fractional derivative* we refer the interested reader to Mainardi, Luchko and Pagnini [9]. Let us now consider the time fractional *Caputo derivative.* Following the original idea by Caputo [1], see also [2,6,12], a proper time fractional derivative of order $\beta \in (m-1, m]$ with $m \in \mathbb{N}$, useful for physical applications, may be defined in terms of the following rule for the Laplace transform:

$$\mathcal{L}\left\{{}_tD_*^\beta f(t);s\right\} = s^\beta\,\widetilde{f}(s) - \sum_{k=0}^{m-1} s^{\beta-1-k} f^{(k)}(0^+), \quad m-1<\beta\le m, \tag{1.6}$$

where $s \in \mathbb{C}$ and $\widetilde{f}(s) = \mathcal{L}\{f(t);s\} = \int_0^\infty \mathrm{e}^{-st} f(t)\,dt$. Then the *Caputo fractional derivative* of $f(t)$ turns out to be defined as

$${}_tD_*^\beta f(t) := \begin{cases} \dfrac{1}{\Gamma(m-\beta)}\displaystyle\int_0^t \frac{f^{(m)}(\tau)\,d\tau}{(t-\tau)^{\beta+1-m}}, & m-1<\beta<m, \\[2ex] \dfrac{d^m}{dt^m} f(t), & \beta = m. \end{cases} \tag{1.7}$$

[a]Let us recall that a generic linear pseudo-differential operator A, acting with respect to the variable $x \in \mathbb{R}$, is defined through its Fourier representation, namely $\int_{-\infty}^{+\infty} \mathrm{e}^{i\kappa x} A[f(x)]\,dx = \widehat{A}(\kappa)\,\widehat{f}(\kappa)$, where $\widehat{A}(\kappa)$ is referred to as symbol of A, given as $\widehat{A}(\kappa) = \left(A\,\mathrm{e}^{-i\kappa x}\right)\mathrm{e}^{+i\kappa x}$.

In order to formulate and solve the Cauchy problems for (1.2) we have to select explicit initial conditions concerning $u(x,0^+)$ if $0 < \beta \le 1$ and $u(x,0^+)$, $u_t(x,0^+)$ if $1 < \beta \le 2$. If $\phi_1(x)$ and $\phi_2(x)$ denote two given real functions of $x \in I\!R$, the Cauchy problems consist in finding the solution of (1.2) subjected to the additional conditions:

$$u(x,0^+) = \phi_1(x)\,, \quad x \in I\!R\,, \quad \text{if} \quad 0 < \beta \le 1\,; \tag{1.8a}$$

$$\begin{cases} u(x,0^+) = \phi_1(x)\,, \\ u_t(x,0^+) = \phi_2(x)\,, \end{cases} \quad x \in I\!R\,, \quad \text{if} \quad 1 < \beta \le 2\,. \tag{1.8b}$$

2 The Green functions

The Cauchy problems can be conveniently treated by making use of the most common integral transforms, *i.e.* the Fourier transform (in space) and the Laplace transform (in time). Indeed, the composite Fourier-Laplace transforms of the solutions of the two Cauchy problems:

(a) $\{(1.2) + (1.8a)\}$ if $0 < \beta \le 1$, (b) $\{(1.2) + (1.8b)\}$ if $1 < \beta \le 2$,

turn out to be, by using (1.4) and (1.6) with $m = 1, 2$,

$$\widehat{\widetilde{u}}(\kappa, s) = \frac{s^{\beta-1}}{s^\beta + \psi_\alpha^\theta(\kappa)}\,\widehat{\phi_1}(\kappa)\,, \quad 0 < \beta \le 1\,, \tag{2.1a}$$

$$\widehat{\widetilde{u}}(\kappa, s) = \frac{s^{\beta-1}}{s^\beta + \psi_\alpha^\theta(\kappa)}\,\widehat{\phi_1}(\kappa) + \frac{s^{\beta-2}}{s^\beta + \psi_\alpha^\theta(\kappa)}\,\widehat{\phi_2}(\kappa)\,, \quad 1 < \beta \le 2\,. \tag{2.1b}$$

By fundamental solutions (or Green functions) of the above Cauchy problems we mean the (generalized) solutions corresponding to the initial conditions:

$$G_{\alpha,\beta}^{\theta\,(1)}(x,0^+) = \delta(x)\,, \quad 0 < \beta \le 1\,; \tag{2.2a}$$

$$\begin{cases} G_{\alpha,\beta}^{\theta\,(1)}(x,0^+) = \delta(x)\,, \\ \dfrac{\partial}{\partial t}\,G_{\alpha,\beta}^{\theta\,(1)}(x,0^+) = 0\,, \end{cases} \quad \begin{cases} G_{\alpha,\beta}^{\theta\,(2)}(x,0^+) = 0\,, \\ \dfrac{\partial}{\partial t}\,G_{\alpha,\beta}^{\theta\,(2)}(x,0^+) = \delta(x)\,, \end{cases} \quad 1 < \beta \le 2\,. \tag{2.2b}$$

We have denoted by $\delta(x)$ the delta-Dirac generalized function, whose (generalized) Fourier transform is known to be 1, and we have distinguished by the apices (1) and (2) the two types of Green functions. From Eqs (2.1a)-(2.1b) the Fourier-Laplace transforms of these Green functions turn out to be

$$\widehat{\widetilde{G}}{}_{\alpha,\beta}^{\theta\,(j)}(\kappa, s) = \frac{s^{\beta-j}}{s^\beta + \psi_\alpha^\theta(\kappa)}\,, \quad 0 < \beta \le 2\,, \quad j = 1, 2\,. \tag{2.3}$$

Furthermore, by recalling the Fourier convolution property, we note that the Green functions allow us the represent the solutions of the above two Cauchy problems through the relevant integral formulas:

$$u(x,t) = \int_{-\infty}^{+\infty} G_{\alpha,\beta}^{\theta\,(1)}(\xi,t)\,\phi_1(x-\xi)\,d\xi\,, \quad 0<\beta\le 1\,; \qquad (2.4a)$$

$$u(x,t) = \int_{-\infty}^{+\infty} [G_{\alpha,\beta}^{\theta\,(1)}(\xi,t)\phi_1(x-\xi)+G_{\alpha,\beta}^{\theta\,(2)}(\xi,t)\phi_2(x-\xi)]\,d\xi,\ 1<\beta\le 2. \quad (2.4b)$$

We recognize from (2.3) that the function $G_{\alpha,\beta}^{\theta\,(2)}(x,t)$ along with its Fourier-Laplace transform is well defined also for $0<\beta\le 1$ even if it loses its meaning of being a fundamental solution of (1.2), resulting

$$G_{\alpha,\beta}^{\theta\,(2)}(x,t) = \int_0^t G_{\alpha,\beta}^{\theta\,(1)}(x,\tau)\,d\tau\,, \quad 0<\beta\le 2\,. \qquad (2.5)$$

By using the known scaling rules for the Fourier and Laplace transforms, and introducing the *similarity variable* $x/t^{\beta/\alpha}$, we infer from (2.3) (thus without inverting the two transforms) the *scaling properties* of the Green functions,

$$G_{\alpha,\beta}^{\theta\,(j)}(x,t) = t^{-\beta/\alpha+j-1}\,K_{\alpha,\beta}^{\theta\,(j)}\left(x/t^{\beta/\alpha}\right)\,, \qquad (2.6)$$

where the one-variable functions $K_{\alpha,\beta}^{\theta\,(1)}(x)$, $K_{\alpha,\beta}^{\theta\,(2)}(x)$ are called the *reduced Green functions*. We also note the *symmetry relation*:

$$G_{\alpha,\beta}^{\theta\,(j)}(-x,t) = G_{\alpha,\beta}^{-\theta\,(j)}(x,t)\,, \quad j=1,2\,, \qquad (2.7)$$

so for the determination of the Green functions we can restrict our attention to $x>0$. Extending the method illustrated in [4,9], where only the Green function of type (1) was determined, we first invert the Laplace transforms getting

$$\widehat{G_{\alpha,\beta}^{\theta\,(j)}}(\kappa,t) = t^{j-1}\,E_{\beta,j}[-\psi_\alpha^\theta(\kappa)t^\beta],\ \widehat{K_{\alpha,\beta}^{\theta\,(j)}}(\kappa) = E_{\beta,j}[-\psi_\alpha^\theta(\kappa)],\ j=1,2, \quad (2.8)$$

where $E_{\beta,j}$ denotes the two-parameter Mittag-Leffler function[b]. We note the normalization property $\int_{-\infty}^{+\infty} K_{\alpha,\beta}^{\theta\,(j)}(x)\,dx = E_{\beta,j}(0) = 1/\Gamma(j) = 1$ for $j=1,2$.

[b]The Mittag-Leffler function $E_{\beta,\mu}(z)$ with $\beta,\mu>0$ is an entire transcendental function of order $\rho=1/\beta$, defined in the complex plane by the power series

$$E_{\beta,\mu}(z) := \sum_{n=0}^{\infty} \frac{z^n}{\Gamma(\beta n+\mu)}\,, \quad \beta,\mu>0\,, \quad z\in\mathbb{C}\,.$$

For information on the Mittag-Leffler-type functions the reader may consult *e.g.* [6,12].

Following [9] we invert the Fourier transforms of $K_{\alpha,\beta}^{\theta\,(j)}(x)$ by using the convolution theorem of the Mellin transforms, arriving at the Mellin-Barnes integral representation[c]:

$$K_{\alpha,\beta}^{\theta\,(j)}(x) = \frac{1}{\alpha x}\frac{1}{2\pi i}\int_{\gamma-i\infty}^{\gamma+i\infty} \frac{\Gamma(\frac{s}{\alpha})\,\Gamma(1-\frac{s}{\alpha})\,\Gamma(1-s)}{\Gamma(j-\frac{\beta}{\alpha}s)\,\Gamma(\rho\,s)\,\Gamma(1-\rho\,s)}\,x^{\,s}\,ds\,, \qquad (2.9)$$

where $0 < \gamma < \min\{\alpha, 1\}$, and $\rho = (\alpha-\theta)/(2\,\alpha)$.

We note that the Mellin-Barnes integral representation $(2.9)^d$ allows us to construct computationally the fundamentals solutions of Eq. (1.2) for any triplet $\{\alpha, \beta, \theta\}$ by matching their convergent and asymptotic expansions, as shown in [9] for the first Green function. For the particular cases that allow simplifications in the integrand of Eq. (2.9), we obtain relevant expressions of the corresponding Green functions. This occurs in the following cases:
(a) for $j = 1$ and $\{0 < \alpha < 2,\ \beta = 1\}$ (*strictly space fractional diffusion*) where we have $K_{\alpha,1}^{\theta\,(1)}(x) = L_\alpha^\theta(x)$, *i.e.* the class of the strictly stable (non-Gaussian) densities[3] exhibiting fat tails (with the algebraic decay $\propto |x|^{-(\alpha+1)}$) and infinite variance;
(b) for $j = 1, 2$ and $\{\alpha = 2,\ 0 < \beta < 2\}$ (*time fractional diffusion* including *standard diffusion*), where we have $K_{2,\beta}^{0\,(j)}(x) = M_{\beta/2}^{(j)}(x)/2$, *i.e.* the class of the Wright-type densities[5,7,8,9,11] exhibiting stretched exponential tails and finite variance proportional to $t^{\beta+j-1}$;
(c) for $j = 1$ and $\{0 < \alpha = \beta < 2\}$ (*neutral fractional diffusion*), where we have $K_{\alpha,\alpha}^{\theta\,(1)}(x) = N_\alpha^\theta(x)$, *i.e.* the class of the Cauchy-type densities[9].

Based on the arguments outlined in[9], we extend the meaning of probability density to the cases $\{0 < \alpha < 2\,,\ 0 < \beta < 1\}$ and $\{1 < \beta \le \alpha < 2\}$ by proving the following composition rules of the Mellin convolution type:

$$K_{\alpha,\beta}^{\theta\,(j)}(x) = \begin{cases} \alpha \displaystyle\int_0^\infty \left[\xi^{\alpha-1}\, M_\beta^{(j)}\,(\xi^\alpha)\right]\, L_\alpha^\theta\,(x/\xi)\,\frac{d\xi}{\xi}\,, & 0 < \beta < 1\,, \\ \displaystyle\int_0^\infty M_{\beta/\alpha}^{(j)}(\xi)\, N_\alpha^\theta(x/\xi)\,\frac{d\xi}{\xi}\,, & 0 < \beta/\alpha < 1\,. \end{cases} \qquad (2.10)$$

---

[c] The names refer to the two authors, who in the beginning of the past century developed the theory of these integrals using them for a complete integration of the hypergeometric differential equation. However, as revisited in [10], these integrals were first introduced in 1888 by S. Pincherle (Professor of Mathematics at the University of Bologna from 1880 to 1928).
[d] Readers acquainted with Fox H functions can recognize in (2.9) the representation of a certain function of this class, see *e.g.* [13]. Unfortunately, as far as we know, computing routines for this general class of special functions are not yet available.

References

1. M. Caputo, Linear models of dissipation whose Q is almost frequency independent, Part II, *Geophys. J. R. Astr. Soc.* **13** (1967) 529–539.
2. M. Caputo and F. Mainardi, Linear models of dissipation in anelastic solids, *Riv. Nuovo Cimento* (Ser. II) **1** (1971) 161–198.
3. W. Feller, *An Introduction to Probability Theory and its Applications*, Vol. 2 (Wiley, New York, 1971).
4. R. Gorenflo, A. Iskenderov and Yu. Luchko, Mapping between solutions of fractional diffusion-wave equations, *Fractional Calculus and Applied Analysis* **3** No 1 (2000) 75–86.
5. R. Gorenflo, Yu. Luchko and F. Mainardi, Wright functions as scale-invariant solutions of the diffusion-wave equation, *J. Computational and Applied Mathematics* **118** No 1-2 (2000) 175-191.
6. R. Gorenflo and F. Mainardi, Fractional calculus: integral and differential equations of fractional order, in: A. Carpinteri and F. Mainardi (Editors), *Fractals and Fractional Calculus in Continuum Mechanics* (Springer Verlag, Wien, 1997), 223–276.
7. F. Mainardi, On the initial value problem for the fractional diffusion-wave equation, in: S. Rionero and T. Ruggeri (Editors), *Waves and Stability in Continuous Media, VII* (World Scientific, Singapore, 1994) 246-251.
8. F. Mainardi, Fractional calculus: some basic problems in continuum and statistical mechanics, in: A. Carpinteri and F. Mainardi (Editors), *Fractals and Fractional Calculus in Continuum Mechanics* (Springer Verlag, Wien and New-York, 1997), 291–248.
9. F. Mainardi, Yu. Luchko and G. Pagnini, The fundamental solution of the space-time fractional diffusion equation, *Fractional Calculus and Applied Analysis* 4 No 2 (2001) 153-192.
10. F. Mainardi and G. Pagnini, Salvatore Pincherle: the pioneer of the Mellin-Barnes integrals, *J. Computational and Applied Mathematics*, submitted.
11. F. Mainardi and G. Pagnini, The Wright functions as solutions of the time-fractional diffusion equation, *Applied Mathematics and Computation*, submitted.
12. I. Podlubny, *Fractional Differential Equations* (Academic Press, San Diego, 1999).
13. H.M. Srivastava, K.C. Gupta and S.P. Goyal, *The H-Functions of One and Two Variables with Applications* (South Asian Publishers, New Delhi and Madras, 1982).

BIFURCATION ANALYSIS OF EQUILIBRIA WITH A MAGNETIC ISLAND IN TWO-DIMENSIONAL MHD

L. MARGHERITI AND C. TEBALDI

Department of Mathematics
University of Lecce, 73100 Lecce - Italy

In this work we have analyzed the sequence of bifurcated equilibria in two-dimensional reduced resistive magnetohydrodynamics. We have considered classes of symmetric equilibria and shown that, as in a previous work[15], after a first symmetry-breaking bifurcation, leading to an equilibrium with a small size magnetic island, the system undergoes a tangent bifurcation for a critical value of the aspect ratio. Above this value no equilibrium with a small island exists and the system rapidly develops an island of macroscopic size. This general property can be proposed as a basic ingredient of the intermittent events observed in magnetically confined plasmas in more general situations. The case of equilibria with a symmetric magnetic field plus a small island perturbation is also considered. The bifurcations in the system are investigated as a function of the aspect ratio ϵ and of the "error" parameter δ measuring the perturbation. When ϵ is decreased from one, also in this case equilibria with a small island disappear because of a tangent bifurcation. The range in ϵ for which the small island equilibria exists becomes smaller and smaller when δ is increased. The possibility to extend investigation to the case of different boundary conditions is also discussed.

1 Introduction

The high performances reached by fusion devices during the last years as well as enhanced astrophysical observations reveal more and more complicated behavior in plasmas, connected to strongly non-linear regimes accessed after destabilization of (axi-)symmetric equilibria.

Often the presence of a slowly growing magnetic island inside the plasma can be identified as a "precursor" of such regimes. This is the case for phenomena of intermittent enhancement of magnetohydrodynamic (MHD) activity as, in the case of Tokamaks, the sawtooth relaxations, the so-called edge-localized modes (ELMs) occurring in high confinement regimes and, as extreme events, disruptions that lead to sudden termination of the current discharge [6].

A theoretical analysis of this type of phenomenology in Tokamaks usually shows that the excitations which break the symmetry of the initially axisymmetric state are of the "soft" type. For values of some control parameter p slightly above the stability threshold p_c of the axisymmetric state, the system settles in a neighboring state of slightly broken symmetry, for example an equilibrium with a small saturated magnetic island.

The difficulty in understanding the frequently observed "hard" excitation on the basis of perturbation analysis around the initially axisymmetric state has become known as the "trigger" problem[20].

In general, abrupt changes of observable quantities are conveniently described in the conceptual framework of bifurcation theory[4]. In a previous paper[15] it has been explored the possibility of overcoming the difficulty posed by the trigger problem by investigating the occurrence of a second bifurcation of the non symmetric saturated state. This was done for a slab model of reduced resistive MHD (RRMHD), by varying the aspect ratio of the slab ϵ, or equivalently, the tearing mode stability parameter Δ'. The main result was that, after a first bifurcation (at $\Delta' \sim 0$) leading to a saturated tearing mode with a magnetic island of small amplitude, the system undergoes a tangent bifurcation at $L\Delta' \sim 1$, where L is a macroscopic scale length. Above this value of Δ' no equilibrium with a small island exists. The system jumps to a state where the island width is of order of the system size.

On the basis of that analysis it was proposed that sudden reconnection events observed in laboratory plasmas would occur from a state of already broken symmetry when the second "hard" bifurcation takes place. This state of slightly broken symmetry can be identified as the precursor state in which a small stable island can in principle be observed.

A more detailed analysis of the parametric dependence of the first bifurcated equilibrium was done[16] by analytic perturbation theory and numerical methods. In that paper the model was also extended to include the diamagnetic effect through the equilibrium ion drift velocity v^* under the assumption of constant (frozen) pressure gradients. The symmetric equilibrium becomes unstable because of a supercritical Hopf bifurcation, giving rise to a time-periodic solution, i.e. one with a small rotating island. The further analysis of the sequence of bifurcations has confirmed qualitatively the bifurcation diagram in Ref.[15], up to a critical value of the magnetic velocity v_c^* (of the order of one tenth of the Alfven speed). Thus the disappearance of small size island because of a tangent bifurcation as a possible mechanism for "hard" excitations is confirmed also in the case of island rotation.

In the following we briefly discuss how the above results relate to previous work .

Two coalescent small size island solutions are found using different perturbation techniques[17,19]. Those results are essentially based on expansions in the parameter Δ', which are only appropriate to treat the region around the first bifurcation point. Thus any conclusion about the nature of a subsequent bifurcation can only be speculative, when based on those methods.

The analysis of Ref.[12] is valid beyond the first bifurcation, but only situations with constant slab aspect ratio were considered, using the Lundquist number S as a bifurcation parameter. The bifurcation sequence in RRMHD was studied in Ref.[8] with an initial value code, using a variable aspect ratio but at a constant S ($S \sim 10^2$). No tangent bifurcation was detected in either work, in which however also the choice of rigid boundary conditions (zero radial flows at the walls) strongly limits the island growth, as the radial flow far from the reconnecting region would tend to grow with the island.

Whereas in the experiments great care is taken to produce axisymmetric plasmas, in some cases the symmetry is established only in an approximate form: one talks of the presence of "error fields". The result is of importance, particularly in connection with large experimental devices. One purpose of this paper is to start to address this problem, considering an equilibrium that for an aspect ratio $\epsilon = 1$ has already a small island, whose size is measured by a new, asymmetry parameter δ. The theory of bifurcation predicts the changes from the symmetric case ($\delta = 0$) at the first (symmetry breaking) bifurcation when $\delta \ll$, but nothing more. In particular nothing can be said a priori on the occurrence of the tangent bifurcation leading to disappearance of the small island equilibrium, which was the main result in Ref. [15].

This work is organized as follows. Sec. 2 is devoted to the illustration of the MHD model and Sec. 3 to the numerical techniques. In Sec. 4 we extend the stability analysis[15] to classes of symmetric equilibria. The case of non-symmetric equilibria is discussed in Sec. 5. In Sec. 6 we introduce a model that allows to consider boundary conditions different from the periodic ones. Conclusions and a discussion of possible developments are given in Sec. 7.

2 The MHD Model

The MHD model used in this work is based on single helicity reduced resistive magneto-hydrodynamics (RRMHD)[5]. The normalized model equations can be written as

$$\partial_t U + [\phi, U] = [J, \psi] + \mu \nabla^2 U \tag{1}$$

$$\partial_t \psi + [\phi, \psi] = -\eta (J - J_e) \tag{2}$$

The equations are defined on a two-dimensional domain with coordinates x and y. With reference to the magnetic geometry of a tokamak, x can be thought of as a radial coordinate and y as a poloidal coordinate. The third direction is considered ignorable. The model equations describe the evolution of the plasma vorticity $U = \nabla^2 \phi$, where ϕ is the electric potential playing the role of the stream function, and of the magnetic flux function ψ associated with the magnetic field in the plasma (a constant magnetic field is assumed in the ignorable direction). The other fields are the current density $J = -\nabla^2\psi$ and a driving current density J_e, associated with equilibrium. Moreover for any two fields A and B, $[A,B] \equiv \partial_x A \partial_y B - \partial_y A \partial_x B$, so that $[\phi,\cdot] = \underline{v}\cdot\nabla$ is the usual advection operator.

Lengths are normalized to a macroscopic length L, which is either a measure of the size of the system or of the scale length of the equilibrium magnetic field. Times are normalized to the Alfven time $\tau_A = L/v_A$, where $v_A = B_\rho/\rho^{1/2}$ is the poloidal Alfven speed associated with the equilibrium field B_ρ (ρ is the mass density).

The dissipation is measured by the viscosity μ and by the resistivity η, which in these units are respectively the inverse of the Reynolds number R, $\mu = 1/R$, and of the Lundquist number S, $\eta = 1/S$.

The domain is taken to be a square box (slab) $[-L_x, L_x] \times [-L_y, L_y]$, where the normalized lengths are of order one. It is convenient to take $L_x = \pi$, $L_y = \pi/\epsilon$, where the slab aspect ratio ϵ has been introduced.

The boundary conditions are taken periodic in both directions. Although periodicity is natural in the y direction, it seems somewhat odd to use it in the radial direction x. In reality when the island width w is sufficiently smaller than the system size, the only effect of periodicity is to add a duplicate island with the same characteristics.

The model is controlled by three dimensionless parameters, ϵ, S and the magnetic Prandtl number $P = S/R$, fixed in the rest of the paper ($P = 0.2$).

3 Numerical techniques

Various numerical techniques were employed. When possible, fixed point methods were used as an efficient way to track the sequence of bifurcated equilibria. In addition, a spectral and a pseudospectral initial value codes were employed, the latter in particular to benchmark the fixed point calculations with higher space resolution.

3.1 Fixed point method

In order to solve the system of Eqs. 1-2 a spectral decomposition is adopted for the unknowns, choosing the eigenfunctions of the laplacian as the complete orthogonal set for the expansion. One has

$$(\phi, \psi) = \sum_{\mathbf{k}} (\phi_{\mathbf{k}}, \psi_{\mathbf{k}}) e^{i\mathbf{k}\cdot\mathbf{x}} \qquad \mathbf{k} = (l, m\epsilon) \quad l, m \quad \text{integers} \tag{3}$$

We truncate the expansion to a finite set L of $2N$ wave vectors ("modes") such that if $\mathbf{k}$ belongs to L also $-\mathbf{k}$ belongs to L. This gives $4N$ ordinary differential equations for $4N$ real unknowns. Moreover, a $2N$ invariant subspace exists, characterized by imaginary amplitudes for the magnetic and velocity fields, which allows to reduce the system to $2N$ real unknowns. Different sets L have been considered, starting from a "ball" around the origin with $N = 100$ and adding modes in a slab centered at $m = 0$ up to $N = 348$.

The time independent version of Eqs. 1-2 can be cast in the form $\mathbf{F}(\mathbf{a}, \mathbf{p}) = \mathbf{0}$ where $\mathbf{a}$ are the unknowns and $\mathbf{p}$ the set of control parameters. A suitable tool to find the solutions is Newton's method, used in connection with the theorems of bifurcation theory[14]. One should stress that essential features of the method are the capability of finding both stable and unstable equilibria and its efficiency in following the sequence of equilibria when parameters are varied. Unstable equilibria are also essential to obtain the bifurcation diagram and then to fully understand the dynamics in the nonlinear regimes.

3.2 Spectral and pseudo-spectral codes

Since the model equations are supplemented with periodic boundary conditions, a suitable initial value code is a spectral code, which advances in time the Fourier amplitudes of the relevant fields. In some cases a direct truncation of the model equations to the relevant degrees of freedom was used either to compute stable equilibria or to study transients. In cases of heavier computations, the code employed takes advantage of the pseudospectral method to compute the non-linear terms, which in Fourier space take the form of convolutions. After computing the derivative in Fourier space, the fields are transformed to real space where the non-linearity are simply products at each grid point. The result is then transferred back to Fourier space. In this way the number of operations to compute the non-linearities scales like $NlogN$ instead of the unfavorable scaling $\sim N^2$ one would get with direct evaluation of correlations.

4 The symmetric case

We report the results obtained when the driving current J_e is taken symmetric, i.e. in normalized form

$$J_e(x) = \cos x \tag{4}$$

It is known[3] that a symmetric magnetic configuration can be unstable to symmetry breaking perturbations under certain conditions.

The stability boundary of the reference equilibrium can be obtained with linear theory. Upon writing

$$(\phi(x,y,t),\psi(x,y,t)) = (\tilde{\phi}(x)e^{-i\omega t+iky}, \psi_e(x) + \tilde{\psi}(x)e^{-i\omega t+iky}) \qquad (5)$$

where $k = m\epsilon$, with m integer, one obtains the linearized version of Eqs. 1-2. Linearized equations can be solved by asymptotic matching. In the limit of large Lundquist number, dissipation is only important in a narrow region around the $x = 0$ and $x = \pm\pi$ lines, where the magnetic field vanishes. Far from this region, outer solutions are exponentials when $k > 1$, which is therefore a sufficient condition for the stability of the reference equilibrium. As ϵ decreases, the first instability occurs for mode number $m = 1$ when $\epsilon = 1$. The outer solutions for $k \leq 1$ are $\tilde{\psi}_{out} = \bar{\psi}\cos[\kappa(|x| - \pi/2)]$ where $\kappa \equiv 1 - k^2$. The parameter $\Delta' \equiv \lim_{x\to 0^+}\partial_x \log\tilde{\psi}_{out} - \lim_{x\to 0^-}\partial_x\log\tilde{\psi}_{out} = 2\kappa\tan(\kappa\pi/2)$ is usually introduced to discuss the stability condition. We note that, for the given equilibrium and for a given mode number m, there is a one-to-one correspondence between Δ' and the slab aspect ratio ϵ. Thus, these two quantities can be used interchangeably as a control parameter and the stability boundary of the zero viscosity case is $\Delta' = 0$ or $\epsilon = 1$. The role of the finite viscosity is to shift the stability boundary to some $\epsilon = \epsilon_c \lesssim 1$ or $\Delta' = \Delta'_c \gtrsim 0$ (Ref. [15,16]).

4.1 Bifurcation diagram

The sequence of equilibria with magnetic islands was studied for $1 > \epsilon > 0.80$ and S up to 1000 using the fixed point code. The spatial resolution in this code was improved with respect to Refs. [15,16] and the investigation has been more refined in checking critical values of the parameter, in order to make comparisons with the asymmetric case of Sec. 5. The results were found to be essentially unaffected by the truncation procedure. The main result is shown in Fig. 1 where the island width w for the different equilibria is plotted against ϵ.

As expected, the initially symmetric equilibrium P_0 with $w = 0$ becomes unstable to tearing-like perturbations when $\epsilon = \epsilon_c = 0.975$ or $\Delta' = 0.19$. This value is consistent with the estimate given in Ref. [16]. The bifurcation is a pitchfork and two new stable equilibria with a small magnetic island, denoted by P_+ and P_-, appear, related by the system symmetry T. For this reason P_+ and P_- have the same behavior and, differently from Ref. [15], they both are represented in the bifurcation diagram (Fig. 1) as a reference to Sec. 5.

When $\epsilon = \epsilon_Q = 0.977$ two pairs of equilibria, Q_+, Q_+^* and Q_-, Q_-^* appear via tangent bifurcation, stable the Q's and unstable the Q^*'s. Furthermore $Q_- = T(Q_+)$ and $Q_-^* = T(Q_+^*)$. At a smaller value $\epsilon = \epsilon_P = 0.896$ the coales-

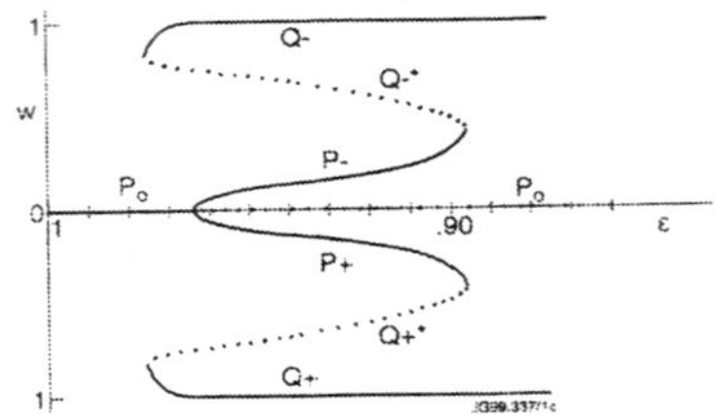

Figure 1. Normalized island width w for the equilibria P_0, P's, Q's versus ϵ for $S = 1000$. Solid lines denote stability, dotted lines instability.

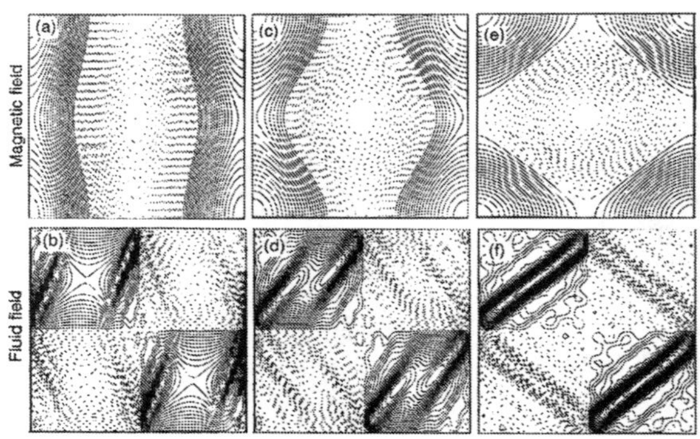

Figure 2. Contour plots of ψ and ϕ at ϵ slightly larger than ϵ_c for: P_+, a) and b), Q_+^*, c) and d), and Q_+, e) and f).

cence of Q_-^* with P_- and of Q_+^* with P_+ takes place, by tangent bifurcation, making the small island stable equilibria disappear. Above ϵ_P the only stable solutions are the Q's, with an island width of the order of the system size. For completeness, in Fig. 2 we show the contour plots of ψ and ϕ for the three equilibria P_+, Q_+^* and Q_+ at a value of ϵ just above ϵ_P. One can see that in the case of P_+ the magnetic island retains approximately its linear shape. It is less so for the velocity field, which however is still organized in four main convective cells. By comparison the island width of the Q-equilibria is comparable to the equilibrium scale length. The corresponding velocity field is more complicated, with four main elongated vortices aligned along the separatrices. We also checked that, in the ϵ range under study, the bifurcation diagram is stable to a further increase of S.

Regarding the behavior of the island width in the small size regime, one can distinguish two phases. The former, which occurs when the control parameter ϵ is just below the threshold of the symmetry-breaking bifurcation, is characterized by a square-root dependence of the amplitude on the departure from threshold, as it is generally the case with this type of bifurcation. For moderately higher values of the amplitude, but still such that the island width is much smaller than the system size, the island width scales linearly.

These results have been confirmed analytically by simultaneous expansion of the deviation from the symmetric equilibrium ψ and ϕ and of the control parameter ϵ in terms of a new smallness parameter[16].

4.2 *Different classes of symmetric equilibria*

The previous results are also confirmed in the case of different symmetric equilibria. To show that, we have considered classes of equilibria, obtained by adding terms in $J_e(x)$ in the form of cosines in such a way to preserve the null lines of the equilibrium magnetic field.
First we considered normalized equilibria with two terms

$$J_e(x;a,b) = a\cos x + b\cos 2x \tag{6}$$

and in particular the one for a=b=0.77, for which the field gradient at $x = 0$ is maximum.

The bifurcation diagram is qualitatively unchanged. It is found that after a first symmetry-breaking bifurcation for $\epsilon_c = 1.027$, giving rise to an equilibrium with a small size magnetic island, a tangent bifurcation occurs for $\epsilon_P = 0.899$. The first bifurcation occurs at a higher value of the aspect ratio ϵ, in agreement with the increase of the gradient in the magnetic field at $x = 0$. On the contrary the value of the tangent bifurcation, at which the small island equilibrium disappears, is essentially unaffected by the change of the starting symmetric equilibrium. For comparison, we recall that for $J_e(x) = \cos x$ the critical values were $\epsilon_c = 0.975$ and $\epsilon_P = 0.896$. Concerning the topology of the fields, we notice that at bifurcation the island width is also essentially unchanged and the fluid field has only some weakening near $x = \pm L_x$.
Analogous results have been obtained with one more term in $J_e(x)$:

$$J_e(x;a,b,c) = a\cos x + b\cos 2x + c\cos 3x \tag{7}$$

The bifurcation diagram has been confirmed, as well as the monotonic dependence of the symmetry-breaking threshold from the magnetic field gradient at $x = 0$.

The phenomenology, in particular the tangent bifurcation, is found largely independent of the specific symmetric equilibrium.

5 The non-symmetric case

Whereas in the experiments great care is taken to produce (axi-)symmetric plasmas, in some cases the symmetry is established in an approximate form only: one talks of the presence of "error fields". The subject is of importance, particularly in connection with large experimental devices.

As a first step to study this problem we consider a perturbation of (4), namely

$$J_e = \cos x + \delta \cos \epsilon y \tag{8}$$

where δ is a small "error" parameter and for $\epsilon = 1$ the associated equilibrium has already a magnetic island, whose normalized width is

$$w = 2\delta^{1/2}/\pi$$

The system of Eqs. 1-2 is now controlled by four dimensionless parameters: the Lundquist number S, the Prandtl number P, the aspect ratio ϵ and the "error" parameter δ. The first two parameters were kept fixed, $S = 1000$ and $P = 0.2$, as in the previous investigation, and the study was performed in the two dimensional parameter space (δ, ϵ).

Bifurcation theory predicts in this case that the symmetry breaking at $\epsilon_c \lesssim 1$ leaves place, generically, to a tangent bifurcation for any δ arbitrarily close but different from zero. Nothing can be said however about the behavior when δ is increased, in particular about the tangent bifurcation for smaller ϵ, responsible in the symmetric case for the loss of island saturation and "explosive" growth.

In the following we present the bifurcation diagram for the equilibria when ϵ is decreased from 1 and for different values of δ, obtained using the fixed point code. The results were found to be essentially unaffected by the truncation procedure.

5.1 *Results when asymmetry is very small:* $\delta = 0.002$

For this value of δ, the magnetic island size at the equilibrium P_- for $\epsilon = 1$, stable, is $w = 0.03$. Decreasing ϵ, the island size slowly increases. When $\epsilon = \epsilon_{Q_-} = 0.9778$ a pair of equilibria, one stable (denoted by Q_-) and one unstable (Q^*_-) appear via tangent bifurcation. At $\epsilon = \epsilon_{Q_+} = 0.9777$ one more pair of equilibria, Q_+ and Q^*_+ , stable the former, unstable the latter, appear

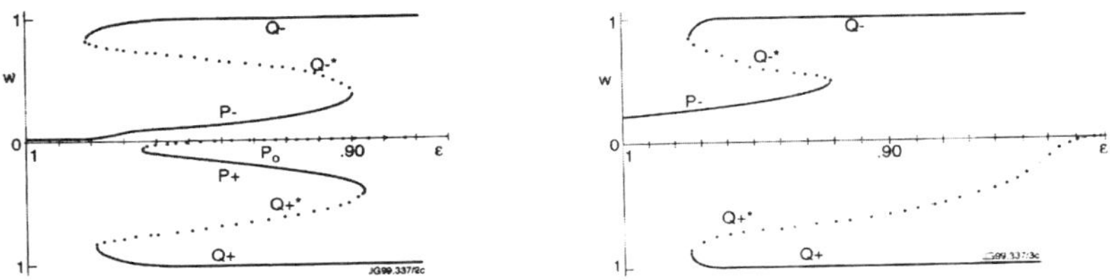

Figure 3. Normalized island width versus ϵ for the equilibria: a) P_0, P's, Q^*'s, Q's for $\delta = 0.002$; b) P_-, Q's, Q^*'s for $\delta = 0.1$. Solid lines denote stability, dotted lines instability.

via tangent bifurcation. At $\epsilon = \epsilon_c = 0.964$ another tangent bifurcation occurs and two more equilibria, P_+ stable and P_0 unstable, appear, with a small island ($w = 0.05$).

Further decreasing ϵ, the island size in the stable equilibria P_- and P_+ increases, while the one for P_0 decreases. At $\epsilon = \epsilon_{P_-} = 0.897$ the small island stable equilibrium P_- and the unstable equilibrium Q^*_- coalesce and disappear in another tangent bifurcation. The same happens at $\epsilon = \epsilon_{P_+} = 0.896$ for P_+ and Q^*_+. Below this value of ϵ only stable solutions of large island sizes exist. Fig. 3a), where the island width w for the different equilibria is plotted against ϵ, shows the bifurcation diagram. When compared with Fig. 1 , Fig. 3a) shows that for this value of δ one has essentially the same phenomenology of the symmetric case.

For $1 \geq \epsilon \geq \epsilon_c$ the stable equilibrium P_-, with a very small magnetic island, plays the role of the symmetric P_0. P_- does not change stability, but the island width increase is strongly enhanced when ϵ is further diminished from ϵ_c, like in P_- for the symmetric case, down to ϵ_P. The analogy for the equilibrium P_+ in the two cases is even more striking and the one for the Q's equilibria is complete.

5.2 Results for increasing asymmetry: $0.02 \leq \delta \leq 0.1$

In order to study the effects of increasing asymmetry, the sequence of equilibria, their stability and their bifurcations have been investigated for a number of values of δ in the range $0.02 \leq \delta \leq 0.1$. The island widths for the equilibria P_- at $\epsilon = 1$ are proportional to $\delta^{1/2}$ in the all range of δ considered and the actual values are given in Table 1.

Up to the value $\delta = 0.06$ the bifurcation diagram shown in Fig. 3a) is

Table 1. As function of δ: island widths w_{P_-} at $\epsilon = 1$, w_{P_+} at $\epsilon = \epsilon_c$; critical values of ϵ.

δ	w_{P_-}	$w_{P_+} = w_{P_0}$	ϵ_c	ϵ_{P_+}	ϵ_{P_-}	ϵ_{Q_+}	ϵ_{Q_-}
0.	0.		0.975	0.896	0.896	0.977	0.977
0.002	0.003	0.05	0.964	0.896	0.897	0.977	0.978
0.02	0.09	0.11	0.925	0.889	0.903	0.976	0.978
0.04	0.13	0.15	0.900	0.882	0.909	0.976	0.979
0.06	0.16	0.17	0.881	0.874	0.914	0.975	0.979
0.08	0.18	0.25	0.863	0.864	0.919	0.975	0.980
0.1	0.2				0.924	0.974	0.980

qualitatively unchanged, with the same equilibria and bifurcations. However the values of ϵ at which the bifurcations take place have quite different dependence on δ, this fact leading to relevant changes in the behavior of the system. The bifurcation points for different values of δ are shown in Table 1.

While ϵ_{Q_+} and ϵ_{Q_-} are left practically unchanged by the increase of δ, ϵ_c strongly decreases. At the same time, ϵ_{P_+} decreases slowly and ϵ_{P_-} increases with increasing δ. The overall result is that the parameter region where saturation to an equilibrium with a small island size takes place becomes smaller and smaller with δ. In particular the role of the equilibrium P_+ becomes less and less relevant, since it exists in the range $\epsilon_c \geq \epsilon \geq \varepsilon_P$. For $\delta = 0.08$ ϵ_c and ϵ_P almost coincide and for $\delta = 0.1$ P_+ has disappeared from the system and Q_+^* , unstable, takes the place of P_0 also at small ϵ.

Fig. 3b) shows the bifurcation diagram for $\delta = 0.1$. When compared with Fig. 3a) one can notice that also the equilibrium P_- exists in a narrower region of the parameter space and with a slightly larger island size.

The change of ϵ_c with δ has also effects on the time dependent behavior of the system. Even if the equilibrium P_0 is unstable, it can play a role through existence of its stable manifold[4] in slowing down the first phase of the island growth. This becomes very crucial when saturation to a small island equilibrium does not take place and the further regime is the one of "explosive" island growth[15]. In any case the island width for the equilibria P_0, P_+ at $\epsilon = \epsilon_c$ is also strongly increasing with δ, as it is shown in Table 1.

6 A finite differences-spectral model

In the previous results the choice of the boundary conditions, taken periodic in both space directions, has the effect to duplicate magnetic islands.

As a general guideline, one expects that whenever the island size is somewhat smaller than the system size, the phenomenology observed in our model should be universal. In this respect we consider the choice of periodic boundary conditions, originally dictated by numerical convenience, to be acceptably close to the more realistic, free-boundary, conditions that one should, in principle, employ in cilindrical or toroidal geometry.

To investigate in this line, we have changed the model in order to relax the periodicity in the x direction and to allow different boundary conditions. This can be obtained by using an expansion in Fourier series only in y and a finite differences scheme in the x-direction. In order to be able to take advantage in our investigation of bifurcation theory, we want however to retain the set up as a dynamical system, without discretizing time derivatives.
The Fourier expansion for the functions ϕ and ψ can be written as

$$\phi(x,y,t) = \sum_k \phi_k(x,t)e^{iky} \tag{9}$$

$$\psi(x,y,t) = \psi_e(x) + \sum_k \psi_k(x,t)e^{iky} \tag{10}$$

where $k = m\epsilon$, m integer and ψ_e is the equilibrium flux function.
We have truncated the expansion to a finite set L of m's, containing zero and $2N$ integers, such that if m belongs to L, also $-m$ belongs to L. We obtained $4N+2$ partial differential equations for the real and imaginary parts of the coefficients. The existence of the invariant subspace $\psi_k^I = \phi_k^R = \phi_0^I = 0$ allows to reduce the system to $2N+1$ equations.

In order to test the code and the efficiency of this approach, the dynamical system obtained has been studied for $\psi_e(x) = \cos x$ taking periodic boundary conditions on the slab $[-\pi,\pi]\times[-\pi/\epsilon,\pi/\epsilon]$. We have fixed $S = 10^2$, $P = 0.2$ and chosen $m = 0,1,2$ and $n = 19$. The results of Ref. [15] have been obtained.

As relevant application, now we can consider initial symmetric equilibria which have zero magnetic field at $x = 0$, but no periodicity in x. An interesting case, $B_e(x) = \tanh(x/\lambda)$, is under study.

7 Final Discussion and Conclusion

In this work, we have analyzed the sequence of bifurcations in two-dimensional resistive MHD for classes of symmetric equilibria. The case of equilibria which have a small magnetic island, ("error fields"), at an aspect ratio $\epsilon = 1$ has also been considered extensively.

The results have confirmed qualitatively the bifurcation diagram obtained in Ref.[15]. This is also the case for small values of the asymmetry parameter

δ. Therefore the disappearance of small size islands because of a tangent bifurcation as a possible mechanism for "hard" excitation is confirmed.

In the latter case, when δ is increased, but still much less than one, equilibria with a small island size exist for smaller and smaller ranges of the aspect ratio since the tangent bifurcation takes place for larger and larger values of ϵ. The possibility of "hard" excitations is then strongly enhanced.

We recall that the generic equation in normal form near the tangent bifurcation is

$$d\xi/dt = \xi^2 + \delta p \tag{11}$$

where ξ is a coordinate on the centre manifold and δp is the deviation of the control parameter from the bifurcation point. Just above the critical value ($\delta p > 0$) no local equilibrium exists and ξ grows as $(\delta p)^{1/2} \tan[(\delta p)^{1/2} t]$. Since the island width and the other observable quantities are a growing function of the generic coordinate ξ, one expects "explosive" growth of the observable quantities as well.

We stress that our approach, being based on a fixed point method, differs from most of the past non-linear analysis, which was traditionally carried out as an initial value problem, typically by deriving an approximate equation for the island dynamics. However, when the comparison is possible we find that our results are consistent, at least qualitatively, with the information provided by such analysis. This is the case in the island saturation regime[10,21,2] and also when the island tends to macroscopic scale[13,7]. Our result also accounts for a transition between Rutherford and Sweet-Parker regimes[18]. In Ref.[16] is given a detailed discussion.

We stress that the occurrence of a tangent bifurcation in dynamical systems is generic and therefore persistent at least to small modifications of the system. On the other hand one should be aware that additional phenomena can in principle appear in fully 3D geometry where new degrees of freedom are allowed. A 3D stability analysis of the 2D saturated island equilibria is required to address this question.

Concerning the intermittent MHD phenomena observed in tokamaks, a natural question is what happens in reality when those hard events take place. On the basis of our analysis, one would be tempted to conclude that a true disruption takes place[19], since we find that after the tangent bifurcation the final state has an island of macroscopic size.

However, one should also consider that RRMHD gives only a partial description of the plasma. In particular the fast transport process occurring during the crash are not described by our model. These processes could introduce a feedback in the system that limits the growth of the magnetic island.

If this were the case, intermittent relaxation events like ELMs would occur. In this respect it is worth reminding that a tangent bifurcation is the basic ingredient of a common intermittency scenario in fluids[9].

Acknowledgments

C.T. acknowledges the kind hospitality of DRFC (CEA),Cadarache (France), where some of the work was carried out under EURATOM mobility program. This paper was partially supported by Murst P.R.I.N. *N.L. Mathematical Problems of Propagation and Stability in Continuous Models*, 2000.

References

1. S.I. Braginskii, in *Rev. of Plasma Phys.* (edited by M.A. Leontovich), Vol.1, Consultants Bureau, New York, 1985.
2. R.Y. Dagazian and R.B. Paris, *Phys. Fluids* **29**, 1986.
3. H.P. Furth *et al*, *Phys. Fluids* **6**, 1963.
4. J. Guckenheimer and P. Holmes, *Nonlinear Oscillations, Dynamical Systems and Bifurcations of Vector Fields*, Springer, New York, 1986.
5. B.B. Kadomtsev and O.P. Pogutse, *Sov. Phys. JETP* **38**, 1974.
6. B.B. Kadomtsev, *Tokamak Plasma: a Complex System*, Institute of Physics Publishing, Bristol, 1992.
7. E.N. Parker, *J. Geophys. Res* **62**, 1957.
8. R.D. Parker *et al*, *Phys. Fluids* **B2**, 1990.
9. Y. Pomeau and P. Manneville, *Comm. Math. Phys.* **74**, 1980.
10. P.H. Rutherford, *Phys. Fluids* **16**, 1973.
11. A. Samain, *Plasma Phys. and Contr. Fusion* **26**, 1984.
12. B. Saramito and E.K. Maschke, in *Magnetic Turbulence and Transport* (P. Hennequin and M.A. Dubois eds.), Editions de Physique, Orsay, 1993.
13. P.A. Sweet, in *Electromagnetic Phenomena in Cosmic Physics* (edited by B. Lehnert), Cambridge University Press, Cambridge, 1958.
14. C. Tebaldi, in *Nonlinear Dynamics*, World Scientific, Singapore, 1989.
15. C. Tebaldi *et al*, *Plasma Phys. and Contr. Fusion* **38**, 1996.
16. C. Tebaldi and M. Ottaviani, *Plasma Physics* **62**, 1999.
17. A. Thyagaraja, *Phys. Rev. Lett.* **24**, 1981.
18. F.L. Waelbroeck, *Phys. Rev. Lett.* **70**, 1993.
19. J.A. Wesson *et al*, in *Proc. of the Tenth Int. Conf. on Plasma Phys. and Controlled Nuclear Fusion Research*, Vol.2, IAEA, 1985.
20. J.A. Wesson *et al*, *Nucl. Fusion* **31**, 1991.
21. R.B. White *et al*, *Phys. Fluids* **20**, 1977.

EXISTENCE OF UNBOUNDED SOLUTIONS IN THERMOELASTICITY

S. MARTIN

Departament Matematica Aplicada IV, Universitat Politecnica de Catalunya, Jordi Girona, 1-3. 08034 Barcelona. Spain.
E-mail: sebasm@mat.upc.es

R. QUINTANILLA

Departament Matematica Aplicada II, Universidad Politecnica de Catalunya, Colom, 11, Terrassa. Barcelona. Spain.
E-mail: ramon@ma2.upc.es

The aim of this paper is to study the solutions of the thermoelasticity equations in the case of unbounded domains and unbounded solutions. We obtain existence, uniqueness and qualitative behaviour results of the solutions in the themoelasticity of type III. Thermoelasticity without energy dissipation is also considered.

1 Introduction

The usual theory of heat conduction based on the Fourier law allows the phenomena of "infinite velocity diffusion" that is not acceptable from a physical point of view. This kind of facts has launched an intense activity in the field of heat propagation. Extensive reviews on the second sound theories are in the work of Chandrasekharaiah [1] and in the book of Muller and Ruggeri [9]. In the more recent surveys of Chandrasekharaiah [2] and Hetnarski and Ignazack [8] the theory proposed by Green and Naghdi [3,4,5,6,7] is considered as an alternative way to model heat propagation. This model is developed in a rational way to produce a fully consistent theory that allows the incorporation of thermal pulse transmission in a very logical manner. They make use of a general entropy balance rather than an entropy inequality. The development is quite general and the characterization of material response for the thermal phenomena is based on three types of constitutive functions, labeled of type I, II and III. When the theory of type I is linearised the parabolic equation of the heat conduction arises. The theory of type II (a limiting case of the type III) does not admit energy dissipation. This theory is usually called "without energy dissipation".

The aim of this paper is to study the solutions of the thermoelasticity equations in the case of unbounded domains and unbounded solutions. We obtain existence, uniqueness and qualitative behaviour results of the solutions in the themoelasticity of type III. We set down suitable conditions under which the problem of the thermoelasticity is well posed. Thermoelasticity without energy dissipation is also considered [10]. Our main tool is the semigroup of linear operators theory.

2 Notation and basic equations

The system of equations that governs the thermoelasticity of type III for a homogeneous and centrosymmeric body are

$$\rho \ddot{u}_i = (a_{ijkh} u_{h,k})_{,j} - (a_{ij}\theta)_{,j} + f_i, \tag{1}$$

$$c\ddot{\alpha} = -a_{ij}\dot{u}_{i,j} + (b_{ij}\theta_{,j})_{,i} + (k_{ij}\alpha_{,j})_{,i} + S. \tag{2}$$

Here u_i is the displacement vector, θ is the temperature field and α is a variable that is usual in this theory and satisfies $\dot{\alpha} = \theta$, f_i and S are the supply terms. If we recall the point of view of Green and Naghdi α is regarded as representing some "mean" thermal displacement magnitude and for brevity is usually referred to as "thermal displacement".

Concerning the constitutive coefficients, we recall that a_{ijkh} is the elasticity tensor, b_{ij} is the thermal conductivity tensor, a_{ij} is the coupling tensor that in the isotropic case is related to the thermal expansion coefficient, ρ is the mass density, c is the coefficient of thermal capacity and k_{ij} is a tensor which is typical of this theory.

Thermoelasticity without energy dissipation can be seen as a limiting case of thermoelasticity of type III and corresponds to the case that $b_{ij} = 0$.

We assume that

(i) $\rho > 0, \qquad c > 0.$

(ii) There are two positive constant a_0, k_0 such that

$$a_{ijrs}\xi_{ij}\xi_{sr} \geq a_0\xi_{ij}\xi_{ij}, \tag{3}$$

$$k_{ij}\xi_i\xi_j \geq k_0, \xi_i\xi_i, \quad b_{ij}\xi_i\xi_j \geq 0, \tag{4}$$

for every tensor ξ_{ij} and every vector ξ_i.

In this paper we obtain an existence theorem of solutions in the case

$$\mathbf{u} = 0, \quad \alpha = 0 \text{ on } \partial B. \tag{5}$$

To define the problem we impose the initial conditions:

$$\mathbf{u}(0,\mathbf{x}) = \mathbf{u}_0(\mathbf{x}),\ \mathbf{v}(0,\mathbf{x}) = \mathbf{v}_0(\mathbf{x}),\ \alpha(0,\mathbf{x}) = \alpha_0(\mathbf{x}),\ \theta(0,\mathbf{x}) = \theta_0(\mathbf{x}),\ \text{in } B. \tag{6}$$

3 Functional statement

We now transform the boundary-initial-value problem determined by the system (1), (2), boundary conditions (5) and initial conditions (6), into an abstract problem on a suitable Hilbert space. We denote

$$\mathcal{Z} = \{(\mathbf{u},\mathbf{v},\alpha,\theta), \mathbf{u} \in \mathbf{W}_0^{1,2}, \alpha \in W_0^{1,2}, \mathbf{v} \in \mathbf{L}^2, \theta \in L^2\},$$

where $W_0^{1,2}$ and L^2 are the usual Sobolev spaces and $\mathbf{W}_0^{1,2} = [W_0^{1,2}]^3, \mathbf{L}^2 = [L^2]^3$.

Let ω be a positive real number and $r^2 = x_k x_k$ the square of the distance to the origin. We define the Hilbert space

$$\mathcal{Z}_\omega = \{(\mathbf{u},\mathbf{v},\alpha,\theta), e^{-\omega r}\mathbf{u} \in \mathbf{W}_0^{1,2}, e^{-\omega r}\alpha \in W_0^{1,2}, e^{-\omega r}\mathbf{v} \in \mathbf{L}^2, e^{-\omega r}\theta \in L^2\},$$

We denote by

$$\begin{array}{ll} A_i\mathbf{u} = \rho^{-1}(a_{ijrs}u_{r,s})_{,j}, \ B_i\theta = -\rho^{-1}(a_{ij}\theta)_{,j}, \ E\mathbf{v} = -c^{-1}a_{ij}v_{i,j}, \ , \\ F\alpha = c^{-1}(k_{ij}\alpha_{,j})_{,i}, \quad G\theta = -c^{-1}(b_{ij}\theta_{,j})_{,i}, \ \mathbf{A} = (A_i), \mathbf{B} = (B_i) \ , \end{array}$$

and

$$\mathcal{A} = \begin{pmatrix} \mathbf{0} & \mathbf{I} & \mathbf{0} & \mathbf{0} \\ \mathbf{A} & \mathbf{0} & \mathbf{0} & \mathbf{B} \\ 0 & 0 & 0 & I \\ 0 & E & F & G \end{pmatrix},$$

where $\mathbf{I}$ and I are the identity operator in the respective spaces.

The problem (1), (2), (5), (6) can be transformed into the following abstract equation in the Hilbert space $\mathcal{Z}_\omega$:

$$\frac{d\varpi}{dt} = \mathcal{A}\varpi(t) + \mathcal{F}(t), \ \ \varpi(0) = \varpi_0, \tag{7}$$

where

$$\mathcal{F}(t) = (\mathbf{0}, \mathbf{f}, 0, S), \ \ \varpi_0 = (\mathbf{u}_0, \mathbf{v}_0, \alpha_0, \theta_0). \tag{8}$$

We denote by $L^2_\omega(B)$ the Hilbert space of the functions u such that $\exp(-\omega r)u \in L^2(B)$. We have the inner product

$$< u, \breve{u} >= \int_B \exp(-2\omega r) u \breve{u} dv.$$

We also define the Hilbert spaces $W^{p,2}_\omega(B)$ of the functions of $L^2_\omega(B)$ such that the p-first derivatives lie in $L^2_\omega(B)$. The inner product is

$$< u, \breve{u} >= \sum_{j \le p} \int_B \exp(-2\omega r) u_{,i_1 \ldots i_j} \breve{u}_{,i_1 \ldots i_j} dv.$$

We also define

$$W^{p,2}_{0,\omega}(B) = \{u \in W^{p,2}_{\omega}(B); \exp(-\omega r)u \in W^{p,2}_{0}(B)\},$$

$$\mathbf{W}^{p,2}_{\omega} = \left[W^{p,2}_{\omega}\right]^3 \text{ and } \mathbf{W}^{p,2}_{0,\omega} = \left[W^{p,2}_{0,\omega}\right]^3$$

Lemma 3.1 . *There exists two positive constants C, d such that*

$$\int_B \exp(-2\omega r)\Big(du_iu_i + a_{ijsr}u_{i,j}u_{r,s}\Big)dv \geq C\int_B \exp(-2\omega r)\Big(u_iu_i + u_{i,j}u_{i,j}\Big)dv$$

Proof. The proof is a direct consequence of the divergence theorem and the use of arithmetic-geometric mean inequality.

Similarly we may prove the existence of two positive constants E, f such that

$$\int_B \exp(-2\omega r)(f\alpha^2 + k_{ij}\alpha_{,i}\alpha_{,j})dv \geq E\int_B (-2\omega r)(\alpha^2 + \alpha_{,i}\alpha_{,i})dv. \quad (9)$$

Now, we introduce in $\mathcal{Z}_\omega$ the inner product:

$$< (\mathbf{u}, \mathbf{v}, \alpha, \theta), (\mathbf{u}^*, \mathbf{v}^*, \alpha^*, \theta^*) > \quad (10)$$

$$= \frac{1}{2}\int_B \exp(-2\omega r)\Big(\rho v_iv_i^* + du_iu_i^* + a_{ijsr}u_{i,j}u^*_{r,s} + c\theta\theta^* + f\alpha\alpha^* + k_{ij}\alpha_i\alpha_j^*\Big)dv.$$

It is also easy to prove that the operators $\mathbf{A}, F$ generate a quasi-contractive semigroup in the spaces $\mathbf{L}^2_\omega(B)$ and $L^2_\omega(B)$ respectively .

4 The existence result

First, we recall that the domain $\mathcal{D}$ of the operator $\mathcal{A}$ is dense in the Hilbert space $\mathcal{Z}_\omega$.

Lemma 4.1 . *There exists a positive constant δ such that for all $\varpi \in \mathcal{D}$, we have:*

$$< \mathcal{A}(\varpi)\varpi, \varpi > \leq \delta||\varpi||^2. \quad (11)$$

Proof. By using the evolution equations and the divergence theorem, we obtain:

$$\begin{aligned}
&< (\mathcal{A}(\mathbf{u}, \mathbf{v}, \alpha, \theta), (\mathbf{u}, \mathbf{v}, \alpha, \theta) > \\
&= \tfrac{1}{2}\textstyle\int_B \exp(-2\omega r)\Big(du_iv_i + f\alpha\theta + a_{ijrs}v_{s,r}u_{i,j} + (a_{ijrs}u_{s,r})_{,j}v_i \\
&-(a_{ij}\theta)_{,j}v_i - a_{i,j}v_{,j}\theta + (b_{ij}\theta_{,j})_{,i}\theta + (k_{ij}\alpha_{,i})_{,j}\theta + k_{ij}\theta_{,j}\alpha_{,i}\Big)dv \\
&= \tfrac{1}{2}\textstyle\int_B \exp(-2\omega r)\Big(du_iv_i + f\alpha\theta + \frac{2\omega x_j}{r}a_{ijrs}u_{s,r}v_i - \frac{2\omega x_j}{r}a_{ij}v_i\theta \\
&-\frac{2\omega x_j}{r}k_{ij}\alpha_{,i}\theta - \frac{2\omega x_j}{r}b_{ij}\theta_{,i}\theta - b_{ij}\theta_{,i}\theta_{,j}\Big)dv.
\end{aligned} \quad (12)$$

After the systematic use of the arithmetic-geometric mean inequality on the right hand side of (12) and in view of the assumptions, we can select a positive constant δ such that (11) is satisfied.

Lemma 4.2 . *There exists a positive constant* λ_0 *such that* $\mathcal{A}$ *satisfies the range condition:*

$$\overline{Range(\lambda_0 \mathcal{ID} - \mathcal{A})} \supset (\mathbf{W}^{2,2}_{\omega})^2 \times (W^{2,2}_{\omega})^2. \tag{13}$$

Proof. Let $\varpi^* = (\mathbf{u}^*, \mathbf{v}^*, \alpha^*, \theta^*) \in (\mathbf{W}^{2,2}_{\omega})^2 \times (W^{2,2}_{\omega})^2$. We must prove that the equation

$$\lambda_0 \varpi - \mathcal{A}\varpi = \varpi^*, \tag{14}$$

has a solution $\varpi = (\mathbf{u}, \mathbf{v}, \alpha, \theta) \in \mathcal{D}(\mathcal{A}))$ for λ_0 sufficiently large. From the definition of $\mathcal{A}$, we obtain the system:

$$\lambda_0 \mathbf{u} - \mathbf{v} = \mathbf{u}^*, \ \lambda_0 \mathbf{v} - \mathbf{A}\mathbf{u} - \mathbf{B}\theta = \mathbf{v}^*, \tag{15}$$

$$\lambda_0 \alpha - \theta = \alpha^*, \ \lambda_0 \theta - E\mathbf{v} - F\alpha - G\theta = \theta^*. \tag{16}$$

By substitution, we get

$$\lambda_0^2 \mathbf{u} - \mathbf{A}\mathbf{u} - \mathbf{B}\theta = \lambda_0 \mathbf{u}^* + \mathbf{v}^*, \tag{17}$$

$$\lambda_0^2 \alpha - F\alpha - \lambda_0 G\alpha - \lambda_0 E\mathbf{u} = \lambda_0 \alpha^* - G\alpha^* - E\mathbf{u}^*. \tag{18}$$

To study system (17),(18) we introduce the following bilinear form

$$\mathcal{B}_{\lambda_0}\Big[(\mathbf{u}, \alpha), (\tilde{\mathbf{u}}, \tilde{\alpha})\Big] = < \begin{pmatrix} \lambda_0^2 - \mathbf{A} & -\mathbf{B} \\ -\lambda_0 E & \lambda_0^2 - F - \lambda_0 G \end{pmatrix} \begin{pmatrix} \mathbf{u} \\ \alpha \end{pmatrix}, (\tilde{\mathbf{u}}, \tilde{\alpha}) > . \tag{19}$$

on $\mathbf{W}^{1,2}_{0,\omega} \times W^{1,2}_{0,\omega}$. After several uses of the divergence theorem an the arithmetic-geometric mean inequality we can prove that $\mathcal{B}_{\lambda_0}$ (for λ_0 sufficiently large) is bounded and coercive. The right had side of (17), (18) lies in $\mathbf{W}^{-1}_{\omega} \times W^{-1}_{\omega}$. Hence Lax- Milgram théorem implies the existence of a solution of the system (15),(16).

The use of lemmas (4.1), (4.2) and the Lummer-Phillips corollary to the Hille-Yosida theorem lead to prove that the operator $\mathcal{A}$ generates a quasi-contractive semigroup in $\mathcal{Z}_{\omega}$.

Thus, we conclude the next theorem:

Theorem 4.1 . *Let us assume that the conditions (3),(4) are satisfied and the supply terms* $\mathbf{F} \in C^1([0,T], \mathbf{L}^2_{\omega}(B)) \cap C^0([0,T], \mathbf{W}^{1,2}_{0,\omega}(B))$, $S \in C^1([0,T], L^2_{\omega}(B))$ $\cap C^0([0,T], W^{1,2}_{0,\omega}(B))$. *Then, for any* $(\mathbf{u}_0, \mathbf{v}_0, \alpha_0, \theta_0)$ *in* $\mathcal{D}$, *there exists a unique solution of the evolution equations; namely, there exists a unique* $(\mathbf{u}(t), \mathbf{v}(t), \alpha(t)$ $, \theta(t)) \in C^1([0,T], \mathcal{Z}_{\omega}) \cap C^0([0,T], \mathcal{D})$.

This theorem states that our problem is well posed. In particular we have the following estimate of the solutions

$$||(\mathbf{u}(t),\mathbf{v}(t),\alpha(t),\theta(t))|| \\ \leq \exp(\delta t)\Big(||(\mathbf{u}_0,\mathbf{v}_0,\alpha_0,\theta_0)|| + \int_0^t \Big(\int_B \exp(-2\omega r)(f_i f_i + S^2)dv\Big)^{1/2}\Big). \qquad (20)$$

Acknowledgments

This work is supported by the project BFM2000-0809.

References

1. D. S.Chandrasekharaiah, Thermoelasticity with second sound: A review. *Appl. Mech. Rev.*, **39**, (1986), 355-376.
2. D. S.Chandrasekharaiah, Hyperbolic thermoelasticity: A review of recent literature. *Appl. Mech. Rev.*,**51**, (1998), 705-729.
3. A.E. Green and P. M. Naghdi, A unified procedure for contruction of theories of deformable media.I. Classical continuum physics,II. Generalized continua, III. Mixtures of interacting continua, *Proc. Royal Society London A*, **448**, (1995a), 335-356, 357-377, 379-388.
4. A.E. Green and P. M. Naghdi, A re-examination of the basic postulates of themomechanics. *Proc. Royal Society London A*, **432**, (1991), 171-194.
5. A.E. Green and P. M. Naghdi, On undamped heat waves in an elastic solid. *J. Thermal Stresses*, **15**, (1992), 253-264.
6. A.E. Green and P. M. Naghdi, Thermoelasticity without energy dissipation. *J. Elasticity*, **31**, (1993), 189-208.
7. A.E. Green and P. M. Naghdi, A new thermoviscous theory for fluids. *Journal of Non-Newtonian Fluid Mech.*, **56**, (1995), 289-306.
8. R. B. Hetnarski and J. Ignazack, Generalized thermoelasticity. *J. Thermal Stresses*, **22**, (1999), 451-470.
9. I. Muller and T. Ruggeri, *Rational and extended thermodynamics.* Springer-Verlag, New-York. 1998
10. R. Quintanilla, Existence in thermoelasticity without energy dissipation. *J. Thermal Stresses*, **25**, (2002), 195-202.

COMPTON COOLING OF A RADIATING FLUID

GIOVANNI MASCALI
Department of Mathematics and Computer Science,
University of Catania,
Viale A. Doria 6, 95125 Italy,
E-mail: mascali@dmi.unict.it

In this paper the Compton cooling of a radiating fluid is studied using the method of moments and the maximum entropy principle. Numerical comparisons are presented which show agreement between the maximum entropy results and Monte Carlo calculations.

1 Radiative fluids

Radiative fluid dynamics plays an important role in the description of many astrophysical phenomena such as supernova explosions, the gravitational collapse, the accretion of matter into black holes or the the flare up of thermal x-ray sources, as well as in plasma physics. In most of these cases a careful relativistic treatment of radiative transfer through matter is essential.
The mathematical model [1 2 3], which describes a radiative fluid, consists of coupled equations for radiation and matter, precisely of the radiative transfer equation, which is an integro-differential equation in seven independent variables, and the conservation laws for the number of material particles and the total energy momentum:

$$L[F] = C[F] \tag{1}$$

$$N^{\mu}_{;\mu} = 0 \qquad T^{\mu\nu}_{tot\ ;\nu} = 0 \tag{2}$$

where $F(x, k)$, x space-time event and k photon 4-momentum, is the photon distribution function and L the Liouville operator which applied to F gives

$$L[F] = k^{\alpha}\frac{\partial F}{\partial x^{\alpha}} - \Gamma^{\alpha}_{\beta\gamma}k^{\beta}k^{\gamma}\frac{\partial F}{\partial k^{\alpha}},$$

with $\Gamma^{\alpha}_{\beta\gamma}$ the Christoffel symbols associated to the metric tensor $g_{\mu\nu}$, and $C[F]$ is the collision term.
N^{μ} is the material particle 4-current density and $T^{\mu\nu}_{tot}$ the total energy-momentum. Semicolon indicates covariant derivative.
To these the equations of state of the material medium and, if required, a dynamical equation for the metric tensor must be added.
The effort to solve this system directly by numerical methods seems prohibitive also for the present computing resources. Therefore it is necessary to

resort to simplified models. The most common among these models is that in which the transfer equation is solved by means of the method of moments.

2 The method of moments

The method of moments consists of expanding the photon distribution function as a series of polynomials whose coefficients, whether integrated or not with respect to frequency, are solution of an infinite set of partial differential equations obtained by considering the moments of the transfer equation.
If we decompose the photon 4-momentum into pieces along and orthogonal to the mean 4-velocity u of the material particles

$$k^{\lambda} = \omega(u^{\lambda} + n^{\lambda})$$

with n the direction of photon motion and ω the photon energy [a], and introduce the projected symmetric trace-free (PSTF) spherical harmonics [b] [1]

$$\begin{aligned} \Phi^{A_k} &= 1 \qquad \text{for } k = 0, \\ \Phi^{A_k} &= n^{<\alpha_1} \ldots n^{\alpha_k>} \quad \text{for } k = 1, 2, \ldots\ldots \end{aligned}$$

the following expansion of the photon distribution function comes after

$$\begin{aligned} F(x,k) &= F(x^{\mu};\omega,n) = \mathcal{F}(x,\omega) + \mathcal{F}_{\mu}(x,\omega)\Phi^{\mu} + \mathcal{F}_{\mu\nu}(x,\omega)\Phi^{\mu\nu} + \cdots\cdots \\ \mathcal{F}^{A_k} &= \frac{(2k+1)!!}{4\pi k!} \int_{\Omega} \Phi^{A_k} F(x;\omega,n) d\Omega \,. \end{aligned}$$

where Ω is the unit sphere of the projected tangent space. From this expansion the definition of the PSTF moments of F of order (k, r) naturally follows

$$\mathcal{M}_r^{A_k}(x) \equiv \int_{P_{ph}} \frac{k^{<\alpha_1} \ldots . k^{\alpha_k>}}{(-k_{\lambda}u^{\lambda})^{k-r-1}} F\,\pi_{ph} = \tag{3}$$

$$= \frac{k!}{\pi^2(2k+1)!!} \int_{P_{ph}} (-k_{\lambda}u^{\lambda})^{r+2}\, \mathcal{F}^{A_k}\, d(-k_{\lambda}u^{\lambda})$$

$$k = 0, 1, \ldots \text{ and } r = 1, 2, \ldots \tag{4}$$

where P_{ph} is the photon momentum space and π_{ph} its volume element.
Only the lower order moments have an immediate physical interpretation

- $\mathcal{M}_1$ ($k = 0$, $r = 1$), energy density

[a] Units such that $c = \hbar = k = 1$ are used, c, $\hbar$, k being the light velocity, the reduced Planck's constant and Boltzmann's constant, respectively. We also work in the framework of classical General Relativity and adopt for the metric tensor the signature +2 [4].
[b] The symbol < > means the projected symmetric trace-free part of a tensor.

- $\mathcal{M}_1^{\mu}$ $(k = 1,\, r = 1)$, energy flux
- $\mathcal{M}_1^{\mu\nu}$ $(k = 2,\, r = 1)$, shear stress tensor.

Tensors with higher k and r respectively take into account the anisotropy and spectral deviations of the photon distribution function from the Planckian. The above-written moments satisfy the following infinite hierarchy of equations obtained starting from the transfer equation [1] [5]

$$\Big\{\mathcal{M}_{r;\gamma}^{A_k\gamma} + \mathcal{M}_{r;\gamma}^{A_k}u^{\gamma} + \frac{k}{2k+1}\mathcal{M}_r^{A_{k-1};\alpha_k} - (k-r)\mathcal{M}_r^{A_k\beta\gamma}\sigma_{\beta\gamma}$$
$$-(k-r)\mathcal{M}_r^{A_k\beta}a_{\beta} + \frac{3+r}{3}\mathcal{M}_r^{A_k}\Theta + \frac{(2r+3)k}{(2k+3)}\mathcal{M}_r^{A_{k-1}\gamma}\sigma_{\gamma}^{\alpha_k}$$
$$-k\mathcal{M}_r^{A_{k-1}\gamma}\omega_{\gamma}^{\alpha_k} + \frac{k(k+r+2)}{(2k+1)}\mathcal{M}_r^{A_{k-1}}a^{\alpha_k} +$$
$$\frac{k(k-1)(k+r+1)}{(2k-1)(2k+1)}\mathcal{M}_r^{A_{k-2}}\sigma^{\alpha_{k-1}\alpha_k}\Big\}^{PSTF} = \mathcal{S}_r^{A_k} \tag{5}$$

where

- $a_{\alpha} = u^{\beta}u_{\alpha;\beta}$ is the 4-acceleration
- $\Theta = u^{\beta}_{;\beta}$ the expansion
- $\sigma_{\alpha\beta} = (u_{\alpha;\beta})^{PSTF}$ the shear
- $\omega = \frac{1}{2}(u_{\alpha;\beta} - u_{\beta;\alpha})^{P}$ the observers' rotation

and

- $\mathcal{S}_r^{\alpha_1\ldots.\alpha_k} \equiv \int_{P_{ph}} \frac{k^{<\alpha_1}\ldots.k^{\alpha_k>}}{(-k_{\lambda}u^{\lambda})^{k-r-1}}\, C[F]\,\pi_{ph}$ are the moments of the collision operator.

In practical physical situations, one has to choose a finite number of moments, $k = 0, \ldots, k_{max}$ and $r = 1, \ldots, r_{max}$, as fundamental variables which describe the physical system, but in so doing the resulting truncated system of equations turns out to be not closed because of the presence of the higher order fluxes and the source moments.
Therefore constitutive relations are needed for these extra-variables.
Using the maximum entropy principle [6] [3], it is possible to obtain an approximation of the photon distribution function, which explicitly depends on the fundamental variables and allows one to get closure relations for the extra-fluxes and the source terms. To compute the latter quantities it is necessary to specify the types of scattering mechanisms occurring between material particles and photons [5].

3 Compton cooling of a plasma

Here we consider the simple physical situation of a radiative field interacting with a static plasma only through Compton scattering [8]. We also suppose that the effects of the gravitational field are neglegible and the system is homogeneous and isotropic. In these conditions the moment equations become much less complicated and in the reference frame at rest with respect to the material medium, read [c]

$$\frac{\partial \mathcal{M}_r}{\partial t} = \mathcal{S}_{r\,C}, \quad r = 1, \ldots, r_{max} \tag{6}$$

with $\mathcal{S}_{r\,C}$ scalar moments of the Compton source term.
In order to describe the time evolution of the system, these equations have to be coupled to the following ones

$$\frac{\partial \mathcal{N}_{ph}}{\partial t} = 0 \qquad \frac{\partial \mathcal{E}_{tot}}{\partial t} = 0 \tag{7}$$

which respectively represent the conservation of the photon number $\mathcal{N}_{ph} \equiv \mathcal{M}_0$ [d] and the total energy $\mathcal{E}_{tot}$, which is the sum of the electron energy $\mathcal{E}_{el} = n_e e$ and the photon energy $\mathcal{M}_1$, $e = m_e + \frac{3}{2} T \left(1 + \frac{5}{4} \frac{T}{m_e}\right)$ being the first relativistic approximation of the energy per electron, with m_e and T electron mass and temperature.
Actually it is more convenient to use the following set of fundamental variables T, $z \equiv e^{\frac{\tilde{\mu}_{ph}}{T_{ph}}}$, T_{ph}, $\mathcal{M}_2, \ldots, \mathcal{M}_{r_{max}}$, where $\tilde{\mu}_{ph}$ and T_{ph} are respectively the photon chemical potential and temperature.
As said, the source terms $\mathcal{S}_{r\,C}$, $\quad r = 1, \ldots, r_{max}$ may be expressed as functions of the above-mentioned variables by means of the maximum entropy principle [5].
So doing the system of equations (6)-(7) (FD model) results to be a completely closed system of ordinary differential equations which can be easily solved by means of standard numerical methods.
We have solved it in corrispondence with the following two sets of initial conditions

- $F(0;\omega) = \left[z(0)^{-1} e^{\frac{\omega}{T_{ph}(0)}} - 1\right]^{-1}$ with $T_{ph}(0) = 1\,keV$ and $z(0) = 1$, $T(0) = 50\,keV$

[c] Only the scalar moments are non-zero because the system is isotropic.
[d] The Compton scattering does not change the number of photons.

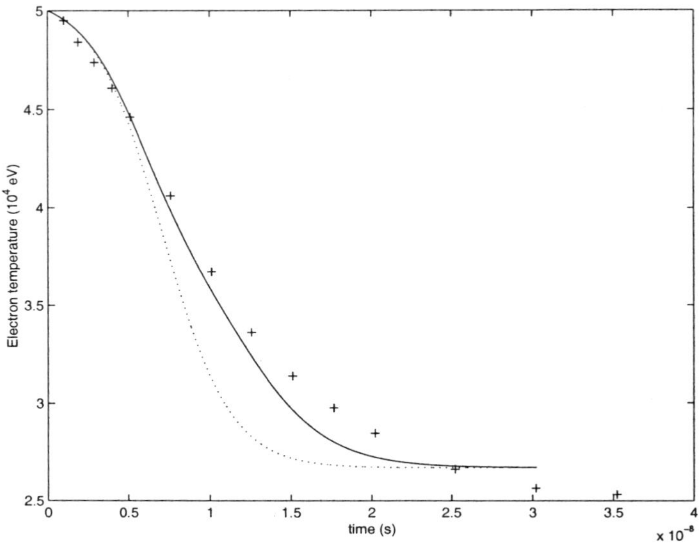

Figure 1. Electron temperature versus time. Crosses: Monte Carlo results. Dotted line: FD model with 2 moments. Continuous line: FD model with 3 moments.

- $F(0;\omega) = \left[z(0)^{-1}e^{\frac{\omega}{T_{ph}(0)}} - 1\right]^{-1}$ with $T_{ph}(0) = 1\,keV$ and $z(0) = 50^3$, $T(0) = 100\,keV$

In both cases we have used $r_{max} = 1,2$. The results relative to the first set of initial condition, see Fig. 1, have been compared with Monte Carlo simulations [8]. Those obtained by using $r_{max} = 2$ are in very good agreement with the Monte Carlo ones. Using a greater number of moments is not convenient because the improvement is not worth the computational cost.
We underline that the equilibrium temperature of our model is different from the Monte Carlo one, however the first is the correct one as may be figured out on the basis of the conservation of the photon number and the total energy.

Acknowledgements

This work was supported by Italian Consiglio Nazionale delle Ricerche (Prg. N.98.01041.CT01, N.99.01714.01 and N.00.00128.ST74), by TMR (Progr. n. ERBFMRXCT970157) and Italian MURST (Prot. n. 9801169828-005).

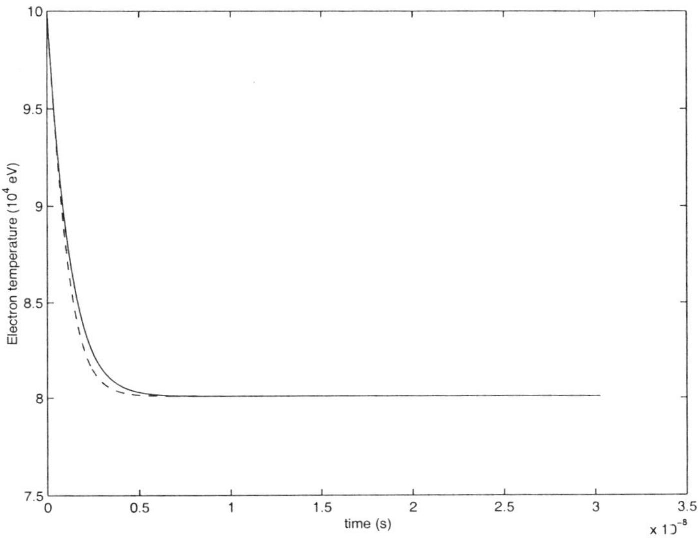

Figure 2. Electron temperature versus time. Dotted line: FD model with 2 moments. Continuous line: FD model with 3 moments.

References

1. K. S. Thorne, Mon.Not.R.Soc. **194**, 439 (1981).
2. M. A. Schweizer, Ann. Phys. **183**, 80 (1988).
3. G. Mascali, V. Romano, Ann. Inst. H. Poincare **67**, 123 (1997).
4. C. W. Misner, K. S. Thorne and J. A. Wheeler in *Gravitation,* ed. Freeman (S. Francisco, 1973).
5. G. Mascali, submitted , (2001).
6. I. Müller and T. Ruggeri in *Rational Extended Thermodynamics,* Springer-Verlag (Berlin, 1998).
7. H. Struchtrup, Ann. Phys. **257**, 111 (1996).
8. G. Cooper, Phys. Rev. D **3**, 2312 (1970).

FAST RELAXATION PHENOMENA IN EXTENDED THERMODYNAMICS OF SUPERFLUIDS

M.S. MONGIOVÌ AND R.A. PERUZZA

Dipartimento di Matematica ed Applicazioni, Università di Palermo, Italy
E-mail: mongiovi@unipa.it, peruzza@math.unipa.it

In previous works a macroscopic monofluid model of liquid helium II, which is based on *Extended Thermodynamics*, was formulated,where the time evolution of the non equilibrium stress tensor was neglected, putting zero the relaxation times τ_0 and τ_2 of the non equilibrium pressure and of the stress deviator. In this work, the complete model with 14 fields is studied in the linear approximation: a dispersion relation is obtained and the solutions of this equation are determined perturbing the solutions obtained in the cases $\tau_0 = \tau_2 = 0$. The corresponding modes are also discussed.

1 Introduction

It is known that liquid helium II shows very odd effects, related to its quantum nature: it possesses very small viscosity and very high thermal conductivity; further temperature waves are propagated in it. In order to describe the behavior of this quantum liquid, *Titza* and *Landau* proposed a phenomenological theory known as the *two-fluid model* [1], which treats helium II as a two component mixture; but this model is not completely satisfactory because of the inseparability of the two components of the mixture. Macroscopic monofluid models of liquid helium II which are based on *Extended Thermodynamics* (E.T.)[2] have been formulated by various Authors, both in the presence and in the absence of dissipative phenomena[3,4,5]. These models are able to describe, in accord with experimental data, the propagation of first sound[4,5] (the normal sound wave), of second sound[3,4,5] (the temperature wave) and of fourth sound[4,5] (a wave propagation in helium II when the latter is forced to flow in very thin capillaries or porous media). In all these studies the time evolution of the stress tensor (trace and deviator) was neglected. In this work this time evolution is not neglected and the complete model with 14 fundamental fields is studied. Introducing new vector variables, the wave propagation is studied and the fast relaxation phenomena are analyzed.

* Supported by MIUR under grant *Nonlinear Mathematical Problems of Wave Propagation and Stability in Models of Continuous Media*

2 Wave propagation in a Superfluid

The fundamental fields of E.T. with 14 fields are the density ρ, the velocity $\mathbf{v} = (v_i)$, the temperature T, the non equilibrium pressure p_V, the stress deviator $m_{<ij>}$ and the heat flux $\mathbf{q} = (q_i)$. The linearized system of field equations for these fields is [2,4]:

$$\begin{cases} \frac{\partial \rho}{\partial t} + \rho \frac{\partial v_j}{\partial x_j} = 0, \\ \rho \frac{\partial v_i}{\partial t} + \frac{\partial}{\partial x_j} \left[\left(p_E + p_V \right) \delta_{ij} + m_{<ij>} \right] = 0, \\ \rho \frac{\partial \epsilon}{\partial t} + \frac{\partial q_j}{\partial x_j} + \left[\left(p_E + p_V \right) \delta_{ij} + m_{<ij>} \right] \frac{\partial v_i}{\partial x_j} = 0, \\ \tau_0 \frac{\partial p_V}{\partial t} + \xi \frac{\partial v_j}{\partial x_j} - \beta' T \xi \frac{\partial q_j}{\partial x_j} = -p_V, \\ \tau_2 \frac{\partial m_{<ik>}}{\partial t} + 2\eta \frac{\partial v_{<i}}{\partial x_{k>}} - 2\beta T \eta \frac{\partial q_{<i}}{\partial x_{k>}} = -m_{<ik>}, \\ \tau_1 \frac{\partial q_i}{\partial t} + \kappa \frac{\partial T}{\partial x_i} - \beta' T^2 \kappa \frac{\partial p_V}{\partial x_i} - \beta T^2 \kappa \frac{\partial m_{<ij>}}{\partial x_j} = -q_i. \end{cases} \tag{2.1}$$

In these equations $\epsilon = \epsilon(p, T)$ is the internal specific energy, $p_E = p_E(\rho, T)$ the equilibrium pressure, τ_0, τ_2 and τ_1 the relaxation times of p_V, $m_{<ij>}$ and q_i respectively; ξ, η and k bulk viscosity, shear viscosity and thermal conductivity respectively; β, β' phenomenological coefficients linked to the thermal dissipation. In a Superfluid the thermal conductivity κ and the relaxation time τ_1 of the stress deviator are extremely high; we suppose that the ratio $\frac{k}{\tau_1} = \zeta$ is finite and non zero[4]. As in He II the thermal dilatation is very small, for sake of simplicity, we will put it equal to zero in this work.

In order to study the propagation of plane harmonic waves of small amplitude we put $U = (p, v_i, T, p_V, m_{<ij>})$ and we look for solutions of (2.1) of the form: $U = U_0 + \overline{U} e^{i(K n_j x_j - \omega t)}$ where $U_0 = (p_0, 0, T_0, 0, 0, 0)$ denotes the unperturbed state, $\overline{U}$ small amplitudes whose products can be neglected, $K = k_r + i k_s$ the wave number, $\omega = \omega_r + i \omega_s$ the frequency, $\mathbf{n} = (0, 0, 1)$ the unit vector orthogonal to the wave front, $t_1 = (1, 0, 0)$ and $t_2 = (0, 1, 0)$ unit vectors tangent to the wave front. We obtain a homogeneous algebraic linear system for the small amplitudes, which we can write in the following matrix form:

$$\begin{pmatrix} A & 0 & 0 & 0 \\ 0 & B & 0 & 0 \\ 0 & 0 & C & 0 \\ 0 & 0 & 0 & D \end{pmatrix} \begin{pmatrix} U_1 \\ U_2 \\ U_3 \\ U_4 \end{pmatrix} = \begin{pmatrix} 0 \\ 0 \\ 0 \\ 0 \end{pmatrix} \tag{2.2}$$

where A is the matrix:

$$\begin{pmatrix} -\omega & \rho K & 0 & 0 & 0 & 0 & 0 \\ \frac{p_\rho}{\rho}K & -\omega & \frac{p_T}{\rho}K & 0 & \frac{K}{\rho} & 0 & 0 \\ 0 & \frac{Tp_T}{\rho c_V}K & -\omega & 0 & 0 & \frac{K}{\rho c_V} & 0 \\ 0 & \lambda_0 K & 0 & -(\omega\tau_0+i) & 0 & -\lambda_0 K\beta' T & 0 \\ 0 & \frac{4}{3}\lambda_2 K & 0 & (\omega\tau_2+i) & -(\omega\tau_2+i) & -\frac{4}{3}\lambda_2 K\beta T & 0 \\ 0 & 0 & \zeta K & (\beta-\beta')T^2\zeta K & -\beta T^2\zeta K & -\left(\omega+\frac{i}{\tau_1}\right) & 0 \\ 0 & -\frac{2}{3}\lambda_2 K & 0 & (\omega\tau_2+i) & 0 & \frac{2}{3}\lambda_2 K\beta T & -(\omega\tau_2+i) \end{pmatrix},$$

and

$$B=-(\omega\tau_2+i), \qquad C=D=\begin{pmatrix} -\omega & 0 & \frac{K}{\rho} \\ 0 & -(\omega+\frac{i}{\tau_1}) & -\beta T^2\zeta K \\ \lambda_2 K & -\lambda_2 K\beta T & -(\omega\tau_2+i) \end{pmatrix},$$

$$U_1=\left(\overline{\rho}\;\overline{v}_3\;\overline{T}\;\overline{p}_V\;\overline{m}_{33}\;\overline{q}_3\;\overline{m}_{11}\right)^T, U_2=\left(\overline{m}_{12}\right),$$
$$U_3=\left(\overline{v}_1\;\overline{q}_1\;\overline{m}_{13}\right)^T, U_4=\left(\overline{v}_2\;\overline{q}_2\;\overline{m}_{23}\right)^T.$$

¿From (2.2) we can see that waves longitudinal and transversal in velocity and heat flux propagate independently.

2.1 *Study of the modes longitudinal in the vector fields*

We consider the first eight equations of (2.2). This subsystem can be written:

$$\begin{cases} -\omega\overline{\rho}+\rho K\left(\frac{R-1}{R}\overline{u}_3^{(s)}+\frac{1}{R}\overline{u}_3^{(n)}\right)=0 \\ -\omega\overline{T}+\frac{K}{\rho c_V}\frac{R-1}{R}\frac{1}{\beta T}u_3^{(s)}-\frac{K}{c_V}\frac{1}{R}\beta T^2\zeta u_3^{(n)}=0 \\ -\left(\omega+\frac{i}{\tau_1}\frac{1}{R}\right)\overline{u}_3^{(s)}-\frac{i}{\tau_1}\frac{1}{R}\overline{u}_3^{(n)}+K\frac{p_\rho}{\rho}\overline{\rho}+K\frac{1}{\rho\beta T^2}\overline{T}+\frac{K}{\rho}\left(1-\frac{\beta'}{\beta}\right)\overline{p}_V=0 \\ -\left(\omega-\frac{i}{\tau_1}\frac{R-1}{R}\right)\overline{u}_3^{(n)}+\frac{i}{\tau_1}\frac{R-1}{R}\overline{u}_3^{(s)}-K\beta T\zeta\overline{T}+\frac{KR}{\rho}\left(\overline{m}_{33}-\overline{p}_V\right)+\frac{KR'}{\rho}\overline{p}_V=0 \\ -\left(\omega\tau_0+i\right)\overline{p}_V+\xi K\left(\frac{R-R'}{R}\overline{u}_3^{(s)}+\frac{R'}{R}\overline{u}_3^{(n)}\right)=0 \\ -\left(\omega\tau_2+i\right)\left(\overline{m}_{33}-\overline{p}_V\right)+\frac{4}{3}\eta K\overline{u}_3^{(n)}=0 \\ -\left(\omega\tau_2+i\right)\left(\overline{m}_{11}-\overline{p}_V\right)-\frac{2}{3}K\eta\overline{u}_3^{(n)}=0 \\ -(\omega\tau_2+i)\overline{m}_{12}=0 \end{cases} \tag{2.1.1}$$

where $R=1+\rho\beta^2T^3\zeta$, $R'=1+\rho\beta\beta' T^3\zeta$, $\mathbf{u}^{(n)}=\mathbf{v}-\beta T\mathbf{q}$ and $\mathbf{u}^{(s)}=\mathbf{v}+\frac{1}{\rho\beta T^2\zeta}\mathbf{q}$.

We observe first that $\overline{m}_{11}=\overline{m}_{22}=3\overline{p}_V-\frac{\overline{m}_{33}}{2}$; then from the two latter equations we obtain $\omega_{1,2}=-\frac{1}{\tau_2}$: we deduce immediately that there are two modes where only the quantities $\overline{m}_{12}$ and $\overline{m}_{11}-\overline{p}_V$ vibrate independently.

We now consider the first six equations of the subsystem (2.1.1). This system admits non trivial solutions if and only if its determinant vanishes; this condition gives rise to the following equation of the sixth degree in ω.

$$\left(-\omega^2 + K^2 V_1^2\right)\left[-\left(\omega + \frac{i}{\tau_1}\right)\omega + K^2 V_2^2\right](\omega\tau_0 + i)(\omega\tau_2 + i) +$$

$$+ \beta^2 T^3 \zeta K^2 \omega\left[-\omega^2 + K^2 V_1^2\right]\left(\xi\frac{\beta'^2}{\beta^2} + \frac{4}{3}\eta\right) + \frac{4}{3}\frac{K^4}{\rho}\beta^2 T^3 \zeta \omega^2 \left(1 - \frac{\beta'}{\beta}\right)^2 \xi\eta$$

$$+ \frac{K^2}{\rho}\left[-\omega^2\left(\omega + \frac{i}{\tau_1}\right) + K^2 V_2^2 \omega\right]\left(\xi + \frac{4}{3}\eta\right) = 0 \qquad (2.1.2)$$

where we have put ${V_1}^2 = p_\rho,\ {V_2}^2 = \frac{\zeta}{\rho c_V}$.

In a superfluid[3,4,5] the relaxation times of viscous pressure and of stress deviator are very small; on the contrary, the evolution time of the heat flux is extremely high.

We will solve dispersion relation (2.1.2) using a perturbative method. We determine first its solutions under the limiting hypothesis $\tau_0 = 0,\ \tau_2 = 0$ and $\frac{1}{\tau_1} = 0$. Under these hypotheses, the solutions $\omega_3^{(0)}$ and $\omega_4^{(0)}$ go to infinity; the other ones $\omega_{5,6}^{(0)}$ and $\omega_{7,8}^{(0)}$ correspond to first sound, where density and velocity vibrate, and to second sound, where temperature and heat flux vibrate.

In this section, we will study in detail the solutions ω_3 and ω_4 when τ_0 and τ_2 are very small, but different from zero. To this purpose we carry out a change of variable in the relation dispersion, putting $\omega = \frac{1}{x}$. We perturb the degenerate solution $x = 0$, putting $x = \xi\epsilon$, $\tau_0 = \alpha_0\epsilon$, $\tau_2 = \alpha_2\epsilon$, $\frac{1}{\tau_1} = \alpha_1\epsilon$, where ϵ is an adimensional coefficient; to first order in ϵ, we obtain the solutions $\omega_3^{(1)}$ and $\omega_4^{(1)}$ and the following corresponding associated modes :

$\omega_3^{(1)} = -\frac{i}{\tau_0} + O(\epsilon^2)$	$\omega_4^{(1)} = -\frac{i}{\tau_2} + O(\epsilon^2)$
$\overline{\rho}^{(1)} = 0,$	$\overline{\rho}^{(1)} = 0,$
$\overline{T}^{(1)} = 0,$	$\overline{T}^{(1)} = 0,$
$\overline{p}_V^{(1)} = p_E\psi,$	$\overline{p}_V^{(1)} = 0,$
$\overline{m}_{33}^{(1)} = p_E\psi,$	$\overline{m}_{33}^{(1)} = p_E\psi,$
$\overline{m}_{11}^{(1)} = p_E\psi$	$\overline{m}_{11}^{(1)} = -\frac{p_E}{2}\psi$
$[\overline{u}_3^{(n)}]^{(1)} = iK\tau_0\frac{p_E}{\rho}R'\psi,$	$[\overline{u}_3^{(n)}]^{(1)} = iK\tau_2\frac{p_E}{\rho}R\psi,$
$[\overline{u}_3^{(s)}]^{(1)} = iK\tau_0\frac{p_E}{\rho}\left(1 - \frac{\beta'}{\beta}\right)\psi$	$[\overline{u}_3^{(s)}]^{(1)} = 0$

The mode with frequency $\omega_3^{(1)}$ corresponds to a very rapid relaxation phenomenon involving essentially p_V, m_{33}, m_{11}, $\mathbf{u}^{(n)}$, $\mathbf{v}$ and $\mathbf{q}$. In particular, if $\beta = \beta'$, in this mode the field $\mathbf{u}^{(\mathbf{s})}$ is not involved. Also the mode with frequency $\omega_4^{(1)}$ corresponds to a rapid relaxation phenomenon, but in this mode the pressure p_V and the vector field $\mathbf{u}^{(s)}$ are not involved, even if β' is different from β.

2.2 Study of the modes transversal in the vector fields

In this section we study the modes transversal in the velocity and in the heat flux; to this purpose we consider, for example, the subsystem $CU_3 = 0$, which in the new variables becomes:

$$\begin{cases} -\omega\left(\frac{R-1}{R}\overline{u}_1^{(s)} + \frac{1}{R}\overline{u}_1^{(n)}\right) + \frac{K}{\rho}\overline{m}_{13} = 0 \\ -\left(\omega + \frac{i}{\tau_1}\right)\frac{1}{\beta T}\frac{R-1}{R}\left(\overline{u}_1^{(s)} - \overline{u}_1^{(n)}\right) - \beta T^2\zeta K\overline{m}_{13} = 0 \\ -\left(\omega\tau_2 + i\right)\overline{m}_{13} + \eta K\overline{u}_1^{(n)} = 0 \end{cases} \tag{2.2.1}$$

This system admits nontrivial solutions if and only if its determinant vanishes; we obtain:

$$\tau_2\omega^3 + i\omega^2\left(1 + \frac{\tau_2}{\tau_1}\right) - \left(\frac{1}{\tau_1} + \frac{\eta}{\rho}K^2R\right)\omega + \frac{i}{\tau_1}\frac{\eta}{\rho}K^2 = 0 \tag{2.2.2}$$

Identical relation is obtained considering the subsystem $DU_4 = 0$. Under the hypotheses that $\tau_2 = 0$ and $\frac{1}{\tau_1} = 0$, we obtain:

$$\omega_9^{(0)} = \infty \quad \omega_{10}^{(0)} = 0, \quad \omega_{11}^{(0)} = -i\frac{\eta}{\rho}K^2R. \tag{2.2.3}$$

Supposing that τ_2 and $\frac{1}{\tau_1}$ are not zero, these solutions correspond to very different modes in the superfluid. Here, we study in detail the solutions ω_9 and ω_{10}. We put $\frac{1}{\tau_1} = \alpha_1\epsilon$, $\tau_2 = \alpha_2\epsilon$, where ϵ is an adimensional coefficient. To the first order in ϵ, we obtain:

$\omega_9^{(1)} = -\frac{i}{\tau_2} + O(\epsilon^2)$	$\omega_{10}^{(1)} = -\frac{i}{R}\frac{1}{\tau_1} + O(\epsilon^2)$
$\overline{m}_{13}^{(1)} = \psi$, $[\overline{u}_1^{(n)}]^{(1)} = i\frac{K}{\rho}\tau_2R\psi$, $[\overline{u}_1^{(s)}]^{(1)} = 0$	$\overline{m}_{13}^{(1)} = 0$, $[\overline{u}_1^{(n)}]^{(1)} = 0$, $[\overline{u}_1^{(s)}]^{(1)} = \frac{R-1}{R}\psi$

As we see, also in the mode with the frequency ω_9 the field $\mathbf{u}^{(\mathbf{s})}$ is not involved, while the mode with frequency ω_{10} corresponds to an extremely slow relaxation phenomenon involving only $\mathbf{u}^{(\mathbf{s})}$; when $\frac{1}{\tau_1}$ goes to zero, any stationary transversal distribution of $\mathbf{u}^{(\mathbf{s})}$ is allowed. Obviously, the system $DU_4 = 0$

yields the solutions $\omega_{12} = \omega_9$, $\omega_{13} = \omega_{10}$ and $\omega_{14} = \omega_{11}$ with similar associated modes.

3 Conclusions

In the work the wave propagation in the extended monofluid theory of a Superfluid is studied. The results of this work suggest us to interpret the fields $\mathbf{u}^{(n)}$ and $\mathbf{u}^{(s)}$ as the velocities of the normal and superfluid components of liquid helium II. Consequently, this theory is in accord with the microscopic theory of liquid helium II and with experimental observations.

References

1. L. Tisza, *Nature*, **141**,913 (1938); L.D. Landau, *J. Phys.*, **5**, 71 (1941)
2. G.M. Kremer in *Extended Thermodynamics Systems*, ed. Sieniutycz S. and Salamon P., (Taylor and Francis, New York, 1992);I. Müller and T. Ruggeri, in *Rational Extended Thermodynamics*, ed. Springer-Verlag, (1996); D. Jou, J. Casas-Vázquez and G. Lebon in *Extended Irreversible Thermodynamics*, ed. Springer-Verlag, (2001).
3. A. Greco and I. Müller, *Arch.Rat.Mech.Anal.*, **85**, 279 (1984)
4. M.S. Mongiovì, *Phys. Rev.* B**48**, 6276 (1993); *Physica A*, **291**, 518 (2001).
5. M.S. Mongiovì and R.A. Peruzza, *Atti AIMETA 1999*, Como, 6-9 Oct.1999; *submitted to Z.A.M.P.*

ON THE REACTION STRESS IN BODIES WITH LINEAR INTERNAL CONSTRAINTS

ADRIANO MONTANARO

Dipartimento di Metodi e Modelli Matematici per le Scienze Applicate
Università di Padova

Via Belzoni, 7   35131 Padova, Italy

E-Mail: montanaro@dmsa.unipd.it

Summary The existence of a material internal constraint in a continuous body generally implies the existence of an indetermination in the reaction stress. One may think that it can be eliminated by the prescriptions of the external contact force and body force: 'The precise value of **N** ('the reaction stress') depends on the external body forces and the boundary tractions.' ([1], p. 133). In [2] the aforementioned indetermination is studied for a linearly elastic, prestressed body B subject to any internal material constraint, with regard to the initial/boundary value-problem; it is pointed out that different external actions can sustain body processes with the same motion and different reaction stresses; hence the notion of *physically equivalent processes* is introduced. A suitable vector *space*, l_τ, *of the reaction stresses that cannot be detected by means of processes which are solution of some ibvp* for B *related with a given portion* τ *of* ∂B, is defined. This space coincides with the class of indetermination of r in the solution (u,r) of any initial/boundary value-problem related with τ, where u is the displacement field and r is the list of the reaction stress multipliers. If the internal constraints are explicitly defined, this indetermination can be computed. Here we show that in some cases it fails to vanish; thus the aforementioned assertion of [1] is not true.

1 Linearized constitutive equations for internally constrained materials with initial stress

We consider a body B composed of a material which is subject to k, $1 \le k \le 6$, internal material constraints.
These constraints are specified by the prescription of k symmetric tensors V_i, $i = 1,2,\ldots,k$. The linear space

$$Model = \{H \in Lin : H \cdot V_i = 0, \quad i = 1,2,\ldots,k\}$$

is called *the space of admissible displacement gradients*. The body B may be kept at rest in the configuration κ under the external action of a suitable body force field $\overset{o}{b}$ in $\mathrm{B} = \kappa(B)$ and a contact force field $\overset{o}{c}$ on $\partial\mathrm{B}$. In the initial state B the body has prestress $\overset{o}{S} = \overset{o}{T} + \overset{o}{N}$, where $\overset{o}{N} = \sum_{i=1}^{k} q_i V_i$ is the initial reaction stress and $\overset{o}{T}$ is the active initial stress. The constitutive equations for the linearized first Piola-Kirchhoff stress have the form

$$S = \overset{o}{S} + \bar{C}[H] + R \ , \tag{1}$$

where $\bar{C}$ is a 4-th order tensor, $\bar{C}[H]$ is the incremental active stress and $R = \sum_{i=1}^{k} r_i V_i$ is the incremental reaction stress. For more details about law (1) see [2], [3]; we point out that any reasoning developed here also holds for internally constrained bodies referred to a natural configuration, considered e.g. in [4], [5].

2 External action for B

An *external (purely mechanical) action* on B in the time interval $Int = [t_0, t_1]$, consists in a triple (b, c_τ, x_π) such that

(i) b is a body force field (per unit mass) defined in $\mathrm{B} \times Int$,
(ii) c_τ is a contact force field defined on $\tau \times Int$, where τ is any given measurable portion of the boundary $\partial\mathrm{B}$ of B,
(iii) x_π is a motion defined on $\pi \times Int$, where $\pi = \partial\mathrm{B} \setminus \tau$.

3 Initial/boundary value-problem for B

Let (b, c_τ, x_π) be a prescribed external action in Int, relative to the complementary parts τ and π of $\partial\mathrm{B}$, and let (u_o, v_o) be prescribed initial conditions for position and velocity in B.

The initial/boundary value-problem - briefly *ibvp* - for B, *related with* (τ, π), (u_o, v_o) and sustained by (b, c_τ, x_π), consists in the problem to find a displacement field $u = u(X, t)$ and a vector field $r = r(X, t)$, with $r = (r_i)_{i=1,..,k}$, such that $(\nabla u, r) \in Model \times \Re^k$ at every $(X,t) \in \mathrm{B} \times Int$, and satisfying the equations

$$H = \nabla u \ , \qquad E = (H + H^T)/2 \ , \qquad \nabla \cdot S + b = \rho \ddot{u} \quad \text{in} \quad \mathrm{B} \times Int \ ,$$

where $\rho =$ initial mass-density,

$$Sn + (n \cdot En - trE)\overset{o}{S} n = c_\tau \quad \text{on} \quad \tau \times Int \ ,$$

$$u = u_\pi \quad \text{on} \quad \pi \times Int$$

and

$$u(., t_o) = u_o(.), \quad \dot{u}(., t_o) = v_o(.) \quad \text{in } \mathrm{B} \ .$$

A *process* of B is a solution (u, r) of an *ibvp* of B.

4 Physically equivalent external actions

Two processes (u, r) and (u, r'), with the same displacement u, are called *kinematically equivalent*. Two *external actions* are said to be *kinematically equivalent* if they can sustain two kinematically equivalent processes. Two kinematically equivalent processes which are sustained by the same external action are said to be *physically equivalent*. In [2] the following proposition is proved.

PROPOSITION Let (u, r) be a process for B in Int and let $r' : \mathrm{B} \times Int \to \Re^k$ be a smooth field. Then, the process (u, r') is physically equivalent to (u, r) if and only if $\nabla \cdot (R' - R) = 0$ in $\mathrm{B} \times Int$ and $(R' - R)n = 0$ on $\tau \times Int$, where $R = \sum_{i=1}^{k} r_i V_i$ and $R = \sum_{i=1}^{k} r'_i V_i$.

5 Indeterminate space in the solution of an ibvp and space of physically detectable reaction stresses

Let $BC^1(S, F)$ be the space of maps, bounded and of class C^1, from a given subset S of an Euclidean space into a Banach space F, endowed with the *sup norm*. Consider the vector space

$$\ell_\tau \equiv \left\{ \lambda \in BC^1(\mathrm{B} \times Int, \Re^k) : \nabla \cdot L = 0 \quad in \quad B \times Int, \quad Ln = 0 \quad on \quad \tau \times Int, \ L = \sum_{i=1}^{k} \lambda_i V_i \right\}$$

The above proposition implies that *two kinematically equivalent processes are physically equivalent if and only if they differ by a (ghost) process* (o, s) *with* $s \in l_\tau$.

That is, for $s \in l_\tau$ the process (o, s) cannot be physically detected. The indetermination of r in the solution (u, r) of any *ibvp* related with τ is just given by l_τ; thus, when u is unique, we say that l_τ is the (physically) *indeterminate space in the solution* (u, r) *of any ibvp related with* τ. Moreover, the quotient vector space

$$V_\tau \equiv BC^1(\mathrm{B} \times Int, \Re^k) / l_\tau ,$$

of $BC^1(\mathrm{B} \times Int, \Re^k)$ modulo l_τ, is called the *space of reaction stresses that are physically detectable* by processes which are solution of some *ibvp* related with τ.

If the k internal constraints are explicitly known, then the indetermination l_τ can be computed and one may see that in some cases it fails to vanish; thus the aforementioned assertion of [1] is not true.

We point out that the indetermination of the reaction stress does not depend on the state of prestress; in particular, it coincides with the indetermination that appears in the case of natural configuration, *i.e.*, when $\overset{o}{T} = 0 = \overset{o}{N},\ \overset{o}{S} = 0$.

6 Examples

Example 1. The constraint of *incompressibility* is specified by $k = 1,\quad V = I$.
Hence, $Model = \{ H \in Lin : trH = 0 \},\quad R = rI$; thus
$\nabla \cdot R = grad\, r = 0 \quad \Leftrightarrow \quad r = constant \quad in \quad \mathrm{B} \times Int$ and
$Rn = rn = 0 \Leftrightarrow r = 0$. As a consequence, we have
$\ell_\tau \equiv \{\lambda \in BC^1(\mathrm{B} \times Int, \Re) : \quad \lambda \quad constant \} \neq \{0\} \Leftrightarrow \tau = \varnothing$ and
$\ell_\tau = \{0\} \Leftrightarrow \tau \neq \varnothing$.

Example 2. The constraint of *inextensibility along, e.g., the direction* e_1 is specified by $k = 1,\quad V = e_1 \otimes e_1$. Hence, $Model = \{ H \in Lin : H^{11} = 0 \}$,
$R = r(e_1 \otimes e_1)$; thus
$\{\nabla \cdot R = \dfrac{\partial r}{\partial x_1} e_1 = 0 \quad \Leftrightarrow \quad r = r(x_2, x_3, t)\}$ and
$Rn = r(e_1 \cdot n)e_1 = 0 \quad on \quad \tau \times Int \quad \Leftrightarrow \quad r(e_1 \cdot n) = 0 \quad on \quad \tau \times Int$.
As a consequence, we have
$\ell_\tau \equiv \{\lambda \in BC^1(\mathrm{B} \times Int, \Re) : \quad \lambda = \lambda(x_2, x_3, t) \quad and \quad \lambda(e_1 \cdot n) = 0 \quad on \quad \tau \times Int \}$.
Hence, if $e_1 \cdot n = 0 \quad on \quad \tau \times Int$, then
$\ell_\tau \equiv \{\lambda \in BC^1(\mathrm{B} \times Int, \Re) : \quad \lambda = \lambda(x_2, x_3, t) \}$;
instead, if $e_1 \cdot n \neq 0 \quad on \quad \tau \times Int$, then
$\ell_\tau \equiv \{\lambda \in BC^1(\mathrm{B} \times Int, \Re) : \quad \lambda = \lambda(x_2, x_3, t) \quad and \quad \lambda = 0 \quad on \quad \tau \times Int \}$.

Example 3. The constraint of *preservation of orthogonality of two directions, e.g.* , e_1 and e_2, is specified by $k = 1,\quad V = e_1 \otimes e_2 + e_2 \otimes e_1$. Hence,
$Model = \{ H \in Lin \ : \ H^{12} + H^{21} = 0 \},\quad R = r(e_1 \otimes e_2 + e_2 \otimes e_1)$; thus
$\{ \nabla \cdot R = \dfrac{\partial r}{\partial x_2} e_1 + \dfrac{\partial r}{\partial x_1} e_2 = 0 \quad \Leftrightarrow \quad r = r(x_3, t)\}$ and
$Rn = r(n_2 e_1 + n_1 e_2) = 0 \quad on \quad \tau \times Int \quad \Leftrightarrow \quad r = 0 \quad on \quad \tau \times Int \quad or \quad n = e_3$.
As a consequence, if $n = e_3 \quad on \quad \tau \times Int$, then
$\ell_\tau \equiv \{\lambda \in BC^1(\mathrm{B} \times Int, \Re) : \quad \lambda = \lambda(x_3, t) \}$.

Instead, if $n = e_1$ *or* $n = e_2$ *on* τ , then
$\ell_\tau \equiv \{\lambda \in BC^1(\mathrm{B}\times Int, \mathfrak{R}):\ \lambda = \lambda(x_3, t)\ \ and\ \ \lambda = 0\ \ on\ \ \tau \times Int\}$.

Example 4. The constraint of *inextensibility along, e.g., the direction* e_1 *and incompressibility* is specified by $k = 2$, $V_1 = e_1 \otimes e_1$, $V_2 = I$. Hence, we have
$Model = \{H \in Lin\ :\ H^{11} = 0 = H^{22} + H^{33}\}$, $\quad R = r_1(e_1 \otimes e_1) + r_2 I$;
thus
$\nabla \cdot R = (\frac{\partial r_1}{\partial x_1} + \frac{\partial r_2}{\partial x_1})e_1 + \frac{\partial r_2}{\partial x_2} e_2 + \frac{\partial r_2}{\partial x_3} e_3 = 0 \Leftrightarrow r_2 = r_2(x_1, t)\ \ and\ \ \frac{\partial (r_1 + r_2)}{\partial x_1} = 0$,
$Rn = (r_1(e_1 \otimes e_1) + r_2 I)n = r_1 n_1 e_1 + r_2 n = (r_1 n_1 + r_2 n_2)e_1 + r_2 n_2 e_2 + r_2 n_3 e_3$.
As a consequence,
$\nabla \cdot R = 0 \ \Leftrightarrow\ r_2 = r_2(x_1, t)\ \ and\ \ r_1 = -r_2(x_1, t) + f(x_2, x_3, t)$ *for any smooth functio*
and $\quad Rn = 0 \ \Leftrightarrow\ r_1 n_1 + r_2 n_2 = 0 = r_2 n_2 = r_3 n_3$.
Hence, if $n = e_1$ *on* $\tau \times Int$, then $r_1 = 0$, $r_2 = r_2(t)$ and
$\ell_\tau \equiv \{\lambda \in BC^1(\mathrm{B}\times Int, \mathfrak{R}^2):\ \lambda_1 = 0\ \ and\ \ \lambda_2 = \lambda_2(t)\ \ on\ \ \tau \times Int\}$;
if $n = e_2$ *or* $n = e_3$ *on* $\tau \times Int$, then $r_2 = 0$, $r_1 = f(x_2, x_3, t)$ and thus
$\ell_\tau \equiv \{\lambda \in BC^1(\mathrm{B}\times Int, \mathfrak{R}^2):\ \lambda_2 = 0\ \ on\ \ \tau \times Int\}$.

Example 5. The internal constraint of *kirchhoff plate-like bodies* is specified by $k = 3$, $V_1 = e_1 \otimes e_3 + e_3 \otimes e_1$, $V_2 = e_2 \otimes e_3 + e_3 \otimes e_2$, $V_3 = e_3 \otimes e_3$. Hence, we have
$Model = \{H \in Lin\ :\ 0 = H^{13} + H^{31} = H^{23} + H^{32} = H^{33}\}$, $\quad R = r_1 V_1 + r_2 V_2 + r_3 V_3$;
thus $\nabla \cdot R = (\frac{\partial r_1}{\partial x_3} + \frac{\partial r_1}{\partial x_1})e_1 + (\frac{\partial r_2}{\partial x_3} + \frac{\partial r_2}{\partial x_2})e_2 + \frac{\partial r_3}{\partial x_3} e_3$ and
$Rn = r_1(n_3 e_1 + n_1 e_3) + r_2(n_3 e_2 + n_2 e_3) + r_3 n_3 e_3$.
As a consequence,
$\nabla \cdot R = 0 \ \Leftrightarrow\ r_3 = r_3(x_1, x_2, t),\ \ \frac{\partial r_1}{\partial x_3} + \frac{\partial r_1}{\partial x_1} = 0\ \ and\ \ \frac{\partial r_2}{\partial x_3} + \frac{\partial r_2}{\partial x_2} = 0$,
$Rn = 0 \ \Leftrightarrow\ r_1 n_3 = 0,\ r_2 n_3 = 0\ \ and\ \ r_1 n_1 + r_2 n_2 + r_3 n_3 = 0$.
Hence, if, *e.g.*, $n = e_3$ *on* $\tau \times Int$, then

$$\ell_\tau \equiv \left\{ \lambda \in BC^1(\mathrm{B}\times Int, \mathfrak{R}^3):\ \lambda_3 = \lambda_3(x_1, x_2, t),\ \frac{\partial \lambda_1}{\partial x_3} + \frac{\partial \lambda_1}{\partial x_1} = 0,\ \frac{\partial \lambda_2}{\partial x_3} + \frac{\partial \lambda_2}{\partial x_2} \right.$$

$$\left. in\ \ \mathrm{B}\times Int\ \ and\ \ 0 = \lambda_1 = \lambda_2 = \lambda_3\ \ on\ \ \tau \times Int \right\}.$$

.1 References

1. Wang C.C., Mathematical Principles of Mechanics and Electromagnetism, Part A: Analytical and Continuum Mechanics, Plenum Press-New York and London, 1976.

2. Montanaro A., Equivalence theorems on the propagation of small-amplitude waves in prestressed linearly elastic materials with internal constraints, *J. Elasticity* **57** (1999) pp. 25-53.

3. Marlow R.S., On the stress in an internally constrained elastic material, *J. Elasticity* **27** (1992) pp. 97-131.

4. Podio-Guidugli P. and Vianello M., The representation problem of constrained linear elasticity, *J. Elasticity* **28** (1992) pp. 271-276.

5. Lembo M. and Podio-Guidugli, P., Plate theory as an exact consequence of three-dimensional linear elasticity, *Eur. J. Mech., A/Solids* **10** (1991) pp. 485-516.

Integration and Segregation in a Population - a Short Account of Socio-Thermodynamics

Ingo Müller
Technical University Berlin

The tools of game theory for the determination of the expected gain from a competition of hawks and doves for a resource are employed to determine the conditions under which the population is integrated or segregated. The birds are supposed to have two contest strategies to choose from and the price of the resource determines which strategy they prefer. The resulting strategy diagram bears a strong resemblance to the phase diagrams of a binary solution or an alloy in different phases and with a miscibility gap in the liquid phase.

1 Introduction

The comportment of large systems of individual elements is governed by the laws of thermodynamics provided that the elements perform a random motion superposed upon their strife for a specific goal.

Thus we may formulate a thermodynamic theory of sociology, – here called socio-thermodynamics – by studying a population of birds, hawks and doves, who search for a resource and compete for it. In doing so both species may employ different strategies, called **A** and **B**, the first one being mildly competitive, while the second one is fiercely competitive.

We allow the price of the resource, – and therefore its value for the birds –, to change, depending on its availability, and we assume that the birds will adopt *the* strategy that maximizes the gain. It turns out that the two species mix freely when the resource is cheap, because it is abundant, and that the species separate when the resource is scarce and therefore dear.

The situation is reminiscent of the comportment of two fluids which mix freely in the vapour phase under low pressure p, while under high pressure, – in the liquid phase –, they separate into drops rich in one constituent or the other. In chemical thermodynamics that behaviour is summarized in a (p, x)-phase diagram, where x is the mol fraction of one constituent. Essential features of such a diagram are:

- full miscibility at low pressure
- partial demixing at intermediate pressure
- miscibility gap at pressures beyond the eutectic pressure.

In the present paper we duplicate these features in a (τ, z)-strategy diagram of a population of hawks and doves; τ is the price of the resource and z is the fraction of hawks.

1.1 Strategies

We consider a mixed population of N birds, hawks and doves, who all compete for a resource whose value is $M\tau$. We call τ the price of the resource. The birds may choose between two strategies, denoted by **A** or **B**.

Strategy A

If two hawks meet over the resource, they fight until one is gravely injured. The winner gains the value, worth $M\tau$ points, while the loser will be punished with $-P^H$ points for being injured and thus losing time for recovery. Two doves do not fight. They merely engage in a symbolic conflict, posturing and threatening but not actually fighting. One of them will eventually win the resource, – always with $M\tau$ points – but they will both be fined $-P^D$ points for wasted time. If a hawk meets a dove, the hawk will always win the resource, gaining $M\tau$ points and the dove will walk away, gaining nothing and avoiding penalty.

Assuming winning and losing equally probable we conclude that the expectation values for the gain per encounter are given by

$$e_A^{HH} = \frac{1}{2}\left(M\tau - P^H\right), \ \mathrm{e}_\mathrm{A}^\mathrm{HD} = \mathrm{M}\tau, \ \ \mathrm{e}_\mathrm{A}^\mathrm{DH} = 0, \ \mathrm{e}_\mathrm{A}^\mathrm{DD} = \frac{1}{2}\mathrm{M}\tau - \mathrm{P}^\mathrm{D} \qquad (2.1)$$

for the four possible encounters HH, HD, DH and DD.

Strategy B

The hawks adjust the gravity of the injury and thus the severity of the punishment to the price τ of the resource. If the price is higher than 1, they fight less and therefore reduce the punishment to a mere $P^H\left(1 - a^H(\tau-1)\right)$ points. Likewise the doves adjust the duration of the posturing – and therefore the penalty – to the price τ. For a given τ they will be fined $P^D\left(1 - a^D(\tau-1)\right)$ points, so that for $\tau > 1$ the penalty is reduced. Doves will still not fight when they find themselves competing with hawks, but they will try to snatch the resource and run. Let them be successful 4 out of 10 times; but, successful or not, they risk injury with a penalty of $W\left(1 + \beta^{DH}(\tau-1)\right)$.

Thus the expected gains for the HH, HD, DH, DD encounters read

$$e_B^{HH} = \tfrac{1}{2}\left(M\tau - P^H\left(1 - a^H(\tau-1)\right)\right), \ \mathrm{e}_\mathrm{B}^\mathrm{HD} = \tfrac{6}{10}\mathrm{M}\tau,$$

$$e_B^{DH} = \tfrac{4}{10}M\tau - W\left(1 + \beta^{DH}(\tau-1)\right), \ \mathrm{e}_\mathrm{B}^\mathrm{DD} = \tfrac{1}{2}\mathrm{M}\tau - \mathrm{P}^\mathrm{D}\left(1 - \mathrm{a}^\mathrm{D}(\tau-1)\right). \qquad (2.2)$$

The expected gain per encounter of a hawk with either another hawk or a dove, and of a dove with either a hawk or another dove are

$$e_i^H = z e_i^{HH} + (1-z) e_i^{HD} \quad \text{and} \quad e_i^D = z e_i^{DH} + (1-z) e_i^{DD}, \tag{2.3}$$

where i stands for strategy **A** or **B** and where $z = N^H/N$ is the fraction of hawks. Finally the expected gain per encounter per bird reads

$$e_i = z e_i^H + (1-z) e_i^D. \tag{2.4}$$

2 Expected gains for the two strategies

For the parameters in (2.1) and (2.2) we choose the values given in Table 1. Therefore we may write e_A and e_B in (2.4) in the more specific forms

$$\begin{aligned} e_A(z;\tau) &= -6z^2 + 2z + \tfrac{5}{2}\tau - 1, \\ e_B(z;\tau) &= \left(\tfrac{24}{5}\tau - \tfrac{24}{5}\right) z^2 - \left(\tfrac{18}{5}\tau + \tfrac{2}{5}\right) z + \left(\tfrac{14}{5}\tau - \tfrac{13}{10}\right). \end{aligned} \tag{3.1}$$

Obviously e_A is a concave parabola as a function of z, while e_B is a convex parabola, at least for $\tau > 1$ which we consider the high-price range.

species	strategy	M	P	α	β^{DH}	W
hawks	A	5	10	$\frac{3}{10}$		
	B	5				
doves	A	5	1	$\frac{3}{10}$	$\frac{1}{2}$	6
	B	5				

Table 1: Parameter values for gains and penalties.

3 Strategy diagram

We shall take it for granted that the birds will employ *the* strategy that provides the largest gain per bird and per encounter.

Fig.1 shows the two parabolae for various values of τ between 0.6 and 3.4. Inspection shows that for $\tau > 1$ strategy **A** provides bigger gains per

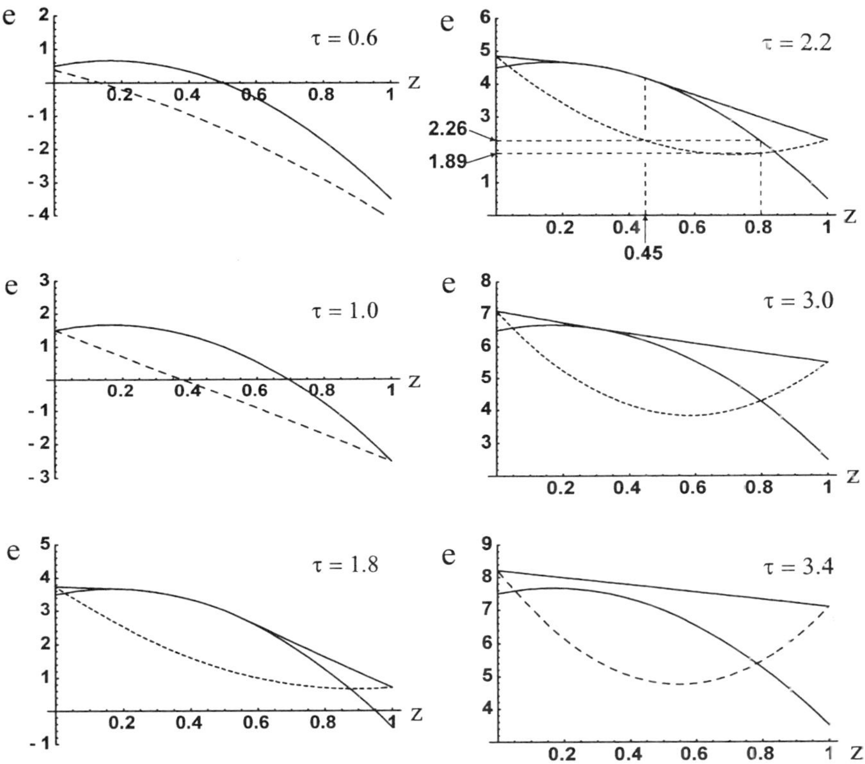

Figure 1: Expected gains as functions of hawk fraction z. Solid lines: Strategy **A**, dashed lines: Strategy **B**.

bird per encounter than strategy **B**. We regard $\tau < 1$ as the low-price range and conclude that the birds will adopt strategy **A**, independent of the hawk fraction, when the price is low.

When the price is higher than $\tau = 1$, the gain curves of Fig.1 intersect twice such that the gains of strategy **B** are higher than those of strategy **A** for small and large hawk fractions. Therefore we might expect that dove-rich and hawk-rich populations employ strategy **B** while populations with a more even distribution of hawks and doves still employ strategy **A**.

However, there is an alternative and the alternative means segregation. Let us consider for instance the case of a population with hawk fraction $z = 0.8$ and a price of τ =2.2 , see one of the graphs of Fig.1. Under those circumstances a homogeneous mixture of hawks and doves would prefer strategy **A** over strategy **B**, since $e_A(0.8; 2.2) = 2.26$ is bigger than $e_B(0.8; 2.2) = 1.89$. A larger gain may be obtained, however, by a segregated population with colonies of pure-hawks employing strategy **B** and colonies with hawks and doves mixed in the proportion $z = 0.45$ which employ strategy **A**. Indeed, if x and $(1 - x)$ are the fractions of birds in either colonies with

$$x \cdot 1 + (1 - x)\, 0.45 = 0.8, \text{ hence x} = 0.64$$

we see that

$$e = x e_{\text{B}}(1; 2.2) + (1 - \text{x})\, e_{\text{A}}(0.45; 2.2) = 3.00$$

is bigger than either $e_A(0.8; 2.2)$ or $e_B(0.8; 2.2)$

Rather obviously this means that the maximum gain per bird per encounter may be realized on the concave envelope of the two curves $e_A(z;\tau)$ and $e_B(z;\tau)$ for each τ. Such concave envelopes are indicated in the graphs of Fig.1. Along their straight parts we always have segregation into either pure hawk colonies and a mixture of hawks and doves or pure dove colonies and a mixture. The pure colonies employ strategy **B** and the mixtures employ strategy **A**.

For $\tau > 3.0$ the mixtures disappear and the segregation of hawks and doves is complete. All birds employ strategy **B** in this high-price-range.

The situation may be summarized by projecting the straight parts of the concave envelopes of Fig. 1 onto the appropriate lines $\tau = const.$ Thus we obtain the (τ, z)-strategy diagram of Fig.2 with its four regions denoted by **I** through **IV**. We have in

- **I:** a homogeneously mixed population of hawks and doves employing strategy **A**.
- **II.** pure hawk colonies employing strategy **B** and separate colonies of mixed populations of hawks and doves employing strategy **A**.
- **III:** Same as **II**, except that the pure colonies are dove colonies.
- **IV:** Complete unmixing into pure hawk and pure dove colonies both employing strategy **B**. We call this region the miscibility gap.

4 Review, outlook and implication

The educated reader may be reminded by all this of the thermodynamic theory of solutions and alloys and, indeed, that theory has originally motivated the author to develop socio-thermodynamics on the basis of strategies discussed in game theory. The simple strategy A seems to go back to an article by

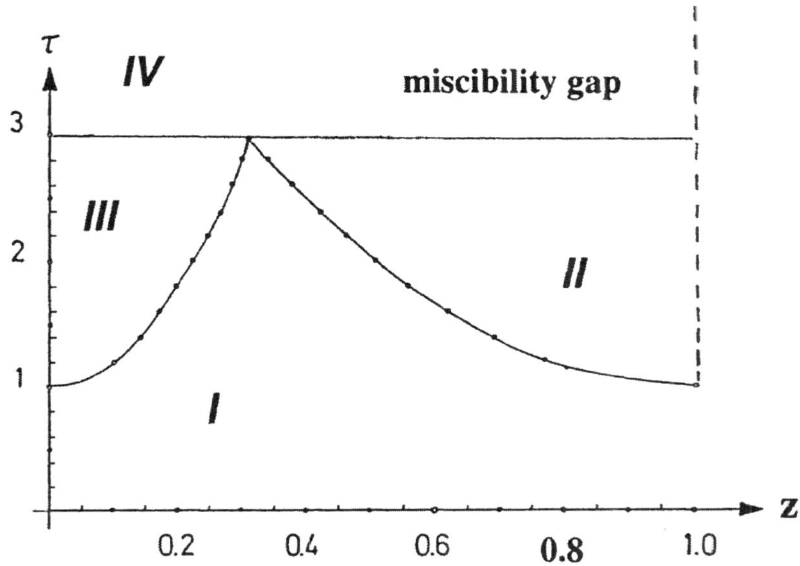

Figure 2: Strategy diagram

Maynard-Smith & Price [1]; it is quoted by Dawkins [2] and also in the book [3] by Straffin.

The theory of socio-thermodynamics, such as it is at this time, is fully described by the author in [4] and a preliminary version, much like the present one was presented in [5].

We must realize that the present argument relies on the axiom that a population strives to maximize its gain. This may be a plausible motive but the axiom is not in the proper spirit of thermodynamics. Thermodynamics bases its growth assumption on the second law which states that heat cannot pass from cold to warm, at least not "by itself". In [4] this law is paraphrased by the statement that a value transfer cannot occur spontaneously from cheap to dear. In this manner socio-thermodynamics can be developed in complete analogy to thermodynamics of mixtures in different phases. Also the trend to a maximum gain appears as a corollary of the axiom on value transfer.

A possible criticism can be levelled at the axiom of maximum gain, whether explicit as in this paper or of a corollary nature as in [4]. Indeed, we assume here that the population will attempt to realize the maximal gain per bird and encounter irrespective of whether it is integrated or segregated. For this to be so the birds must feel a high degree of solidarity – either volontary or enforced – with the whole population and not only within their own colonies. Thus for instance for $\tau = 3.4$ in the pure-dove-colonies the birds would have a higher gain than in the pure-hawk-colonies, if they did not share.

The implication of all this is obvious: If a king in the kingdom of birds want to keep his population integrated, he better keeps the price of the resources low. Otherwise there will be segregation. The gentle reader is invited to speculate on the interpretation of these results outside the kingdom of birds.

References

1. J. Maynard-Smith, G.R. Price. The logic of animal conflict. Nature **246** (1973).
2. R. Dawkins. The Selfish Gene. Oxford University Press (1989).
3. P.D. Straffin. Game Theory and Strategy. New Mathematial Library. The Math. Assoc. of America **36** (1993).
4. I. Müller, Socio-thermodynamics - Integration and Segregation in a Population. Cont. Mech. & Thermodyn. (in press) 2002.
5. I. Müller, Integration and segregation in a population – a thermodynamicist's view. In: Continuum Mechanics and Applications in Geophysics and the Environment. B. Straughan, R. Greve, H. Ehrentraut, Y. Wang (eds.), Springer, Heidelberg, Berlin (2001).

SECOND SOUND PROPAGATION IN SUPERFLUID HELIUM VIA EXTENDED THERMODYNAMICS

A. MURACCHINI, T. RUGGERI & L. SECCIA

Department of Mathematics and Research Center of Applied Mathematics. University of Bologna, C.I.R.A.M. Via Saragozza 8, 40123 - Bologna, Italy.
E-mail: muracchi@ciram.ing.unibo.it

In this paper we study the second sound propagation in superfluid helium using a model performed in the case of a rigid conductor. The theoretical basis of our arguments is given by a recent paper [1] in which is proved that the differential system of a binary Euler's fluid can be written as a system for a single heat conducting fluid. Then the phenomenon of second sound propagation in crystals and in Helium II and, more in general the phenomenology arising in these materials at low temperatures, can be described within a unique framework.

1 Introduction

The unusual properties of liquid He II (i.e. second sound, high conductivity, very small viscosity and shape changes in wave propagation) can be explained by considering a mixture of two non interacting fluids [2,3,4,5]. In [6,7] it was proved that such model is a particular case of a binary mixture of Euler's fluids and, in this framework, the superfluidity effects of He II and the second sound propagation have been explained by a macroscopic point of view. Nevertheless, due to an intuition of Peshkov [8], the second sound phenomenon is not typical of He II appearing also in many others materials at very low temperature. In fact, it is observed in pure crystals of ^{3}He [9], ^{4}He [10], NaF [11] and Bi [12].
In this context some theories, very different from the two-fluids model, have been developed in order to describe the main features of the phenomenon. In particular, in [13], a model based on the kinetic theory of a phonon gas was established.
To explain the thermal disturbances propagating with finite velocity many other macroscopic theories have been proposed [14] starting, as a rule, by a modification of the Cattaneo's equation [15]. As a consequence, recently, using the universal principles of Extended Thermodynamics [16], Ruggeri and coworkers [17] studied the second sound propagation in crystals introducing a thermal inertia factor and so were able to prove the existence of a characteristic temperature $\widetilde{\theta}$ typical of the material. This temperature plays an important role on the shape changes of the second sound wave: in particular, in the papers [17,18], the existence of cold and hot shocks is pointed out. The model is

based on a differential non linear system with an energy balance conservation equation and an evolution equation for the heat flux

$$\frac{\partial \rho e}{\partial t} + \operatorname{div} \mathbf{q} = 0 \quad ; \quad \frac{\partial (\alpha \mathbf{q})}{\partial t} + \operatorname{grad} \nu = -b\,\mathbf{q} \tag{1}$$

Here ρ is the constant mass density, $e = e(\theta)$ the internal energy density, $\mathbf{q}$ the heat flux vector; α (thermal inertia), ν and b are constitutive scalars depending on the temperature θ.
The equation $(1)_2$ reduces to Fourier law when $\alpha = 0$, provided that $b = \nu'/\kappa$ (κ is the heat conductivity), while the Cattaneo equation is obtained if α is constant.
Although this model gives satisfactory results about the phenomenology of second sound propagation when the thermal inertia α is a function of the temperature, nevertheless it is not easy to understand the physical meaning of α. Recently, this problem has been cleared by T. Ruggeri [1] in the context of binary mixtures of simple fluids and so it is possible to construct a unified macroscopic theory of second sound valid for both He II and crystals.
In this paper, assuming this theoretical background, the second sound propagation in superfluid helium is studied by using the model performed in the case of a rigid conductor [17,18].

2 The model

The differential system for a binary mixture of Euler's fluids (i.e. fluids that are neither viscous nor heat conducting and without chemical reactions) is [16]

$$\begin{aligned}
&\frac{\partial \rho}{\partial t} + \operatorname{div}(\rho \mathbf{v}) = 0\\
&\frac{\partial \rho_1}{\partial t} + \operatorname{div}(\rho_1 \mathbf{v}_1) = \tau_1\\
&\frac{\partial \rho \mathbf{v}}{\partial t} + \operatorname{div}(\rho \mathbf{v} \otimes \mathbf{v} - \mathbf{t}) = \mathbf{0}\\
&\frac{\partial \rho_1 \mathbf{v_1}}{\partial t} + \operatorname{div}(\rho_1 \mathbf{v}_1 \otimes \mathbf{v}_1 - \mathbf{t}_1) = \mathbf{m}_1\\
&\frac{\partial \left(\frac{1}{2}\rho v^2 + \rho e\right)}{\partial t} + \operatorname{div}\left\{\left(\frac{1}{2}\rho \mathrm{v}^2 + \rho e\right)\mathbf{v} - \mathbf{t}\mathbf{v} + \mathbf{q}\right\} = 0.
\end{aligned} \tag{2}$$

Here τ_1 is the chemical production density and

$$\rho = \sum_{\beta=1}^{2} \rho_\beta \quad , \quad \mathbf{v} = \sum_{\beta=1}^{2} \frac{\rho_\beta}{\rho} \mathbf{v}_\beta$$

The heat flux $\mathbf{q}$ and the stress tensor $\mathbf{t}$ are given by

$$\mathbf{q} = \sum_{\beta=1}^{2} \left\{ \rho_\beta \left(e_\beta + \frac{1}{2}\mathbf{u}_\beta^2 \right) + p_\beta \right\} \mathbf{u}_\beta \quad ; \quad \mathbf{t} = -p\mathbf{I} - \sum_{\beta=1}^{2} \rho_\beta \mathbf{u}_\beta \otimes \mathbf{u}_\beta$$

where $p = \sum_{\beta=1}^{2} p_\beta$ and the diffusion velocity $\mathbf{u}_\beta = \mathbf{v}_\beta - \mathbf{v}$ has been introduced. The eqs. (2) are nine balance laws in correspondence with the nine unknown fields (ρ, ρ_1, $\mathbf{v}$, $\mathbf{v}_1$, θ). Ruggeri [1] rewrites $\mathbf{v}_1$ and $\mathbf{v}_2$ as a linear combination of $\mathbf{v}$ and $\mathbf{q}$

$$\mathbf{v}_1 = \mathbf{v} + \frac{\alpha}{\rho_1}\mathbf{q} \quad , \quad \mathbf{v}_2 = \mathbf{v} - \frac{\alpha}{\rho_2}\mathbf{q}$$

with α given by

$$\alpha = \frac{1}{\left(e_1 + \frac{p_1}{\rho_1} + \frac{1}{2}\mathbf{u}_1^2\right) - \left(e_2 + \frac{p_2}{\rho_2} + \frac{1}{2}\mathbf{u}_2^2\right)}$$

and the system (2) yields

$$\begin{aligned}
&\frac{\partial \rho}{\partial t} + \operatorname{div}(\rho \mathbf{v}) = 0 \\
&\frac{\partial (\rho c)}{\partial t} + \operatorname{div}(\rho c \mathbf{v} + \alpha \mathbf{q}) = \tau \\
&\frac{\partial \rho \mathbf{v}}{\partial t} + \operatorname{div}\left(\rho \mathbf{v} \otimes \mathbf{v} + \mathrm{p}\mathbf{I} + \frac{\alpha^2}{\rho \mathrm{c}(1-\mathrm{c})}\mathbf{q} \otimes \mathbf{q}\right) = 0 \\
&\frac{\partial (\rho c \mathbf{v} + \alpha \mathbf{q})}{\partial t} + \operatorname{div}\left\{\rho c \mathbf{v} \otimes \mathbf{v} + \frac{\alpha^2}{\rho c}\mathbf{q} \otimes \mathbf{q} + \alpha(\mathbf{v} \otimes \mathbf{q} + \mathbf{q} \otimes \mathbf{v}) + \nu \mathbf{I}\right\} = -\mathrm{b}\mathbf{q} \\
&\frac{\partial \left(\frac{1}{2}\rho v^2 + \rho e\right)}{\partial t} + \operatorname{div}\left\{\left(\tfrac{1}{2}\rho \mathrm{v}^2 + \rho \mathrm{e} + \mathrm{p}\right)\mathbf{v} + \left(\frac{\alpha^2 \mathbf{v}\cdot\mathbf{q}}{\rho \mathrm{c}(1-\mathrm{c})} + 1\right)\mathbf{q}\right\} = 0
\end{aligned} \tag{3}$$

Therefore, the binary mixture can be seen as a single fluid with a heat conduction in the spirit of an Extended Thermodynamics with nine fields (in the eqs. (3) we put $\nu = p_1, \tau = \tau_1$, and, as $\mathbf{m}_1$ is an objective vector, $\mathbf{m}_1 = -b\mathbf{q}$). Moreover, $c = \rho_1/\rho$ is a concentration variable. At low temperatures the concentration c is a function of the temperature and $(3)_2$ is out of the system and is used in order to calculate the mass production τ.

To isolate the thermal wave (second sound) we consider the body as rigid putting $\rho =$const, $\mathbf{v} = 0$. So it is possible to prove that only the last two equations of the system (3), i.e. the energy equation and the one describing the evolution of the heat flux, must be considered

$$\frac{\partial(\rho e)}{\partial t} + \operatorname{div}\mathbf{q} = 0 \quad ; \quad \frac{\partial(\alpha\mathbf{q})}{\partial \mathrm{t}} + \operatorname{div}\left(\nu\mathbf{I} + \frac{\alpha^2}{\rho c}\mathbf{q}\otimes\mathbf{q}\right) = -\mathrm{b}\,\mathbf{q}$$

These equations, neglecting the quadratic terms in $\mathbf{q}$, coincide with the ones given by (1). Now the advantage is that the physical meaning of the thermal inertia α and the function ν are well explained (ν is a "pressure" and α is the inverse of a difference of enthalpies and kinetic energies) and moreover the model (1), used in the case of a rigid conductor, is helpful also for the Helium II provided we concentrate our attention only to the propagation of the second sound wave.

3 Outlines of shock waves theory for hyperbolic systems

Let's consider a generic first order quasi-linear system of N balance laws

$$\partial_t \mathbf{u} + \partial_i \mathbf{F}^i(\mathbf{u}) = \mathbf{f}(\mathbf{u}), \qquad (i = 1, 2, 3) \tag{4}$$

($\mathbf{F}^i$, $\mathbf{f}$, $\mathbf{u} = \mathbf{u}(t, x^i) \in R^N$) with a supplementary conservation law (entropy principle)

$$\partial_t h^o + \partial_i h^i = g \leq 0 \tag{5}$$

where -h^o, -h^i and -g are, respectively, the entropy density, the entropy flux and the entropy production. The Rankine-Hugoniot equations, i.e. the compatibility conditions ensuring the existence of the shocks for a system of type (4), are

$$-s\,[\![\mathbf{u}]\!] + [\![\mathbf{F}^i]\!]\, n_i = 0 \quad \Longrightarrow \quad -s\mathbf{u}_1 + \mathbf{F}^i(\mathbf{u}_1)n_i = -s\mathbf{u}_o + \mathbf{F}^i(\mathbf{u}_o)n_i \tag{6}$$

where $\mathbf{n} \equiv (n_i)$ is the unit normal to the shock surface, s denotes the normal velocity of the shock and $[\![\beta]\!] = \beta_1 - \beta_o$ $(\forall\beta)$ is the jump between the limit values of a generic quantity β on the left (β_1) and on the right (β_o) of the shock wave front. If we consider the supplementary law (5) we obtain

$$\eta = -s\,[\![h^o]\!] + [\![h^i]\!]\, n_i$$

which is, generally, non vanishing [19,20]. The condition $\eta > 0$ is often referred in literature as "*entropy growth condition*" and is assumed as a criterion to pick up physical shocks among the solutions of the Rankine-Hugoniot equations. Another criterion, instead the entropy growth, for selecting physical shocks among all solutions of eqs.(6) is furnished, as well known, by the *Lax conditions* which, in general way, can be written as follows

$$\lambda_o < s < \lambda_o^{(k+1)}, \quad \lambda_1^{(k-1)} < s < \lambda_1^{(k)}; \qquad \forall k = 1, ..., N \tag{7}$$

or

$$\lambda_o < s < \lambda_1 \tag{8}$$

where λ_o, λ_1 are the characteristic velocities evaluated, respectively, in the states ahead and behind the shock front. The so called *Equivalence Theorem* holds: For each k-shock [a] and waves satisfying the so called genuine non-linearity [19]

$$\nabla\lambda \cdot \mathbf{d} \neq 0\,, \tag{9}$$

there exists a neighbourood of the null shock such that

$$\eta > 0 \quad \Longleftrightarrow \quad \lambda_o < s < \lambda_1$$

In (9), $\nabla \equiv \partial/\partial\mathbf{u}$ while λ, $\mathbf{d}$ are, respectively, a characteristic velocity and the corresponding right eigenvector of the characteristic eigenvalue problem.

4 Shock waves in superfluid helium

In [17] we have seen that the characteristic temperatures in a crystal are the ones for which the genuine non linearity is lost in equilibrium state. This means that they are solutions of

$$(\nabla\lambda \cdot \mathbf{d})_{\mathbf{q}=0} = 0 \tag{10}$$

(here λ, $\mathbf{d}$ are, respectively, the positive characteristic velocity and the corresponding right eigenvector of the system (1)).
In [17] it was also proved that the solutions of (10) are the extrema of the shape function

$$\Phi(\theta) = U_E \theta^{1/3} c_v^{1/6} \tag{11}$$

where $U_E\,(\theta)$ is the second sound velocity corresponding to the characteristic eigenvalue λ evaluated in an equilibrium state and c_v is the specific heat.
In our case the shock stability is given by the Liu conditions [21] instead of the Lax and entropy conditions. However, it is possible to show that here the Liu condition coincides with the Lax conditions and so, in the following, we shall refer always to these last ones for simplicity. Then, starting from this theoretical point of view, it is possible to study the second sound propagation in superfluid He II. For this aim let us consider eqs. (1) using the constitutive relations [17] and the experimental data of $U_E\,(\theta)$ and c_v available for superfluid

[a] A k-shock is a shock for which $\lim_{s \to \lambda_o^{(k)}} \mathbf{u}_1 = \mathbf{u}_o$ holds.

helium [22,23]. In this way, at equilibrium, from (11) the following three values for the critical temperature are found

$$\widetilde{\theta}_1 \simeq 0.55K, \quad \widetilde{\theta}_2 \simeq 1K, \quad \widetilde{\theta}_3 \simeq 1.8K, \tag{12}$$

corresponding to the extrema of the shape function (11). It is very important to underline that the temperatures (12) coincide with those experimentally identified in He II [24]. Some important features of the function (11) are deducible from the figure 1. In particular, we see that:

- on the contrary of the case examined in [17], the function Φ shows two maxima and a minimum too;
- the existence of a critical temperature $\widetilde{\theta}_2$, corresponding to the minimum point of $\Phi(\theta)$, leads to shape changes of second sound wave very different from those obtained in [17].

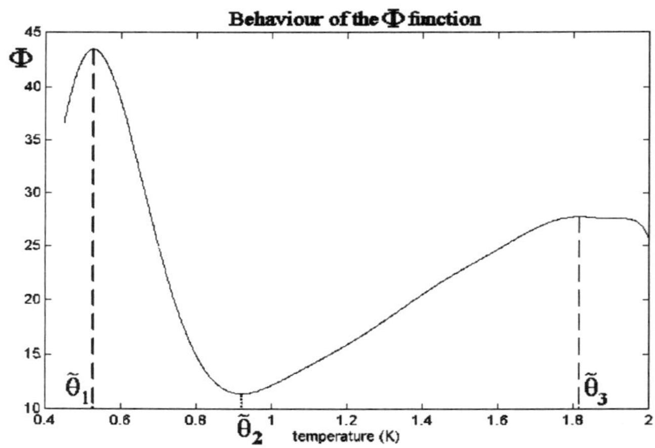

Figure 1. The plot of the function $\Phi(\theta)$.

To confirm the physical difference concerning the processes that happen in this temperature range, note that the second sound velocity is not longer approximable with a function of the type [17]

$$U_E^{-2} = A + B\theta^n \tag{13}$$

which, in the present case, is valid only in a neighbourood of $\widetilde{\theta}_1$ or $\widetilde{\theta}_3$.

4.1 The numerical evaluation

In order to deduce the curves of $s(\theta)$, $\lambda(\theta)$ and $\eta(\theta)$ we take into account the experimental data given in [22,23], namely the second sound velocity and the specific heat as functions of the temperature θ. Furthermore we need to obtain some interpolating functions for $U_E(\theta)$ and $c_v(\theta)$. Because there exist the three characteristic temperatures (12) changing the shock structures, the interesting temperature range can be restricted between $0.45K$ and $2K$, so as to include these ones and exclude the particular situations near $0K$ and the "lambda point" corresponding to $2.2K$. As regard to the specific heat, a polynomial function of $5th$ degree furnishes an enough good approximating function. On the contrary, the problem of finding a curve fitting the second sound values is more difficult. In fact, if we want to state an unique function, for example of polynomial type, we are forced to accept a $12th$ order polynomial, presenting some little numerical oscillations close to the extreme points of the temperature range. In any way, this approach is useful since it permits us to study globally the trend of the shock phenomena, deducing their main features. On the other side, if one wishes to improve the numerical evaluation, it is necessary to divide the second sound data at least in three different temperature intervals: in this way it becomes possible to choose three different approximations, fitting very well the experimental values. In particular, the decreasing behaviour of U_E between both $0.45K$ and $0.75K$ and $1.6K$ and $2K$ permits us to use as the second sound speed the empirical equation (13) already used for the case of high-purity crystals [25]; but, now, we have different numerical values of A, B, n namely

$$A = 5.27 \cdot 10^{-5} \qquad B = 2.54 \cdot 10^{-2} \qquad n = 13.20\ ; \qquad 0.45K \leq \theta \leq 0.75K$$

$$A = 2.38 \cdot 10^{-3} \qquad B = 5.10 \cdot 10^{-9} \qquad n = 18.05\ ; \qquad 1.60K \leq \theta \leq 2.00K$$

This fact highlights the physical resemblance among the conditions for the second sound propagation in crystals and in the superfluid helium inside the previous temperature ranges. In the interval between $0.8K$ and $1.5K$, the wave velocity has a minimum and this implies the request for a different curve fitting and suggests the existence of a different physical situation for the propagating wave. Now a $5th$ order polynomial represents a best fit to the data.

5 Shape changes of second sound wave: two possible interpretations

Usually a temperature pulse in the superfluid helium is generated putting a heater first on, so to increase the temperature of the helium bath supposed at equilibrium, and then rapidly off, so to decrease the temperature of the non-equilibrium region until the equilibrium one. In a similar -also if not usual- way, we could perform first a temperature decrease with respect to the helium bath at equilibrium, and then come back heating. Both these procedures lead to an initial rectangular wave, that in a case we can define "positive" and in the other one "negative". This rectangular profile (positive or negative) is formed joining two successive shock fronts. At this point, in order to analyse their stability, one must choose between two different interpretations.
Following the first one, that we have applied in [17], if the duration of the pulse is short (tipically some micro-seconds), it is possible to consider all the upper (lower) region of the rectangular pulse in a non-equilibrium state. So, the first shock corresponds to an hot (cold) shock, while the second one -in the rear part of the pulse- is not a cold (hot) shock, because at present the right side of the shock (unperturbed region) is the non equilibrium state ($\theta_1, q_1 \neq 0$), while the left side (perturbed region) is the equilibrium state ($\theta_o, q_o = 0$). Since in the previous shock analysis we have chosen always the right side coincident with the equilibrium state, then to study this second shock it is necessary to change in (6) the state $\mathbf{u}_o$ with $\mathbf{u}_1$ and vice versa. However it is a simple matter, in general, to show [17] that the Lax conditions (7) become now, instead of (8), the complementary ones

$$\lambda_1 < s < \lambda_o \quad ; \quad \lim_{s \to \lambda_o} \theta_1 = \theta_o,$$

and moreover, for the entropy growth condition, η must be negative.
Let us resume all the results by means of the table which follows (here we call ${\theta_1}^L$ the temperature limit, depending on θ_o, for the Lax region).
Instead, following the second interpretation, that in our opinion appears more in accordance with the experimental data available for the superfluid helium, if the duration of the pulse is of some ten of micro-seconds, it is possible to consider the upper (lower) region of the rectangular pulse divided in two sub-regions: one of these, near the front part of the pulse, in a non-equilibrium state, while the other one, near the back part of the pulse, in an equilibrium state. So, in this case, the first shock corresponds again to an hot (cold) shock, while the second one -in the rear part of the wave- is actually a true cold (hot) shock, because at present the right side of the shock (unperturbed region) represents also an equilibrium state ($\theta_1, q_1 = 0$), while the left side

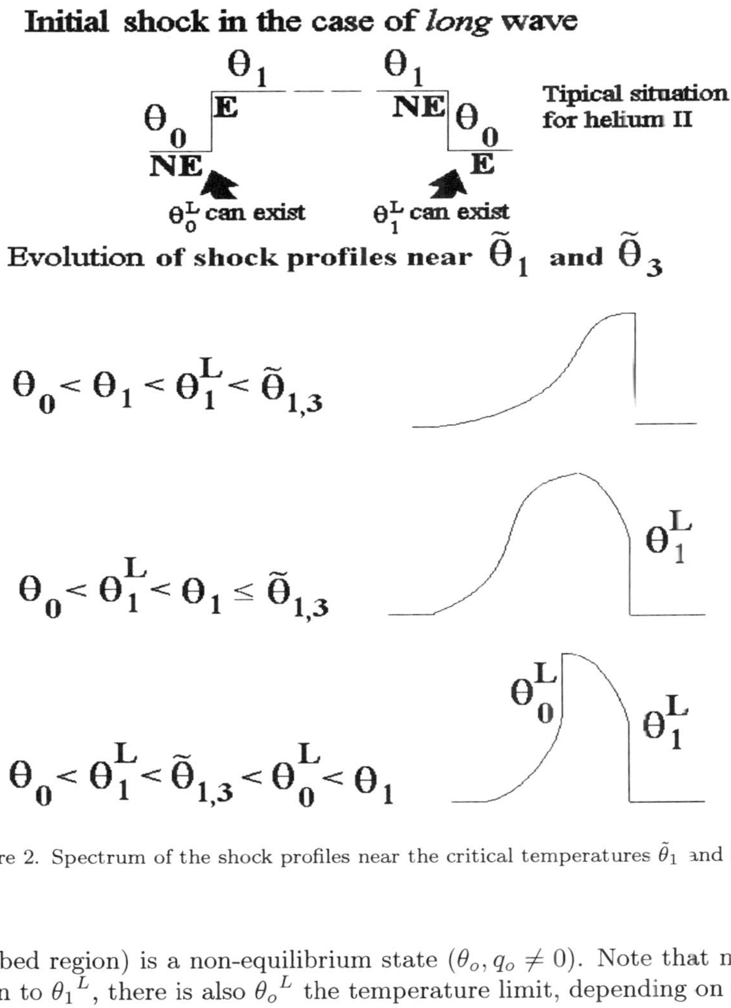

Figure 2. Spectrum of the shock profiles near the critical temperatures $\tilde{\theta}_1$ and $\tilde{\theta}_3$.

(perturbed region) is a non-equilibrium state $(\theta_o, q_o \neq 0)$. Note that now, in addition to ${\theta_1}^L$, there is also ${\theta_o}^L$ the temperature limit, depending on θ_1, for the Lax region with respect to the rear part of the wave (see the top part of the figure 2).

The full analysis of the results originated from this second interpretation will be showed in a forthcoming paper. In the fig. 2 we present only the spectrum of possible modifications for an initial positive rectangular pulse, when the temperature of the helium bath is near to the critical temperatures $\tilde{\theta}_1$ or $\tilde{\theta}_3$. Both the cases present the formation of a double shock pulse, even if this very interesting phenomenon has been clearly identified [26] by shock wave

Range of the initial temperature	Temperatures of the initial shock	HOT SHOCK 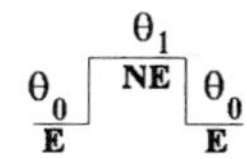	Temperatures of the initial shock	COLD SHOCK 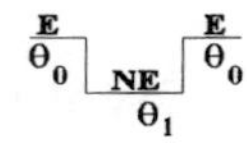
$\theta_0 < \widetilde{\theta}_1$	$\theta_0 < \theta_1 < \theta_1^L$ weak shock		$\theta_1 < \theta_0$ weak shock	
	$\theta_0 < \theta_1^L < \theta_1$ strong shock	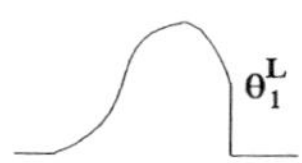		
$\theta_0 = \widetilde{\theta}_1$	$\theta_0 < \theta_1$ weak shock	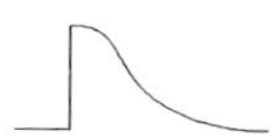	$\theta_1 < \theta_0$ weak shock	
$\widetilde{\theta}_1 < \theta_0 < \widetilde{\theta}_2$	$\theta_0 < \theta_1 < \theta_1^L$ weak shock	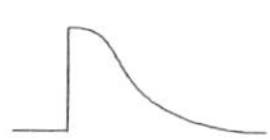	$\theta_1^L < \theta_1 < \theta_0$ weak shock	
	$\theta_0 < \theta_1^L < \theta_1$ strong shock		$\theta_1 < \theta_1^L < \theta_0$ strong shock	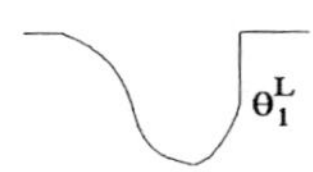

$\theta_0 = \widetilde{\theta}_2$	$\theta_0 < \theta_1 < \theta_1^L$ weak shock		$\theta_1^L < \theta_1 < \theta_0$ weak shock	
	$\theta_0 < \theta_1^L < \theta_1$ strong shock		$\theta_1 < \theta_1^L < \theta_0$ strong shock	
$\widetilde{\theta}_2 < \theta_0 < \widetilde{\theta}_3$	$\theta_0 < \theta_1 < \theta_1^L$ weak shock		$\theta_1^L < \theta_1 < \theta_0$ weak shock	
	$\theta_0 < \theta_1^L < \theta_1$ strong shock	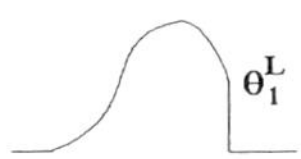	$\theta_1 < \theta_1^L < \theta_0$ strong shock	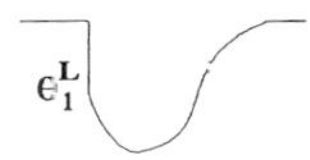
$\theta_0 = \widetilde{\theta}_3$	$\theta_0 < \theta_1$ weak shock	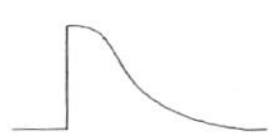	$\theta_1^L < \theta_1 < \theta_0$ weak shock	
			$\theta_1 < \theta_1^L < \theta_0$ strong shock	
$\theta_0 > \widetilde{\theta}_3$	$\theta_0 < \theta_1$ weak shock	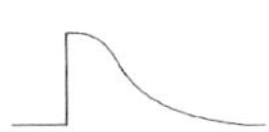	$\theta_1^L < \theta_1 < \theta_0$ weak shock	
			$\theta_1 < \theta_1^L < \theta_0$ strong shock	

experiments only in a neighborhood of $\tilde{\theta}_3 \sim 1.8K$.

6 Conclusions

In this paper our previous model for the second sound propagation in the case of rigid conductors at low temperatures [17] has been furthermore explained with respect to a binary Euler's fluid. In He II the study of the wave propagation has pointed out the presence of three characteristic temperatures $\tilde{\theta}$, instead of only one as for the rigid conductor, in a neighbourhood of whom the pulse profiles, experimentally revealed, dramatically change. We have given an explanation of these modifications on the ground of a stability analysis starting from an initial shock pulse (positive and negative). Moreover, since in the Helium II the duration of the generated thermal pulse is greatly increased with respect to the conditions for a rigid conductor, we have suggested a new extended interpretation for the initial wave signal, that has permitted us to describe in a coherent way the experimental evidence of double shocks near $\tilde{\theta}_3$.
In a forthcoming paper the consequences of this new way of considering the initial rectangular pulse will be completely developed, both for positive and negative shocks.

References

1. T. Ruggeri, in: *Continuum Mechanics and Applications in Geophysics and Environment.* B. Straughan, R. Greve, H. Ehrentraut and Y. Wang Editors. p. 79, Springer-Verlag, Berlin (2001).
2. F. London, Nature **141**, 643 (1938) ; Phys. Rev. **54**, 947 (1938).
3. L. Landau, J. Phys. U.S.S.R. **5**, 71 (1941).
4. L. Landau, E. Lifsits , *Mècanique des fluides.* (MIR, Moscow, 1971).
5. S. J. Putterman, *Superfluid Hydrodynamics.* (North Holland, Amsterdam, 1974).
6. W. Dreyer, Dissertation Technische Universität Berlin (1983).
7. I. Müller, *Thermodynamics.* (Pitman, New York, 1985).
8. V. Peshkov, in: *Report on an International Conference on Fundamental Particles and Low Temperature Physics*, Vol. II (The Physical Society of London, 19 (1947)).
9. C.C. Ackerman, W.C. Overton Jr, Phys. Rev. Lett. **22** (15), 764 (1969).
10. C.C. Ackerman, B. Bertman, H.A. Fairbank, R.A. Guyer, Phys. Rev. Lett. **16**(18), 789 (1966).

11. H.E. Jackson, C.T. Walker and T.F. McNelly, Phys. Rev. Lett. **25**, 26 (1970).
12. V. Narayanamurti, R.C. Dynes, Phys. Rev. Lett. **28**, 1461 (1972).
13. R.A. Guyer, J.A. Krumhansl, Phys. Rev. Lett. **148**, 766 (1966).
14. M.E. Gurtin, A.C. Pipkin, Arch. Rat. Mech. Anal. **31**, 113 (1968); A.E. Green, K.A. Lindsay, J.Elasticity **2**(1), 1 (1972); G. Grioli, Atti Acc. Naz. Lincei, **67**, (1979); D.D. Joseph, L. Preziosi, Rev. Mod. Phys. **61**, (1989); D.D. Joseph, L. Preziosi, Rev. Mod. Phys. **62**, (1990).
15. C. Cattaneo, Atti Sem. Mat. Fis. Univ. Modena **3**, 83 (1948).
16. I. Müller, T. Ruggeri, *Rational Extended Thermodynamics*. Springer Tracts in Natural Philosophy vol. 37, new edition (Springer-Verlag, New York, 1999).
17. T. Ruggeri, A. Muracchini and L. Seccia, Phys. Rev. Lett. **64**, 2640 (1990); Il Nuovo Cimento **16**D, N.1 15 (1994); Phys. Rev. B **54**(1), 332 (1996).
18. A. Muracchini, L. Seccia, Journ. Math. Anal. Appl. **240**, 383 (1999).
19. P.D. Lax, in: *Contribution to Non Linear Functional Analysis*, edited by E.H. Zarantonello, (Academic Press, New York 1971).
20. G. Boillat, C. R. Acad. Sci. Paris, **283**A, 409 (1976) ; T. Ruggeri, A. Strumia, Ann. Inst. H. Poincarè (1) **34**, 65 (1981).
21. T.P. Liu, J. Differential Equations, **18**, 218 (1975) ; T.P. Liu, T. Ruggeri *Entropy Production and Admissibility of Shocks.* To appear.
22. D. De Klerk, R.P. Hudson and J.R. Pellam, Phys. Rev. **93**, 28 (1954).
23. H.C. Kramers, J.D. Wasscher and J. Gorter, Physica **18**, 5 (1952).
24. J. Cummings, D.W. Schmidt and W.J. Wagner, Phys. Fluids **21**, (5) (1978).
25. B.D. Coleman, D.C. Newman, Phys. Rev. B **37**, 1492 (1988).
26. T. N. Turner, Physica **107**B, 701 (1981).

BGK MODEL FOR SIMULATING ELECTRON TRANSPORT IN SEMICONDUCTOR DEVICES

O. MUSCATO

Dipartimento di Matematica e Informatica
viale A. Doria 6, 95125 Catania , Italy
E-mail: muscato@dmi.unict.it

A BGK approximation of the Boltzmann Transport Equation is used for simulating carrier transport in silicon semiconductors. An analytic solution has been obtained in the stationary and homogeneous case. Some properties of this solution have been discussed and its validity have been assessed by Monte Carlo simulations. The BGK model has been also used for simulating a $n^+ - n - n^+$ silicon diode.

1 Introduction.

Recent advances in submicron semiconductor devices led to regimes of non equilibrium hot electron transport in which traditional drift diffusion models break down. This failure is due to the small size of the devices and manifests itself with relatively large carrier mean free path and stronger electric fields and gradients. The semiclassical Boltzmann Transport Equation (BTE hereafter) coupled with the Poisson equation provides a natural framework for modeling this transport. To solve the BTE is a difficult task because it is an integro-differential equation with six dimensions in the phase space and one in time: also the band structure of the material and the electron-phonon coupling introduce complex kernel in the collisional operator. The most popular numerical methods used to solve the BTE are Monte Carlo simulations (MC hereafter) in which one generates typical histories of many electrons and then averages over the ensemble to find the macroscopic quantities. This approach suffers from the considerable computation time needed to acquire enough statistics in order to reduce the noise.

Alternative, simpler expressions have been proposed for the BTE collisional operator: the most widely known collision model is usually called Bhatnagar, Gross and Krook (hereafter BGK) model. That has been proved useful in many other areas of physics , including kinetic theory of gases, Fermi liquid theory etc. In this case it is assumed that the distribution function relaxes to its equilibrium value determined by the local density and the lattice temperature and that this process can be characterized by a constant relaxation time [1 2]. This approach is certainly not exact for scattering in semiconductors, because neglects the energy and angular dependence of the scattering rate as well as the discrete amounts of energy lost in the scattering from optical phonons. However we guess that this approximation is adequate for studying submicron structures because they contain many ballistic electrons, for which the details of the scattering process are irrelevant. The main effect is determined by the magnitude of the scattering rate which can be correctly incluted by a relaxation term. In this paper we will investigate the case in which the relaxation time is not a constant but a function of the electric field, in such a way to fit the mobility obtained by Monte Carlo simulations in the homogeneous, stationary regime.

2 Basic equations.

The BTE for electrons and one conduction band writes [3]:

$$\frac{\partial f}{\partial t} + \vec{v}(\vec{k}) \cdot \nabla_{\vec{x}} f - \frac{q}{\hbar} \vec{E} \cdot \nabla_{\vec{k}} f = Q(f) \tag{1}$$

where the unknown function $f(t, \vec{x}, \vec{k})$ represents the probability density of finding an electron at time t in the position $\vec{x} \in \mathbb{R}^3$, with the wave-vector $\vec{k}$; q is the absolute value of the electron charge and the electric field $\vec{E}$ is related to the electron distribution by Poisson's equation

$$\vec{E} = -\nabla_{\vec{x}} \phi, \quad \epsilon \Delta \phi = q(N_D(\vec{x}) - n(t, \vec{x}))$$

where ϕ is the electric potential, N_D is the donor density (which is a positive function) , ϵ the dielectric constant and n the particle density, which is related to f by

$$n(t, \vec{x}) = \int_{\mathbb{R}^3} f(t, \vec{x}, \vec{k}) \, d\vec{k} \quad .$$

The electron group velocity $\vec{v}(\vec{k})$ is given by

$$\vec{v}(\vec{k}) = \frac{1}{\hbar} \nabla_{\vec{k}} \varepsilon$$

where $\varepsilon(\vec{k})$ is the energy of the considered conduction band structure of the crystal measured from the band minimum and $\hbar$ the Planck constant. In the neighborhood of the band minimum a good dispersion relation is given by the parabolic approximation :

$$\varepsilon(\vec{k}) = \frac{\hbar^2 \vec{k}^2}{2m^\star}, \quad \vec{k} \in \mathbb{R}^3, \tag{2}$$

where $m^\star$ the effective electron mass (which is $0.32\, m_e$ in silicon). The collision operator can be schematically written as

$$Q(f) = \int_{\mathbb{R}^3} \left[S(\vec{k}', \vec{k}) f(\vec{k}') - S(\vec{k}, \vec{k}') f(\vec{k}) \right] d\vec{k}' \tag{3}$$

where $S(\vec{k}, \vec{k}')$ is the transition probability, per unit time, from a state $\vec{k}$ to a state $\vec{k}'$. The first term in (3) represents the gain and the second one the loss. In the limit of low electron density the electron-electron scattering can be neglected in the collisional operator and consequently in our MC simulator we will consider the main scattering mechanism, i.e. the electron-phonon interactions (see Jacoboni and Reggiani [3] for details) .
In the BGK approximation the collisional operator writes:

$$Q(f) = -\frac{f(\vec{x}, \vec{k}, t) - n(t, \vec{x}) f_M}{\tau} \tag{4}$$

where

$$f_M = \sqrt{\frac{m^\star}{2\pi k_B T_0}} \quad exp\left[-\frac{m^\star v^2}{2 k_B T_0} \right]$$

is the maxwellian , τ is the characteristic relaxation time and T_0 the lattice temperature. This BGK model satisfies the particle number conservation equation but not the energy and momentum conservation equations. We remark that collisions are completely effective in cancelling any information about non equilibrium configuration that electrons may be carrying. This overestimates the efficacy of collisions in restoring equilibrium which suggests that non equilibrium effects will be underestimated. On the other hand , the use of a single τ underestimates the rate of relaxation of high energy electrons and so tends to overestimate the non equilibrium effects. The result of the competition between these two effects is not clear.
In the stationary , homogeneous regime by multipling eqs.(1),(4) by the velocity component along the electric field direction and by integrating over $d\vec{k}$ one obtains the drift velocity

$$v_d = \mu E, \quad \mu = -\frac{e}{\hbar}\tau \tag{5}$$

where μ is the mobility. It is clear that from eq.(5) no velocity saturation is obtained with constant a τ, despite to physical experiments. For this reason we use a physical field dependent relaxation time given by:

$$\tau(E) = \frac{m^\star}{e}\mu(E) = \frac{m^\star}{e} \frac{2\mu_0}{1 + \sqrt{1 + 4\left(\frac{\mu_0 E}{v_0}\right)^2}} \tag{6}$$

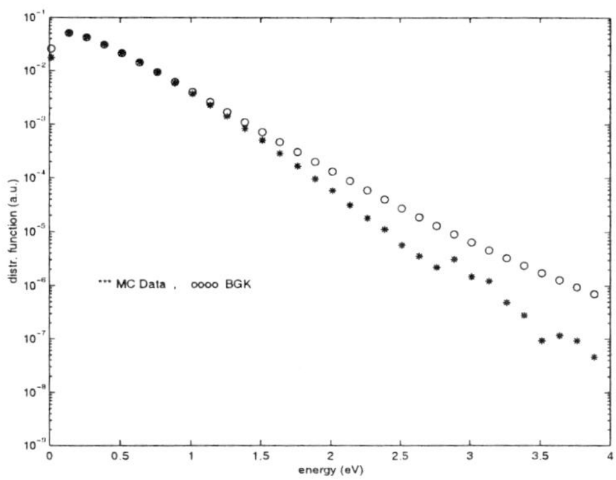

Figure 1. The distribution function $F_B(w)$ obtained by the BGK solution eq.(7) and that obtained by MC simulations, for an electric field of 53570 V/cm.

where the parameters μ_0 and v_0 are obtained as fitting parameters with MC data in bulk silicon.

3 Stationary , homogeneous solution of the BGK model

In this case the BGK equation reduces to a linear first order partial differential equation and for, an electric field direction along the x axis, its solution is [1] :

$$f_B(\vec{k}) = \sqrt{\frac{\pi}{2}} exp\left(\frac{1}{2}\eta^2 - \eta k_x\right) exp\left(-\frac{k_y^2 + k_z^2}{2}\right) erfc\left[-\frac{1}{\sqrt{2}}(k_x - \eta)\right] \quad (7)$$

where

$$\eta = \frac{\sqrt{m^\star k_B T_0}}{m^\star \mu(E) E} \quad .$$

From this distribution function it is possible to calculate the distribution function as function of the average energy w, i.e.

$$F_B(w) = \int f_B(\vec{k}) \delta\left(\varepsilon(\vec{k}) - w\right) d\vec{k} \quad .$$

In figure 1 we compare the distribution function $F_B(w)$ with that obtained by MC simulations: the agreement is very good for energies less then 1 eV whereas, in the tail, there are some differences.

4 Simulation of a 1D $n^+ - n - n^+$ silicon diode and conclusions.

The BGK equation can be solved numerically by using finite difference scheme **ENO** and **WENO** [4] [5]. Time discretization is via **TVD** (Total-variation-diminishing) third ordee explicit Runge-Kutta method. Uniform high order accuracy can be achieved without introducing any oscillations near discontinuities or sharp gradient regions: these algorithms are extremely stable and robust in all the numerical simulations [6]. The $n^+ - n - n^+$ diode has been intensively studied as a simple, prototypical submicron structure. It is formed by a layer of lightly doped material (n=2 $10^{15} cm^{-3}$) called channel between two slabs (0.1 μm wide each) of highly doped material ($n^+ = 10^{18} cm^{-3}$) called cathode and anode and the device is at room temperature (T_0 =300 ^0K) . Baranger and Wilkins [7] used a 1D stationary BGK model with a constant relaxation time to study transport phenomena in a $n^+ - n - n^+$ Gallium Arsenide structure. In this paper we will study a 1D silicon diode in the more complex time-depending case, by using the BGK model with the electric field depending mobility given by eq.(6) . The results will be compared with the MC simulations obtained by the `Damocles` code [8].

Simulation results for the mean velocity are shown in figure 2 for two channel lengths. The comparison for the longer channel (i.e. 0.4 μm) correspond to a length scale that balances the strength of collisions with the drift strength. As a result, we have remarkable coincidence of the results for the 1D velocity BGK model with respect to the MC simulations. The comparison of the energy is much worse than the one for the mean velocity because, in order to compare the energies of the 1D BGK model, we need to assume that the corresponding distribution function in 3-D corresponds to the 1D distribution function multiplied by a maxwellian with zero mean velocity and background temperature in the orthogonal directions. As a consequence the 1D BGK model fails in capturing the energy of the MC simulations. For the shorter device (0.05 μm channel length) the 1D field dependent BGK model fails to produce a good approximation for the mean velocity (see figure 2, right column) and also for any of the other moments. In this case the BGK model fails to capture the high energy effects coming from stronger scattering rates. This means that the full acoustic and non-polar optical phonons scattering collision operator must be used in the BTE .

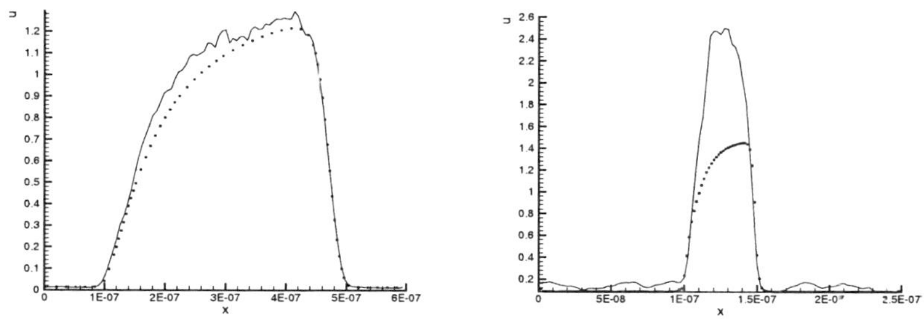

Figure 2. Velocity profiles obtained by the 1D BGK model (circles) and by Monte Carlo simulation (continuos line) with a bias voltage of 1 Volt. On the left column it is shown the case of the longer channel (0.4 μm) , on the right column it is reported the case of the shorter channel (0.05 μm). The velocity u is in $10^7 cm/sec$ units and the position x in m.

Acknowledgments

This work has been supported by the C.N.R. project No. 98.03630.ST74 *Modelli Matematici per i semiconduttori* , MURST Cof 2000 *Mathematical problems of Kinetic theories* and the *Laboratory for Computational Astrophysics* at Catania Astrophysical Observatory.

References

1. S.A. Trugman and A.J. Taylor, Phys. Rev. B **33**, 5575 (1986)
2. N.S. Wingreen, C.J. Stanton and J.W. Wilkins, Phys. Rev. Lett. **57**, 1086 (1986).
3. C. Jacoboni and L. Reggiani , Rev. Modern Phys. **55**, 645 (1983)
4. C.-W. Shu and S. Osher, J. Comp. Phys. **83**, 32 (1989)
5. G. Jiang and C.-W. Shu, J. Comp, Phys. **126**, 202 (1998)
6. J.A. Carrillo, I.M. Gamba, O. Muscato and C.-W. Shu, *Comparison of Monte Carlo and deterministic simulations of a silicon diode*, preprint (2001)
7. H.U. Baranger and J. W. Wilkins ,Phys. Rev. B **36**, 1436 (1987).
8. S. E. Laux, M. V. Fischetti and D. J. Frank, IBM J. Res. Develop. **34**, 466 (1990)

NONLINEAR WAVES IN CONTINUA WITH SCALAR MICROSTRUCTURE

FRANCESCO OLIVERI

Department of Mathematics, University of Messina
Contrada Papardo, Salita Sperone 31, 98166 Sant'Agata, Messina, Italy
E-mail: oliveri@mat520.unime.it

We consider the equations governing a continuum with scalar microstructure, and look for solutions having the features of progressive waves by means of a suitable asymptotic expansion of the field variables. The result of the procedure shows that the wave process can be studied by means of nonlinear evolution equations that capture the complexity of the considered models. Some examples are provided.

1 Introduction

In the classical theory of continua, a body is thought of as a set of material elements each having a distinct identity and occupying at each instant of time an exclusive place. It is assumed that nothing of geometrically interesting could be perceived by a finer observation of the element. On the other hand, when we consider the case of continua with microstructure[1], a closer look at the element reveals a sort of microscopic order that is described by a certain number of order parameters. The simplest models are those involving only one order parameter: typical examples are given by continua vith voids, bubbly liquids, granular materials. Despite of the variety of materials described as continua with microstructure, it has been shown that a unified approach for their study is possible[1].

It has to be stressed that the kinetic energy per unit mass of a continuum with microstructure is the sum of the classical term and of a term, related to the microstructure, given by the non-negative function $\kappa(\nu, \dot{\nu})$:

$$\frac{1}{2}\mathbf{v}^2 + \kappa(\nu, \dot{\nu}), \quad \kappa(\nu, 0) = 0, \quad \frac{\partial^2 \kappa}{\partial \dot{\nu}^2} \neq 0,$$

where ν is the order parameter. The local description of continua with microstructure consists of a system of partial differential equations stating the balance laws for mass, momentum and micromomentum[1]:

$$\begin{aligned}
&\dot{\rho} + \rho \nabla \cdot \mathbf{v} = 0, \\
&\rho \dot{\mathbf{v}} = \rho \mathbf{f} + \nabla \cdot T, \\
&\rho \left[\overline{\left(\frac{\partial \chi}{\partial \dot{\nu}} \right)}^{\,\cdot} - \left(\frac{\partial \chi}{\partial \nu} \right) \right] = \rho \beta - \zeta + \nabla \cdot \mathbf{s},
\end{aligned} \tag{1}$$

where ρ is the mass density of the continuum, $\mathbf{v}$ the velocity, $\mathbf{f}$ the external force per unit volume, T the stress tensor, and χ the density of kinetic coenergy related to the function κ by the relation

$$\kappa = \left(\frac{\partial \chi}{\partial \dot{\nu}}\right) \dot{\nu} - \chi;$$

furthermore, β is the density per unit mass of the external actions acting on the microstructure (for instance electric actions on the molecules of a liquid crystal), ζ the density per unit volume of the internal microstructural actions, which do not necessarily sum to zero as internal forces do, $\mathbf{s}$ represents the actions exerted on the microstructure through the element, and the dot stands for lagrangian time derivative. Finally, stress T, microstress ζ and $\mathbf{s}$ must be assigned as constitutive functions of the geometric and kinematic variables (e.g., the deformation gradient F, $D = \mathrm{sym}\nabla \mathbf{v}$, ν, $\nabla \nu$, ...). The stress tensor T is symmetric if the parameter ν is not affected by rigid rotations, whereas κ and χ coincide if κ is quadratic in $\dot{\nu}$.

If thermal effects are relevant then the balance equation for the energy

$$\rho \dot{e} = T \cdot \nabla \mathbf{v} + \zeta \cdot \dot{\nu} + \mathbf{s} \cdot \nabla \dot{\nu} - \nabla \cdot \mathbf{q} + \rho h,$$

where e is the internal energy per unit mass, $\mathbf{q}$ the heat flux vector, h the heat supply per unit mass, and the entropy inequality in the classical Clausius-Duhem form

$$\rho \dot{\eta} \geq -\nabla \cdot \left(\frac{\mathbf{q}}{\theta}\right) + \frac{\rho h}{\theta},$$

where η is the specific entropy and θ is the absolute temperature, need to be postulated.

In the following we shall consider the case in which the order parameter ν is a scalar real variable.

2 Progressive waves

Here attention is devoted to the search of solutions of the field equations having the features of a *progressive wave*[2]. We have a progressive wave when there is a family of propagating surfaces such that the magnitude of the rate of change of the field variables (or, eventually, of their derivatives) when we move with such surfaces is small compared with the one observed when we move across the surfaces (this concept is a generalization of the simple waves).

The starting point of the procedure we will use consists in writing the field equations in dimensionless form. As a consequence, in a natural way, a "small" parameter ϵ (usually a ratio between a quantity related to the microstructure and a macroscopic quantity) enters the equations. If we restrict ourselves to the one-dimensional case, and introduce the notation

$$\mathbf{U}^T = [\rho, v, \ldots] \in \mathbb{R}^N, \quad \mathbf{V}^T = \left[\mathbf{U}^T, \nu\right] \in \mathbb{R}^{N+1},$$

i.e., the N-vector of "macroscopic" (dimensionless) field variables (mass density, velocity, ...) and the $(N+1)$-vector of "all" the (dimensionless) field variables, the balance equations of continua with microstructure usually appear in the form:

$$\begin{aligned}
&\frac{\partial \mathbf{U}}{\partial t} + \mathcal{A}(\mathbf{U})\frac{\partial \mathbf{U}}{\partial x} = \\
&\quad = \epsilon^{\alpha_0}\mathbf{B}^{(0)}(\mathbf{V}) + \epsilon^{\alpha_1}\mathbf{B}^{(1)}\left(\mathbf{V}, \frac{\partial \mathbf{V}}{\partial x}\right) + \sum_{i=2}^{p} \epsilon^{\alpha_i} B^{(i)}\left(\mathbf{V}, \frac{\partial \mathbf{V}}{\partial x}\right)\frac{\partial^i \mathbf{V}}{\partial x^i}, \\
&\epsilon^2 \rho \left[\overline{\left(\frac{\partial \chi}{\partial \dot{\nu}}\right)}^{\cdot} - \left(\frac{\partial \chi}{\partial \nu}\right)\right] = \\
&\quad = \epsilon^{\beta_0} F^{(0)}(\mathbf{V}) + \epsilon^{\beta_1} F^{(1)}\left(\mathbf{V}, \frac{\partial \mathbf{V}}{\partial x}\right) + \sum_{i=2}^{q} \epsilon^{\beta_i} F^{(i)}\left(\mathbf{V}, \frac{\partial \mathbf{V}}{\partial x}\right)\frac{\partial^i \mathbf{V}}{\partial x^i},
\end{aligned}$$

where various functions of the indicated arguments appear. Moreover, α_i and β_j are constants subjected to the constraints

$$\alpha_i \geq i, \quad (i = 0, \ldots, p), \qquad \beta_j \geq j, \quad (j = 0, \ldots, q).$$

The last step of the procedure involves the use of an asymptotic expansion of the field variables, whereupon we are able to get the wave speeds (that are the eigenvalues, assumed to be real, of the matrix $\mathcal{A}$) and a transport equation ruling the time evolution of the wave amplitudes.

3 Asymptotic solutions

To investigate the propagation of progressive waves by means of the asymptotic expansion of the field variables, it is convenient to introduce an extra "fast" variable $\xi = \epsilon^{-1}\varphi(x,t)$, $\varphi(x,t)$ being the phase function, and assume that the field variables depend upon x, t and ξ.

Therefore, let us consider the following asymptotic solution[3,4]:

$$\mathbf{V} = \mathbf{V}_0(x,t) + \epsilon \mathbf{V}_1(x,t,\xi) + \epsilon^2 \mathbf{V}_2(x,t,\xi) + O(\epsilon^3), \tag{2}$$

where $\mathbf{V}_0$ is the unperturbed state (solution of the system at hand when $\epsilon = 0$). If $\mathbf{V}_0$ is constant it follows:

$$\frac{\partial \mathbf{V}}{\partial t} = \varphi_t \frac{\partial \mathbf{V}_1}{\partial \xi} + \epsilon\left(\frac{\partial \mathbf{V}_1}{\partial t} + \varphi_t \frac{\partial \mathbf{V}_2}{\partial \xi}\right) + O(\epsilon^2),$$

$$\frac{\partial \mathbf{V}}{\partial x} = \varphi_x \frac{\partial \mathbf{V}_1}{\partial \xi} + \epsilon \left(\frac{\partial \mathbf{V}_1}{\partial x} + \varphi_x \frac{\partial \mathbf{V}_2}{\partial \xi} \right) + O(\epsilon^2),$$
$$\frac{\partial^2 \mathbf{V}}{\partial x^2} = \epsilon^{-1} \varphi_x^2 \frac{\partial^2 \mathbf{V}}{\partial \xi^2} + \varphi_{xx} \frac{\partial \mathbf{V}_1}{\partial \xi} + 2\varphi_x \frac{\partial^2 \mathbf{V}_1}{\partial \xi \partial x} + O(\epsilon)$$
$$\dots\dots,$$

where the subscripts x and t denote partial derivatives. If we insert the asymptotic expansion (2) into the dimensionless field equations, and set the coefficients of the resulting series in ϵ termwise to zero, we get:

$$(\varphi_t + \mathcal{A}_0 \varphi_x) \frac{\partial \mathbf{U}_1}{\partial \xi} = 0, \tag{3}$$

$$(\varphi_t + \mathcal{A}_0 \varphi_x) \frac{\partial \mathbf{U}_2}{\partial \xi} + \frac{\partial \mathbf{U}_1}{\partial t} + \mathcal{A}_0 \frac{\partial \mathbf{U}_1}{\partial x} + ((\nabla_{\mathbf{U}} \mathcal{A})_0 \, \mathbf{U}_1) \, \varphi_x \frac{\partial \mathbf{U}_1}{\partial \xi} =$$
$$= \alpha_0^0 \left(\nabla_{\mathbf{V}} \mathbf{B}^{(0)} \right)_0 \mathbf{V}_1 + \alpha_1^1 \mathbf{B}^{(1)} \left(\mathbf{V}_0, \varphi_x \frac{\partial \mathbf{V}_1}{\partial \xi} \right) +$$
$$+ \alpha_i^i \sum_{i=2}^{p} B^{(i)} \left(\mathbf{V}_0, \varphi_x \frac{\partial \mathbf{V}_1}{\partial \xi} \right) \frac{\partial^i \mathbf{V}_1}{\partial \xi^i}, \tag{4}$$

$$\rho_0 \left(\frac{\partial^2 \chi}{\partial \nu^2} \right)_0 (\varphi_t + v_0 \varphi_x)^2 \frac{\partial^2 \nu_1}{\partial \xi^2} = \beta_0^0 \left(\nabla_{\mathbf{V}} F \right)_0 \mathbf{V}_1 +$$
$$+ \beta_1^1 F^{(1)} \left(\mathbf{V}_0, \varphi_x \frac{\partial \mathbf{V}_1}{\partial \xi} \right) + \beta_i^i \sum_{i=2}^{q} F^{(i)} \left(\mathbf{V}_0, \varphi_x \frac{\partial \mathbf{V}_1}{\partial \xi} \right) \frac{\partial^i \mathbf{V}_1}{\partial \xi^i}, \tag{5}$$

where α_i^i and β_i^i represent suitable constants belonging to the set $\{0, 1\}$, and the subscript $_0$ means that a quantity is evaluated for $\mathbf{V} = \mathbf{V}_0$. From (3), by introducing the normal wave speed $\lambda = -\dfrac{\varphi_t}{\varphi_x}$, it follows

$$\det(\mathcal{A}_0 - \lambda I) = 0,$$

i.e., $\lambda(\mathbf{U}_0)$ is an eigenvalue of the matrix $\mathcal{A}_0$, whereas the corresponding right eigenvector $\mathbf{d}_0 = \mathbf{d}(\mathbf{U}_0)$ allows us to write:

$$\mathbf{U}_1 = \pi(x, t, \xi) \mathbf{d}_0,$$

π being the so-called *wave amplitude factor*. Thus, the wave speeds are determined by the subsystem (assumed to be hyperbolic), involving only the "macroscopic" field variables, obtained from the complete system for $\epsilon \to 0$.

By considering the propagation of waves in a constant state, from the definition of λ and the initial condition $\varphi(x, 0) = x$, it follows $\varphi(x, t) = x - \lambda t$.

Finally, the evolution equation for the leading terms of the asymptotics will arise from the remaining conditions and take into account the effects due to the microstructure on the wave process. In fact, by left multiplying condition (4) by the left eigenvector $\mathbf{l}_0$, we get:

$$(\mathbf{l}_0 \cdot \mathbf{d}_0)\frac{\partial \pi}{\partial \sigma} + \mathbf{l}_0\left(\left(\nabla_{\mathbf{U}}\mathcal{A}\right)_0 \mathbf{d}_0\right)\mathbf{d}_0 \pi \frac{\partial \pi}{\partial \xi} =$$

$$= \mathbf{l}_0 \left[\alpha_0^0 \left(\nabla_{\mathbf{V}}\mathbf{B}^{(0)}\right)_0 \mathbf{V}_1 + \alpha_1^1 \mathbf{B}^{(1)}\left(\mathbf{V}_0, \frac{\partial \mathbf{V}_1}{\partial \xi}\right) + \right.$$

$$\left. + \alpha_i^i \sum_{i=2}^{p} B^{(i)}\left(\mathbf{V}_0, \frac{\partial \mathbf{V}_1}{\partial \xi}\right)\frac{\partial^i \mathbf{V}_1}{\partial \xi^i}\right],$$

where $\dfrac{\partial}{\partial \sigma} = \dfrac{\partial}{\partial t} + \lambda \dfrac{\partial}{\partial x}$ is the time derivative along the characteristic rays.

Ultimately, the leading terms π and ν_1 of the asymptotics are ruled by:

$$\frac{\partial \pi}{\partial \sigma} + \kappa_1 \pi \frac{\partial \pi}{\partial \xi} + \kappa_2 \pi + \kappa_3 \nu_1 + \chi_1\left(V_0, \frac{\partial \pi}{\partial \xi}, \frac{\partial \nu_1}{\partial \xi}\right) +$$

$$+ \sum_{i=2}^{p} \chi_i^{(1)}\left(V_0, \frac{\partial \pi}{\partial \xi}, \frac{\partial \nu_1}{\partial \xi}\right)\frac{\partial^i \pi}{\partial \xi^i} + \sum_{i=2}^{p} \chi_i^{(2)}\left(V_0, \frac{\partial \pi}{\partial \xi}, \frac{\partial \nu_1}{\partial \xi}\right)\frac{\partial^i \nu_1}{\partial \xi^i} = 0,$$

$$\frac{\partial^2 \nu_1}{\partial \xi^2} + \kappa_4 \pi + \kappa_5 \nu_1 + +\psi_1\left(V_0, \frac{\partial \pi}{\partial \xi}, \frac{\partial \nu_1}{\partial \xi}\right) +$$

$$+ \sum_{i=2}^{q} \psi_i^{(1)}\left(V_0, \frac{\partial \pi}{\partial \xi}, \frac{\partial \nu_1}{\partial \xi}\right)\frac{\partial^i \pi}{\partial \xi^i} + \sum_{i=2}^{q} \psi_i^{(2)}\left(V_0, \frac{\partial \pi}{\partial \xi}, \frac{\partial \nu_1}{\partial \xi}\right)\frac{\partial^i \nu_1}{\partial \xi^i} = 0,$$

where various constants and functions appear. By eliminating ν_1, when possible, we obtain the evolution equation for π that in general allows for various descriptions of the wave process according to the orders of magnitude of the different coefficients involved (wave hierarchies problem)[5,6].

4 Examples

4.1 Immiscible mixtures of two perfect fluids

We assume[7] the continuum composed of spherical elements of radius ϑ_1 of a perfect fluid containing a concentrical spherical inclusion of variable radius ϑ_2 of another perfect fluid that expands and contracts homogeneously (no diffusion). The small parameter is $\epsilon = \left(\dfrac{\vartheta_1}{L}\right)^{2/3}$, where L is a macroscopic length, and the evolution equation for the wave amplitude factor is

$$\frac{\partial \pi}{\partial \sigma} + \pi \frac{\partial \pi}{\partial \xi} + \kappa_1 \frac{\partial \pi}{\partial \xi} + \kappa_2 \frac{\partial^3 \pi}{\partial \xi^3} + \kappa_3 \frac{\partial^2}{\partial \xi^2}\left(\frac{\partial \pi}{\partial \sigma} + \pi \frac{\partial \pi}{\partial \xi}\right) = 0, \tag{6}$$

with κ_1, κ_2 and κ_3 constants, *i.e.*, a modified KdV equation.

4.2 Bubbly liquids

In this case[8,9] the small parameter is

$$\epsilon = \frac{R_0}{L}, \quad \text{or, equivalently,} \quad \epsilon = \frac{\Phi_{2_0}}{\Phi_{1_0}}.$$

where R_0 is a typical bubble radius, L a macroscopic length, Φ_{1_0} and Φ_{2_0} the volume fractions of the liquid and gaseous phases, respectively. If diffusion of the bubbles is allowed, we have the transport equation

$$\frac{\partial \pi}{\partial \sigma} + \pi \frac{\partial \pi}{\partial \xi} + \kappa_1 \frac{\partial^2 \pi}{\partial \xi^2} + \kappa_2 \frac{\partial^3 \pi}{\partial \xi^3} + \kappa_3 \frac{\partial^4 \pi}{\partial \xi^4} + \\ + \kappa_4 \frac{\partial}{\partial \xi}\left(\frac{\partial \pi}{\partial \sigma} + \pi \frac{\partial \pi}{\partial \xi}\right) + \kappa_5 \frac{\partial^2}{\partial \xi^2}\left(\frac{\partial \pi}{\partial \sigma} + \pi \frac{\partial \pi}{\partial \xi}\right) + \kappa_6 \frac{\partial^3}{\partial \xi^3}\left(\frac{\partial \pi}{\partial \sigma} + \pi \frac{\partial \pi}{\partial \xi}\right) = 0,$$

i.e., a fourth order equation with higher order nonlinear terms; on the contrary, if diffusion of bubbles is neglected, we get

$$\frac{\partial \pi}{\partial \sigma} + \pi \frac{\partial \pi}{\partial \xi} + \kappa_1 \left(\kappa_3 \frac{\partial^2 \pi}{\partial \xi^2} + \kappa_4 \frac{\partial^3 \pi}{\partial \xi^3}\right) + \kappa_2 \left[\left(\kappa_3 \frac{\partial}{\partial \xi} + \kappa_4 \frac{\partial^2}{\partial \xi^2}\right)\left(\frac{\partial \pi}{\partial \sigma} + \pi \frac{\partial \pi}{\partial \xi}\right)\right] = 0,$$

i.e., a modified KdVB equation (for very large Reynolds numbers it results $\kappa_3 = 0$, *i.e.*, we have a modified KdV equation).

4.3 Granular materials

A granular material is thought of as a suspension of spheres in a compressible fluid of negligible density with respect to that of the suspended particles. An element of the continuum is modeled as a spherical envelope of radius ϑ_1 containing some spherical inclusions (of radius ϑ_2) called grains (in particular, an element of the continuum is thought of as a grain surrounded by its 12 immediate neighbours). The allowed motions are expansions (or contractions) of the inclusions and radial motions of the spherical crust due to the displacements of the grains (neither diffusion of the grains through the envelope, nor relative rotations of the elements or of the grains are considered). The small parameter ϵ is proportional to $\left(\frac{\vartheta_1}{L}\right)^{2/3}$, with L a macroscopic length.

If we consider compressible grains and neglect dissipation[10], the evolution equation for the wave amplitude factor is:

$$\frac{\partial \pi}{\partial \sigma} + \pi \frac{\partial \pi}{\partial \xi} + \kappa_1 \frac{\partial^3 \pi}{\partial \xi^3} + + \kappa_2 \frac{\partial^2}{\partial \xi^2}\left(\frac{\partial \pi}{\partial \sigma} + \pi \frac{\partial \pi}{\partial \xi} + \kappa_3 \frac{\partial^3 \pi}{\partial \xi^3}\right) = 0. \qquad (7)$$

On the contrary, if the grains are incompressible but we include dissipative effects[11], the evolution equation for the wave amplitude factor becomes:

$$\frac{\partial \pi}{\partial \sigma} + \pi \frac{\partial \pi}{\partial \xi} - \left(\kappa_1 + \kappa_2 \left(\frac{\partial \pi}{\partial \xi} \right)^2 \right) \frac{\partial^2 \pi}{\partial \xi^2} + \kappa_3 \frac{\partial^3 \pi}{\partial \xi^3} = 0, \tag{8}$$

that is, a modified KdVB equation.

Acknowledgments

Work supported by the Research Project "Problemi matematici non lineari di propagazione e stabilità nei modelli del continuo", Cofin. MURST 2000.

References

1. G. Capriz. Continua with microstructure. Springer Tracts in Natural Phylosophy, **35**, 1989.
2. P. Germain. Progressive waves. Jahrbuch der DGLR, 1971.
3. Y. Choquet-Bruhat. Ondes asymptotiques et approchées pour des systèmes d'équations aux dérivées partielles non linéaires. J. Math. Pure et Appliquée, **48**, 117-158, 1969.
4. G. Boillat. Ondes asymptotiques non linéaires. Ann. Mat. Pura e Applicata, **61**, 31-44, 1976.
5. G.B. Whitham. Linear and nonlinear waves. John Wiley & Sons, New York, 1974.
6. D. Fusco. Some comments on wave motion described by non-homogeneous quasilinear first order hyperbolic systems. Meccanica, **17**, 128-137, 1982.
7. P. Giovine & F. Oliveri. Wave features related to a model of compressible immiscible mixtures of two perfect fluids. Acta Mechanica, **96**, pp. 85-96, 1993.
8. D. Fusco & F. Oliveri. Derivation of a non-linear model equation for wave propagation in bubbly liquids. Meccanica, **24**, pp. 15-25, 1989.
9. F. Oliveri. Nonlinear wave propagation in a non-diffusive model of bubbly liquids. Acta Mechanica, **83**, pp. 135-148, 1990.
10. P. Giovine & F. Oliveri. Dynamics and wave propagation in dilatant granular materials. Meccanica, **30**, pp. 341-357, 1995.
11. C. Godano & F. Oliveri. Nonlinear seismic waves: a model for site effects. Int. J. Non-linear Mech., **34**, 457-468, 1999.

A NAVIER-STOKES MODEL FOR CHEMICALLY REACTING GASES

M. PANDOLFI BIANCHI
Dipartimento di Matematica, Politecnico di Torino, Corso Duca degli Abruzzi, 24
10129 Torino, Italy
E-mail: miriam@polito.it

A. J. SOARES
Departamento de Matemática, Universidade do Minho, Campus de Gualtar
4710-057 Braga, Portugal
E-mail: ajsoares@math.uminho.pt

The purpose of this work is to deduce the system of Navier-Stokes equations specialized for a quite general discrete kinetic model of a gas mixture with a general reversible chemical reaction. A perturbation scheme of Chapman-Enskog type is applied to the kinetic equations leading to the constitutive laws of the model. In this frame the reactive Euler and Navier-Stokes equations of the model have been formally deduced in dependence on the macroscopic variables.

1 Introduction

In recent papers[1,2,3], kinetic models of the Boltzmann equation extended to chemically reacting gases have shown a good capability to describe multi–particle interactions through an accurate microscopic description of the gas mixture. Such models are also able to recover, in good agreement with the laws of chemical kinetics, fluid dynamical processes as interaction between a gas and a radiation field, detonation waves with the related problem of linear stability, and sound propagation in different flow regimes. In order to study further gas phenomena including transport effects within kinetic theory, it is necessary to derive the Navier-Stokes equations of the model. A first approach to such a topic goes back in time respectively for an extended full Boltzmann equation[4] and for a class of discrete models of inert gases[5]. More recently[6], the Navier-Stokes equations have been dealt with for a gas with a bimolecular reaction. Aim of the paper is to deduce the reactive Navier-Stokes equations starting from a discrete kinetic model with a quite general velocity discretization and general reversible chemical reaction. A perturbation scheme of Chapman-Enskog type is applied to the reactive kinetic equations of such model, under the assumption that the mechanical relaxation time is short compared with the chemical relaxation time. Accordingly, the closed systems of both reactive Euler and Navier-Stokes equations of the model are then obtained in dependence on the macroscopic variables.

2 The mathematical model

With reference to a recent paper by the authors [7], where all mathematical aspects and chemical kinetic properties of a general discrete model of a reacting gas have been investigated, consider a gas of $r+s$ species $A_1, \ldots, A_r, B_1, \ldots, B_s$, $r, s \in \mathbb{N}$, whose particles of species M move with $p \in \mathbb{N}$ selected velocities

$$\vec{v}_i^{\,M}, \quad i = 1, \ldots, p, \quad M = A_1, \ldots, A_r, B_1, \ldots, B_s\,,$$

and undergo chemical interactions obeying the reversible law

$$A_1 + \cdots + A_r \rightleftharpoons B_1 + \cdots + B_s\,. \tag{1}$$

Moreover, multiple elastic collisions are admitted involving h particles, at most, h being the maximum between the number of reactants, r, and products, s, of the chemical reaction (1), i.e. $h = \max\{r, s\}$.

Kinetic equations. The microscopic description of the gas mixture is given by the kinetic equations in the unknown number densities N_i^M of particles of species M with velocity $\vec{v}_i^{\,M}$. Their matrix form is

$$\frac{\partial \mathbf{N}}{\partial t} + \mathcal{A}\mathbf{N} = \mathbf{Q}(\mathbf{N}) + \mathbf{R}(\mathbf{N})\,, \tag{2}$$

where the vector functions are defined by

$$\begin{aligned}
\mathbf{N} &= (N_1^{A_1}, \ldots, N_p^{A_1}, \ldots, N_1^{A_r}, \ldots, N_p^{A_r}, N_1^{B_1}, \ldots, N_p^{B_1}, \ldots, N_1^{B_s}, \ldots, N_p^{B_s})\\
\mathbf{Q} &= (Q_1^{A_1}, \ldots, Q_p^{A_1}, \ldots, Q_1^{A_r}, \ldots, Q_p^{A_r}, Q_1^{B_1}, \ldots, Q_p^{B_1}, \ldots, Q_1^{B_s}, \ldots, Q_p^{B_s})\\
\mathbf{R} &= (\mathcal{R}_1^{A_1}, \ldots, \mathcal{R}_p^{A_1}, \ldots, \mathcal{R}_1^{A_r}, \ldots, \mathcal{R}_p^{A_r}, \mathcal{R}_1^{B_1}, \ldots, \mathcal{R}_p^{B_1}, \ldots, \mathcal{R}_1^{B_s}, \ldots, \mathcal{R}_p^{B_s})
\end{aligned}$$

and $\mathcal{A}$ is the diagonal matrix of order $p(r+s)$ with elements $\mathcal{A}_{ii} = \vec{v}_i^{\,M} \cdot \overrightarrow{\text{grad}}$. The nonlinear collision terms Q_i^M, due to elastic scattering, and $\mathcal{R}_i^M$, due to inelastic interactions, have been already evaluated in explicit form [7]. On the other hand, the collision operator $\mathbf{Q}$ can be decomposed on the basis of multiple collisions in the more convenient form, recalling the meaning of h,

$$\mathbf{Q}(\mathbf{N}) = \mathbf{Q}_2(\mathbf{N},\mathbf{N}) + \mathbf{Q}_3(\mathbf{N},\mathbf{N},\mathbf{N}) + \ldots + \mathbf{Q}_h(\mathbf{N},\ldots,\mathbf{N})\,, \tag{3}$$

each term $\mathbf{Q}_j$, $j = 2, 3, \ldots, h$, being the contribution due to all j–encounters.

Introduce the space $\mathcal{F}$ of mechanical collision invariants, with dimension q, and the space $\mathcal{M}$ of mechanical-reactive collision invariants, with dimension $q-1$, and assume the orthonormal basis $\mathcal{B}_{\mathcal{F}} \subset \mathcal{F}$ and $\mathcal{B}_{\mathcal{M}} \subset \mathcal{M}$, defined by

$$\mathcal{B}_{\mathcal{M}} = \left\{\Upsilon^{(1)}, \ldots, \Upsilon^{(q-1)}\right\}, \ \mathcal{B}_{\mathcal{F}} = \left\{\Upsilon^{(1)}, \ldots, \Upsilon^{(q-1)}, \Upsilon^{(q)}\right\}, \ \mathcal{B}_{\mathcal{M}} \subset \mathcal{B}_{\mathcal{F}}\,. \tag{4}$$

Reminding the meaning of the collision invariants [7], one has

$$< \mathbf{R}(\mathbf{N})\,,\,\Upsilon^{(k)} >= 0,\ k = 1,\ldots,q-1 \ \text{ and } \ < \mathbf{R}(\mathbf{N})\,,\,\Upsilon^{(q)} >= r_f - r_b\,, \qquad (5)$$

where $r_f - r_b$ represents the reaction rate of the model, r_f and r_b being the forward and backward reaction rate, respectively, depending on both number densities of the corresponding reactants and reactive collision frequencies.

Remark. If the kinetic model admits only one velocity modulus for each M–species, then it can be proven that the dimension of $\mathcal{F}$ is $q = r + s + 2$. Conversely, if the model admits more than one velocity modulus for at least one of the M–species, then the dimension is $q = r + s + 3$. ■

Conservation laws. Projecting Eq.(2) onto the $q-1$ invariants of $\mathcal{B}_{\mathcal{M}}$, and recalling the first equality in Eq. (5), one obtains the $q-1$ conservation laws

$$\frac{\partial}{\partial t} < \mathbf{N}\,,\,\Upsilon^{(k)} > + < \mathcal{A}\mathbf{N}\,,\,\Upsilon^{(k)} > = \ 0\,, \quad \Upsilon^{(k)} \in \mathcal{B}_{\mathcal{M}}\,, \quad k = 1,\ldots,q-1\,. \ (6)$$

For each k, such equations express the conservation of the corresponding macroscopic variables a_k, defined by

$$a_k \ = < \mathbf{N}\,,\,\Upsilon^{(k)} >\,, \qquad k = 1,\ldots,q-1\,. \qquad (7)$$

Rate equation. Analogously, projecting Eq.(2) onto the mechanical invariant $\Upsilon^{(q)} \in \mathcal{B}_{\mathcal{F}} \setminus \mathcal{B}_{\mathcal{M}}$ and recalling the second equality in Eq. (5), the rate equation of the model is deduced in the form

$$\frac{\partial}{\partial t} < \mathbf{N}, \Upsilon^{(q)} > \ + \ < \mathcal{A}\mathbf{N}, \Upsilon^{(q)} > = \ r_f - r_b\,, \qquad (8)$$

which describes the evolution of the last macroscopic variable a_q defined by

$$a_q \ = < \mathbf{N}\,,\,\Upsilon^{(q)} > \ . \qquad (9)$$

The quantity a_q identifies the progress variable of the model and specifies the gas chemical composition.

Equilibrium. The chemical equilibrium condition is represented by the equality

$$r_f - r_b = 0\,,$$

which means that the forward and backward reactions proceed at equal rates. Moreover, the mechanical equilibrium of the gas corresponds to Maxwellian number densities, say $\hat{N}_i^M(a_1,\ldots,a_q)$, which are characterized by

$$\mathbf{N} \ = \ \hat{\mathbf{N}}(a_1,\ldots,a_q) \qquad \Longleftrightarrow \qquad \mathbf{Q}(\mathbf{N}) \ = \ 0\,. \qquad (10)$$

The governing equations of the model (6), (8) constitute a non closed system in $p(r+s)$ unknowns N_i^M which must be rewritten in a closed form.

3 Chapman–Enskog expansion

The dimensionless kinetic equations can be written in the form

$$\varepsilon\left[\frac{\partial \mathbf{N}}{\partial t} + \mathcal{A}\mathbf{N} - \mathbf{R}(\mathbf{N})\right] = \mathbf{Q}(\mathbf{N}), \tag{11}$$

where $\varepsilon = \ell/d$ is the Knudsen number, ℓ and d being the mean free path and a typical length, respectively. $\mathbf{N}$ is now the vector function of the dimensionless number densities and $\mathbf{Q}(\mathbf{N})$ obeys the decomposition (3). The presence of the reactive term at the left hand side of Eq. (11) agrees with the assumption that the effects due to convection and chemical reaction are considered small in comparison with the ones due to elastic scattering. In order to derive the reactive Navier-Stokes equations of the model, a first order Chapman-Enskog procedure is applied, assuming the following expansions for the number densities and time derivatives of the macroscopic variables

$$\begin{aligned}
&\mathbf{N} = \mathbf{N}^{(0)}(a_1,\ldots,a_q) + \varepsilon\,\mathbf{N}^{(1)}(a_1,\ldots,a_q,\frac{\partial a_1}{\partial x},\ldots,\frac{\partial a_q}{\partial x})\\
&\frac{\partial a_k}{\partial t} = \frac{\partial a_k}{\partial t}^{(0)}(a_1,\ldots,a_q,\frac{\partial a_1}{\partial x},\ldots,\frac{\partial a_q}{\partial x}) +\\
&\qquad +\varepsilon\frac{\partial a_k}{\partial t}^{(1)}(a_1,\ldots,a_q,\frac{\partial a_1}{\partial x},\ldots,\frac{\partial a_q}{\partial x}), \qquad k = 1,\ldots,q.
\end{aligned} \tag{12}$$

Inserting expansions (12) into Eq. (11), and neglecting terms with ε^2 yield

$$\begin{aligned}
&\varepsilon\left[\frac{\partial \mathbf{N}}{\partial t}^{(0)} + \mathcal{A}\mathbf{N}^{(0)} - \mathbf{R}(\mathbf{N}^{(0)})\right] = \\
&\qquad = \mathbf{Q}(\mathbf{N}^{(0)}) + 2\varepsilon\mathbf{Q}_2(\mathbf{N}^{(0)},\mathbf{N}^{(1)}) + \cdots + h\varepsilon\mathbf{Q}_h(\mathbf{N}^{(0)},\ldots,\mathbf{N}^{(0)},\mathbf{N}^{(1)})\,.
\end{aligned} \tag{13}$$

It may be noted that the right-hand side of Eq. (13) represents the linearized elastic collision operator about $\mathbf{N}^{(0)}$ and that it preserves the orthogonality property of $\mathbf{Q}$ with respect to the mechanical invariants $\Upsilon^{(k)} \in \mathcal{F}$, namely

$$<\Upsilon^{(k)},\,\mathbf{Q}(\mathbf{N}^{(0)})+\cdots+h\varepsilon\mathbf{Q}_h(\mathbf{N}^{(0)},\ldots,\mathbf{N}^{(0)},\mathbf{N}^{(1)})>= 0\,,\ k = 1,\ldots,q. \tag{14}$$

Equation (13) leads to the characterization of both zero and first order approximations, when the corresponding terms in ε are equated and the time derivative of $\mathbf{N}^{(0)}$ is evaluated recalling the dependence of $\mathbf{N}^{(0)}$ on the macroscopic variables. Accordingly, one obtains

$$\mathbf{Q}(\mathbf{N}^{(0)}) \;=\; 0\,; \tag{15}$$

$$2\mathbf{Q}_2(\mathbf{N}^{(0)},\mathbf{N}^{(1)}) + \cdots + h\mathbf{Q}_h(\mathbf{N}^{(0)},\ldots,\mathbf{N}^{(0)},\mathbf{N}^{(1)}) = \\ = \sum_{\ell=1}^{q} \frac{\partial \mathbf{N}^{(0)}}{\partial a_\ell}\frac{\partial a_\ell}{\partial t}^{(0)} + \mathcal{A}\mathbf{N}^{(0)} - \mathbf{R}(\mathbf{N}^{(0)}). \tag{16}$$

Constitutive conditions. The unique solution to Eq. (15), as assured by condition (10), is the function $\hat{\mathbf{N}}$ of the Maxwellian number densities, i.e.

$$\mathbf{N}^{(0)} \;=\; \hat{\mathbf{N}}\,(a_1,\ldots,a_q)\,.$$

On the other hand, Eq. (16) involves the time derivatives, at the zero-order, of variables $a_1,\ldots,a_q$. These derivatives can be evaluated projecting Eq. (16) on the orthonormal basis $\mathcal{B}_{\mathcal{F}}$ and recalling the orthogonality properties (14). The constitutive conditions then follow as

$$\frac{\partial a_k}{\partial t}^{(0)} = - < \mathcal{A}\mathbf{N}^{(0)}\,,\,\Upsilon^{(k)} > + < \mathbf{R}(\mathbf{N}^{(0)})\,,\,\Upsilon^{(k)} >\,,\;\; k=1,\ldots,q\,. \tag{17}$$

Zero-order approximation: reactive Euler equations. Having in mind that invariants $\Upsilon^{(k)}$ satisfy Eqs. (5), the reactive Euler equations of the model can be recovered straightforward from conditions (17) in the form

$$\begin{aligned} &\frac{\partial a_k}{\partial t}^{(0)} + < \mathcal{A}\mathbf{N}^{(0)}\,,\,\Upsilon^{(k)} > = \;0\,,\quad k=1,2,\ldots,q-1,\\ &\frac{\partial a_k}{\partial t}^{(q)} + < \mathcal{A}\mathbf{N}^{(0)}\,,\,\Upsilon^{(q)} > = \;r_f - r_b\,. \end{aligned} \tag{18}$$

First-order approximation: reactive Navier–Stokes equations. The vector function $\mathbf{N}^{(1)}$ is obtained from Eq. (16), taking into account the constitutive conditions (17). Accordingly, $\mathbf{N}^{(1)}$ still depends on the macroscopic variables a_k and their space derivatives only. Therefore, the first-order perturbed number density, say $\mathbf{N}^{①}$, can be expressed as

$$\mathbf{N}^{①} \;=\; \hat{\mathbf{N}}\,(a_1,\ldots,a_q) \;+\; \varepsilon\,\mathbf{N}^{(1)}\,(a_1,\ldots,a_q,\frac{\partial a_1}{\partial x},\ldots,\frac{\partial a_q}{\partial x}). \tag{19}$$

The Navier-Stokes equations can now be derived, writing the system of conservation and rate equations (6), (8) in closed form, when $\mathbf{N}$ is substituted by the perturbed density $\mathbf{N}^{①}$ as defined by Eq. (19). They formally read

$$\begin{aligned}
&\frac{\partial a_1}{\partial t} + < \mathcal{A}\,\mathbf{N}^{①}\,,\,\Upsilon^{(1)} > = 0 \\
&\dots \qquad \dots \qquad \dots \\
&\frac{\partial a_{q-1}}{\partial t} + < \mathcal{A}\,\mathbf{N}^{①}\,,\,\Upsilon^{(q-1)} > = 0 \qquad (20)\\
&\frac{\partial a_q}{\partial t} + < \mathcal{A}\mathbf{N}^{①}, \Upsilon^{(q)} > = r_f - r_b .
\end{aligned}$$

The formal structure of system (20) evidentiates the conservative character of the first $q-1$ equations and the evolution character of the last equation which specifies the chemical process.

Further extensions, in view of the computation of transport coefficients, are possible once system (20) has been specialized for a particular reversible reaction and velocity geometry.

Fluid dynamical processes with chemical reactions including effects of viscosity, heat conduction and diffusion, can then be studied within discrete kinetic models starting from the related reactive Navier–Stokes equations.

Acknowledgments

The present paper has been partially supported by Progetto Nazionale Cofin 2000-2001 (Coord. Prof. T. Ruggeri) and Minho University Mathematics Centre (CMAT – Programa de Financiamento Plurianual da FCT – Portugal).

References

1. M. Groppi, G. Spiga, *J. Math. Chem* **26**, 197 (2000).
2. F. Hanser, R. Monaco, F. Schurrer, *Transp. Theory Stat. Phys.* **30**, 537 (2001).
3. M. Pandolfi Bianchi, A. J. Soares, *Phys. Fluids* **8**, 3423 (1996).
4. J. Ross, P. Mazur *J. Chem. Phys.* **35**, 19 (1961).
5. R. Gatignol, *Théorie cinétique des gaz à répartition discrète des vitesses* (Springer–Verlag, Berlin, 1975).
6. R. Monaco, M. Pandolfi Bianchi, A. Rossani, *Math. Models and Methods in Appl. Sci.* **4**, 355 (1994).
7. M. Pandolfi Bianchi, A. J. Soares, *Continuum Mech. Thermodyn.* **12**, 53 (2000).

SLOW EIGENMODES OF THE SHALLOW-WATER EQUATIONS

FRANCESCO PAPARELLA
Dipartimento di Matematica E. de Giorgi
Università di Lecce
Via Arnesano - 73100 Lecce - Italy
E-mail: francesco.paparella@unile.it

We present a survey of recent work on the lowest end of the eigenmode spectrum of the shallow water equations in a rotating reference frame. The results are complemented with numerical simulations of the fully nonlinear equations. Having care of using physically correct, mass-conserving boundary conditions, a description in terms of slow eigenmodes seems to be a key component in the explanation of climate's decadal variability.

1 Introduction

The shallow-water equations are a conceptual model for a multitude of physical phenomena. In particular, in the field of geophysical fluid dynamics, their modeling power ranges from the description of near-shore surface waves, to the general circulation of the oceans. To study large-scale flows, it is convenient to use a Cartesian rotating reference frame attached to the Earth's surface. By convention, x spans the zonal direction (positive eastward), and y spans the meridional direction (positive northward). We consider the idealized basin sketched in fig.1. A shallow layer of incompressible fluid, having total depth h and density ρ_1, lies on top of an infinitely deep, incompressible layer of fluid having density $\rho_2 > \rho_1$. The interface depth is decomposed as $h = h_0 + h'$, where h_0 is the average value of h. In a rotating reference frame, the shallow-water equations describing the dynamics of the interface depth, are:

$$\begin{aligned} u_t + uu_x + vu_y - fv &= -g'h_x + \mathcal{D}^x + \mathcal{F}^x \\ v_t + uv_x + vv_y + fu &= -g'h_y + \mathcal{D}^y + \mathcal{F}^y \\ h_t + (hu)_x + (hv)_y &= 0 \end{aligned} \tag{1}$$

where t is time; u and v are, respectively, the zonal and meridional components of the velocity; $g' = g(1 - \rho_1/\rho_2)$ is the reduced gravity, where g is the acceleration of gravity; $f = 2\Omega\cos\theta$ is the Coriolis parameter, where Ω is the modulus of Earth's angular velocity and θ is the latitude; for our purposes it will be sufficient to express the Coriolis parameter as $f = f_0 + \beta y$, where β is a constant which depends on the latitude of origin of the reference frame. The

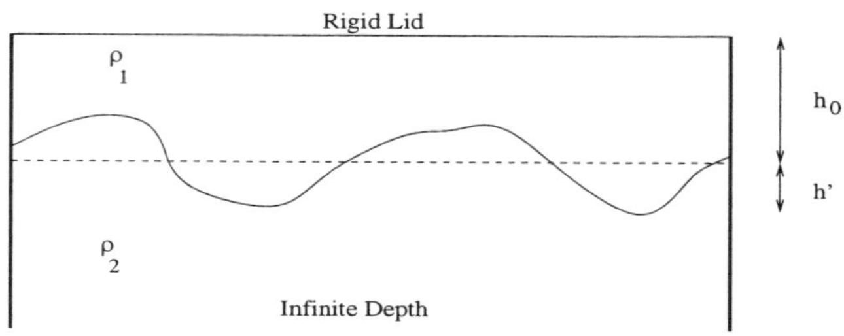

Figure 1. Side view of an idealized basin with a shallow fluid layer of depth $h = h_0 + h'$ and density ρ_1, lying on a motionless, infinitely deep, fluid layer of density ρ_2.

dissipative terms $\mathcal{D}^x$ and $\mathcal{D}^y$ are suitable functions of, or differential operators acting on, the velocity. The forcing terms $\mathcal{F}^x$, $\mathcal{F}^y$ are prescribed functions of time and space. For the moment, we shall enforce no-normal flow conditions at the rigid boundaries located at $x = 0, L$ and $y = 0, L$.

Equations (1) are a simple model for ocean dynamics, which still retains an extremely rich phenomenology. Among other applications, it has successfully been used in climate studies, such as Cane and Zebiak's famous first predictive model of El Niño[1]. One of the reasons for this success, is the large variety of waves which can be derived as special cases of the linearized version of eq. (1). A remarkable class of waves, known as Rossby waves (or "planetary" waves) arises when one considers perturbations to the so-called *geostrophic* flows, defined as the steady flows which satisfies

$$\begin{aligned} -f_0 v &= -g' h_x \\ f_0 u &= -g' h_y \end{aligned} \tag{2}$$

for a given $h(x, y)$. Rossby waves are peculiar: at wavelengths shorter than a critical value, their propagation is always westward, while at longer wavelengths their propagation is always eastward. Their dynamics is much slower than that of any other family of shallow water waves, and, in the long wavelength limit, the group velocity reaches the constant value of $-\beta g h_0 / f_0$.

2 The importance of the boundary conditions

Slow, long waves have a special appeal, because they have the potential for modifying the sea temperature patterns with magnitudes, and on time scales, which are of climatological interest. This leads to the question whether, and in which form, they contribute to the dynamics of the oceans. Such a complex question, which involves many different disciplines, for the applied mathematician boils down to two distinct problems. Since we know that wave-like solutions exist, we would like to know how quickly they are damped (or, conversely, if a realistic intensity of the forcing is enough to maintain sustained oscillations). The second point is to investigate the nature of the oscillations once they have reached an intensity large enough to excite the nonlinear terms of the governing equations.

For some time those questions have been considered somewhat settled by appealing to the behavior of a simpler version of the shallow water equation, called the *quasi-geostrophic* approximation. The same perturbative approach which reveals the nature of the Rossby waves, leads also, at the first order, to a self-contained equation for the fluctuations h' of the interface height, acting as streamfunction of an incompressible two-dimensional fluid. This approximation effectively filters out all kinds of fast waves, retaining only the mechanism leading to Rossby waves. The resulting quasi-geostrophic equation has been much studied because it lends itself to analytical manipulation. In particular, for rectangular domains, the eigenfunctions of its linearization can be expressed as simple combinations of trigonometric functions[2]. Almost all the studies employing the quasi-geostrophic equations have, unfortunately, imposed the no-normal flow boundary conditions by assuming $h' = 0$ at the boundaries of the domain Ω. While this yields a mathematically well-posed problem, it is physically unacceptable, because it violates the principle of mass conservation. The physically correct, mass-conserving boundary condition is expressed through the following integral constraint:

$$\frac{d}{dt} \int_{\Omega} h'(x,y)dxdy = 0. \tag{3}$$

A note about this quirk has long been available in the literature[3], however practitioners in the field have quietly dismissed the issue by assuming, without too an extensive checking, that both boundary conditions would give similar results[4]. While this is true for some aspects of quasi-geostrophic dynamics (such as the theory of western boundary currents), the same assumption gives misleading results when applied to the study of the eigenfunction spectrum of the linearized quasi-geostrophic equations. In fact, the boundary condition $h' = 0$ gives supercritically damped eigenfunctions, and this fact has generated the widespread belief that, even at very low frequencies, basin-scale, mid-latitude, planetary waves dissipates far too quickly to have any climatological impact.

Very recently this belief has been challenged[5]. It has been shown that, by employing the physically correct boundary conditions, the low frequencies eigenmodes of the linearized quasi-geostrophic equations are so slightly damped that they can easily resonate upon stochastic forcing[6]. The intuitive analogy is that between a large metallic shield (a *gong*) which is clamped at its rim, and a similar object having free edges. The latter will oscillate for a much longer time than the second! Thus, the question of the climatic relevance of waves at the lowest end of the frequency spectrum is again wide open.

3 Slow eigenmodes of the shallow water equations in enclosed basins

The next logical step is that to investigate the linearized shallow water equations with semi-realistic forcing and dissipation. For simplicity, we shall consider rectangular basins, and use a linear Eckman damping, in the form $\mathcal{D}^x = -ru$ and $\mathcal{D}^y = -ry$, with constant r. We assume that the forcing has no meridional component: $\mathcal{F}^y = 0$. The zonal component of the forcing is set to be: $\mathcal{F}^x = \alpha \cos(\pi y/L) < h' > /h_0$, where α is a coupling constant, L is the basin height, and the angular brackets denote the average over a restricted region of the domain, adjacent to the western boundary. The rationale behind this choice becomes more clear by recalling that temperature fluctuations at the surface of the ocean, which feedback on the wind forcing, are proportional to h' (for details see Cessi and Paparella[7]). Assuming $[u, v, h'](x, y, t) = e^{-i\omega t}[\hat{u}, \hat{v}, \hat{h}](x, y)$, the linear problem is written as the eigensystem:

$$\begin{aligned} -i\omega\hat{u} - f\hat{v} &= -g'\hat{h}_x - r\hat{u} + \alpha\cos(\pi y/L) < \hat{h} > /h_0 \\ -i\omega\hat{v} + f\hat{u} &= -g'\hat{h}_y - r\hat{v} \\ -i\omega\hat{h} &= -h_0\hat{u}_x - h_0\hat{v}_y \end{aligned} \tag{4}$$

Figure 2 (left panel) shows the slowest, least damped portion of the eigenfunction spectrum of eq. 4, for five different values of the coupling constant α. It is evident the bifurcation which leads to sustained (growing) oscillations of the lowest eigenmode.

The nonlinear equilibration is observed by means of numerical simulation of the fully nonlinear equations (1). Our second-order numerical code staggers the variables on an Arakawa C-grid with an Asselin-corrected leap-frog time integrator. A slightly perturbed initial condition is evolved for 500 years. Growth rate and period (4.4 years) agree with the linear calculations. In the equilibrated regime the observed period is 8.8 years, and the alternance of high and low maxima suggests the presence of a period doubling bifurcation. This

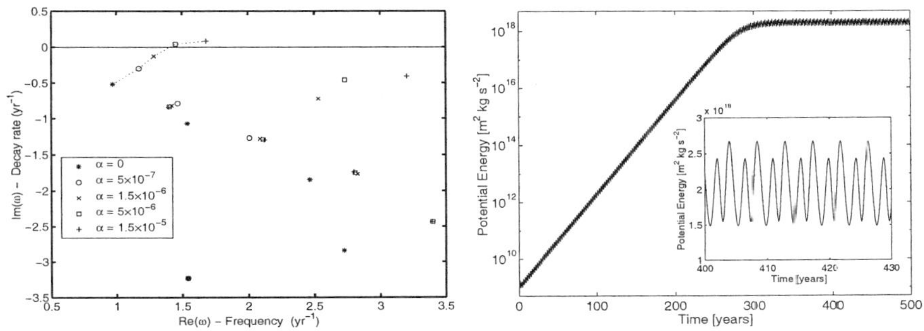

Figure 2. Left: lowest, least damped portion of the eigenspectrum of eq. (4) for five different strengths of the ocean-atmosphere coupling. Right: available potential energy vs. time for $\alpha = 1.5 \cdot 10^{-6}$; the inset reveals period doubling. All parameters are chosen to be representative of the North Atlantic.

is illustrated in fig. 2 (right panel). Note that, being the available potential energy a quadratic quantity, it oscillates at half the period.

4 Slow eigenmodes in periodic topologies

All the results discussed above apply to closed basins only. It is well known that with open, or periodic, topologies, the shallow water equations (and their quasi-geostrophic relative) prefer jet-like solutions[8]. Such is the nature, for example, of the Stratospheric Jet-Stream, and of the Antarctic Circumpolar Current.

Seen on a global scale, the world's oceans girdle the Earth. It is then legitimate the question of how robust are our results upon a change of topology. A complete answer to this question is well beyond the scope of the present work, and shall be presented elsewhere. As a preliminary investigation, we set-up our numerical code with a complex coastline geometry, vaguely reminiscent of the actual land-mass distribution. A periodic channel is open on the southernmost end of the simulation domain. On the northern and southern edges of the domain, and on all the coastlines, the boundary conditions are of no-normal flow. No forcing is imposed, and dissipation is the same as in the previous section, so any initial condition freely decays to the asymptotic state of quiet.

We initialize the code by setting up a height anomaly of Gaussian shape in the middle of the "southern pacific". Figure 3 shows the variable h' after

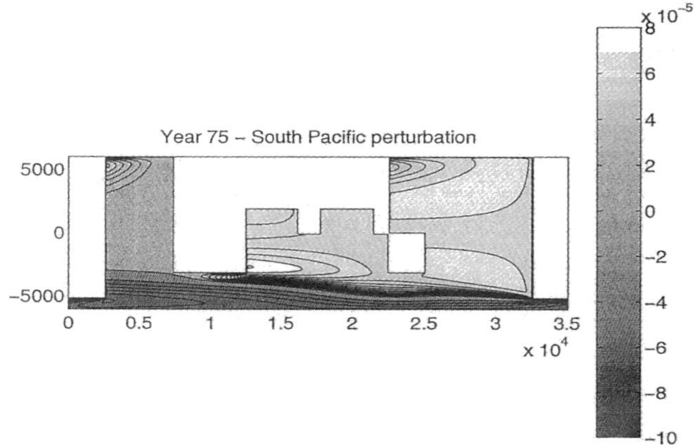

Figure 3. Interface height h' in a freely-decaying simulation with realistic topology. Lengths are expressed in kilometers.

75 years. The most notable feature, due to the periodic topology of the basin, is the eastward-going, jet-like current at the southern end of the domain, very reminiscent of the Antarctic Circumpolar Current. Yet in the northern hemisphere, westward-propagating, wave-like activity is still very evident, with qualitative features highly similar to those of the slowest linear eigenmode of the shallow water equations in enclosed basins.

Acknowledgments

The author wishes to thank the organizers of the XI International Conference on Waves and Stability in Continuous Media, and prof. Claudio Tebaldi for continuous encouragement and support.

References

1. M.A. Cane *et al*, *Nature* **321**, 827 (1986)
2. J. Pedlosky, *Geophysical Fluid Dynamics*, (Springer-Verlag, New York, 1987).
3. J.C. McWilliams, *Dyn. Atmos. Oceans* **1**, 427 (1977).
4. G.R. Flierl, *Dyn. Atmos. Oceans* **1**, 443 (1977).
5. P. Cessi, F. Primeau *J. Phys. Oceanogr.* **31**, 127 (2001).
6. P. Cessi, S. Louazel *J. Phys. Oceanogr.* , in press (2001).
7. P. Cessi, F. Paparella *Geophys. Res. Lett.* **28**, 2437 (2001).
8. A.E. Gill, *Atmosphere-Ocean Dynamics*, (Academic Pr, San Diego, 1982).

A COMPARISON BETWEEN RELATIVISTIC EXTENDED THERMODYNAMICS WITH 14 FIELDS AND THAT WITH 30 FIELDS

S. PENNISI

Dip. Matematica, Via Ospedale 72, Cagliari. E-mail: spennisi@unica.it

A further extension of Relativistic Extended Thermodynamics is here investigated by considering 30 independent variables, instead of 14. The two models are compared by seeing their implications on a well known iterative procedure.

1 Introduction.

Relativistic Extended Thermodynamics, as exposed in the pioneering work [1] by Liu, Müller and Ruggeri, deals with 14 independent variables and with the first 3 equations of the following system

$$\partial_\alpha V^\alpha = 0\,,\ \partial_\alpha T^{\alpha\beta} = 0\,,\ \partial_\alpha A^{\alpha\beta\gamma} = I^{\beta\gamma}\,,\ \partial_\alpha A^{\alpha\beta\gamma\delta} = I^{\beta\gamma\delta}\,, \tag{1}$$

The further extension presented in this work is to consider also the independent components of $A^{\alpha\beta\gamma}$ as independent variables, and to add the new last equation of the system (1). In this way, we have 30 independent equations in 30 independent variables. Moreover, we know that $I^\alpha_\alpha = 0$, $I^{\beta\gamma}_\gamma = 0$. The entropy principle, via the Liu Theorem [2], says that a 4-vector h'^α and the Lagrange Multipliers ξ , λ_β , $\Sigma_{\beta\gamma}$, $\Sigma_{\beta\gamma\delta}$ exist, such that $\Sigma^\alpha_\alpha = 0$, $\Sigma^{\beta\gamma}_\gamma = 0$ and

$$V^\alpha = \frac{\partial h'^\alpha}{\partial \xi}\,,\ T^{\alpha\beta} = \frac{\partial h'^\alpha}{\partial \lambda_\beta}\,,\ \left(A^{\alpha\beta\gamma} - \frac{\partial h'^\alpha}{\partial \Sigma_{\mu\nu}}\right)\left(g^\beta_\mu g^\gamma_\nu - \frac{1}{4} g_{\mu\nu} g^{\beta\gamma}\right) = 0\,, \tag{2}$$

$$\left(A^{\alpha\beta\gamma\delta} - \frac{\partial h'^\alpha}{\partial \Sigma_{\mu\nu\lambda}}\right)\left(g^\beta_\mu g^\gamma_\nu g^\delta_\lambda - \frac{1}{2} g_{\mu\nu} g^{(\delta}_\lambda g^{\beta\gamma)}\right) = 0\,,\ \Sigma_{\beta\gamma} I^{\beta\gamma} + \Sigma_{\beta\gamma\delta} I^{\beta\gamma\delta} \geq 0\,,$$

where the Lagrange multipliers have been taken as independent variables. The exploitation of eqs. $(2)_{1-4}$ is here omitted, for the sake of brevity, and will be reported in a subsequent work [3]. Here, in the next section, we will investigate the residual inequality $(2)_5$. Now the equations of ordinary thermodynamics are usually derived from those of extended thermodynamics by an iterative procedure that is akin to the Maxwellian iteration used in the kinetic theory of gases. In sect. 4 this methodology will be applied to this 30 fields theory, showing that, if some arbitrary functions are chosen to be zero, the results coincide exactly with those obtained starting with the 14 fields one. A similar methodology will be applied in sect. 3 to recover the constitutive functions appearing in the 14 fields approach from the results in the 30 fields one.

Therefore, the 2 approaches confirm one another. I conclude this section reporting the equilibrium values of V^α, $T^{\alpha\beta}$, $A^{\alpha\beta\gamma}$, $A^{\alpha\beta\gamma\delta}$, i.e.,

$$V^\alpha = nU^\alpha \quad , \quad T^{\alpha\beta}_{eq.} = ph^{\alpha\beta} + eU^\alpha U^\beta \quad , \quad (3)$$

$$A^{\alpha\beta\gamma}_{eq.} = 4\pi y m^5 \left(J_{4,1} h^{(\alpha\beta} U^{\gamma)} + J_{2,3} U^\alpha U^\beta U^\gamma \right) \quad ,$$

$$A^{\alpha\beta\gamma\delta}_{eq.} = 4\pi y m^6 \left(2J_{4,2} h^{(\alpha\beta} U^\gamma U^{\delta)} + J_{2,4} U^\alpha U^\beta U^\gamma U^\delta + \frac{1}{5} J_{6,0} h^{(\alpha\beta} h^{\gamma\delta)} \right) \quad ,$$

$$\text{with} \quad n = 4\pi y m^3 J_{2,1}(\alpha,\gamma) \, , \, p = \frac{4}{3}\pi y m^4 J_{4,0}(\alpha,\gamma) \, , \, (4)$$

$$e = 4\pi y m^4 J_{2,2}(\alpha,\gamma) \, , \, s = \alpha + \frac{k\gamma}{m}\frac{e+p}{n} \, , \, \gamma = \frac{m}{kT} \, , \, h^{\alpha\beta} = g^{\alpha\beta} + U^\alpha U^\beta \, ,$$

$$J_{m,n}(\alpha,\gamma) = \int_0^\infty \frac{\sinh^m \rho \cosh^n \rho}{\exp\left(\frac{\alpha}{k} + \gamma \cosh\rho\right) \pm 1} d\rho \, , \, I_{m,n}(\alpha,\gamma) = \frac{\partial}{\partial\alpha} J_{m,n}(\alpha,\gamma) \, .$$

2 The residual inequality.

We have now to write the expressions of the productions terms $I^{\beta\gamma}$, $I^{\beta\gamma\delta}$; with the methods of the theory of representations theorems [4], we obtain

$$I^{\beta\gamma} = B_1(h^{\beta\gamma} + 3U^\beta U^\gamma) U^\mu U^\nu \Sigma_{\mu\nu} + 2B_2 U^{(\beta} h^{\gamma)\mu} \Sigma_{\mu\nu} U^\nu + \quad (5)$$
$$B_3(h^{\beta\mu}h^{\nu\gamma} - \tfrac{1}{3}h^{\beta\gamma}h^{\mu\nu})\Sigma_{\mu\nu} + C_1(h^{\beta\gamma} + 3U^\beta U^\gamma)U^\mu U^\nu U^\phi \Sigma_{\mu\nu\phi} +$$
$$2C_2 U^{(\beta} h^{\gamma)\mu} \Sigma_{\mu\nu\phi} U^\nu U^\phi + C_3(h^{\beta\mu}h^{\nu\gamma} - \tfrac{1}{3}h^{\beta\gamma}h^{\mu\nu})\Sigma_{\mu\nu\phi}U^\phi \, ,$$
$$I^{\beta\gamma\delta} = 3A_1(h^{(\beta\gamma}U^{\delta)} + U^\beta U^\gamma U^\delta)U^\mu U^\nu \Sigma_{\mu\nu} + 3A_2(h^{(\beta\gamma}h^{\delta)\mu} +$$
$$5U^{(\beta}U^\gamma h^{\delta)\mu})\Sigma_{\mu\nu}U^\nu + A_3(3h^{(\beta}_\mu h^\gamma_\nu U^{\delta)} - h^{(\beta\gamma}U^{\delta)}h_{\mu\nu})\Sigma^{\mu\nu} + 3E_1(h^{(\beta\gamma}U^{\delta)} +$$
$$U^\beta U^\gamma U^\delta)U^\mu U^\nu U^\phi \Sigma_{\mu\nu\phi} + 3E_2(h^{(\beta\gamma}h^{\delta)\mu} + 5U^{(\beta}U^\gamma h^{\delta)\mu})\Sigma_{\mu\nu\phi}U^\nu U^\phi +$$
$$E_3(3h^{(\beta}_\mu h^\gamma_\nu U^{\delta)} - h^{(\beta\gamma}U^{\delta)}h_{\mu\nu})\Sigma^{\mu\nu\phi}U_\phi + E_4(5h^\beta_\mu h^\gamma_\nu h^\delta_\phi - 3h^{(\beta\gamma}h^{\delta)}_\mu h_{\nu\phi})\Sigma^{\mu\nu\phi} \, .$$

I have chosen to write these representations in terms of $\Sigma_{\beta\gamma}$ and $\Sigma_{\beta\gamma\delta}$ in order to abbreviate the following exploitation of the residual inequality $(2)_5$. In any case, by using the expressions of $\Sigma_{\beta\gamma}$ and $\Sigma_{\beta\gamma\delta}$ in the previous section, it becomes expressed in terms of the old variables. We have to impose also that $I_{\beta\gamma}$, $I_{\beta\gamma\delta}$ are invertible functions of $\Sigma_{\beta\gamma}$ and $\Sigma_{\beta\gamma\delta}$; this fact, together with $(2)_5$, is equivalent to say that the quadratic form $\Sigma_{\beta\gamma}I^{\beta\gamma} + \Sigma_{\beta\gamma\delta}I^{\beta\gamma\delta}$ is positive-definite. We can now see that $\Sigma^{\alpha\beta} = \sigma^{<\alpha\beta>} + \sigma(h^{\alpha\beta} + 3U^\alpha U^\beta) + 2U^{(\alpha}\sigma^{\beta)}$,
$\Sigma^{\alpha\beta\gamma} = S(U^\alpha U^\beta U^\gamma + U^{(\alpha}h^{\beta\gamma)}) + 3S^{(\alpha}(U^\beta U^{\gamma)} + \frac{1}{5}h^{\beta\gamma)}) + 3S^{(<\alpha\beta>}U^{\gamma)} + S^{<\alpha\beta\gamma>}$,
with $\sigma = \frac{1}{3}\Sigma^{\alpha\beta}h_{\alpha\beta}$, $\sigma^\mu = -\Sigma^{\alpha\beta}h^\mu_\alpha U_\beta$, $\sigma^{<\mu\nu>} = \Sigma^{\alpha\beta}(h^\mu_\alpha h^\nu_\beta - \frac{1}{3}h^{\mu\nu}h_{\alpha\beta})$,
$S = -\Sigma^{\alpha\beta\gamma}U_\alpha U_\beta U_\gamma$, $S^\mu = \Sigma^{\alpha\beta\gamma}U_\alpha U_\beta h^\mu_\gamma$, $S^{<\mu\nu>} = -\Sigma^{\alpha\beta\gamma}(h^\mu_\alpha h^\nu_\beta - \frac{1}{3}h^{\mu\nu}h_{\alpha\beta})U_\gamma$,
$S^{<\mu\nu\delta>} = \Sigma^{\alpha\beta\gamma}(h^\mu_\alpha h^\nu_\beta h^\delta_\gamma - \frac{3}{5}h^{(\mu\nu}h^{\delta)}_\alpha h_{\beta\gamma})$.
By using this decompositions, the above mentioned quadratic form becomes

$$36B_1\sigma^2 - 6(2C_1 + 3A_1)\sigma S + 6E_1 S^2 + 2B_2\sigma_\mu\sigma^\mu - 2(C_2 + 9A_2)\sigma_\mu S^\mu + 18E_2 S_\mu S^\mu +$$
$$B_3\sigma^{<\mu\nu>}\sigma_{<\mu\nu>} - (C_3 + 3A_3)\sigma^{<\mu\nu>}S_{<\mu\nu>} + 3E_3 S^{<\mu\nu>}S_{<\mu\nu>} + 5E_4 S^{<\mu\nu\phi>}S_{<\mu\nu\phi>} \, ,$$

which is positive-definite iff $B_1 > 0\,,\ B_2 > 0\,,\ B_3 > 0\,,\ E_4 > 0\,,$ (6)

$$\begin{vmatrix} 12B_1 & -2C_1 - 3A_1 \\ -2C_1 - 3A_1 & 2E_1 \end{vmatrix} > 0\,, \begin{vmatrix} 2B_2 & C_2 + 9A_2 \\ C_2 + 9A_2 & 18E_2 \end{vmatrix} > 0\,, \begin{vmatrix} 2B_3 & -C_3 - 3A_3 \\ -C_3 - 3A_3 & 6E_3 \end{vmatrix} > 0\,,$$

from which it follows also $E_1 > 0\,,\ E_2 > 0\,,\ E_3 > 0\,.$ (7)

3 Transition to the Thermodynamics with 14 independent fields.

The equations of ordinary thermodynamics are usually derived from those of extended thermodynamics by an iterative procedure that is akin to the Maxwellian iteration used in the kinetic theory of gases. This will be performed in the next section. Here I want to propose a similar approach to derive the equations found by Liu, Müller and Ruggeri in the theory with 14 independent fields. The iterative scheme proceeds as follows: Let us substitute in the left-hand side of eq. $(1)_4$ the expression of $A^{\alpha\beta\gamma}$, which has been obtained in the theory with 14 independent fields, and let us find the first iterates from the right-hand side. It will depend on the fields and on their derivatives with respect to x_α; these last ones have to be dropped, because they have not been considered in the theory with 14 independent fields. This is equivalent to find the first iterate from $I^{\beta\gamma\delta} = 0$ or, by using eq $(5)_2$, from $U^\mu U^\nu U^\phi \Sigma_{\mu\nu\phi} = -\frac{A_1}{E_1} U^\mu U^\nu \Sigma_{\mu\nu}\,,\ h^{\delta\mu}\Sigma_{\mu\nu\phi}U^\nu U^\phi = -\frac{A_2}{E_2} h^{\delta\mu}\Sigma_{\mu\nu}U^\nu\,,$
$(h^\beta_\mu h^\gamma_\nu - \frac{1}{3}h^{\beta\gamma}h_{\mu\nu})\Sigma^{\mu\nu\phi}U_\phi = -\frac{A_3}{E_3}(h^\beta_\mu h^\gamma_\nu - \frac{1}{3}h^{\beta\gamma}h_{\mu\nu})\Sigma^{\mu\nu}\,,$
$(5h^\beta_\mu h^\gamma_\nu h^\delta_\phi - 3h^{(\beta\gamma}h^{\delta)}_\mu h_{\nu\phi})\Sigma^{\mu\nu\phi} = 0\,.$ But, in ref. [3] it is observed that the closure for $A^{\alpha\beta\gamma}$ in the theory of 14 independent fields, is a consequent of the present results if and only if $\Sigma^{\alpha\beta\gamma} = 0$; consequently our aim in this iterative procedure is achieved if and only if $A_1 = 0\,,\ A_2 = 0\,,\ A_3 = 0$. We note also that these values for A_i are compatible with the inequalities (6). After that, we see that the expression $(5)_1$ for $I^{\beta\gamma}$ calculated in $\Sigma^{\mu\nu\phi} = 0$ becomes the same of the theory of 14 independent fields. Let us now perform the most significative iterative procedure, leading to ordinary thermodynamics, and see that the result coincides with the corresponding one in the theory of 14 independent fields.

4 Transition to Ordinary Thermodynamics.

It is here convenient to rewrite the representations (5) in terms of the original variables; by defining $\pi = \frac{1}{3}(T^{\alpha\beta} - T^{\alpha\beta}_{eq.})h_{\alpha\beta}\,,\ q^\mu = -(T^{\alpha\beta} - T^{\alpha\beta}_{eq.})h^\mu_\alpha U_\beta\,,\ t^{<\mu\nu>} = (T^{\alpha\beta} - T^{\alpha\beta}_{eq.})(h^\mu_\alpha h^\nu_\beta - \frac{1}{3}h^{\mu\nu}h_{\alpha\beta})\,,$
$a^\mu = (A^{\alpha\beta\gamma} - A^{\alpha\beta\gamma}_{eq.})U_\alpha U_\beta h^\mu_\gamma\,,\ a^{<\mu\nu>} = -(A^{\alpha\beta\gamma} - A^{\alpha\beta\gamma}_{eq.})(h^\mu_\alpha h^\nu_\beta - \frac{1}{3}h^{\mu\nu}h_{\alpha\beta})U_\gamma\,,$
$a^{<\mu\nu\delta>} = (A^{\alpha\beta\gamma} - A^{\alpha\beta\gamma}_{eq.})(h^\mu_\alpha h^\nu_\beta h^\delta_\gamma - \frac{3}{5}h^{(\mu\nu}h^{\delta)}_\alpha h_{\beta\gamma})\,,\ a = -(A^{\alpha\beta\gamma} - A^{\alpha\beta\gamma}_{eq.})U_\alpha U_\beta U_\gamma\,,$

they become $I^{\beta\gamma} = \tilde{B}_1(h^{\beta\gamma} + 3U^\beta U^\gamma)\pi + 2\tilde{B}_2 U^{(\beta}q^{\gamma)} + \tilde{B}_3 t^{<\beta\gamma>} +$ (8)

$\tilde{C}_1(h^{\beta\gamma} + 3U^\beta U^\gamma)a + 2\tilde{C}_2 U^{(\beta}a^{\gamma)} + \tilde{C}_3 a^{<\beta\gamma>}\,,$
$I^{\beta\gamma\delta}m^{-1} = 3\tilde{A}_1(h^{(\beta\gamma}U^{\delta)} + U^\beta U^\gamma U^\delta)\pi + 3\tilde{A}_2(h^{(\beta\gamma}q^{\delta)} + 5U^{(\beta}U^\gamma q^{\delta)})+$
$3\tilde{A}_3 t^{(<\beta\gamma>}U^{\delta)} + 3\tilde{E}_1(h^{(\beta\gamma}U^{\delta)} + U^\beta U^\gamma U^\delta)a+$
$3\tilde{E}_2(h^{(\beta\gamma}a^{\delta)} + 5U^{(\beta}U^\gamma a^{\delta)}) + 3\tilde{E}_3 a^{(<\beta\gamma>}U^{\delta)} + \tilde{E}_4 a^{<\beta\gamma\delta>}\,.$

The expressions of the coefficients in terms of the previous one are reported in the

Appendix. Now we can perform the iterative procedure. Let us substitute in the left-hand side of eqs. (1) the equilibrium values of the variables, and let us find the first iterates of $T^{\alpha\beta} - T^{\alpha\beta}_{eq.}$ and $A^{\alpha\beta\gamma} - A^{\alpha\beta\gamma}$ from the right-hand side. In other words, let us consider the system $\partial_\theta(nU^\theta) = 0$,
$\partial_\theta[pg^{\theta\beta} + (e+p)U^\theta U^\beta] = 0$, $4\pi y m^5 \partial_\theta(J_{4,1}h^{(\theta\beta}U^{\gamma)} + J_{2,3}U^\theta U^\beta U^\gamma) = I^{\beta\gamma}$,
$4\pi y m^6 \partial_\theta(2J_{4,2}h^{(\theta\beta}U^\gamma U^{\delta)} + J_{2,4}U^\theta U^\beta U^\gamma U^\delta + \frac{1}{5}J_{6,0}h^{(\theta\beta}h^{\gamma\delta)}) = I^{\beta\gamma\delta}$.
By using the representations (8) and denoting $\partial_\theta\chi$ with $\chi_{,\theta}$, it becomes

$$I_{2,1}\frac{1}{k}U^\theta\alpha_{,\theta} + I_{2,2}U^\theta\gamma_{,\theta} = I_{4,0}\frac{\gamma}{3}U^\theta_{,\theta}\,,\; I_{2,2}\frac{1}{k}U^\theta\alpha_{,\theta} + I_{2,3}U^\theta\gamma_{,\theta} = I_{4,1}\frac{\gamma}{3}U^\theta_{,\theta}\,, \tag{9}$$

$$h^\beta_\mu U^\theta U^\mu_{,\theta} = \frac{I_{4,0}}{k\gamma I_{4,1}}h^{\beta\theta}\alpha_{,\theta} + \frac{1}{\gamma}h^{\beta\theta}\gamma_{,\theta}\,,\; I_{2,3}\tfrac{1}{k}U^\theta\alpha_{,\theta} + I_{2,4}U^\theta\gamma_{,\theta} - I_{4,2}\tfrac{\gamma}{3}U^\theta_{,\theta} = \frac{3(\tilde{B}_1\overset{1}{\pi} + \tilde{C}_1\overset{1}{a})}{k4\pi y m^5},$$

$$I_{4,1}h^{\delta\theta}\alpha_{,\theta} + kI_{4,2}h^{\delta\theta}\gamma_{,\theta} - kI_{4,2}\gamma U^\theta h^\delta_\mu U^\mu_{,\theta} = \frac{3(\tilde{B}_2\overset{1}{q}{}^{\delta} + \tilde{C}_2\overset{1}{a}{}^{\delta})}{4\pi y m^5}\,,$$

$$I_{6,0}(h^{\theta(\delta}h^{\psi)\phi} - \tfrac{1}{3}h^{\theta\phi}h^{\delta\psi})U_{\phi,\theta} = \frac{-15(\tilde{B}_3\overset{1}{t}{}^{<\delta\psi>} + \tilde{C}_3\overset{1}{a}{}^{<\delta\psi>})}{8\pi y m^5 k\gamma}\,,$$

$$I_{2,4}\tfrac{1}{k}U^\theta\alpha_{,\theta} + I_{2,5}U^\theta\gamma_{,\theta} - I_{4,3}\tfrac{\gamma}{3}U^\theta_{,\theta} = \frac{3(\tilde{A}_1\overset{1}{\pi} + \tilde{E}_1\overset{1}{a})}{k4\pi y m^5}\,,$$

$$I_{4,2}h^{\delta\theta}\alpha_{,\theta} + kI_{4,3}h^{\delta\theta}\gamma_{,\theta} - kI_{4,3}\gamma U^\theta h^\delta_\mu U^\mu_{,\theta} = \frac{15(\tilde{A}_2\overset{1}{q}{}^{\delta} + \tilde{E}_2\overset{1}{a}{}^{\delta})}{4\pi y m^5}\,,$$

$$I_{6,1}(h^{\theta(\delta}h^{\psi)\phi} - \tfrac{1}{3}h^{\theta\phi}h^{\delta\psi})U_{\phi,\theta} = \frac{-15(\tilde{A}_3\overset{1}{t}{}^{<\delta\psi>} + \tilde{E}_3\overset{1}{a}{}^{<\delta\psi>})}{8\pi y m^5 k\gamma}\,, \quad \overset{1}{a}{}^{<\beta\gamma\delta>} = 0\,.$$

We see that the above eqs. $(9)_{1,2,4,7}$ can be written in the form

$$\begin{pmatrix} I_{2,1} & I_{2,2} & 0 & 0 \\ I_{2,2} & I_{2,3} & 0 & 0 \\ I_{2,3} & I_{2,4} & \tilde{B}_1 & \tilde{C}_1 \\ I_{2,4} & I_{2,5} & \tilde{A}_1 & \tilde{E}_1 \end{pmatrix} \begin{pmatrix} \frac{1}{k}U^\theta\alpha_{,\theta} \\ U^\theta\gamma_{,\theta} \\ \frac{-3\overset{1}{\pi}}{k4\pi y m^5} \\ \frac{-3\overset{1}{a}}{k4\pi y m^5} \end{pmatrix} = \begin{pmatrix} I_{4,0} \\ I_{4,1} \\ I_{4,2} \\ I_{4,3} \end{pmatrix} \tfrac{\gamma}{3}U^\theta_{,\theta}\,,$$

a system in the unknowns $U^\theta\alpha_{,\theta}$, $U^\theta\gamma_{,\theta}$, $\overset{1}{\pi}$, $\overset{1}{a}$. In particular $\overset{1}{\pi} = \nu U^\theta_{,\theta}$,
is one of its consequences, with ν (bulk viscosity) given by

$$\nu = \frac{4\pi y m^6}{9T}\left(\tilde{E}_1\begin{vmatrix} I_{2,0} & I_{2,1} & I_{2,2} \\ I_{2,1} & I_{2,2} & I_{2,3} \\ I_{2,2} & I_{2,3} & I_{2,4}\end{vmatrix} - \tilde{C}_1\begin{pmatrix} I_{2,0} & I_{2,1} & I_{2,2} \\ I_{2,1} & I_{2,2} & I_{2,3} \\ I_{2,3} & I_{2,4} & I_{2,5}\end{pmatrix}\right)\begin{vmatrix} I_{2,1} & I_{2,2} \\ I_{2,2} & I_{2,3}\end{vmatrix}^{-1}\begin{vmatrix} \tilde{B}_1 & \tilde{C}_1 \\ \tilde{A}_1 & \tilde{E}_1\end{vmatrix}^{-1}.$$

We see that this expression coincides with the corresponding one in the theory of 14 independent fields if $\tilde{C}_1 = 0$. Similarly, the above eqs. $(9)_{3,5,8}$ can be written as

$$\begin{pmatrix} I_{4,0} & 0 & 0 \\ I_{4,1} & \tilde{B}_2 & \tilde{C}_2 \\ I_{4,2} & 5\tilde{A}_2 & 5\tilde{E}_2 \end{pmatrix} \begin{pmatrix} h^{\delta\theta}\alpha_{,\theta} \\ \frac{-3\overset{1}{q}{}^{\delta}}{4\pi y m^5} \\ \frac{-3\overset{1}{a}{}^{\delta}}{4\pi y m^5} \end{pmatrix} = \begin{pmatrix} I_{4,1} \\ I_{4,2} \\ I_{4,3} \end{pmatrix} \left(-kh^{\delta\theta}\gamma_{,\theta} + k\gamma U^\theta h^\delta_\mu U^\mu_{,\theta}\right)\,,$$

which yields $\overset{1}{q}{}^{\delta} = Kh^{\delta\mu}\left(T_{,\mu} + TU^\theta U^\mu_{,\theta}\right)$ with K (heat conductivity)
given by $K = -\frac{4\pi y m^6}{15T^2 I_{4,0}}\left(5\tilde{E}_2\begin{vmatrix} I_{4,0} & I_{4,1} \\ I_{4,1} & I_{4,2}\end{vmatrix} - \tilde{C}_2\begin{vmatrix} I_{4,0} & I_{4,1} \\ I_{4,2} & I_{4,3}\end{vmatrix}\right)\begin{vmatrix} \tilde{B}_2 & \tilde{C}_2 \\ \tilde{A}_2 & \tilde{E}_2\end{vmatrix}^{-1}$.
This expression coincides with the corresponding one in the theory of 14 indepen-

dent fields if $\tilde{C}_2 = 0$. Finally, the eqs. $(9)_{6,9}$ can be written in the form

$$\begin{pmatrix} \tilde{B}_3 & \tilde{C}_3 \\ \tilde{A}_3 & \tilde{E}_3 \end{pmatrix} \begin{pmatrix} \overset{1}{t}{}^{<\delta\psi>} \\ \overset{1}{a}{}^{<\delta\psi>} \end{pmatrix} = \begin{pmatrix} I_{6,0} \\ I_{6,1} \end{pmatrix} \frac{-8\pi y m^5 k\gamma}{15} \left(h^{\theta(\delta} h^{\psi)\phi} - \tfrac{1}{3} h^{\theta\phi} h^{\delta\psi} \right) U_{\phi,\theta} \,,$$

from which, in particular, $\overset{1}{t}{}^{<\delta\psi>} = \mu \left(h^{\theta(\delta} h^{\psi)\phi} - \frac{1}{3} h^{\theta\phi} h^{\delta\psi} \right) U_{\phi,\theta}$,

with μ (viscosity) given by $\mu = -\frac{8\pi y m^6}{15T} \begin{vmatrix} I_{6,0} & \tilde{C}_3 \\ I_{6,1} & \tilde{E}_3 \end{vmatrix} \begin{vmatrix} \tilde{B}_3 & \tilde{C}_3 \\ \tilde{A}_3 & \tilde{E}_3 \end{vmatrix}^{-1}$.

This expression coincides with the corresponding one in the theory of 14 independent fields if $\tilde{C}_3 = 0$. Regarding the sign of ν, K, μ we see that the expressions in the Appendix yield

$$\begin{vmatrix} \tilde{B}_3 & \tilde{C}_3 \\ \tilde{A}_3 & \tilde{E}_3 \end{vmatrix} = 3 \begin{vmatrix} -B_3 & C_3 \\ -A_3 & E_3 \end{vmatrix} \begin{vmatrix} -I_{6,2} - \frac{7A_2}{\gamma^8} & I_{6,1} \\ -I_{6,1} - \frac{A}{\gamma^7} & I_{6,0} \end{vmatrix} \left(\frac{5}{8\pi y m^8 D_3} \right)^2 m = \frac{-75}{64\pi^2 y^2 m^{15} D_3} \begin{vmatrix} -B_3 & C_3 \\ -A_3 & E_3 \end{vmatrix},$$

from which it follows that

$$\mu = \frac{-64\pi^2 y^2 m^{12}}{75 T E_3} \frac{[2E_3 I_{6,0} - m I_{6,1}(C_3 + 3A_3)]^2 + m^2 (I_{6,1})^2 \begin{vmatrix} 2B_3 & -C_3 - 3A_3 \\ -C_3 - 3A_3 & 6E_3 \end{vmatrix}}{(C_3 - 3A_3)^2 + \begin{vmatrix} 2B_3 & -C_3 - 3A_3 \\ -C_3 - 3A_3 & 6E_3 \end{vmatrix}} \,;$$

from the inequalities (6) it follows that $\mu < 0$. Similarly, we have

$$\begin{vmatrix} \tilde{B}_2 & \tilde{C}_2 \\ \tilde{A}_2 & \tilde{E}_2 \end{vmatrix} = \frac{5 I_{4,0}}{64\pi^2 y^2 m^{15} D_2} \begin{vmatrix} B_2 & C_2 \\ A_2 & E_2 \end{vmatrix},$$

from which it follows that

$$K = \frac{-32\pi^2 y^2 m^{18}}{225 T^2 B_2 (I_{4,0})^2} \left(\left[18 B_2 m \begin{vmatrix} I_{4,0} & I_{4,2} \\ I_{4,1} & I_{4,3} \end{vmatrix} - 5(C_2 + 9A_2) \begin{vmatrix} I_{4,0} & I_{4,1} \\ I_{4,1} & I_{4,2} \end{vmatrix} \right]^2 + 25 \begin{vmatrix} I_{4,0} & I_{4,1} \\ I_{4,1} & I_{4,2} \end{vmatrix}^2 \begin{vmatrix} 2B_2 & C_2 + 9A_2 \\ C_2 + 9A_2 & 18A_2 \end{vmatrix} \right) \left[(-C_2 + 9A_2)^2 + \begin{vmatrix} 2B_2 & C_2 + 9A_2 \\ C_2 + 9A_2 & 18E_2 \end{vmatrix} \right]^{-1};$$

from the inequalities (6) it follows that $K < 0$. At last, we have

$$\begin{vmatrix} \tilde{B}_1 & \tilde{C}_1 \\ \tilde{A}_1 & \tilde{E}_1 \end{vmatrix} = \frac{-9}{32\pi^2 y^2 m^{15} D_1} \begin{vmatrix} B_1 & C_1 \\ A_1 & E_1 \end{vmatrix} \begin{vmatrix} I_{2,1} & I_{2,2} \\ I_{2,2} & I_{2,3} \end{vmatrix}$$

from which it follows that

$$\nu = \frac{-16\pi^2 y^2 m^{12}}{81 T B_1} \begin{vmatrix} I_{2,1} & I_{2,2} \\ I_{2,2} & I_{2,3} \end{vmatrix}^{-2} \left[(2C_1 - 3A_1)^2 + \begin{vmatrix} 12B_1 & -2C_1 - 3A_1 \\ -2C_1 - 3A_1 & 2E_1 \end{vmatrix} \right]^{-1} \cdot$$

$$\left\{ \left[6mB_1 \begin{vmatrix} I_{2,0} & I_{2,1} & I_{2,3} \\ I_{2,1} & I_{2,2} & I_{2,4} \\ I_{2,2} & I_{2,3} & I_{2,5} \end{vmatrix} - (3A_1 + 2C_1) \begin{vmatrix} I_{2,0} & I_{2,1} & I_{2,2} \\ I_{2,1} & I_{2,2} & I_{2,3} \\ I_{2,2} & I_{2,3} & I_{2,4} \end{vmatrix} \right]^2 + \begin{vmatrix} I_{2,0} & I_{2,1} & I_{2,2} \\ I_{2,1} & I_{2,2} & I_{2,3} \\ I_{2,2} & I_{2,3} & I_{2,4} \end{vmatrix}^2 \begin{vmatrix} 12B_1 & -2C_1 - 3A_1 \\ -2C_1 - 3A_1 & 2E_1 \end{vmatrix} \right\} < 0 \,.$$

Appendix. Relation between (5) and (8).
The representations (5), by using the expressions of $\Sigma_{\beta\gamma}$ and $\Sigma_{\beta\gamma\delta}$ in ref. [3], become (8) with

$$\tilde{B_1} = \frac{-9B_1}{16\pi y m^6 D_1} \begin{vmatrix} I_{2,1} & I_{2,2} & I_{2,4} \\ I_{2,2} & I_{2,3} & I_{2,5} \\ I_{2,3} & I_{2,4} & I_{2,6} + \frac{7A_2}{\gamma^8} \end{vmatrix} + \frac{3C_1}{8\pi y m^7 D_1} \begin{vmatrix} I_{2,1} & I_{2,2} & I_{2,3} \\ I_{2,2} & I_{2,3} & I_{2,4} \\ I_{2,3} & I_{2,4} & I_{2,5} + \frac{A}{\gamma^7} \end{vmatrix},$$

$$\tilde{C_1} = \frac{-3B_1}{16\pi y m^7 D_1} \begin{vmatrix} I_{2,0} & I_{2,1} & I_{2,3} \\ I_{2,1} & I_{2,2} & I_{2,4} \\ I_{2,2} & I_{2,3} & I_{2,5} \end{vmatrix} + \frac{C_1}{8\pi y m^8 D_1} \begin{vmatrix} I_{2,0} & I_{2,1} & I_{2,2} \\ I_{2,1} & I_{2,2} & I_{2,3} \\ I_{2,2} & I_{2,3} & I_{2,4} \end{vmatrix},$$

$$\tilde{B_2} = \frac{3B_2}{8\pi y m^6 D_2} \begin{vmatrix} I_{4,0} & I_{4,2} \\ I_{4,2} & I_{4,4} + \frac{7A_2}{\gamma^8} \end{vmatrix} + \frac{-5C_2}{24\pi y m^7 D_2} \begin{vmatrix} I_{4,0} & I_{4,1} \\ I_{4,2} & I_{4,3} + \frac{A}{\gamma^7} \end{vmatrix},$$

$$\tilde{C_2} = \frac{-3B_2}{8\pi y m^7 D_2} \begin{vmatrix} I_{4,0} & I_{4,2} \\ I_{4,1} & I_{4,3} \end{vmatrix} + \frac{5C_2}{24\pi y m^8 D_2} \begin{vmatrix} I_{4,0} & I_{4,1} \\ I_{4,1} & I_{4,2} \end{vmatrix},$$

$$\tilde{B_3} = \frac{15B_3}{8\pi y m^6 D_3} \left(I_{6,2} + \frac{7A_2}{\gamma^8} \right) + \frac{-5C_3}{8\pi y m^7 D_3} \left(I_{6,1} + \frac{A}{\gamma^7} \right),$$

$$\tilde{C_3} = \frac{-15B_3}{8\pi y m^7 D_3} I_{6,1} + \frac{5C_3}{8\pi y m^8 D_3} I_{6,0},$$

$$3m\tilde{A_1} = \frac{-27A_1}{16\pi y m^6 D_1} \begin{vmatrix} I_{2,1} & I_{2,2} & I_{2,4} \\ I_{2,2} & I_{2,3} & I_{2,5} \\ I_{2,3} & I_{2,4} & I_{2,6} + \frac{7A_2}{\gamma^8} \end{vmatrix} + \frac{9E_1}{8\pi y m^7 D_1} \begin{vmatrix} I_{2,1} & I_{2,2} & I_{2,3} \\ I_{2,2} & I_{2,3} & I_{2,4} \\ I_{2,3} & I_{2,4} & I_{2,5} + \frac{A}{\gamma^7} \end{vmatrix},$$

$$3m\tilde{E_1} = \frac{-9A_1}{16\pi y m^7 D_1} \begin{vmatrix} I_{2,0} & I_{2,1} & I_{2,3} \\ I_{2,1} & I_{2,2} & I_{2,4} \\ I_{2,2} & I_{2,3} & I_{2,5} \end{vmatrix} + \frac{3E_1}{8\pi y m^8 D_1} \begin{vmatrix} I_{2,0} & I_{2,1} & I_{2,2} \\ I_{2,1} & I_{2,2} & I_{2,3} \\ I_{2,2} & I_{2,3} & I_{2,4} \end{vmatrix},$$

$$3m\tilde{A_2} = \frac{9A_2}{8\pi y m^6 D_2} \begin{vmatrix} I_{4,0} & I_{4,2} \\ I_{4,2} & I_{4,4} + \frac{7A_2}{\gamma^8} \end{vmatrix} + \frac{-5E_2}{8\pi y m^7 D_2} \begin{vmatrix} I_{4,0} & I_{4,1} \\ I_{4,2} & I_{4,3} + \frac{A}{\gamma^7} \end{vmatrix},$$

$$3m\tilde{E_2} = \frac{-9A_2}{8\pi y m^7 D_2} \begin{vmatrix} I_{4,0} & I_{4,2} \\ I_{4,1} & I_{4,3} \end{vmatrix} + \frac{5E_2}{8\pi y m^8 D_2} \begin{vmatrix} I_{4,0} & I_{4,1} \\ I_{4,1} & I_{4,2} \end{vmatrix},$$

$$3m\tilde{A_3} = \frac{45A_3}{8\pi y m^6 D_3} \left(I_{6,2} + \frac{7A_2}{\gamma^8} \right) + \frac{-15E_3}{8\pi y m^7 D_3} \left(I_{6,1} + \frac{A}{\gamma^7} \right),$$

$$3m\tilde{E_3} = \frac{-45A_3}{8\pi y m^7 D_3} I_{6,1} + \frac{15E_3}{8\pi y m^8 D_3} I_{6,0}, \quad m\tilde{E_4} = \frac{175}{8\pi y m^8} \frac{E_4}{I_{8,0} + \frac{7A_2}{\gamma^8}}.$$

References

1. I-S. Liu, I. Müller, T. Ruggeri, Ann.of Phys. **169**, 191 (1986).
2. I-S. Liu, Arch.Rational Mech.Anal. **456**, 131 (1972).
3. S. Pennisi, In preparation. , (2001).
4. S. Pennisi and M. Trovato, LE MATEMATICHE. **44**, 173 (1989).

SOME NUMERICAL RESULTS ON THE DEVELOPMENT OF SINGULARITIES IN THE DYNAMICS OF HARMONIC MAPS

F. PISTELLA, V. VALENTE

Istituto per le Applicazioni del Calcolo, CNR, Viale del Policlinico, 137, 00161 Roma, Italy

E-mail: pistella@iac.rm.cnr.it

The paper deals with the numerical study of the appearance of several blow-ups in time which may appear in the evolution of ferromagnets and the heat flow of harmonic maps. By the implementation of a semi-implicit numerical scheme the role of the dissipative term and the effects of the boundary condition on the development of singular solutions are shown.

1 Introduction

Let Ω be an open set of $\mathbb{R}^n$ with $n \geq 2$. We consider the evolution equations

$$\mathbf{m}_t = \Delta\mathbf{m} + |\nabla\mathbf{m}|^2\mathbf{m} - \alpha(\mathbf{m}_t \times \mathbf{m}), \quad |\mathbf{m}| = 1, \tag{1}$$

depending on a parameter $\alpha \geq 0$, where the vector $\mathbf{m} : \Omega \times \mathbb{R}^+ \to \mathrm{S}^2$, with S^2 the unit sphere in $\mathbb{R}^3$. For $\alpha = 0$, the equations (1) reduce to the well known equations for the heat flow of harmonic maps

$$\mathbf{m}_t = \Delta\mathbf{m} + |\nabla\mathbf{m}|^2\mathbf{m}, \qquad |\mathbf{m}| = 1,$$

studied by many authors (see for example the paper by M. Struwe [7] and the references therein). For $\alpha > 0$ the equations (1) represent an alternative formulation to study the dynamical behavior of ferromagnets. Indeed, the cross product of $(1)_1$ by the vector $\mathbf{m}$ leads to the classical Gilbert equation

$$\alpha\,\mathbf{m}_t = \mathbf{m} \times (-\mathbf{m}_t + \Delta\mathbf{m}). \tag{2}$$

The equation (2) is the simplest one to describe the ferromagnet dynamics; no magnetoelastic interaction is taken into account and the effects of the anisotropy energy and the demagnetization field are supposed negligible. For an exhaustive formulation of the dynamics of deformable ferromagnets we refer to the papers [1,3]. The existence of weak solutions to the initial boundary value problem for the equation (2) is proved in [1,6] with different techniques. Here we undertake the numerical study of these equations and present numerical experiments with the aim to exhibit some phenomena which may occur. An open question in the mathematical study of ferromagnets concerns with the appearance of singularities. A partial regularity result has been proved [7] for the heat flow of harmonic maps in the two-dimensional case. The problem is to find a smooth initial datum $\mathbf{m}^0$ for the evolution equations (1) such that the solution of the corresponding initial-boundary value problem does not belong to $W^{1,\infty}(\mathbb{R}^n)$ for N (finite) values of the time t, i.e.

$$\lim_{t \to t_s^-} ||\nabla \mathbf{m}||_\infty^2 \to \infty, \qquad s = 1, 2, ..., N$$

The problem has been partially solved by Chang-Ding-Ye [2], by constructing an example of finite-time blow-up, in the case N=1, for the solutions of the heat flow for harmonic maps ($\alpha = 0$) on a unit disk. Starting from this example we proved [4], by a penalty method, the numerical existence of a singular solution also for the dynamics of ferromagnets ($\alpha > 0$). Here we adopt a more efficient procedure, based on a direct semi-implicit finite difference approximation of (2), and discuss some computational results concerning the development of several singularities in finite time for $\alpha \geq 0$.

2 The numerical scheme

By defining a space grid of steps h_x, h_y, h_z, denoting the time step by δt and adopting centered second order approximation for the Laplacian operator, we construct a semi-implicit finite difference scheme for the approximation of the Gilbert equation on an interval of the space

$$\alpha \frac{\mathbf{m}_{i,j,l}^{n+1} - \mathbf{m}_{i,j,l}^{n}}{\delta t} = \frac{\mathbf{m}_{i,j,l}^{n+1} + \mathbf{m}_{i,j,l}^{n}}{2} \times \left(-\frac{\mathbf{m}_{i,j,l}^{n+1} - \mathbf{m}_{i,j,l}^{n}}{\delta t} + \Delta^h \mathbf{m}_{i,j,l}^{n}\right) \qquad (3)$$

where $i = 1, ..., imax$, $j = 1, ..., jmax$, $l = 1, ..., lmax$ and

$$\mathbf{m}_{i,j,l}^{n} = \mathbf{m}(ih_x, jh_y, lh_z, n\delta t)$$

$$\Delta^h \mathbf{m}_{i,j,l}^{n} = (\Delta_x^h + \Delta_y^h + \Delta_z^h)\mathbf{m}_{i,j,l}^{n}$$

with

$$\Delta_x^h \mathbf{m}_{i,j,l}^{n} = \frac{\mathbf{m}_{i+1,j,l}^{n} - 2\mathbf{m}_{i,j,l}^{n} + \mathbf{m}_{i-1,j,l}^{n}}{h_x^2}$$

and analogously in the directions y and z.

We call (3) semi-implicit due to the term on the right hand side which is averaged at time levels n and $n+1$. This choice allows to preserve the unitarity of the solution, as can be easily proved by scalar multiplication of both sides of (3) by this averaged term. So the modulus of the numerical solution remains the same as time iterations go on.

First order approximations of the gradient are also defined

$$\nabla^h \mathbf{m}_{i,j,l}^n = (\nabla_x^h \mathbf{m}_{i,j,l}^n, \nabla_y^h \mathbf{m}_{i,j,l}^n, \nabla_z^h \mathbf{m}_{i,j,l}^n)$$

where

$$\nabla_x^h \mathbf{m}_{i,j,l}^n = \frac{\mathbf{m}_{i,j,l}^n - \mathbf{m}_{i-1,j,l}^n}{h_x}$$

and analogously in y and z. Zero normal derivative at the boundary, as required in the case of ferromagnets, or assignement of the solution on the boundary are assumed to carry out the numerical tests in the next section.

The scheme is stable. Indeed a discrete analogous of the energy inequality holds, by which the energy does not grow in time. We prove[5] that if the time step satisfies

$$\delta t \leq \frac{h^2}{2}, \qquad h = \frac{\min(h_x, h_y, h_z)}{\sqrt{3}} \tag{4}$$

then, for each $n > 0$,

$$\|\nabla^h \mathbf{m}^n\|_0^2 = \sum_{i,j,l} \left|\nabla^h \mathbf{m}_{i,j,l}^n\right|^2 \leq \sum_{i,j,l} \left|\nabla^h \mathbf{m}_{i,j,l}^0\right|^2 = \|\nabla^h \mathbf{m}^0\|_0^2$$

where

$$\left|\nabla^h \mathbf{m}_{i,j,l}^n\right|^2 = \left|\nabla_x^h \mathbf{m}_{i,j,l}^n\right|^2 + \left|\nabla_y^h \mathbf{m}_{i,j,l}^n\right|^2 + \left|\nabla_z^h \mathbf{m}_{i,j,l}^n\right|^2.$$

In order to monitor the occurrence of singularities in the next section we also use the descrete analogous of the L^∞norm defined by

$$\|\nabla^h \mathbf{m}^n\|_\infty^2 = \max_{i,j,l} \left|\nabla^h \mathbf{m}_{i,j,l}^n\right|^2$$

Note that the stability condition (4) does not depend on α and moreover is the same required by a classical explicit approximation for the simplest heat equation.

The proposed scheme (3) gives directly the sense of the approximation in time and is useful to prove the stability result. But for the implementation we reformulate it by providing an explicit formulation of the solution at the new level

$$\mathbf{m}_{i,j,l}^{n+1} = \mathbf{m}_{i,j,l}^n + \delta t \frac{\alpha \mathbf{u}_{i,j,l}^n + \mathbf{u}_{i,j,l}^n \times \mathbf{w}_{i,j,l}^n}{\alpha^2 + |\mathbf{w}_{i,j,l}^n|^2}$$

where $\mathbf{u}_{i,j,l}^n = \mathbf{m}_{i,j,l}^n \times \Delta^h \mathbf{m}_{i,j,l}^n$ and $\mathbf{w}_{i,j,l}^n = \mathbf{m}_{i,j,l}^n + \frac{\delta t}{2} \Delta^h \mathbf{m}_{i,j,l}^n$.

Note that if the stability condition is strictly verified, that is $\delta t < \frac{h^2}{2}$, the scheme is consistent also for $\alpha = 0$ since in that case $|\mathbf{w}_{i,j,l}^n|$ is not equal to zero.

3 Numerical experiments

In this section we present some numerical tests of our scheme in the bidimensional case. The initial datum which has been proved by Chang, Ding and Ye to evolve into a singular solution for the heat flow of harmonic maps ($\alpha = 0$) in two space dimensions is

$$\mathbf{m}^0 = [\frac{x}{r} sin(f(r)), \frac{y}{r} sin(f(r)), cos(f(r))]$$

in the disk $D = x^2 + y^2 = r^2 \leq r_0^2$ when the smooth function $f(r)$ satisfies the boundary conditions $f(0) = 0$, $|f(r_0)| > \pi$.

Note that due to the radial symmetry the system in $\mathbf{m}$ reduces to only one equation in f. In our tests we assume $f(r) = 8k\pi r(r-1)$ for $r \leq r_0 = 1/2$ and consider two different values of k, that is $k = 1, 2$.

To reveal the appearance of a blow-up, we monitor the norms $||\nabla^h \mathbf{m}^n||_\infty^2$ and $||\nabla^h \mathbf{m}^n||_0^2$ as functions of time.

So our tests evolve in this way: firstly we reconstruct the results of Chang, Ding and Ye by means of a simple explicit approximation, just to produce plots of the predicted blow-up. Then we explore the possibility of several blow-ups, still in the case $\alpha = 0$ and radial symmetry, by changing the condition at $r = 0$. Then we neglect the radial symmetry and integrate the equation for $\mathbf{m}$ on a square for both the assignments on the boundary. Finally, the effect of $\alpha > 0$ is analyzed.

In Fig.1 we report the evolution in time of the two norms in the case of radial symmetry, $\alpha = 0$ and $f(0,t) = 0$, $f(r_0, t) = f(r_0, 0)$ for $k = 1, 2$. We see that the maximum modulus of the gradient suddenly reaches a plateau and remains at that level. This level is a function of the space grid (it increases for decreasing h, runs not reported) and of k. The corresponding plots of the L_2norm of the gradient has a steep descent, which represents a quick loss of energy. To investigate the possibility of multiple blow-ups still in the case $\alpha = 0$, we weaken the assignment at $r = 0$ (which is not a physical boundary) by setting $f'(0) = 0$. It causes the infinite norm of the gradient to reduce its maximum value and go decrease quickly. Several peaks appear, each one corresponding to a jump in the energy.

In Fig.2 we have again $\alpha = 0$ but the radial symmetry hypothesis is neglected, that is we solve the system for $\mathbf{m}$ in a square $Q = [-r_0, r_0] \times [-r_0, r_0]$, with a grid of steps $h_x = h_y = 1/300$, assuming $f(r) = f(r_0)$ for $r > r_0$ and implementing the semi-implicit scheme (3). Now to afford the ferromagnet problem we need only to modify the condition on the boundary of the square, by setting the normal derivative equal to 0, and to introduce the term in α.

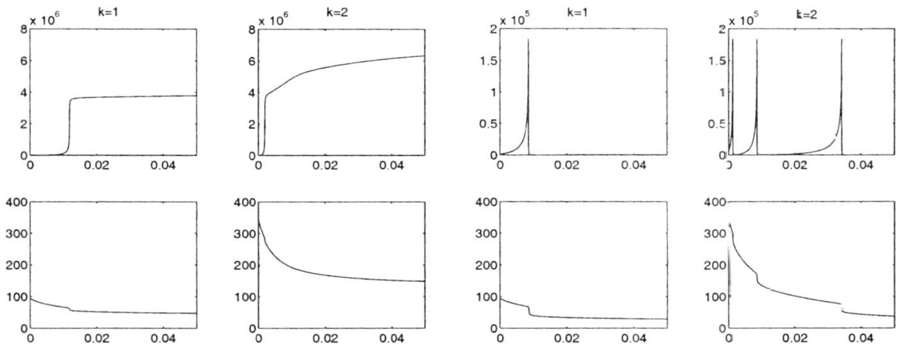

Figure 1: Computed solution in the disk $D = x^2 + y^2 \leq 0.25$ for $\alpha = 0$. The first row of plots shows the behaviour of the norm $||\nabla^h \mathbf{m}^n||_\infty^2$ versus time when the Dirichet boundary conditions for $k = 1, 2$ (the first two plots) and the Neumann boundary conditions for $k = 1, 2$ (the last two plots) are assumed at the centre of the disk ($r = 0$). The second row of plots shows the behaviour of the energy norm $||\nabla^h \mathbf{m}^n||_0^2$ at the same conditions.

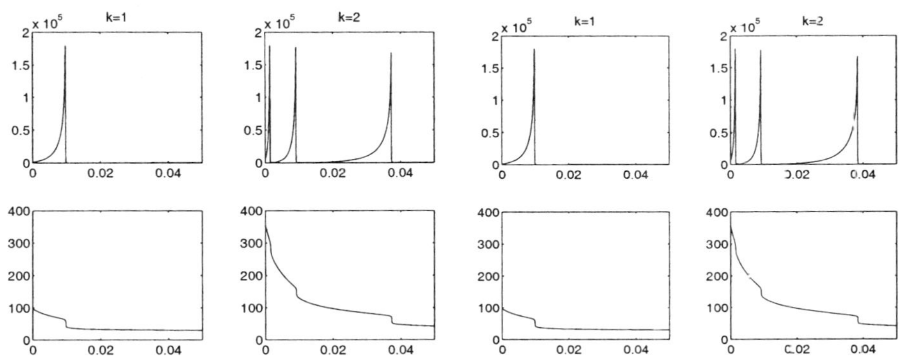

Figure 2: Computed solution in the square $Q = [-0.5, 0.5] \times [-0.5, 0.5]$ for $\alpha = 0$. The first row of plots shows the behaviour of the norm $||\nabla^h \mathbf{m}^n||_\infty^2$ versus time when the Dirichet boundary conditions for $k = 1, 2$ (the first two plots) and the Neumann boundary conditions for $k = 1, 2$ (the last two plots) are assumed at the boundary of Q. The second row of plots shows the behaviour of the energy norm $||\nabla^h \mathbf{m}^n||_0^2$ at the same conditions.

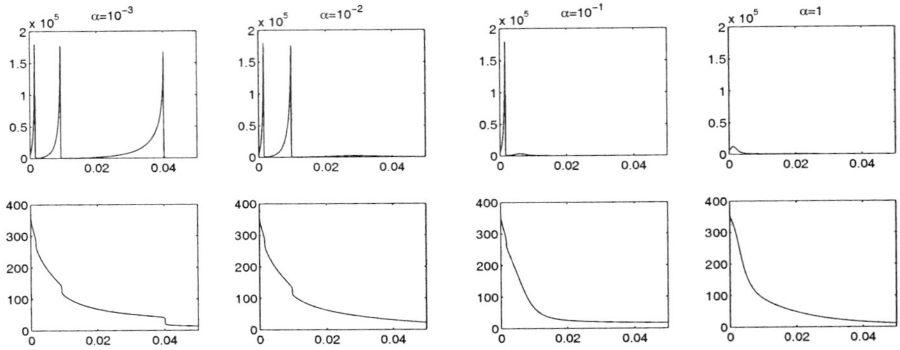

Figure 3: Computed solution in the square $Q = [-0.5, 0.5] \times [-0.5, 0.5]$ for fixed $k = 2$. The first row of plots shows the behaviour of the norm $||\nabla^h \mathbf{m}^n||^2_\infty$ versus time for $\alpha = 10^{-3}, 10^{-2}, 10^{-1}, 1$. The second row of plots shows the behaviour of the energy norm $||\nabla^h \mathbf{m}^n||^2_0$ at the same conditions.

The effects of the first step are not relevant: the plots show only a little delay in the appearance of the last peak.

On the contrary, the term in α produces increasing effects. We focalize on $k = 2$, corresponding to three peaks when $\alpha = 0$. For $\alpha = 10^{-3}$ we have only a little delay in the appearance the last peak, but by increasing α a regularizing effect is evident: only two peaks for $\alpha = 10^{-2}$, just one peak for $\alpha = 10^{-1}$ and a very flat trend for $\alpha = 1$ (Fig.3).

References

1. M. Bertsch, P. Podio-Guidugli and V. Valente, *Ann. Mat. Pura Appl. (IV)*, **CLXXIX**, 331 (2001).
2. K.–C. Chang , W.–Y. Ding and R. Ye, *J. Diff. Geom.*, **36**, 507 (1992).
3. A. DeSimone and P. Podio-Guidugli , *O Arch. Rational Mech. Anal.*, **136**, 201 (1996).
4. F. Pistella and V. Valente, *Numer. Methods Partial Differential Equations* **15** , no. 5, 544 (1999).
5. F. Pistella and V. Valente, (2001), submitted.
6. P. Podio-Guidugli, and V. Valente, *Nonlinear Anal. Theory, Meth. Appl.*, **47**, 147 (2001).
7. M. Struwe, *Nonlinear Partial Differential Equations in Differential Geometry,* (Park City UT 1992) 257 IAS/Park City Math. Ser. 2, Amer. Math. Soc. Providence, RI (1996).

ON BIFURCATIONS IN MULTILATTICES

MARIO PITTERI

Dipartimento di Metodi e Modelli Matematici per le Scienze Applicate
Università di Padova
Via Belzoni 7, 35131 Padova, Italy

We give an example of how a kinematics of multilattices, constructed to keep track in a detailed way of the symmetry changes in deformable complex crystals and of describing correctly the invariance of their constitutive equations, are useful in the analysis of certain weak phase changes, either configurational or structural.

1 Introduction

A fair amount of work has been recently devoted to studying the geometry and kinematics of multilattices, in view of constructing a nonlinear model of the thermomechanical behavior of complex crystals. Some details and references are given by P&Z[9], where some accomplishments as well as still unsolved problems are outlined. Among the accomplishments we mention a systematic classification scheme for crystalline structures, the simplest instances of which, besides the known 14 Bravais classes of simple lattices, have been shown by F&Z[2] for planar 2-lattices, and by F&Z[3] for 3-dimensional 2-lattices.

Here we show in a simple case how the knowledge of the full kinematics of a multilattice can help in classifying all the possibilities for their weak (that is, involving suitably small distortions) symmetry-breaking thermoelastic phase changes. The motivation for this procedure can be already seen in the case of simple lattices. In the literature one usually finds the description of the so-called reduced problems – for instance in Go[5] or T&D[11] – which are well known and only depend on the structure of the original symmetry group. These are the right object to consider for classification purposes, but they contain only part of the information present in the original problem. This may be the reason why, in spite of the reduced problems having been known for some time, their application to the analysis of weak symmetry-breaking transitions in simple lattices, as given by Er[1], does not seem to be as well established. For instance, one and the same reduced problem in $\mathbb{R}^2$ endowed with the symmetries of an equilateral triangle describes the following transformations of simple 3-dimensional lattices: (1) rhombohedral to (base) centered monoclinic; (2) hexagonal to base-centered orthorhombic; (3) primitive [face- or body-centered] cubic to primitive [centered] tetragonal. This information on the transition cannot be obtained from the reduced problem alone. Even

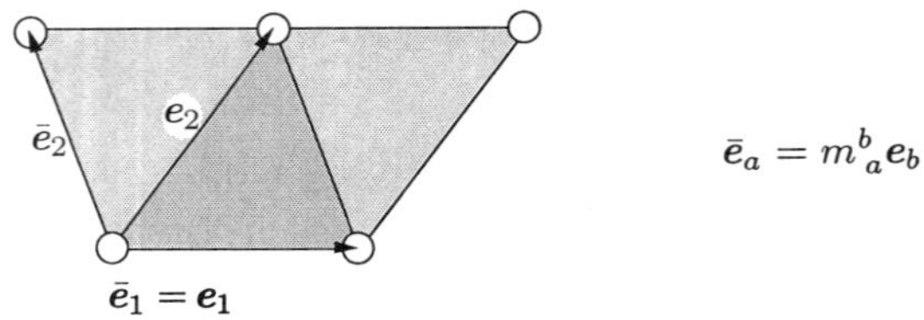

Figure 1. Equivalent bases for a given (planar) lattice

more so for the transitions in complex crystals, where one usually focusses on the kinematic variables that are relevant, thus addressing directly the reduced problem. This procedure allows one to avoid the complexities of starting from the full kinematics; but only the latter can describe all the possibilities, some of which may be otherwise overlooked.

The full kinematic description of the thermoelastic symmetry breaking in simple lattices allows T&Z[12] to describe the tetragonal-orthorhombic-monoclinic transitions in zirconia (ZrO_2), including the triple point, by choosing a different monoclinic variant than the one proposed earlier. The resulting phase diagram is in qualitative agreement with the experimental one, and a quantitative agreement is under investigation.

2 Preliminaries

Let $\mathbb{Z}$ and $\mathbb{R}$ denote the integral and real numbers, respectively. Consider first the simplest triply-periodic structures, that is, *simple lattices* (or 1-*lattices*):

$$\mathcal{L} = \{N^a \boldsymbol{e}_a,\ a = 1, 2, 3, N^a \in \mathbb{Z}\} = \mathcal{L}(\boldsymbol{e}_a). \tag{1}$$

The *lattice vectors* (or *lattice basis*) $\boldsymbol{e}_a$ are linearly independent in $\mathbb{R}^3$.

The basis $\boldsymbol{e}_a$ uniquely determines the 1-lattice $\mathcal{L}(\boldsymbol{e}_a)$, but not vice versa:

$$\mathcal{L}(\bar{\boldsymbol{e}}_a) = \mathcal{L}(\boldsymbol{e}_a) \Leftrightarrow \bar{\boldsymbol{e}}_a = m_a^b \boldsymbol{e}_b\,, \quad m \in GL(3, \mathbb{Z}). \tag{2}$$

Here $GL(3, \mathbb{Z})$ is the group of 3 by 3 integral matrices with determinant ± 1.

2.1 'Geometric' and 'arithmetic' symmetries of simple lattices

In the study of crystal symmetry various aspects have been considered classically: Kepler (1611-1619), followed by Hooke (1665) and, later, Haüy (1822), studied periodic structures, and tiling and packing problems. A major early advance was the classification of the different 'kinds' of subgroups of the orthogonal group $O(3)$, by various authors, circa 1830. These are the well known 'point groups', i.e., the groups of symmetries of bounded objects. The name relates to the fact that any affine isometry leaving a bounded object invariant has a fixed point. Here, 'kind' means *conjugacy class in* $O(3)$. This criterion rests on the idea that the same basic symmetry is shared by rotated objects. For the group $SO(3)$ of *proper* orthogonal tensors (*rotations*), we have:

Theorem 2.1 *The finite subgroups of* $SO(3)$ *are all orthogonally equivalent to one in the following list:* $\mathbf{1}$, $\mathcal{Z}_n$, $\mathcal{D}_n$, $\mathcal{T}$, $\mathcal{O}$, $\mathcal{I}$, *which represent the rotational symmetries of an asymmetric body, of the n-pyramid, of the n-prism, and of the five Platonic solids: the tetrahedron, the octahedron (and the cube), the icosahedron (and the penthagon-dodecahedron), respectively.*

When also $-\mathbf{1}$ is taken into account, the possible groups are obtained by suitably 'patching together' subgroups of the groups above, possibly multiplied by $-\mathbf{1}$. An important result is the following:

Corollary 2.1 *Among the infinite ones above, 32 are the* crystal classes, *that is, the classes of groups leaving some simple lattice invariant; their elements can only have order 1, 2, 3, 4 or 6 (the* crystallographic restriction*). Groups belonging to these 32 crystal classes are called* crystallographic point groups. The crystallographic point group, $P(\boldsymbol{e}_a)$, of the lattice $\mathcal{L}(\boldsymbol{e}_a)$ is given by:

$$P(\boldsymbol{e}_a) = \{\boldsymbol{Q} \in O(3) : \boldsymbol{Q}\mathcal{L}(\boldsymbol{e}_a) = \mathcal{L}(\boldsymbol{e}_a) \text{ or, the same, } \boldsymbol{Q}\boldsymbol{e}_a = m^b{}_a \boldsymbol{e}_b\}. \quad (3)$$

Notice that the basis $\boldsymbol{e}_a$ satisfies (3) if and only if the *lattice metric*

$$C = (C_{ab}), \quad C_{ab} = \boldsymbol{e}_a \cdot \boldsymbol{e}_b \quad (4)$$

is a fixed point of the map

$$C \mapsto m^t C m. \quad (5)$$

Only the groups in 7 of the 32 crystal classes are maximal for the property of leaving a 1-lattice invariant; these are the familiar *crystal systems*: triclinic, monoclinic, orthorhombic, rhombohedral, tetragonal, hexagonal, cubic.

Let us interpret the basic relation (3) differently. By looking at the left-hand side we obtained the crystallographic point groups; we now focus on the right-hand side, introducing the *lattice group* $L(\boldsymbol{e}_a)$ of a lattice $\mathcal{L}(\boldsymbol{e}_a)$:

$$L(\boldsymbol{e}_a) = \{m \in GL(3, \mathbb{Z}) : m^b{}_a \boldsymbol{e}_b = \boldsymbol{Q}\boldsymbol{e}_a, \, \boldsymbol{Q} \in P(\boldsymbol{e}_a)\}; \quad (6)$$

it is the group of integral matrices representing the lattice symmetries *in the lattice basis.* Consider now two simple lattices $\mathcal{L} = \mathcal{L}(\boldsymbol{e}_a)$ and $\bar{\mathcal{L}} = \mathcal{L}(\bar{\boldsymbol{e}}_a)$.

Definition 2.1 $\mathcal{L}$ *and* $\bar{\mathcal{L}}$ *are of the same Bravais type if and only if for suitable choices of their bases* $\mathbf{e}_a$ *and* $\bar{\mathbf{e}}_a$ *they have the same lattice group. Equivalently:* $\mathcal{L}$ *and* $\bar{\mathcal{L}}$ *are of the same Bravais type if and only if their lattice groups are* arithmetically equivalent, *that is, conjugate in* $GL(3, \mathbb{Z})$ – *not in* $O(3)$.

With this definition one analyzes the (Bravais) lattice types by studying the *conjugacy classes* of lattice groups in $GL(3, \mathbb{Z})$.

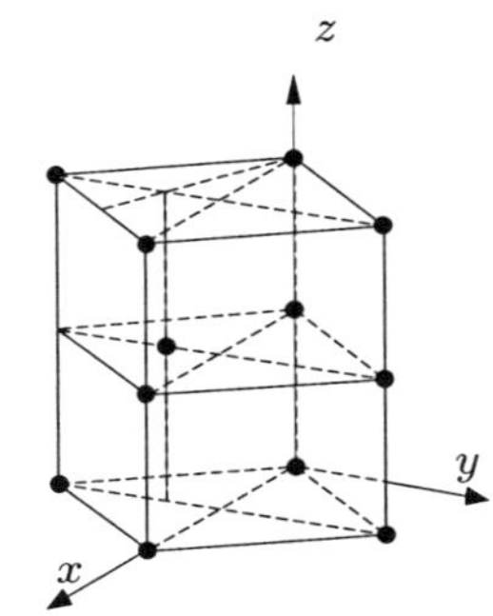

Figure 2. The hexagonal close-packed 2-lattice

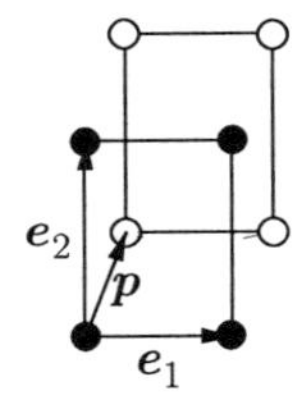

Figure 3. Descriptors for a planar 2-lattice

2.2 Outline of the extension to n-lattices

Real crystals (hcp metals, alloys, etc.) are not in general 1-lattices. We need to describe their geometry and kinematics by means of *multilattices*, which are the union of interpenetrating translates of a 1-lattice. A simple, well known example of 2-lattice is the hexagonal close-packed (hcp) structure in Fig. 2.

Following Pi[6], an n-lattice $\mathcal{M}$ in 3-dimensional affine space can be defined as follows, by choosing the origin O at one of the lattice points and setting $\boldsymbol{p}_0 = \boldsymbol{0}$ for convenience:

$$\mathcal{M} = \mathcal{M}(\boldsymbol{e}_a, \boldsymbol{p}_1, \ldots, \boldsymbol{p}_{n-1}) = \cup_0^{n-1}\{O + \mathcal{L}(\boldsymbol{e}_a) + \boldsymbol{p}_i). \tag{7}$$

Fig. 3 is a schematic picture of a 2-dimensional (planar) 2-lattice; if the atoms represented by filled circles are phisically indistinguishable from the ones represented by open circles, the 2-lattice is called *monatomic*, otherwise *diatomic*. The *multilattice descriptors* $(\boldsymbol{e}_a, \boldsymbol{p}_i) =: \boldsymbol{\varepsilon}_\sigma$, $\sigma = 1, \ldots, n+2$, satisfy the following conditions, which express the three-dimensionality of the multilattice and the non-overlap of the constituent 1-lattices:

$$\boldsymbol{e}_1 \cdot \boldsymbol{e}_2 \times \boldsymbol{e}_3 \neq 0, \quad \boldsymbol{p}_i \neq l_i^a \boldsymbol{e}_a, \quad \boldsymbol{p}_i \neq \boldsymbol{p}_j + l_{ij}^a \boldsymbol{e}_a, \quad i,j = 1, \ldots, n-1, l_i^a, l_{ij}^a \in \mathbb{Z}. \tag{8}$$

The geometric symmetry of n-lattices is classically studied through the *space groups*, i.e. the groups of affine isometries leaving some multilattice invariant. The classification of space groups was obtained at the turn of the century (Jordan, Schönfliess, Fedorov, Frobenius): *there exist 217 affine conjugacy classes of space groups in three dimensions.* Equivalence of the classifications by group isomorphism and by affine conjugacy constitutes a famous theorem by Bieberbach solving the second part of Hilbert's 18th problem.

The use of space groups in the analysis of multilattice symmetry has the same drawback as the use of crystal systems for 1-lattices: it is *too coarse.* Indeed, many distinct 'types' of multilattices may be compatible with the same space group. To mineralogists and metallurgists this is known; yet in crystallography there is no general notion of what is meant by *distinct multilattice types.* Metallurgists refer to empirical catalogues of structures, the St[10] being perhaps the best known compilation, very widely used but lacking a theoretical basis. Our viewpoint is that the study of the *arithmetic symmetry of multilattices*, which is introduced by mimicking the procedure for 1-lattices, gives the needed group-theoretical framework. Support to this view is given by Pi[7], P&Z[8], F&Z[2], F&Z[3], F&Z[4]. We now give the basic information in steps, referring to monatomic multilattices for simplicity and to *essential descriptors*: the lattice $\mathcal{L}(\boldsymbol{e}_a)$ should contain *all* the translations mapping the multilattice to itself. This is a delicate assumption, related to the still open problem of a global kinematic description of multilattices of different complexity; some comments can be found in P&Z[8], P&Z[9].

First step. We replace the descriptors $\boldsymbol{\varepsilon}_\sigma$ introduced above by the $O(3)$-invariant *multilattice metric*

$$K = (K_{\sigma\tau}), \quad K_{\tau\sigma} = K_{\sigma\tau} = \boldsymbol{\varepsilon}_\sigma \cdot \boldsymbol{\varepsilon}_\tau, \quad \sigma, \tau = 1, \dots, n-1, \tag{9}$$

which is an analog of (4). The manifold of multilattice metrics $\mathcal{Q}^m_{n+2}$ is a *submanifold* of the vector space $\mathcal{Q}_{n+2}$ of all symmetric matrices in $\mathbb{R}^{n+2}$; $\mathcal{Q}^m_{n+2}$ is a 'state space' for n-lattices analogous to the set $\mathcal{Q}^>_3$ of positive-definite quadratic forms in $\mathbb{R}^3$ (the lattice metrics in (4)) for 1-lattices.

Second step. We identify the 'global symmetry group' of n-lattices, which must act on $\mathcal{Q}^m_{n+2}$ (recall the action of $GL(3, \mathbb{Z})$ on $\mathcal{Q}^>_3$):

Proposition 2.1 *Let $\mathcal{M}(\boldsymbol{\varepsilon}_\sigma)$ be a (monatomic) n-lattice. Then $\bar{\boldsymbol{\varepsilon}}_\sigma$ are new descriptors for $\mathcal{M}$ up to a translation (that is, $\mathcal{M}(\bar{\boldsymbol{\varepsilon}}_\sigma) = \mathcal{M}(\boldsymbol{\varepsilon}_\sigma) + \boldsymbol{t}$) if and only if there exists a suitable matrix μ such that $\bar{\boldsymbol{\varepsilon}}_\sigma = \mu^\tau{}_\sigma \boldsymbol{\varepsilon}_\tau$; the matrix μ has the following form, for $a, b = 1, 2, 3$, $i, j = 1, \dots, n-1$, $(m^b{}_a) \in GL(3, \mathbb{Z})$, $l^b{}_i \in \mathbb{Z}$, and $\alpha = (\alpha^j{}_i)$ belonging to the finite non-commutative group of matrices generated by the permutation matrices of the set $\{1, \dots, n-1\}$ and by the*

$n-1$ by $n-1$ matrices of the form $\bar{\alpha}$ below, which are obtained from the identity by replacing one of its rows, say the i-th, by a row of -1s:

$$(\mu^{\tau}{}_{\sigma}) = \left(\begin{array}{ccc|c} & m^{b}{}_{a} & & l^{b}_{1}\cdots l^{b}_{n-1} \\ \hline 0 & 0 & 0 & \\ & \vdots & & \alpha^{j}_{i} \\ 0 & 0 & 0 & \end{array}\right), \quad \bar{\alpha} = \begin{pmatrix} 1 & 0 & \cdots & 0 & 0 \\ 0 & 1 & \cdots & 0 & 0 \\ \vdots & \vdots & \cdots & \vdots & \vdots \\ -1 & -1 & \cdots & -1 & -1 \\ \vdots & \vdots & \cdots & \vdots & \vdots \\ 0 & 0 & \cdots & 1 & 0 \\ 0 & 0 & \cdots & 0 & 1 \end{pmatrix}, \tag{10}$$

The matrix μ uniquely determines the new descriptors $\bar{\varepsilon}_{\sigma}$, and vice versa.

The group of the above matrices μ is denoted by Γ_{n+2}, and it acts in a natural way, analogous to (5), on the manifold of multilattice metrics:

$$K \mapsto \mu^{t} K \mu. \tag{11}$$

Third step. Among all changes of descriptors, particular importance have those which produce an affine transformation of the multilattice onto itself. If, with respect to the chosen origin O (which is one of the lattice points) we represent the isometry as a pair $(\boldsymbol{t}, \boldsymbol{Q})$, $\boldsymbol{t} \in \mathbb{R}^{3}$, $\boldsymbol{Q} \in O(3)$, it must be, for the same index i appearing in the matrix α – see (10):

$$\boldsymbol{Q}\boldsymbol{\varepsilon}_{\sigma} = \mu^{\tau}{}_{\sigma}\boldsymbol{\varepsilon}_{\tau}, \qquad \boldsymbol{t} = \boldsymbol{p}_{i} + n^{a}_{0}\boldsymbol{e}_{a},\ n^{a}_{0} \in \mathbb{Z}. \tag{12}$$

In terms of the multilattice metric K in (9) the equality $(12)_1$ is equivalent to K being a fixed point for the corresponding map (11):

$$\mu^{t} K \mu = K. \tag{13}$$

For any essentially described multilattice $\mathcal{M} = \mathcal{M}(\boldsymbol{\varepsilon}_{\sigma})$ the solutions $(\boldsymbol{t}, \boldsymbol{Q})$ of (12) depend only on $\mathcal{M}$ and not on its specific descriptors $\boldsymbol{\varepsilon}_{\sigma}$, and constitute the *space group* $S(\mathcal{M})$ of $\mathcal{M}$. The corresponding matrices $\mu \in \Gamma_{n+2}$ form the *lattice group* $\Lambda(\boldsymbol{\varepsilon}_{\sigma}) < GL(3(n+2), \mathbb{Z})$ of $\mathcal{M}(\boldsymbol{\varepsilon}_{\sigma})$. This finite group depends on the $\boldsymbol{\varepsilon}_{\sigma}$, actually on the corresponding metric K (so, $\Lambda(\boldsymbol{\varepsilon}_{\sigma}) =: \Lambda(K)$), and changes to a conjugate group in Γ_{n+2} under a change of descriptors. One can reconstruct the space group of a multilattice from its lattice group:

Proposition 2.2 *The lattice group $\Lambda(\boldsymbol{\varepsilon}_{\sigma})$ of $\mathcal{M}(\boldsymbol{\varepsilon}_{\sigma})$ is isomorphic to the quotient group $S(\mathcal{M}(\boldsymbol{\varepsilon}_{\sigma}))\backslash\mathcal{L}(\boldsymbol{e}_{a})$.*

The three steps above show how studying the conjugacy classes of the subgroups of Γ_{n+2} formalizes the notion of distinct (arithmetic) *n-lattice types.* Analyzing the action of Γ_{n+2} on the state space $\mathcal{Q}^{m}_{n+2}$ gives information on the kinematics of deformable n-lattices. This analysis is considerably simplified if one restricts the attention to suitably small distortions of a given multilattice:

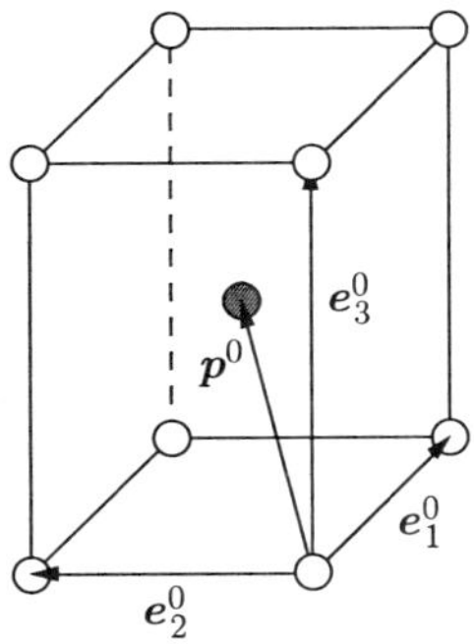

Figure 4. A diatomic tetragonal 2-lattice

Proposition 2.3 *Any multilattice metric K admits a $\Lambda(K)$-invariant neighborhood $\mathcal{N}$ in $\mathcal{Q}_{n+2}$, to be called a* wt-nbhd *of K, such that*

$$\mu^t \mathcal{N} \mu \cap \mathcal{N} \neq \emptyset \;\Leftrightarrow\; \mu \in \Lambda(K) \;\Leftrightarrow\; \mu^t K \mu = K. \tag{14}$$

Therefore, in any neighborhood of K contained in $\mathcal{N}$ the global Γ_{n+2}-invariance reduces to the invariance under the lattice group $\Lambda(K)$:

$$\text{for any } \bar{K} \in \mathcal{N} \quad \Lambda(\bar{K}) \leq \Lambda(K) \qquad (\leq \text{ means 'subgroup of'}). \tag{15}$$

This result allows us to efficiently reduce the description of the invariance in the wt-nbhds, and to greatly simplify the classification of generic elastic bifurcations for essential multilattices.

3 A simple example of bifurcation

Consider the diatomic tetragonal 2-lattice illustrated above (the open and the filled circles represent physically different atoms), with descriptors $\boldsymbol{\varepsilon}^0_{\jmath} = (\boldsymbol{e}^0_1, \boldsymbol{e}^0_2, \boldsymbol{e}^0_3, \boldsymbol{p}^0)$ such that

$$\boldsymbol{e}^0_r \cdot \boldsymbol{e}^0_s = 0, r \neq s, \;\; a = \|\boldsymbol{e}^0_1\| = \|\boldsymbol{e}^0_2\| \neq \|\boldsymbol{e}^0_3\| = c, \;\; \boldsymbol{p}^0 = \frac{1}{2}(\boldsymbol{e}^0_1 + \boldsymbol{e}^0_2 + \boldsymbol{e}^0_3). \tag{16}$$

The space group is symmorphic (semidirect product), based on the lattice $\mathcal{L}(\boldsymbol{e}^0_a)$ and on the tetragonal holohedry

$$D_{4h} = \mathcal{T}_{\boldsymbol{k}} \cup -\mathcal{T}_{\boldsymbol{k}}, \quad \mathcal{T}_{\boldsymbol{k}} = \{\mathbf{1}, \boldsymbol{R}^{\pi/2}_{\boldsymbol{k}}, \boldsymbol{R}^{\pi}_{\boldsymbol{k}}, \boldsymbol{R}^{3\pi/2}_{\boldsymbol{k}}, \boldsymbol{R}^{\pi}_{\boldsymbol{i}}, \boldsymbol{R}^{\pi}_{\boldsymbol{j}}, \boldsymbol{R}^{\pi}_{\boldsymbol{i}+\boldsymbol{j}}, \boldsymbol{R}^{\pi}_{\boldsymbol{i}-\boldsymbol{j}}\}; \tag{17}$$

here $(\boldsymbol{i}, \boldsymbol{j}, \boldsymbol{k}) = (\boldsymbol{e}^0_1/a, \boldsymbol{e}^0_2/a, \boldsymbol{e}^0_3/c)$ and, for any vector $\boldsymbol{v}$ and angle ω, $\boldsymbol{R}^{\omega}_{\boldsymbol{v}}$ is the rotation by the angle ω about the direction of $\boldsymbol{v}$.

Since this 2-lattice is diatomic, all the matrices $\mu \in \Lambda(\varepsilon^0_\sigma)$, which describe the invariance in a wt-nbhd of the multilattice metric K^0, have the form

$$(\mu^\tau{}_\sigma) = \left(\begin{array}{ccc|c} & & & l^1 \\ & m^b{}_a & & l^2 \\ & & & l^3 \\ \hline 0 & 0 & 0 & 1 \end{array}\right), \quad m \in L(e^0_a); \tag{18}$$

here the lattice group $L(e^0_a)$ is the integral representation of the tetragonal holohedry D_{4h} in the lattice basis. We do not need to compute the elements of $\Lambda(\varepsilon^0_\sigma)$; as an example, the one representing the central inversion of D_{4h} has the form (18) with $m = -1$ and $l^1 = l^2 = l^3 = -1$.

Consider now the potential energy of a diatomic 2-lattice per unit lattice cell, to be denoted by ϕ. Since the position of all the points is determined by the multilattice descriptors, we can assume ϕ to depend on (e_a, p). Also, any change of descriptors leaves the multilattice unaffected, hence does not change the energy. Therefore we have the global invariance

$$\phi(e_a, p) = \phi(m^b{}_a e_b, p + l^a e_a) \quad \text{for all } m \in GL(3, \mathbb{Z}), l^a \in \mathbb{Z}. \tag{19}$$

If we consider the multilattice configuration described by the vectors (e^0_a, p^0) introduced above, or by the related multilattice metric K^0, and restrict the domain of ϕ to all the descriptors in a neighborhood N of (e^0_a, p^0) such that the related metric K belongs to the wt-nbhd $\mathcal{N}$ of Proposition 2.3, we have for the energy the (local) invariance given by $(19)_1$ for all μ in $\Lambda(\varepsilon^0_\sigma)$ whose m component has positive determinant; these represent the rotation group $\mathcal{T}_k$. The pair (e^0_a, p^0) will later represent the equilibrium configuration of the 2-lattice at the phase change.

It is reasonable to assume the energy function ϕ to be also Galilean invariant. This amounts to requiring that the restricted domain N be $SO(3)$-invariant and that, for any rotation R and any $(e_a, p) \in N$,

$$\phi(e_a, p) = \phi(Re_a, Rp). \tag{20}$$

Thus, in N, the energy can be regarded as a function of K; or, better,

$$\phi(e_a, p) = \Phi(C_{ab}, p_a), \quad C_{ab} = e_a \cdot e_b, \quad p_a = p \cdot e_a. \tag{21}$$

Recall that, for all $\mu \in \Lambda(\varepsilon^0_\sigma)$ corresponding to rotations $Q \in \mathcal{T}_k$,

$$m^b{}_a e^0_b = Qe_a, \quad p^0 + l^a e_a = Qp^0. \tag{22}$$

To normalize the orientation of the deformed lattice, consider only 'stretched' pairs $(e_a, p) \in N$ such that $e_a = Ue^0_a$, U positive definite symmetric, and define the vector π and the tensors C and E (nonlinear strain) by

$$p = U(p^0 + \pi), \quad U^2 = C = 1 + 2E. \tag{23}$$

Then $C_{ab} = \boldsymbol{e}_a^0 \cdot \boldsymbol{e}_b^0$, $p_a = \boldsymbol{e}_a^0 \cdot \boldsymbol{C}(\boldsymbol{p}^0 + \boldsymbol{\pi})$, and

$$\Phi(C_{ab}, p_a) = \bar{\Phi}(\boldsymbol{C}, \boldsymbol{\pi}) = \tilde{\Phi}(\boldsymbol{E}, \boldsymbol{\pi}). \tag{24}$$

The invariance of the function $\tilde{\Phi}(\boldsymbol{E}, \boldsymbol{\pi})$ now follows from (19)–(24):

$$\tilde{\Phi}(\boldsymbol{E}, \boldsymbol{\pi}) = \tilde{\Phi}(\boldsymbol{Q}^t \boldsymbol{E} \boldsymbol{Q}, \boldsymbol{Q}\boldsymbol{\pi}) \quad \text{for all } \boldsymbol{Q} \in \mathcal{T}_{\boldsymbol{k}}. \tag{25}$$

By identifying the potential energy per unit lattice cell with the free energy density at zero absolute temperature θ, and extending the considerations above to positive temperatures, we can interpret (25), in which now $\tilde{\Phi}$ is also a function of θ, as giving the invariance of the free energy function.

For the reduced problems below it will be convenient to introduce in $\mathbb{R}^3$ the orthonormal basis $(\boldsymbol{i}, \boldsymbol{j}, \boldsymbol{k})$ above, and for symmetric tensors the basis $\boldsymbol{V}_1, \ldots, \boldsymbol{V}_6$ represented in the basis $(\boldsymbol{i}, \boldsymbol{j}, \boldsymbol{k})$ by the matrices in (34)–(36) below. Then any pair $(\boldsymbol{E}, \boldsymbol{\pi})$ is represented by nine coordinates:

$$\boldsymbol{E} = \sum_{r=1}^{6} y^r \boldsymbol{V}_r, \ \boldsymbol{\pi} = \sum_{\alpha=7}^{9} y^\alpha \boldsymbol{c}_\alpha, \ \ (\boldsymbol{E}, \boldsymbol{\pi}) \approx (y^1, \ldots, y^9) = (y^i) = (y^r, y^\alpha), \tag{26}$$

and any transformation $(\boldsymbol{E}, \boldsymbol{\pi}) \mapsto (\boldsymbol{Q}^t \boldsymbol{E} \boldsymbol{Q}, \boldsymbol{Q}\boldsymbol{\pi})$ by a 9×9 matrix:

$$y^i \mapsto A^i_{\ j} y^j, \quad \text{with } A^r_{\ \alpha} = 0 = A^\alpha_{\ r}, \tag{27}$$

so that the matrix A is a block matrix, with a 6×6 and a 3×3 blocks, each one itself orthogonal. We denote by G the group of such orthogonal matrices A corresponding to $\boldsymbol{Q} \in \mathcal{T}_{\boldsymbol{k}}$. We also introduce the representation of the free energy density in the basis above, and of its invariance based on (27):

$$\tilde{\Phi}(\boldsymbol{E}, \boldsymbol{\pi}, \theta) = \Psi(y^i, \theta), \quad \Psi(y^i, \theta) = \Psi(A^i_{\ j} y^j, \theta). \tag{28}$$

In terms of Ψ the equilibrium equations are, in a convenient notation,

$$\Psi_i := \frac{\partial \Psi}{\partial y^i}(y^j, \theta) = 0, \ i = 1, \ldots, 9. \tag{29}$$

We assume these conditions to hold for $\theta = \bar{\theta}$ and $y^i = 0$, the latter giving the (reference) multilattice $\mathcal{M}(\varepsilon_\sigma^0)$. Consider now the second-order symmetric tensor of moduli at the transition:

$$\mathsf{L}_{ij} = \Psi_{ij}(0, \bar{\theta}); \tag{30}$$

if this is invertible, then, by the implicit function theorem, the equilibrium equations (29) have one solution $y^i = \bar{y}^i(\theta)$ for θ near $\bar{\theta}$, such that $\bar{y}^i(\bar{\theta}) = 0, i = 1, \ldots, 9$. Also, by continuity and uniqueness, all points on this equilibrium branch have the same symmetry as $\mathcal{M}(\varepsilon_\sigma^0)$. Therefore symmetry breaking can only occur if the tensor of moduli has a nontrivial kernel. For any matrix $A \in G$, the invariance $(28)_2$ forces, by differentiation, the following identity among the second derivatives of Ψ at $(0,\bar{\theta})$:

$$\mathsf{L} = A^t \mathsf{L} A, \tag{31}$$

which implies that the eigenspaces of L are invariant under the action of G:

$$A^t \mathsf{L} A y = \mathsf{L} y = \lambda y \quad \Leftrightarrow \quad \mathsf{L} A y = \lambda A y. \tag{32}$$

Equivalently, invariance forces certain eigenvalues of L to be equal. We focus on the case, that we may call *generic*, in which the only conditions to be imposed on derivatives of Ψ are those guaranteeing that $(0,\bar{\theta})$ is a stable equilibrium at which bifurcation occurs, and those forced by invariance, for instance (31). In particular, the only eigenvalues of L that are equal are the ones that are forced to be so by invariance; or, the eigenspaces of L are *irreducible invariant* (i.i.) subspaces of $\mathbb{R}^9$ under the action (27) of G, and exactly one of them is the kernel of L. Then, the condition that a stable phase exists, say, for $\theta > \bar{\theta}$ forces all the other eigenvalues to be strictly positive.

Since the action of G is factorized (the first 6 coordinates and the last 3 do not mix), the i.i. subspaces of $\mathbb{R}^9$ are necessarily of either one of the forms

$$\mathcal{V}_1 \times \{\mathbf{0}\} \quad \text{or} \quad \{\mathbf{0}\} \times \mathcal{V}_2, \tag{33}$$

where $\mathcal{V}_1$ $[\mathcal{V}_2]$ is an i.i. subspace of $\mathbb{R}^6$ [of $\mathbb{R}^3$]. Case $(33)_1$ corresponds to *configurational transitions*: the center atom follows the deformation of the sheleton, at least in the beginning. Case $(33)_2$ describe *structural transitions*, which are driven by the displacement of the central atom off center, followed by a suitable consequent deformation of the skeleton. From now on we recall what follows from the classical procedure of determining the i.i. subspaces of *Sym* [of $\mathbb{R}^3$] for case $(33)_1$ [$(33)_2$], and then considering the related *reduced problem*, a description of which can be found in Go[5], Er[1], T&D[11], P&Z[9].

Case $(33)_1$ *(i)* The set $\hat{C}(\boldsymbol{e}_a)$ of symmetric tensors left fixed by the transformation $\boldsymbol{E} \mapsto \boldsymbol{Q}^t \boldsymbol{E} \boldsymbol{Q}$ forms a 2-dimensional subspace, an orthonormal basis for which is given by the tensors $\boldsymbol{V}_1$, $\boldsymbol{V}_2$ represented in the basis $(\boldsymbol{i},\boldsymbol{j},\boldsymbol{k})$ by matrices of the form

$$\boldsymbol{V}_1 = \frac{1}{\sqrt{2}} \begin{pmatrix} 1 & 0 & 0 \\ 0 & 1 & 0 \\ 0 & 0 & 0 \end{pmatrix} \quad \text{and} \quad \boldsymbol{V}_2 = \begin{pmatrix} 0 & 0 & 0 \\ 0 & 0 & 0 \\ 0 & 0 & 1 \end{pmatrix}. \tag{34}$$

The i.i. subspaces contained in $\hat{C}(\boldsymbol{e}_a)$ are those and only those that are 1-dimensional. In each one of them the reduced action is trivial, the bifurcation point is a *turning point*, with change of stability but no change of symmetry.

(ii) In the orthogonal complement $\hat{C}(\boldsymbol{e}_a)^{\perp}$ of $\hat{C}(\boldsymbol{e}_a)$ there are two mutually orthogonal 1-dimensional i.i. subspaces $\mathcal{V}_1$ and $\mathcal{V}_2$ generated by $\boldsymbol{V}_3$ and $\boldsymbol{V}_4$, respectively, and a 2-dimensional subspace $\mathcal{V}_3$ generated by $\boldsymbol{V}_5$, $\boldsymbol{V}_6$, with

$$\boldsymbol{V}_3 = \frac{1}{\sqrt{2}}\begin{pmatrix} 1 & 0 & 0 \\ 0 & -1 & 0 \\ 0 & 0 & 0 \end{pmatrix}, \quad \boldsymbol{V}_4 = \frac{1}{\sqrt{2}}\begin{pmatrix} 0 & 1 & 0 \\ 1 & 0 & 0 \\ 0 & 0 & 0 \end{pmatrix}, \tag{35}$$

$$\boldsymbol{V}_5 = \frac{1}{\sqrt{2}}\begin{pmatrix} 0 & 0 & 0 \\ 0 & 0 & 1 \\ 0 & 1 & 0 \end{pmatrix}, \quad \boldsymbol{V}_6 = \frac{1}{\sqrt{2}}\begin{pmatrix} 0 & 0 & 1 \\ 0 & 0 & 0 \\ 1 & 0 & 0 \end{pmatrix}. \tag{36}$$

(iii) The reduced group $\mathcal{P}$ on $\mathcal{V}_1$ and $\mathcal{V}_2$ is $\mathcal{P} = \{1, -1\}$. The action of $\mathcal{T}_{\boldsymbol{k}}$ on the typical element of $\mathcal{V}_1$ [$\mathcal{V}_2$] thus produces two variants which have *orthorhombic* symmetry, and primitive [base-centered] centering. The reduced bifurcation diagram is the standard pitchfork.

(iv) The reduced group $\mathcal{P}$ on the 2-dimensional i.i. subspace $\mathcal{V}_3$ has 8 elements and, $r(\omega)$ being the rotation by ω in $\mathbb{R}^2$, is generated by

$$f := \begin{pmatrix} -1 & 0 \\ 0 & 1 \end{pmatrix} \approx \boldsymbol{R}_{\boldsymbol{i}}^{\pi} \quad \text{and} \quad r(\frac{\pi}{2}) := \begin{pmatrix} 0 & -1 \\ 1 & 0 \end{pmatrix} \approx \boldsymbol{R}_{\boldsymbol{k}}^{\frac{3\pi}{2}}. \tag{37}$$

The action of $\mathcal{T}_{\boldsymbol{k}}$ on the typical tensor of $\mathcal{V}_3$ thus produces eight variants; these have *triclinic* symmetry, except on suitable 1-dimensional *monoclinic* subspaces. These correspond to subspaces of $\mathbb{R}^2$ that are left invariant by some subgroup of $\mathcal{P}$. Since the latter is the group of symmetries of a square centered at the origin and edges parallel to the coordinate axes, one can see easily that invariant subspaces are: the origin itself (under the central inversion representing the monoclinic group $\mathcal{M}_{\boldsymbol{k}} = \{\boldsymbol{1}, \boldsymbol{R}_{\boldsymbol{k}}^{\pi}\}$ but also, of course, under the whole of $\mathcal{P}$); the two axes x and y in $\mathbb{R}^2$, corresponding to the tetragonally conjugate monoclinic groups $\mathcal{M}_{\boldsymbol{j}}$ and $\mathcal{M}_{\boldsymbol{i}}$, respectively, in obvious notation; and the two diagonals $x = y$ and $x = -y$, corresponding to the conjugate monoclinic groups $\mathcal{M}_{\boldsymbol{i}-\boldsymbol{j}}$ and $\mathcal{M}_{\boldsymbol{i}+\boldsymbol{j}}$, respectively. The variants on the x and y axes are primitive monoclinic, while those on $x = \pm y$ are (base-) centered.

The bifurcating curves consist in two pairs of pitchforks, the first in the (x, θ) and (y, θ) planes, and the second in the $(x = y, \theta)$ and $(x = -y, \theta)$ planes. Assuming the high-symmetry phase $x = 0 = y$ to be stable for $\theta > \bar{\theta}$, a pair of pitchforks is stable only if it exists for $\theta \leq \bar{\theta}$ (supercritical); then also the other pair has to be supercritical, and exactly one of the pairs is stable.

Case $(33)_2$ T&D[11] have considered these structural bifurcations in their introductory chapter. The complete analysis presented here shows that not only these are possible, but also the configurational ones of case $(33)_1$. Moreover, the configurational and structural transitions here analyzed are alternative (in the sense of (33)) and indipendent bifurcation mechanisms.

There are two i.i. subspaces of $\mathbb{R}^3$ under the action $\boldsymbol{\pi} \mapsto \boldsymbol{Q}\boldsymbol{\pi}$:
(a) the z axis; the reduced problem is the one in *(iii)* above. The resulting two variants have in this case C_{4v} symmetry: a fourfold z axis, and four vertical (containing the z axis) symmetry planes: $x = 0$, $y = 0$, $x = y$, and $x = -y$.
(b) the (x, y) plane; the reduced problem is the one in *(iv)* above. For each pair of pitchforks the four variants have C_{2v} symmetry: a twofold z axis, and the vertical plane containing $\boldsymbol{\pi}$; this is one of the four planes in *(a)*.

Acknowledgments

This work is part of the research activities of the TMR Network "Phase Transitions in Crystalline Solids", and is also partially supported by the Italian Cofinanziamento 2000 "Modelli Matematici per la Scienza dei Materiali".

References

1. J. L. Ericksen in *Microstructure and phase transition*, IMA Volumes in Mathematics and its Applications 54, eds. J.L. Ericksen, R.D. James, D. Kinderlehrer and M. Luskin (Springer-Verlag, New York, etc. 1993).
2. G. Fadda and G. Zanzotto, *Acta Cryst.* A**56**, 36 (2000).
3. Fadda, G. and G. Zanzotto, *Int. J. of Nonl. Mech.* **36**, 527 (2001).
4. G. Fadda and G. Zanzotto, *Acta Cryst.*, to appear (2001).
5. M. Golubitsky *et al.*, *Singularities and groups in bifurcation theory, Vol II* (Springer Verlag, New York, etc. 1988).
6. M. Pitteri, *J. of Elasticity* **15**, 3 (1985).
7. M. Pitteri, *Int. J. of Plasticity* **14**, 139 (1998).
8. M. Pitteri and G. Zanzotto, *Acta Cryst.* **A54**, 359 (1998).
9. M. Pitteri and G. Zanzotto, *Continuum models for phase transitions and twinning in crystals* (CRC/Chapman & Hall, London 2001, to appear).
10. *Strukturberichte* (Akademische Verlagsgesellschaft, Leipzig 1915–1940).
11. P. Tolédano and V. Dmitriev, *Reconstructive phase transformations: in crystals and quasicrystals* (World Scientific, Singapore etc. 1996)
12. L. Truskinovsky and G. Zanzotto, Elastic crystals with a triple point, to appear (2001).

AN H-THEOREM IN A SIMPLE MODEL OF CHEMICALLY REACTING DENSE GASES

JACEK POLEWCZAK

Department of Mathematics, California State University, Northridge, CA 91330-8313, USA

E-mail: jacek.polewczak@csun.edu

Existence of an H-theorem for a simple model of chemically reacting dense gases is obtained. The model amounts to a *coloring* process of the four component mixture of hard-spheres with probability $0 < \alpha_R \leq 1$.

1 Introduction

In this note I show that a simplified version the simple reacting spheres (SRS) kinetic model possess an H-functional.

The simple reacting spheres (SRS) model has been proposed by Marron [1] and Xystris and Dahler [2]. The reacting molecules behave as if they were single mass points with two internal states of excitation. Collisions may alter the internal states (this occurs when the kinetic energy associated with the reactive motion exceeds the activation energy) but can not transfer mass from one molecule to another. Reactive and non-reactive collision events are considered to be hard spheres–like. In a four component mixture A, B, A^*, B^*, and the chemical reaction of the type

$$A + B \rightleftharpoons A^* + B^*. \tag{1}$$

A^* and B^* are the distinct species from A and B. Indices $1, 2, 3, 4$ are for the particles A, B, A^*, B^*, respectively. There is **no** net mass transfer in reactive collisions. This implies $m_1 = m_3$ and $m_2 = m_4$, where m_i denotes the mass of the i-th particle, $i = 1, \ldots, 4$. Reactions take place when the reactive particles are separated by a distance $\sigma_{12} = \frac{1}{2}(d_1 + d_2)$, where d_i denotes the diameter of the i-th particle. In the SRS model, reactions do not change diameters of the particles, $d_1 = d_3$ and $d_2 = d_4$, thus $\sigma_{34} = \sigma_{12}$. By not allowing the hard sphere diameter to change upon reaction one avoids complications of producing overlapping configurations (see, Bose-Ortoleva [3]).

The SRS, being a natural extension of the hard–sphere collisional model, reduces itself to the (revised) Enskog theory when the chemical reactions are turned off. Furthermore, in the dilute–gas limit it provides an interesting kinetic model of chemical reactions considered by Polewczak [4].

2 The simple reacting spheres (SRS) kinetic model

The collisional encounters of SRS model consists of two types: elastic and reactive. In the case of elastic collisions between a pair of particles from species i and s, the initial velocities v, w take post–collisional values

$$v' = v - 2\frac{\mu_{is}}{m_i}\epsilon\langle\epsilon, v-w\rangle, \quad w' = w + 2\frac{\mu_{is}}{m_s}\epsilon\langle\epsilon, v-w\rangle. \tag{2}$$

Here, $\langle\cdot\,,\cdot\rangle$ is the inner product in R^3, ϵ is a vector along the line passing through the centers of the spheres at the moment of impact, i.e.,

$$\epsilon \in S_+^2 = \{\epsilon \in R^3 : |\epsilon| = 1, \langle\epsilon, v-w\rangle \geq 0\},$$

and $\mu_{is} = m_i m_s(m_i + m_s)$ is the reduced mass of the colliding pair, where m_i and m_s are the masses of particles from i-th and s-th species, respectively $(i, s = 1, 2, 3, 4)$. Finally, the assumptions $m_1 = m_3$ and $m_2 = m_4$ imply $\mu_{12} = \mu_{34}$.

For the reactive collision between particles of species i and s to occur $(i, s = 1, \ldots, 4)$, the kinetic energy associated with the relative motion along the line of centers must exceed the activation energy γ_i (defined below),

$$\frac{1}{2}\mu_{is}\big(\langle\epsilon, v-w\rangle\big)^2 \geq \gamma_i, \tag{3}$$

with ϵ having the same meaning as above. In the case of the (endothermic) reaction $A + B \rightarrow A^* + B^*$ the velocities v, w take their post–reactive values

$$v^\ddagger = v - \tfrac{\mu_{12}}{m_1}\epsilon\left[\langle\epsilon, v-w\rangle - \alpha^-\right], \tag{4}$$

$$w^\ddagger = w + \tfrac{\mu_{12}}{m_2}\epsilon\left[\langle\epsilon, v-w\rangle - \alpha^-\right], \tag{5}$$

with $\alpha^- = \sqrt{\big(\langle\epsilon, v-w\rangle\big)^2 - 2E_{abs}/\mu_{12}}$ and, E_{abs}, the energy absorbed by the internal degrees of freedom. The absorbed energy E_{abs} has the property

$$E_{abs} = E_3 + E_4 - E_1 - E_2 > 0, \tag{6}$$

where $E_i > 0$, $i = 1, \ldots 4$, is the energy of i-th particle associated with its internal degrees of freedom.

Now, in order to complete the definition of the model, the activation energies γ_1, γ_2 for A and B are chosen to satisfy $\gamma_1 \geq E_{abs} > 0$, and by symmetry, $\gamma_2 = \gamma_1$.

For the inverse (exothermic) reaction, $A^* + B^* \to A + B$, the post–reactive velocities are given by

$$v^\dagger = v - \tfrac{\mu_{34}}{m_3}\epsilon\left[\langle\epsilon, v-w\rangle - \alpha^+\right], \tag{7}$$

$$w^\dagger = w + \tfrac{\mu_{34}}{m_4}\epsilon\left[\langle\epsilon, v-w\rangle - \alpha^+\right], \tag{8}$$

with $\alpha^+ = \sqrt{\left(\langle\epsilon, v-w\rangle\right)^2 + 2E_{abs}/\mu_{34}}$, and the activation energies for A^* and B^*, $\gamma_3 = \gamma_1 - E_{abs}$ and, $\gamma_4 = \gamma_3$. I note that the model allows for the exothermic reaction to have nonzero activation energy.
Post- and pre-collisional velocities of reactive pairs satisfy conservation of the momentum

$$m_1 v + m_2 w = m_3 v^\ddagger + m_4 w^\ddagger,\ m_3 v + m_4 w = m_1 v^\dagger + m_2 w^\dagger. \tag{9}$$

A part of kinetic energy is exchanged with the energy absorbed by the internal states. The following equalities hold:

$$m_1 v^2 + m_2 w^2 = m_3 v^{\ddagger 2} + m_4 w^{\ddagger 2} + 2E_{abs}, \tag{10}$$

$$m_3 v^2 + m_4 w^2 = m_1 v^{\dagger 2} + m_2 w^{\dagger 2} - 2E_{abs}. \tag{11}$$

For $i = 1,2,3,4$, $f_i(t,x,v)$ denotes the one-particle distribution function of the ith component of the reactive mixture. The SRS kinetic system has the form

$$\frac{\partial f_i}{\partial t} + v\frac{\partial f_i}{\partial x} = J_i^E + J_i^R, \quad i = 1,2,3,4, \tag{12}$$

where J_i^E is the non-reactive (hard-sphere) collision operator

$$J_i^E = \sum_{s=1}^{4}\left\{\sigma_{is}^2 \int_{R^3\times S_+^2}\left[f_{is}^{(2)}(t,x,v',x-\sigma_{is}\epsilon,w') - f_{is}^{(2)}(t,x,v,x+\sigma_{is}\epsilon,w)\right]\langle\epsilon, v-w\rangle\, d\epsilon dw\right\}, \tag{13}$$

where $S_+^2 = \{\epsilon \in R^3 : |\epsilon| = 1, \langle\epsilon, v-w\rangle \ge 0\}$, and $f_{ij}^{(2)}$ approximates the density of pairs of particles in collisional configurations.
For $i = 1,2,3,4$, the reactive terms $J_i^R =$

$$\beta\sigma_{ij}^2 \int_{R^3\times S_+^2}\left[f_{kl}^{(2)}(t,x,v_{ij}^{\odot},x-\sigma_{ij}\epsilon,w_{ij}^{\odot}) - f_{ij}^{(2)}(t,x,v,x+\sigma_{ij}\epsilon,w)\right]\times$$

$$\Theta\left(\langle\epsilon, v-w\rangle - \Gamma_{ij}\right)\langle\epsilon, v-w\rangle d\epsilon dw$$

$$-\beta\sigma_{ij}^2 \int\limits_{R^3\times S_+^2} \left[f_{ij}^{(2)}(t,x,v',x-\sigma_{ij}\epsilon,w') - f_{ij}^{(2)}(t,x,v,x+\sigma_{ij}\epsilon,w) \right] \times$$

$$\Theta(\langle\epsilon,v-w\rangle - \Gamma_{ij})\,\langle\epsilon,v-w\rangle d\epsilon dw, \tag{14}$$

Here, $0 \le \beta \le 1$ is the steric factor, $\Gamma_{ij} = \sqrt{2\gamma_i/\mu_{ij}}$, and Θ is the Heaviside step function. The pairs of velocities $(v_i^{\odot}, v_j^{\odot})$ refer to post-reactive velocities, i.e., $(v_{ij}^{\odot}, w_{ij}^{\odot}) = (v^{\ddagger}, w^{\ddagger})$ for $i,j = 1,2$, and $(v_{ij}^{\odot}, w_{ij}^{\odot}) = (v^{\dagger}, w^{\dagger})$ for $i,j = 3,4$. The pairs of indices (i,j) and (k,l) are associated with the following set of quadruples (i,j,k,l):

$$\{(1,2,3,4),\ (2,1,4,3),\ (3,4,1,2),\ (4,3,2,1)\}. \tag{15}$$

In the dilute-gas case of (12),

$$f_{ij}^{(2)}(t,x_1,v_1,x_2,x_2) = f_i(t,x_1,v_1)f_j(t,x_2,v_2), \tag{16}$$

and the expressions $x \pm \sigma_{is}$ and $x \pm \sigma_{ij}$ appearing as the arguments of the distribution functions in (13) and (14), respectively, are replaced by x. In the dilute-gas case, Polewczak [4] showed that for a nonnegative solutions f_i of (12) the convex function $H(t)$ defined by

$$H(t) = \sum_{i=1}^{4} \int\limits_{\Omega\times R^3} f_i(t,x,v)\log f_i(t,x,v)\,dvdx \tag{17}$$

is non-increasing in t, with Ω being a spatial domain. Furthermore, the following equivalent statements characterize the equilibrium states [4]:

1. $f_i = n_i(t,x)\left(\frac{m_i}{2\pi kT(t,x)}\right)^{3/2} \exp\left(-\frac{m_i(v-u(t,x))^2}{2kT(t,x)}\right)$ and

   $n_1(t,x)n_2(t,x) = n_3(t,x)n_4(t,x)\exp\left(\frac{E_{abs}}{kT(t,x)}\right)$,
2. $J_i^E(\{f_i\}) = 0$ and $J_i^R(\{f_i\}) = 0$, $i = 1,\ldots,4$,
3. $\sum\limits_{i=1}^{4}\int\limits_{R^3} \left[J_i^E(\{f_i\}) + J_i^R(\{f_i\})\right]\log f_i\,dv = 0$,

where $n_i(t,x)$ is the number density of specie i, $u(t,x)$ is the macroscopic velocity, and $T(t,x)$ is the macroscopic temperature. In addition, Polewczak [4] obtained global existence and convergence to equilibrium results in the dilute-gas case.

The case of dense gases is obtained when

$$f_{ij}^{(2)}(t,x,v,x\pm\sigma_{ij},w) = g_2^{(ij)}(x,x\pm\sigma_{ij})f_i(t,x,v)f_j(t,x\pm\sigma_{ij},w), \tag{18}$$

with $g_2^{(ij)}(r_1,r_2) = g_2^{(ij)}(r_1,r_2,\{\rho_i\})$ being a pair correlation function for the system of hard-sphere mixture at non-uniform equilibrium with $\rho_i = m_i n_i$. It's Mayer expansion has the form:

$$g_2^{(ij)}(r_1,r_2) = \exp(-\beta\phi_{ij}^{HS}(|r_1-r_2|))\Big\{1+\textstyle\sum_{s=1}^4 \int V(12\,|\,3)\rho_s(t,r_3) \\ +\frac{1}{2}\sum_{s=1}^4\sum_{l=1}^4 \int\int V(12\,|\,34)\rho_s(t,r_3)\,\rho_l(t,r_4)\,dr_3 dr_4 + \cdots \quad (19) \\ +\frac{1}{(k-2)!}\sum_{i_1,\ldots,i_k=1}^4 \int dr_3 \cdots \int dr_k\,\rho_{i_1}(3)\cdots\rho_{i_k}(k)\,V(12\,|\,3\cdots k)+\cdots\Big\}$$

where $\rho_{i_j}(j) = \rho_{i_j}(t,r_j)$, $\beta = 1/k_B T$, and $V(12\,|\,3\cdots k)$ (Husimi function) is the sum of all graphs of k labeled points which are biconnected when the Mayer factor $f^{ij}(r_1,r_2) = \exp(-\beta\phi_{ij}^{HS}(|r_1-r_2|))-1$ is added. In the case of the hard-spheres mixture, the Mayer factors $f^{ij}(r_1,r_2) = \Theta(|r_1-r_2|-\sigma_{ij})-1$, where Θ is the step function.
It is an open problem whether the system (12) with $f_{ij}^{(2)}$ given in (18) has an H-function.

3 H-functional for the simplified version of SRS

Consider the following caricature model of the SRS: assume that when a hard-sphere A collides with another hard-sphere B, there is a probability α_R such that

$$A + B \to A^* + B^*,$$

i.e., A^* and B^* are formed. Similarly, there is a probability $(1-\alpha_R)$ that the pair A and B will scatter elastically. This simple dynamic model, considered by Kapral [5], amounts to a *coloring* process with the probability $0 \le \alpha_R \le 1$. The model by can be formally derived from the SRS kinetic system by setting the activation energies $\gamma_i = 0$ in (3), for $i = 1,2,3,4$. The corresponding system has the form of (12) with J_i^E given by

$$J_i^E = \sum_{s=1}^4 \Big\{\sigma_{is}^2 \lambda_{is}^E \int_{R^3\times S_+^2} \Big[f_{is}^{(2)}(t,x,v',x-\sigma_{is}\epsilon,w') \\ -f_{is}^{(2)}(t,x,v,x+\sigma_{is}\epsilon,w)\Big]\,\langle\epsilon, v-w\rangle\,d\epsilon dw\Big\}, \quad (20)$$

where $\lambda_{is}^E = (1-\alpha_R)m_i m_s$ for $(i,s) \in \{(1,2),(2,1),(3,4),(4,3)\}$ and $\lambda_{is}^E = m_i m_s$ for all other pairs (i,s). The corresponding reactive terms are

$$J_i^R = \alpha_R m_i m_j \sigma_{ij}^2 \int_{R^3\times S_+^2} \Big[f_{kl}^{(2)}(t,x,v',x-\sigma_{ij}\epsilon,w') - f_{ij}^{(2)}(t,x,v,x+\sigma_{ij}\epsilon,w) \Big] \langle \epsilon, v-w\rangle d\epsilon dw. \tag{21}$$

The pairs (i,j) and (k,l) are associated with the set of quadruples (i,j,k,l) given in (15), while the pair distribution function $f_{ij}^{(2)}$ in (20-21) is given by (18). I remark that in contrast to a non-reactive four species revised Enskog equation, there is no mass conservation for the individual species in the above model. We have

Theorem

The kinetic model (12) with J_i^E and J_i^R given by (20) and (21), respectively, possesses an H-function $H(t)$ of the form

$$H(t) = \sum_{i=1}^{4} \int_{\Omega\times R^3} f_i(t,x,v)\log f_i(t,x,v)\, dvdx - H_{corr}(t),$$

with

$$H_{corr}(t) = \tfrac{1}{2}\sum_{i=1}^{4}\sum_{j=1}^{4} \int \rho_i(t,r_1)\rho_j(t,r_2)V(12)\, dr_1 dr_2$$
$$+\sum_{k=3}^{\infty}\frac{1}{k!}\sum_{i_1,\dots,i_k=1}^{4} \int dr_1 \cdots \int dr_k\, \rho_{i_1}(1)\cdots\rho_{i_k}(k)\, V(12\cdots k),$$

where $\rho_{i_j}(j) = \rho_{i_j}(t,r_j) = m_{i_j} n_{i_j}(t,r_j)$ and $V(12\cdots k)$ is the sum of all irreducible Mayer graphs which doubly connect k particles.

For example,

$$\rho_i(t,r_1)\rho_j(t,r_2)V(12) = \rho_i(t,r_1)\rho_j(t,r_2)f^{ij}(r_1,r_2)$$

and

$$\rho_i(t,r_1)\rho_j(t,r_2)\rho_k(t,r_3)V(123) =$$
$$\rho_i(t,r_1)\rho_j(t,r_2)\rho_k(t,r_3)f^{ij}(r_1,r_2)f^{ik}(r_1,r_3)f^{jk}(r_2,r_3),$$

where the Mayer factors $f^{ij}(r_1,r_2) = \Theta(|r_1-r_2|-\sigma_{ij})-1$, with Θ being the step function. As in the case of non-reactive hard-sphere mixture one expects

$$-\beta H_{corr} = \mathcal{A}^{excess}$$

to be the excess free energy for the mixture of hard-spheres in the **non-uniform** equilibrium.

The proof of the Theorem follows almost the same line of arguments as in the non-reactive, one specie, revised Enskog equation [6], and will be provided in a forthcoming paper.

Finally, if f_i $(i = 1, 2, 3, 4)$ are nonnegative solutions of (12) with J_i^E and J_i^R given by (20) and (21), the following equivalent statements characterize the equilibrium states:

1. $f_i = n_i(t,x)\left(\frac{m_i}{2\pi kT(t)}\right)^{3/2} \exp\left(-\frac{m_i(v-u(t))^2}{2kT(t)}\right)$ and
   $n_1(t,x)n_2(t,x) = n_3(t,x)n_4(t,x)$, (the equilibrium reaction rate)

2. $\sum\limits_{i=1}^{4} \int\limits_{\Omega\times R^3} \left[J_i^E(\{f_i\}) + J_i^R(\{f_i\})\right] \log f_i \, dv dx = 0.$

References

1. M. T. Marron, J. Chem. Phys. **52**, 4060 (1970) .
2. N. Xystris and J. S. Dahler, *Kinetic theory of simple reacting spheres*, J. Chem. Phys. **68**, 387 (1978).
3. S. Bose and P. Ortoleva, *Reacting hard sphere dynamics: Liouville equation for condensed media*, J. Chem. Phys. **70**, 3041 (1979).
4. J. Polewczak, *The kinetic theory of simple reacting spheres: I. Global existence results in a dilute-gas limit*, J. Stat. Physics **100**, 327 (2000).
5. R. Kapral, *Kinetic theory of chemical reactions in liquids*, in **Advances in Chemical Physics**, eds. I. Prigogine and S. A. Rice, vol. 48, J. Wiley and Sons, New York, 1981, pp. 71–181.
6. M. Mareschal, J. Bławzdziewicz, and J. Piasecki, *Local entropy production from the revised Enskog equation: General formulation for inhomogeneous fluids*, Phys. Rev. Lett. **52**, 1169 (1984).

FREE BOUNDARY IN RADIAL SYMMETRIC CHEMOTAXIS

MARIO PRIMICERIO
Dipartimento di Matematica, "U.Dini", Viale Morgagni 67/a, 50134, Firenze, Italia
E-mail: primicer@math.unifi.it

AND

BORIS ZALTZMAN
DEEP, Blaustein Institute for Desert Research, Ben-Gurion University of the Negev,
Sede-Boqer Campus, 84990, Israel
E-mail: boris@bgumail.bgu.ac.il

We consider a parabolic-elliptic system of partial differential equations modelling the chemotaxis. We assume that the concentration of the organisms cannot exceed a limit value $\overline{A}$. Consequently, a free boundary can exist separating a region where $A = \overline{A}$ from the region where $A < \overline{A}$. In this paper we generalize the results of our privious study of the one-dimensional free boundary problem to the two- and three-dimensional radial symmetric cases.

1 Introduction

In this paper we study a free boundary problem arising in chemotaxis. As it is well known, chemotaxis is the motion of the individuals of a (cellular) population induced by the gradient of a chemical substance, often secreted by the members of the same population. The phenomenon plays as important role in ecology, in morphogenesis and in many other biological processes (see e.g. Ref.[1]).

Mathematical modelling of chemotaxis was first proposed in the pioneer paper by E. F. Keller and L. A. Segel[2]. In dimensionless variables the respective governing equations read as follows

$$\frac{\partial a}{\partial t} = \nabla\left(\nabla a - a\nabla p\right), \tag{1}$$

$$\varepsilon\frac{\partial p}{\partial t} = ka - \eta p + \Delta p, \tag{2}$$

which hold for $\underline{x}$ in a regular domain Ω and for $t > 0$, and which are to be solved once proper initial and boundary conditions are prescribed. The unknown functions a and p correspond to the dimensionless concentrations of cells and the attractant, respectively. The coefficient ε, which is typically very small, is the ratio of the diffusivities of the cellular population and attractant, and the coefficients $k = O(1)$ and $\eta = O(\varepsilon)$ are dimensionless rates of production and spontaneous decay of the chemoattractant.

Attention of mathematicians has been mainly devoted to the possible occurrence of the so-called chemotactic collapse, i.e. the concentration of the whole population in a single point in finite time. Problem (1), (2) was considered in Refs.[3−9] assuming that total flux of $\frac{\partial}{\partial n}a - a\frac{\partial}{\partial n}p$ vanishes on $\partial\Omega$, together with the flux of chemoattractant $\frac{\partial}{\partial n}p$, and that the initial data

$$a(\underline{x},0) = a_0(\underline{x}), \quad p(\underline{x},0) = p_0(\underline{x}), \quad \underline{x} \in \Omega$$

satisfy standard regularity conditions. It was proved that if the problem is one-dimensional, then a solution (a,p) exists for all time and moreover $\int_\Omega a^2(\underline{x},t)d\underline{x}$ is finite for any t. On the other hand, it was also shown that singularity can actually develop in a two- and three-dimensional radial symmetric problem.

In what follows we assume that zero total flux condition for $a(x,t)$ holds on fixed boundary $S = \partial\Omega$ together with the constant Dirichlet boundary condition on the chemoattractant concentration.

Letting ε tend to zero we obtain the following limiting problem

$$\frac{\partial a}{\partial t} = \nabla\left(\nabla a - a\nabla p\right), \quad \text{in } \Omega \tag{3}$$

$$\Delta p = -ka, \tag{4}$$

$$\frac{\partial a}{\partial n} - a\frac{\partial p}{\partial n} = 0 \quad \text{on } S, \tag{5}$$

$$p = p^1 = \text{const} \quad \text{on } S. \tag{6}$$

The aim of the present paper is to introduce in the model the concept of "maximum packing" of the cells: in other words, we assume that cells concentration A cannot exceed a value $\overline{A}$ corresponding to the situation where the intercellular spacing essentially reduces to zero. To "cast" this constraint – which appears to be quite natural – into the model we assume that the sensitivity of cells to chemotactic stimulus vanishes for $A > \overline{A}$ and, consequently, in dimensionless variable we obtain

$$\frac{\partial a}{\partial t} = \nabla\left(\nabla a - a\chi(a)\nabla p\right), \quad \text{in } \Omega \tag{7}$$

$$\Delta p = -ka, \tag{8}$$

$$\frac{\partial a}{\partial n} - \chi(a)a\frac{\partial p}{\partial n} = 0 \quad \text{on } S, \tag{9}$$

$$p = p^1 = \text{const} \quad \text{on } S, \tag{10}$$

with

$$\chi(a) = \begin{cases} 1, & 0 \le a < 1, \\ \in [0,1], & a = 1, \\ 0, & a > 1. \end{cases} \tag{11}$$

The problem (7)–(11) was studied in one-dimensional case in Ref.[10]. In this paper we generalize the results obtained in Ref.[10] to the two- and three-dimensional radial-symmetric cases. The plan of the paper is the following. In Sec.2 we give the definition of weak solution to problem (14)–(19). We then restrict ourselves by the consideration of monotonically decreasing initial data and prove that monotonicity is preserved for all times. Moreover, continuous and monotone dependence upon the initial datum is proved, implying uniqueness. In Sec.3 we consider stationary solutions and discriminate situations in which a region $a = 1$ (aggregation case) exists or not. Then, we prove that every weak solution converge in $C^\alpha(0,1)$ to the stationary solution corresponding to the same initial datum. In Sec.4 we study the regularity of the solution i.e. the smoothness of the free boundary $x = R(t)$ separating the aggregated region $a = 1$ from the region where $a < 1$. We consider the situations $R(0) > 0$ and $R(0) = 0$ separately giving some additional information on the free boundary.

2 Radial-symmetric free boundary chemotaxis problem

Hereinafter we restrict ourselves to the consideration of the problem (7)–(11) in the radial–symmetric case. Rewriting this problem in polar ($n = 2$) and spherical ($n = 3$) coordinates, respectively, we obtain

$$\frac{\partial a}{\partial t} = \frac{1}{r^{n-1}} \frac{\partial}{\partial r} \left(r^{n-1} \frac{\partial a}{\partial r} - a\chi(a) r^{n-1} \frac{\partial p}{\partial r} \right) \quad \text{for } r < 1, \tag{12}$$

$$\frac{1}{r^{n-1}} \frac{\partial}{\partial r} \left(r^{n-1} \frac{\partial a}{\partial r} \right) = -ka \quad \text{for } r < 1, \tag{13}$$

$$\frac{\partial a}{\partial r} - \chi(a) a \frac{\partial p}{\partial r} = 0 \qquad \text{for } r = 1, \tag{14}$$

$$p = p^1 \qquad \text{for } r = 1. \tag{15}$$

To define the weak solution to the problem (12)–(15) we integrate the function a in $\underline{x}$

$$G(r,t) \stackrel{def}{=} \int_{|\underline{x}|<r} a(\underline{x},t) d\underline{x} \Big/ \begin{cases} 2\pi, & n = 2 \\ 4\pi^2, & n = 3 \end{cases} = \int_0^r r^{n-1} a dr, \tag{16}$$

where $r = |\underline{x}|$. This integration reduces the parabolic–elliptic system (12)–(15) to a parabolic initial-boundary value problem for the function $G(r,t)$

$$G_t = r^{n-1}\frac{\partial}{\partial r}\left(\frac{1}{r^{n-1}}\frac{\partial G}{\partial r}\right) + k\chi\left(\frac{1}{r^{n-1}}\frac{\partial G}{\partial r}\right)\frac{1}{r^{n-1}}\frac{\partial G}{\partial r}G, \quad r<1,\ 0<t<T, \tag{17}$$

$$G(1,t) = M \stackrel{def}{=} \int_0^r r^{n-1}a_0 dr,\ 0<t<T, \tag{18}$$

$$G(r,0) = G_0(r), \quad r<1. \tag{19}$$

Definition *The function $a(x,t)$ is a weak solution to the problem (12)–(15) if the function $G(r,t)$, defined by (16), belongs to the space $W_2^{2,1}((0,1)\times(0,T))$, satisfies the conditions (18), (19) and the equation (17) almost everywhere in $(0,1)\times(0,T)$.*

Similarly to Theorem 1 of Ref.[10], we establish

Theorem 1 *Let us assume that $a_0 \in L^\infty(0,1)$ and $0 \le a_0 \le 1$ a.e. in $(0,1)$. Then there exists a weak solution to the problem (17)–(19) such that*

$$G(r,t) \in W_p^{2,1}((0,1)\times(0,T)) \ \textit{for all}\ p>1, \tag{20}$$

$$0 \le G(r,t) \le M, \qquad 0<r<1,\ 0<t<T, \tag{21}$$

$$0 \le G_r \le r^{n-1}, \qquad 0<r<1,\ 0<t<T. \tag{22}$$

If a_0 is decreasing in r, then

$$\frac{\partial}{\partial r}\left(\frac{1}{r^{n-1}}\frac{\partial G}{\partial r}\right) \le 0 \ \textit{a.e. in}\ (0,1)\times(0,T). \tag{23}$$

The uniqueness is obtained only for weak solutions monotonic in r:

Theorem 2 *The decreasing in r weak solution ($a_r(x,t) \le 0$ and $\frac{\partial}{\partial r}\left(\frac{1}{r^{n-1}}\frac{\partial G}{\partial r}\right) \le 0$) to the problem (17)–(19) is unique.*

Proof. First, we prove the stability of any decreasing weak solution. We consider the difference $\overline{G}(r,t) = G^2(r,t) - G^1(r,t)$ of two different solutions to the problem (17)–(19) corresponding to the initial conditions $G^1(r,0) = G_0^1(r)$ and $G^2(r,0) = G_0^2(r)$, respectively, such that $\frac{\partial}{\partial r}\left(\frac{1}{r^{n-1}}\frac{\partial G^1}{\partial r}\right) \le 0$, $\frac{\partial}{\partial r}\left(\frac{1}{r^{n-1}}\frac{\partial G^2}{\partial r}\right) \le 0$ a.e. in $(0,1)\times(0,T)$. Taking the difference of the respective equations for functions G^1 and G^2 we obtain

$$\overline{G}_t = r^{n-1}\frac{\partial}{\partial r}\left(\frac{1}{r^{n-1}}\frac{\partial \overline{G}}{\partial r}\right) + k\chi\left(\frac{1}{r^{n-1}}\frac{\partial G^1}{\partial r}\right)\frac{1}{r^{n-1}}\frac{\partial G^1}{\partial r}\overline{G} +$$
$$k\chi\left(\frac{1}{r^{n-1}}\frac{\partial G^1}{\partial r}\right)\frac{1}{r^{n-1}}\frac{\partial \overline{G}}{\partial r}G^2 +$$
$$k\left[\chi\left(\frac{1}{r^{n-1}}\frac{\partial G^1}{\partial r}\right) - \chi\left(\frac{1}{r^{n-1}}\frac{\partial G^2}{\partial r}\right)\right]\frac{1}{r^{n-1}}\frac{\partial G^2}{\partial r}G^2, \quad r<1, \ \ 0<t<T. \tag{24}$$

Since $\frac{\partial}{\partial r}\left(\frac{1}{r^{n-1}}\frac{\partial G^i}{\partial r}\right) \le 0$, $i=1,2$, then

$$\chi\left(\frac{1}{r^{n-1}}\frac{\partial G^i}{\partial r}\right) = \begin{cases} 1, & G^i - \frac{r^n}{n+1} < 0 \\ \in [0,1], & G^i - \frac{r^n}{n+1} = 0 \end{cases}, \quad i=1,2$$

and

$$\left[\chi\left(\frac{1}{r^{n-1}}\frac{\partial G^1}{\partial r}\right) - \chi\left(\frac{1}{r^{n-1}}\frac{\partial G^2}{\partial r}\right)\right](G^1 - G^2) \le 0 \quad \text{in} \quad (0,1)\times(0,\tau). \tag{25}$$

Multiplying equation (24) by $\overline{G}/r^{n-1}$ and using inequality (25) we obtain

$$\frac{d}{dt}\int_0^1 \overline{G}^2/r^{n-1}dr \le \kappa\int_0^1 \overline{G}^2/r^{n-1}dr, \tag{26}$$

for some positive constant $\kappa > 0$. The inequality (26) yields

$$\int_0^1 \overline{G}^2/r^{n-1}dr \le e^{\kappa t}\int_0^1 \left(G_0^1(r) - G_0^2(r)\right)^2/r^{n-1}dx. \tag{27}$$

The stability estimate (27) yields the uniqueness of the decreasing weak solution to the problem (17)–(19) and completes the proof. ■

Hereinafter, we restrict ourselves to consider decreasing weak solutions $a(r,t)$. Similarly to Ref.[10] we establish the following results

Lemma 3 *If $G_0^1(r) \ge G_0^2(r)$ and the functions $a_o^i(r)$ decrease in r, then $G^1(r,t) \ge G^2(r,t)$ for $0<r<1$, $0<t<T$, where G^1 and G^2 are the solutions to the problem (17)–(19) with the initial conditions $G^i(r,0) = G_0^i(r)$, $i=1,2$, respectively.*

Lemma 4 *If $G_t|_{t=0} = r^{n-1}\frac{\partial}{\partial r}\left(\frac{1}{r^{n-1}}\frac{\partial G_0}{\partial r}\right) + \kappa\chi\left(\frac{1}{r^{n-1}}\frac{\partial G_0}{\partial r}\right)\frac{1}{r^{n-1}}\frac{\partial G_0}{\partial r}G_0 \ge 0$ a.e. in $(0,1)$ then $G_t(r,t) \ge 0$ for a.e. $(r,t)\in(0,1)\times(0,T)$. If $G_t|_{t=0} \le 0$ a.e. in $(0,1)$ then $G_t(r,t) \le 0$ for a.e. $(r,t)\in(0,1)\times(0,T)$.*

3 Stationary solutions to the chemotaxis free boundary problem

We start with a consideration of the stationary solution to the problem (17)–(19) in two-dimensional case, $n = 2$. First, we consider a stationary solution assuming $a = G_r/r$ being less than threshold $a = 1$. Then the respective stationary boundary-value problem reads

$$r\frac{\partial}{\partial r}\left(\frac{1}{r}\frac{\partial G}{\partial r}\right) + k\frac{1}{r}\frac{\partial G}{\partial r}G = 0, \qquad r < 1, \tag{28}$$

$$G(0) = 0, \quad G(1) = M. \tag{29}$$

Integrating the problem (28), (29) we find

$$G(r) = \frac{r^2}{\frac{1}{M} + \frac{k}{4}\left(r^2 - 1\right)}, \ r \leq 1. \tag{30}$$

The function $a = G_r/r$ tends to infinity at $r = 0$ as $M \to 4/k$ and the solution (30) breaks down for $M \geq 4/k$. This corresponds to the chemotaxis blowup in the time-dependent problem (17)–(19) (see Ref.[11]). The function a reaches threshold 1 at $r = 0$ for $M = 4/\left(k+8\right)$ and the problem (28), (29) ceases to hold for $M > 4/\left(k+8\right)$. The respective free boundary stationary problem reads

$$r\frac{\partial}{\partial r}\left(\frac{1}{r}\frac{\partial G}{\partial r}\right) + k\frac{1}{r}\frac{\partial G}{\partial r}G = 0, \qquad r_0 < r < 1, \tag{31}$$

$$G(r_0) = r_0^2/2, \tag{32}$$

$$\frac{\partial G}{\partial r}(r_0) = r_0, \tag{33}$$

$$G(1) = M. \tag{34}$$

Integrating the problem (31)–(34) we obtain that

$$G = \left(A_1\left(\frac{r}{r_0}\right)^{-kR_0} + \frac{1}{2R_0}\right)^{-1} + \frac{2}{k} - R_0, \qquad r_0 \leq r \leq 1, \quad \text{where} \tag{35}$$

$$A_1 = \frac{1}{R_0 + \frac{r_0^2}{2} - \frac{2}{k}} - \frac{1}{2R_0}, \tag{36}$$

$$R_0 = \sqrt{\frac{4}{k^2} + \frac{r_0^4}{4}}; \tag{37}$$

and the free boundary r_0 is the solution of the following algebraic equation

$$G(1) = M. \tag{38}$$

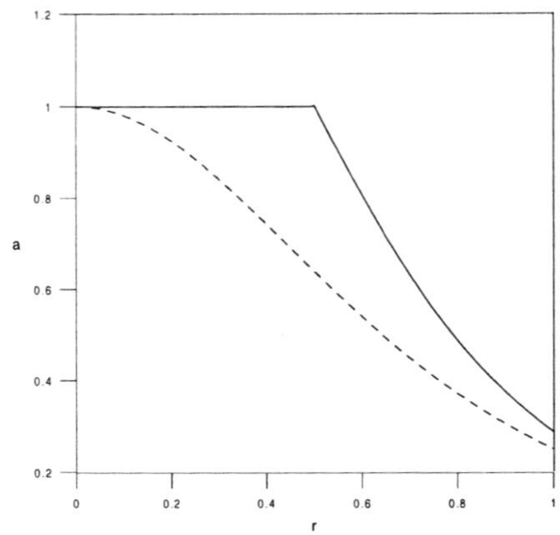

Figure 1. Two-dimensional stationary solutions: $k = 8$, $M = 0.25$ (dashed line) and $M = \frac{1}{3}$ (continuous line).

In Fig.1 we present the graphs of stationary solutions for $k = 8$ corresponding to the critical value of $M = 0.25$ and $M = \frac{1}{3}$, respectively.

In three-dimensional case the stationary problem reads as follows

$$r^2 \frac{\partial}{\partial r}\left(\frac{1}{r^2}\frac{\partial G}{\partial r}\right) + k\chi\left(\frac{1}{r^2}\frac{\partial G}{\partial r}\right)\frac{1}{r^2}\frac{\partial G}{\partial r}G = 0, \quad 0 < r < 1, \tag{39}$$

$$G(1) = M. \tag{40}$$

Considering the possible solution to the following stationary problem

$$r^2 \frac{\partial}{\partial r}\left(\frac{1}{r^2}\frac{\partial G}{\partial r}\right) + k\frac{1}{r^2}\frac{\partial G}{\partial r}G = 0, \quad 0 < r < 1,$$

$$G(1) = M,$$

in the vicinity of $r = 0$ and comparing it with the solution to the problem (28)–(29) we find that there is no such solution with a finite and non-negative function $a = (\partial G/\partial r)G/r^2$ for all values of $M > 0$. This corresponds to the occurence of the chemotaxis blowup in the time-dependent problem (17)–(19) for all values of $M > 0$ in three dimensions (see Ref. [11]). The solution to the respective free boundary problem

$$r^2 \frac{\partial}{\partial r}\left(\frac{1}{r^2}\frac{\partial G}{\partial r}\right) + k\frac{1}{r^2}\frac{\partial G}{\partial r}G = 0, \quad r_0 < r < 1,$$
$$G(r_0) = r_0^3/3,$$
$$G'(r_0) = r_0^2,$$
$$G(1) = M,$$

is unique and may be found using the numerical integration of this problem.

Lemma 5 *Every weak solution $G(r,t)$ converges to the respective stationary solution $G^{\infty}(r)$, corresponding to the same values of κ and M, as $t \to \infty$, in the following sense*

$$G(r,t) \underset{t\to\infty}{\longrightarrow} G^{\infty}(r) \quad \textit{in} \;\; C^{1+\alpha}[0,1], \quad 0 < \alpha < 1 \tag{41}$$

Proof. The proof is based on the comparison of the solution $G(r,t)$ with the upper barrier function $\overline{G}(r,t)$ and the lower barrier $\underline{G}(r,t)$, which are, respectively, decreasing and increasing in time solutions to the problem (7)–(11) with the initial conditions

$$\underline{G}(r,0) = Mr^n, \;\; r \leq 1$$

and

$$\overline{G}(r,0) = \begin{cases} r^n/n, & r \leq \sqrt[n]{Mn}; \\ M, & \sqrt[n]{Mn} \leq r \leq 1, \end{cases}$$

respectively. ■

4 Classical solution to the chemotaxis free boundary problem

In this Section we study the regularity of the solution to the problem (12)–(15), i.e. the smoothness of the free boundary $x = R(t)$ separating the aggregated region $a = 1$ from the region where $a < 1$. The respective chemotaxis free boundary problem reads as follows

$$a(r,t) = \frac{1}{r^{n-1}}\frac{\partial G}{\partial r} = 1, \;\; p_r = -k\frac{r^{n-1}}{n}, \quad r \leq R(t), \tag{42}$$

$$G_t = r^{n-1}\frac{\partial}{\partial r}\left(\frac{1}{r^{n-1}}\frac{\partial G}{\partial r}\right) + k\frac{1}{r^{n-1}}\frac{\partial G}{\partial r}G, \quad R(t) < r < 1, \;\; 0 < t < T, \tag{43}$$

$$G(R(t),t) = \frac{R^n(t)}{n}, \quad G_r(R(t),t) = R^{n-1}(t)a(R(t),t) = R^{n-1}(t), \quad 0 < t < T, \tag{44}$$

$$\left(a(R(t),t) = 1, \;\; a_r(R(t)+,t) = -k\frac{R(t)}{n}\right), \quad 0 < t < T$$

$$G(1,t) = M, \quad 0 < t < T, \tag{45}$$

$$G(x,0) = G_0(x), \quad 0 < x < 1. \tag{46}$$

In what follows we address two typical situations $R(0) = 0$ and $R(0) > 0$, studying, respectively, the smoothness of the free boundary in the case when aggregated zone appears at the initial instant and for initially existing aggregated zone.

Theorem 6 *If the following conditions on the initial datum $a_0(r) = a(r,0)$ hold*

$$\begin{aligned} a_0(0) = 1, \;\; a_0'(0) = 0,& \\ a_0(r) < 1, \;\; a_0'(r) \leq 0,& \quad x \in (0,1], \\ \gamma \stackrel{def}{=} a_0''(0) \in (-k/n, 0) \quad \text{and}& \quad a_0 \in C^2[0,1], \end{aligned}$$

then the free boundary appears and has the following initial asymptotic behavior

$$\lim_{t\to 0+} \frac{R(t)}{\sqrt{t}} = \alpha > 0, \tag{47}$$

where α is a root of the following algebraic equation

$$e^{\alpha^2/4}\frac{k\alpha^4}{8}\int_\alpha^\infty \frac{1}{\xi}e^{-\xi^2/4}d\xi - \frac{k\alpha^2}{4} = -\frac{k}{2} - \gamma, \quad n = 2; \tag{48}$$

$$e^{\alpha^2/4}\frac{k\alpha^5}{12}\int_\alpha^\infty \frac{1}{\xi^2}e^{-\xi^2/4}d\xi - \frac{k\alpha^2}{6} = -\frac{k}{3} - \gamma, \quad n = 3. \tag{49}$$

Proof. To prove Theorem 6 we use the following similarity solution

$$\widetilde{G}(r,t) = \int_0^r \int_0^t r^{n-1}\theta(r,t)drdt + \frac{r^n}{n} + \widetilde{\gamma}\frac{r^{n+2}}{n+2},$$
$$\theta(r,t) = \Theta(\xi), \quad \xi = \frac{r}{\sqrt{t}},$$
$$\widetilde{R}(t) = \widetilde{\alpha}\sqrt{t},$$

where Θ and $\alpha > 0$ are a solution to the following boundary-value problem

$$\Theta'' + \Theta'\left(\frac{1}{2}\xi + \frac{n-1}{\xi}\right) = 0,$$
$$\Theta(\alpha) = k\frac{\alpha^2}{2n},$$
$$\Theta'(\alpha) = -k\frac{\alpha^3}{4n},$$
$$\Theta(\infty) = \widetilde{\gamma} + \frac{k}{n}.$$

Comparing the solution to the problem (42)–(46) with $\widetilde{G}(r,t)$, we find that in the sufficiently small vicinity of the point $r = 0$, $t = 0$, the following estimates hold

$$G(r,t) \le \widetilde{G}(r,t) \quad \text{if } \widetilde{\gamma} > \gamma,$$
$$G(r,t) \ge \widetilde{G}(r,t) \quad \text{if } \widetilde{\gamma} < \gamma.$$

Taking the limit $\widetilde{\gamma} \to \gamma$ completes the proof. ∎

Theorem 7 *Let the following conditions hold*

$$R(0) = R_0 > 0,$$
$$a_0(r) = 1 \quad \textit{for } 0 \le r \le R_0 \quad \textit{and} \quad a_0(r) < 1, \; a_0'(r) \le 0 \textit{ for } r \in (R_0, 1]$$
$$a_0(r) \in C^1[R_0, 1].$$

Then there exists a solution to the problem (42)–(46) for all $t \in [0, t_0]$, *where* t_0 *is an instant of a possible degeneration of the free boundary* $\lim_{t \to t_0-} R(t) = 0$.

Proof. The proof of the local existence of the classical solution is rather standard. The more interesting is the proof of the existence of the global in time classical (smooth) solution to the problem (42)–(46) . Here we only sketch the main ideas of the proof. We introduce the new unknown function $v = a_r + kr/n$ and consider the respective free boundary problem on it. At first, comparing this function v with the respective similarity solutions, used as upper and lower barrier functions, we obtain the following Hölder continuity of the free boundary $R(t)$

$$|R(t_1) - R(t_2)| \leq \eta\,|t_1 - t_2|\,, \quad \eta > 0,$$

for all t_1, $t_2 \in [0, t_\delta]$, where

$$t_\delta = \inf\{t > 0,\ R(t) \leq \delta\}.$$

At the next stage we use this estimate in order to find more appropriate upper and lower barrier solutions. The comparison of the function v with these barrier functions yields the boundedness of the free boundary velocity $R'(t)$ on the interval $[0, t_\delta]$ and completes the proof. The techniques used on the last stage is quite similar to that used in Ref.[12] for the analysis of the global solvability of the undercooled Stefan problem. ∎

Finally, in Fig.2a,b we present the graphs of the solution $a(r,t)$ to the three-dimensional free-boundary problem (12)–(15), with $k = 1$ and $a_0(r) = \ln(1 + r^3)/10$. The graph of the free boundary $R(t)$ is presented in Fig.2a and in Fig.2b we present the long-time asymptotic of the solution $a(r,t)$: $A(r) = \lim_{t\to\infty} a(r,t)$.

References

1. W. Alt, *Lecture Notes in Biomath.* **38**, 353 (1980).
2. E.F. Keller, L.A. Segel, *J. Theor. Biol.* **26**, 399 (1970).
3. S. Childress, J.K. Percus, *Math. Biosc.* **56**, 217 (1981).
4. M.A. Herrero, *in Applied and Industrial Mathematics*, (R. Spigler ed. Kluwer 2000), 89.
5. M.A. Herrero, E. Medina, J.J.L. Velazquez, *J. Comp. Appl. Math.* **97**, 99 (1998).
6. M.A. Herrero, M.A., J.J.L. Velazquez, *J. Math. Biol.* **35**, 177 (1996).
7. M.A. Herrero, J.J.L. Velazquez, *Math. Ann.* **306**, 583 (1996).
8. M.A. Herrero, J.J.L. Velazquez, *Ann. Scuola Normale Sup. Pisa* **XXIV**, 633 (1997).
9. W. Jager, S. Luckhaus, *Trans. Amer. Math. Soc.* **329**, , (819) 1992
10. M. Primicerio, B. Zaltzman, *Interfaces and Free Boundaries* , (subm. publ.).
11. T. Nagai, *Adv. Math. Sci. Appl.* **5**, 1 (1995).
12. A. Fasano, M. Primicerio, *Math. Methods Appl. Sci.* **5**, 84 (1983).

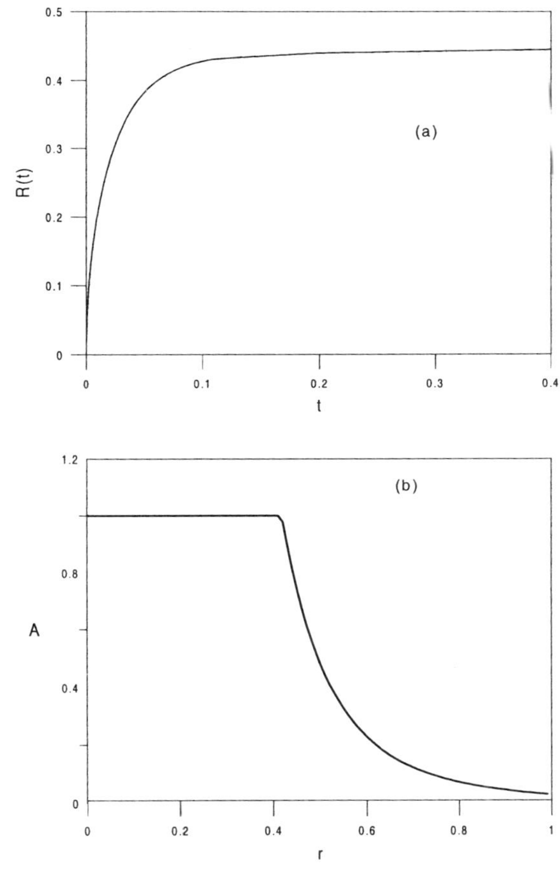

Figure 2. Three-dimensional radial symmetric stationary solution to the 'obstacle' chemotaxis problem: (a) free boundary $R(t)$, (b) stationary solution $A(r)$

MODELING OF DISSIPATIVE PROCESSES

K.R. RAJAGOPAL
Department of Mechanical Engineering & Mathematics
Texas A&M University
College Station, TX 77840-3123
USA

1 Introduction

Recently, a thermodynamic frame-work has been developed that has been used to study a variety of dissipative phenomena: traditional plasticity, twinning, solid to solid phase transitions, response of multi-network polymers, viscoelastic response of materials, crystallization of polymers and growth and adaptation of biological materials. The salient features of the theory will be discussed here.

In the case of a classical elastic solid, it is assumed that there is only one stress-free configuration of the body, modulo rigid body motion. As the body is deformed, its underlying stress-free state (natural configuration) does not change. From a microstructural perspective, the underlying micro-mechanism does not change due to the deformation. However, there are many real materials which can exist stress-free in a variety of configurations, these configurations not being related to each other by a rigid motion. An example of the same is the response exhibited by a traditional "plastic" material. We see such bodies have an infinity of natural configurations (stress-free configurations).

To describe the response of such bodies, it is necessary to know (or to infer) how the natural configurations evolve as the body undergoes a thermodynamic process. In addition to knowing how the natural configurations evolve, it is necessary to know the response of the materials from these configurations. In general, it is not necessary to use a natural configuration but any number of an equivalence class of configurations to which this natural configuration belongs (see Rajagopal (2002)). For instance, in the case of traditional plasticity, it is necessary to know the stress response from these various natural configurations (i.e., the shape of the response curves OP, AP', BP," etc). Thus, traditional plasticity can be considered as a multi-parameter (in general three-dimensional case) family of elastic responses from evolving natural configurations. From a three-dimensional perspective, it is not just the shape of this curve but also the invariance properties of the response functions. It is possible that with regard to these various natural configurations, the material's symmetries are not related by Noll's rule. This allows us to model the change of symmetries that accompany solid to solid phase transitions as well as crystallization of polymers and in general solidification and melting.

Here, we shall discuss briefly, through an example, the basic notions associated with materials that have more than one natural configuration.

When energy is supplied to a body, a part of the energy could be stored, a part of it could be converted to energy in its thermal form (it is said to be dissipated) a part of the energy could go towards the latent energy[1], a part towards the latent heat etc. Here, we will restrict ourselves to a purely isothermal problem and we will not consider phase transitions. Of the energy that is stored, a part could be recovered in a purely mechanical process while another part might be recovered only in a thermal process (e.g. Energy stored in the dislocation structure can be recovered by annealing but not in a purely mechanical process). To describe a real material we need to make all these specifications concerning the body. The general framework is discussed elsewhere with regard to the multi-network response of polymers (Rajagopal and Wineman (1992), Wineman and Rajagopal (1990), Huntley, Rajagopal and Wineman (2000)), twinning (Rajagopal and Srinivasa (1995), (1997), Lapcyzk, Rajagopal and Srinivasa (2000)), traditional plasticity (Rajagopal and Srinivasa (1998), (1998)) solid to solid phase transitions (Rajagopal and Srinivasa (2001)), viscoelasticity (Rajagopal and Srinivasa (2000), Muralikrishnan and Rajagopal (2002)), Anisotropic liquids (Rajagopal and Srinivasa (2001))), crystallization in polymers (Rao and Rajagopal (1999), (2001), (2002)), Growth and adaptation in biological materials (Humphrey and Rajagopal (2002), Rao, Humphrey and Rajagoapl (2002)). Here, we shall illustrate our ideas by considering the response of viscoelastic fluids.

We shall consider viscoelastic fluids that have an instantaneous elastic response. The elastic response is determined by how the material stores its energy, namely the stored energy function (Helmholtz potential) and the rate of dissipation. Different forms for storing the energy and the rate of dissipation lead to different types of viscoelastic fluids. Such an approach to modeling in fact mirrors the original ideas of Maxwell for describing the response of viscoelastic fluids and it also reflects the mechanical analogs that have been used to describe such fluids via springs and dashpots.

2 Kinematics

Let $k_R(B)$ and $k_t(B)$ denote the reference and current configuration of the body B. The motion of the body χ_{k_R} assigns to each point $\mathbf{X} \in k_R(B)$, a point $\mathbf{x} \in k_t(B)$, at each instant of time t, i.e.,

$$\mathbf{x} = \chi_{k_R}(\mathbf{X}, t). \tag{1}$$

[1] Latent energy is not to be confused with latent heat. Latent energy is the difference in the internal energy between the phases and plays a very important role in phase transitions.

We shall denote $k_R(B)$ and $k_t(B)$, by k_R, k_t etc for the sake of convenience. The velocity $\mathbf{v}$ and the deformation gradient $\mathbf{F}_{k_R}$ are defined through

$$\mathbf{V} = \frac{\partial \chi_{k_R}}{\partial t}, \tag{2}$$

$$\mathbf{F}_{k_R} = \frac{\partial \chi_{k_R}}{\partial \mathbf{X}}. \tag{3}$$

Let $k_{p(t)}$ denote the natural configuration corresponding to the configuration at time t .

Now, the Cauchy-Green stretch tensor $\mathbf{B}_{k_R}$ and $\mathbf{C}_{k_R}$ are defined through

$$\mathbf{B}_{k_R} = \mathbf{F}_{k_R} \mathbf{F}_{k_R}^T, \qquad \mathbf{C}_{k_R} = \mathbf{F}_{k_R}^T \mathbf{F}_{k_R}. \tag{4}$$

We introduce the upper convected derivative of a tensor $\mathbf{A}$ is defined through

$$\overset{\nabla}{\mathbf{A}} = \frac{d\mathbf{A}}{dt} - \mathbf{LA} - \mathbf{AL}^T, \tag{5}$$

where $\frac{d}{dt}$ denotes the usual material time derivative and $\mathbf{L}$ denotes the velocity gradient

$$\mathbf{L} = \frac{\partial \mathbf{v}}{\partial \mathbf{x}}. \tag{6}$$

The symmetric part of the velocity gradient $\mathbf{D}$ and the Rivlin-Ericksen tensor $\mathbf{A}_1$ are defined through

$$\mathbf{D}=\frac{1}{2}\left[\mathbf{L}+\mathbf{L}^T\right], \qquad \mathbf{A}_1=2\mathbf{D}. \tag{7}$$

In the case of a body that has been deformed homogeneously, from the current configuration $\mathbf{k}_t$ we can instantaneously elastically unload to the natural configuration $k_{p(t)}$. We can then define the deformation gradient associated with the motion from $k_{p(t)}$ to k_t, i.e., $\mathbf{F}_{k_{p(t)}}$. This notion can be extended in the case of inhomogeneous deformations also and the reader referred to Rajagopal and Srinivasa (1998) where this issue is discussed at length. In general $\mathbf{F}_{k_{p(t)}}$ is not the gradient of a mapping but is a mapping from the tangent space at the point $\mathbf{X}_{k_{p(t)}} \in k_{p(t)}$ to the tangent space containing the point $x \in k_t$.

Having defined $\mathbf{F}_{k_{p(t)}}$, we then define the tensor $\mathbf{G}$ through

$$\mathbf{G} := \mathbf{F}_{k_{p(t)}}^{-1}\mathbf{F}_{k_R}. \tag{8}$$

We can then define

$$\mathbf{B}_{k_{p(t)}}=\mathbf{F}_{k_{p(t)}}\mathbf{F}_{k_{p(t)}}^T, \quad \mathbf{C}_{k_{p(t)}}=\mathbf{F}_{k_{p(t)}}^T\mathbf{F}_{k_{p(t)}}, \tag{9}$$

and

$$\mathbf{L}_{k_{p(t)}}=\dot{\mathbf{G}}\mathbf{G}^{-1}. \tag{10}$$

3 Thermodynamic Preliminaries

The stored energy ψ for the class of materials we are interested in is given by

$$\psi=\psi_{k_t}\left(\mathbf{F}_{k_{p(t)}}\right), \tag{11}$$

thus reflecting the fact that the elastic response depends on the deformation gradient from the natural configuration corresponding to the current state of the body at time t. The suffix k_t for the stored energy reflects the fact that the form of the elastic response can be different. Of course, a special example of the same is

$$\psi = \psi\left(\mathbf{F}_{k_{p(t)}} \right), \tag{12}$$

that is the form of the stored energy does not change with time.

The rate of dissipation ξ for the class of materials being considered is

$$\xi = \xi_{k_t}\left(\mathbf{F}_{k_{p(t)}}, \mathbf{G}, \dot{\mathbf{G}} \right). \tag{13}$$

Once again, a special case is

$$\xi = \xi\left(\mathbf{F}_{k_{p(t)}}, \mathbf{G}, \dot{\mathbf{G}} \right). \tag{14}$$

The rate of dissipation is given by

$$\xi = \mathbf{T} \cdot \mathbf{D} - \rho\dot{\psi} \tag{15}$$

when ζ is the density.

Since we shall be concerned with an incompressible fluid, it can undergo only isochoric motions and thus

$$tr\mathbf{D} = 0\,, \tag{16}$$

$$tr\mathbf{D}_{k_{p(t)}} = 0\,. \tag{17}$$

We will require that the fluid meet the above constraints in all processes.

Also, since the fluid is isotropic, the stored energy ψ depends on $\mathbf{F}_{k_{p(t)}}$ through the principal invariants of $\mathbf{B}_{k_{p(t)}}$, i.e., through

$$I_1 = tr\mathbf{B}_{k_{p(t)}}, \quad I_2 = \frac{1}{2}\left[\left(tr\mathbf{B}_{k_{p(t)}} \right)^2 - tr\mathbf{B}^2_{k_{p(t)}} \right], \tag{18}$$

and

$$I_3 = \det \mathbf{B}_{k_{p(t)}}. \tag{19}$$

However, as the material is incompressible $I_3 = 1$, and thus

$$\psi = \psi(I_1, I_2). \tag{20}$$

Also, as there is no dissipation if the natural configuration does not change, the choice of the rate of dissipation has to be such that

$$\xi = \hat{\xi}\left(\mathbf{F}_{k_{p(t)}}, \mathbf{L}_{k_{p(t)}}\right). \tag{21}$$

Frame Indifference and isotropy then imply that

$$\xi = \tilde{\xi}\left(\mathbf{B}_{k_{p(t)}}, \mathbf{D}_{k_{p(t)}}\right), \tag{22}$$

We shall assume that

$$\tilde{\xi}\left(\mathbf{B}_{k_{p(t)}}, \mathbf{0}\right) = 0, \tag{23}$$

that is if the underlying natural configuration does not change, there is no dissipation.

Unlike the usual procedure of using the Clausius-Duhem inequality as the second law of thermodynamics to place restrictions on the forms of the constitutive relations, we shall assume a constitutive form for the rate of dissipation that automatically meets the non-negativity of the rate of dissipation, thereby ensuing that the second law is met.

We shall determine the evolution of these natural configurations by requiring that the rate of dissipation be maximized. We shall require that the energy-dissipation equation be met in all processes, that is, we shall treat is as a constraint while maximizing the dissipation. Also, as we are interested in incompressible viscoelastic fluids, we need to enforce (16) and (17) as constraints.

4 Viscoelastic Fluid Models

We shall now show how the framework can be used to obtain generalizations of two very popular models in viscoelasticity, the Maxwell fluid and the Burger fluid. We start with a discussion of the Maxwell fluid.

We shall suppose that the stored energy has the special form

$$\psi = \frac{\mu}{2}(I_1 - 3), \tag{24}$$

where

$$I_1 = tr\mathbf{B}_{k_{p(t)}}, \tag{25}$$

i.e., the material stores energy like a neo-Hookean solid, but the strain is measured from the evolving set of configuration k_t. The constant μ is the sheer modulus associated with the elastic response.

Next, we suppose that the rate of dissipation given by

$$\xi = \eta \mathbf{B}_{k_{p(t)}} \cdot \mathbf{D}_{k_{p(t)}} \mathbf{B}_{k_{p(t)}}, \tag{26}$$

that is we have a dashpot which take into account the natural configuration is changing. The choice of the stored energy and rate of dissipation of the above forms in tantamount to a spring-dashpot model that is in series, in a one-dimensional model.

Carrying out the maximization of the rate of dissipation subject to the constraints (15), (16) and (17) leads to (see Rajagopal and Srinivasa (2000) for details):

$$\mathbf{T} = -p\mathbf{1} + \mathbf{B}_{k_{p(t)}}, \tag{27}$$

$$\overset{\nabla}{\mathbf{B}}_{k_{p(t)}} = \frac{\mu}{\varsigma}\left[\mathbf{B}_{k_{p(t)}} - \lambda \mathbf{1}\right], \tag{28}$$

where

$$\lambda = \frac{3}{\left[trB_{k_{p(t)}}^{-1} \right]}. \tag{29}$$

The classical Maxwell fluid is given through the following constitutive representation:

$$\mathbf{T} = -\hat{p}\mathbf{1} + \mathbf{S} \tag{30}$$

$$\mathbf{S} + \lambda_1 \overset{\nabla}{\mathbf{S}} = \eta \mathbf{D}\,. \tag{31}$$

Though the models (27), (28) and (29), (30) do not look similar, it is easy to show that the two are equivalent (Rajagopal and Srinivasa (2000)), provided we allow for small elastic displacement gradients, i.e.,

$$\left\| \mathbf{B}_{k_{p(t)}} - \mathbf{1} \right\| = 0(\varepsilon), \quad \varepsilon << 1\,. \tag{32}$$

Thus, the Maxwell fluid stores energy like a linearized elastic spring and dissipates like a viscous fluid, the reason $\mathbf{B}_{k_{p(t)}}$ appears in the expression for the dissipation is to account for the natural configuration changing or in the parlance of a spring-dashpot model, the spring and dashpot are in series.

The next model we document is the Burger's model (Burger (1939)) which is used to describe the response of asphalt. Materials such as asphalt have two characteristic relaxation times and it is best to model it as co-existing components with two natural configurations which move together. The stored energy ψ, are given through

$$\psi_i = \psi_i\left(\theta, (\mathrm{I}_1)_i, (\mathrm{I}_2)_i\right), \quad i = 1,2, \tag{33}$$

and the rate of dissipation given through

$$\xi_i = \xi_i\left(\theta, \mathbf{B}_{k_{p_i(t)}}, \mathbf{D}_{k_{p_i(t)}}\right), \quad i = 1,2. \tag{34}$$

As the material is incompressible we will have to impose the constraints (16) and (17).

Maximizing the rate of dissipation subject to (15), (16) and (17) as constraints leads to

$$\mathbf{T} = -\mathrm{p}\mathbf{1} + \mu_1 \mathbf{B}_{\mathrm{k}_{\mathrm{p(t)}}} + \mu_2 \mathbf{B}_{\mathrm{k}_{\mathrm{p(t)}}}, \tag{35}$$

$$\frac{1}{2}\overset{\nabla}{\mathbf{B}}_{\mathrm{k}_{\mathrm{p_i(t)}}} = \frac{\mu_{\mathrm{i}}}{\eta_1}\left[\frac{3}{\left(\mathrm{tr}\mathrm{B}^{-1}_{\mathrm{k}_{\mathrm{p_i(t)}}}\right)}\mathbf{1} - \mathbf{B}_{\mathrm{k}_{\mathrm{p_i(t)}}}\right] \tag{36}$$

where

$$\eta_1 = \overline{\eta}_1\left(\left(\mathrm{Ntr}\left(\mathbf{B}_{k_{p_i(t)}}\right) - 3\right)^m + 1\right), \tag{37}$$

$$\eta_2 = \overline{\eta}_2, \tag{38}$$

where $\overline{\eta}_1$ and $\overline{\eta}_2$ are constraints.

It can be shown that the above model is equivalent to

$$\mathbf{T} = -p\mathbf{1} + \overline{\mathbf{S}}, \tag{39}$$

$$\mathbf{S} = \mu_1 \mathbf{B}_{k_{p_i(t)}} + \mu_2 \mathbf{B}_{k_{p_i(t)}}, \tag{40}$$

$$\mathbf{S} + \alpha \overset{\nabla}{\mathbf{S}} + \beta \overset{\nabla\nabla}{\mathbf{S}} = \left[\gamma_1 \mathbf{D} + \gamma_2 \overset{\nabla}{\mathbf{D}}\right], \tag{41}$$

A very important aspect to describing these models for viscoelastic fluids as elastic responses from an evolving set of configurations is that in the form (35), (36) does not pose any difficulties one usually encounters with the prescription of boundary conditions: These issues are discussed elsewhere (see Rajagopal (2002)).

Thus, by picking different forms for the manner in which the materials stores energy and how it dissipates energy we can develop a variety of models for describing the behavior of viscoelastic fluids. As mentioned in the introduction, the framework provides for a great deal more than that, it can model a variety of processes in which entropy is produced (see references in the introduction by Rajagopal and his co-workers).

Acknowledgement: I thank the National Science Foundation for its support of this work.

Bibliography

1. J.C. Maxwell, On the dynamical theory of gases, *Phil. Trans. Roy. Soc.* London A **157**, 26-78 (1866).
2. K.R. Rajagopal and A.R. Srinivasa, On the inelastic behavior of Solids. Part I
3. K.R. Rajagopal and A.R. Srinivasa, On the inelastic behavior of Solids. Part II. Energetics of deformation twinning, *Intl. J. Plasticity*, **13**, 1-35 (1997).
4. K.R. Rajagopal and A.R. Srinivasa, Mechanics of the inelastic behavior of materials. Part I. *Intl. J. Plasticity*, **14**, 945-966 (1998).

5. K.R. Rajagopal and A.R. Srinivasa, Mechanics of the inelastic behavior of materials. Part II. *Intl. J. Plasticity*, **14**, 967-995 (1998).
6. K.R. Rajagopal and A.R. Srinivasa, A thermodynamic framework for rate type fluid models, *J. Non-Newtonian Fluid Mechanics*, **88**, 207-227 (2000).
7. K.R. Rajagopal, Report 6. Multiple Natural Configurations in Continuum Mechanics, Institute for Computational and Applied Mechanics, University of Pittsburgh (1995).
8. K.R. Rajagopal and A.S. Wineman, A constitutive equation for non-linear elastic materials which undergo deformation induced microstructural changes, *Intl. J. of Plasticity*, **8**, 385-395 (1992).
9. I.J. Rao and K.R. Rajagopal, A study of strain-induced crystallization in polymers, *Intl. J. Solids and Structures*, **38**, 1149-1169 (2001).
10. I.J. Rao and K.R. Rajagopal, A thermomechanical framework for the crystallization of polymers, *ZAMP*, **53**, 1-41 (2001).
11. I.J. Rao and K.R. Rajagopal, Phenomenological modeling of polymer crystallization using the notion of multiple natural configurations, *Interfaces and Free Boundaries*, **2**, 73-94 (2000).
12. K.R. Rajagopal and A.R. Srinivasa, Thermomechanical modeling of shape memory alloys, *ZAMP*, **50**, 459-496 (1999).
13. K.R. Rajagopal and N. Chandra, A Thermodynamic framework for the superplastic response of materials, Superplast. Adv. Mater., ICSAM-2000, *Material Sci. Forum*, **357**, 261-271 (2001).
14. I.J. Rao, J.D. Humphrey and K.R. Rajagopal, Biological growth and remodeling: An uniaxial example with application to tendons and ligaments, In Press, *Compt. Model. Eng. Science.*
15. A.S. Wineman and K.R. Rajagopal, On a constitutive theory for materials undergoing microstructural changes, *Arch. Mechanics*, **42**, 53-75 (1990).
16. J.D. Humphrey and K.R. Rajagopal, A constrained mixture model for growth and remodeling of soft tissues, *Math. Models and Methods in the Appl. Sciences*, In Press.
17. C. Eckart, The thermodynamics of irreversible processes IV, the theory of elasticity and inelasticity, *Phys. Rev.*, **73**, 373-382 (1948).
18. J. Murali Krishnan and K.R. Rajagopal, A thermodynamic framework for the constitutive modeling of asphalt concrete, To appear in *J. Mater. Civ. Engineering*.
19. J.G. Oldroyd, On the formulation of the rheological equations of state, *Proc. Roy-Soc. London*, **A200**, 523-591 (1950).
20. J. M. Burgers, Mechanical Considerations- model systems- phenomenological theories of relaxation and viscosity, First report on viscosity and plasticity, *Nordemann Publishing Company Inc., New York*, 2nd edition. Prepared by the committee for the study of viscosity of the Academy of Sciences, Amsterdam (1939).
21. K.R. Rajagopal, Art, Craft and Philosophy of Modeling, To appear.

ON A TRANSPORT PROBLEM IN A TIME-DEPENDENT DOMAIN

R. RIGANTI, F. SALVARANI
Dipartimento di Matematica, Politecnico di Torino,
C.so Duca degli Abruzzi 24, 10129 Torino, Italy
E-mail: riganti@polito.it

In this note we study a transport linear integro-differential equation in a time-dependent domain with slab geometry. A singular perturbations technique is applied in order to test the error between the exact solution and its ***quasi-static approximation***, which satisfies a simpler equation.

1 Introduction

We study a transport equation for photons in a slab of given thickness, divided in three time-dependent regions. Each region is characterized by different total and scattering cross sections and by different sources. The boundaries of the central region are moving with an assigned law, depending on t. Such problem, theoretically studied by Belleni-Morante, Monaco, Salvarani and Roach,[1,2] has interesting applications, for instance in cardiology[3] and in astrophysics[4].

In a non-dimensional setting, the model under investigation is the following. We consider a slab $\Omega = [-1, 1]$, divided in three time-dependent regions Ω_1, Ω_2 and Ω_3 such that

$$\Omega_1(t) = \{x \in \Omega \; : \; -1 \leq x < a(t)\}, \quad \Omega_2(t) = \{x \in \Omega \; : \; a(t) \leq x < a(t) + h\},$$

$$\Omega_3(t) = \{x \in \Omega \; : \; a(t) + h \leq x \leq 1\},$$

where $a(t)$ and $a(t) + h$ are the known positions of two moving boundaries and h is a positive constant. Moreover it is assumed that $a(t)$ is a continuous and differentiable function of $t \geq 0$, $a(t) \in (-1, 1 - h)$, for all $t \geq 0$, and $\sup|\dot{a}(t)| << c$, where c is the speed of light.

We shall assume that the photon transport phenomenon is one-dimensional in the variable $x \in \Omega$, so that the photon velocity is specified by $\mu \in [-1, 1]$ (μ is the cosine of the angle between the x-axis and the velocity).

Let $n(t, x, \mu)$ be the photon number density, and $q_i \geq 0$ $i = 1, \ldots, 3$ a given photon source relative to the region Ω_i. The quantities $\sigma_i > 0$ and $\sigma_{si} \geq 0$ are the total and scattering cross-sections respectively. We assume that q_i, σ_i and σ_{si} are constant in each Ω_i. Moreover, we impose non-reentry boundary conditions, that is:

$$n(t, -1, \mu) = 0 \quad \mu \in (0, 1], \quad n(t, 1, \mu) = 0 \quad \mu \in [-1, 0) \quad t \geq 0.$$

The equation under investigation, written as an abstract problem in the Banach space $X = L^1(\Omega \times (-1,1))$, is the following:

$$\begin{cases} \dfrac{dn}{dt}(t) = \left[A - \dfrac{\sigma(t)}{\varepsilon} I + \dfrac{\sigma_s(t)}{\varepsilon} K\right] n(t) + Q(t), & t > 0 \\ n(0) = n_0, \end{cases} \tag{1}$$

where $n(t) : \mathbf{R}^+ \to X$; $\varepsilon = \sup|\dot{a}(t)|/c$ is a dimensionless parameter, and assuming $\sup|\dot{a}(t)| = 1$ (hence $1/\varepsilon = c$), the following operators have been defined:

$$Af(x,\mu) = -\frac{\mu}{\varepsilon}\frac{\partial f}{\partial x}, \qquad \mathcal{R}(A) \subseteq X,$$

$$\mathcal{D}(A) = \{f \in X : Af \in X, f(-1,\mu) = 0, \text{ if } \mu > 0, \ f(1,\mu) = 0, \text{ if } \mu < 0\},$$

$$Kf = \frac{1}{2}\int_{-1}^{1} f(x,\mu')d\mu', \qquad \mathcal{D}(K) = X, \ \mathcal{R}(K) \subseteq X.$$

Moreover:

$$Q(t) = q_1 + [q_2 - q_1]H(x - a(t)) + [q_3 - q_2]H(x - a(t) - h),$$

where $H(y)$ is a Heaviside-like function of class $C^2(\mathbf{R})$, with the following properties: *i)* $H(y) = 0$ for all $y \leq 0$, $H(y) = 1$ for all $y \geq \xi$; *ii)* $0 \leq H'(y) \leq \beta \in \mathbf{R}^+$, for all $y \in [0,\xi]$, where ξ is given, and it is such that $0 < \xi << 1$. The quantities $\sigma(t)$ and $\sigma_s(t)$ are defined in a similar way.

Moreover, we assume that the initial photon density n_0 satisfies the following stationary problem, obtained from (1) at $t = 0$:

$$\left[A - \frac{\sigma(0)}{\varepsilon} I + \frac{\sigma_s(0)}{\varepsilon} K\right] n_0 + Q(0) = 0.$$

This means that the time-independent photon distribution n_0 corresponds to the structure that the three regions have at $t = 0$, with the two boundaries placed at $x = a(0)$ and $x = a(0) + h$ respectively.

A simpler *ordinary* integro-differential equation:

$$\left[A - \frac{\sigma(t)}{\varepsilon} I + \frac{\sigma_s(t)}{\varepsilon} K\right] n_Q(t) + Q(t) = 0, \tag{2}$$

obtained by deleting the time derivative term in the original problem (1), was also studied [1]. Its solution $n_Q(t)$, called *quasi-static approximation*, is yet depending on time because of the time dependence of the source and of the cross sections. Since ε is a very small parameter it may be considered as a good approximation of the solution to (1) at least outside an initial layer with thickness of order ε.

In this paper, in order to develop a good procedure for the numerical solution of the singularly perturbed equation (1), we will use Hilbert expansions in a composite form, which will permit to take into account also the initial layer of the solution.

The rest of the paper is organized as follows: in the next section we present the Hilbert expansion technique used to solve problem (1) in composite form. In Section 3 we show both the numerical strategy adopted to obtain the uniformly-valid solution and some numerical results concerning also the comparison with the *quasi-static* approximation.

2 Expansion of the solution

In this section we collect the main results in the study of problems (1) and (2). In what follows, we shall use the following notations: $R_M^{i,j} = |R_i - R_j|$, where $i, j = 1, 2, 3$ and R_i is a function defined on the region Ω_i. Furthermore: $\sigma_M = \max\{\sigma_1, \sigma_2, \sigma_3\}$, $\sigma_m = \min\{\sigma_1, \sigma_2, \sigma_3\}$, $\sigma_{s,M} = \max\{\sigma_{s1}, \sigma_{s2}, \sigma_{s3}\}$. We shall assume that $\sigma_m > \sigma_{s,M}$; this assumption is realistic from a physical viewpoint because σ_m is a *total* cross section, and $\sigma_{s,M}$ is a *scattering* cross section.

We recall that existence and uniqueness results for problems (1) and (2) have been proved in the paper by Belleni-Morante et al.[1] In particular, the following estimate of the error $E_Q(t) = n(t) - n_Q(t)$ has been proved:

Theorem 1: *The norm of the absolute error E_Q between the exact and the quasi-static approximation of problem (1) satisfies the inequality*

$$\|E_Q(t)\| \le \frac{2\varepsilon^2}{(\sigma_m - \sigma_{s,M})^2}\left[(q_M^{2,1} + q_M^{3,2}) + \frac{q_M(\sigma_M^{2,1} + \sigma_M^{3,2} + \sigma_{s,M}^{2,1} + \sigma_{s,M}^{3,2})}{(\sigma_m - \sigma_{s,M})}\right].$$

The solution of the singularly perturbed problem (1), which is characterized by non-uniform convergence with respect to the time t as the small parameter ε goes to zero, can be determined by using Hilbert expansions in a composite form, as formerly proposed by Tikhonov, Vasil'eva and coworkers[6] and more recently by O'Malley[7]. Such a solution is written as the sum

$$n(t, x; \mu, \varepsilon) = N(t, x; \mu, \varepsilon) + \eta(\tau, x; \mu, \varepsilon) \tag{3}$$

of an *outer (bulk)* approximation N representing the solution in the space domain Ω at times which are outside a (time) initial layer with a size of order

ε, and of an *inner* approximation η which represents the initial layer correction term depending on the stretched independent variable $\tau = t/\varepsilon$.

In a similar way, the initial datum satisfies the condition:

$$n(t=0,x;\mu,\epsilon) = n_Q(t=0,x;\mu,\epsilon) = N(0,x;\mu,\epsilon) + \eta(0,x;\mu,\varepsilon).$$

In equation (3) the inner solution η, which must vanish outside the initial layer, and the outer solution N are subjected to the conditions:

$$\begin{array}{llll} \mu \in (0,1]: & \eta(\tau,-1;\mu,\varepsilon) = 0, & N(t,-1;\mu,\varepsilon) = 0, & \tau \geq 0 \\ \mu \in [-1,0): & \eta(\tau,1;\mu,\varepsilon) = 0, & N(t,1;\mu,\varepsilon) = 0, & \tau \geq 0. \end{array} \quad (4)$$

Thanks to these initial and boundary conditions, the unknown N satisfies the following evolution equation in the Banach space $L^1(\Omega \times (-1,1))$:

$$\begin{cases} \dfrac{dN}{dt}(t) = \left[A - \dfrac{\sigma(t)}{\varepsilon}I + \dfrac{\sigma_s(t)}{\varepsilon}K\right] N(t) + Q(t), & t > 0 \\ N(0) = n_Q(0) - \eta(\tau = 0). \end{cases} \quad (5)$$

and $\eta(\tau,x;\mu,\varepsilon)$ must be the solution of the equation

$$\frac{d\eta}{d\tau} = [\varepsilon A - \sigma(\varepsilon\tau)I + \sigma_s(\varepsilon\tau)K]\,(n(\varepsilon\tau) - N(\varepsilon\tau)), \quad (6)$$

with boundary conditions given by (4).

A suitable matching of the outer and inner approximations at the ends of their respective regions of validity allows to determine a uniformly valid asymptotic expansion for the solution $n(t,x;\mu,\varepsilon)$. In order to obtain this solution, let us consider formal power series expansions in ε of N, η and n_Q:

$$N \sim \sum_k N^{(k)}(t,x;\mu)\varepsilon^k;\ \eta \sim \sum_k \eta^{(k)}(\tau,x;\mu)\varepsilon^k;\ n_Q \sim \sum_k n_Q^{(k)}(x;\mu,t)\varepsilon^k \quad (7)$$

where the terms in the expansions must satisfy, following (3) and (4):

$$k = 0,1,\ldots:\ \eta^{(k)}(0,x;\mu) = n_Q^{(k)}(x;\mu,t=0) - N^{(k)}(0,x;\mu) \quad (8)$$

$$\eta^{(k)}(\tau,-1;\mu) \equiv N^{(k)}(\varepsilon\tau,-1;\mu) = 0,\ \mu \in (0,1],\ \tau \geq 0 \quad (9)$$

$$\eta^{(k)}(\tau,1;\mu) \equiv N^{(k)}(\varepsilon\tau,1;\mu) = 0,\ \mu \in [-1,0),\ \tau \geq 0. \quad (10)$$

Substituting into equations (5) and (6) and equating the terms with equal powers in ε, we obtain a sequence of equations in the unknown terms $N^{(k)}$, $\eta^{(k)}$, which are to be solved with the conditions (8), (9) and (10). By indicating with B the operator εA, this sequence of equations is:

$$[B - \sigma(t)I + \sigma_s(t)K]N^{(0)} = 0 \qquad t > 0 \tag{11}$$

$$[B - \sigma(t)I + \sigma_s(t)K]N^{(1)} + Q(t) = \frac{\partial N^{(0)}}{\partial t} \qquad t > 0 \tag{12}$$

$$[B - \sigma(t)I + \sigma_s(t)K]N^{(k)} = \frac{\partial N^{(k-1)}}{\partial t} \qquad t > 0, \quad k \in \mathbf{N}, \quad k \geq 2 \tag{13}$$

$$\begin{cases} \dfrac{d\eta^{(k)}}{d\tau} = [B - \sigma(\varepsilon\tau)I + \sigma_s(\varepsilon\tau)K]\eta^{(k)} \qquad k \in \mathbf{N}, \quad \tau \geq 0 \\ \eta^{(k)}(0) = n_Q^{(k)}(0) - N^{(k)}(0). \end{cases} \tag{14}$$

Our main results concerning the solutions of the above sequence are summarized by the lemmas which follow; their proofs are reported in a previous paper [5].

Lemma 2: *Let the power series expansion of n_Q given in equation (7) be the solution of problem (2). Then* $n_Q^{(k)}(x;\mu,t) \equiv 0$ for all $k \neq 1$.

By this lemma it follows that the quasi-static solution n_Q is of order ε, and namely

$$n_Q(x;\mu,\varepsilon,t) = \varepsilon n_Q^{(1)}(x;\mu,t). \tag{15}$$

Lemma 3: *The solution of order zero in ε for equation (1), $N^{(0)} + \eta^{(0)}$, is identically zero.*

Lemma 4: *The term $\eta^{(1)}$ of the initial layer solution is identically zero, and the ε-order term of the composite solution (3) is coincident with the quasi-static solution (15) of problem (2), i.e. $n_Q = \varepsilon N^{(1)}$.*

As a consequence of the above lemmas, the exact solution $n(t)$ of problem (1) differs from the quasi-static one $n_Q(t)$ by terms of order ε^2 and higher ones; in the composite form (3) it is given by

$$n(t,x;\mu,\varepsilon) = n_Q(x;\mu,t,\varepsilon) + \sum_{k\geq 2}[N^{(k)}(t,x;\mu) + \eta^{(k)}(t/\varepsilon,x;\mu)]\,\varepsilon^k, \tag{16}$$

where $N^{(k)}$ and $\eta^{(k)}$ are solutions, respectively, of equations (13) and (14) with $k \geq 2$.

Some results on the errors introduced by truncating the power series (16) are given by the following two lemmas.

Lemma 5: *Let $\eta^{(k)}$ be the solution of problem (14). Then*

$$\left\|\eta^{(k)}(\tau)\right\| \leq \left\|n_Q^{(k)}(0) - N^{(k)}(0)\right\| e^{-(\sigma_m - \sigma_{s,M})\tau}.$$

Lemma 6: *Let $n(t)$ be the solution of problem (1). Then the error obtained by truncating the power expansion (16) to the order ε^k, is of order ε^{k+1}, i.e. $\|E^{(k)}(t)\| = O(\varepsilon^{k+1})$.*

3 Numerical techniques and results

In this Section the numerical treatment of the approximated solution of order two is presented.

For the numerical study of the quasi-static approximation, the well known discrete-ordinate method[8,9] has been applied. It consists in replacing equation (2) by a system of $2M$ coupled equations, obtained by discretization of the angular variable μ. By replacing the operator K with a $2M$-points Gauss quadrature formula, the following equations have been obtained:

$$\left[\mu_k \frac{d}{dx} - \sigma(t)I\right] n_Q(\mu_k) + \frac{1}{2}\sigma_s(t)\sum_{i=1}^{2M} \alpha_i n_Q(\mu_i) + \varepsilon Q(t) = 0, \quad k = 1, \ldots, 2M$$

where μ_k are the nodes in Gauss' formula, and α_k are the respective weights. Moreover the boundary conditions are

$$\begin{aligned} n_Q(t, 1, \mu_k) &= 0 \quad \text{for } k = 1, \ldots, M \\ n_Q(t, -1, \mu_k) &= 0 \quad \text{for } k = M+1, \ldots, 2M. \end{aligned}$$

These equations have been solved by a standard Runge-Kutta routine.

With regard to the numerical treatment of the external solution, we used again the discrete-ordinate method, owing to the analogy between equation (13) and the quasi-static equation (2), and applied it to the second-order term $N^{(2)}$ which satisfies equation (13) with k=2. In order to determine its right side term $\partial N^{(1)}/\partial t = \dot{n}_Q/\varepsilon$ (see Lemma 4), a suitable numerical procedure has been used to evaluate the time derivative of the quasi static solution $n_Q(t)$. It satisfies the following equation:

$$-B\dot{n}_Q = -\sigma\dot{n}_Q + \sigma_s K\dot{n}_Q + \left[\varepsilon\dot{Q} - \dot{\sigma}n_Q + \dot{\sigma}_s K n_Q\right] \tag{17}$$

with a source term given by the quantity between square brackets. Hence a reasonable approximation for this source term has been computed[5] and the discrete-ordinate method has been applied again to solve equation (17) for $\dot{n}_Q(t)$.

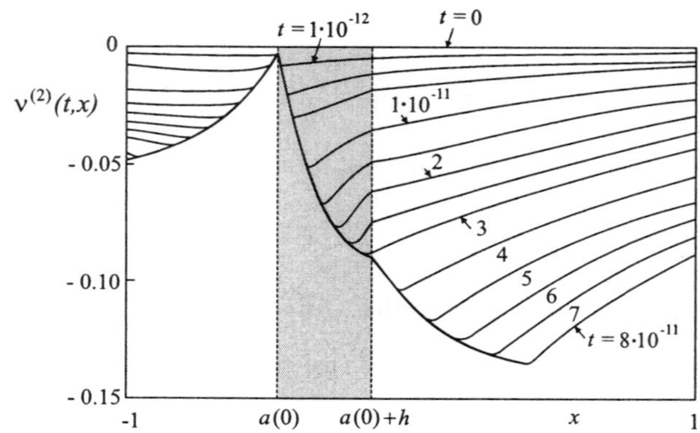

Figure 1: Total density in the initial layer: case of low absorption and scattering.

The initial-boundary value problem for equation (14) has been approached by means of a splitting technique in the version proposed by Strang [10], which allows to achieve a second order accuracy.

Numerical simulations of the time-space evolution of the second-order terms $N^{(2)}$ and $\eta^{(2)}$ in the solution (16) have been made by considering, in dimensionless variables, a source $q_1 = 9 \times 10^{-2}$ in Region 1, and a constant speed $\dot{a}(t) = 1$ for the motion of the intermediate Region 2.

The source term $\dot{n}_Q/\varepsilon$ in the evolution equation of $N^{(2)}$ has been previously determined at fixed values of time $t \in [0, 0.5]$ by solving equation (17) as explained above, and the results $\dot{n}_Q(t, x, \mu)$ were used to obtain the external solution $N^{(2)}(x, t, \mu)$ by a further application of the discrete-ordinate method.

Then, the numerical solution $N^{(2)}$ obtained at $t = 0$ has been used as the initial condition for equation (14) with $k = 2$, in order to determine the initial layer correction term $\eta^{(2)}(\tau, x, \mu)$ by means of the above outlined Strang's splitting technique.

By simulating a low absorption and scattering cross sections in the intermediate Region 2 we assumed: $\sigma_1 = 0.6, \sigma_2 = 1, \sigma_3 = 0.4, \sigma_{si} = \sigma_i/10,\ i = 1, 2, 3.$ The results obtained in this case for the total density second-order term

$$\nu^{(2)}(t, x) = \int_{-1}^{1} [N^{(2)}(t, x, \mu) + \eta^{(2)}(t/\varepsilon, x, \mu)]d\mu$$

are plotted in Figure 1 versus space and for very small times, i.e. of the order of the initial layer with thickness $\varepsilon = 1/3 \times 10^{-10}$. Figure 2 shows the total density outside this time initial layer.

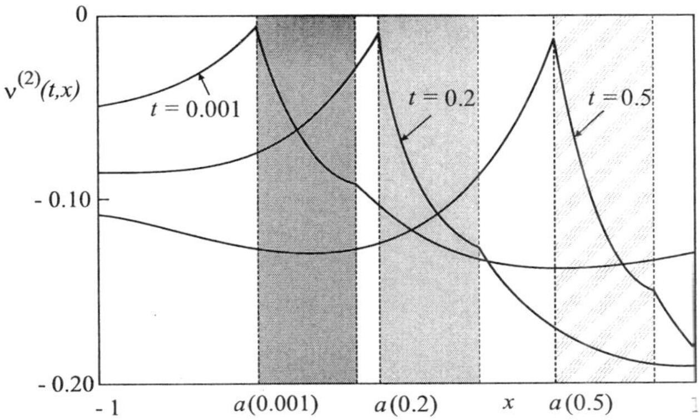

Figure 2: Total density outside the initial layer.

Assuming that the truncated ε^2-order composite solution is a reasonably good approximation of the exact one $n(t,x)$, it is possible to determine the time evolution of the relative error $e_Q(t)$ of the quasi-static approximation n_Q by computing the ratio

$$e_Q(t) = \frac{\|n(t,x) - n_Q(t,x)\|}{\|n(t,x)\|} \simeq \frac{\varepsilon^2 \|N^{(2)} + \eta^{(2)}\|}{\|n_Q + \varepsilon^2(N^{(2)} + \eta^{(2)})\|},$$

with $N^{(2)}$ and $\eta^{(2)}$ obtained by the above numerical simulation.

The results are shown in Figure 3, where the relative error of n_Q is plotted versus time t both in the case of low cross sections shown by Figures 1,2 and by assuming as a second numerical case: $\sigma_1 = 0.8, \sigma_2 = 5, \sigma_3 = 0.6, \sigma_{si} = \sigma_i/10$, $i = 1, 2, 3$, which may simulate high values of the absorption and scattering cross sections in Region 2. After a monotonic growing in the initial layer, the error clearly decreases with time in the case of low absorption and scattering, while in the simulation with high absorbtion effects it shows very small oscillations about an approximately constant value. In both cases, the maxima of the computed errors are lower than the corresponding theoretical values, computed by using the bound given by Theorem 1.

Acknowledgments

This work has been partially supported by the CNR "Progetto Coordinato" No. 99.01706 and by PRIN 2000, "Problemi matematici non lineari di propagazione e stabilità nei modelli del continuo" (Prof. T. Ruggeri, co-ordinator).

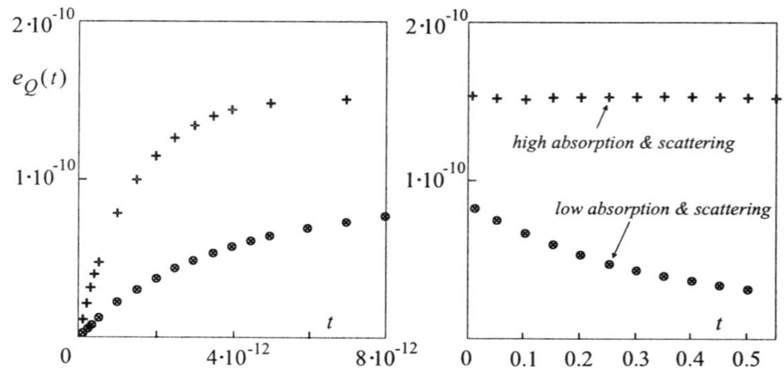

Figure 3: Relative error of n_Q in the initial layer (*left*) and at higher times (*right*).

References

1. A. Belleni-Morante, R. Monaco and F. Salvarani, Approximated solutions of photon transport in a time dependent region, *Transp. Theory Stat. Phys.* **30**, 2001, 421-438.
2. A. Belleni-Morante and G.F. Roach, A mathematical model for gamma-ray transport in the cardiac region, *J. Math. Anal. and Appl.* **244**, 2000, 498-514.
3. F.J. Wackers, R. Soufer and B.L. Zaret, *Heart Disease, a textbook of Cardiovascular Medicine*, W. B. Saunders and Co., Philadelphia, 1977.
4. S. Chandrasekhar, *Radiative Transfer*, Dover, New York, 1960.
5. R. Riganti and F. Salvarani, A singularly perturbed problem of photon transport in a time-dependent region. In print on *Math. Methods in Applied Sciences.*
6. A.B. Vasil'eva A.B. V. Butuzov and L. Kalachev, The boundary function method for singular perturbation problems, *SIAM Studies in Appl. Math.* **14**, Philadelphia, 1995.
7. R.E. O'Malley, *Singular Perturbation Methods for Ordinary Differential Equations*, Springer-Verlag, New York, 1991.
8. H.B. Keller, Approximated solutions of transport problems. II: Convergence and applications of the discrete-ordinate method, *J. Soc. Indust. Appl. Math.* **8**, 1960, 43-73.
9. H.B. Keller, On the pointwise convergence of the discrete-ordinate method, *J. Soc. Indust. Appl. Math.* **8**, 1960, 560-567.
10. G. Strang, On the construction and comparison of difference schemes, *SIAM J. Numer. Anal.* **5**, 1968, 506-517.

ON THE LONG TIME BEHAVIOUR OF THE SOLUTIONS OF NON-LINEAR PARABOLIC EQUATIONS IN UNBOUNDED DOMAINS

S. RIONERO
Department of Mathematics, University of Naples "Federico II",
Complesso Universitario Monte S. Angelo,
Via Cinzia, 80126 Naples - ITALY
E-mail: rionero@unina.it

The nonlinear reaction-diffusion equation $u_t = \Delta F(\mathbf{x}, u) - g(\mathbf{x}, u)$ in an unbounded domain is considered. In the case of Dirichlet boundary data and for solutions which – a priori – can grow at large spatial distances matters relating to the asymptotic behaviour as $t \to \infty$ are studied. Conditions ensuring the existence of absorbing sets in a suitably weighted L^2 space are obtained.

1 Introduction

Let u be a scalar valued function from $I\!R^3$ into $I\!R$, Ω a domain in $I\!R^3$ and $\Omega_T = \Omega \times (0, T)$ with T=const.> 0. We consider – under prescribed initial and Dirichlet boundary data – the non-linear reaction-diffusion equation of the type

$$u_t = \Delta F(\mathbf{x}, u) - g(\mathbf{x}, u), \qquad \mathbf{x} \in I\!R^3. \tag{1}$$

As it is well known, when Ω is bounded, equations of kind (1) have been deeply and extensively investigated by many authors (see, for instance,[1]-[6] and the references therein).

On the contrary – when Ω is unbounded and u can grow at large spatial distances or does not belong to some $L^p(\Omega)$ – there are only few authors who have investigated equations of kind (1) (we recall here the papers [8]-[14] and the references therein). But this case is of big relevance both mathematically and physically [a].

---

[a]Let us consider for instance the steady solutions of the heat equation in the half space $z \geq 0$ and in the exterior of sphere of radius $r_0 > 0$, under the boundary data $u(z = 0) = u_0 =$ const. and $u(r_0) = \bar{u}_0 =$ const., respectively. One obtains the solutions

$$u = u_0 + \lambda z, \qquad \lambda = \text{const.} \in I\!R$$

$$u = \bar{u}_0 + \mu\left(\frac{1}{r} - \frac{1}{r_0}\right), \qquad \mu = \text{const.} \in I\!R$$

which are or unbounded or not belonging to $L^2(\Omega)$.
Another relevant difference between the Ω bounded case and the Ω unbounded case is that the embedding of $H^1(\Omega)$ in $L^2(\Omega)$ is compact only when Ω is bounded. If Ω is bounded, the existence of a bounded absorbing set in $H^1(\Omega)$ yields the existence of the maximal attractor

Our aim is to obtain stability results for the equations of type (1) when Ω is unbounded, letting u grow at large spatial distances. Precisely our principal goal is to study matters relating to the asymptotic behaviour of solutions of (1). In particular we are interested in obtaining conditions on F and g under which each solution tends towards a steady state when $t \to \infty$ or is definitively bounded. Smooth solutions are assumed throughout, but most of the arguments are easily adapted for weak solutions.

In Section 2 we obtain a weighted energy relation which is used for obtaining conditions for the asymptotic convergence, as $t \to \infty$, to a steady state (Section 3) and conditions allowing the existence of absorbing sets in a suitably weighted L^2 (Section 4).

2 A weighted energy relation

Let Ω be the exterior of a bounded regular domain $\Omega_0 \subset I\!R^3$ or the half space $z \geq 0$ and let us consider the initial boundary value problem (i.b.v.p.)

$$\begin{cases} u_t = \Delta F(\mathbf{x}, u) - g(\mathbf{x}, u), & (\mathbf{x}, t) \in \Omega \times I\!R^+ \\ u(\mathbf{x}, 0) = u_0(\mathbf{x}) & \mathbf{x} \in \Omega \\ u(\mathbf{x}, t) = u_1(\mathbf{x}, t) & \mathbf{x} \in \partial\Omega_0 \end{cases} \tag{2}$$

and the associated boundary value problem (b.v.p.)

$$\begin{cases} \Delta F(\mathbf{x}, U) = g(\mathbf{x}, U) & \mathbf{x} \in \Omega \\ U(\mathbf{x}) = u_1(\mathbf{x}) & \mathbf{x} \in \partial\Omega_0. \end{cases} \tag{3}$$

Let (3) be solvable and let U denote a solution. On setting

$$\begin{cases} u = U + v \\ \mathcal{L} = F(\mathbf{x}, U + v) - F(\mathbf{x}, U) \\ \mathcal{L}^* = g(\mathbf{x}, U + v) - g(\mathbf{x}, U), \end{cases} \tag{4}$$

---

in $L^2(\Omega)$. When Ω is unbounded in order to overcome the lack of compactness of $H^1(\Omega)$ into $L^2(\Omega)$ it is usually necessary to introduce L^p-weighted spaces.

one obtains the i.b.v.p. for the perturbation v to the basic state U

$$\begin{cases} v_t = \Delta\mathcal{L} - \mathcal{L}^* & (\mathbf{x}, t) \in \Omega \times I\!R^+ \\ v(\mathbf{x}, 0) = v_0 & \mathbf{x} \in \Omega \\ v = 0 & \mathbf{x} \in \partial\Omega_0. \end{cases} \tag{5}$$

Now let

$$\begin{cases} w: \ \mathbf{x} \in I\!R^3 \Rightarrow w(\mathbf{x}) \in I\!R^+, & \lim_{|\mathbf{x}|\to\infty} w(\mathbf{x}) = 0 \\ w \in C^2(I\!R^3), & G(U, v) = \int_0^v \mathcal{L}(U, \bar{v}) d\bar{v} \end{cases} \tag{6}$$

and let $I_w(\Omega)$ denote the class of perturbations v such that

$$v \in I_w(\Omega) \Rightarrow \begin{cases} \lim_{|\mathbf{x}|\to\infty} |w\mathcal{L}\nabla\mathcal{L}| = 0 \\ |wG| + |\mathcal{L}^2\nabla w| + w(\nabla\mathcal{L})^2 + w|\mathcal{L}\mathcal{L}^*| \in L(\Omega). \end{cases} \tag{7}$$

On setting

$$V = \int_\Omega wG d\Omega \tag{8}$$

the following theorem holds

Theorem 1 - *Let* (3) *be solvable and* $v \in I_w(\Omega)$. *Then along the solutions of* (5) *it turns out that*

$$\frac{dV}{dt} = \int_\Omega \left(\frac{1}{2}\mathcal{L}^2\Delta w - w(\nabla\mathcal{L})^2 - w\mathcal{L}\mathcal{L}^* \right) d\Omega \tag{9}$$

and hence

$$V(t) = V(t_0) + \int_{t_0}^t d\tau \int_\Omega \left(\frac{\mathcal{L}^2\Delta w}{2} - w(\nabla\mathcal{L})^2 - w\mathcal{L}\mathcal{L}^* \right) d\Omega. \tag{10}$$

Proof. On multiplying $(5)_1$ by $w\mathcal{L}$ it turns out

$$\frac{d}{dt} wG = w\mathcal{L}\Delta\mathcal{L} - \mathcal{L}\mathcal{L}^* \qquad (\mathbf{x}, t) \in \Omega \times I\!R^+. \tag{11}$$

But

$$\begin{cases} w\mathcal{L}\Delta\mathcal{L} = \nabla \cdot (w\mathcal{L}\nabla\mathcal{L}) - \frac{1}{2}\nabla w \cdot \nabla\mathcal{L}^2 - w(\nabla\mathcal{L})^2 \\ -\nabla w \cdot \nabla\mathcal{L}^2 = -\nabla \cdot (\mathcal{L}^2\nabla w) + \mathcal{L}^2\Delta w \\ v = 0 \Rightarrow \mathcal{L} = 0 \end{cases} \tag{12}$$

hence, letting $v \in I_w(\Omega)$ and integrating (11) on Ω, (10) immediately follows.

Remark 1 - *For the sake of simplicity in the sequel we shall consider only the case*

$$\Omega = \textit{half space } z \geq 0 \tag{13}$$

3 On the asymptotic convergence to a steady state

Let Ω be the half space $z \geq 0$ and let (3) be solvable. In this section our goal is to obtain conditions ensuring that, as $t \to \infty$, the solutions of (2) converge to a solution of (3). The following theorem holds.

Theorem 2 - *Let*

$$\frac{\partial \mathcal{L}}{\partial v} = \frac{\partial \mathcal{L}}{\partial u} = \frac{\partial F}{\partial u} \geq m = const. > 0, \qquad \forall \mathbf{x} \in \Omega, \forall v \in I\!R. \tag{14}$$

Then one has

$$G(U,0) = \left[\frac{\partial G}{\partial v}\right]_{v=0} = 0, \quad \frac{\partial^2 G}{\partial v^2} \geq m, \quad G(U,v) \geq \frac{1}{2}mv^2 \tag{15}$$

$$\frac{\mathcal{L}}{v} \geq m \quad \forall v \neq 0, \quad G < v\mathcal{L} \quad \forall v \neq 0 \tag{16}$$

$$\mathcal{L}^2 \geq mG \qquad \forall v \in I\!R. \tag{17}$$

Proof. See Theorem 4 and Remark 3 in [12].

Theorem 3 - *Let* (14) *holds and*

$$\frac{\partial \mathcal{L}}{\partial v} \leq m^* = const. > 0, \qquad (\mathbf{x}, v) \in \Omega \times I\!R. \tag{18}$$

Then, setting $\mu = m^{*2}/m$, it follows that

$$\mathcal{L}^2 \leq 2\mu G. \tag{19}$$

Proof. From $\{m \leq \frac{\partial \mathcal{L}}{\partial v} \leq m^*,\ G \geq \frac{m}{2}v^2\}$, one obtains $\mathcal{L}^2 \leq m^{*2}v^2 \leq \frac{2(m^*)^2}{m}G$.

Theorem 4 - *Let* (14)$-$(18) *hold, U be a fixed solution of (3), $\{w = e^{-\alpha z},\ \alpha =$ positive constant$\}$*, $w[\mathcal{L}^2 + (\nabla\mathcal{L})^2 + |\mathcal{L}\mathcal{L}^*|] \in L(\Omega)$, $\forall \alpha > 0$

$$\int_\omega w\mathcal{L}\mathcal{L}^* d\Omega \geq 0, \qquad t \geq t_0. \tag{20}$$

Then U is the asymptotic state of any solution $u = U + v$ of (2) such that

$$\exists t_0(v):\ v_0 = [v]_{t=t_0} \Rightarrow G_0 = G(\mathbf{x}, U, v_0) \in L(\Omega) \tag{21}$$

according to

$$E(t) + \int_{t_0}^t \int_\Omega (\nabla\mathcal{L})^2 d\Omega d\tau \leq E(t_0) \tag{22}$$

with

$$E = \int_\Omega G(\mathbf{x}, U, v) d\Omega, \qquad E(t_0) = \int_\Omega G_0 d\Omega. \tag{23}$$

Proof. Let us notice that $(15)_3$ implies that G is a positive definite functional of v and so is V, i.e. V is a Liapunov functional. From

$$w = e^{-\alpha z} \qquad \alpha = \text{const.} > 0 \tag{24}$$

it turns out that

$$\Delta w = \alpha^2 w \tag{25}$$

therefore (9) and (21) imply

$$\frac{dV}{dt} - \mu\alpha^2 V \leq -\int_\Omega w[(\nabla\mathcal{L})^2 + \mathcal{L}\mathcal{L}^*] d\Omega \tag{26}$$

and hence

$$V + \int_{t_0}^t e^{\mu\alpha^2(t-\tau)} d\tau \int_\Omega w[(\nabla\mathcal{L})^2 + \mathcal{L}\mathcal{L}^*] d\Omega \leq V(t_0) e^{\mu\alpha^2 t}. \tag{27}$$

Letting $\alpha \to 0$, the right hand side of (26) - in view of (21) - converges to $\int_\Omega G(\mathbf{x}, U, v_0) d\Omega$, hence the monotone convergence theorem implies (22).

Remark 2 *Let $\|\cdot\|_2$ be the L^2-norm. Because of $(15)_3, (22)$ and the weighted Poincaré inequality {Cfr. [7], theorem 6.11, p. 182}*

$$m^2 \int_\Omega \frac{v^2}{(1+z)^2} d\Omega \leq \int_\Omega \frac{\mathcal{L}^2}{(1+z)^2} d\Omega \leq 4 \int_\Omega (\nabla \mathcal{L})^2 d\Omega \tag{28}$$

from (22) it turns out that

$$\frac{1}{2} m \|v\|_2^2 + \frac{m^2}{4} \int_{t_0}^t \Psi(\tau) d\tau \leq \frac{m^*}{2} \|v_0\|_2^2 \tag{29}$$

with

$$\Psi(t) = \int_\Omega \frac{v^2}{(1+z)^2} d\Omega = \left\| \frac{v}{1+z} \right\|^2. \tag{30}$$

Therefore (29) implies the stability of U with respect to to L^2-norm of the perturbation v and the "weak" asymptotic stability with respect to the L^2-weighted norm (30) according to

$$\int_{t_0}^t \Psi(\tau) d\tau \leq \frac{2m^*}{m^2} \|v_0\|^2 \qquad \forall t \geq t_0. \tag{31}$$

Remark 3. *Let $\{g \equiv 0,\ F = F(u)\}$. Then (3) reduces to*

$$\begin{cases} \Delta F = 0 & \mathbf{x} \in \Omega \\ F = F_1 & \mathbf{x} \in \partial\Omega \end{cases} \tag{32}$$

where $F_1 = \varphi(U)$, φ being the inverse of the function Ψ such that $F = \Psi(U)$. Therefore if $\bar{F}$ is a solution of (32), $F = \bar{F} + \lambda z$ is a solution too, $\forall \lambda \in I\!R$. We remark that (21) selects uniquely the attracting steady solution. In fact let

$$\begin{cases} U = \Psi(\bar{F} + \lambda z) = \bar{U} + \left(\dfrac{d\Psi}{dF} \right)_{\bar{F} + \theta \lambda z} \lambda z, \\ \bar{U} = \Psi(\bar{F}), \qquad \theta \in]0, 1[. \end{cases}$$

Then it follows that

$$\begin{cases} u = \bar{U} + v = \bar{U} + \left(\dfrac{d\Psi}{dF}\right)_{\bar{F}+\theta\lambda z} \lambda z + v_1, \\ v_1 = v - \left(\dfrac{d\Psi}{dF}\right)_{\bar{F}+\theta\lambda z} \lambda z. \end{cases}$$

Therefore

$$G(\mathbf{x}, \bar{U}, v_1) = \int_0^{v_1} \mathcal{L}(\mathbf{x}, U, \bar{v}_1) d\bar{v}_1 = G(\mathbf{x}, \bar{U}, v) + \int_v^{v_1} \mathcal{L}(\mathbf{x}, U, \bar{v}) d\bar{v} \geq$$

$$> G(\mathbf{x}, \bar{U}, v) - m|v| \left|\frac{d\Psi}{dF}\right|_{\bar{F}+\theta\lambda z} \lambda z$$

with $\dfrac{1}{m^*} \leq \left|\dfrac{d\Psi}{dF}\right| \leq \dfrac{1}{m}$. *Although* $v_0 \in L^2(\Omega),\ v_0 z \notin L^2(\Omega),\ \forall v_0$.

Remark 4. *Let us notice that*

i) (22) *and* $v_0 \in L^2(\Omega)$ *imply that* $|v|$ *and* $|\nabla \mathcal{L}|$ *belong to* $L^2(\Omega),\ \forall t \geq t_0$*;*

ii) $\mathcal{L}\mathcal{L}^* \geq \varphi(x, y) e^{\varepsilon \alpha z}$ *with* $\{\varepsilon \in (0,1), \varphi(x, y) \in L(\mathbb{R}^2), \int_{\mathbb{R}^2} \varphi(x, y) dx dy \geq 0\}$ *imply* (22)*;*

iii) $g(\mathbf{x}, u) = \varphi(x, y) f(z) F(\mathbf{x}, u) \Rightarrow \mathcal{L}\mathcal{L}^* = \varphi(x, y) f(z) \mathcal{L}^2$.

4 Absorbing sets

Let us consider the weighted space

$$L^2_w(\Omega) = \{v \text{ measurable}, \|\sqrt{w}\, v\|_2^2 = \int_\Omega w u^2 d\Omega < \infty\}$$

with w given by (24) and notice that because of

$$\frac{1}{2} m \|\sqrt{w}\, v\|_2^2 \leq V \leq \frac{1}{2} m^* \|\sqrt{w}\, v\|_2^2$$

$V^{1/2}$ is a norm equivalent to $\|\sqrt{w}\, v\|_2$. In this Section we shall adopt the $V^{1/2}$-weighted norm and denote by $L^2_V(\Omega)$ the space $L^2_w(\Omega)$ normed with $V^{1/2}$.

We begin by recalling the weighted Poincaré inequality for the half space $z \geq 0$ {Cfr. [7], theorem 6.9, p. 180}

$$\int_\Omega e^{-\alpha z}\mathcal{L}^2 d\Omega \leq \frac{4}{\alpha^2}\int_\Omega e^{-\alpha z}(\nabla\mathcal{L})^2 d\Omega. \tag{33}$$

From (9) and (21) then it turns out that

$$\frac{dV}{dt} + \gamma V \leq \int_\Omega w\mathcal{L}\left(\frac{\alpha^2}{2}\mathcal{L} - \mathcal{L}^*\right) d\Omega \tag{34}$$

with

$$\gamma = \frac{\alpha^2}{2}\mu. \tag{35}$$

Theorem 5 - *Let*

$$\int_\Omega \left(\frac{\alpha^2}{2}\mathcal{L}^2 - \mathcal{L}\mathcal{L}^*\right) w d\Omega \leq c = const. > 0 \tag{36}$$

and set

$$r_0^2 = \frac{c}{\gamma}. \tag{37}$$

Then

$$\lim_{t\to\infty} \sup V^{1/2} \leq r_0 \tag{38}$$

and there exists an absorbing set B_0 *in* $L_V^2(\Omega)$*, namely, any ball of* $L_V^2(\Omega)$ *centered at* O *of radius* $V^{1/2} = r > r_0$.

Proof. (34) implies

$$V \leq V_0 e^{-\gamma t} + \frac{c}{\gamma}(1 - e^{-\gamma t}) \tag{39}$$

and hence (38). Let $\bar{B}$ is a bounded set of $L_V^2(\Omega)$and $B(O,R)$ the ball of $L_V^2(\Omega)$ centered at O of radius R such that $\bar{B} \subset B$. From (39) with $V = R^2$ we infer that at any instant

$$t > t_0 = \frac{1}{\gamma}\log\frac{R^2 - r_0^2}{r^2 - r_0^2}$$

the orbit of $\bar{B}$ is $B(O,r)$. We end remarking that $R \to r$ implies $t_0 = 0$, i.e. the sphere $B(O,r)$ is *positively invariant.*

Let $\{O, \rho, \theta, z\}$, with $\{\rho = \sqrt{x^2+y^2}$, $\theta \in (0, 2\pi)$, $z \geq 0\}$ be a cylindrical frame of reference with the origin at $z = 0$ and let w_1 denote the weight function

$$w_1 = e^{-(\alpha z + \alpha_1 \rho)} \tag{40}$$

with α and α_1 positive constants and I_{w_1} the class of perturbation v such that (7) is verified with w substituted by w_1. It turns out that $I_{w_1} \supset I_{\upsilon}$. On taking into account that

$$\Delta w_1 = (\alpha^2 + \alpha_1^2) w_1 - \frac{\alpha_1}{\rho} w_1 \tag{41}$$

then (9) implies

$$\frac{dV_1}{dt} \leq \frac{\alpha^2 + \alpha_1^2}{2} \|\sqrt{w_1}\ \mathcal{L}\|_2^2 - \|\sqrt{w_1}\ \nabla\mathcal{L}\|_2^2 - \int_\Omega w_1 \mathcal{L}\mathcal{L}^* d\Omega \tag{42}$$

with

$$V_1 = \int_\Omega w_1 G d\Omega. \tag{43}$$

On the other hand from the weighted Poincaré inequality for the half space $z \geq 0$ {Cfr. [7], p. 180}

$$\int_0^\infty e^{-\alpha z} \mathcal{L}^2 dz \leq \frac{4}{\alpha^2} \int_0^\infty e^{-\alpha z} \left(\frac{\partial \mathcal{L}}{\partial z} \right)^2 dz$$

it follows that

$$\int_\Omega w_1 \mathcal{L}^2 d\Omega \leq \frac{4}{\alpha^2} \int_\Omega w_1 (\nabla \mathcal{L})^2 d\Omega \tag{44}$$

hence (42) implies

$$\frac{dV_1}{dt} + \gamma V_1 \leq \int_\Omega w_1 \mathcal{L} \left(\frac{\alpha^2 + \alpha_1^2}{2} \mathcal{L} - \mathcal{L}^* \right) d\Omega. \tag{45}$$

Since (45) can be obtained from (34) through the substitution

$$\begin{pmatrix} V_1 & w_1 & \alpha^2 + \alpha_1^2 \\ \\ V & w & \alpha^2 \end{pmatrix} \tag{46}$$

and denoting by $L^2_{V_1}(\Omega)$ the space $L^2_w(\Omega)$ normed with $V_1^{1/2}$, theorem 5 implies the following

Theorem 6 - *Let*

$$\int_\Omega w_1 \mathcal{L} \left(\frac{\alpha^2 + \alpha_1^2}{2} \mathcal{L} - \mathcal{L}^* \right) d\Omega \le c_1(\alpha, \alpha_1) = const. > 0 \tag{47}$$

and set

$$r_0 = \frac{c_1}{\gamma}. \tag{48}$$

Then

$$\lim_{t\to\infty} \sup V_1^{1/2} \le r_0 \tag{49}$$

and there exists an absorbing set B_0 in $L^2_{V_1}(\Omega)$, namely, any ball of $L^2_{V_1}(\Omega)$ centered at point $O \equiv \{V_1 \equiv 0\}$ of radius $V_1^{1/2} = r > r_0$.

Acknowledgments

This work has been performed under the auspicies of the G.N.F.M. of I.N.D.A.M. and M.U.R.S.T. (P.R.I.N.): "Nonlinear mathematical problems of wave propagation and stability in continuous media".

References

1. J. Smoller in *Shock waves and reaction-diffusion equation*, (Springer-Verlag, n.258 of "A Series of Comprehensive Studies in Mathematics", 1983).
2. J.K. Hale in *Asymptotic behaviour of dissipative systems*, (Math. Surveys and Monographs, n. 25, Amer. Math. Soc. Providence, Rhode Islands, 1988).
3. R. Temam in *Infinite-Dimensional Dyamical Systems in Mechanics and Physics*, (Appl. Math. Sc. 68, Springer-Verlag, n.26, 1988).
4. A.V. Babin and M.I. Visik in Attractors of evolution equations (Holland, n.25 of " Studies in Math. and its Appl., 1992)
5. J.D. Murray in *Mathematical Biology*, (Biomathematics Texte, Springer-Verlag, n.19, 1989).
6. O.A. Ladyzenshaja and V.A. Solonnikov *Linear and quasilinear equations of parabolic type*, (Transl. Math. Monogr. A.M.S., Providence Rhode Islands, n. 23, 1968).
7. J.N. Flavin and S. Rionero in *Qualitative estimates for partial differential equations. An introduction*, (CRC Press, Boca Raton, Florida, 1996).
8. S. Rionero in *Continuum mechanics and applications in geophysics and environment*, (Springer (Physics and Astronomy), eds. B. Straughan, R. Greve, H. Ehrentrant, Y. Wang, 2001), pp. 56-78.

9. G.P. Galdi and S. Rionero in *Weighted energy methods in Fluid Dynamics and elasticity*, (Springer-Verlag, Berlin Heidelberg New York Tokyo 1134 of "Lecture Notes in Mathematics", 1985).
10. S. Rionero and G.P. Galdi *Ricerche Mat.* **VXXX**, 85 (1981)
11. R. Russo and P. Maremonti *Ricerche Mat.* **41**, 11 (1993)
12. E. Feireisl and P. Laurencot and F. Simondon and H. Touré *C.R. Acad. Sci. Paris* **319**, 147 (1994)
13. S. Rionero and F. Perrini "On the nonlinear diffusion in a half space" (to appear)
14. I. Torcicolo and M. Vitiello "On the nonlinear diffusion in the exterior of a sphere" *Rend. Acc. Sc. fis. mat. Napoli* **LXVIII**, (2001)

ON THE GEOMETRY OF SPATIAL HYDRODYNAMIC MOTIONS. SOLITONIC CONNECTIONS

COLIN ROGERS

School of Mathematics, The University of New South Wales, Sydney, NSW 2052, Australia

E-mail: c.rogers@unsw.edu.au

It is established that in spatial hydrodynamic motions with velocity $\mathbf{q} = q\mathbf{t}$ subject to the geometric restriction $\operatorname{div} \mathbf{t} = 0$, the $\mathbf{t}$-field is constrained by the integrable Heisenberg spin equation on individual constant pressure surfaces. The same geometric formulation is then applied to the kinematics of ideal fibre-reinforced fluids to obtain necessary geometric constraints on the motions and again a connection to the Heisenberg spin equation.

1 Introduction

Howard[1], Wasserman[2] and Marris[3] all investigated, in detail, steady hydrodynamic motions subject to the condition $\operatorname{div} \mathbf{t} = 0$ where $\mathbf{t}$ is the unit tangent to the streamline. It has been recently shown by Rogers[4] that, remarkably, for such motions the $\mathbf{t}$-field necessarily satisfies the Heisenberg spin equation on individual constant pressure surfaces. It is recalled that, the Heisenberg spin equation is equivalent to the celebrated nonlinear Schrödinger (NLS) equation. Indeed, the latter equation results from the Gauss-Mainardi-Codazzi equations for the constant pressure surfaces.

The NLS equation was derived by Hasimoto[5] in nonsteady hydrodynamics in an approximation to the motion of an inextensible vortex filament in an unbounded medium. The components of the NLS equation in that context describe the temporal evolution of the curvature and torsion of the vortex filament and were originally set down by Da Rios[6] in 1906. These and subsequent results by Da Rios on what is essentially the localised induction approximation (LIA) were subsequently collected in a survey by Levi-Civita[7]. The re-discovery of the work of Da Rios, a student of Levi-Civita, has an interesting history recounted by Ricca[8]. The intrinsic geometry of the NLS equation and its auto-Bäcklund transformation has been described in[9]. It is remarked that the analysis presented here is valid 'mutatis mutandis' both for spatial motion of a Prim gas with $\operatorname{div} \mathbf{t} = 0$ and for an equivalent magnetohydrodynamic system[4]. In the latter context, the Heisenberg spin equation holds on total magnetic pressure surfaces. Geometric configurations wherein these surfaces comprise nested toroids which are foliated in accordance with the condition $\operatorname{div} \mathbf{t} = 0$ have recently been constructed by Schief[10].

In[11], Spencer investigated kinematic aspects of a continuum model of the motion of a fibre-reinforced fluid wherein a fibre direction characterised by a unit vector $\mathbf{a}$ is attributed to each fluid particle. In particular, fibre divergence free motions with $\operatorname{div} \mathbf{a} = 0$ were considered. Here, it is shown that the necessary kinematic conditions for a particular class of such motions again lead to the Heisenberg spin equation.

2 The geometric formulation

Here, we adopt a characterisation of a three-dimensional vector field $\mathbf{\Phi}$ in terms of anholonomic coordinates, as introduced by Vranceanu[12]. This formalism was subsequently employed by Marris and Passman[13] to derive kinematic properties in hydrodynamics. Therein, the orthonormal basis $\mathbf{t}, \mathbf{n}, \mathbf{b}$ is introduced along the tangent, principal normal and binormal directions to the vector lines of nonvanishing $\mathbf{\Phi}$. The system governing the directional derivatives of the orthonormal triad may then be shown to be[13]:

$$\frac{\delta}{\delta s}\begin{pmatrix}\mathbf{t}\\ \mathbf{n}\\ \mathbf{b}\end{pmatrix} = \begin{pmatrix} 0 & \kappa & 0 \\ -\kappa & 0 & \tau \\ 0 & -\tau & 0 \end{pmatrix}\begin{pmatrix}\mathbf{t}\\ \mathbf{n}\\ \mathbf{b}\end{pmatrix}, \tag{1}$$

$$\frac{\delta}{\delta n}\begin{pmatrix}\mathbf{t}\\ \mathbf{n}\\ \mathbf{b}\end{pmatrix} = \begin{pmatrix} 0 & \theta_{ns} & \Omega_b + \tau \\ -\theta_{ns} & 0 & -\operatorname{div} \mathbf{b} \\ -(\Omega_b + \tau) & \operatorname{div} \mathbf{b} & 0 \end{pmatrix}\begin{pmatrix}\mathbf{t}\\ \mathbf{n}\\ \mathbf{b}\end{pmatrix}, \tag{2}$$

$$\frac{\delta}{\delta b}\begin{pmatrix}\mathbf{t}\\ \mathbf{n}\\ \mathbf{b}\end{pmatrix} = \begin{pmatrix} 0 & -(\Omega_n + \tau) & \theta_{bs} \\ \Omega_n + \tau & 0 & \kappa + \operatorname{div} \mathbf{n} \\ -\theta_{bs} & -(k + \operatorname{div} \mathbf{n}) & 0 \end{pmatrix}\begin{pmatrix}\mathbf{t}\\ \mathbf{n}\\ \mathbf{b}\end{pmatrix}. \tag{3}$$

Here, $\delta/\delta s, \delta/\delta n$ and $\delta/\delta b$ denote directional derivatives in the tangential, principal normal and binormal directions respectively. Thus, (1) represents the usual Serret-Frenet relations while (2) and (3) provide the directional derivatives of the orthonormal triad $\{\mathbf{t}, \mathbf{n}, \mathbf{b}\}$ in the $\mathbf{n}$- and $\mathbf{b}$-directions respectively. Accordingly,

$$\boldsymbol{\nabla} = \mathbf{t}\frac{\delta}{\delta s} + \mathbf{n}\frac{\delta}{\delta n} + \mathbf{b}\frac{\delta}{\delta b}, \tag{4}$$

while θ_{bs} and θ_{ns} are geometric quantities originally introduced by Bjørgum in [14] via

$$\theta_{ns} = \mathbf{n} \cdot \frac{\delta \mathbf{t}}{\delta n}, \quad \theta_{bs} = \mathbf{b} \cdot \frac{\delta \mathbf{t}}{\delta b}\,. \tag{5}$$

It is noted that

$$\operatorname{div}\mathbf{t} = (\mathbf{t}\frac{\delta}{\delta s} + \mathbf{n}\frac{\delta}{\delta n} + \mathbf{b}\frac{\delta}{\delta b}) \cdot \mathbf{t} = \theta_{ns} + \theta_{bs} \tag{6}$$

and similarly

$$\operatorname{div}\mathbf{n} = -\kappa + \mathbf{b} \cdot \frac{\delta \mathbf{n}}{\delta b}\,, \tag{7}$$

$$\operatorname{div}\mathbf{b} = -\mathbf{b} \cdot \frac{\delta \mathbf{n}}{\delta n}. \tag{8}$$

Moreover,

$$\operatorname{curl}\mathbf{t} = \left(\mathbf{t} \times \frac{\delta}{\delta s} + \mathbf{n} \times \frac{\delta}{\delta n} + \mathbf{b} \times \frac{\delta}{\delta b}\right) \mathbf{t} = \Omega_s \mathbf{t} + \kappa \mathbf{b}, \tag{9}$$

where

$$\Omega_s = \mathbf{t} \cdot \operatorname{curl}\mathbf{t} = \mathbf{b} \cdot \frac{\delta \mathbf{t}}{\delta n} - \mathbf{n} \cdot \frac{\delta \mathbf{t}}{\delta b} \tag{10}$$

is termed the abnormality of the **t**-field. The relation (9) is of particular importance. It was originally obtained in 1927 by Masotti[15] and rediscovered by Bjørgum[14]. In a similar manner, we obtain the companion results

$$\operatorname{curl}\mathbf{n} = -(\operatorname{div}\mathbf{b})\mathbf{t} + \Omega_n \mathbf{n} + \theta_{ns}\mathbf{b} \tag{11}$$

where

$$\Omega_n = \mathbf{n} \cdot \operatorname{curl}\mathbf{n} = \mathbf{t} \cdot \frac{\delta \mathbf{n}}{\delta b} - \tau \tag{12}$$

and

$$\operatorname{curl}\mathbf{b} = (\kappa + \operatorname{div}\mathbf{n})\mathbf{t} - \theta_{bs}\mathbf{n} + \Omega_b \mathbf{b} \tag{13}$$

where

$$\Omega_b = \mathbf{b} \cdot \operatorname{curl}\mathbf{b} = -\tau - \mathbf{t} \cdot \frac{\delta \mathbf{b}}{\delta n}. \tag{14}$$

Addition of the abnormalities Ω_s, Ω_n and Ω_b as given by (10), (12) and (14) respectively, produces the relation

$$\Omega_s + \Omega_n + \Omega_b = 2(\Omega_s - \tau). \tag{15}$$

This result appears in another guise in the treatise of Weatherburn[16]. Therein, Ω_s, Ω_n and Ω_b are called the total moments of the **t**, **n** and **b** fields respectively.

The identity curl grad $\phi = \mathbf{0}$ yields

$$\begin{aligned}\frac{\delta\phi}{\delta s}\operatorname{curl}\mathbf{t} + \operatorname{grad}\left(\frac{\delta\phi}{\delta s}\right)\times\mathbf{t} + \frac{\delta\phi}{\delta n}\operatorname{curl}\mathbf{n} + \operatorname{grad}\left(\frac{\delta\phi}{\delta n}\right)\times\mathbf{n} \\ + \frac{\delta\phi}{\delta b}\operatorname{curl}\mathbf{b} + \operatorname{grad}\frac{\delta\phi}{\delta b}\times\mathbf{b} = \mathbf{0}\end{aligned} \tag{16}$$

whence, on use of the relations (10), (12) and (14), it follows that[13]

$$\begin{aligned}\frac{\delta^2\phi}{\delta n\delta b} - \frac{\delta^2\phi}{\delta b\delta n} &= -\frac{\delta\phi}{\delta s}\Omega_s + \frac{\delta\phi}{\delta n}\operatorname{div}\mathbf{b} - \frac{\delta\phi}{\delta b}(\kappa + \operatorname{div}\mathbf{n}),\\ \frac{\delta^2\phi}{\delta b\delta s} - \frac{\delta^2\phi}{\delta s\delta b} &= -\frac{\delta\phi}{\delta n}\Omega_n + \frac{\delta\phi}{\delta b}\theta_{bs},\\ \frac{\delta^2\phi}{\delta s\delta n} - \frac{\delta^2\phi}{\delta n\delta s} &= -\frac{\delta\phi}{\delta s}\kappa - \frac{\delta\phi}{\delta n}\theta_{ns} - \frac{\delta\phi}{\delta b}\Omega_b\ .\end{aligned} \tag{17}$$

The compatibility of the system (1)-(3) imposes, on use of the relations (17), the following set of nine conditions on the eight geometric parameters $\kappa, \tau, \Omega_s, \Omega_n, \operatorname{div}\mathbf{n}, \operatorname{div}\mathbf{b}, \theta_{ns}$ and θ_{bs}[13,17]:

$$\begin{aligned}\frac{\delta}{\delta b}\theta_{ns} + \frac{\delta}{\delta n}(\tau + \Omega_n) = (\kappa + \operatorname{div}\mathbf{n})(\Omega_s - 2\Omega_n - 2\tau)\\ + \operatorname{div}\mathbf{b}(\theta_{bs} - \theta_{ns}) + \Omega_s\kappa,\end{aligned} \tag{18}$$

$$\begin{aligned}\frac{\delta}{\delta b}(\tau + \Omega_n - \Omega_s) + \frac{\delta}{\delta n}\theta_{bs} = (\kappa + \operatorname{div}\mathbf{n})(\theta_{ns} - \theta_{bs})\\ + \operatorname{div}\mathbf{b}(\Omega_s - 2\Omega_n - 2\tau)\ ,\end{aligned} \tag{19}$$

$$\begin{aligned}\frac{\delta}{\delta b}(\operatorname{div}\mathbf{b}) + \frac{\delta}{\delta n}(\kappa + \operatorname{div}\mathbf{n}) = (\tau + \Omega_n)(\tau + \Omega_n - \Omega_s) - \theta_{ns}\theta_{bs}\\ - \tau\Omega_s - (\operatorname{div}\mathbf{b})^2 - (\kappa + \operatorname{div}\mathbf{n})^2\ ,\end{aligned} \tag{20}$$

$$\frac{\delta}{\delta s}(\tau + \Omega_n) + \frac{\delta\kappa}{\delta b} = -\Omega_n\theta_{ns} - (2\tau + \Omega_n)\theta_{bs}\ , \tag{21}$$

$$\frac{\delta}{\delta s}\theta_{bs} = -\theta_{bs}^2 + \kappa(\kappa + \operatorname{div}\mathbf{n}) - \Omega_n(\tau + \Omega_n - \Omega_s) + \tau(\tau + \Omega_n), \tag{22}$$

$$\frac{\delta}{\delta s}(\kappa + \operatorname{div}\mathbf{n}) - \frac{\delta\tau}{\delta b} = -\Omega_n\operatorname{div}\mathbf{b} - \theta_{bs}(2\kappa + \operatorname{div}\mathbf{n}), \tag{23}$$

$$\begin{aligned}\frac{\delta\kappa}{\delta n} - \frac{\delta}{\delta s}\theta_{ns} = \kappa^2 + \theta_{ns}^2 + (\tau + \Omega_n)(3\tau + \Omega_n)\\ - \Omega_s(2\tau + \Omega_n)\ ,\end{aligned} \tag{24}$$

$$\begin{aligned}\frac{\delta}{\delta s}(\tau + \Omega_n - \Omega_s) = -\theta_{ns}(\Omega_n - \Omega_s)\\ + \kappa\operatorname{div}\mathbf{b} + \theta_{bs}(-2\tau - \Omega_n + \Omega_s)\ ,\end{aligned} \tag{25}$$

$$\frac{\delta\tau}{\delta n} + \frac{\delta}{\delta s}(\operatorname{div}\mathbf{b}) = -\kappa(\Omega_n - \Omega_s) - \theta_{ns}\operatorname{div}\mathbf{b} + (\kappa + \operatorname{div}\mathbf{n})(-2\tau - \Omega_n + \Omega_s) \; . \tag{26}$$

3 The case $\mathbf{\Omega}_n = 0$

In the sequel, our concern will be with geometries such that

$$\Omega_n = 0 \tag{27}$$

Magnetohydrodynamic motions subject to this restriction have been investigated by Rogers and Kingston[18]. The condition (27) guarantees the existence of functions Σ, Ψ such that

$$\mathbf{n} = \Psi\nabla\Sigma \; . \tag{28}$$

As $\mathbf{n}$ is parallel to the normal to the surfaces $\Sigma =$ constant, the vector lines of $\mathbf{t}$ (commonly termed s-lines) constitute geodesics on these surfaces. Accordingly, the vector lines of $\mathbf{b}$ (the b-lines) are necessarily parallels thereon[16].

If the s-lines and b-lines are taken as parametric curves on the surfaces $\Sigma =$ constant then the surface metric of an individual such surface may be brought into the form

$$\mathrm{I}_{\Sigma} = ds^2 + g(s,b)db^2 \; , \tag{29}$$

and the two-parameter surface gradient for $\Sigma =$ constant is given by

$$\boldsymbol{\nabla}_{\Sigma} = \mathbf{t}\frac{\delta}{\delta s} + \mathbf{b}\frac{\delta}{\delta b} = \mathbf{t}\frac{\partial}{\partial s} + \frac{\mathbf{b}}{g^{1/2}}\frac{\partial}{\partial b} \tag{30}$$

where

$$\theta_{bs} = \frac{\partial}{\partial s}\ln g^{1/2} \; . \tag{31}$$

The Gauss-Weingarten equations for an individual surface $\Sigma =$ constant become

$$\begin{aligned} \frac{\partial}{\partial s}\begin{pmatrix}\mathbf{t}\\ \mathbf{n}\\ \mathbf{b}\end{pmatrix} &= \begin{pmatrix}0 & \kappa & 0\\ -\kappa & 0 & \tau\\ 0 & -\tau & 0\end{pmatrix}\begin{pmatrix}\mathbf{t}\\ \mathbf{n}\\ \mathbf{b}\end{pmatrix} \; , \\ \frac{1}{g^{1/2}}\frac{\partial}{\partial b}\begin{pmatrix}\mathbf{t}\\ \mathbf{n}\\ \mathbf{b}\end{pmatrix} &= \begin{pmatrix}0 & -\tau & \theta_{bs}\\ \tau & 0 & \kappa + \operatorname{div}\mathbf{n}\\ -\theta_{bs} & -(\kappa + \operatorname{div}\mathbf{n}) & 0\end{pmatrix}\begin{pmatrix}\mathbf{t}\\ \mathbf{n}\\ \mathbf{b}\end{pmatrix} \; . \end{aligned} \tag{32}$$

Moreover, if $\mathbf{r}$ denotes the generic position vector to a surface Σ = constant then

$$\frac{\partial \mathbf{r}}{\partial s} = \mathbf{t} \ , \quad \frac{\partial \mathbf{r}}{\partial b} = g^{1/2}\mathbf{b} \tag{33}$$

so that compatibility requires the $\mathbf{t}$-field to satisfy the Heisenberg spin-type equation

$$\frac{\partial \mathbf{t}}{\partial b} = \frac{\partial}{\partial s}\left(h\mathbf{t} \times \frac{\partial \mathbf{t}}{\partial s}\right) \ , \quad h = g^{1/2}/\kappa \ . \tag{34}$$

Importantly, if

$$g^{1/2} = \nu\kappa \ , \ \partial\nu/\partial s = 0 \tag{35}$$

so that

$$\theta_{bs} = \kappa_s/\kappa \tag{36}$$

then, on absorption of ν into the coordinate b, the integrable Heisenberg spin equation

$$\frac{\partial \mathbf{t}}{\partial b} = \mathbf{t} \times \frac{\partial^2 \mathbf{t}}{\partial s^2} \tag{37}$$

results. In this case, the compatibility conditions for the Gauss-Weingarten system (32), namely (22)-(24) reduce to the nonlinear Schrödinger (NLS) equation.

4 The hydrodynamics system with $\operatorname{div} \mathbf{t} = 0$

Here, we investigate the steady classical hydrodynamics system

$$\operatorname{div} \mathbf{q} = 0 \ , \tag{38}$$

$$\rho(\mathbf{q} \cdot \nabla)\mathbf{q} + \nabla p = \mathbf{0} \ , \tag{39}$$

where $\mathbf{q}$ is the fluid velocity and p, ρ denote the pressure and constant density respectively.

If we now set $\mathbf{q} = q\,\mathbf{t}$ where $\mathbf{t}$ is the unit tangent to the streamline then decomposition of the continuity and momentum equations (38), (39) yields

$$\frac{\delta q}{\delta s} + q\operatorname{div} \mathbf{t} = 0 \ , \tag{40}$$

together with

$$\frac{\delta p}{\delta s} = \rho q^2 \operatorname{div} \mathbf{t} \ , \ \frac{\delta p}{\delta n} = -\rho q^2 \kappa \ , \ \frac{\delta p}{\delta b} = 0 \ . \tag{41}$$

On use of the commutator relations (17) it is seen that the compatibility conditions for the pressure field p require that

$$\begin{aligned} 2\left(\frac{\delta}{\delta n}\ln q\right)\operatorname{div}\mathbf{t} &= -\frac{\delta\kappa}{\delta s}+\kappa\theta_{bs}-\frac{\delta}{\delta n}\operatorname{div}\mathbf{t}+2\kappa\operatorname{div}\mathbf{t}\ , \\ 2\left(\frac{\delta}{\delta b}\ln q\right)\operatorname{div}\mathbf{t} &= \kappa\Omega_n-\frac{\delta}{\delta b}\operatorname{div}\mathbf{t}\ , \\ 2\kappa\left(\frac{\delta}{\delta b}\ln q\right) &= -\operatorname{div}(\kappa\mathbf{b})-\Omega_s\operatorname{div}\mathbf{t}\ . \end{aligned} \tag{42}$$

The relations (42) show that, if $\operatorname{div}\mathbf{t}\neq 0$ then the hydrodynamic motions are completely determined by the streamline pattern. The result is of long standing and was obtained in another manner by Prim[19]. On the other hand, if $\operatorname{div}\mathbf{t}=0$ and $\kappa\neq 0$ then the compatibility conditions (42) yield

$$\theta_{bs}=\kappa_s/\kappa\ ,\quad \Omega_n=0\quad . \tag{43}$$

These relations are augmented by

$$\frac{\delta}{\delta s}\ln q=-\frac{1}{2\kappa}\operatorname{div}\kappa b\ ,\quad q\neq 0 \tag{44}$$

which is readily shown to be compatible with the relation

$$\frac{\delta q}{\delta s}=0 \tag{45}$$

It is noted that the latter is equivalent, by virtue of the continuity equation (40) to the geometric condition $\operatorname{div}\mathbf{t}=0$ if stagnation points are excluded. It was with hydrodynamic motions subject to the condition (43) that the doctoral thesis of Howard in 1953 was concerned[1]. The relations (43) establish the remarkable result originally obtained in[4] that for such hydrodynamic motions the $\mathbf{t}$-field satisfies the integrable Heisenberg spin equation. The result is encapsulated in the following:

Theorem 1

In steady hydrodynamic motions with $\delta q/\delta s=0$, the unit tangent $\mathbf{t}$ to the streamlines on individual constant pressure surfaces obeys the Heisenberg spin-type equation

$$\frac{\partial\mathbf{t}}{\partial b}=\frac{\partial}{\partial s}\left(\mathbf{t}\times\frac{\partial\mathbf{t}}{\delta s}\right)\ ,$$

where s denotes arc length of the streamlines and b parametrises the binormal lines.

5 The geometry of the spatial motion of fibre-reinforced fluids

Spencer in[11] examined kinematic aspects of the motion of an ideal fibre-reinforced fluid. Therein, a fibre-direction is characterised by a unit vector **a** attributed to each fluid particle. The fibres are the trajectories of the vector **a**. The basic kinematic conditions were shown to reduce to consideration of the incompressibility condition

$$\operatorname{div} \mathbf{v} = 0 \ , \tag{46}$$

together with the requirement that the fibres cónvect with the flow, namely

$$\frac{\partial \mathbf{a}}{\partial t} + \mathbf{v} \cdot \nabla \mathbf{a} - \mathbf{a} \nabla \mathbf{v} = \mathbf{0} \ . \tag{47}$$

It was noted in[11], that (47) incorporates, on scalar multiplication by **a**, the condition of fibre inextensibility. Moreover, it was observed that (46) and (49) combine to show that

$$\left(\frac{\partial}{\partial t} + \mathbf{v} \cdot \nabla \right) \operatorname{div} \mathbf{a} = 0 \ . \tag{48}$$

Thus, if

$$\operatorname{div} \mathbf{a} = 0 \tag{49}$$

initially then this divergence remains zero throughout the motion. Such flows were termed *fibre-divergence free* by Spencer. It will be with the geometry of this important class that we shall now be concerned.

We proceed with the assumption that the fibre direction at a point is constant in time so that $\dfrac{\partial \mathbf{a}}{\partial t} = 0$. Accordingly, in view of (46) and (47), the convection condition (47) yields

$$\operatorname{curl} (\mathbf{v} \times \mathbf{a}) = \mathbf{0} \tag{50}$$

whence the fibre and streamlines lie on a family of surfaces Σ.

If the fluid velocity **v** is now decomposed according to

$$\mathbf{v} = v_a \mathbf{a} + v_n \mathbf{n} + v_b \mathbf{b} \tag{51}$$

in terms of the fibre direction **a**, its principal normal **n** and binormal **b** then the continuity and convection conditions yield, in turn

$$\frac{\delta v_a}{\delta s} + v_a \operatorname{div} \mathbf{a} + \frac{\delta v_n}{\delta n} + v_n \operatorname{div} \mathbf{n} + \frac{\delta v_b}{\delta b} + v_b \operatorname{div} \mathbf{b} = 0 \ , \tag{52}$$

together with

$$v_n\kappa - \frac{\delta v_a}{\delta s} = 0, \tag{53}$$

$$v_n\theta_{ns} - v_b\Omega_n - \frac{\delta v_n}{\delta s} = 0, \tag{54}$$

$$v_n\Omega_b + v_b\theta_{bs} - \frac{\delta v_b}{\delta s} = 0\ , \tag{55}$$

while the condition (49) is equivalent to

$$\theta_{ns} + \theta_{bs} = 0\ . \tag{56}$$

In what follows, attention is restricted to the case $v_n = 0,\ v_b \neq 0$. The system (52)-(55) then reduces, in view of (49) to

$$\frac{\delta v_b}{\delta b} + v_b \operatorname{div} \mathbf{b} = 0, \tag{57}$$

$$\frac{\delta v_a}{\delta s} = 0\ , \tag{58}$$

$$\Omega_n = 0\ , \tag{59}$$

$$v_b\theta_{bs} - \frac{\delta v_b}{\delta s} = 0\ . \tag{60}$$

The condition (59) shows that the normal to the surfaces Σ is parallel to the principal normal $\mathbf{n}$ to the fibre direction $\mathbf{a}$. Hence the fibres are geodesics on the surfaces Σ.

The relations (57) and (60) have compatibility condition

$$\frac{\delta\theta_{bs}}{\delta b} + \frac{\delta}{\delta s}\operatorname{div}\mathbf{b} = -(\operatorname{div}\mathbf{b})\theta_{bs}\ . \tag{61}$$

However, the geometric conditions (18) and (26) reduce, in view of (56) and (59) to

$$-\frac{\delta}{\delta b}\theta_{bs} + \frac{\delta\tau}{\delta n} = (\kappa + \operatorname{div}\mathbf{n})(\Omega_s - 2\tau) + 2\theta_{bs}\operatorname{div}\mathbf{b} + \Omega_s\kappa\ ,$$

and

$$\frac{\delta\tau}{\delta n} + \frac{\delta}{\delta s}\operatorname{div}\mathbf{b} = \Omega_s\kappa + \theta_{bs}\operatorname{div}\mathbf{b} + (\kappa + \operatorname{div}\mathbf{n})(\Omega_s - 2\tau)\ ,$$

whence, on subtraction, (61) results. Accordingly, the compatibility requirement on the equation (57) and (60) for the binormal component v_b of the fluid velocity imposes no additional constraint.

If the fibre lines and their orthogonal trajectories are now taken as parametric curves on the surfaces Σ so that the surface metric I_Σ adopts the form (29) then the condition (60) shows that

$$v_b = \alpha g^{1/2} , \quad (\delta\alpha/\delta s = 0) \tag{62}$$

while (57) requires that

$$\operatorname{div}(\alpha g^{1/2}\mathbf{b}) = 0 . \tag{63}$$

In particular, for geometries with

$$g^{1/2} = \kappa \tag{64}$$

(60) shows that $\theta_{bs} = \kappa_s/\kappa$. The latter relation together with $\Omega_n = 0$ establishes the following result:

Theorem 2

In spatial motions of an ideal fibre-reinforced fluid with $\mathbf{v} = v_a\mathbf{a} + \alpha\kappa\mathbf{b}$, *where* v_a *and* α *are constant on individual fibres and the steady fibre direction* $\mathbf{a}$ *is such that* $\operatorname{div}\mathbf{a} = 0$, *the unit vector* $\mathbf{a}$ *necessarily obeys the integrable Heisenberg spin equation*

$$\frac{\partial\mathbf{a}}{\partial b} = \frac{\partial}{\partial s}\left(\mathbf{a} \times \frac{\partial\mathbf{a}}{\delta s}\right)$$

on the individual NLS surfaces Σ *containing the fibre and streamlines. The binormals* $\mathbf{b}$ *to the fibres are such that* $\operatorname{div}(\alpha\kappa\mathbf{b}) = 0$.

In view of the relation[4]

$$\frac{\partial\Omega_s}{\delta s} + \Omega_s \operatorname{div}\mathbf{a} + \operatorname{div}\kappa\mathbf{b} = 0$$

it is seen that

$$\frac{\delta\Omega_s}{\delta s} = \frac{\delta}{\delta b}\ln\alpha . \tag{65}$$

It is remarked that the foliation of NLS surfaces subject to the condition $\operatorname{div}\mathbf{a} = 0$ recently constructed by Schief[10] has $\delta\Omega_s/\delta s = 0$.

6 Orthogonal families of fibres

Spencer, in his monograph[20] on fibre-reinforced materials considered kinematically admissible deformations involving surfaces reinforced by two families of continuously distributed fibres. In the case of motions of fibre-reinforced fluids with

$$\mathbf{v} = v_a\mathbf{a} + \alpha g^{1/2}\mathbf{b} .$$

if the **b**-lines are also fibres and div **b** vanishes initially so that

$$\operatorname{div}\mathbf{b} = 0 \tag{66}$$

throughout the motion then the convection condition yields

$$\operatorname{curl}(v_a\mathbf{n}) = 0 \tag{67}$$

whence

$$\frac{\delta v_a}{\delta b} = 0\ , \tag{68}$$

$$\theta_{ns} = 0\ . \tag{69}$$

Thus,

$$\operatorname{curl}\mathbf{n} = \mathbf{0} \tag{70}$$

while, in view of (56), the condition (69) is equivalent to

$$\theta_{bs} = 0\ . \tag{71}$$

But, the Gaussian curvature $\mathcal{K}$ of the surfaces Σ bearing the fibre and streamlines in given by

$$\mathcal{K} = |\mathbf{N}, \operatorname{curl}_\Sigma\mathbf{t},\ \operatorname{curl}_\Sigma\mathbf{b}| = -\frac{\delta\theta_{bs}}{\delta s} - \theta_{bs}^2\ . \tag{72}$$

Accordingly, $\mathcal{K} = 0$ and, in view of (70), the surfaces Σ are parallel developables. Moreover, θ_{ns} is the geodesic curvature of the **b** fibres so that (69) shows that these also are geodesics on the surfaces Σ. Thus, we obtain the following result:

Theorem 3

In spatial motion of an ideal fibre-reinforced fluid with $\mathbf{v} = v_a\mathbf{a} + v_b\mathbf{b}$, *wherein both the* **a** *and* **b** *lines are fibres with* $\operatorname{div}\mathbf{a} = \operatorname{div}\mathbf{b} = 0$, *the surfaces* Σ *on which this double family of fibres and the streamlines lie constitute parallel developables. The* **a** *and* **b** *fibres are geodesics thereon. The velocity components* v_a *and* v_b *are constant on individual developables.*

It is interesting to record that Spencer discussed in[20], kinematically possible classes of deformations of doubly fibre-reinforced materials in which the fibres lie on parallel developables.

References

1. L. Howard, *Constant Speed Flows*, PhD Thesis, Princeton University (1953).
2. R.H. Wasserman, *On a class of three-dimensional compressible fluid flows*, J. Math. Anal. **5**, 119-135 (1962).
3. A.W. Marris, *On motions with constant speed and streamline parameters*, Arch. Rat. Mech. Anal. **90**, 1-14 (1985).
4. C. Rogers, *On the Heisenberg spin equation in hydrodynamics*, Preprint. Instituto de Matematica Pura & Applicada, Rio de Janeiro, Série B-127 (2000).
5. H. Hasimoto, *A soliton on a vortex filament*, J. Fluid. Mech. **51**, 477-485 (1972).
6. L.S. Da Rios, *Sul moto d'un liquido indefinito con un filetto vorticoso di forma qualunque*, Rend. Circ. Mat. Palermo **22**, 117-135 (1906).
7. T. Levi-Civita, *Attrazione Newtoniana dei Tubi Sottili e Vortici Filiformi*, Annali R Scuola Norm Pisa, Zanichelli, Bologna (1932).
8. R.L. Ricca, *Rediscovery of Da Rios equation*, Nature **352**, 561-562 (1991).
9. C. Rogers and W.K. Schief, *Intrinsic geometry of the NLS equation and its auto-Bäcklund transformation*, Stud. Appl. Math. **101**, 267-287 (1998).
10. W.K. Schief, Nested toroidal surfaces in magnetohydrostatics. Generation via soliton theory, in preparation (2001).
11. A.J.M. Spencer, *Fibre-streamline flows of fibre-reinforced viscous fluids*, Euro. Jnl. Appl. Math. **8**, 209-215 (1997)
12. M.G. Vranceanu, *Les éspaces non holonomes et leurs applications mécaniques*, Mém. Sci. Math. **76** (1936).
13. A.W. Marris and S.L. Passman, *Vector fields and flows on developable surfaces*, Arch. Rat. Mech. Anal. **32**, 29-86 (1969).
14. O. Bjørgum, *On Beltrami vector fields and flows*, Part I, Universitat I, Bergen, Arbok Naturvitenskapelig rekke n1 (1951).
15. M. Massoti, *Decomposizione intrinseca del vortice a sue applicazioni*, Instituto Lombardo di Scienze a Lettere Rendiconti (2) **60**, 869-879 (1927).
16. C.E. Weatherburn, *Differential Geometry of Three Dimensions* Vol II, Cambridge University Press (1930).
17. W.L. Yin and A.C. Pipkin, *Kinematics of viscometric flow*, Arch. Rat. Mech. Anal. **37**, 115-135 (1970).
18. C. Rogers and J.G. Kingston, *Non-dissipative magneto-hydrodynamic flows with magnetic and velocity field lines orthogonal geodesics on a normal congruence*, Soc. Ind. Appl. Math. Jl. Appl. Math. **26**, 183-195

(1974).
19. R. Prim, *On the uniqueness of flows with given streamlines*, J. Math. Phys. **28**, 50-53 (1949).
20. A.J.M. Spencer, *Deformations of Fibre-reinforced Materials*, Oxford University Press (1972).

RAYLEIGH WAVES IN HORIZONTALLY STRATIFIED MEDIA: RELEVANCE OF RESONANT FREQUENCIES

ROMA V.[*] LANCELLOTTA R.[*], RIX G.J.[‡]

1 Abstract

The interest of researchers towards Rayleigh waves is justified by the importance that they have in engineering applications such as geotechnical soil characterization and dynamic response of structures[1,2]. In this work the frequencies of resonance of a layered half-space will be presented.i.e the frequencies for which the vertical displacements due to travelling Rayleigh waves are maximised. Successively a sensitivity analysis for a simple system has allowed for an approximate relationship to be found among the frequencies of resonance and the properties of the system.

2 Theoretical Considerations

The geometry of the system is a set of n horizontally infinite layers overlaying an indefinite half-space. For both layers and half-space the hypotheses of homogeneity, isotropy, linear elastic mechanical behaviour will be assumed, so that the theory of linear elasticity and the principle of superposition of effects hold. The geometrical-mechanical properties of the system are (see fig.1): thickness h_i, shear wave velocities V_{S_i}, Poisson ratios v_i, mass densities ρ_i, in which the sub-indeces i and ∞ refer to the generic i-th layer and to the half-space respectively. A coordinate system (z, r) is introduced as it is shown in fig.1. If we consider the wave equation of motion in plane strain conditions for each layer and if we assume a solution in terms of displacements u and stresses σ of the following form:

$$u_l(x,z,t) = u_{0l}(z)e^{i(\omega t - kx)}, \sigma_l(x,z,t) = \sigma_{0l}(z)e^{i(\omega t - kx)} \tag{1}$$

in which ω is the circular frequency of excitation, k is the wave number in x direction, $u_{0l}(z)$ and $\sigma_{0l}(z)$ depend on the depth z only, then it has been shown [3] that the equation of equilibrium of forces for each layer is:

$$F_l(\omega, k) = K_l(\omega, k) \cdot u_l(\omega, k) \tag{2}$$

*POLITECNICO DI TORINO (ITALY)
‡GEORGIA INSTITUTE OF TECHNOLOGY, ATLANTA (USA)

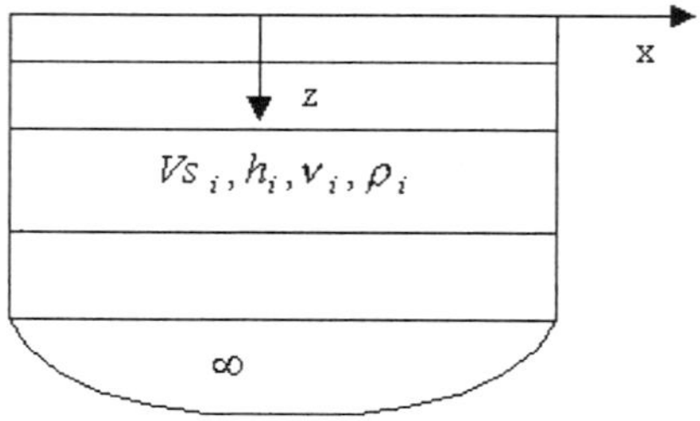

Figure 1. Model of the site.

where $F_l(\omega, k)$ is the vector of external loads, $u_l(\omega, k)$ is the displacements vector and $K_l(\omega, k)$ is the stiffness matrix of the layer and l refers to the layer. The matrices of all the layers can be properly assembled, by imposing the equilibrium of forces and the congruence of displacements at the interfaces of the layers, so that the equation of dynamical equilibrium of the global system can be written as [3]:

$$F(\omega, k) = K(\omega, k) \cdot u(\omega, k) \tag{3}$$

in which $F(\omega, k)$ and $u(\omega, k)$ represent the external loads and the displacements at the interfaces of the layers, $K(\omega, k)$ is the global stiffness matrix of the system and depends on the geometrical and mechanical properties of the system and on the circular frequency ω and the wave number k. In order to study the free vibrations of the system the external loads are set equal to zero:

$$F(\omega, k) = K(\omega, k) \cdot u(\omega, k) = 0 \tag{4}$$

and nontrivial solutions are searched solving the eigenvalue problem:

$$det[K(\omega, k)] = 0 \tag{5}$$

Since the terms of the stiffness matrix K are transcendental, (5) has in general infinitely many solutions and it must be solved by search techniques. The (5) represents the *Geometrical Rayleigh Dispersion Relation*. For a fixed frequency ω, the eigenvalues k that satisfy (5) are the wave numbers corresponding to the generalised Rayleigh waves. Each solution is named Rayleigh mode of vibration and it physically represents a simple wave that can propagate through the system under the condition of free vibrations. The dispersion phenomenon exists when the several components of a general disturbance travel with different velocities because of their different wavelengths. As a consequence after a certain time and space the original disturbance will change its shape and will spread into a long wave train with the wavelength

Layer	Thickness h(m)	Vp(m/s)	Vs(m/s)	Mass density (Kg/m^3)
1	5	1500	1000	1800
2	5	750	500	1800
Half-space	∞	1500	1000	1800

Table 1. Geometrical and mechanical characteristics of the layered system taken as example

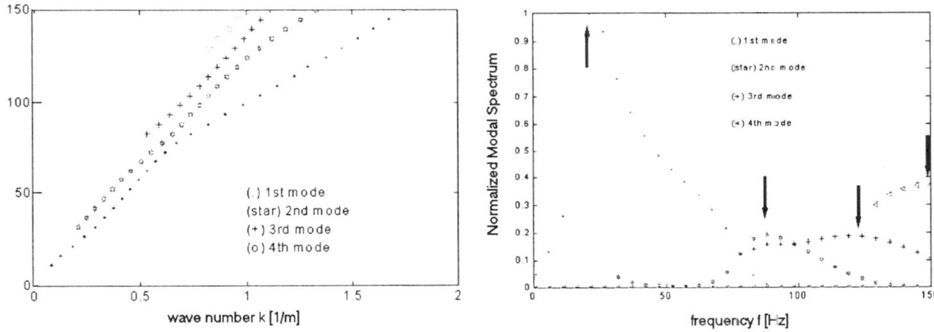

Figure 2. (a)Solutions of the geometrical dispersion relation in the [f,k] domain for the numerical example: Rayleigh modes of vibration.(b)Normalised spectrum of vertical modal displacements on the surface of the layered half-space for the numerical example.

varying rather gradually along it [2]. The *Geometrical Dispersion* is intrinsically due to the propagation of the Rayleigh waves through a stratified wave guide. A clear picture of this phenomenon is given by the principle of constructive interference [5].

3 Numerical Example

Consider a system where n=2, i.e. two layers over an infinite half-space whose characteristics are summarised in the table 1.

Since the stiffness does not always increase with depth, but a softer layer is trapped between two stiffer ones the system is an inversely dispersive profile, i.e. the group velocity is greater than the phase velocity for some frequencies. The example is just too complicate to analytically express the dispersion relation (5), so only the numerical solutions of (5) are presented in the (f, k) domain (see fig.2(a)). It is visible that (5) has multiple branches,i.e. for a fixed frequency more than one wave number may exist and vice versa for a fixed wave number more than one frequency may satisfy (5). Each branch represents a Rayleigh mode that follows an independent path in the f-k domain, an aspect that will be stressed in the following. A typical characteristic of the higher modes is that for each one of them a cut-off frequency exists, below which the mode cannot propagate. In fact only real values of the wave number k correspond to propagating Rayleigh waves, when no dissipative mechanisms are encountered in the model and complex values of the wave number k have no physical meaning. If a vertical harmonic source of the type:

$$F_z(t) = F_0 \cdot e^{i\omega t} \tag{6}$$

is applied in a point at the free surface of the layered half-space then different types of waves are generated: body and Rayleigh waves that travel with spherical and cylindrical wave fronts respectively. Because of their geometrical attenuation and the distribution of energy among them, in the far field the main contribution to the response of the system is given by Rayleigh waves, so the global displacement field at a certain position in space for a given frequency of excitation ω can be expressed as a proper superposition of all the Rayleigh modes [4]:

$$u_\beta(x, z, \omega, t) = \sum_{j=1}^{\infty} [A_\beta(x, z, \omega)]_j \cdot e^{i(\omega t - k_j x + \varphi_\beta)} \tag{7}$$

$$[A_\beta(x, z, \omega)]_j = \frac{F_0 \cdot r_2(z_{source}, k_j, \omega)}{4v_j \cdot U_j \cdot I_j \cdot \sqrt{2\pi x \cdot k_j}} \cdot r_{1or2}(z_source, k_j, \omega) \tag{8}$$

$$I_j(z, k_j, \omega) = \frac{1}{2} \int_0^\infty \rho(z) \cdot [r_1^2(z, k_j, \omega) + r_2^2(z, k_j, \omega)] \cdot dz \tag{9}$$

in which β indicates either the vertical or the horizontal component of motion, j denotes the generic j-th mode, M is the total number of modes, k_j is the wave number corresponding to the j-th mode for the given frequency of excitation ω,z_{source} is the source depth, $\varphi_\beta$is a phase shift equal to $\pm\frac{p}{4}$ according to β, $A_j$is the amplitude of the modal displacement, v_j and U_j are the modal phase and group velocities, F_0 is the amplitude of the vertical harmonic force, I_j is the first energy integral and $r_1, r_2$are the eigenvectors as they are defined in [6]. Thanks to the linearity of the equations of motion each modal component represents a solution, so, instead of considering the global response of the system with all the modes combined together, it is insightful to take every single mode and study its behaviour independently from the others [7]. This way will enable us to understand which is the relative importance of each mode compared to the others and how it can be correlated to the properties of the system under investigation [2]. By Fourier transforming each modal component of the displacements from the (f, x) domain to the (f, k) domain a perspective look can be given to the normalised spectrum of the displacements 2(b). In the (f, k) domain the spectrum of displacements with all the modes together forms a surface with ridges and valleys, but if each modal component is followed along its path in the (f, k) domain, then along this modal branch both the frequency and the wave number change and there will be a particular couple of values of frequency and wave number for which the response of the system will be maximised. Such peaks in the spectrum represent the positions in which resonance is reached. In fig.2(b) the spectra of all the modes have been projected upon the $k = 0$ plane so that a representation in the frequency domain is available. The arrows indicate the positions of the peaks of the modal spectra and it is worthy to note that in cases like this, where the site is inversely dispersive, different modes are predominant in different ranges of frequency and the higher modes become important in evaluating the system response [2]. So for the example it can be observed that the 1st mode drops down after a frequency of about 75 Hz and the 2nd mode at about 130 Hz. The positions of these peaks strictly depend on the characteristics of the system they are referred to, but it is not easy to analytically establish an explicit relationship, even for the simplest case of a single layer over an infinite half-space. For this latter simple case a sensitivity analysis has been conducted. The parameters that have been considered are:V_{S_1}, h_1, V_{S_∞}, $\upsilon_1, \upsilon_\infty$ and the number j of the Rayleigh mode under investigation. The strategy that has been adopted is to hold all the parameters constant except the one of interest, so that its influence on the position of the peaks can be easily put in evidence. As results it has been observed that a linear dependence of the frequency of resonance f_R exists from the number j of the Rayleigh mode and the shear wave velocity of the layer and an inverse proportionality between the frequency of resonance and the thickness of the layer. It can also be said that the shear wave velocity of the half-space influences only the amplitude of the peak in the spectrum, not its position. Finally the results show that the frequency of resonance f_R is practically independent from the Poisson ratios in a wide range of variation. By summurizing all the results obtained herein an approximate relationship can be written among the parameters of the system under consideration and the frequency of resonance f_R for the generic j-th Rayleigh mode:

$$f_R = (A + Bj)\frac{V_{S1}}{h_1} \tag{10}$$

in which the two constants $A = -0.09$ and $B = 0.65$ have been determined with a relative error less than 5% by means of all the simulations that have been performed in terms of the dimensionless variable

$$y = f_R \frac{h_1}{V_{S1}} \tag{11}$$

4 Conclusions

At first the geometrical dispersion phenomenon of Rayleigh waves in a layered half-space has been introduced and an example has been illustrated when no external forces are applied on the system. Secondly the superposition of several modes of vibration has been presented, when dealing with a harmonic vertical source on the free surface of the half-space. By studying the relative importance of each mode respect to the others it has been possible to find out the existence of relative peaks in the frequency-wavenumber spectrum of displacements.

References

1. Lai C.G. (1998) *Simultaneous inversion of Rayleigh phase velocity and attenuation for near-surface site characterization*, PhD Diss., Georgia Inst. of Technology, Atlanta (Georgia, USA)
2. Roma V. (2001) *Soil Properties and Site Characterization through Rayleigh waves*, PhD Thesis, Politecnico di Torino, Torino (Italy)
3. Kausel, Roesset (1981) Stiffness matrices for layered soils, *Bullettin of seismological Society of America*, vol.71 (6), pp.1743-1761
4. Achenbach J.D. (1984) *Wave propagation in elastic solids*, North-Holland, Amsterdam, Netherlands
5. Tolstoy (1960) *Wave propagation*, McGraw-Hill, New York
6. Aki , Richards (1980) *Quantitative Seismology I and II*, Freeman, S.Francisco
7. Roma V. Rix G., Lai C.G.,Hebeler G.L.(2002) Geotechnical Soil Characterization using Fundamental and Higher Rayleigh Modes in Layered Media *12th European Conference on Earthquake Engineering* September 9-13(accepteded for publication)
8. Kramer (1996) *Geotechnical Earthquake Engineering*, Prentice Hall, New York

SYMMETRY CLASSIFICATION FOR A CLASS OF ENERGY-TRANSPORT MODELS

V. ROMANO, A. VALENTI
Dipartimento di Matematica e Informatica, Università di Catania, viale A. Doria, 6, 95125 Catania, Italy
E-mail: romano@dmi.unict.it, valenti@dmi.unict.it

The symmetry classification of a class of energy-transport models, arising in hydrodynamical modeling of charge transport in semiconductors, is presented. Optimal systems of one dimensional Lie subalgebras are obtained.

1 Introduction

Continuum models for the description of charge carrier transport in semiconductors have attracted in the last years the attention of applied mathematicians and engineers on account of their applications in the design of electron devices. The *energy transport* models for semiconductors (hereafter ET models) are macroscopic models that take into account also the thermal effects related to the electron flow through the crystal at variance with the popular *drift-diffusion* [1,2,3] models that are based on the assumption of isothermal motion.

The evolution equations are given by the balance equations for density and energy of the charge carriers, coupled to the Poisson equation. The pioneering models were proposed in [4,5] on the basis of heuristic argument.

Here the symmetry classification [6,7,8,9,10] for a class of energy transport models is investigated in the one dimensional case. We give the functional form of the constitutive functions (mobilities, energy relaxation time and doping profile) so that the balance equations admits symmetries. In particular one recovers for the fluxes the same expressions as those of the energy transport model of Chen et el. [4]. Optimal systems of one dimensional Lie subalgebras are given.

2 The symmetry classification

The energy transport models comprise the balance equations for density and energy of the charge carriers, coupled to the Poisson equation for the electric potential. In the one dimensional case they are given by the following class $\mathcal{C}$ of PDE's

$$n_t + J_x = 0, \tag{1}$$
$$\frac{3}{2}(nT)_t + S_x + JE + \frac{3}{2}n\frac{(T - T_L)}{\tau_W(T)} = 0, \tag{2}$$
$$\lambda^2 E_x + n - c(x) = 0. \tag{3}$$

n is the electron density, T is the electron temperature, T_L the lattice temperature, assumed to remain constant. E is the electric field, which is related to the potential V in the usual way $E = -V_x$, $\tau_W(T)$ is the energy relaxation time and $c(x)$ is the doping profile. All the previous variables are to be intended in a scaled form.

J and S are the relevant components of the electron momentum and energy flux. They are related to n, T and E through the constitutive relations

$$J = -\left[\left(\mu^{(1)}(T)Tn\right)_x + \mu^{(1)}(T)nE\right], \tag{4}$$
$$S = -\left[\left(\mu^{(2)}(T)T^2 n\right)_x + \mu^{(2)}(T)TnE\right]. \tag{5}$$

$\mu^{(1)}(T)$, $\mu^{(1)}(T)$ are the mobilities.

According to the specific form of $\mu^{(1)}(T)$, $\mu^{(1)}(T)$ and $\tau_W(T)$ one has different energy transport models (see [11] for a complete review).

We discuss the symmetry classification of the systems belonging to the class $\mathcal{C}$ of PDE's by the infinitesimal Lie method. This latter allows us to find the infinitesimal generator of the symmetry transformations and, at the same time, gives the functional dependence of the constitutive functions $\mu^{(1)}(T)$, $\mu^{(1)}(T)$, $\tau_W(T)$ and $c(x)$ for which the system does admit symmetries.

We consider the one-parameter Lie group of infinitesimal transformations in (x, t, n, T, E)-space given by

$$\hat{t} = t + \varepsilon\, \xi^1(x, t, n, T, E) + \mathcal{O}(\varepsilon^2), \tag{6}$$
$$\hat{x} = x + \varepsilon\, \xi^2(x, t, n, T, E) + \mathcal{O}(\varepsilon^2), \tag{7}$$
$$\hat{n} = n + \varepsilon\, \eta^1(x, t, n, T, E) + \mathcal{O}(\varepsilon^2), \tag{8}$$
$$\hat{T} = T + \varepsilon\, \eta^2(x, t, n, T, E) + \mathcal{O}(\varepsilon^2), \tag{9}$$
$$\hat{E} = E + \varepsilon\, \eta^3(x, t, n, T, E) + \mathcal{O}(\varepsilon^2), \tag{10}$$

where ε is the group parameter and the associated Lie algebra $\mathcal{L}$ is the set of vector fields of the form

$$X = \xi^1\frac{\partial}{\partial t} + \xi^2\frac{\partial}{\partial x} + \eta^1\frac{\partial}{\partial n} + \eta^2\frac{\partial}{\partial T} + \eta^3\frac{\partial}{\partial E}. \tag{11}$$

One then requires that the transformation (6)-(10) leaves invariant the set of solutions of the system (1)-(3). In others words, one requires that the transformed system has the same form as the original one. This yields to an over determined linear system of partial differential equations for the infinitesimals ξ^1, ξ^2, η^1, η^2 and η^3, called determining system, which leads to the following conditions

$$\xi^1 = -[2ma_1 + (1+m)b_1]t + b_0, \quad \xi^2 = a_1 x + a_0, \tag{12}$$

$$\eta^1 = b_1 n, \quad \eta^2 = (2a_1 + b_1)T, \quad \eta^3 = (a_1 + b_1)E, \tag{13}$$

$$(2a_1 + b_1)\left(T\mu_T^{(1)} + m\mu^{(1)}\right) = 0, \tag{14}$$

$$(2a_1 + b_1)\left(T\mu_T^{(2)} + m\mu^{(2)}\right) = 0, \tag{15}$$

$$(a_1 x + a_0)c_x - b_1 c = 0, \tag{16}$$

$$(2a_1 + b_1)\tau_{WT} + \frac{1}{T}[(2a_1 + b_1)\frac{T_L}{(T_L - T)}$$

$$+2ma_1 + (1+m)b_1]\tau_W = 0, \tag{17}$$

where a_0, a_1, b_0, b_1 and m are constants.

For $\mu^{(1)}$, $\mu^{(2)}$, τ_W and c arbitrary, from (12)-(17), we have that the Principal Lie Algebra $\mathcal{L}_\mathcal{P}$ of the system (1)-(3) is one-dimensional and it is spanned by the operator

$$X_1 = \frac{\partial}{\partial t}. \tag{18}$$

Otherwise, we obtain

$$\mu^{(1)} = \mu_0^{(1)} T^m, \qquad \mu^{(2)} = \mu_0^{(2)} T^m, \tag{19}$$

with $\mu_0^{(1)}$ and $\mu_0^{(2)}$ constants. In this case the Lie algebras extend of one dimension $\mathcal{L}_\mathcal{P}$. The complete Lie group classification for the system (1)-(3) is reported in table 1.

3 Optimal systems

In general, when a system of differential equations admits a Lie group $\mathcal{G}_r$ and its Lie algebra $\mathcal{L}_r$ is of dimension $r > 1$, one desires to minimize the search for invariant solutions by finding the nonequivalent branches of solutions. This leads to the concept of the optimal systems.

In fact if two subalgebras are similar, i.e. they are connected by a transformation belonging to the symmetry group (with Lie algebra $\mathcal{L}_r$), then their corresponding invariant solutions are connected by the same transformation.

case	Forms of $\tau_W(T)$ and $c(x)$	Extensions of $L_{\mathcal{P}}$
I	τ_W arbitrary $c = c_0$	$X_2 = \frac{\partial}{\partial x}$
II	$\tau_W = \frac{\tau_0(T-T_L)}{T^{(m+2)}}$ $c = c_0 e^{px}$ $p \neq 0$	$X_2 = -p(1+m)t\frac{\partial}{\partial t} + \frac{\partial}{\partial x}$ $+pn\frac{\partial}{\partial n} + pT\frac{\partial}{\partial T} + pE\frac{\partial}{\partial E}$
III	$\tau_W = \frac{\tau_0(T-T_L)}{T^{m+\frac{2(p+1)}{p+2}}}$ $c = c_0(x+q)^p,$ $p \neq -2$	$X_2 = -[2m + p(1+m)]t\frac{\partial}{\partial t}$ $+(x+q)\frac{\partial}{\partial x} + pn\frac{\partial}{\partial n}$ $+(2+p)T\frac{\partial}{\partial T} + (1+p)E\frac{\partial}{\partial E}$

Table 1. Lie group classification. $\mu^{(1)} = \mu_0^{(1)}T^m$, $\mu^{(2)} = \mu_0^{(2)}T^m$. c_0, τ_0, p and q are constitutive constants.

Therefore, it is sufficient to put into one class all similar subalgebras of a given dimension, say s, and select a representative from each class. The set of the representatives of all these classes is called an optimal system of order s [6]. In order to find all invariant solutions with respect to s-dimensional subalgebras, it is sufficient to construct invariant solutions for the optimal system of order s. The set of invariant solutions obtained in this way is called an optimal system of invariant solutions.

For one-dimensional subalgebras, the problem of finding an optimal system of subgroups is essentially the same as the problem of classifying the orbits of the adjoint transformations. In [6] in constructing the one-dimensional optimal system the global matrix of the adjoint transformation is used. In [7] a slightly different technique is employed: it consists in constructing a table (adjoint table) showing the separate adjoint actions of each element in $\mathcal{L}_r$ as it acts on all the other elements.

We recall that if X and Y are two element of the Lie Algebra, the adjoint action of the group generated by X on Y is given by the Lie serie

$$Ad(\exp(\varepsilon X))Y = Y - \varepsilon\,[X,Y] + \frac{\varepsilon^2}{2}\,[X,[X,Y]] - \frac{\varepsilon^3}{3!}\,[X,[X,[X,Y]]] + \cdots (1)$$

where ε is the group parameter and $[X,Y] = YX - XY$ is the commutator.

Here we show the details of the analysis only of some cases. The other cases can be investigated in a straightforward way.

CASE II with $m \neq -1$, $p \neq 0$. The only non trivial commutator is

$$[X_1, X_2] = (1+m)pX_1.$$

and the following adjoint table 2 has been obtained

Ad	X_1	X_2
X_1	$X_1 - \varepsilon(1+m)pX_1 + X_2$	
X_2	$e^{\varepsilon(1+m)p}$	X_2

Table 2. the (i,j)-th entry indicates $Ad(\exp(\varepsilon X_i))X_j$

Now let us consider an element of the Lie symmetry algebra: $X = a_1X_1 + a_2X_2$, $a_i \in \Re$. First we investigate the case $a_2 \neq 0$. One may set $a_2 = 1$ without loss of generality. By using the adjoint table 2, one finds

$$Ad\,(\exp\,(\varepsilon X_1))\,X_2 = [a_1 - \varepsilon(1+m)p]\,X_1 + X_2.$$

With the choice $\varepsilon = \frac{a_1}{(1+m)p}$, the coefficient of X_1 in $Ad\,(\exp\,(\varepsilon X_1))\,X_2$ vanishes. Therefore the elements of the Lie algebra of the type $X = a_1X_1 + a_2X_2$, $a_2 \neq 0$, are conjugate with X_2.

If $a_2 = 0$ no further reduction is possible.

CASE III with $2m + (1+m)p \neq 0$. The only non trivial commutator is

$$[X_1, X_2] = [2m + (1+m)p]\,X_1.$$

and the following adjoint table 3 has been obtained

Ad	X_1	X_2
X_1	X_1	$-\varepsilon\,[2m + (1+m)p1]\,X_1 + X_2$
X_2	$e^{\varepsilon[2m+(1+m)p]}$	X_2

Table 3. the (i,j)-th entry indicates $Ad(\exp(\varepsilon X_i))X_j$

As in the previous case we consider a generic element of the Lie symmetry algebra: $X = a_1X_1 + a_2X_2$, $a_i \in \Re$. In the case $a_2 \neq 0$, setting $a_2 = 1$ without loss of generality and by using the adjoint table 3, one finds

$$Ad\,(\exp(\varepsilon X_1))\,X_2 = [a_1 - \varepsilon(2m + (1+m)p)]\,X_1 + X_2.$$

With the choice $\varepsilon = \frac{a_1}{2m+(1+m)p}$, the coefficient of X_1 in $Ad\,(\exp(\varepsilon X_1))\,X_2$ vanishes. Therefore the element of the Lie algebra of the type $X = a_1X_1 + a_2X_2$, $a_2 \neq 0$, are conjugate with X_2.

If $a_2 = 0$ no further reduction is possible.

The complete results on the optimal system are summarized in table 4.

case	Generators of the optimal systems
I	$X_o = a\frac{\partial}{\partial t} + \frac{\partial}{\partial x}$
II_a	$m = -1$, $X_o = a\frac{\partial}{\partial t} + \frac{\partial}{\partial x} + pn\frac{\partial}{\partial n} + pT\frac{\partial}{\partial T} + pE\frac{\partial}{\partial E}$
II_b	$m \neq -1$, $X_o = -p(1+m)t\frac{\partial}{\partial t} + \frac{\partial}{\partial x}$ $+pn\frac{\partial}{\partial n} + pT\frac{\partial}{\partial T} + pE\frac{\partial}{\partial E}$
III_a	$2m + p(1+m) = 0$, $X_o = a\frac{\partial}{\partial t} + (x+q)\frac{\partial}{\partial x} + pn\frac{\partial}{\partial n}$ $+(2+p)T\frac{\partial}{\partial T} + (1+p)E\frac{\partial}{\partial E}$
III_b	$2m + p(1+m) \neq 0$, $X_o = -[2m + p(1+m)]t\frac{\partial}{\partial t} + (x+q)\frac{\partial}{\partial x}$ $+pn\frac{\partial}{\partial n} + (2+p)T\frac{\partial}{\partial T} + (1+p)E\frac{\partial}{\partial E}$

Table 4. Non trivial generators of the optimal systems. a is a real parameter.

Acknowledgments

The author V. R. acknowledges the support from CNR (program Agenzia2000, grant n. CNRG000DB7) and from TMR (program Asymptotic Methods in Kinetic Theory, grant n. ERBFMRXCT970157). The author A. V. acknowledges the support from CNR through G.N.F.M. and by M.U.R.S.T. Project: *Non Linear Mathematical Problems of Wave Propagation and Stability in Models of Continuous Media.*

References

1. S. Selberherr, *Analysis and simulation of semiconductor devices*(Wien - New York, Springer-Verlag, 1984).
2. W. Hänsch, *The drift-diffusion equation and its applications in MOSFET modeling*(Wien, Springer-Verlag, 1991).
3. P. Markowich, C. A. Ringhofer and C. Schmeiser, *Semiconductor equations* (Wien, Springer-Verlag, 1990).
4. D. Chen, E. C. Kan, U. Ravaioli, C-W. Shu, R. Dutton, *IEEE on Electron Device Letters* **13**, 26 (1992).
5. E. Lyumkis, B. Polsky, A. Shir and P. Visocky, *Compel* **11**, 311 (1992).
6. L. V. Ovsiannikov, *Group Analysis of Differential Equations* (Academic Press, New York, 1982).
7. P. J. Olver, *Applications of Lie Groups to Differential Equations* (Springer-Verlag, New York, 1986).
8. G. W. Bluman and S. Kumei, *Symmetries and Differential Equations* (Springer-Verlag, New-York, 1989).
9. N. H. Ibragimov, *CRC Hanbook of Lie Group Analysis of Differential Equations* (CRC Press, Boca Raton, FL, 1994).
10. W. I. Fushchych and W. M. Shtelen, *Symmetry Analysis and Exact Solutions of Nonlinear Equations of Mathematical Physics* (Kluwer, Dordrecht, 1993).
11. N. Ben Abdallah and P. Degong, *J. Math. Phys.* **37**, 205 (1996).

THERMODYNAMICS AND BALANCE LAWS FOR PROCESSES OF INELASTIC DEFORMATIONS

E. I. ROMENSKI

Sobolev Institute of Mathematics, Novosibirsk 630090, Russia

E-mail: evrom@math.nsc.ru

Governing differential equations in conservative form are formulated for the process of inelastic deformation. Three different definitions of the rate of inelastic deformation are analysed. One of them defines the rate of inelastic deformation by the nonlinear kinetics of Maxwell relaxation model. The other one is based on the introducing the stress diffusion. The third one assumes that the rate of inelastic deformation is proportional to the vector of density of defects of structure. This vector is managed by a conservation law in which the flux of defects of structure is combined with another balance law. In this connection the field of defects of structure can exchange by an energy with the stress field and generate stress waves, but the total energy is not changed in such a process. Thermodynamical properties and symmetric form of governing equations are discussed.

1 Introduction

Well-posed models of complicated continuous media should take into account fundamental laws of physics, and laws of nonequilibrium thermodynamics in particular. Thermodynamically correct governing equations for processes of inelastic deformation based on the Maxwell relaxation model, have been formulated in the book [1]. These equations are widely used to analyse various concrete problems of continuum mechanics [2-5].

It is necessary to note that the succesful applications of above mentioned equations are conditioned by their properties, which are the hyperbolicity and Galilean invariance. A possibility to derive the conservative form of these equations was unknown so far, and we present it in this paper using idea of [6]. A conservative form of equations can be useful to develop efficient numerical methods, and to analyse mathematical properties of equations (such as solvability, definition of discontinuous solutions, etc.)

Several ways to define constitutive equations for inelastic processes are studied in this paper. To formulate thermodynamically compatible systems of balance laws the methods of extended thermodynamics are used [1,6-9]. These methods have been succefully applied to derive governing equations for various complicated media, such as multipase flows with different velocities of phase moving [1,6,9].

The thermodynamically compatible system of conservation laws (overdetermined as a rool) is a basis to derive equations for any model. Equations obtained can be transformed to a symmetric hyperbolic system. An additional energy conservation law for such the system guarantees the validity of thermodynamics laws.

Here, within a framework of the theory of thermodynamically compatible systems, three different ways to define the rate of inelastic deformation tensor are analysed. This tensor appears in balance equations for the elastic distortion tensor [1,10]. As was mentioned, one of possible definitions of the inelastic deformation tensor is well known and based on the Maxwell relaxation model for the kinetics of inelastic deformation. Another one is based on the introducing the stress diffusion into balance equations for the distortion tensor [11]. The third one consits of the introducing an additional field of defects of structure, which vector flux generates an inelastic deformation. This field interacts with the stress field by means of source terms in balance equations for distortion tensor and in field's equations for the vector flux of defects density. Above interaction can generate the stress waves caused by oscillations of the field of the density of defects without variation of total energy of a medium. The new "two-field" governing equations for the stress waves propagation with internal field of defects of structure can be considered as a phenomenological continual model of wave diffraction and scattering in the medium with inclusions of other medium. The "internal defects field" is a heuristic interpretation of interior continuum of oscillators, such as inclusions in elastic medium.

The three mentioned models of inelastic deformation processes have a different physical meaning and different range of behaviour. Naturally, all these ways can be coupled together.

2 Conservation laws of nonlinear elasticity theory in Euler coordinates

Consider the moving elastic medium and suppose that the parameters of state are ρ - mass density, S - specific entropy, u^i - velocity vector, c^i_j - distortion (the gradient of deformation) tensor. The closed system of differential equations consists of mass, entropy, momentum, and distortion conservation laws [1,7].

$$\begin{aligned}
&\frac{\partial \rho}{\partial t} + \frac{\partial \rho u^\alpha}{\partial x^\alpha} = 0, \\
&\frac{\partial \rho S}{\partial t} + \frac{\partial \rho S u^\alpha}{\partial x^\alpha} = 0, \qquad (1) \\
&\frac{\partial \rho u_i}{\partial t} + \frac{\partial (\rho u_i u^\alpha - \sigma^\alpha_j)}{\partial x^\alpha} = 0,
\end{aligned}$$

$$\frac{\partial \rho c_j^i}{\partial t} + \frac{\partial(\rho c_j^i u^\alpha - \rho c_j^\alpha u^i)}{\partial x^\alpha} = 0.$$

Here σ_j^i - the stress tensor connected with the equation of state E (specific internal energy) by the formula

$$\sigma_i^\alpha = \rho c_j^\alpha E_{c_j^i} - \rho^2 E_\rho \delta_j^\alpha. \tag{2}$$

The equation of state is the function of the density, entropy, and distortion tensor:

$$E = E(\rho, S, c_j^i).$$

Note, that the density is defined by the formula $\rho = \rho_0 det(c_j^i)$, where ρ_0 is the initial density, and the temperature is defined by the formula $T = E_S$.

The system (1) has the remarkable properties, analogs of which will be used in the further construction. Namely, the system (1) admits an additional stationary conservation law for the compatibility of deformation

$$\frac{\partial \rho c_j^i}{\partial x^i} = 0. \tag{3}$$

This equation can be easily derived from the distortion tensor conservation law of the system (1) by the differentiating this equation with respect to x^i. It leads directly to the equality

$$\frac{\partial}{\partial t}\frac{\partial \rho c_j^i}{\partial x^i} = 0,$$

and if the equality (3) holds for $t = 0$, then it holds for $t > 0$. Note, that in the case of inelastic deformation the right hand side arises in the compatibility equation (3).

The other important property of the system (1) is the implementation of additional energy conservation law for its solution:

$$\frac{\partial \rho(E + u_i u^i/2)}{\partial t} + \frac{\partial[\rho(E + u_i u^i/2)u^\alpha - u^i \sigma_i^\alpha]}{\partial x^\alpha} = 0. \tag{4}$$

To prove this, it is necessary to sum all equations of the system (1) and the stationary equation (3) multiplied by $q_0 = E + \rho E_\rho - S E_S - c_j^i E_{c_j^i} - u^i u_i/2$, $q_\omega = E_S$, u_i, $r_j^i = E_{c_j^i}$, $u_i E_{c_j^i}$ respectively.

With the help of the methods of extended thermodynamics it is possible to transform the system (1) to a symmetric form. To do this, it is necessary to use already determined new variables $q_0, q_\omega, u_i, r_j^i$ and define the generating potential

$$L = p = \rho^2 E_\rho.$$

Then the system (1) can be written in the form

$$
\begin{aligned}
&\frac{\partial L_{q_0}}{\partial t} + \frac{\partial (u_k L)_{q_0}}{\partial x^k} = 0,\\
&\frac{\partial L_{q_\omega}}{\partial t} + \frac{\partial (u_k L)_{q_\omega}}{\partial x^k} = 0,\\
&\frac{\partial L_{u_i}}{\partial t} + \frac{\partial [(u_k L)_{u_i} - r^i_\alpha L_{r^\alpha_k}]}{\partial x^k} = 0,\\
&\frac{\partial L_{r^j_i}}{\partial t} + \frac{\partial [(u_k L)_{r^j_i} - u^i L_{r^j_k}]}{\partial x^k} = 0,
\end{aligned} \tag{5}
$$

The stationary compatibility equation (3) can be written in the form

$$
\frac{\partial L_{r^j_i}}{\partial x^i} = 0, \tag{6}
$$

and the energy conservation law (4) takes the form

$$
\frac{\partial}{\partial t}\left(q_0 L_{q_0} + q_\omega L_{q_\omega} + u_i L_{u_i} + r^i_j L_{r^i_j} - L\right) +
$$

$$
\frac{\partial}{\partial x^k}\left(q_0 (u^k L)_{q_0} + q_\omega (u_k L)_{q_\omega} + u_i (u_k L)_{u_i} + r^i_j (u_k L)_{r^i_j} - u_k L - u^\alpha r^\beta_\alpha L_{r^\beta_\alpha}\right) = 0.
$$

The symmetric form of the system (5) can be derived after transformation of its third and fourth equations with the use of the stationary conservation law (6), and can be written in the following equivalent form

$$
\begin{aligned}
&\frac{\partial L_{q_0}}{\partial t} + \frac{\partial (u_k L)_{q_0}}{\partial x^k} = 0,\\
&\frac{\partial L_{q_\omega}}{\partial t} + \frac{\partial (u_k L)_{q_\omega}}{\partial x^k} = 0,\\
&\frac{\partial L_{u_i}}{\partial t} + \frac{\partial (u_k L)_{u_i}}{\partial x^k} - L_{r^\alpha_k} \frac{\partial r^i_\alpha}{\partial x^k} = 0,\\
&\frac{\partial L_{r^j_i}}{\partial t} + \frac{\partial (u_k L)_{r^j_i}}{\partial x^k} - L_{r^j_k} \frac{\partial u^i}{\partial x^k} = 0.
\end{aligned} \tag{7}
$$

It is obvious, that two first terms in all equations of the system can be written with the help of symmetric matrices at $\partial/\partial t$, $\partial/\partial x^k$, and the matrices forming other terms in the third and fourth equations are also symmetric. The symmetry of matrices of the system (7) together with the convexity of generating potential L guarantee the hyperbolicity of the system (7), and hence the system (5). The convexity of the equation of state in the nonlinear elasticity theory is discussed in [1,7].

3 Modelling of inelastic deformation by the source terms in equations for the distortion tensor

A concept of unstressed reference state is explored usually in models of inelastic deformations [1], and introduced in order to define an effective elastic deformation of element of a medium. This deformation is realized by the isentropic transformation of the element of a medium in the unstressed state to the actual state with the given stress tensor. Above concepts allow us to assume that governing equations for the effective elastic distortion tensor can be written in the conservative form:

$$\frac{\partial \rho c^i_j}{\partial t} + \frac{\partial(\rho c^i_j u^\alpha - \rho c^\alpha_j u^i)}{\partial x^\alpha} = \Phi^i_j. \tag{8}$$

Here the tensor Φ^i_j defines the rate of inelastic variation of the distortion tensor. This tensor must satisfy to some restrictions which are connected with the properties of nonlinear elasticity equations described in the previous section. These properties are the implementation of energy conservation law and the implementation of distortion compatibility equations.

Consider a few different methods to define the tensor Φ^i_j. They describe a different physical types of inelastic deformation process and can be coupled together, if it is necessary.

The first of them is based on the assumption that Φ^i_j is a nonlinear function of parameters of state (stresses, temperature, and other variables for complicated processes). The definition of the rate of inelastic deformation by such a way is connected with the formation of defects of structure of a medium (for example, dislocations in metals) [1,4]. The corresponding model has been developed on the base of nonlinear Maxwell relaxation model and used for the analysis of high rate deformation of metals in particularly. But the possibility to reduce equations of this model to the conservative form is unknown till now.

So, consider the system which is derived from the generalization of conservation laws of the elasticity theory:

$$\begin{aligned}
&\frac{\partial \rho}{\partial t} + \frac{\partial \rho u^\alpha}{\partial x^\alpha} = 0,\\
&\frac{\partial \rho S}{\partial t} + \frac{\partial \rho S u^\alpha}{\partial x^\alpha} = Q = \frac{E_{c^i_j}\phi^i_j}{E_S},\\
&\frac{\partial \rho u_i}{\partial t} + \frac{\partial(\rho u_i u^\alpha - \sigma^\alpha_j)}{\partial x^\alpha} = 0,\\
&\frac{\partial \rho c^i_j}{\partial t} + \frac{\partial(\rho c^i_j u^\alpha - \rho c^\alpha_j u^i)}{\partial x^\alpha} = -(u^i\beta_j + \phi^i_j),
\end{aligned} \tag{9}$$

$$\frac{\partial \rho c^i_j}{\partial x^i} = \beta_j,$$
$$\frac{\partial \beta_j}{\partial t} + \frac{\partial (u^i \beta_j + \phi^i_j)}{\partial x^i} = 0,$$

The new variables β_j, ϕ^i_j arise in the system of governing equations, but it is supposed that the internal energy does not depend on these parameters:

$$E = E(\rho, S, c^i_j).$$

Emphasize, that to close the system it is enough to define the rate of inelastic deformation tensor ϕ^i_j as the function of parameters of state, such as the density, stresses, and temperature, or of the equivalent set of variables ρ, S, c^i_j. Further it is possible to choose a closed subsystem for variables ρ, S, u^i, c^i_j in order to find a solution of the system (9). Such system has the form:

$$\begin{aligned}
&\frac{\partial \rho}{\partial t} + \frac{\partial \rho u^\alpha}{\partial x^\alpha} = 0,\\
&\frac{\partial \rho S}{\partial t} + \frac{\partial \rho S u^\alpha}{\partial x^\alpha} = \frac{E_{c^i_j} \phi^i_j}{E_S},\\
&\frac{\partial \rho u_i}{\partial t} + \frac{\partial (\rho u_i u^\alpha - \sigma^\alpha_j)}{\partial x^\alpha} = 0,\\
&\frac{\partial \rho c^i_j}{\partial t} + \frac{\partial \rho c^i_j u^\alpha}{\partial x^\alpha} - \rho c^\alpha_j \frac{\partial u^i}{\partial x^\alpha} = -\phi^i_j,
\end{aligned} \tag{10}$$

After finding a solution of the system (10) it is possible to find β_j from the stationary conservation law of the system (9) if necessary.

The conservative form of equations (9) can be useful to develop numerical algorithms and to analyse discontinuous solutions.

Let us prove that the system (9) is compatible. It is necessary to prove that the right hand sides in the fourth and fifth equations of the system (9) are in accordance with its last equation. Actually, the differentiation of the fourth equation with respect to x^i leads to the equality

$$\frac{\partial}{\partial t}\left(\frac{\partial \rho c^i_j}{\partial x^i}\right) = -\frac{\partial (u^i \beta_j + \phi^i_j)}{\partial x^i}.$$

Its straightforward consequence is

$$\frac{\partial}{\partial t}\left(\frac{\partial \rho c^i_j}{\partial x^i} - \beta_j\right) = 0,$$

which can be obtained with the help of the last equation of the system (9). One can conclude that if for $t = 0$ the equality

$$\frac{\partial \rho c^i_j}{\partial x^i} - \beta_j = 0,$$

holds, then it holds for $t > 0$. Hense, the last three equations of the system (9) are compatible.

To prove that the (9) admits an additional energy conservation law it is nesessary to sum all equations of the system multiplied by $q_0 = E + \rho E_\rho - SE_S - c^i_j E_{c^i_j}$, $q_\omega = E_S$, u_i, $r^i_j = E_{c^i_j}$, $u_\alpha E_{c^\alpha_j}$ respectively.

As a result of transformation and elimination of right hand terms we obtain the following equality

$$\frac{\partial \rho(E + u_i u^i/2)}{\partial t} + \frac{\partial[\rho(E + u_i u^i/2) - u^i \sigma^\alpha_i]}{\partial x^\alpha} = 0.$$

It is the differential form of energy conservation law in the model of inelastic deformation under consideration.

Note, that the source term in the balance equation for entropy arises as a result of inelastic behaviour of a medium. This term is called the entropy production, and must be nonnegative due to the second principle of thermodynamics. The nonegativity requirement leads to some restrictions for the dependence of the tensor ϕ^i_j on the parameters of state. Different ways to determine ϕ^i_j in the framework of the Maxwell relaxation model are discussed in [1,3,4,10].

Consider the formalization of the system (9) which is based on the extended thermodynamics theory and can be realized by the generating potential $L(q_0, q_\omega, u^i, r^i_j) = \rho^2 E_\rho$ depending on the set of variables $q_0 = E + \rho E_\rho - SE_S - c^i_j E_{c^i_j}, q_\omega = E_S, u_i, r^i_j = E_{c^i_j}$. Using formulas $L_{q_0} = \rho, L_{q_\omega} = \rho S, L_{u_i} = \rho u_i, L_{r^i_j} = \rho c^i_j$, the system (9) can be written in the form

$$\begin{aligned}
&\frac{\partial L_{q_0}}{\partial t} + \frac{\partial (u_k L)_{q_0}}{\partial x^k} = 0,\\
&\frac{\partial L_{q_\omega}}{\partial t} + \frac{\partial (u_k L)_{q_\omega}}{\partial x^k} = \frac{r^i_j \phi^i_j}{q_\omega},\\
&\frac{\partial L_{u_i}}{\partial t} + \frac{\partial[(u_k L)_{u_i} - r^i_\alpha L_{r^\alpha_k}]}{\partial x^k} = 0,\\
&\frac{\partial L_{r^j_i}}{\partial t} + \frac{\partial[(u_k L)_{r^j_i} - u^i L_{r^j_k}]}{\partial x^k} = -(u^i \beta_j + \phi^i_j),\\
&\frac{\partial L_{r^j_i}}{\partial x^i} = \beta_j,
\end{aligned} \tag{11}$$

$$\frac{\partial \beta_j}{\partial t} + \frac{\partial (u^i \beta_j + \phi^i_j)}{\partial x^i} = 0.$$

The hyperbolicity of the system (11) can be proved by the calculation of its characteristcs. But this system can not be written in a symmetric form directly because of the presence of additional variables β_j. A possibility to write the system similar to (11) in the symmetric form using additional equations for derivatives of variables is discussed in [12]. It is interesting to try to construct such a symmetrization for the system (11) in terms of generating potential.

4 Stress diffusion

Consider the other model of inelastic deformation which is based on the introducing the spacial second order derivatives into equations for effective elastic distortion tensor. A physical meaning of such a definition of the rate of inelastic deformation can be motivated by phenomenon of the moving of defects of structure of a medium [11]. This model is discussed in [12] without requirements of conservativeness of governing equations.

The governing equations of the model are the same as the system (9):

$$\begin{aligned}
&\frac{\partial \rho}{\partial t} + \frac{\partial \rho u^\alpha}{\partial x^\alpha} = 0,\\
&\frac{\partial \rho S}{\partial t} + \frac{\partial \rho S u^\alpha}{\partial x^\alpha} = Q,\\
&\frac{\partial \rho u_i}{\partial t} + \frac{\partial (\rho u_i u^\alpha - \sigma^\alpha_j)}{\partial x^\alpha} = 0,\\
&\frac{\partial \rho c^i_j}{\partial t} + \frac{\partial (\rho c^i_j u^\alpha - \rho c^\alpha_j u^i)}{\partial x^\alpha} = -(u^i \beta_j + \phi^i_j),\\
&\frac{\partial \rho c^i_j}{\partial x^i} = \beta_j,\\
&\frac{\partial \beta_j}{\partial t} + \frac{\partial (u^i \beta_j + \phi^i_j)}{\partial x^i} = 0,
\end{aligned} \tag{12}$$

The distinctions are the other formulas for the tensor ϕ^i_j

$$\phi^i_j = -\frac{\partial}{\partial x^\alpha} \kappa^{i\alpha m}_{j\beta n} \frac{\partial E_{c^m_n}}{\partial x^\beta}$$

and for the entropy production

$$Q = \frac{1}{E_S} \kappa^{i\alpha m}_{j\beta n} \frac{\partial E_{c^i_j}}{\partial x^\alpha} \frac{\partial E_{c^m_n}}{\partial x^\beta}.$$

Note, that the overdetermined system (12) is compatible, the proof of this fact is the same as was for the system (9) in the previous section. But the energy conservation law for the system (12) differs from (4), because of the presence of additional terms with spacial derivatives in the fluxes:

$$\frac{\partial}{\partial t}\rho(E + u_i u^i/2) + \frac{\partial}{\partial x^\alpha}\left[\rho(E + u_i u^i/2)u^\alpha - u^i \sigma_i^\alpha - E_{c_j^i} \kappa_{j\beta n}^{i\alpha m} \frac{\partial E_{c_n^m}}{\partial x^\beta}\right] = 0.$$

The tensor $\kappa_{j\beta n}^{i\alpha m}$ must be choosen in accordance with the second principle of thermodynamics by such a way that the entropy production is nonnegative. This implies that the quadratic form

$$\kappa_{j\beta n}^{i\alpha m} Z_{i\alpha m} Z^{j\beta n}$$

is nonnegative for each nonzero tensor Z_{ijk} (we deal with the Cartesian coordinates, hence $Z_{ijk} = Z^{ijk}$)

5 Inelastic deformation caused by the field of internal parameters

Consider a process of inelastic deformation caused by the action of internal variables field. This field is generated by the action of stresses, has a finite velocity of propagation, and generates the stress field by itself. In other words, above process is the result of interaction and energy exchange between stress field and internal parameters field.

The formalization of this interaction is based on the ideas of the previous sections. It is supposed that the tensor ϕ_j^i and the vector β_j can be parameters of state of a medium. Hence, they give a contribution to the internal energy and it is necessary to derive an additional governing equations.

Assume that the internal energy depends on the density ρ, the entropy S, the distortion tensor c_j^i, and on the new additional parameters of state. These are the vector θ_j and the tensor ψ_j^i characterizing the defect structure of medium. The closed system of balance laws for such a medium can be written in the following form

$$\begin{aligned}
&\frac{\partial \rho}{\partial t} + \frac{\partial \rho u^k}{\partial x_k} = 0,\\
&\frac{\partial \rho u_i}{\partial t} + \frac{\partial(\rho u_i u^k - \sigma_i^k)}{\partial x^k} = 0,\\
&\frac{\partial \rho c_j^i}{\partial t} + \frac{\partial(\rho u^k c_j^i - \rho u^i c_j^k)}{\partial x^k} = -\alpha(\rho u^i \theta_j + E_{\psi_j^i}),
\end{aligned}$$

$$\frac{\partial\rho\theta_i}{\partial t}+\frac{\partial(\rho u^k\theta_i+E_{\psi^i_k})}{\partial x_k}=0, \tag{13}$$
$$\frac{\partial\rho\psi^i_j}{\partial t}+\frac{\partial(\rho u^k\psi^i_j+\delta^{kj}E_{\theta_j})}{\partial x_k}=\alpha E_{c^j_i}-\chi E_{\psi^j_i},$$
$$\frac{\partial\rho S}{\partial t}+\frac{\partial\rho u^k S}{\partial x_k}=Q=\frac{\chi}{E_S}E_{\psi^i_j}E_{\psi^j_i}$$
$$\frac{\partial\rho c^i_j}{\partial x^i}=\alpha\rho\theta_j$$

Here σ^k_i – is the stress tensor, determined by the formula (2):

$$\sigma^i_i=\rho c^i_j E_{c^i_j}-\rho^2 E_\rho\delta^i_j.$$

The rate of inelastic deformation tensor in the system (13) is proportional to the flux of vector $\rho\theta_j$ with the constant coefficient of proportionality $-\alpha$. In this connection the source term arises in the equation for ψ^i_j as the function of stresses (the coefficient of proportionality is α) By virtue of antisymmetry of above two source terms the total energy of the system does not vary, but the exchange between fields of stresses and internal parameters is admissible.

The rate of relaxation of the tensor ψ^i_j to the equilibrium state is determined by the coefficient χ. The equilibrium state is realized for $E_{\psi^j_i}=0$ in the unsteressed state, when $E_{c^j_i}=0$. Naturally, χ can be a function of the parameters of state.

It is easy to prove that the last stationary equation of the system (13) is compatible with other equations of the system. For this purpose it is necessary to differentiate the third equation of (13) with respect to x^i and then to use the fourth equation. As a result we obtain

$$\frac{\partial}{\partial t}\left(\frac{\partial\rho c^i_j}{\partial x^i}-\alpha\theta_j\right)=0.$$

It is obvious that if for $t=0$ the equality

$$\frac{\partial\rho c^i_j}{\partial x^i}-\alpha\theta_j=0,$$

holds, then it holds for $t>0$, and hence the stationary equation of the system (13) is compatible with other its equations.

It is easy to show that the system (13) is thermodynamically compatible, that is its solutions fulfil to the additional energy conservation law. To prove it is necessary to sum all equations multiplied by $q_0=E+\rho E_\rho-SE_S-c^i_J E_{c^i_j}-\theta_i E_{\theta_i}-\psi^i_j E_{\psi^i_j}-u_i u^i/2$, u^i, $r^j_i=E_{c^i_j}$, $n^i=E^i_\theta$, $J^j_i=E_{\psi^i_j}$, $q_\omega=E_S$,

$u_iE_{c^i_j}$ respectively. As a reult all source terms are annihilated, and the energy conservation law has the form

$$\frac{\partial\rho(E+u^\alpha u^\alpha/2)}{\partial t}+\frac{\partial(\rho u^k(E+u^\alpha u^\alpha/2)-u^\alpha\sigma^k_\alpha+E_{\theta^\alpha}E_{\psi_{k\alpha}})}{\partial x^k}=0, \qquad (14)$$

Note, that inelastic processes lead to the entropy production caused by the relaxation of the field ψ^i_j:

$$Q=\frac{\chi}{E_S}E_{\psi^i_j}E_{\psi^j_i}\geq 0$$

Assume that the generating potential L, depending on the variables $q_0, u_i, r^j_i, n^i, J^j_i, q_\omega$ exists. Using the following identification of variables $\rho = L_{q_0}, \rho u^i = L_{u_i}, \rho c^i_j = L_{r^j_i}, \rho\theta_i = L_{n^i}, \rho\psi^i_j = L_{J^j_i}, \rho S = L_{q_\omega}$, the system (13) can be written in the form:

$$\begin{aligned}
&\frac{\partial L_{q_0}}{\partial t}+\frac{\partial(u_kL)_{q_0}}{\partial x^k}=0,\\
&\frac{\partial L_{u_i}}{\partial t}+\frac{\partial[(u_kL)_{u_i}-r^i_\alpha L_{r^\alpha_k}]}{\partial x^k}=0,\\
&\frac{\partial L_{r^j_i}}{\partial t}+\frac{\partial[(u_kL)_{r^j_i}-u^iL_{r^j_k}]}{\partial x^k}=-\alpha(u^iL_{n^j}+J^j_i),\\
&\frac{\partial L_{n^i}}{\partial t}+\frac{\partial((u_kL)_{n^i}+J^k_i)}{\partial x^k}=0,\\
&\frac{\partial L_{J^j_i}}{\partial t}+\frac{\partial((u_kL)_{J^j_i}+\delta^k_jn^i)}{\partial x^k}=\alpha r^i_j-\chi J^i_j,\\
&\frac{\partial L_{q_\omega}}{\partial t}+\frac{\partial(u_kL)_{q_\omega}}{\partial x^k}=\frac{\chi}{q_\omega}J^j_iJ^i_j,\\
&\frac{\partial L_{r^j_i}}{\partial x^i}=\alpha L_{n_j},
\end{aligned}$$

This system can be rewritten in the symmetric form. To prove it is necessary to transform the second and third equations using the last stationary conservative equation. As the result we obtain

$$\begin{aligned}
&\frac{\partial L_{q_0}}{\partial t}+\frac{\partial(u_kL)_{q_0}}{\partial x^k}=0,\\
&\frac{\partial L_{u_i}}{\partial t}+\frac{\partial(u_kL)_{u_i}}{\partial x^k}-L_{r^\gamma_k}\frac{\partial r^i_\gamma}{\partial x^k}=\alpha r^i_\gamma L_{n_\gamma},\\
&\frac{\partial L_{r^j_i}}{\partial t}+\frac{\partial(u_kL)_{r^j_i}}{\partial x^k}-L_{r^j_k}\frac{\partial u^i}{\partial x^k}=-\alpha J^j_i,
\end{aligned}$$

$$\frac{\partial L_{n^i}}{\partial t} + \frac{\partial((u_k L)_{n^i}}{\partial x^k} + \frac{\partial J_i^k}{\partial x^k} = 0,$$
$$\frac{\partial L_{J_i^j}}{\partial t} + \frac{\partial((u_k L)_{J_i^j}}{\partial x^k} + \frac{\partial n^i}{\partial x^j} = \alpha r_j^i - \chi J_j^i,$$
$$\frac{\partial L_{q_\omega}}{\partial t} + \frac{\partial(u_k L)_{q_\omega}}{\partial x^k} = \frac{\chi}{q_\omega} J_i^j J_j^i,$$

Obvious, that the system is symmetric, because the conservative terms can be written by the symmetric matrices of the second derivatives of potentials L, $u^k L$. The other nonconservative terms are symmetric too.

References

1. S.K. Godunov and E.I. Romenski, *Elements of Continuum Mechanics and Conservation Laws*, (Scientific Books, Novosibirsk, 1998).
2. L.A. Merzhievsky *et al, Sov. Phys. Dokl.* V **31(10)**, 812 (1986)
3. V.N. Dorovsky *et al, Zh. Prikl. Mekh. Tekh. Fiz.* V **24**, No 4, 10 (1983)
4. L.A. Merzhievsky and A.D. Resnyansky *J. de Physique* V **46** 67 (1985)
5. A.D. Resnyansky and E.I. Romensky in *Proc. 11th Int. Conf. on Composite Materials*, ed. Murray, L.Scott (Woodhead Publishing Ltd, 1997).
6. E.I. Romensky in *Proc. Int. Conf. "Godunov Methods: Theory and Applications"* ed. E.F. Toro (Kluwer/Plenum, 2001).
7. S.K. Godunov and E.I. Romensky in *Computational Fluid Dynamics REVIEW95*, ed. A. Hafez, S. Oshima (John Wiley, NY, 1995)
8. E.I. Romensky *Math. Comput. Modelling* V **28**, No 10, 115 (1998).
9. I. Mueller and T. Ruggeri, *Extended Thermodynamics* (Springer-Verlag, 1993).
10. E.I. Romensky *Siberian Math. Journal* V **30**, No 4, 135 (1988).
11. L.D. Landau and Ye.M. Lifshiz *Elasticity Theory* (Nauka, Moscow, 1987).
12. E.I. Romenski *Siberian Advances in Mathematics* V **5**, No 1, 120 (1995).

EULER EQUATIONS ARISING FROM EXTENDED KINETIC THEORY: SOUND WAVE PROPAGATION

Alberto Rossani
Istituto Nazionale di Fisica della Materia
Dipartimento di Fisica, Politecnico di Torino
C.so Duca degli Abruzzi, 24 - 10129 Torino, Italy

1 Introduction

In the last years, fluid dynamic problems, at a molecular level, have been considered by taking into account chemical reactions and/or gas-radiation interaction. From such Extended Kinetic Theory, fluid dynamic models can be derived by adopting suitable closure procedures on the moment equations. One of the classical fluid dynamic problems to deal with is the study of sound wave propagation and the relevant dispersion relation. In the frame of classical Fluid Dynamics such a study can be found in well known textbooks [1], based on macroscopic considerations. However, starting from a kinetic model, all the fluid dynamic properties relevant to this problem can be derived from microscopic inputs only.

In the present paper, starting from some mesoscopic model of Extended Kinetic Theory, we utilize fluid dynamic approximations at the Euler level to study sound wave propagation. In particular, we derive explicitly the dispersion relations for a mixture of reacting gases and for a gas costituted by many-level atoms. In both the cases, a relationship like $k^2P(i\omega)+\omega^2Q(i\omega) = 0$, where P and Q are polynomials (with real coefficients) of $i\omega$, with the same degree, is found.

2 Reacting gases

Consider a mixture of four gases A, B, C, D, which, besides all the elastic collisions, can interact according to the following reversible bimolecular reaction: $A + B \rightleftharpoons C + D$. Particles A, B, C, D, to be labeled in the sequel by 1, 2, 3, 4, are endowed with internal energies E_i $i = 1, ..., 4$, respectively. In ref. [2] kinetic and moment equation for such a mixture are provided. Balance equations for particles i reads

$$\frac{\partial n_i}{\partial t} + \nabla \cdot (n_i \mathbf{u}_i) = S_i, \tag{1}$$

where n_i is the number density, $\mathbf{u}_i$ is the mean velocity, and S_i is the source term due to chemical interactions. Balance of momentum and kinetic energy are expressed as

$$\frac{\partial}{\partial t}(\rho\mathbf{u}) + \nabla\cdot(\rho\mathbf{u}\otimes\mathbf{u}+\mathbf{P}) = \mathbf{0}, \tag{2}$$

$$\frac{\partial}{\partial t}(e_{th}+\rho u^2/2) + \nabla\cdot\{[(e_{th}+\rho u^2/2)\mathbf{I}+\mathbf{P}]\cdot\mathbf{u}+\mathbf{q}_{th}\} = S_1\Delta E, \tag{3}$$

where $\rho = \sum_i m_i n_i$ is the total density; $\mathbf{u} = (1/\rho)\sum_i m_i n_i \mathbf{u}_i$ is the mean velocity of the mixture; $\mathbf{P}$ is the stress tensor; $e_{th} = (3/2)n\chi T$ is the thermal energy density ($n = \sum_i n_i$ is the total number density, χ is the Boltzmann constant, T is the absolute temperature); $\mathbf{q}_{th}$ is the thermal energy flux; $\Delta E = \sum_i \lambda_i E_i$ is the molecular heat of reaction ($\lambda_i = 1$ for $i = 1, 2$, $\lambda_i = -1$ for $i = 3, 4$).

Mass and total energy conservation read as follows:

$$\frac{\partial\rho}{\partial t} + \nabla\cdot(\rho\mathbf{u}) = 0, \tag{4}$$

$$\frac{\partial e}{\partial t} + \nabla\cdot[(e\mathbf{I}+\mathbf{P})\cdot\mathbf{u}+\mathbf{q}_{th}+\mathbf{q}_{int}] = 0, \tag{5}$$

where $e = (1/2)\rho u^2 + e_{th} + e_{int}$ ($e_{int} = \sum_i E_i n_i$ is the internal energy density); $\mathbf{q}_{int}$ is the internal energy flux.

A closed set of equations for n_i, $\mathbf{u}$, T, is obtained, at the Euler level, by setting

$$\mathbf{u}_i = \mathbf{u}, \quad \mathbf{P} = n\chi T\mathbf{I}, \quad \mathbf{q}_{th} = \mathbf{q}_{int} = \mathbf{0}, \tag{6}$$

and

$$S_i = \lambda_i[\nu_{34}(T)n_3n_4 - \nu_{12}(T)n_1n_2].$$

The effective collision frequencies ν_{ij} satisfy Arrhenius law:

$$\frac{\nu_{34}(T)}{\nu_{12}(T)} = \exp\left(\frac{\Delta E}{\chi T}\right)\left(\frac{m_1m_2}{m_3m_4}\right)^{3/2}. \tag{7}$$

Now, starting from an homogeneous equilibrium state

$$n_i = n_{i0}, \quad \mathbf{u} = \mathbf{0}, \quad T = T_0,$$

with $S_i = S_{i0} = 0$, we consider the propagation of an harmonic plane wave

$$n_i = n_{i0} + N_i\exp[i(kx-\omega t)], \qquad u = u_0 + W\exp[i(kx-\omega t)],$$

$$T = T_0 + \Theta\exp[i(kx-\omega t)],$$

where $u = \mathbf{u} \cdot \mathbf{i}$.

By inserting this ansatz, the linearization with respect to N_i, W, and Θ gives:

$$-i\omega N_1 + ikn_{i0}W = S_1^*,$$

$$-i\omega(N_1 - N_2) + ik(n_{10} - n_{20})W = 0,$$

$$-i\omega(N_1 + n_3) + ik(n_{10} + n_{30})W = 0,$$

$$-i\omega(N_1 + N_4) + ik(n_{10} + n_{40})W = 0,$$

$$-i\omega\rho_0 W + ik\chi\left(n_0\Theta + T_0\sum_i N_i\right) = 0,$$

$$-i\omega(3/2)\chi\left(n_0\Theta + T_0\sum_i N_i\right) - i\omega\sum_i E_i N_i +$$

$$+ik\left((5/2)n_0\chi T_0 + \sum_i E_i n_{i0}\right)W = 0,$$

where

$$S_1^* = \nu_{34}(T_0)(n_{30}N_4 + n_{40}N_3) - \nu_{12}(T_0)(n_{10}N_2 + n_{20}N_1) +$$

$$+\Theta[\nu_{34}'(T_0)n_{30}n_{40} - \nu_{12}'(T_0)n_{10}n_{20}].$$

Observe that, from (7), we have

$$\frac{\nu_{34}'(T)}{\nu_{34}(T)} - \frac{\nu_{12}'(T)}{\nu_{12}(T)} = -\frac{\Delta E}{\chi T^2}. \tag{8}$$

Since

$$\nu_{34}(T_0)n_{30}n_{40} = \nu_{12}(T_0)n_{10}n_{30} = R,$$

eq. (8) gives

$$\nu_{34}'(T_0)n_{30}n_{40} - \nu_{12}'(T_0)n_{10}n_{40} = -\frac{R\Delta E}{\chi T_0^2}.$$

Non trivial solutions to the system for N_i, W, Θ, are obtained if and only if the determinant of the coefficients vanishes. Such a condition gives the dispersion relationship, which links k and ω:

$$-\frac{\omega^2}{a_e^2} + k^2 - i\omega\tau\left(-\frac{\omega^2}{a_f^2} + k^2\right) = 0,$$

where the coefficients a_e and a_f are known in the literature as equilibrium flow and frozen flow speed of sound, respectively, while τ is a characteristic time of the reaction. Explicitly we have

$$a_f^2 = \frac{5}{3}\,\frac{n_0 \chi T_0}{\rho_0},$$

$$a_e^2 = a_f^2 \left[\sum_i \frac{1}{n_{i0}} + \frac{2}{5n_0}\left(\frac{\Delta E}{\chi T_0}\right)^2\right] \Big/ \left[\sum_i \frac{1}{n_{i0}} + \frac{2}{3n_0}\left(\frac{\Delta E}{\chi T_0}\right)^2\right]$$

$$\frac{1}{\tau} = R\left[\sum_i \frac{1}{n_{i0}} + \frac{2}{5n_0}\left(\frac{\Delta E}{\chi T_0}\right)^2\right].$$

It is easy to verify that $a_f > a_e$, according to non-equilibrium thermodynamics [3].

Observe that the dispersion relation can be written as $k^2 P(i\omega) + \omega^2 Q(i\omega) = 0$, where P and Q are first order polynomials of $i\omega$. We shall show that this is only the simplest case of a more general dispersion relation, where P and Q are polynomials of the same order $L \geq 1$.

3 Atoms with internal energy levels

Consider a gas constituted by atoms A_ℓ ($\ell = 1, 2, ..., N$) endowed with a finite number N of internal energy levels

$$E_1 = 0 < E_2 < \ldots < E_N,$$

with transitions from one state to another made possible by inelastic scattering. Here we restrict ourselves to the simplest case:

$$A_k + A_i \rightleftharpoons A_k + A_j,$$

which include both the elastic ($i = j$) and the inelastic ($i \neq j$) interactions.

In ref. [4] kinetic and macroscopic equations for such a mixture are provided.

Balance equation for particles A_i reads

$$\frac{\partial n_\ell}{\partial t} + \nabla \cdot (n_\ell \mathbf{u}_\ell) = \sum_{j>\ell} S_{\ell j} - \sum_{i<\ell} S_{i\ell}. \tag{9}$$

Momentum and kinetic energy balance are written as follows:

$$\frac{\partial}{\partial t}(\rho \mathbf{u}) + \nabla \cdot (\rho \mathbf{u} \otimes \mathbf{u} + \mathbf{P}) = \mathbf{0}, \tag{10}$$

$$\frac{\partial}{\partial t}(e_{th} + \rho u^2/2) + \nabla \cdot \{[(e_{th} + \rho u^2/2)\mathbf{I} + \mathbf{P}] \cdot \mathbf{u} +$$
$$+\mathbf{q}_{th}\} = \sum_j \sum_{i<j} S_{ij}\Delta_{ij}.$$

Conservation of mass and total energy read

$$\frac{\partial n}{\partial t} + \nabla \cdot (n\mathbf{u}) = 0, \tag{11}$$

$$\frac{\partial e}{\partial t} + \nabla \cdot [(e\mathbf{I} + \mathbf{P}) \cdot \mathbf{u} + \mathbf{q}_{th} + \mathbf{q}_{int}] = 0. \tag{12}$$

At the Euler level, a closed set of equations for the unknowns n_ℓ, $\mathbf{u}$, T, is found by setting (6) and

$$S_{ij} = G_{ij}[n_j - n_i \exp(-\Delta_{ij}/\chi T)],$$

where

$$G_{ij} = \sum_k n_k \nu_{kj}^{ki}(T).$$

The set of equations (11), (9) with $\ell > 1$, (10), and (12) is now closed. By considering the propagation of an harmonic wave we are led to the following system:

$$-i\omega \sum_\ell N_\ell + ikW \sum_\ell n_{\ell 0} = 0,$$

$$-i\omega N_\ell + ikn_{\ell 0}W = \sum_i \alpha_{\ell i} N_i + \beta_\ell \Theta,$$

$$-i\omega \rho_0 W + i\chi k(T_0 \sum_\ell N_\ell + n_0 \Theta) = 0,$$

$$-i\omega[(3/2)\chi(n_0\Theta + T_0 \sum_\ell N_\ell) + \sum_\ell E_\ell N_\ell] +$$
$$+ik((5/2)n_0\chi T_0 + \sum_\ell n_{\ell 0})W = 0,$$

where $\alpha_{\ell i}$ and β_ℓ are defined by

$$\sum_i \alpha_{\ell i} N_i + \beta_\ell \Theta = \sum_{i>\ell} S^*_{\ell i} - \sum_{i<\ell} S^*_{i\ell},$$

and

$$S^*_{ij} = \sum_k [\nu_{kj}^{ki}(T_0)(n_{k0}N_j + n_{j0}N_k) +$$

$$-\nu_{ki}^{kj}(T_0)(n_{k0}N_i + n_{i0}N_k)] - \frac{R_{ij}\Delta_{ij}}{\chi T_0^2}\Theta$$

with

$$R_{ij} = \sum_k \nu_{kj}^{ki}(T_0)n_{j0}n_{k0} = \sum_k \nu_{ki}^{kj}(T_0)n_{i0}n_{k0}.$$

By solving the first equation with respect to W we have

$$W = \frac{\omega}{k}\frac{\sum_\ell N_\ell}{\sum_\ell n_{\ell 0}}. \tag{13}$$

Now, by inserting this expression, the third equation gives

$$((5/2)n_0\chi T_0 + \sum_\ell E_\ell n_{\ell 0})\sum_\ell N_\ell =$$

$$= [(3/2)\chi(n_0\Theta + T_0\sum_\ell N_\ell) + \sum_\ell E_\ell N_\ell]\sum_\ell n_{\ell 0},$$

that is

$$\Theta = \sum_\ell \eta_\ell N_\ell. \tag{14}$$

The last equation, thanks to (13) and (14), becomes

$$-\omega^2\rho\sum_\ell N_\ell + \chi k^2(T_0\sum_\ell N_\ell + n_0\sum_\ell \eta_\ell N_\ell)\sum_\ell n_{\ell 0} = 0$$

or

$$\sum_\ell (a_\ell k^2 - \rho_0\omega^2)N_\ell = 0. \tag{15}$$

By inserting (13) and (14), the second equation gives

$$i\omega(-N_\ell\sum_i n_{i0} + n_{\ell 0} + n_{\ell 0}\sum_i N_i) =$$

$$= (\sum_i \alpha_{\ell i}N_i + \beta_\ell\sum_i \eta_i N_i)\sum_i n_{i0},$$

that is

$$\sum_i (C_{\ell i} + i\omega D_{\ell i})N_i = 0 \quad \ell > 1. \tag{16}$$

Now the solvability condition for the system of equations (15) and (16) reads as follows

$$\begin{vmatrix} a_1k^2-\rho_0\omega^2 & a_2k^2-\rho_0\omega^2 & \dots & a_Nk^2-\rho_0\omega^2 \\ C_{21}+i\omega D_{21} & C_{22}+i\omega D_{22} & \dots & C_{2N}+i\omega D_{2N} \\ \vdots & \vdots & \ddots & \vdots \\ C_{N1}+i\omega D_{N1} & C_{N2}+i\omega D_{N2} & \dots & C_{NN}+i\omega D_{NN} \end{vmatrix} = 0,$$

that is $k^2P(i\omega)+\omega^2Q(i\omega)=0$, where P and Q are $(N-1)$th order polynomials of $i\omega$, which is the more general dispersion relationship we anticipated.

References

1. Landau L., and Lifchitz E., *Mecanique des Fluides*, Editions Mir, Moscou (1971)
2. Rossani A., and Spiga G., A note on the Kinetic Theory of chemically reacting gases, *Physica A* **272**, 563-573 (1999).
3. S.R. de Groot, and P. Mazur, *Non-equilibrium thermodynamics*, Dover Publications, New York (1984)
4. Rossani A., and Spiga G., On a dynamical system arising from the kinetic theory of atoms and photons, *Rend. del Circ. Mat. di Palermo* **57**, 439-445 (1997).

CENTRAL SCHEMES FOR BALANCE LAWS

G. RUSSO
Dipartimento di Matematica e Informatica
Università di Catania
Viale A. Doria 6
95125 Catania, Italy
E-mail: russo@dmi.unict.it

In this talk a review is given of some modern shock capturing schemes for the numerical approximation of hyperbolic systems of balance laws. The focus is on finite volume central schemes, which can be easily applied to a variety of problems. After a brief introduction on conservative schemes and on second order central schemes, a way to construct high order schemes is presented. The treatment of systems with a stiff source term is also considered, and an application to Extended Thermodynamics is presented.

1 Introduction

The numerical solution of hyperbolic conservation laws has been an active field of research in the last two decades. The reasons for this high interest are due to the challenge of the problem, from a mathematical point of view, and to the numerous important applications of hyperbolic systems of conservation laws in several fields.

The conceptual basis for the construction of shock capturing schemes rely on the mathematical theory of hyperbolic systems of conservation laws. Solutions to quasilinear hyperbolic systems may develop singularities in finite time, even with smooth initial data. When singularities develop, the system has to be written in an appropriate weak form (conservation form), and the solution have to be considered in the weak sense. The propagation speed of such discontinuities is related to the jump across them by the so called jump relations (also known as Rankine-Hugoniot relations), which naturally appear when the system is written in conservation form. Furthermore, uniqueness of the solution is in general lost after singularity formation, and new selection criteria (such as, for example, the entropy condition) are needed to restore uniqueness.

Using these concepts as guideline, numerical schemes can be constructed which satisfy a discrete conservation form (which guarantees the correct propagation speed of discontinuities) and a discrete entropy condition, which guarantees, at least in some cases, that the numerical solution converges to the unique entropy solution of the system.

The motivation that comes from the applications is evident. Physical systems such as waves in gas dynamics, or in magneto-hydrodynamics, are described by hyperbolic systems of conservation laws, other models, such as, for example, discrete velocity models in kinetic theory, hydrodynamical models describing charge transport in semiconductors[1], shallow water equations with variable bottom profile, gas in Extended Thermodynamics[14], and many more, are described by systems of hyperbolic equations of balance laws.

The numerical solution of such systems is important both in the case of well-established systems, such as Euler equations, MHD equations or shallow water equations, for the immediate applications in Aeronautic Engineering, Astrophysics and Oceanography or river flow problems, as well as in the case of new physical models, for example traffic flow models, semiconductor models, kinetic models, because accurate numerical solutions can be used to test the models against experiments, to validate or to void them, and therefore to help improving them.

As we said, quasilinear hyperbolic systems develop jump discontinuities (shocks) in finite time. The most commonly used schemes for the numerical approximation of conservation laws are the so called *shock capturing schemes*, for which no information is required about the structure or position of the shock, and the latter appears as a region of steep gradients in the numerical solution.

Among the several shock capturing schemes for conservation laws, we mention finite element, finite difference and finite volume methods. We shall focus on finite volume methods, without going into details about the relative merit of the different approaches.

We can distinguish two main categories of shock capturing finite volume schemes, namely schemes that are based on the solution of (exact or approximate) Riemann problems, and schemes that do not require such knowledge.

An introductory text on shock capturing finite volume schemes for conservation laws is the book by LeVeque[6], while a good review on modern shock capturing schemes can be found, for example, in the lecture notes of a CIME course[4], where several different kinds of shock-capturing schemes, including finite element, are reviewed.

Usually, schemes based on Riemann solvers are able to give sharper numerical solutions in some specific cases, such as in the detection and propagation of contact and linearly degenerate discontinuities. On the other hand, these schemes require the knowledge of the solution of the Riemann problem and therefore they can be quite expensive, especially in those cases for which no simple analytical solution of the Riemann problem is known.

A family of schemes that belongs to the second category is the family of central schemes. The basic first order prototype scheme is the Lax-Friedrichs scheme (exactly as the basic first order Godunov scheme is the prototype of Riemann-based upwind schemes).

These schemes do not require the knowledge of eigenvalues and eigenvectors of the system, and therefore they are particularly useful when such information is not available, of when characteristic decomposition is expensive to compute.

The first second order central scheme in one space dimension was proposed independently by Nessyahu and Tadmor [15] and by Sanders and Weiser [17], and applied to gas dynamics.

High order central schemes have been developed by several authors (see for example the papers [12,3,7]). Central schemes have been extended in several dimensions. Second order central schemes on rectangular grids, for example, have been introduced in [2,5], and high order central schemes in 2D are derived in [8,9]. Because of their simplicity and robustness, central schemes have been used to solve a large variety of problems. We mention here only a few significant cases, namely hyperbolic systems with source term[13], Hamilton-Jacobi equation[11], incompressible flows[10], hydrodynamical models of semiconductors[1,16].

2 Central schemes for conservation laws

Let us consider a system of equations of the form

$$\frac{\partial u}{\partial t} + \frac{\partial f(u)}{\partial x} = 0 \tag{1}$$

where $u(x,t) \in \mathbb{R}^d$ is the unknown vector field, and $f : \mathbb{R}^d \to \mathbb{R}^a$ is assumed to be a smooth function. The system is hyperbolic in the sense that for any $u \in \mathbb{R}^d$, the Jacobian matrix $A = \nabla_u f(u)$ has real eigenvalues and its eigenvectors span $\mathbb{R}^d$.

Such system is linear if the Jacobian matrix does not depend on u, otherwise it is called quasilinear.

Finite volumes schemes for system (1) are obtained by discretizing space into cells $I_j = [x_{j-1/2}, x_{j+1/2}]$, and time in discrete levels t_n. For simplicity we assume that the cells are all of the same size $\Delta x = h$, so that the center of cell j is $x_j = x_0 + jh$. This assumption is not necessary for finite volume schemes.

Integrating the conservation law on a cell in space-time $I_j \times [t_n, t_{n+1}]$ (see left picture in Fig. (1)) one has

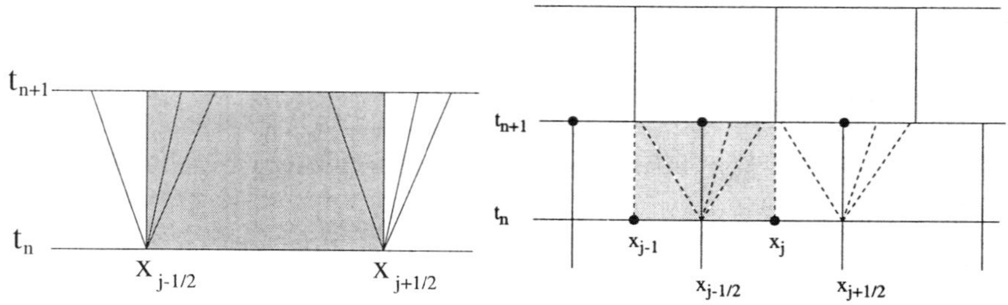

Figure 1. Integration over a non staggered cell and Godunov methods (left), and on staggered cell and central schemes (right)

$$\int_{x_{j-1/2}}^{x_{j+1/2}} u(x,t_{n+1}) = \int_{x_{j-1/2}}^{x_{j+1/2}} u(x,t_n) - \int_{t_n}^{t_{n+1}} (f(u(x_{j+1/2},t)) - f(u(x_{j-1/2},t)))\, dt \tag{2}$$

This (exact) relation suggests the use of numerical scheme of the form

$$\bar{u}_j^{n+1} = \bar{u}_j^n - \frac{\Delta t}{h}(F_{j+1/2} - F_{j-1/2}), \tag{3}$$

where $\bar{u}_j^n$ denotes an approximation of the cell average of the solution on cell j at time t_n, and $F_{j+1/2}$, which approximates the integral of the flux on the boundary of the cell, is the so called *numerical flux*, and depends on the cell average of the cells surrounding point $x_{j+1/2}$.

In a Godunov scheme the solution is approximated by a piecewise constant function, and the numerical flux at the edge between cell j and cell $j+1$ is given by

$$F_{j+1/2} = F(u^*(\bar{u}_j, \bar{u}_{j+1})),$$

where $u^*(u_-, u_+)$ is the solution of the Riemann problem at $x_{j+1/2}$ with values u_- and u_+ on the two sides of the discontinuity. This scheme is first order accurate and, when applied to a linear hyperbolic system, it reduces to the well-known first order upwind scheme. Such scheme can be extended to higher order, if instead of a piecewise constant approximation, a higher order reconstruction is used for the solution at each time step (and, of course, a high order scheme is used for the integration in time). The reconstruction step is a crucial point in the development of high order shock capturing schemes.

In order to derive central schemes, let us integrate the equation (1) on a staggered grid, as shown in the right picture of Fig. (1), obtaining

$$\int_{x_j}^{x_{j+1}} u(x,t_{n+1}) = \int_{x_j}^{x_{j+1}} u(x,t_n) - \int_{t_n}^{t_{n+1}} (f(u(x_{j+1},t)) - f(u(x_j,t)))\,dt \quad (4)$$

Once again, this formula is *exact*. In order to convert it into a numerical scheme one has to approximate the staggered cell average at time t_n, and the time integral of the flux on the border of the cells.

Let us assume that the function $u(x,t_n)$ is reconstructed form cell averages as a piecewise polynomial function. Then the function is smooth at the center of the cell and its discontinuities are at the edge of the cell. If we integrate the equation on a staggered cell, then there will be a fan of characteristics (and maybe shocks) propagating from the center of the staggered cell, while the function on the edge (dashed vertical lines in the figure) will remain smooth, provided the characteristic fan does not intersects the edge of the cell, i.e. provided a suitable CFL condition of the form

$$\Delta t < \frac{\Delta x}{2\rho(A)} \quad (5)$$

is satisfied.

The simplest central scheme is obtained by piecewise constant reconstruction of the function, and by using a first order quadrature rule in the evaluation of the integrals. The resulting scheme is the Lax-Friedrichs scheme on a staggered grid.

A second order scheme is obtained by using a piecewise linear approximation for the reconstruction of the function, and a second order quadrature rule (for example the midpoint rule) for the computation of the time integral of the flux on the edges of the cell.

Such scheme, proposed by Nessyahu and Tadmor[15], has the following simple expression

$$\bar{u}^{n+1}_{j+1/2} = \bar{u}^n_{j+1/2} - \lambda(f(u^{n+1/2}_{j+1}) - f(u^{n+1/2}_j)) \quad (6)$$

where $\lambda = \Delta t/\Delta x$ is the mesh ratio, $\bar{u}^n_{j+1/2}$ denotes the staggered cell average and is computed as

$$\bar{u}^n_{j+1/2} = \frac{1}{h}\int_{x_j}^{x_{j+1}} R(x;\bar{u}^n) = \frac{1}{2}(\bar{u}^n_j + \bar{u}^n_{j+1}) - \frac{1}{8}(u'_{j+1} - u'_j), \quad (7)$$

where R denotes the piecewise linear reconstruction of the function u, and u'_j/h denotes a first order approximation of the derivative of the function in the cell. The value of the field u at the node of the midpoint rule, $u^{n+1/2}_j$, can be computed by first order Taylor expansion, which is equivalent to forward Euler scheme,

$$u_j^{n+1/2} = \bar{u}_j^n - (\lambda/2) f_j'.$$

Note that the cell average $\bar{u}_j^n$ can be used rather than the pointwise value u_j^n, since their difference is only second order in h. Here f_j'/h denotes a first order approximation of the space derivative of the flux.

In order to prevent spurious oscillations in the numerical solution, it is essential that these derivatives are computed by using a suitable *slope limiter*. Several choices are possible for the slope limiter. The simplest one is the MinMod limiter, defined according to

$$\mathrm{MM}(a,b) = \frac{1}{2}(\mathrm{sign}(a) + \mathrm{sign}(b)) \min(|a|,|b|).$$

Such simple limiter, however, degrades the accuracy of the scheme near extrema. The use of other limiters is described, for example, in the paper[15].

3 High order central schemes

In most applications, second order schemes have been successfully used. There is, however, a great interest in constructing and using high order schemes.

It may seem strange to look for high order schemes for the numerical solution of quasilinear hyperbolic systems, since the solution itself is not regular, and in general it presents jump discontinuities. However, there are several reasons for using high order schemes. They provide good accuracy of the solution in the smooth regions. They also provide sharp resolution of the discontinuities. These two benefits justify the use of high order schemes (see the example illustrated in Fig.(3.1) below).

In the construction of high order central schemes, let us start from relation (4). We assume that the solution at time t_n is reconstructed as a piecewise polynomial function. Then, by approximating the integral of the flux by a quadrature formula, one obtains the general scheme

$$\bar{u}_{j+1/2}^{n+1} = \frac{1}{\Delta x}\int_{x_j}^{x_{j+1}} R(x;\bar{u}^n)\,dx \tag{8}$$

$$+\lambda \sum_{i=0}^{s} \gamma_l [f(\hat{u}(x_j, t_n + \beta_l \Delta t)) - f(\hat{u}(x_{j+1}, t_n + \beta_l \Delta t))] \tag{9}$$

The parameters γ_l and β_l are the weights and the nodes of the particular quadrature formula, and $\hat{u}$ are the intermediate values of the field at the nodes, predicted by Runge-Kutta method. For a fourth-order method one can use, for example, Simpson's rule.

The staggered cell-averages at time t_n, $\bar{u}^n_{j+1/2}$, are given by

$$\bar{u}^n_{j+1/2} = \frac{1}{h}\int_{x_j}^{x_{j+1}} R(x;\bar{u}^n)dx = \frac{1}{h}\left[\int_{x_j}^{x_{j+1/2}} R_j(x)dx + \int_{x_{j+1/2}}^{x_{j+1}} R_{j+1}(x)dx\right]. \tag{10}$$

The pointwise values of the solution at the nodes of the quadrature formula can be computed by using Natural Continuous Extension of Runge-Kutta Schemes. A detailed description of such procedure can be found in [3] and reference therein.

3.1 *The Central-Weighted Essentially Non Oscillatory (CWENO) Reconstruction*

Here we briefly describe Central-WENO (CWENO) piecewise-parabolic reconstruction which can be utilized to construct a fourth-order method.

Let u be a given function, and $\bar{u}_j$ its cell average on the cell I_j. Starting from the data $\{\bar{u}_j\}$ we apply the reconstruction scheme, obtaining the function $R(x;\bar{u})$. By construction, $R(x;\bar{u}) = \sum_j R_j(x)\chi_j(x)$, where in this case $R_j(x)$ is a polynomial of degree 2.

This polynomial has the following form

$$R_j(x) = \sum_{k=-1}^{1} w_k P_{j+k}(x) \tag{11}$$

where $P_j(x)$ are polynomials of second degree, obtained imposing that their cell averages on cells $j-1$, j, and $j+1$ are given, respectively, by $\bar{u}_{j-1}$, $\bar{u}_j$, and $\bar{u}_{j+1}$. The weights w_k, $k=-1,0,1$ are chosen according to the following criteria:

1. Accuracy. For smooth solutions the reconstruction should provide an accurate approximation of i) half cell average, ii) pointwise values of the function $u(x_j)$, iii) pointwise values of the space derivative of the flux f. Because the three requirements are different, three different sets of weights w will in general be found.

2. Non-oscillatory Reconstruction. Avoid spurious oscillations reconstruction for non smooth functions u[7].

The last requirement imposes that the weights depend on the smoothness of the function. Their effect can be summarized as follows: if the function to

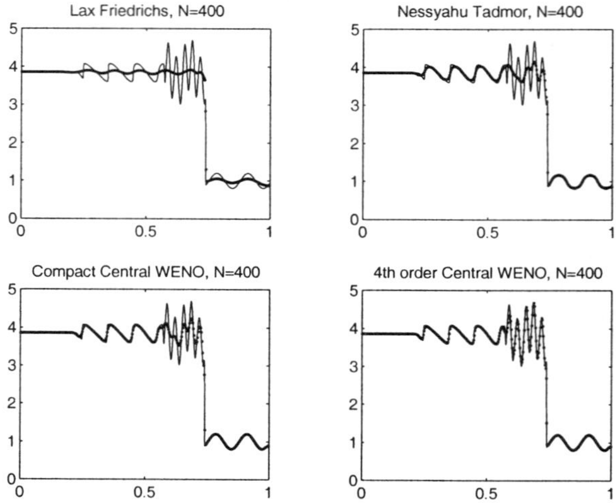

Figure 2. Numerical solution of gas dynamics: shock-density wave interaction. Density profiles obtained by first, second, third and fourth order central schemes

be reconstructed is smooth, then the weights should be close to the constant values that guarantees high accuracy, and if the function has some discontinuity near the point x_j then the weights will adjust in such a way that only the values belonging to the smooth side will enter in the reconstruction. For example, if the discontinuity is just on the right of x_j, then the weight w_1 will be close to zero.

As an example to illustrate the importance of high order schemes, we consider the interaction between a shock wave and a density wave in gas dynamics. The density profiles obtained by central schemes of different orders (first to fourth) are shown in Fig. (3.1). As it is evident from the figures, high order schemes are required to capture the fine detail of the solution.

3.2 Multidimensional schemes

Central schemes can be extended to problems in several dimensions. Second order central schemes on rectangular grids have been considered by Jiang and Tadmor[5], Sanders and Weiser[17], and by Arminjon et al.[2]. High order central schemes in two dimensions have been considered in the several papers[8,9].

The extension to two dimensions is also based on piecewise polynomial reconstruction, used to compute the staggered cell average, pointwise values, and integral of the flux. The latter requires the use of a suitable quadrature formula on rectangles in space-time.

As in the one dimensional case, the reconstruction step is crucial. In several dimensions it is very important to use a compact stencil, in order to have an efficient scheme.

4 Central schemes for systems with stiff source

Several models in mathematical physics are described by quasilinear hyperbolic systems with source term. The source may be of relaxation type, or it may represent a diffusion term. In these cases it is more appropriate to talk about balance laws, rather than conservation laws.

Hyperbolic systems with relaxation appear in discrete velocity models in kinetic theory, gas with vibrational degrees of freedom, gas in Extended Thermodynamics [14], hydrodynamical models for semiconductors [1], shallow water equations, and in several other physical models.

In this section we consider systems of balance laws of the form

$$\frac{\partial u}{\partial t} + \frac{\partial f(u)}{\partial x} = \frac{1}{\epsilon} g(u) \tag{12}$$

The parameter ϵ represents the relaxation time. If it is very small than we say the the relaxation term is *stiff*. Integrating Eq.(12) over the space-time (see right picture in Fig. 1) one has

$$\begin{aligned} \int_{x_j}^{x_{j+1}} u(x, t_{n+1}) = & \int_{x_j}^{x_{j+1}} u(x, t_n) + \frac{1}{\epsilon} \int_{t_n}^{t_{n+1}} \int_{x_j}^{x_{j+1}} g(x, t) \, dx \, dt \\ & - \int_{t_n}^{t_{n+1}} (f(u(x_{j+1}, t)) - f(u(x_j, t))) \, dt. \end{aligned} \tag{13}$$

Numerical schemes are obtained by a suitable discretization of the integrals. A scheme which is stable and accurate even for very stiff source has been proposed in [13]. It is obtained by using two predictor stages, one for the flux, and one which is needed for the source. The scheme can be written in the form

$$\begin{aligned} u_j^{n+1/2} &= \bar{u}_j^n - \frac{\lambda}{2} f_j' + \frac{\Delta t}{2} g(u_j^{n+1/2}) \\ u_j^{n+1/3} &= \bar{u}_j^n - \frac{\lambda}{3} f_j' + \frac{\Delta t}{3} g(u_j^{n+1/3}) \\ u_{j+1/2}^{n+1} &= \bar{u}_{j+1/2}^n - \lambda(f(u_{j+1}^{n+1/2}) - f(u_j^{n+1/2})) \\ & \quad + \frac{\Delta t}{8} (3g(u_j^{n+1/3}) + 3g(u_{j+1}^{n+1/3}) + 2g(u_{j+1/2}^{n+1})) \end{aligned} \tag{14}$$

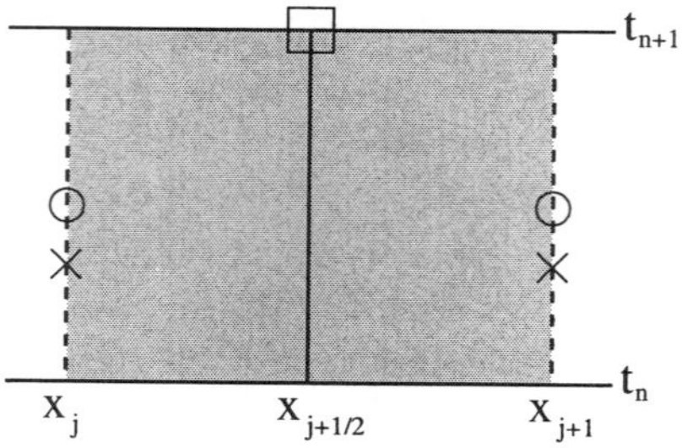

Figure 3. Nodes in space time for the second order Uniform Central Scheme

where $\bar{u}^n_{j+1/2}$ is computed by

$$\bar{u}^n_{j+1/2} = \frac{1}{2}(u^n_j + u^n_{j+1}) + \frac{1}{8}(u'_j - u'_{j+1}). \tag{15}$$

This scheme has been called Uniformly accurate Central Scheme of order 2 (UCS2)[13], and has the following properties. When applied to hyperbolic systems with relaxation, it is second order accurate in space and time both in the non stiff case (i.e. $\epsilon = 1$) and in the stiff limit (i.e. $\epsilon = 0$). A small degradation of accuracy is observed for intermediate values of the relaxation parameter ϵ.

The time discretization used in the above scheme is a particular case of Runge-Kutta Implicit-Explicit (IMEX) scheme. Such schemes are particularly important when one has to solve systems that contain the sum of e non stiff (possibly expensive to compute) and a stiff term. These systems may be convection-diffusion equations, or hyperbolic systems with stiff relaxation. In these cases it is highly desirable to use a scheme which is explicit in the non stiff term, and implicit in the stiff term.

As an application of the scheme UCS2, we consider the a Riemann problem for a gas in Extended Thermodynamics[14].

In one dimension the gas is described by a system of five balance equations, the usual three conservation equations for mass, momentum, and energy, and two additional equations for the balance of stress σ (just one component in one dimension) and heat flux q.

Such system is hyperbolic in a suitable region of the field vector u, which contains the equilibrium manifold $\sigma = 0$, $q = 0$. As $\epsilon \to 0 \Rightarrow \sigma \to 0, q \to 0$ and the equations reduce to the Euler equations for monoatomic gas.

No simple expression for the eigenvalues and eigenvector is known for such system, therefore it would be expensive to use characteristic based methods.

The numerical solution to a typical Riemann problem for a monoatomic gas in Extended Thermodynamics is shown in Fig. (4) (consinuous line). By comparison, the solution to the Euler equations corresponding to the same initial conditions is reported (dashed line). The two solutions are very close for density and velocity, because the value of the relaxation parameter is very small ($\epsilon = 2 \times 10^{-4}$). Stress tensor and heat flux are mainly concentrated at the shock.

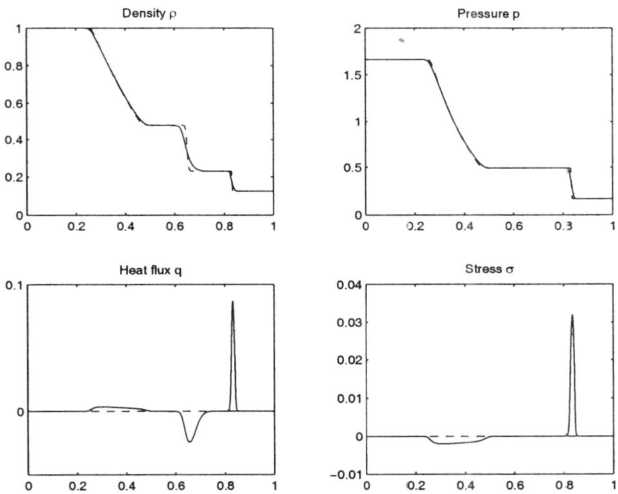

Figure 4. Solutions of the Riemann problem in Extended Thermodynamics model with $\varepsilon = 2 \times 10^{-4}$ and comparison with the solution of the Euler equations of gas dynamics. Top pictures: density (left) and pressure (right) computed with 800 grid points with UCS2 for Extended Thermodynamics (continuous line) and NT scheme for Euler equation (dashed line). Bottom pictures: Heat flux (left) and shear stress (right).

References

1. A. M. Anile, V. Romano and G. Russo, *Extended hydrodynamical model of carrier transport in semiconductors*, SIAM Journal Applied Mathematics **61**, 74-101 (2000).
2. P. Arminjon and M.C. Viallon, *Généralisation du schéma de Nessyahu-Tadmor pour une équation hyperbolique à deux dimensions d'espace*, C.

R. Acad. Sci. Paris Sr. I Math. **320**, 85-88 (1995).

3. F. Bianco, G. Puppo and G. Russo, *High Order Central Schemes for Hyperbolic Systems of Conservation Laws*, SIAM J. Sci. Comput. **21**, 294–322 (1999).
4. B. Cockburn, C. Johnson, C.-W. Shu and E. Tadmor, *Advanced Numerical Approximation of Nonlinear Hyperbolic Equations*, Lecture Notes in Mathematics 1697 (editor: A. Quarteroni), Springer, Berlin, 1998.
5. G.-S. Jiang and E. Tadmor, *Nonoscillatory Central Schemes for Multidimensional Hyperbolic Conservation Laws*, SIAM J. Sci. Comput. **19**, 1892–1917 (1998).
6. R. J. LeVeque, *Numerical Methods for Conservation Laws*, Lectures in Mathematics, Birkhäuser Verlag, 1992.
7. D. Levy, G. Puppo and G. Russo, *Central WENO Schemes for Hyperbolic Systems of Conservation Laws*, Math. Model. and Numer. Anal. **33**, 547–571 (1999).
8. D. Levy, G. Puppo and G. Russo, *A Third Order Central WENO Scheme for 2D Conservation Laws*, Appl. Num. Math. **33**, 407–414 (2000).
9. D. Levy, G. Puppo and G. Russo, *Compact Central WENO Schemes for Multidimensional Conservation Laws*, SIAM J. Sci. Comp. **22**, 656–672 (2000).
10. D. Levy and E. Tadmor, *Non-oscillatory central schemes for the incompressible 2-D Euler equations*, Mathematics Research Letters **4**, 1–20 (1997).
11. C.-T. Lin and E. Tadmor, *High-resolution non-oscillatory central scheme for Hamilton-Jacobi equations*, SIAM Journal on Scientific Computation **21**, 2163–2186 (2000).
12. X.-D. Liu and E. Tadmor, *Third Order Nonoscillatory Central Scheme for Hyperbolic Conservation Laws*, Numer. Math. **79**, 397–425 (1998).
13. S. F. Liotta, V. Romano and G. Russo, *Central schemes for balance laws of relaxation type*, SIAM J. Numer. Anal. **38**, 1337–1356 (2000).
14. I. Müller and T. Ruggeri, *Rational extended thermodynamics*, Springer-Verlag, Berlin, 1998.
15. H. Nessyahu and E. Tadmor, *Non-oscillatory Central Differencing for Hyperbolic Conservation Laws*, J. Comput. Phys. **87**, 408–463 (1990).
16. V. Romano and G. Russo, *Numerical solution for hydrodynamical models of semiconductors*, Mathematical Models & Methods in Applied Sciences **10**, 1099-1120 (2000).
17. R. Sanders and W. Weiser, *High resolution staggered mesh approach for nonlinear hyperbolic systems of conservation laws*, J. Comput. Phys. **10**, 314 (1992).

LONG TIME BEHAVIOR OF A SHALLOW WATER MODEL FOR A BASIN WITH VARYING BOTTOM TOPOGRAPHY

MARCO SAMMARTINO AND VINCENZO SCIACCA

Department of Mathematics and Applications, University of Palermo, Via Archirafi 34, 90123 Palermo (Italy)

E-mail: marco@dipmat.math.unipa.it *and* sciacca@dipmat.math.unipa.it

We study the long time behavior of a shallow water model introduced by Levermore and Sammartino to describe the motion of a viscous incompressible fluid confined in a basin with topography. Here we prove the existence of a global attractor and give an estimate on its Hausdorff and fractal dimension.

1 Introduction

In a recent paper[1] Levermore and Sammartino (see also the references therein) derived a model to describe the motion of a viscous incompressible fluid confined in a basin with a slowly varying bottom topography and with a free upper boundary. In deriving their model the two main assumptions were the following: First, the basin is shallow, i.e. the depth is much smaller than the tipical orizzontal length. Second, the tipical velocity of the fluid is much smaller than the velocity of the gravity waves. This is equivalent to consider the fluid motion on time scales much longer than the period of the gravity waves so that averaging on time suppresses gravity waves.

Under these hypotheses, and after the appropriate scaling the following 2D model equations were derived[1] :

$$
\begin{aligned}
\frac{\partial \boldsymbol{u}}{\partial t} + \boldsymbol{u}\cdot\boldsymbol{\nabla}_{\boldsymbol{x}}\boldsymbol{u} + \boldsymbol{\nabla}_{\boldsymbol{x}} p + \eta \boldsymbol{u} &= \nu b^{-1}\boldsymbol{\nabla}_{\boldsymbol{x}}\cdot\left[b(\boldsymbol{\nabla}_{\boldsymbol{x}}\boldsymbol{u} + (\boldsymbol{\nabla}_{\boldsymbol{x}}\boldsymbol{u})^T - \mathbb{I}\,\boldsymbol{\nabla}_{\boldsymbol{x}}\cdot\boldsymbol{u})\right] + \boldsymbol{f} \\
\boldsymbol{\nabla}_{\boldsymbol{x}}\cdot(b\boldsymbol{u}) &= 0 \\
\boldsymbol{u}(\boldsymbol{x}, t=0) &= \boldsymbol{u}_0 \\
\boldsymbol{\nu}\cdot\boldsymbol{u} &= 0 \qquad \boldsymbol{x}\in\partial\Omega \\
\boldsymbol{\tau}\cdot(\boldsymbol{\nabla}_{\boldsymbol{x}}\boldsymbol{u} + (\boldsymbol{\nabla}_{\boldsymbol{x}}\boldsymbol{u})^T)\cdot\boldsymbol{\nu} &= -\beta\boldsymbol{u}\cdot\boldsymbol{\tau} \qquad \boldsymbol{x}\in\partial\Omega \quad .
\end{aligned}
\tag{1.1}
$$

In the above equations $\boldsymbol{u}(\boldsymbol{x}, t)$ is the horizontal component of the fluid velocity, $\Omega \subset \mathbb{R}^2$ is a bounded domain, ν is the viscosity, $b(\boldsymbol{x})$ represents the topography of bottom of the basin, $\mathbb{I}$ is the 2×2 identity matrix, η is a positive regular function representing the effects of the friction at the bottom, $\boldsymbol{f}(\boldsymbol{x}, t)$ is a forcing term e.g. due to the presence of the wind; $\boldsymbol{\nu}$ and $\boldsymbol{\tau}$ are the normal and the tangent at the boundary $\partial\Omega$. The incompressibility condition is weighted by the presence of the bottom topography, and would reduce to the usual divergence free condition for a flat bottom. The boundary conditions are the usual no-flux condition, and the Navier condition which seems the most appropriate when the typical fluid particle size is much bigger than the boundary layer size. Notice also the complicated structure of the diffusion term that takes into account the non isotropy induced by the presence of the bottom topography.

The starting point in deriving[1] the above equations were the 3D Navier-Stokes equations with the appropriate boundary conditions and with the consistent expression for the stress tensor. The asymptotic analysis using the Froude number and the aspect ratio as smallness parameters led to the model equations Eqs. (1.1). Similar models were previously derived[2,3], without taking into account the effect of the viscosity of the fluid and without considering the interaction of the fluid with the wind and with the boundary.

It can be proved that Eqs. (1.1) are well posed globally in time[1]; for the precise statement see Section 2 below.

In this paper we shall be concerned with the long time behavior of the solutions of the system (1.1). We shall prove the existence of a global attractor in the function space H (for the definition of H see Section 2), and give an estimate on the dimension of this attractor. The main result of this paper is the following Theorem:

Theorem 1 *The dynamical system associated to Eqs. (1.1) admits an attractor $\mathcal{A}$ that is compact, connected and maximal in H. Moreover, if we define the integer* m *by:*

$$\mathrm{m} - 1 < \frac{C^3}{c^{7/2}} \left(\frac{\Pi \Lambda (1 + \gamma \mu)}{d \lambda_1} \right)^{1/2} \frac{|\boldsymbol{f}|_b}{\nu^2} \leq \mathrm{m} ,$$

then the Hausdorff dimension and the fractal dimension of $\mathcal{A}$ are bounded by m *and* 2m *respectively.*

The constants introduced in the definition of m will be formally defined in the rest of the paper. Here we just notice that that Λ, d, γ and Π depends on the domain Ω, λ_1 is the first eigenvalue of the Stokes operator associated to Eqs. (1.1), C and c depends on the topography, μ depends on the curvature of $\partial\Omega$ and on the friction at the boundary.

The rest of the paper will be devoted to proving the above Theorem. Our treatment will be necessarily sketchy, and we leave the presentation of the details of our proofs to a forthcoming paper.

2 Well posedness of the model and the existence of an attractor

In this Section we shall introduce the appropriate functional setting for Eqs. (1.1), recall an existence and uniqueness Theorem proved in[1], and give the proof of the existence of an attractor.

Let us define the following Hilbert spaces:

$$H = \{\boldsymbol{u} : \boldsymbol{u} \in L^2_b, \quad \boldsymbol{\nabla_x} \cdot (b\boldsymbol{u}) = 0, \quad \boldsymbol{\nu} \cdot \boldsymbol{u} = 0 \quad \boldsymbol{x} \in \partial\Omega\}$$
$$V = \{\boldsymbol{u} : \boldsymbol{u} \in H^1_b, \quad \boldsymbol{\nabla_x} \cdot (b\boldsymbol{u}) = 0, \quad \boldsymbol{\nu} \cdot \boldsymbol{u} = 0 \quad \boldsymbol{x} \in \partial\Omega\}$$

where L^2_b, H^1_b are the usual Sobolev spaces with weighted norms. These norms will be denoted with $|\cdot|_b$ and $||\cdot||_b$ respectively. One can prove[1] the following Theorem, estabilishing the well posedness of the model equations:

Theorem 2 *Let Ω be smooth. Suppose that $b(\boldsymbol{x})$, $\nu(\boldsymbol{x})$, and $\eta(\boldsymbol{x})$ are non-negative functions over $\overline{\Omega}$. Suppose moreover that b and ν are smooth, that $b\nu \geq C > 0$, and that $\beta(\boldsymbol{x}) \geq \kappa(\boldsymbol{x})$ on $\partial\Omega$, where $\kappa(\boldsymbol{x})$ is the curvature of $\partial\Omega$ at $\boldsymbol{x}$. Let $\boldsymbol{u}_0 \in H^2 \cap V$ and $\boldsymbol{f} \in L^2_b$.*

Then the system (1.1) has a unique solution $\boldsymbol{u} \in L^\infty([0,T], H^2) \cap C([0,T], V)$. Moreover, $\partial_t \boldsymbol{u} \in L^\infty([0,T], H) \cap L^2([0,T], V)$.

Let us define $c = \inf_\Omega b$ e $C = \sup_\Omega b$. Then the following generalized Poincaré inequality, with $\Lambda = \Lambda(\Omega)$ depending on Ω[9], holds:

$$|\boldsymbol{u}|^2_b \leq \frac{C\Lambda}{c} ||\boldsymbol{u}||^2_b \tag{2.1}$$

It will be useful to write Eqs. (1.1) in the weak form[1]:

$$\frac{d}{dt}\boldsymbol{u} + \nu A_b \boldsymbol{u} + B(\boldsymbol{u}, \boldsymbol{u}) + \eta \boldsymbol{u} = \boldsymbol{f} \tag{2.2}$$

where the bilinear operator $B(\cdot,\cdot)$ is defined as $B(\boldsymbol{u}, \boldsymbol{v}) = \boldsymbol{u} \cdot \boldsymbol{\nabla v}$, and the elliptic operator νA_b is the Stokes operator relative to Eqs. (1.1)[1]. Under the condition $\beta(\boldsymbol{x}) \geq \kappa(\boldsymbol{x})$ one can prove the coercivity of the operator A_b:

$$(\nu A_b \boldsymbol{u}, \boldsymbol{u})_b \geq \frac{c\nu}{C} ||\boldsymbol{u}||^2_b. \tag{2.3}$$

We are now in the position of proving the existence of the attractor $\mathcal{A}$. In fact we prove the following:

(i) the existence in H of a bounded absorbing set;

(ii) the evolution semigroup is uniformly compact.

These properties will lead[4] to the existence of the attractor.

We prove the first property. If one takes the scalar product in L^2_b of Eq. (2.2) with $\boldsymbol{u}$ and uses the Poincaré and Young inequalities, and the coercivity of A_b, one easily gets the following energy estimate:

$$\frac{d}{dt}|\boldsymbol{u}|_b^2 + \frac{c^2\nu}{C^2\Lambda}|\boldsymbol{u}|_b^2 \leq \frac{C^2\Lambda}{c^2\nu}|\boldsymbol{f}|_b^2 .$$

Using the Gronwall lemma one gets the existence in H of an absorbing set.

To prove (ii) it will be enough (given that V is compactly imbedded in H) to prove the existence in V of an absorbing set.

If one takes the scalar product of Eq. (2.2) with $A_b\boldsymbol{u}$, and one uses the well known[4] properties of the operator $B(\cdot,\cdot)$, together with the Young inequality, one readily gets:

$$\frac{d}{dt}||\boldsymbol{u}||_b^2 + \frac{2C\nu}{c\Lambda_1}||\boldsymbol{u}||_b^2 \leq c_1|\boldsymbol{u}|_b^2 + c_2|\boldsymbol{u}|_b^2\,||\boldsymbol{u}||_b^4 + \frac{3C}{\nu c}|\boldsymbol{f}|_b^2\,, \tag{2.5}$$

where $\lambda_1 > 0$ is the first eigenvalue of the operator A_b, and the constants c_1 and c_2 depends on Ω. Equation (2.5), together with the integrability with respect to time of $||\boldsymbol{u}||_b$, and the use of the uniform Gronwall lemma[4], gives the uniform compactness of the evolution semigroup. Property (ii) is threfore proved, and the existence of a compact attractor thus achieved.

3 The dimension of the global attractor

In this Section we shall give an estimate on the Hausdorff dimension of the global attractor for the dynamical system associated to the system (1.1). In our analysis we shall follow the general theory[4,6]. Equation (2.2) can be written in the form

$$\frac{d\boldsymbol{u}}{dt} = F(\boldsymbol{u})\,. \tag{3.1}$$

Let $S(t)$ the evolution semigroup generated by the above equation, so that $S(t)\boldsymbol{u}_0$ is the solution of (3.1) with initial condition $\boldsymbol{u}_0$. To estimate the rate of expansion of the volume in H, we linearize Eq. (3.1) and get:

$$\begin{aligned}\frac{d\boldsymbol{U}}{dt} + \nu A_b\boldsymbol{U} + B(\boldsymbol{u},\boldsymbol{U}) + B(\boldsymbol{U},\boldsymbol{u}) + \eta\boldsymbol{U} &= 0\\ \boldsymbol{U}(0) &= \boldsymbol{\xi}\,.\end{aligned} \tag{3.2}$$

We denote with F' the Fréchet derivative of the operator F introduced in (3.1). We can finally define the numbers q_m, with $m \in \mathbb{N}$, as:

$$q_m = \limsup_{t\to\infty} \sup_{\boldsymbol{u}_0\in\mathcal{A}} \sup_{\boldsymbol{\xi}_i\in H,\ |\boldsymbol{\xi}_i|_b\le 1; i=1,\dots,m} \frac{1}{t}\int_0^t \mathrm{Tr}(\mathrm{F}'(\mathrm{S}(\tau)\boldsymbol{u}_0)\circ \mathrm{Q_m}(\tau))d\tau\ , \tag{3.3}$$

where Q_m is the projector on the space spanned by the $\boldsymbol{\xi}_i$, $i = 1, \dots, m$, and where Tr denotes the trace. One can see that if there exists m such that $q_{\mathrm{m}} < 0$, then the Hausdorff dimension of the attractor $\mathcal{A}$ is bounded and $\dim_{\mathrm{H}}(\mathcal{A}) \le \mathrm{m}$. The rest of the paper will be devoted to estimate this m.

Let $\boldsymbol{\phi}_j(\tau)$, $j = 1, \dots, m$, an orthonormal basis of $Q_m(\tau)H$. One has:

$$\mathrm{TrF}'(\mathrm{S}(\tau)\boldsymbol{u}_0)\circ \mathrm{Q_m}(\tau) = \sum_{j=1}^{m}(-\nu A_b\boldsymbol{\phi}_j - B(\boldsymbol{u},\boldsymbol{\phi}_j) - B(\boldsymbol{\phi}_j,\boldsymbol{u}) - \eta\boldsymbol{\phi}_j,\boldsymbol{\phi}_j)_b.$$

With the use of the Cauchy-Schwarz inequality one gets:

$$\mathrm{Tr}\left(\mathrm{F}'(\mathrm{S}(\tau)\boldsymbol{u}_0)\circ \mathrm{Q_m}(\tau)\right) \le -\frac{c\nu}{C}\sum_{j=1}^{m}\|\boldsymbol{\phi}_j\|_b^2 + \|\boldsymbol{u}\|_b|\boldsymbol{\rho}|_b. \tag{3.4}$$

where we have denoted $\boldsymbol{\rho}(\tau) = \sum_{j=1}^{m}\boldsymbol{\phi}_j(\tau)\cdot\boldsymbol{\phi}_j(\tau)$. It is easy to prove that there exists a constant Π (depending on Ω), such that:

$$|\boldsymbol{\rho}|_b^2 \le \frac{C^2\Pi}{c^3}\sum_{j=1}^{m}\|\boldsymbol{\phi}_j\|_b^2\ . \tag{3.5}$$

Using this inequality in (3.4) one has:

$$\mathrm{Tr}\left(\mathrm{F}'(\mathrm{S}(\tau)\boldsymbol{u}_0)\circ \mathrm{Q_m}(\tau)\right) \le -\frac{\nu c}{2C}\sum_{j=1}^{m}\|\boldsymbol{\phi}_j\|_b^2 + \frac{C\Pi}{2c^2\nu}\|\boldsymbol{u}\|_b^2. \tag{3.6}$$

The following Lemma will be crucial in our estimates:

Lemma 1 *Let $\boldsymbol{\phi}_1, \dots, \boldsymbol{\phi}_m$ orthonormal in H. Then:*

$$\sum_{j=1}^{m}\|\boldsymbol{\phi}_j\|_b^2 \ge \frac{cd\lambda_1}{C(1+\gamma\mu)}m^2\ . \tag{3.7}$$

The meaning of the constants in the above Lemma is as follows: λ_1 is the first eigenvalue of the Stokes operator; $\mu = \sup_{\partial\Omega}(\beta - \kappa) \ge 0$; d is the constant that gives the following estimates on the sequence of the eigenvalues of the Stokes operator in terms of the first eigenvalue[7]: $\lambda_j \le d\lambda_1 j$; γ is the constant, depending on Ω, appearing in the trace Theorem[8].

Using the above Lemma in the estimate (3.6), one has:

$$\mathrm{TrF}'(\mathrm{S}(\tau)\boldsymbol{u}_0) \circ \mathrm{Q_m}(\tau) \leq -\frac{\nu \mathrm{c}^2 \mathrm{d}\lambda_1}{2\mathrm{C}^2(1+\gamma\mu)}\mathrm{m}^2 + \frac{\mathrm{C}\Pi}{2\mathrm{c}^2\nu}||\boldsymbol{u}||_\mathrm{b}^2 \,. \tag{3.8}$$

Inserting (3.8) in (3.3), and using the classical estimate[5] on the integral with respect to time of $||\boldsymbol{u}(\tau)||_b^2$:

$$\limsup_{t\to\infty} \sup_{\boldsymbol{u}_0 \in \mathcal{A}} \frac{1}{t}\int_0^t ||\boldsymbol{u}(\tau)||_b^2 d\tau \leq \frac{C^3\Lambda}{c^3\nu^2}|\boldsymbol{f}|_b^2 \,,$$

one finally gets:

$$q_m \leq -\frac{\nu c^2 d\lambda_1}{2C^2(1+\gamma\mu)}m^2 + \frac{C^4\Pi\Lambda}{2c^5}\frac{|\boldsymbol{f}|_b^2}{\nu^3} \,. \tag{3.9}$$

The above inequality, and the fact that $q_m < 0$ implies that $\dim_H(\mathcal{A}) \leq \mathrm{m}$. This completes the proof of Theorem 1.

Acknowledgments

This work has been supported by the MURST under the grant *Non linear mathematical problems of wave propagation and stability in models of continuous media.*

References

1. C.D. Levermore and M. Sammartino, *Nonlinearity*, **14**(6), 1493 (2001).
2. R. Camassa, D.D. Holm and C.D. Levermore, *J. Fluid Mech.*, **349**, 173 (1997).
3. C.D. Levermore, M. Oliver and E.S. Titi, *Ind. Univ. Math. J.*, **45**, 479 (1996).
4. R. Temam, *Infinite dimensional dynamical systems in mechanics and physics (second editions)*, AMS 68, Springer Verlag.
5. R. Temam, *Navier-Stokes equations, Theory and numerical analysis*, 3rd rev. ed., 1984, North-Holland Amsterdam.
6. J. M. Ghidaglia and R. Temam, *J. Math. Pures and Appl.*, **66**, 273 (1987).
7. G. Metivier, *J. Math. Pures Appl.*, **57**, 133 (1978).
8. M. E. Taylor, *Partial differential equations*, 1995, Springer-Verlag, New-York.
9. J. Deny and J. L. Lions, *Ann. Inst. Fourier*, **5**, 305 (1954).

A MODEL FOR THE CHEMOTHERAPY OF THE HIV INFECTION WITH ANTIGENIC VARIATION

M. SAMMARTINO
Dip. di Matematica, Università di Palermo, Italia
E-mail: marco@dipmat.math.unipa.it

L. SETA
I.T.D.F. - C.N.R., Palermo, Italia
E-mail: seta@itdf.pa.cnr.it

In this work we propose a new approach to model the drug effectiveness in the chemotherapy of the HIV infection. We introduce the drug resistance as a dynamical variable, and model its dynamics through a mechanism that has a simple biological interpretation in terms of the virus fitness. This new dynamical system is able to reproduce the lack of virus eradication. Stability of the equilibria of the systems is investigated. Numerical simulations have been performed to investigate the global dynamics of the model. Numerical and analytic results show a good agreement with the available clinical data.

1 Introduction: the modelling the HIV infection

In the last decade many mathematical models (see Perelson & Nelson[1] for a review) have been developed to describe the immunological response to the HIV infection and the drugs action. These models have been used to explain different phenomena such as the complicated kinetics of the virus load detected in the blood or the decline of the CD4+T cells, the principal target of the virus attack.

As a representative of the many models appeared in the literature we consider the following model for the dynamic of HIV infection, without drug effect, derived by Culshaw & Ruan[2]:

$$\begin{aligned}
\frac{dT}{dt} &= s - \mu_T T + rT\left(1 - \frac{T+T^*}{T_{\max}}\right) - kVT; \\
\frac{dT^*}{dt} &= kVT - \mu_{T^*} T^*; \\
\frac{dV}{dt} &= N\mu_{T^*} T^* - cV.
\end{aligned} \tag{1}$$

In these equation $T(t)$ represents the concentration of healthy CD4+T cells at time t, $T^*(t)$ represents the concentration of infected CD4+T cells, and

Table 1. Parameters mean values.

s	Source term for CD4+T cells	10 $day^{-1}mm^{-3}$
μ_T	Natural death rate of CD4+T cells	0.02 day^{-1}
r	Growth rate of CD4+T cells	0.03 day^{-1}
$T_{\max}$	Maximal population level of CD4+T cells	1500 mm^3
k	Rate of the infection for CD4+T cells	2.4×10^{-5} $day^{-1}mm^3$
μ_{T^*}	Natural death rate of infected CD4+T cells	0.26 day^{-1}
N	Number of viruses produced by infected T–cells	≈500 (*varies*)
c	Natural death rate of viruses	2.4 day^{-1}

$V(t)$ the concentration of the free HIV. Moreover s is the source of CD4+T cells, r is their growth rate and $T_{\max}$ is this carrying capacity; the parameter k represents the rate of infection of T–cells; μ_T, μ_{T^*} and c are the natural death rate of the uninfected cells, of the the infected cells and of the viruses respectively; finally N represents the number of viruses released by each lysing infected T–cells. This oversimplified model describe the first stage of the HIV infection and in particular the evolution towards the infected state when a low viral load is present in the blood.

The system (1) has two equilibrium state:

- the uninfected steady state: $E_0 = (T_0, 0, 0)$, where

$$T_0 = \frac{r - \mu_T + \left[(r-\mu_T)^2 + 4rsT_{\max}^{-1}\right]^{1/2}}{2rT_{\max}^{-1}}; \tag{2}$$

- the infected steady state: $\widehat{E} = (\widehat{T}, \widehat{T}^*, \widehat{V})$, where

$$\begin{aligned} \widehat{T} &= \frac{c}{kN}; \qquad \widehat{T}^* = \frac{c}{N\mu_{T^*}}\widehat{V}; \\ \widehat{V} &= \frac{\mu_{T^*}\left[kNcT_{\max}(r-\mu_T) - rc^2 + sT_{\max}k^2N^2\right]}{kc\,(rc + T_{\max}k\mu_{T^*}N)} . \end{aligned} \tag{3}$$

The stability of these steady solutions depends crucially on the values of the parameters. These parameters show an individual variability and it is possible to furnish only mean values. The typical values in the literature, see Culshaw & Ruan[2], are shown in Table 1. For these values one easily verifies that the uninfected state E_0 is unstable and the infected state $\widehat{E}$ is stable.

Figure 1 shows a numerical solution of this simple system, when $T(0) = 2000$, $T^*(0) = 0$ and $V(0) = 0$.

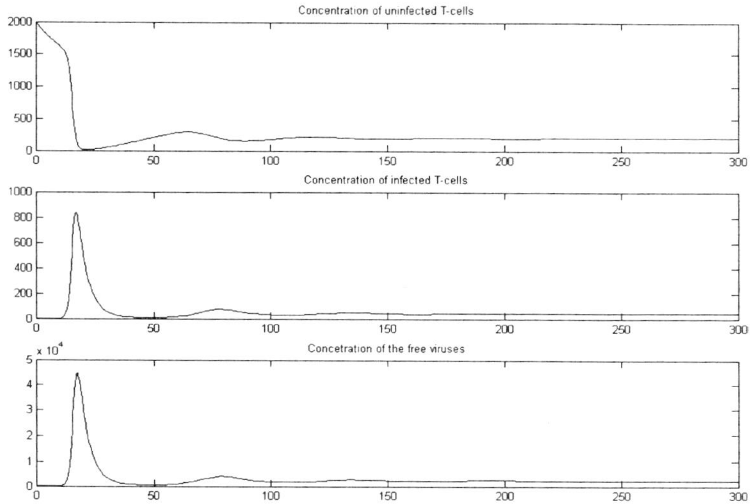

Figure 1. Solution of the basic model: variations of T, T^* and V with respect to the time (in days).

2 The effectiveness

We now consider the situation when a treatment with an antivirus drug is started, e.g. a treatment with a Reverse Transcriptase Inhibitor. When the reverse transcriptase enzyme is inhibited the HIV can enter a cell but it will not successfully infect it: in fact it will not be possible to make a copy of the viral genome and the cell will not produce viral proteins or virus particles.

The system to describe the effect of the RT inhibitor proposed by Perelson & Nelson[1] it is the following :

$$\begin{aligned} \frac{dT}{dt} &= s - \mu_T T + rT\left(1 - \frac{T+T^*}{T_{\max}}\right) - kVT; \\ \frac{dT^*}{dt} &= (1-\eta)kVT - \mu_{T^*}T^*; \\ \frac{dV}{dt} &= N\mu_{T^*}T^* - cV. \end{aligned} \tag{4}$$

In this system η is a constant between 0 and 1 used to describe the effectiveness of the drug. When $\eta = 1$ the inhibition is fully effective, while if

$\eta = 0$ there is no inhibition.

The principal problem with the antiviral therapy is the resistance that the virus is able to mount to particular drugs due to its high rate of mutation. This resistance is also the principal obstacle to successful long term control of the HIV infection. In fact the effectiveness of the drug falls during the treatment, and eventually the viral load increases considerably to reach the level of an untreated infection: this results in a dramatic worsening of the conditions of the patients. Many models have been tested to mimic the viral resistance during the treatment with different type of drugs. These models consider the effectiveness of the drug as a constant and introduce new strains of viruses resistant to a specific drug: see Kirschner & Webb[3]. In general these models seem to be unable to reproduce correctly the dynamics of the developing of the resistance and of the raising of the viral load.

In this paper we propose a new approach to model the effectiveness of the drug. Our approach is based on this simple consideration: resistance occurs due to the selection of the resistant strains of viruses. Therefore resistance is directly related to the number of the viruses which the drug has effectively acted on. This leads to considering the effectiveness η as a new dynamical variable and we have postulated the following law:

$$\eta(t) = \frac{1}{1 + ak \int_0^t \eta(s) V(s) T(s)\, ds} \tag{5}$$

In this formula the integral term represents the number of viruses the drug has inhibited in the period $[0, t]$. The positive constant a represents the rate of growth of the resistance for the specific drug.

Our model equations are Eqs. (4) together with the following equation which is equivalent to Eq. (5):

$$\frac{d\eta}{dt} = -a\eta^3 VT. \tag{6}$$

3 Some results

In this Section we show some of the results obtained using our model. We initialize it imposing as initial conditions the infected state (3) for T, T^* and V, and $\eta(0) = 1$. The equilibrium states are:

- a segment in the phase space (T, T^*, V, η) corresponding to the "*unin-*

fected states":

$$T = T_0, \quad T^* = 0, \quad V = 0, \quad \eta \in [0,1]; \tag{7}$$

- the "*infected state*":

$$T = \widehat{T}, \quad T^* = \widehat{T}^*, \quad V = \widehat{V}, \quad \eta = 0. \tag{8}$$

The stability analysis shows that the uninfected states with $\eta \geq \overline{\eta} = 0.8$ have an attractive character, while the states with $\eta < \overline{\eta}$ are repulsive. What happens is that the system stays close to an uninfected state while is η monotonously decreasing; this behavior persists until η becomes smaller than the critical value $\overline{\eta}$, which causes the system to evolve rapidly towards the infected state. This is confirmed both by the numerics and the analysis. Moreover this behavior is in fact consistent with clinical trials and it is typical of the evolution of the HIV infection under treatment. The period during which the concentration of HIV drops under the detectable level (which is actually observed in clinical trials) is called latency period and it is characterized by the disappearance of the symptoms of the infection. Our model is able to reproduce the latency period; the length of this period is related to the the value of the parameter a. Figure 2 shows a a typical behavior predicted by our model, with $a = 1.5 \times 10^{-7}\text{day}^{-1}\text{mm}^3$. In this case the latency period lasts about three months. Increasing the parameter a the pattern stays the same and latency period increases but remains finite.

4 Conclusions

In this paper we have developed a model to describe the dynamics of the HIV infection under a drug treatment. Treating the drug effectiveness as a dynamic variable, we have been able to reproduce both the occurrence of the apparent eradication of the illness during the first stage of the treatment, and the subsequent developing of the drug resistance with the decline in the count of CD4+T cell and the raising of the viral load. The length of the latency period (the period of the apparent eradication of the infection) is related to a parameter we have introduced (the parameter a in Eq. (6)); this parameter is characteristic of a specific treatment, and should be an important feature of a treatment to be tested because it would give anticipation on the long term effectiveness of a treatment. It should also be interesting to extend our model to the case of multi-therapy where the occurrence cross resistance should also be modelled.

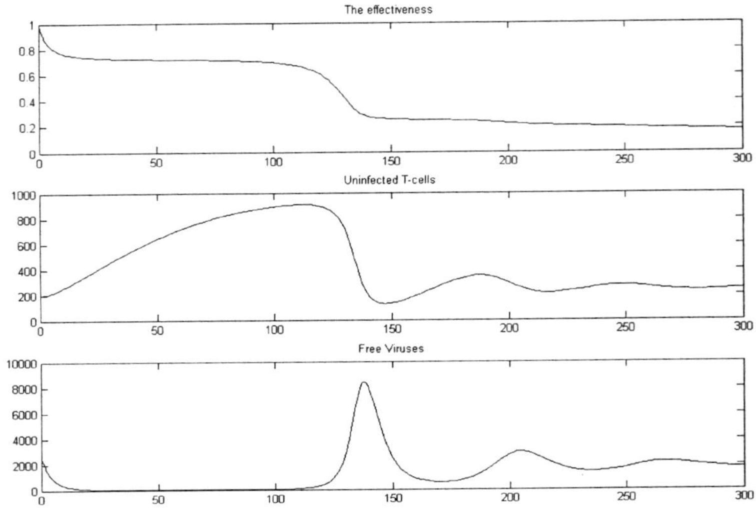

Figure 2. Solution of the model with RT inhibitor: variations of η, T and V with respect the time (in days).

Acknowledgments

This paper was supported in part by the MURST under grant "Problemi matematici non lineari di propagazione e stabilità nei modelli del continuo".

References

1. A.S. Perelson and P.W. Nelson. Mathematical analysis of HIV–1 dynamics in vivo. *SIAM Review*, 1(41):3–44, 1999.
2. R.V. Culshaw and S. Ruan. A delay–differential equation model of HIV infection of CD4+T–cells. *Mathematical Biosciences*, (165):27–39, 2000.
3. D.E. Kirschner and G.F. Webb. Understanding drug resistance for monotherapy treatment of HIV infection. *Bullettin of Mathematical Biology*, 59(4):763–785, 1997.

KINETIC AND FLUID DYNAMIC APPROACHES TO FOUR–WAVE–MIXING AND THERMAL ACOUSTIC PHENOMENA IN QUANTUM OPTICS

F. SCHÜRRER, W. KOLLER AND F. HANSER

Institute for Theoretical Physics, Graz University of Technology, Petersgasse 16, A-8010 Graz, Austria

E-mail: ferdinand.schuerrer@tugraz.at

The degenerate four-wave-mixing technique and laser-induced thermal acoustics represent interesting quantum optic phenomena. Two recently developed kinetic methods based on discrete velocity models and on a semi-continuous formulation of the Boltzmann equation provide deep insights into the dynamics of these phenomena. The results obtained in the fluid dynamic limit of the kinetic equations agree well with those gained from a purely fluid dynamic approach.

1 Introduction

The excitation of acoustic and thermal waves by means of pulsed laser light plays an important role in quantum optics. An interesting application in the field of spectroscopy is the degenerate four-wave-mixing technique (DFWM) [1]. The recently developed method of laser-induced thermal acoustics (LITA) [2] to measure physical and chemical properties of gases both remotely and non-intrusively is also supported by this technique.

So far it has been common to use a linearized fluid dynamic model for the description of such phenomena, although this method fails when the Knudsen numbers are high. From a kinetic point of view, the four-wave-mixing technique is interesting since it addresses a spatially periodic one-dimensional transport problem. It involves several species of gases that interact elastically as well as inelastically via binary collisions and with monochromatic photons. Moreover, kinetic methods are also applicable in the case of high Knudsen numbers corresponding to very small fringe spacings.

After a short presentation of the usual fluid dynamic approach two alternative kinetic treatments of these phenomena are given. First, a discrete model of Shizuta type with 36 velocities is used to describe the dynamics of laser-induced density and thermal gratings. Second, a semi-continuous formulation of the extended Boltzmann equation is applied to this problem. The comparison of the results obtained by these three different methods reveals their assets and drawbacks in dealing with the complicated structure of the dynamics of laser-pulse-driven gratings.

This paper is organized as follows: Sec. 2 provides a short introduction into the physics of degenerate four–wave–mixing and laser-induced thermal acoustics. In Sec. 3 we touch on the fluid dynamic approach, whereas Sec. 4 is devoted to the kinetic treatment of these phenomena.

2 The Physics of Laser-Induced Thermal Acoustics

The process of laser-induced thermal acoustics (left sketch in Fig. 1) can be subdivided into two steps. In the first one, light from a powerful pulsed laser is split into three phase-coherent beams. Two of them, the forward pump and the forward probe beam, intersect at a shallow angle within a strongly absorbing collision dominated gas-phase system. This results in an interference pattern, i.e. a periodically varying photon intensity (right sketch in Fig. 1). The period λ is called fringe spacing.

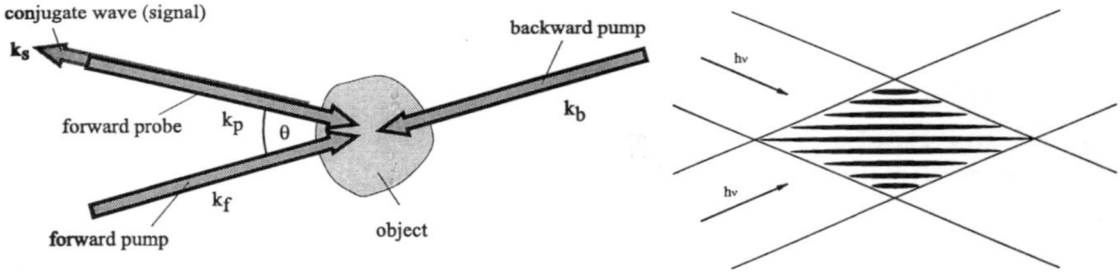

Figure 1. Idealized geometry of LITA experiments (left plot). Region of interference: electric field intensity grating (right plot).

The laser frequency is tuned as to excite electronically a rare species of the mixture. In course of this nonlinear interaction of the electric field intensity grating with the gas mixture, the excited species lose their internal energy due to inelastic collisions with the dominant species. Since this process occurs periodically according to the interference pattern of the laser beams, a stationary acoustic wave is triggered.

The second step involves scattering of the backward pump beam into a coherent signal beam by optoacoustic effects. In this way, it is possible to detect properties of the gas components which influence the formation of the laser induced density grating. Therefore, deep insight and understanding of the dominant processes is very important. This can be obtained, but only under certain conditions, by means of a fluid dynamic approach. More detailed information can by gained by treating it on a kinetic level.

The driver pulse of the laser is assumed to have a temporal variation of the intensity that can be well described by

$$I(t) = \frac{2t}{\tau_L^2} \exp\left[-\left(\frac{t}{\tau_L}\right)^2\right]$$

with a time scale τ_L typically between 5 ns and 10 ns [1]. The spatial variation of the Intensity I governed by the laws of optical interference is proportional to $\cos^2(\pi\, x/2\lambda)$.

It is natural to use the fringe spacing λ in order to characterize this regime by means of Knudsen numbers. Assuming hard sphere particles with diameter a, the Knudsen number reads $Kn = 1/n\pi a^2\lambda$. The number density of gas particles is denoted by n. As long as $Kn \ll 1$, the discrete nature of the gas particles has negligible influence and fluid dynamic equations are well suited for an accurate description of the evolution of the gas. On the other hand, for approx. $Kn > 0.2$, this is no longer true and we have to resort to a kinetic model. However, an overlap exists where both approaches apply and we can compare the results obtained by these different approaches.

3 Fluid dynamic Approach

Starting from a one–dimensional version of the Navier-Stokes closure of the moment equations resulting from the Boltzmann equation [3], it is possible to derive the hydrodynamic equations that Cummings [2] uses to describe LITA and DFWM experiments. Since the perturbations induced by the laser beams are very week, a small departure from equilibrium can be assumed. Consequently, the Navier-Stokes equations can be linearized by means of the ansatz $\psi_1 = \partial v/\partial x$, $\rho = \rho_0 + \rho_1$ and $T = T_0 + T_1$; perturbation terms are indicated by the subscript 1 and unperturbed stationary quantities are labelled by 0. This results finally in

$$\frac{\partial \rho_1}{\partial t} + \rho\, \psi_1 = 0\ , \tag{1a}$$

$$\frac{\partial \psi_1}{\partial t} + \frac{k_B}{m}\frac{\partial^2 T_1}{\partial x^2} - D_V\,\frac{\partial^2 \psi_1}{\partial x^2} + \frac{k_B T}{m\rho}\frac{\partial^2 \rho_1}{\partial^2 x} = \frac{\partial}{\partial x} X\ , \tag{1b}$$

$$\frac{\partial T_1}{\partial t} + \frac{2}{3}\psi_1\, T - \frac{2}{3} D_T\,\frac{\partial^2 T_1}{\partial x^2} = \frac{2}{3}\frac{m}{k_B} v\, X\ . \tag{1c}$$

As usual, ρ, v and T denote the mass density, drift velocity and temperature. External forces acting on the fluid (e.g. the laser pulse) are denoted by capital X. The thermal diffusivity D_T and the longitudinal kinematic viscosity D_V

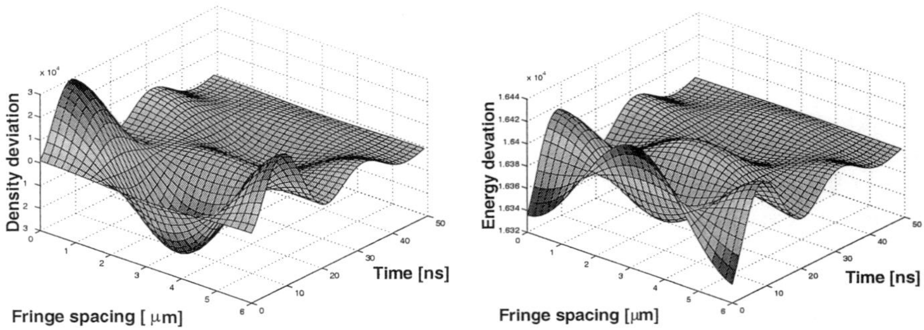

Figure 2. The evolution of density and energy oscillations within a fringe spacing of 10 μm due to the linearized hydrodynamic model for $p = 0.1$ bar.

are given by

$$D_T = \frac{m\kappa}{k_B \rho}, \quad \text{and} \quad D_V = \frac{2\mu + \lambda}{\rho}, \tag{2}$$

where k_B denotes Boltzmann's constant. For the transport coefficients viscosity μ and thermal conductivity κ, we use the first order Chapman–Enskog formulae [4]

$$\mu = 2.6693 \times 10^{-10} \frac{\sqrt{mT}}{\sigma^2} \frac{\text{kg}}{\text{ms}}, \quad \kappa = 8.2949 \times 10^{-2} \frac{\sqrt{T/m}}{\sigma^2} \frac{\text{J}}{\text{msK}}.$$

Furthermore, we neglect the so-called bulk viscosity by setting $\lambda = -(2\mu)/3$.

The obtained equations (1a), (1b) and (1c) form a closed set of three linear partial differential equations, which can easily be solved by resorting to a Fourier expansion in real space.

As initial conditions, we choose constant density and zero flux within a gas of hard sphere particles with atomic mass 28 (diameter 4.4 Å) and a pressure of $p = 0.1$ bar. Instead of exciting the system by an external force X, we assume at $t = 0$ a temperature profile that varies between 292 and 294 K.

The first 50 ns of the time evolutions of the density and energy oscillations within a fringe spacing of 6 μm are shown in Fig. 2. We observe that the high energy density in the center of the slab at t=0 leads to a flow of particles to the borders of the slab. This results in a particle density depletion in the center, whereas the particle density increases at the borders. This, in turn, increases the energy density at the borders, which implies a backflow of the gas to the center of the slab. With increasing energy density in the center, the next strongly damped period of an acoustic wave starts. By comparing experimental measurements with results from the linearized hydrodynamic equations, one is able to determine transport properties of the involved gas mixture [1,2].

4 Kinetic Modeling

To deal with laser-induced thermal acoustics on a kinetic level, the molecular structure of the gas within the region of interference of the laser beams has to be considered. We assume that the gas comprises three different species A, B and B^*. The species B^* denotes an electronically excited state of B. In real experiments the dominant species A stands for N_2, CO_2 or simply the ambient air. The rare species B can, e.g., be a pollutant arising in the course of a combustion such as NO, NO_2 or also OH.

The laser acts as a source of coherent monochromatic photons p of frequency ν that induce the transitions $B + p \rightarrow B^*$. Apart from the dominant elastic interactions, spontaneous and stimulated emission of photons, $B^* \rightarrow B + p$ and $B^* + p \rightarrow B + p + p$, respectively, play an important role. In regions of high laser intensity, local excitations of B result in a higher local excitation–temperature of the rare species B. Binary collisions with the much colder buffer gas A de–excite the molecules B^* according to the inelastic transition $A + B^* \rightarrow A + B$. In order not to offend microreversibility, its (rare) inverse process, $A + B \rightarrow A + B^*$, has also to be included in the picture.

4.1 Discrete Velocity Model

A kinetic model describing a mixture of elastically and inelastically interacting gases excited by monochromatic photons consists of a set of Boltzmann equations [5,6]. These coupled nonlinear integro-differential equations oppose direct analytical and numerical treatment. One way out of this problem is to discretize the velocity space [7]. In our approach, we use for each species the favorable Shizuta type model [8] with 36 velocities and 6 speeds of relative magnitude a, $\sqrt{5}a$, $3a$, $\sqrt{13}a$, $5a$, $\sqrt{29}a$ (Fig. 3). It is characterized by a lot of mixing speed collisions, allowing an efficient energy transfer among particles of different speeds, and an H-theorem assures a stable equilibrium.

Establishing this discrete velocity model demands sophisticated computational techniques. The major problem to overcome is the high number (about 10^4) of admissible collisions inflating the collisional term of the kinetic equations. A special computer program yields the complete collisional scheme and the governing transport equations as a set of coupled nonlinear PDE for the

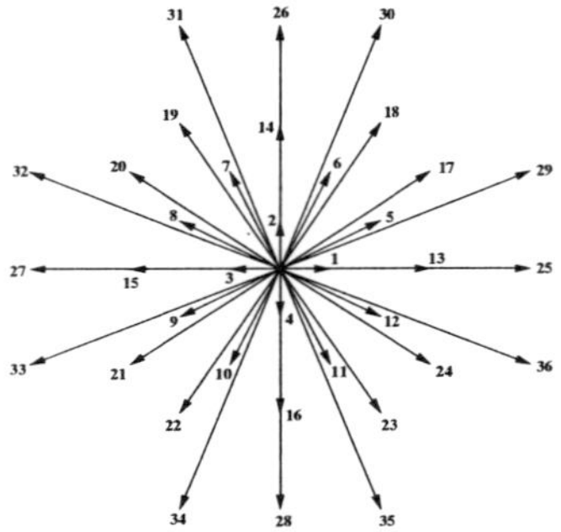

Figure 3. Regular discrete velocity model of Shizuta-type with 36 velocities and six speeds.

number densities of all three species of the gas mixture:

$$\left(\frac{\partial}{\partial t} + \mathbf{v}_i^B \cdot \frac{\partial}{\partial \mathbf{x}}\right) N_i^B = \mathcal{J}_i^B[\underline{N}^M] + \mathcal{R}_i^B[\underline{N}^M, I], \tag{3a}$$

$$\left(\frac{\partial}{\partial t} + \mathbf{v}_i^{B^*} \cdot \frac{\partial}{\partial \mathbf{x}}\right) N_i^{B^*} = \mathcal{J}_i^{B^*}[\underline{N}^M] - \mathcal{R}_i^B[\underline{N}^M, I], \tag{3b}$$

$$\left(\frac{\partial}{\partial t} + \mathbf{v}_i^A \cdot \frac{\partial}{\partial \mathbf{x}}\right) N_i^A = \mathcal{J}_i^A[\underline{N}^M], \tag{3c}$$

where $N_i^M(\mathbf{x}, t)$ is the number density of the molecules M ($M = B, B^*, A$) with velocity $\mathbf{v}_i^M$ ($i = 1, \ldots, 36$) at the point $\mathbf{x}$ at time t. Here, $\underline{N}^M$ is an abbreviation of $(N_1^M, \ldots, N_{36}^M)$. Interactions between the radiation intensity and the particle species B and B^* are taken into account by

$$\mathcal{R}_i^B[\underline{N}^M, I] = (\alpha + \beta I) N_i^{B^*} - \beta I N_i^B,$$

where $I = I(\mathbf{x}, t)$ is the laser intensity and α and β are the Einstein coefficients. The functionals $\mathcal{J}_i^B[\underline{N}^M]$, $\mathcal{J}_i^{B^*}[\underline{N}^M]$ and $\mathcal{J}_i^A[\underline{N}^M]$ are the collision terms. Each collision event contributes an expression of quadratic nonlinearity in the number densities of the molecules. The particular form reads as

$$A_{i,j}^{k,l}(N_k^{M'} N_l^{N'} - N_i^M N_j^N)$$

with $M', N', M, N = B, B^*, A$. The transition rates $A_{i,j}^{k,l}$ are defined by

$$A_{i,j}^{k,l} = \sigma_{MN \to M'N'}^{el(in)}(|\mathbf{v}_i^M - \mathbf{v}_j^N|)|\mathbf{v}_i^M - \mathbf{v}_j^N| a_{i,j}^{k,l}.$$

To acquire more information in this context see [9]. The macroscopic quantities of interest are the number densities n^M, the average gas velocity $\mathbf{u}^M$ and the

the average internal kinetic energy e^M of each species:

$$n^M = \sum_{i=1}^{36} N_i^M, \quad \mathbf{u}^M = \frac{1}{\rho^M} \sum_{i=1}^{36} m^M \mathbf{v}_i^M N_i^M, \tag{4a}$$

$$e^M = \frac{1}{n^M} \sum_{i=1}^{36} \frac{1}{2} m^M \left(\mathbf{v}_i^M - \mathbf{u}^M\right)^2 N_i^M. \tag{4b}$$

To simulate the formation and evolution of the laser-induced gratings, we solve numerically the discrete kinetic equations by applying the fractional step method. This approach divides the evolution of the system into a free–streaming part and a collision part within each time step. For the number density of molecules A, we choose $n^A = 10^{25}\mathrm{m}^{-3}$. Molecules B and B^* are assumed to be of low concentration: $n^B + n^{B*} = 10^{-3} n^A$. It is sufficient to consider only one fringe spacing together with periodic boundary conditions. The scaling parameter a of the Shizuta type model is determined due to the fact that the perturbations propagate with the speed of sound. The gas-mixture is assumed to be in thermodynamic equilibrium before the laser pulse starts interacting with the rare species B and B^*.

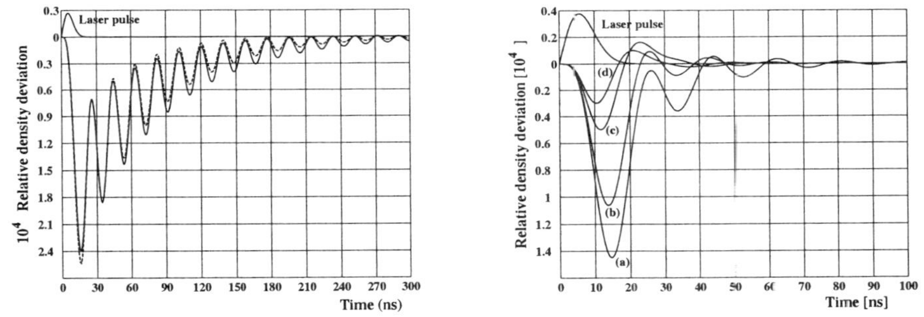

Figure 4. Time evolution of the relative density deviation in the middle of the fringe spacing. In the left plot, the dashed and solid lines show the results obtained by the fluid dynamic and the discrete kinetic model ($Kn = 0.0033$), respectively. In the right plot (a), (b), (c), and (d) correspond to the Kn values 0.021, 0.033, 0.083, and 0.167, respectively.

For small Knudsen numbers the gas behaves as a fluid, and therefore, the results can be compared with those obtained from the fluid dynamic approach. The left plot in Fig. 4 shows such a comparison for the temporal evolution of the relative density deviation in the middle of the fringe spacing for $Kn = 0.0033$. We observe good agreement apart from small discrepancies, which should be attributed to the discrete character of the kinetic model.

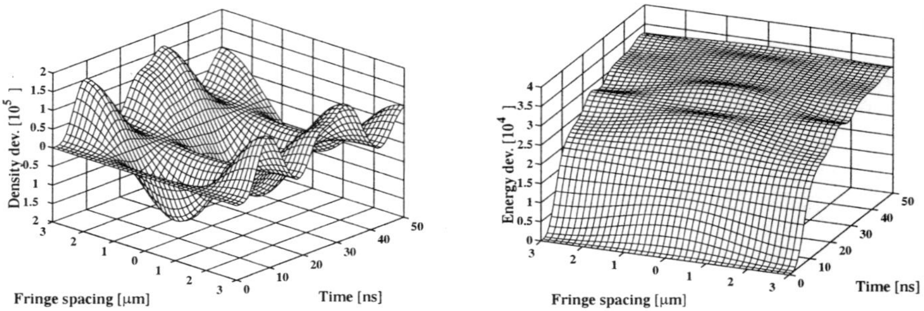

Figure 5. Spatio-temporal evolution of the relative density deviation (left plot) and relative kinetic energy deviation (right plot) inside the fringe spacing for $Kn = 1.67$.

The right plot in Fig. 4 reveals an interesting and unexpected behavior of the gas mixture. The temporal density oscillations disappear almost completely with increasing Knudsen number. At $Kn = 0.083$, the damping effect attains its strongest influence.

In the case of high Knudsen numbers, where only a molecular gas kinetic model applies, we observe a regeneration of the density oscillations as demonstrated in the left plot of Fig. 5. For $Kn = 1.67$, as shown in this figure, the gas mixture behaves as a Knudsen gas, which means that the spatio-temporal evolution of the gas is mainly determined by the periodic boundary conditions. This explains the re-appearance of the density oscillations. Moreover, the oscillations become more and more nonlinear. Even the kinetic energy no longer increases monotonously as can be seen in the right plot of Fig. 5.

4.2 Semi–Continuous Kinetic Model

The semi-continuous approach is based on a recently published paper of Preziosi and Longo [10]. In this paper, they provide a very useful semi-continuous formulation of the nonlinear Boltzmann equation which reduces essentially the complexity of the collision terms. However, their model refers only to elastic collisions in a one-component gas. Our intention is to generalize their ideas in order to cope with - in an alternative way as before in Sec. 4.1 - the problem of laser induced thermal acoustics.

For the purpose of discretizing the Boltzmann equations describing the evolution of an irradiated gas mixture [5,6], we resort to a polar decomposition of the velocity variable $\boldsymbol{v} = v\hat{\boldsymbol{\Omega}}$ with $v = |\boldsymbol{v}|$ and restrict the range of the particle's kinetic energies to the interval $I_\nu = [E_m, E_M)$. The bounds of

I_ν are to be chosen such that all particles with kinetic energies outside of I_ν may be neglected. Next, we introduce on I_ν an arithmetic sequence of energies $E_i = E_m + (i + 1/2)\,\delta, \quad i = 0, 1, \ldots, n$, indicating the centers of the corresponding subintervals I_i in a way that the energy gap for inelastic interactions $\Delta E = q\,\delta$ with $q \in \{1, 2, \ldots, 2n{-}1\}$. Then, any function of kinetic energy (and thus of the speed v) in the continuous Boltzmann equations has to be approximated by a piecewise constant interpolant $g(E) \approx \sum_{i=0}^{n} g_i\,\chi_{I_i}(E)$, where $\chi_B(.)$ denotes the characteristic function of the set B.

The combination of conservation of momentum and total energy implies a restriction of $\hat{\mathbf{\Omega}}_*$ on values which are compatible with a chosen pair of pre-collisional (v, v_*) and post-collisional (v', v'_*) speeds [11]. When fixing $\hat{\mathbf{\Omega}}$, the unit vector pointing in the direction of the incoming test particle, the variation of the solid angle $\hat{\mathbf{\Omega}}_*$, indicating the direction of the incoming field particle, is restricted to the sets:

$$\begin{pmatrix} \hat{D}_* \\ \check{D}_* \end{pmatrix} = \left\{ \hat{\mathbf{\Omega}}_* \in \mathbb{S}^2 \,\middle|\, -\frac{v'v'_*}{v\,v_*} \mp \frac{\epsilon^2}{2vv_*} \le \hat{\mathbf{\Omega}} \cdot \hat{\mathbf{\Omega}}_* \le \frac{v'v'_*}{v\,v_*} \mp \frac{\epsilon^2}{2vv_*} \right\} :$$

with $\epsilon^2 = 2q\delta/m$. For elastic collisions ($\epsilon = 0$), the inner product $\hat{\mathbf{\Omega}} \cdot \hat{\mathbf{\Omega}}_*$ is symmetric with respect to zero. In this case we define $D_* = \hat{D}_* = \check{D}_*$. For inelastic excitation ($\hat{D}_*$) and de–excitation ($\check{D}_*$) processes, this is no longer true.

Integrating the continuous Boltzmann equations over one energy interval I_i leads, after some algebra, to a set of semi-continuous kinetic equations:

$$\frac{\partial f_i}{\partial t} + v_i \hat{\mathbf{\Omega}}\,\frac{\partial f_i}{\partial \boldsymbol{x}} = \mathcal{J}_i + J_i^{AA} + J_i^{AB} + J_i^{AB^*}\,, \tag{5a}$$

$$\frac{\partial \check{f}_i}{\partial t} + v_i \hat{\mathbf{\Omega}}\,\frac{\partial \check{f}_i}{\partial \boldsymbol{x}} = \check{\mathcal{J}}_i - \mathcal{R}_i(I) + J_i^{BB} + J_i^{BA} + J_i^{BB^*}\,, \tag{5b}$$

$$\frac{\partial \hat{f}_i}{\partial t} + v_i \hat{\mathbf{\Omega}}\,\frac{\partial \hat{f}_i}{\partial \boldsymbol{x}} = \hat{\mathcal{J}}_i + \mathcal{R}_i(I) + J_i^{B^*B^*} + J_i^{B^*A} + J_i^{B^*B}\,, \tag{5c}$$

where $f_i \equiv f_i^A$, $\check{f}_i \equiv f_i^B$ and $\hat{f}_i \equiv f_i^{B^*}$ are the discretized distribution functions of the species a, B and B^*. The semi-continuous inelastic collision terms are, e.g., of the form

$$\check{\mathcal{J}}_i[f, \check{f}, \hat{f}] = C_\chi^2 \sum_{j=0}^{n} v_j \sum_{\substack{h,k=0 \\ h+k=i+j-q}}^{n} \int_0^{2\pi} d\vartheta \int_{\hat{D}_*(v_i, v_j, v_h)} d\hat{\mathbf{\Omega}}_*\, \hat{A}_{ij}^{hk}(\hat{\mathbf{\Omega}} \cdot \hat{\mathbf{\Omega}}_*)(\hat{f}'_h\, f'_{*k} - \check{f}_i\, f_{*j})\,, \tag{6a}$$

where $C_\chi = \delta/m$. From a physical point of view, these operators describe the hopping of the gas particles from one energy group to another due to binary collisions. In leaving a continuous set of allowed directions of velocities, semi–continuous models provide a larger and more realistic set of possible outcomes of binary collision processes than discrete velocity models. The kernels are given by

$$\hat{A}^{hk}_{ij}(\hat{\mathbf{\Omega}} \cdot \hat{\mathbf{\Omega}}_*) = \frac{4g}{g^-} \frac{\hat{\sigma}(g)}{R} \ . \tag{7}$$

In this formula, the quantities g, g^- and R have to be evaluated at the speeds corresponding to the energies E_i,

$$g = \sqrt{v_i^2 + v_j^2 - 2v_i v_j \hat{\mathbf{\Omega}} \cdot \hat{\mathbf{\Omega}}_*} \ , \qquad R = \sqrt{v_i^2 + v_j^2 + 2v_i v_j \hat{\mathbf{\Omega}} \cdot \hat{\mathbf{\Omega}}_*} \tag{8}$$

and so on for g^-. The minus sign refers to the relative speed after an excitation process.

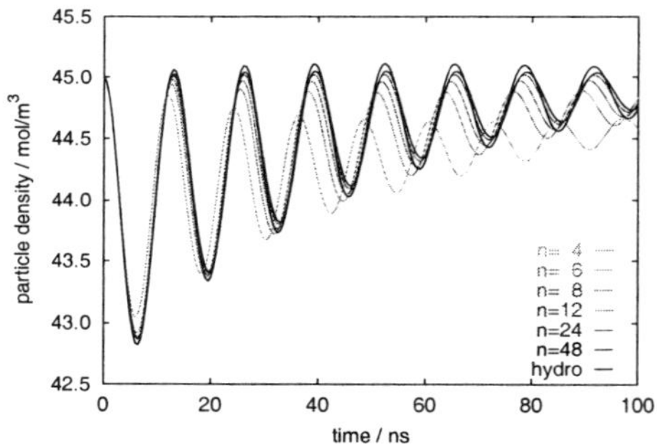

Figure 6. Dependence of the evolution of the particle density in the middle of the fringe spacing on the number of energy groups at a pressure of 1 bar ($Kn = 0.04$).

Dealing with spatially dependent problems requires not only the treatment of the speed variable of the distribution function, but also the direction dependence of the particle velocities. Since we are treating a one–dimensional problem in real space, we expand the distribution functions for each discretized speed in terms of Legendre polynomials (P_N-Approximation [12]) with respect to the angle θ between the velocity and the direction vector in real space:

$$f_i(\hat{\mathbf{\Omega}}, \boldsymbol{x}, t) = f_i(\cos\theta, z, t) = \sum_{l=0}^{N} a_l^{(i)}(z,t) P_l(cos\theta) \ .$$

Inserting it into the semi–continuous Boltzmann equation and projecting it over the Legendre polynomials up to order $N = 3$, results in a closed, coupled set of nonlinear PDEs for the moments of the expansion. The set of truncated moment equations displays conservation of mass, momentum and energy.

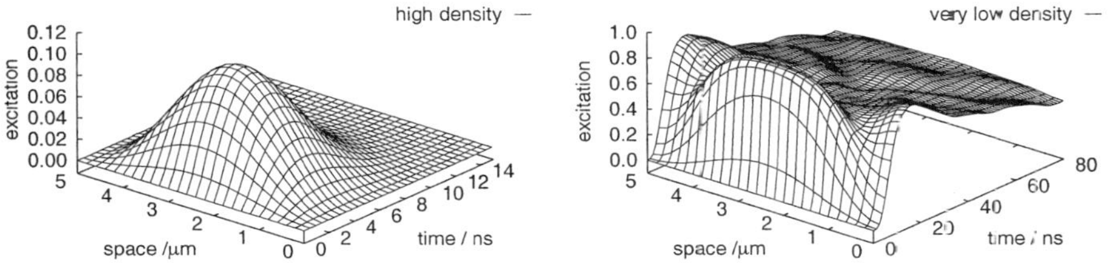

Figure 7. Temporal evolution of the excitation of B^* (fraction B^*/B) within one fringe spacing at a high density of the gas mixture (left plot, $Kn = 0.012$) and at a very low density (right plot, $Kn = 12.2$).

We solve the resulting system numerically by means of the operator splitting method. The first simulation, based on a P_1-Approximation, represents a comparison with results obtained by solving the fluid dynamic equations. Furthermore, we study the influence of the number n of energy groups on the evolution of the gas. Figure 6 shows a striking dependence of the density oscillations in the middle of the fringe spacing on this quantity as long as the number n is small. However, the calculated curves tend towards a common limit when we increase the number of energy groups. We find out that 12 groups yield a fairly good approximation. The solid limit line displays the fluid dynamic solution, which agrees well with the kinetic one.

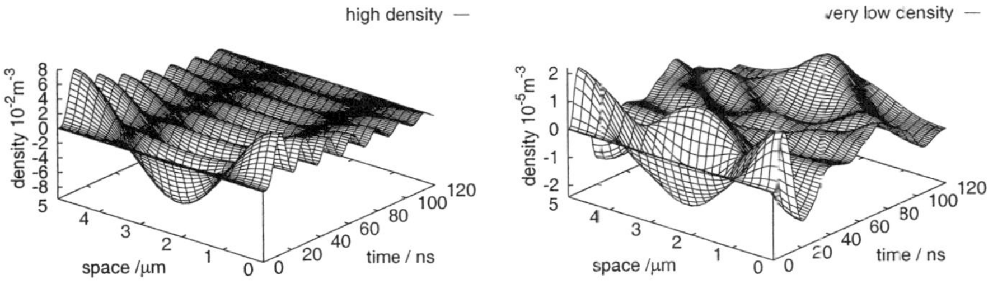

Figure 8. Temporal evolution of the particle density deviation within one fringe spacing at a high density of the gas mixture (left plot, $Kn = 0.012$) and at a very low density (right plot, $Kn = 12.2$).

In order to illustrate different aspects of the dynamics of transient gratings, the following simulations in P_3-approximation have been run at different

Knudsen numbers. Figure 7 shows the excitation (fraction B^*/B) within one fringe spacing in the course of time. For high densities of the gas mixture (left plot) the excitation is less pronounced (up to 12%) and follows exactly the shape of the laser pulse. For low and very low densities the saturation of the transition $B \to B^*$ plays a dominant role. For this reason, a plateau of high excitation (almost 100%) is formed in the center of the fringe spacing.

The oscillation of the particle density is most pronounced at high densities (left plot in Fig. 8) and becomes strongly damped when the gas is rarefied. At very low densities (Knudsen gas), however, the oscillation regenerates because the behavior of the gas is dominated by the cyclic boundary conditions (right plot in Fig. 8).

Recapitulating, we would like to point out that the simulations using discrete velocities or a semi-continuous model provide a valid picture of the physics of thermal gratings for high Knudsen numbers as well as in the fluid dynamic limit.

References

1. P.H. Paul, R.L. Farrow, and P.M. Danehy, *J. Opt. Soc. Am.* B **12**, 384 (1995).
2. E.B. Cummings, *Laser-Induced Thermal Acoustics*, PhD thesis, (California Institute of Technology, Pasadena, 1995).
3. C. Cercignani, *The Boltzmann Equation and Its Applications*, (Springer, New York, 1988).
4. J.O. Hirschfelder, C.F. Curtiss, and R.B. Bird, *Molecular Theory of Gases and Liquids*, (Wiley, New York, 1964).
5. C.R. Garibotti and G. Spiga, *J. Phys.* A **27**, 2709 (1994).
6. A. Rossani, G. Spiga, and R. Monaco, *Mech. Res. Comm.* **24**, 237 (1997).
7. R. Monaco and L. Preziosi, *Fluid Dynamic Application of the Discrete Boltzmann Equation*, (World Scientific, Singapore, 1991).
8. V. Shizuta and S. Kawashima, *Proc. J. Acad.* Ser. A **61**, 252 (1985).
9. F. Hanser, W. Koller, and F. Schürrer, *Phys. Rev.* E **61**, 2065 (2000).
10. L. Preziosi and E. Longo, *Japan J. Indust. Appl. Math.* **14**, 399 (1997).
11. W. Koller, F. Hanser, and F. Schürrer, *J. Phys.* A **33**, 3417 (2000).
12. W. Koller and F. Schürrer, *Trans. Th. Stat. Phys.* **30**, 471 (2001).

LINEARIZATION AND SOLUTIONS OF A SIMPLIFIED MODEL FOR REACTING MIXTURES

M. SENTHILVELAN

School of Physics, The University of Sydney, NSW 2006, Australia

M. TORRISI

Dipartimento di Matematica ed Informatica, Via. A. Doria 6, Catania, Italy
E-mail: torrisi@dmi.unict.it

After recalling some recent results on the potential symmetries of a simplified nonlinear model for a reacting mixture we linearize the system and derive new solutions.

1 Introduction

We consider the following system, say $R\{t, x, u, q\}$,

$$u_t + \left(\frac{u^2}{2} - \alpha q - \beta u_x\right)_x = 0,$$
$$q_x - \gamma q f(u) = 0, \tag{1}$$

which represents a simplified model for a binary unimolecular irreversible exothermic reacting mixture.

- In the above system $R\{t, x, u, q\}$

    1. u denotes a lumped variable with some features of pressure or temperature
    2. q is mass fraction of the reactant
    3. $\alpha > 0$ is heat released during the reaction
    4. $f(u)$ is a structure function derived through asymptotic considerations
    5. $\beta > 0$ is a lumped diffusion coefficient
    6. $\gamma > 0$ is the reaction rate
    7. t can be viewed as a time variable along the characteristics
    8. x is a generalized coordinate which represents space-time of the reaction zone.

- The system $R\{t,x,u,q\}$ also represents an unidimensional specialization of a simplified asymptotic model, see Rosales and Majda[1] and Majda [2], which describes non-linear detonation waves in a chemically reacting fluid mixture which propagate with wave speeds closer to acoustical sound speed. In this regime the chemical and fluid mechanical phenomena interact with each other. The equations of the above model have been derived from Navier -Stokes equations through multiple scaling and matched asymptotic expansions.

- Auxiliary conditions: the initial data of $u(t,x)$ and signalling data of $q(t,x)$ at $x=+\infty$ should satisfy

$$\begin{aligned} u(0,x) &= u_0(x), \qquad -\infty < x < +\infty, \\ \lim_{x\to+\infty} q(t,x) &= q_0(t), \qquad\qquad t \geq 0, \end{aligned} \tag{2}$$

  where $u_0(x)$ and $q_0(t)$ are prescribed functions. Assigning data (2) to (1) leads to a well posed problem [1].

Recently some attempts were made to study Eqs. (1) numerically for some special cases of $R\{t,x,u,q\}$ with the auxiliary conditions given above [3].

Very recently the present authors have carried out a detailed study on the invariance properties of (1) and shown that it admits potential symmetries for a specific form when

$$f(u) = \frac{u}{2\beta\gamma} + k,$$

where k is an arbitrary constant (see [4] and [5]).

We note that the above form of $f(u)$ is consistent with the available forms for rate functions in condensed explosives, see for example [6] and [7].

In this short paper we show some results found in [4] and [5]. The plan of the paper is as follows. In Sec. 2 the potential symmetries are derived and an invertible map is found which put the system in linear form in the case $k=0$. In Secs. 3 and 4, using the usual reduction techniques, we show new solutions for the system (1).

2 Potential symmetries and linearization

Introducing a **potential variable**, v, as a further unknown function, we can rewrite $R\{t,x,u,q\}$ as another system, **auxiliary system**, say $S\{t,x,u,v,q\}$

$$v_x = u,$$

$$v_t = -\frac{u^2}{2} + \alpha q + \beta u_x, \qquad q_x = \gamma q f(u). \tag{3}$$

A point symmetry of the auxiliary system which is not projectable to the space $\{t, x, u, v, q\}$ is a potential symmetry for $R\{t, x, u, q\}$.

These (local) symmetries of $S\{t, x, u, v, q\}$ induce non-local symmetries (i.e. symmetries whose infinitesimal generators depend on integrals of the dependent variables in some specific manner) for the original system $R\{t, x, u, q\}$. These kinds of non-local symmetries have been called, by Bluman and Kumei, as potential symmetries [8,9,10,11].

A potential symmetry of the system under consideration, $R\{t, x, u, q\}$, is a point symmetry of the system $S\{t, x, u, v, q\}$ so one can extend the uses of point symmetries to the potential symmetries. In particular:

i) If $S\{t, x, u, v, q\}$ is linearizable then $R\{t, x, u, q\}$ is also linearizable.

ii) Invariant solutions of $S\{t, x, u, v, q\}$ give solutions of $R\{t, x, u, q\}$. These solutions, of course, are not invariant solutions of any local symmetry admitted by $R\{t, x, u, q\}$.

The invariance of $S\{t, x, u, v, q\}$ under the one-parameter Lie group of infinitesimal transformations [8, 12] leads to the following point symmetry classification [4] :

Case: 1 $f(u) =$ *an arbitrary function.*
In this case the associated Lie algebra is the *principal Lie algebra* $L_{\mathcal{P}}$ which is three-dimensional and spanned by

$$X_0 = \partial_t, \quad X_1 = \partial_x, \quad X_2 = \partial_v.$$

Case: 2 $f(u) =$ *constant* $= f_0$.
In this case we obtain one more generator, in addition to $L_{\mathcal{P}}$

$$X_3 = t\partial_x + \partial_u + x\partial_v.$$

The infinitesimal operators $X_0, X_1, X_2,\ X_3$ are projectable to the space $\{t, x, u, q\}$ and so become point symmetries for the system $R\{t, x, u, q\}$.

Case: 3 $f(u) = \frac{u}{2\beta\gamma} + k$, where k is an arbitrary constant.
In this case we have three more generators in addition to $L_{\mathcal{P}}$ of the form

$$\begin{aligned} X_4 &= \left(\frac{x}{2} - k\beta\gamma t\right)\partial_x + t\partial_t - \left(\frac{u}{2} + k\beta\gamma\right)\partial_u \\ &\quad -k\beta\gamma x\partial_v - q\partial_q, \\ X_{c_1(t,x)} &= \left[c_{1x}(t,x) + \frac{u}{2\beta}c_1(t,x)\right] e^{\frac{v}{2\beta}}\partial_u \end{aligned} \tag{4}$$

$$+c_1(t,x)e^{\frac{v}{2\beta}}\partial_v + \frac{q}{2\beta}c_1(t,x)e^{\frac{v}{2\beta}}\partial_q,$$

$$X_{\chi(t)} = \frac{1}{\alpha}\chi(t)e^{\frac{v}{2\beta}+k\gamma x}\partial_q,$$

where $c_1(t,x)$ is a solution of the following inhomogeneous linear diffusion equation

$$c_{1t}(t,x) - \beta c_{1xx}(t,x) = \chi(t)e^{k\gamma x}, \tag{5}$$

and $\chi(t)$ is an arbitrary function of t.

The infinitesimal operator X_4 is projectable to the space $\{t,x,u,q\}$ while $X_{c_1(t,x)}$ and $X_{\chi(t)}$ are not.

The latter two infinitesimal operators are the desired potential symmetries for (1).

This result allows us to find an invertible mapping which linearizes the system $S\{t,x,u,v,q\}$ through the theorem given in [8]. This theorem (recently considered in [13,14]) gives a necessary and sufficient condition for the existence of an invertible mapping which linearizes a system.

For our following analysis we take $k = 0$. We mention that, this choice matches with some forms of rate functions considered in condensed high explosives [7]. Also it provides a good vehicle for analytic calculations.

In this case the infinitesimal operator generating the infinite parameter group of transformations is

$$\hat{X} = \left[\frac{u}{2\beta}c_1(t,x) + c_{1x}(t,x)\right]e^{\frac{v}{2\beta}}\partial_u$$
$$+c_1(t,x)e^{\frac{v}{2\beta}}\partial_v + \left[\frac{q}{2\beta}c_1(t,x) + \frac{1}{\alpha}\chi(t)\right]e^{\frac{v}{2\beta}}\partial_q. \tag{6}$$

Following the algorithm given in [8] we found the following overdetermined system of PDEs for the unknown $\Phi(t,x,u,v,q)$

$$\frac{u}{2\beta}\Phi_u + \Phi_v + \frac{q}{2\beta}\Phi_q = 0, \quad \Phi_u = 0, \quad \Phi_q = 0 \tag{7}$$

whose two independent solutions, say z_1 and z_2, forms new independent variables for the linearizing transformation. We choose $z_1 = t$ and $z_2 = x$.

The transformation for the remaining variables can be obtained from the solutions of the following system of PDEs for the unknowns $\psi_i(t,x,u,v,q)$, $i = 1,2,3$,

$$e^{\frac{v}{2\beta}}\left(\frac{u}{2\beta}\psi_{1u} + \psi_{1v} + \frac{q}{2\beta}\psi_{1q}\right) = 1 \quad \psi_{1u} = 0, \quad \psi_{1q} = 0,$$

$$\frac{u}{2\beta}\psi_{2u} + \psi_{2v} + \frac{q}{2\beta}\psi_{2q} = 0,\ e^{\frac{v}{2\beta}}\psi_{2u} = 1,\ \psi_{2q} = 0, \tag{8}$$

$$\frac{u}{2\beta}\psi_{3u} + \psi_{3v} + \frac{q}{2\beta}\psi_{3q} = 0, \quad \psi_{3u} = 0, \quad \psi_{3q} = \alpha e^{-\frac{v}{2\beta}}.$$

Particular independent solutions of Eqs (8) can be easily found

$$\psi_1 = -2\beta e^{-\frac{v}{2\beta}}, \quad \psi_2 = ue^{-\frac{v}{2\beta}}, \quad \psi_3 = \alpha qe^{-\frac{v}{2\beta}}. \tag{9}$$

As a consequence we obtain a linearizing transformation

$$z_1 = t, \quad z_2 = x, \quad \psi_1 = -2\beta e^{-\frac{v}{2\beta}}, \quad \psi_2 = ue^{-\frac{v}{2\beta}}, \quad \psi_3 = \alpha qe^{-\frac{v}{2\beta}}. \tag{10}$$

Substituting the transformation (10) in (3) with $f(u) = \frac{u}{2\beta\gamma}$ we get a linear system

$$\psi_{1z_2} - \psi_2 = 0, \qquad \psi_{1z_1} - \beta\psi_{2z_2} - \psi_3 = 0, \qquad \psi_{3z_2} = 0. \tag{11}$$

It is a simple matter to verify that the system $R\{t, x, u, q\}$ can also be linearized.

Further, we note that Eqs. (11) can be rewritten as a single second order PDE

$$\psi_{1t}(t,x) - \beta\psi_{1xx}(t,x) = \mu(t), \tag{12}$$

where $\mu(t)$ is an arbitrary function of t and $z_1 = t, z_2 = x$. By introducing a new function

$$\Psi(t,x) = \psi_1(t,x) - \int \mu(t)dt, \tag{13}$$

Eq. (12) can be brought to the form

$$\Psi_t - \beta\Psi_{xx} = 0. \tag{14}$$

In this way the hunt for solutions of Eqs. (1) with auxiliary conditions (2) is reduced to the search for solutions of Eq. (14) with an initial condition

$$\Psi(0,x) = \Psi_0(x) \tag{15}$$

while $\mu(t)$ is determined from the signalling data of $q(x,t)$.

3 Similarity reductions

The most general operator for Case 3 is given by

$$X = b_0X_0 + b_1X_1 + b_2X_2 + a_1X_4 + X_{c_1(t,x)} + X_{\chi(t)} \tag{16}$$

with b_0, b_1, b_2, a_1 arbitrary constants.

The characteristic equations associated with the infinitesimal operator (16), when $k = 0$, can be written as

$$\frac{dx}{b_1 + \frac{a_1}{2}x} = \frac{dt}{b_0 + a_1 t} = \frac{du}{c_{1x}(t,x)e^{\frac{v}{2\beta}} + [\frac{1}{2\beta}c_1(t,x)e^{\frac{v}{2\beta}} - \frac{a_1}{2}]u}$$

$$= \frac{dv}{c_1(t,x)e^{\frac{v}{2\beta}} + b_2} = \frac{dq}{\frac{1}{\alpha}\chi(t)e^{\frac{v}{2\beta}} + [\frac{1}{2\beta}c_1(t,x)e^{\frac{v}{2\beta}} - a_1]q}. \tag{17}$$

Eq. (5) becomes

$$c_{1t}(t,x) - \beta c_{1xx}(t,x) = \chi(t). \tag{18}$$

We proceed our analysis with the general solution

$$c_1(t,x) = \bar{c}_1(t,x) + \int \chi(t)dt, \tag{19}$$

where $\bar{c}_1(t,x)$ is the solution of the homogeneous part of Eq. (18).

We have solved Eq. (17) in [5] by distinguishing

$$(i)\ a_1 = 0 \quad and \quad (ii)\ a_1 \neq 0.$$

Here we show only the case $a_1 \neq 0$. In this case we get similarity variables

$$\begin{aligned}
z &= \frac{x}{(b_0 + a_1 t)^{(1/2)}} + \frac{2b_1}{a_1(b_0 + a_1 t)^{(1/2)}}, \\
w_1 &= u(b_0 + a_1 t)^{\frac{1}{2}} \left[w_2 - \frac{1}{2\beta}\int \left[\bar{\bar{c}}_1(t,z) + \int \chi(t)dt\right](b_0 + a_1 t)^{A-1}dt\right] \\
&\quad - \int \bar{\bar{c}}_{1z}(t,z)(b_0 + a_1 t)^{A-1}dt, \qquad (20) \\
w_2 &= (b_0 + a_1 t)^A e^{-\frac{v}{2\beta}} + \frac{1}{2\beta}\int \left[\bar{\bar{c}}_1(t,z) + \int \chi(t)dt\right](b_0 + a_1 t)^{A-1}dt, \\
w_3 &= (b_0 + a_1 t)q \left[w_2 - \frac{1}{2\beta}\int \left[\bar{\bar{c}}_1(t,z) + \int \chi(t)dt\right](b_0 + a_1 t)^{A-1}dt\right] \\
&\quad - \frac{1}{\alpha}\int \chi(t)(b_0 + a_1 t)^A dt,
\end{aligned}$$

where $A = \frac{b_2}{2\beta a_1}$ and $\bar{\bar{c}}_1(t,z)$ is a function obtained from $\bar{c}_1(t,x)$, solution of (18), by replacing x in terms of z and t.

Using (20) we can reduce Eqs (3) to the form

$$2\beta w_2' + w_1 = 0, \tag{21}$$

$$a_1\beta z w_2' - \beta w_1' + b_2 w_2 - \alpha w_3 = 0, \tag{22}$$

$$w_3' = 0, \tag{23}$$

where the prime denotes differentiation with respect to the variable z.

4 Solutions of the reduced system

4.1 $b_2 \neq 0$

Substituting $w_3 = I_1$ and Eq. (21) in (22) we get a linear second order ODE in the variable w_2

$$w_2'' + \frac{a_1}{2\beta} z w_2' + \frac{b_2}{2\beta^2} w_2 - \frac{\alpha I_1}{2\beta^2} = 0. \tag{24}$$

Rewriting $\frac{b_2}{2\beta^2} w_2 - \frac{\alpha I_1}{2\beta^2} = \bar{w}_2$, we get

$$\bar{w}_2'' + \frac{a_1}{2\beta} z \bar{w}_2' + \frac{b_2}{2\beta^2} \bar{w}_2 = 0. \tag{25}$$

Introducing $\bar{w}_2(z) = y(s)$, $2s = -\frac{a_1}{2\beta} z^2$ we can rewrite Eq. (25) as

$$s y'' + (\frac{1}{2} - s) y' - A y = 0. \tag{26}$$

where $A = \frac{b_2}{2\beta a_1}$.

Eq. (26) is the well known confluent hypergeometric equation [15,16] whose general solution (in terms of z) can be written in the form :

$$\begin{aligned} w_2 = {} & \frac{\alpha I_1}{b_2} + \frac{2\beta^2 I_2}{b_2} \Phi(A, \frac{1}{2}; (\frac{-a_1 z^2}{4\beta})) \\ & + \frac{2\beta^2 I_3}{b_2} (\frac{-a_1 z^2}{4\beta})^{\frac{1}{2}} \Phi(A + \frac{1}{2}, \frac{3}{2}; (\frac{-a_1 z^2}{4\beta})), \end{aligned} \tag{27}$$

where I_2 and I_3 are integration constants and $\Phi(a, c, s)$ is the function of Pochhammer-Kummer defined as

$$\Phi(a, c, s) = \sum_{n=0}^{\infty} c_n s^n$$

with

$$c_{n+1} = \frac{n+a}{(n+1)(n+c)} c_n, \qquad c_0 = 1. \tag{28}$$

Using (27) in (21), we obtain an explicit form for w_1. Substituting w_1, w_2 and w_3 in (20), we obtain an explicit solution for the system (3) which in turn leads to the solution for (1). For sake of simplicity, we omit the derivations here.

4.2 $b_2 = 0$

Substituting $w_3 = I_1$ and Eq. (21) in (22) we get a linear first order ODE for the unknown w_1

$$2\beta w_1' + a_1 z w_1 + 2\alpha I_1 = 0, \tag{29}$$

whose general solution can be found easily

$$w_1 = e^{\frac{-a_1}{4\beta} z^2} \left(I_2 - \frac{\alpha I_1}{\beta} \int_{z_0}^{z} e^{\frac{a_1}{4\beta}\tau^2} d\tau \right). \tag{30}$$

From (30) we can determine the form of w_2 through Eq. (21). Substituting the expressions in (20) we arrive at finally

$$\begin{aligned} u &= \frac{e^{\frac{-a_1}{4\beta} z^2} \left(I_2 - \frac{\alpha I_1}{\beta} \int_{z_0}^{z} e^{\frac{a_1}{4\beta}\tau^2} d\tau \right) + \int \bar{\bar{c}}_{1z}(t,z)(b_0 + a_1 t)^{A-1} dt}{(b_0 + a_1 t)^{\frac{1}{2}} \left[I_3 - \frac{1}{2\beta} \int_{z_0}^{z} e^{\frac{-a_1}{4\beta}\theta^2} \left(I_2 - \frac{\alpha I_1}{\beta} \int_{\theta_0}^{\theta} e^{\frac{a_1}{4\beta}\tau^2} d\tau \right) d\theta - M(t,x) \right]}, \\ v &= 2\beta log \frac{(b_0 + a_1 t)^A}{\left[I_3 - \frac{1}{2\beta} \int_{z_0}^{z} e^{\frac{-a_1}{4\beta}\theta^2} \left(I_2 - \frac{\alpha I_1}{\beta} \int_{\theta_0}^{\theta} e^{\frac{a_1}{4\beta}\tau^2} d\tau \right) d\theta - M(t,x) \right]}, \\ q &= \frac{I_1 + \frac{1}{\alpha} \int \chi(t)(b_0 + a_1 t)^A dt}{(b_0 + a_1 t) \left[I_3 - \frac{1}{2\beta} \int_{z_0}^{z} e^{\frac{-a_1}{4\beta}\theta^2} \left(I_2 - \frac{\alpha I_1}{\beta} \int_{\theta_0}^{\theta} e^{\frac{a_1}{4\beta}\tau^2} d\tau \right) d\theta - M(t,x) \right]}, \end{aligned} \tag{31}$$

where

$$\begin{aligned} z &= \frac{x}{(b_0 + a_1 t)^{(1/2)}} + \frac{2b_1}{a_1 (b_0 + a_1 t)^{(1/2)}}, \\ M(t,x) &:= \frac{1}{2\beta} \int \left[\bar{\bar{c}}_1(t,z) + \int \chi(t) dt \right] (b_0 + a_1 t)^{A-1} dt. \end{aligned}$$

Acknowledgements: MT was supported by G.N.F.M, University of Catania and MURST Project: *Non Linear Mathematical Problems of Wave Propagation and Stability in Models of Continuous Media.*

References

1. Rosales R and Majda A, 1983, SIAM J. Appl. Math., **43**, 1086.
2. Majda A, 1986, in *Reacting Flows: Combustion and Chemical Reactors* AMS Lectures in Applied Mathematics **24**, 109.
3. Majda A and Roytburd V 1988 in *Computational Fluid Dynamics and Reacting Gas Flows* IMA Volumes in Math. and its Appl., (New York, Springer), **12**, 195.
4. M Senthilvelan and M Torrisi, 2000, J. Phys. A: Math. Gen., **33**, 405.
5. M Senthilvelan and M Torrisi, 2002, Nonlinear Dynamics, *On Certain New Solutions of a Simplified Model for Reacting Mixtures,* (to appear).
6. G Kanel' and A Dremin, 1977, Fizika Goreniya i Vzryva, **13**, 85.
7. Stewart D S 1986 in *Reacting Flows: Combustion and Chemical Reactors* AMS Lectures in Applied Mathematics, **24**, 403.
8. Bluman G W and Kumei S, 1989, *Symmetries and Differential Equations* (New York, Springer).
9. Bluman G W and Kumei S, 1990, Eur. J. Appl. Math. **1**, 189.
10. Bluman G W, 1993, in *Lectures in Applied Mathematics*, American Mathematical Society, Providence, RI, p.97.
11. Bluman G W,1993, in *Modern Group Analysis: Advanced Analytical and Computational Methods in Mathematics and Physics*, Kluwer, Dordrecht, 71.
12. Ames W F and Rogers C (Eds.), 1992, *Nonlinear Equations in the Applied Sciences*, Academic Press, Boston.
13. Donato A and Oliveri F, 1994, J. Math. Anal. Appl. **188**, 552.
14. Donato A and Oliveri F, 1996, Appl. Anal. **58**, 313.
15. Murphy G M, 1969, *Ordinary Differential Equations and Their Solutions*, Affiliated East West, New Delhi.
16. Tricomi F, 1952, *Lezioni sulle funzioni ipergeometriche confluenti*, Editrice Gheroni, Torino.

ELASTIC WAVES IN MATERIALS WITH THIN LAYERS

MARIA PAOLA SPECIALE

Department of Mathematics, University of Messina
Contrada Papardo, Salita Sperone 31, I-98166 Sant'Agata, Messina
E-mail: speciale@mat520.unime.it

MAURIZIO BROCATO

IEI–CNR, Area della Ricerca di Pisa – 1, via G. Moruzzi, I-56124 Pisa
and LAMI–ENPC/LCPC, 1, av. B. Pascal
Cité Descartes, Champs sur Marne – F-77455 Marne la Vallée Cedex
E-mail: brocato@iei.pi.cnr.it

We study a material body made of thin and thick layers of two different elastic constituents, under the assumption that layers in the first family have almost vanishing thickness when compared with those in the second family.
We consider a wave propagating through the layers and derive the consequences of our assumptions. A two dimensional case is studied and a particular explict solution is reported.

1 Introduction

In ref. [1] we proposed the continuum model of an heterogeneous stratified body made of thin and thick regularly alternating layers—beds and courses—of two different elastic constituents, of which the softer fills the thin beds. The model was proposed for applications to the study of masonry structures (made of mortar beds and lapideous courses) or sedimentary rocks (when periods of continuous depositions gave rise to some regularly layered pattern).

The reduced thickness of soft beds (if compared to stiff courses) suggested the idea that stresses can be considered as independent of a coordinate taken along their width. Consequently, a special expression for compatible displacement fields within beds was found, which allowed us to represent any thin bed as a surface endowed with a microstructure. The search for minimal potential energy, under imposed tractions at the boundary, among displacement fields of the kind described above gave us the set of equilibrium equations for the body.

Here, through a procedure similar to the one we used in statics, we write the equations of motion of the system under the assumption of small perturbations (see Section 2), then we study the propagation of elastic waves through the layered continuum (Section 3). In section 4 we restrict attention to a two-dimensional case and write explicitly the form of the elastic waves which can propagate through the layers.

Applications can be sought in seismology or in the dynamics of masonry structures.

Let T be the Cauchy stress, u the displacement, ρ the mass per unit volume, μ and λ the Lame's coefficients, $E :=$ symgrad$u$ the strain, ψ the elastic energy per unit volume, whether on thick courses or on thin beds depending on the suffix A or B.

Let us focus on a representative volume element compound of a thin planar bed and half of the two thick courses next to it. The thin bed has middle surface Γ (the plane $x_3 = 0$) and constant thickness 2δ; the exponent $+$ or $-$ denotes objects at either side of Γ.

2 The model

2.1 Kinematics

Due to the assumption that *the stress state is independent on the x_3 coordinate within B*, to the elastic behaviour and to compatibility requirements, the displacement u_B must be of the form ((x_1, x_2) coordinates on Γ, τ is the time):

$$u_B = \tilde{u}_B + w \; : \quad \begin{cases} \tilde{u}_B = \begin{bmatrix} \alpha_3(\tau)x_2 - \frac{\alpha_1(\tau)}{2}x_3 + \gamma_1(\tau) \\ -\alpha_3(\tau)x_1 - \frac{\alpha_2(\tau)}{2}x_3 + \gamma_2(\tau) \\ \alpha_1(\tau)x_1 + \alpha_2(\tau)x_2 + \varepsilon(\tau) \end{bmatrix} x_3 \,, \\ \\ w = w(x_1, x_2, \tau) \end{cases} \tag{1}$$

(α_1, α_2, and α_3, functions of time, representing relative rotations of the planes parallel to Γ, γ_1 and γ_2 uniform shear of fibers normal to Γ, and ε uniform elongation of the same fibers).

The assumption of *perfect contact* between A and B and (1) allow us to write the jump of u_A across B in terms of the parameters of $\tilde{u}_B$, and thus:

$$u_A = \bar{u}_A + 2\delta \begin{bmatrix} \alpha_3 x_2 + \gamma_1 \\ -\alpha_3 x_1 + \gamma_2 \\ \alpha_1 x_1 + \alpha_2 x_2 + \varepsilon \end{bmatrix} \chi, \quad \chi := \begin{cases} \frac{1}{2} \text{ in } V_{A^+} \\ 0 \text{ in } V_B \\ -\frac{1}{2} \text{ in } V_{A^-} \end{cases} . \tag{2}$$

2.2 Equations of motion

The equations of motion are obtained through application of Euler's principle to the requirement that the sum of the elastic and the kinetic energy of the system minus the potential energy of the external actions must be minimum.

Searching the minimizing displacement among fields verifying (1) and (2) we will obtain the equations of motion of the special system described above. The elastic and the kinetic energy are, for this purpose, written neglecting terms of order greater than one in δ (see ref. [1] for further details); *e.g.* the elastic energy of the whole system is:

$$\int_V \psi\, dv = \int_V \psi_A\, d(vol) + 2\delta \int_\Gamma \left(\psi_B - \frac{1}{2}\, E_A|_\Gamma \cdot T_A|_\Gamma \right) d(area)\,,$$

where ψ_A, ψ_B, T_A can be expressed through the laws of elasticity in terms of $\bar{u}_A$, w, and the parameters α_i, γ_i, and ε.

Thus we have standard Navier equations for $\bar{u}_A$:

$$\begin{cases} \mu_A \Delta \bar{u}_A + (\mu_A + \lambda_A)\,\mathrm{grad}(\,\mathrm{div}\bar{u}_A) = \rho_A \ddot{\bar{u}}_A \ \text{ in } V\,, \\ (2\mu_A\,\mathrm{symgrad}\bar{u}_A + \lambda_A\,\mathrm{div}\bar{u}_A I)n = f \qquad \text{on } \partial V\,; \end{cases} \tag{3}$$

the field equation on Γ for w:

$$\begin{gathered} \mu^* \Delta w + (\mu^* + \lambda^*)\,\mathrm{grad}(\,\mathrm{div}w) + \lambda_B\, s = \\ \rho^* \ddot{w}_A + \mu_A\,\mathrm{div}\left(u_{A,3}|_\Gamma\right) n_\Gamma + \lambda_A\,\mathrm{grad}\left(u_{A3,3}|_\Gamma\right), \end{gathered} \tag{4}$$

where $\lambda^* := \lambda_B - \lambda_A$, $\mu^* := \mu_B - \mu_A$, $\rho^* = \rho_B - \rho_A$, $s := (\alpha_1, \alpha_2, 0)$ and $g := (\gamma_1 + \alpha_3 x_2, \gamma_2 - \alpha_3 x_1, 0)$; the conditions on $\partial\Gamma$ for w ($m \cdot n_\Gamma = 0$):

$$\begin{gathered} 2\mu^* m \cdot (\,\mathrm{symgrad}w)m + \lambda^*\,\mathrm{div}w + \lambda_B(\alpha_1 x_1 + \alpha_2 x_2 + \varepsilon) = \lambda_A\, u_{A3,3}|_\Gamma\,, \\ 2\mu^* n_\Gamma \cdot (\,\mathrm{symgrad}w)m + \mu_B g \cdot m = \mu_A\, u_{A,3}|_\Gamma \cdot m\,; \end{gathered} \tag{5}$$

and the two systems of linear equations for α_i, γ_j, and ε:

$$\begin{aligned} \begin{bmatrix} J_{11} & J_{12} & S_1 \\ J_{12} & J_{22} & S_2 \\ S_1 & S_2 & \Gamma \end{bmatrix} \begin{bmatrix} \alpha_1 \\ \alpha_2 \\ \varepsilon \end{bmatrix} &= -\tfrac{\lambda_B}{\lambda_B + 2\mu_B} \textstyle\int_\Gamma (w_{1,1} + w_{2,2}) \begin{bmatrix} x_1 \\ x_2 \\ 1 \end{bmatrix} d(area) \\ &\quad + \tfrac{1}{\lambda_B + 2\mu_B} \textstyle\int_{\partial V} f_3 \begin{bmatrix} x_1 \\ x_2 \\ 1 \end{bmatrix} \chi d(area)\,, \end{aligned} \tag{6}$$

$$\begin{aligned} \begin{bmatrix} J_{11} + J_{22} & S_2 & -S_1 \\ S_2 & \Gamma & 0 \\ -S_1 & 0 & \Gamma \end{bmatrix} \begin{bmatrix} \alpha_3 \\ \gamma_1 \\ \gamma_2 \end{bmatrix} &= -\textstyle\int_\Gamma \begin{bmatrix} w_{3,1} x_2 - w_{3,2} x_1 \\ w_{3,1} \\ w_{3,2} \end{bmatrix} d(area) \\ &\quad + \tfrac{1}{2\mu_B} \textstyle\int_{\partial V} \begin{bmatrix} f_1 x_2 - f_2 x_1 \\ f_1 \\ f_2 \end{bmatrix} \chi d(area)\,; \end{aligned} \tag{7}$$

where:

$$J_{ij} := \textstyle\int_\Gamma x_i x_j\, d(area)\,, \quad S_i := \int_\Gamma x_i\, d(area)\,, \quad \Gamma := \int_\Gamma d(area)\,.$$

3 Wave propagation

Let us take $\bar{u}_A = v_A \sin(k_A \cdot x - \chi_A \tau)$, with v_A wave amplitude vector, k_A—such that $|k_A| = 1$—direction of propagation, and χ_A velocity of propagation of the wave, such that:

- $\chi_A = \sqrt{\frac{\mu_A}{\rho_A}}$ and $v_A \cdot k_A = 0$ ($\bar{u}_A$ is a transversal wave),
- $\chi_A = \sqrt{\frac{\lambda_A + 2\mu_A}{\rho_A}}$ and $v_A \times k_A = 0$ ($\bar{u}_A$ is a longitudinal wave).

Let us also take $w = w^{(f)}(\omega^{(f)}) + w^{(b)}(\omega^{(b)})$, where $\omega^{(f)} = (k_w \cdot x - \chi_w \tau)$ and $\omega^{(b)} = (k_w \cdot x + \chi_w \tau)$ with k_w—such that $|k_w| = 1$—direction and χ_w velocity of propagation of the wave through Γ. Equation (4) becomes (let $\tilde{k}_A = (k_{A1}, k_{A2}, 0)$)

$$G(w^{(f)\prime\prime} + w^{(b)\prime\prime}) + \lambda_B\, s = -h k_{A3} \sin(\tilde{k}_A \cdot x - \chi_A \tau) \cos(k_{A3}\delta)\,, \tag{8}$$

where

$$G = (\mu^* - \rho^* \chi_w^2) I + (\mu^* + \lambda^*)(k_w \otimes k_w)\,, \quad h = \mu_A (v_A \cdot \tilde{k}_A) n_\Gamma + \lambda_A v_{A3} \tilde{k}_A\,,$$

and one can distinguish three interesting cases (the case $G = 0$ will not be considered here):

1. $k_{A1} \neq 0$, $k_{A2} \neq 0$, $\cos(k_{A3}\delta) \neq 0 \Rightarrow k_{A1}k_{w2} - k_{A2}k_{w1} = 0$ (*i.e.* the waves propagate in A and B along the same direction): we obtain $s = s^{(f)}(\omega^{(f)}) + s^{(b)}(\omega^{(b)})$ with $s^{(f)} = a^{(0)}\omega^{(f)} + a^{(f)}$ and $s^{(b)} = -a^{(0)}\omega^{(b)} + a^{(b)}$, $a^{(0)}$, $a^{(f)}$ and $a^{(b)}$ constant vectors, and
$$\chi_w = \pm \frac{k_{w1}}{k_{A1}} \chi_A\,;$$
2. $\cos(k_{A3}\delta) = 0 \Rightarrow$ (8) becomes $G(w^{(f)\prime\prime} + w^{(b)\prime\prime}) + \lambda_B s = 0$ $(\forall h)$,
3. $k_{A1} = k_{A2} = 0 \Rightarrow$ (8) becomes $G(w^{(f)\prime\prime} + w^{(b)\prime\prime}) + \lambda_B s = 0$ and $h = 0$.

Integration of equation (8) can be performed, in any case, to get $w(\omega^{(f)}, \omega^{(b)})$—for the sake of shortness the result will be shown in two dimensions in §4. Then, all terms at the right-hand side of system (6) and (7) will be known and the parameters α_i, γ_j, and ε can be determined after integration on Γ of w and on ∂V of the force vector f which, due to the boundary condition on u_A, is

$$f = (2\mu_A \operatorname{sym}(v_A \otimes k_A) + \lambda_A (v_A \cdot k_A) I)\, n \cos(k_A \cdot x - \chi_A \tau). \tag{9}$$

4 Two dimensional case

Let us restrict attention to a two-dimensional case (looking for solutions which are independent of x_2), assume the origin of coordinates x_1 to be in the center of Γ, and consider a periodic boundary condition for the continuous field $\bar{u}_A$ on the opposite boundaries of the system parallel to Γ; the three different circumstances listed in §3 admit the following solutions:

1. If $\chi_w = \frac{\chi_A}{k_{A1}}$ it is

$$
\begin{aligned}
w_1 &= \tfrac{1}{G_{11}} \Big(\tfrac{k_{A3}}{k_{A1}^2} h_1 \sin(k_{A1}\omega^{(f)}) \cos(k_{A3}\delta) \\
&\quad - b_1^{(1)} (\omega^{(f)2} - \omega^{(b)2}) + b_1^{(2)} \omega^{(f)} + b_1^{(3)} \omega^{(b)} + b_1^{(4)} \\
&\quad - \lambda_B \big(\tfrac{a_1^{(0)} (\omega^{(f)3} - \omega^{(b)3})}{6} + \tfrac{a_1^{(f)} \omega^{(f)2} + a_1^{(b)} \omega^{(b)2}}{2} \big) \Big) , \qquad (10) \\
w_3 &= \tfrac{1}{G_{33}} \Big(\tfrac{k_{A3}}{k_{A1}^2} h_3 \sin(k_{A1}\omega^{(f)}) \cos(k_{A3}\delta) \\
&\quad - b_3^{(1)} (\omega^{(f)2} - \omega^{(b)2}) + b_3^{(2)} \omega^{(f)} + b_3^{(3)} \omega^{(b)} + b_3^{(4)} \Big) ,
\end{aligned}
$$

where $G_{33} = \mu^* - \rho^* \chi_w$, $G_{11} = 2\mu^* + \lambda^* - \rho^* \chi_w$ and $b^{(i)}$ $(i = 1, \dots 4)$ are arbitrary constant vectors; otherwise, if $\chi_w = -\frac{k_{w1}}{k_{A1}} \chi_A$, w can be written as in (10) by permutation of $\omega^{(f)}$ and $\omega^{(f)}$. Integration of (6) and (7), leads to:

$$
\gamma_1 = \Big(\tfrac{\mu_A}{\mu_B k_{A1}} (v_{A1} k_{A3} + v_{A3} k_{A1}) [\sin(k_{A1} x_1 - \chi_A \tau)]_0^\Gamma - [w_3]_0^\Gamma \Big) \tfrac{1}{\Gamma} ,
$$

$$
a_1^{(0)} = 0 , \qquad a_1^{(f)} = -a_1^{(b)} , \qquad (11)
$$

$$
\epsilon = \tfrac{12}{(\lambda_B + 2\mu_B)\Gamma} \big(\tfrac{\phi_1}{3} - \tfrac{\phi_2}{2\Gamma} \big) ,
$$

where

$$
\begin{aligned}
\phi_1 &= \frac{v_{A1} k_{A1} + (2\mu_A + \lambda_A) v_{A3} k_{A3}}{k_{A1}} [\sin(k_{A1} x_1 - \chi_A \tau)]_0^\Gamma - \lambda_B [w_1]_0^\Gamma \Gamma , \\
\phi_2 &= \frac{v_{A1} k_{A1} + (2\mu_A + \lambda_A) v_{A3} k_{A3}}{k_{A1}^2} [\cos(k_{A1} x_1 - \chi_A \tau) \\
&\quad - k_{A1} x_1 \sin(k_{A1} x_1 - \chi_A \tau)]_0^\Gamma - \lambda_B \int_0^\Gamma w_{1,1} x_1 dx_1 .
\end{aligned}
$$

2. If $\cos(k_3\delta), = 0$ solutions have the form written in equation (10), but with the first term of sums missing.

3. If $k_{A1} = 0$ solutions are again as in the previous case.

Acknowledgments.

Work supported by the Research Project "Problemi matematici non lineari di propagazione e stabilità nei modelli del continuo", Cofin. M.U.R.S.T 2000.

References

1. M. P. Speciale, M. Brocato, *Layered Materials* Proc. Int. Conf. on Mathematical Models for Soils Mechanics, Scilla (Italy), 19/22-9-2000, under press. IEI–CNR Tech. Rep. 06, 2001.
2. M. Gurtin, *An Introduction to Continuum Mechanics* Accademic Press, 1982.

UNCONDITIONAL NONLINEAR STABILITY VIA THE ENERGY METHOD

B. STRAUGHAN

Department of Mathematical Sciences, University of Durham, DH1 3LE, U.K.
E-mail: Brian.Straughan@durham.ac.uk

In this article we review the state of the art concerning recent developments where *unconditional* nonlinear stability is proved in some problems in fluid dynamics. Open questions to fundamental problems are included and we detail several recent advances yielding unconditional stability results via new Lyapunov functionals. The problem of heating a fluid layer from below is highlighted when the viscosity varies with the temperature field.

1 Introduction: an illustration of difficulties with the energy method

The equations governing problems in hydrodynamic stability such as those one encounters in thermal convection are typically of the form

$$Au_t = L_S u + L_A u + N(u), \tag{1}$$

where u is a Hilbert space valued function, u_t is its time derivative, A is a bounded linear operator (typically a matrix with constant entries), $L = L_S + L_A$ is an unbounded, sectorial linear operator, and $N(u)$ represents the nonlinear terms. The operator L_S is the symmetric part of L while L_A denotes the anti-symmetric part. Such abstract equations and examples in fluid dynamics are discussed in e.g. Doering & Gibbon [1], Flavin & Rionero [2], and Straughan [3,4].

The classical theory of linear instability writes

$$u = e^{\sigma t}\phi$$

and discards the $N(u)$ term in (1). One is then faced with solving the eigenvalue problem

$$\sigma A\phi = L_S\phi + L_A\phi, \tag{2}$$

where σ is the eigenvalue and ϕ the eigenfunction.

It is important to note that equation (2) involves both the skew-symmetric operator L_A and the symmetric operator L_S. In general, σ is complex, and one looks for the eigenvalue with largest real part to become positive for instability.

Nonlinear energy stability on the other hand commences by forming the inner product of u with (1). If $(\cdot,\cdot)$ denotes the inner product on the Hilbert space in question then one finds

$$\frac{d}{dt}\frac{1}{2}(u, Au) = (u, L_S u) + (u, N(u)) \tag{3}$$

since

$$(u, L_A u) = 0.$$

Nonlinear energy stability follows from (3) (for details see e.g. Flavin & Rionero [2], Straughan [3]) and it is very important to note that in this way the nonlinear stability boundary does not involve the skew part of L, L_A. Thus, one may expect, in general, that the linear instability and nonlinear stability boundaries are very different.

In fact, there are two fundamental problems arising from (3) when one is faced with deriving *unconditional* nonlinear stability results. These are

(a) the effect of L_A on the nonlinear stability boundary;

(b) what does one do when $\big(u, N(u)\big) \not\geq 0$?

To illustrate some of the difficulties we now outline some fundamental hydrodynamic stability problems which are largely open.

2 Examples where unconditional stability is still open

2.1 The rotating Bénard problem

The equations for a perturbation in the nonlinear rotating Bénard problem are

$$u_{i,t} = -\pi_{,i} + R\theta k_i + \Delta u_i + T(\mathbf{u}\times\mathbf{k})_i - u_j u_{i,j}$$

$$u_{i,i} = 0$$

$$Pr\theta_{,t} = Rw + \Delta\theta - Pru_i\theta_{,i}$$

where u_i, π, θ are velocity, pressure, temperature, T is the Taylor number, Pr is the Prandtl number, and R^2 is the Rayleigh number. Standard indicial notation is employed and the equations hold in the domain $(x, y) \in \mathbf{R}^2, z \in (0, 1)$. The classical energy for this problem is

$$E(t) = \frac{1}{2}||\mathbf{u}||^2 + \frac{1}{2}\lambda Pr||\theta||^2,$$

where $\lambda(> 0)$ is a coupling parameter, and $||\cdot||$ is the norm on $L^2(V)$, V being the period cell for the disturbance. The difficulty here stems from the fact that

$$\int_V \mathbf{u} \cdot (\mathbf{u} \times \mathbf{k}) dV = 0 \equiv (u, L_A u).$$

This means that with the standard energy, the effect of rotation is completely lost. In the linear instability theory rotation has a very strong stabilising effect and so one wishes to incorporate rotation in any nonlinear analysis. This has been done by construction of novel Lyapunov functions. However, to my knowledge there is always a restriction on $E(0)$. This restriction is severe. Full details may be found in Mulone & Rionero [5]. As far as I am aware the nonlinear problem for unconditional stability is still wide open.

2.2 The magnetic Bénard problem

The magnetic Bénard problem is in some ways analogous to the rotating Bénard problem in the nonlinear energy stability sense. Full details of this may be found in Mulone & Rionero [6]. Again, I believe this problem is open.

2.3 Parallel shear flows

Here we have in mind the nonlinear stability of the classical flows of Poiseuille and Couette.

This is another area in which little progress has been made with a nonlinear energy stability analysis. Many recent details may be found in chapter 8 of Straughan [4]. An interesting recent paper dealing with the problem of throughflow imposed on the Couette solution is that of Doering *et al* [7].

While the three classes of problem mentioned in section 2 are relatively untouched from a nonlinear unconditional energy stability viewpoint there are, fortunately, many other problems which have recently proved amenable to analysis. We outline some of these below.

3 Examples where substantial progress has been made with unconditional nonlinear stability

3.1 Penetrative convection (Veronis model)

The problem of penetrative convection is described in e.g. Straughan [3]. The model we use is based on the density

$$\rho = \rho_m \Big[1 - \alpha (T-4)^2 \Big],$$

where ρ_m is a constant and T is temperature. The nonlinear perturbation equations are

$$u_{i,t} = -\pi_{,i} - R\theta(\xi - z)k_i + \Delta u_i - u_j u_{i,j} + Pr k_i \theta^2$$

$$u_{i,i} = 0$$

$$Pr\theta_{,t} = -Rw + \Delta\theta - Pr u_i \theta_{,i}$$

where $\xi = T_u/4$, T_u being the upper temperature. Other notation is as before. In this section $L_A \neq 0$; however, the real problem is the fact that with the standard energy a term of form $Pr(\theta^2, w)$ arises. This being cubic causes difficulty. Payne & Straughan [8] removed this by employing a weighted energy, see also Straughan & Walker [9] and Gentile & Rionero [10]. Other penetrative convection results are given by Chasnov &Tse [11] and Tse & Chasnov [12].

3.2 Double diffusive convection (Heat and salt below)

This is a problem for which the equations look tantalisingly simple. They are

$$u_{i,t} = -\pi_{,i} + R\theta k_i + \Delta u_i - R_C \phi k_i - u_j u_{i,j}$$

$$u_{i,i} = 0$$

$$Pr\theta_{,t} = Rw + \Delta\theta - Pr u_i \theta_{,i}$$

$$Pc\,\phi_{,t} = \Delta\phi + R_C w - Pc\, u_i \phi_{,i}$$

However, the skew effect is associated with R_C, the salting effect. Mulone [13] was the first to recognise the importance of two natural variables in this problem. This has led to a series of important papers establishing unconditional nonlinear stability, Mulone [13,14,15], Mulone & Rionero [16], Lombardo *et al.* [17]. Tracey [18] has established some very interesting results in the triply diffusive problem.

3.3 Rotating porous convection

Conditional nonlinear stability was obtained for the rotating porous Bénard problem by Guo & Kaloni [19] and Qin & Kaloni [20]. Vadasz [21] gave some striking results in the linearised instability and weakly nonlinear problems. Unlike the fluid case, the nonlinear energy stability problem is completely resolved; see Straughan [22].

3.4 Temperature dependent viscosity and temperature dependent thermal diffusivity

The Bénard problem is undoubtedly important and in general both viscosity and thermal diffusivity vary with temperature. Experiments show this effect clearly as is explained by Zhao *et al.* [23].

The theory of nonlinear energy stability with temperature dependent viscosity was first addressed by Richardson [24], Capone & Gentile [25,26]. They employed sophisticated Lyapunov functionals to obtain optimal Rayleigh number thresholds, although their results are conditional.

Flavin & Rionero [27,28,29,30] and Rionero [31] showed how one could cleverly transform the equations to obtain unconditional stability results in the Bénard problem with temperature dependent thermal diffusivity. They also handled the vertical convection problem for a porous slab.

For the porous medium problem, very sharp unconditional results have been obtained by Capone & Rionero [32,33], Rionero [34], and Payne & Straughan [35].

3.5 Thawing subsea permafrost

This is an interesting problem involving motion of a layer of salty sediments beneath the sea off the coast of Alaska, see e.g. Straughan [3], chapter 7. Budu [36] has obtained some sharp unconditional nonlinear stability results when the density is cubic in temperature and salt, employing a realistic density.

3.6 Convection in a dielectric fluid

Mulone *et al.* [37] develop an unconditional nonlinear stability analysis for the problem of convection in a dielectric fluid. This work is interesting in that at first sight the Lyapunov functional they use does not appear to be positive definite. It is, and does lead to unconditional nonlinear stability in a problem where the equations are highly nonlinear.

3.7 Inclined temperature gradient, variable gravity, time dependent boundary conditions

There are many interesting problems under the category of this heading. Interesting stability results have been found for such problems by Alex *et al.* [38], Capone & Rionero [39], Kaloni & Qiao [40,41,42], Néel & Nemrouch [43], Nield [44,45,46,47] and Siddheshwar & Pranesh [48].

4 Variable viscosity convection

Payne & Straughan [35] present an analysis of nonlinear stability of thermal convection in a porous layer when the viscosity depends on temperature. The analysis derives fully unconditional nonlinear stability thresholds, i.e. for *all* initial data, and these are very sharp and hence useful in practical porous media work. The theory is developed with the aid of the porous media theory which includes inertial drag terms as introduced by Forchheimer.

Payne & Straughan [35] allow a linear viscosity of form

$$\mu(T) = \mu_0\big[1 - \gamma(T - T_0)\big],$$

where T_0, μ_0, γ are constants. Their equations take the form

$$\frac{\partial \pi}{\partial x_i} = -(1 + \Gamma z)u_i + \frac{\Gamma}{R}\theta u_i + R\theta\delta_{i3} - b|\mathbf{u}|u_i - c|\mathbf{u}|^2 u_i$$

$$\frac{\partial u_i}{\partial x_i} = 0$$

$$\frac{\partial \theta}{\partial t} + u_i\frac{\partial \theta}{x_i} = Rw + \Delta\theta$$

where the spatial domain is the layer $(x, y) \in \mathbf{R}^2$, $z \in (0, 1)$. The Rayleigh number, Ra, is given by $Ra = R^2$. On the boundaries there holds

$$w = \theta = 0, \qquad z = 0, 1.$$

Additionally (u_i, θ, π) have a periodic plan-form shape which tiles the plane. The period cell which arises is denoted by V.

Payne & Straughan [35] show that a standard energy analysis cannot yield unconditional results because the ratio I/D cannot have a maximum in the natural space. Instead they work with a Lyapunov functional of form

$$E(t) = \frac{1}{2}||\theta||^2 + \frac{a}{3}||\theta||_3^3,$$

where $a(> 0)$ is a positive constant to be judiciously chosen and $||\theta||_3$ denotes the norm of θ on $L_3(V)$.

Payne & Straughan [35] show that for the Forchheimer case $c = 0$ unconditional nonlinear stability holds provided

$$R < R_E = R_L$$

and

$$b > \frac{9\Gamma}{8\pi^2}.$$

This is a very strong result. This paper also studies other viscosity and density effects and derives further unconditional stability results.

Acknowledgments

I am indebted to Professor Salvatore Rionero for helpful discussions on this and related work. Finally, I should like to thank the local organisers of Wascom01, Professors Roberto Monaco and Miriam Pandolfi-Bianchi.

References

1. C.R. Doering and J.D. Gibbon, *Applied Analysis of the Navier-Stokes Equations.* (Cambridge University Press, Cambridge, 1995).
2. J.N. Flavin and S. Rionero, *Qualitative Estimates for Partial Differential Equations.* (CRC Press, Boca Raton, 1995).
3. B. Straughan, *The energy method, stability, and nonlinear convection.* (Springer-Verlag, New York, 1992).
4. B. Straughan, *Explosive Instabilities in Mechanics.* (Springer-Verlag, Heidelberg, 1998).
5. G. Mulone and S. Rionero, *Continuum Mech. Thermodyn.* **9**, 347 (1997).

6. G. Mulone and S. Rionero, *Zeitschrift fur angewandte Mathematik und Mechanik* **73**, 35 (1993).
7. C.R. Doering *et al*, *Phys. Fluids* **12**, 1955 (2000).
8. L.E. Payne and B. Straughan, *Geophys. Astrophys. Fluid Dyn.* **39**, 57 (1987).
9. B. Straughan and D.W. Walker, *Proc. Roy. Soc. London A* **457**, 87 (1996).
10. M. Gentile and S. Rionero, *Rend. Accad. Sci. Fis. Matem. Napoli* **67**, 129 (2000).
11. J.R. Chasnov and K.L. Tse, *Fluid Dyn. Res.* **28**, 397 (2001).
12. K.L. Tse and J.R. Chasnov, *J. Computational Phys.* **142**, 489 (1998).
13. G. Mulone, *Continuum Mech. Thermodyn.* **6**, 161 (1994).
14. G. Mulone, *Rend. Circolo Matem. Palermo* **57**, 347 (1998).
15. G. Mulone in *Proceedings of Wascom 99*, ed. V. Ciancio, A. Donato, F. Oliveri and S. Rionero (World Scientific, Singapore, 2001).
16. G. Mulone and S. Rionero, *Rend. Mat. Acc. Lincei* **9**, 221 (1998).
17. S. Lombardo *et al*, *Math. Methods Appl. Sci.* **23**, 1447 (2000).
18. J. Tracey *Stability analyses of multi-component convection-diffusion problems.* Ph.D. Thesis, Glasgow Univ, (1997).
19. J. Guo and P.N. Kaloni, *J. Math. Anal. Appl.* **190**, 373 (1995).
20. Y. Qin and P.N. Kaloni, *Quart. Appl. Math.* **53**, 129 (1995).
21. P. Vadasz, *J. Fluid Mech.* **376**, 351 (1998).
22. B. Straughan, *Proc. Roy. Soc. London A* **457**, 87 (2001).
23. A.X. Zhao *et al*, *Phys. Fluids* **7**, 1576 (1995).
24. L.L. Richardson *Nonlinear stability analyses for variable viscosity and compressible convection problems.* Ph.D. Thesis, Glasgow Univ, (1993).
25. F. Capone and M. Gentile, *Acta Mechanica* **107**, 53 (1994).
26. F. Capone and M. Gentile, *Continuum Mech. Thermodyn.* **7**, 297 (1995).
27. J.N. Flavin and S. Rionero, *J. Math. Anal. Appl.* **228**, 119 (1998).
28. J.N. Flavin and S. Rionero, *Continuum Mech. Thermodyn.* **11**, 173 (1999).
29. J.N. Flavin and S. Rionero, *Q. Jl. Mech. Appl. Math.* **52**, 441 (1999).
30. J.N. Flavin and S. Rionero, in *Proceedings of Wascom 99*, ed. V. Ciancio, A. Donato, F. Oliveri and S. Rionero (World Scientific, Singapore, 2001).
31. S. Rionero, in *Continuum Mechanics and Applications in Geophysics and the Environment*, ed. B. Straughan, R. Greve, H. Ehrentraut and Y. Wang (Springer, Heidelberg, 2001).
32. F. Capone and S. Rionero, *Rend. Accad. Sci. Fis. Matem. Napoli* **66**, 159 (1999).

33. F. Capone and S. Rionero, in *Proceedings of Wascom 99*, ed. V. Ciancio, A. Donato, F. Oliveri and S. Rionero (World Scientific, Singapore, 2001).
34. S. Rionero, in *Symp. Trends in the Application of Mathematics to Mechanics*, ed. P.E. O'Donoghue and J.N. Flavin. (Elsevier, 2000).
35. L.E. Payne and B. Straughan, *Stud. Appl. Math.* **105**, 59 (2000).
36. P. Budu, Ph.D. Thesis, University of Durham, Forthcoming (2002).
37. G. Mulone *et al*, *Atti Accad. Naz. Lincei* **7**, 241 (1996).
38. S.M. Alex *et al*, *Fluid Dyn. Res.* **29**, 1 (2001).
39. F. Capone and S. Rionero, *Rend. Accad. Sci. Fis. Matem. Napoli* **67**, 119 (2000).
40. P.N. Kaloni and Z.C. Qiao, *Int. J. Heat Mass Transfer* **40**, 1611 (1997).
41. P.N. Kaloni and Z.C. Qiao, *Continuum Mech. Thermodyn.* **12**, 185 (2000).
42. P.N. Kaloni and Z.C. Qiao, *Int. J. Heat Mass Transfer* **44**, 1585 (2001).
43. M.C. Néel and F. Nemrouch, *Continuum Mech. Thermodyn.* **13**, 41 (2001).
44. D.A. Nield, *Int. J. Heat Mass Transfer* **34**, 87 (1991).
45. D.A. Nield, *Int. J. Heat Fluid Flow* **15**, 157 (1994).
46. D.A. Nield, *Int. J. Heat Mass Transfer* **37**, 3021 (1994).
47. D.A. Nield, *Int. J. Heat Mass Transfer* **41**, 241 (1998).
48. P.G. Siddheshwar and S. Pranesh, *J. Magnetism and Magnetic Materials* **192**, 159 (1999).

ON THE NONLINEAR DIFFUSION IN THE EXTERIOR OF A SPHERE

I. TORCICOLLO, M. VITIELLO

Dipartimento di Matematica ed Applicazioni, Università di Napoli Federico II, Complesso Monte S.Angelo, Via Cintia, 80126 Napoli, Italie.
e-mail: torcicol@unina.it; vitiello@matna2.dma.unina.it

The long time behaviour of the solutions of the equation $u_t = \Delta F(u)$ in exterior domains is studied.

1 Introduction

Let $\Omega_0 \subset R^3$ be a sufficiently smooth bounded fixed domain ensuring the validity of divergence-like theorems and denote by Ω the exterior of Ω_0. Let us consider the initial boundary value problem

$$\begin{cases} u_t = \Delta F(u) & (\mathbf{x}, t) \in \Omega \times R^+ \\ u(\mathbf{x}, 0) = u_0(\mathbf{x}) & \mathbf{x} \in \Omega \\ u(\mathbf{x}, t) = u_1(\mathbf{x}) & \mathbf{x} \in \partial\Omega, \quad t \in R^+ \end{cases} \tag{1}$$

where $F \in C^2(R)$, $u_0 \in C(\overline{\Omega})$, $u_1 \in C(\partial\Omega)$ are assigned functions.
The problem (1) can model many physical, chemical or biological phenomena. In fact equation $(1)_1$ can be found in the porous fluid theory, in the filtration theory, in the mathematical biology and anisotropic MHD [5,11]. Problem (1) has been considered in [1] and [2] in the case of a bounded domain where, among other things, matters related to the asymptotic behaviour of solutions have been explored. Some stability and instability results concerning a more general initial boundary value problem are obtained in [12] and [2]. In particular in [2] the long time behaviour of solutions in the case of the bounded domain has been studied. Here we reconsider the problem, but in the case of exterior domains.

Our aim is to obtain, in the class of smooth solutions, conditions on F under which each solution tends toward a steady state when $t \to \infty$, in the case that Ω is an unbounded domain. Precisely, considering the steady boundary value problem

$$\begin{cases} \Delta F(U) = 0 & \mathbf{x} \in \Omega \\ U = u_1(\mathbf{x}) & \mathbf{x} \in \partial\Omega \end{cases} \tag{2}$$

our goal is to show that, when

$$m \leq F'(u) \leq m^\star \quad \forall u \in R, \tag{3}$$

where m and $m^\star$ are positive constants, then the solution U of (2) denotes the asymptotic state of any solution of (1).
Let us underline that, under the assumption

$$F'(u) > 0 \quad \forall u \in R, \tag{4}$$

setting

$$F_1 = F(u_1)$$

the solvability of the problem (2) depends on that of the problem

$$\begin{cases} \Delta F = 0 \ \mathbf{x} \in \Omega \\ F = F_1 \ \ \mathbf{x} \in \partial\Omega. \end{cases} \tag{5}$$

If we denote by F^* the solution of (5), from $F(U) = F^*$, on taking into account the assumption (4) and denoting by φ the (unique) inverse of F, it follows that

$$U = \varphi(F^\star). \tag{6}$$

Then the uniqueness for the problem (5) gives the uniqueness for (2).

The plan of the paper is the following. Section 2, after the introduction of the perturbation problem, is dedicated to some necessary preliminaries. Section 3 deals with the asymptotic stability of solutions in a suitable weighted L^2–norm. The note ends with the application to the case concerning the heat conduction with exponential diffusivity, in the exterior of a sphere (Section 4).

2 Preliminaries

In order to write the perturbation problem let us define

$$u = U + v$$

and

$$L(U, v) = F(U + v) - F(U);$$

then, from (1)-(2), it follows that

$$\begin{cases} v_t = \Delta L \qquad (\mathbf{x}, t) \in \Omega \times R^+ \\ v(\mathbf{x}, 0) = v_0 \ \mathbf{x} \in \Omega \\ v = 0 \qquad\qquad \mathbf{x} \in \partial\Omega. \end{cases} \tag{7}$$

Remark 1. *It is useful to emphasize that* $v = 0 \Rightarrow L = 0$. *In addition (4) implies* $v = 0 \Leftrightarrow L = 0$.

Let us recall here, a suitable Poincaré inequality [4], we will use in the next section and some results proved in [1] and [3], useful throughout the paper.

A weighted Poincaré inequality. *Let Ω the exterior of a sphere Ω_0 of radius r_0, and S the set of functions $u : \Omega \to R^3$ such that*

$$\int_\Omega \left[\frac{u^2}{r^k} + \nabla u : \nabla u\right] d\Omega < \infty, \qquad u|_{\partial\Omega} = 0 \qquad k \in [2, 3[.$$

Then, $\forall u \in S$, the following inequality holds

$$\int_\Omega \frac{u^2}{(\frac{r}{r_0})^k} d\Omega \leq (\frac{2r_0}{3-k})^2 \int_\Omega (\nabla u)^2 d\Omega \tag{8}$$

where $r = (\sum_{i=1}^3 x_i^2)^{\frac{1}{2}}$.

Lemma 1. *Let*

$$G(v, U) = \int_0^v L(\overline{v}, U) d\overline{v} \qquad v \in R. \tag{9}$$

One has

$$G(0, U) = [\frac{\partial G}{\partial v}]_{v=0} = 0, \tag{10}$$

$$L^2(v, U) \geq 2mG(v, U), \tag{11}$$

$$G(v, U) \geq \frac{1}{2} m v^2, \tag{12}$$

$$L^2(v, U) \leq 2m^\star G(v, U). \tag{13}$$

3 Convergence to asymptotic states

In this section we study the asymptotic stability in a suitable weighted L^2–norm of solutions to the system (2) and show that U is the asymptotic state of any solution of (1). We reach our task in the class of perturbations v such that (3) holds and $v_0 \in L^2(\Omega)$. To this end let us introduce the functional

$$V(t) = \int_\Omega g G[v(\mathbf{x}, t), U(\mathbf{x})] d\Omega \tag{14}$$

where $G(U, v)$ is defined as in (9) and the weight is given by $g = e^{-\alpha r}$ $(\alpha > 0)$. Because of (10) and (12), it turns out that V is positive definite .

Evaluating the time derivative of (14) along the solutions of (7) we obtain

$$\dot{V}(t) = \int_\Omega gL\Delta L \, d\Omega = \frac{1}{2}\int_\Omega \Delta g L^2 \, d\Omega - \int_\Omega g(\nabla L)^2 \, d\Omega. \tag{15}$$

Being $\Delta g \leq \alpha^2 g$ and (13) holding, the application of Gronwall's lemma gives

$$V(t) + \int_0^t e^{\alpha^2 m^*(t-\tau)} \int_\Omega g(\nabla L)^2 \, d\Omega d\tau \leq V_0 e^{\alpha^2 m^* t}. \tag{16}$$

Since the right hand side converges to a finite quantity as $\alpha \to 0$, it follows from the monotone convergence theorem, on letting $\alpha \to 0$ that

$$V_1(t) + \int_0^t \int_\Omega (\nabla L)^2 \, d\Omega d\tau \leq V_1(0) \tag{17}$$

where

$$V_1 = \int_\Omega G(v, U) d\Omega. \tag{18}$$

We are now in position to obtain stability on initial data. In fact $V_1(t) \leq V_1(0)$, $\forall t$ implies the stability in the measure of V_1. Precisely, by using (12), from

$$\int_\Omega v^2 d\Omega \leq \frac{2}{m} V_1(0) \tag{19}$$

it turns out the stability in the L^2-norm. On the other hand, from (17), (8), (11) the

$$(\frac{3-k}{2r_0})^2 \int_0^t \int_\Omega \frac{G}{(\frac{r}{r_0})^k} d\Omega d\tau \leq \frac{V_1(0)}{2m}, \tag{20}$$

setting

$$W(t) = \int_\Omega \frac{G(v,U)}{(\frac{r}{r_0})^k} d\Omega,$$

becomes

$$\int_0^t W(t) d\tau \leq V_1(0) \qquad t \in [0, \infty[.$$

From the last inequality it follows that

$$\lim_{t \to \infty} W(t) = 0$$

that is a type of asymptotic stability in the measure of W.

Remark 2. *Concerning the initial value $V_1(0)$ of V_1 let us remark that, by using (3) it follows that*

$$V_1(0) = \int_\Omega \int_0^{v_0} L dv d\Omega = \int_\Omega \int_0^{v_0} F'(U + \sigma v) v dv d\Omega \leq \frac{m^*}{2} \int_\Omega {v_0}^2 d\Omega \qquad [\sigma \in (0,1)$$

4 An application

In the past several years, diffusion systems have attracted a great deal of attention from mathematicians and other scientists. In fact nonlinear diffusion mechanisms are well-known in literature and the equation $(1)_1$, where u denotes a "concentration", is one of the most used models. In this section we show an application of the results obtained in the previous ones.
Let $\Omega \subset R^3$ be the exterior domain of a sphere Ω_R of radius R, centered at the origin in R^3 and let us consider the well-known solution of the problem (5)

$$F^\star(r) = \frac{F_1 R}{r} \qquad (r > R). \tag{21}$$

Let us consider the case concerning the nonlinear heat conduction in solids where u denotes the temperature and $F(u)$ is connected to the diffusivity $k(u)$ by

$$F(u) = \int_0^u k(\overline{u})d\overline{u}. \tag{22}$$

In particular let us consider the case of exponential diffusivity [13]

$$k(u) = k_0 e^{-\gamma u} \tag{23}$$

where k_0 and $\gamma < \dfrac{k_0}{F_1}$ are positive constants. Being

$$F'(u) = k_0 e^{-\gamma u} > 0 \tag{24}$$

in view of (6), it follows that

$$U = \log\left(1 - \frac{\gamma}{k_0}\frac{F_1 R}{r}\right)^{-\frac{1}{\gamma}}; \tag{25}$$

in addition, $0 < u < u^*$ implies

$$m < F'(u) < m^* \tag{26}$$

where $m^* = k_0$ and $m = k_0 e^{-\gamma u^*}$. According to (9), let us introduce

$$G = \int_0^v L d\overline{v} \tag{27}$$

with

$$L(U,v) = F(U+v) - F(U) = \frac{k_0}{\gamma} e^{-\gamma U}(1 - e^{-\gamma v}). \tag{28}$$

Easily it follows that $L(U,0) = 0$ and $G(U,v)$ is positive definite.

We are in position to apply the results obtained in section 3 and hence the solution (25) is the asymptotic state of each solution.

Acknowledgements

This work has been performed under the auspices of the G.N.F.M. of I.N.D.A.M. and M.U.R.S.T. (P.R.I.N.): "Nonlinear mathematical problems of waves propagation and stability in continuous media". The Authors thank gratefully Prof. Salvatore Rionero for having proposed the present research and his helpful suggestions.

References

1. J. N. Flavin and S. Rionero, "Asymptotic and other properties of a nonlinear diffusion model", *J. Math. Analysis Appl.* **228**, 119 (1998).
2. S. Rionero, "Asymptotic and other properties of some nonlinear diffusion models", *Cont. Mech. Appl. Geoph. Env.* , 56 (2001),(Eds. B. Straughan, R. Greve, H. Ehrentraut, Y. Wang, Springer).
3. J. N. Flavin and S. Rionero, "A Free Boundary Value Problem for a nonlinear Heat Equation", (to appear).
4. F. Capone and M. De Angelis, "On the energy stability of fluid motions in exterior of a sphere under free boundary like conditions", *Rend. Acc. Sci. Fis. Mat.* **LX**, (1995).
5. J. N. Flavin and S. Rionero, *Qualitative Estimates for Partial Differential Equation, An Introduction*, (CRC Press, Boca Raton, FL, 1995).
6. G. P. Galdi and S. Rionero, *Weighted energy methods in fluid dynamics and elasticity*, (Lectures Notes in Math., n.1134, Springer- Verlag ,1985).
7. J. Smoller, *Shock waves and Reaction-Diffusion Equations*, (Springer-Verlag, n.258 of "A Series of Comprehensive Studies in Mathematics", 1983).
8. J. D. Murray, *Mathematical Biology*, (Biomathematics Text, n.19, Springer-Verlag, 1989).
9. M. E. Gurtin and R. C. Mac Camy, "On the diffusion of biological populations", *Mathematical Biosciences* **33**, 35 (1977).
10. M. Maiellaro and S. Rionero, "On the stability of Couette-Poiseuille flows in the anisotropic M.H.D. via the Liapunov Direct Method", *Rend. Acc. Sci. Fis. Mat.* , (1995).
11. P. Benilan and M. G. Grandall, "The Continuous dependence of solutions of $u_t - \Delta(\phi(u)) = 0$", *Indiana Un. Mat.* **30**, 1 (1981).
12. F. Capone, S. Rionero and I. Torcicollo, "On the stability of solutions of the remarkable equation $u_t = \Delta F(x, u) - g(x, u)$", *Rend. Cir. Mat. Palermo* **45**, (1996).
13. J. Crank, *The Mathematics of Diffusion*, 2nd ed., (Clarendon Press, Oxford, 1957).

ENTROPY METHODS FOR THE ASYMPTOTIC BEHAVIOUR OF FOURTH-ORDER NONLINEAR DIFFUSION EQUATIONS

G. TOSCANI

Dipartimento di Matematica, via Ferrata 1, 27100 Pavia, ITALY
E-mail: toscani@dimat.unipv.it

We review some recent results on the large-time asymptotics of fourth-order nonlinear parabolic equations in two cases: 1) scalar problems in bounded domains; 2) scalar problems with confinement by a uniformly convex potential The main analytical tool relies on the analysis of the entropy dissipation. A generalized Csizár-Kullback inequality allows for an estimation of the L^1-decay to equilibrium in terms of the relative entropy.

1 Introduction

This note is concerned with a short review of recent results on the large-time asymptotics of solutions of fourth-order (possibly degenerate) scalar parabolic equations. While in the last years noticeable progresses on the these asymptotics have been reached for second-order scalar parabolic convection-diffusion equations[12,25,10,1,21,26], only few results are available for fourth-order diffusion equations. This is a consequence of the fact that, in most of the cases of interest in applications[22], the existence theory for degenerate diffusion equations of higher order is at present not well developed, except in particular cases. One of the main motivations is that maximum principles are in general not available for higher-order equations such that the positivity or non negativity property has to be proved by *ad hoc* techniques. In this paper we will be mainly concerned with two model equations: the so-called thin-film equation[22], and a fourth-order equation which appeared originally as a scaling limit of interface fluctuations in a certain spin system[9].

The first model reads

$$\frac{\partial u}{\partial t} = (|u|^n u_{xxx})_x, \quad (x \in \mathbb{R}, t > 0) \tag{1.1}$$

with $n > 0$. This equation has proved to be rather useful (when $u(x,t)$ remains nonnegative along the evolution) for the modeling of several physical processes: surface–tension–dominated dynamics of thin viscous films, the motion of spreading viscous droplets, oxidation of silicon in semiconductor devices (when $n = 3$) and evolution of thin necks of flows in Hele–Shaw cells (when $n = 1$), among others.

The second model is given by the equation

$$n_t = -(n(\log n)_{xx})_{xx}, \quad (x \in \mathbb{R}, t > 0). \tag{1.2}$$

(1.2) arose originally as a scaling limit in the study of interface fluctuations in a certain spin system[9]. The same equation also arises in the modeling of quantum semiconductor devices[17].

These equations attracted a lot of attention in the mathematical literature in the last ten years. In one space dimension, questions of existence, uniqueness and finite speed of propagation of the support of weak solutions to (1.1) have been addressed in a series of papers [7,5,8,3], which followed the pioneering study of Bernis and Friedman[4]. The majority of these papers refers to the problem of existence in a bounded domain. The asymptotic behavior of (1.1) for the initial–boundary value problem with periodic boundary conditions has been studied in[3,8]. Bernis and Friedman[4] studied the initial boundary value problem for equation (1.1) in a cylinder $Q_T = \Omega \times (0,T)$, where Ω is a bounded interval, assuming that the initial datum $0 \le u_0 \in H^1(\Omega)$, and setting $u_x = u_{xxx} = 0$ on the lateral boundary. They proved a weak positivity property and the mass conservation law. Beretta, Bertsch and Dal Passo[3] proved that, when $0 < n < 3$, the solution converges to its mean value as $t \to \infty$, uniformly in Ω. Furthermore, if $n \ge 3$ and the initial data $u_0 > 0$ the same asymptotic result holds true, provided that some technical assumptions are also fulfilled. More recently[8], the initial boundary value problem has been also dealt with in one space dimension with periodic boundary conditions. There, the authors extend the results of Bernis and Friedman[4] and prove that there exists a weak nonnegative solution for all times which becomes a strong positive solution after some critical finite time in the case $0 < n < 3$. They also analyze the long–time behavior of weak solutions as $t \to \infty$ by introducing a family of regularized convex entropies. This technique allows them to handle the large–times problem when $0 < n < 3$. In addition, for all $n > 0$ they also show exponential decay of the distributional solutions (corresponding to strictly positive initial data) in L^∞–norm. Nonnegative source type solutions to (1.1) in $S = \mathbb{R} \times \mathbb{R}^+$ have been characterized By Bernis and coworkers[6]. If $0 < n < 3$, then there exists precisely one even nonnegative solution to (1.1) such that $u_0(x) = c\delta(x)$, where δ denotes the Dirac mass centered at the origin. This solution has compact support. On the other hand, source–type solutions do not exist if $n \ge 3$.

The initial boundary value problem for equation (1.2), with periodic boundary conditions, has been first considered by Bleher, Lebowitz and Speer[9]. Assuming (strictly) positive initial H^1–data, they showed that there exists a unique positive classical solution, locally in time. For suitably small initial data, the solution is even global in time. However, the problem whether non–negative solutions for general (non–negative) initial data exist globally in time, even in one dimension, remained open. A partial answer to this problem was done by Jüngel and Pinnau[17], provided equation (1.2), in the bounded domain $\Omega = (0,1)$, is subjected to the boundary conditions

$$n(0) = n(1) = 1, \quad n_x(0) = n_x(1) = 0.$$

Unlikely, these results are not enough to give a rigorous treatment to the problem of the asymptotic behavior of equation (1.2) on the whole line.

In all the aforementioned papers, the problem of finding a *sharp* rate of convergence toward the steady state (in a bounded domain), or towards the similarity solution (in the whole space), has not been addressed. Only recently, the application of the so–called entropy methods, so powerful when applied to second-order diffusion equations, appeared to be fruitful for higher order equations. The rate of convergence in L^1 towards the steady state (in a bounded interval with periodic boundary conditions) for the thin–film model has been studied by Lopez, Soler and the author[20]. Subsequently, the study of the intermediate asymptotics of the solution in the case of the exponent $n = 1$ in the whole space has been performed by Carrillo and Toscani[13]. There, an explicit rate of convergence both in L^1 and in L^∞ towards the similarity solution has been found. The rate of convergence towards the steady state for equation (1.2) in the situation considered by Jüngel and Pinnau[17] has been obtained by Jüngel and the author[18]. In the forthcoming sections, we will briefly describe these results.

2 The method

In the last years, the analytical theory for nonlinear convection–diffusion equations of the form

$$\frac{\partial u}{\partial t} = \operatorname{div}(u\nabla V(x) + \nabla f(u)), \qquad (x \in \mathbb{R}^d, t > 0), \tag{2.3}$$

$$u(x, t = 0) = u_0(x) \geq 0, \qquad (x \in \mathbb{R}^d) \tag{2.4}$$

has developed rapidly. In Eq. (2.3) the function f and the confining potential V satisfy suitable assumptions which guarantee the existence of a (unique) stationary solution u_∞. For parabolicity one requires $f'(u) > 0$ for $u > 0$. The break–through came from the Bakry–Emery discovery[2] of the interplay between linear diffusion equations and some functional inequalities (so-called *logarithmic Sobolev inequality*) which had become famous over the last decades for their appearing in various branches of mathematics[16,24]. The strategy in these problems consists essentially in the study of the time-decay of a suitable convex functional. Indeed, if Vu and $\phi(u)$ belong to $L^1(\mathbb{R}^d)$, a direct calculation shows that the entropy functional defined by

$$E(u(t)) := \int_{\mathbb{R}^d} (V(x)u(x,t) + \Phi(u(x,t)))\, dx \tag{2.5}$$

is non increasing in time when evaluated along a solution of (2.3), and

$$\frac{dE(u(t))}{dt} = -\int_{\mathbb{R}^d} u(x,t)\, \left|\nabla V + \Phi''(u)\nabla u\right|^2 (x,t)\, dx. \tag{2.6}$$

Equation (2.6) measures the dissipation of entropy, and the functional

$$I(u(t)) := \int_{\mathbb{R}^d} u(x,t)\, \left|\nabla V + \Phi''(u)\nabla u\right|^2 (x,t)\, dx \tag{2.7}$$

is usually referred as the entropy production functional. Now, one can show that an equilibrium solution of the evolution problem (2.3) is the minimizer of the entropy functional E on the set of all admissible comparison functions. Hence, the convergence of the solution of (2.3) towards the stationary solution can be seen as a consequence of the tendency of the system to evolve towards the state of minimal entropy.

The main tool of the entropy method is to analyze the relationship between the entropy production and its time derivative. In fact, if the system is such that, for some $\lambda > 0$

$$\frac{dI(u(t))}{dt} = -\lambda I(u(t)) - R(t) \tag{2.8}$$

where the remainder $R(t) \geq 0$ on $\mathbb{R}^d$, we obtain coupling equation (2.8) with (2.6)

$$\frac{dE(u(t))}{dt} = \frac{1}{\lambda}\left[\frac{dI(u(t))}{dt} + R(t)\right]$$

which implies at once

$$E(u_0) - E(u_\infty) \leq \frac{1}{\lambda} I(u_0) \tag{2.9}$$

and

$$\frac{d[E(u(t)) - E(u_\infty)]}{dt} \leq -\lambda[E(u(t)) - E(u_\infty)]$$

namely exponential convergence of the energy functional towards its minimum at a rate λ. Then, a Csiszar-Kullback type inequality[14,19] gives exponential convergence in L^1 at a rate $\lambda/2$. It is interesting to remark that (2.9) is a generalized Sobolev type inequality. Thus, in addition to the exponential decay of the solution of the diffusion-advection equation, one obtains as a by-product a proof of differential inequalities which in several cases where unknown. As shown by Carrillo and the author[13], these inequalities are of paramount importance in the study of the large–time asymptotics of higher order equations.

3 Droplet breakup in a Hele–Shaw cell

Let us consider the one–dimensional fourth–order nonlinear degenerate diffusion equation

$$\frac{\partial u}{\partial t} = -(uu_{xxx})_x, \qquad (x \in \mathbb{R}, t > 0), \tag{3.10}$$

This equation, derived from a lubrication approximation, models the surface tension dominated motion of thin viscous films and spreading droplets[22] On the real line $\mathbb{R}$, the problem of the asymptotic behavior of equation (3.10) has been recently considered by Carrillo and the author[13]. The basic remark is that (3.10) can be written as

$$u_t = -2\left[u^{\frac{3}{2}}\left(u^{\frac{1}{2}}\right)_{xx}\right]_{xx}. \tag{3.11}$$

Let us set

$$\alpha(t) = e^t \quad \text{and} \quad \beta(t) = \left(e^{5t} - 1\right)/5.$$

Thanks to the standard time dependent change of variables

$$v(x,t) = \alpha(t)u(\alpha(t)x, \beta(t)), \tag{3.12}$$

equation (3.11) becomes

$$v_t = -2\left[v^{\frac{3}{2}}\left(v^{\frac{1}{2}}\right)_{xx}\right]_{xx} + (xv)_x. \tag{3.13}$$

Equation (3.13) has a unique $C^1(\mathbb{R})$ compactly supported steady state of given mass M,

$$v_\infty(x) = \frac{1}{24}(C^2 - x^2)_+^2 \tag{3.14}$$

with $C = C(M)$, and, as usual, g_+ indicates the positive part of g. The steady state (3.14) is nothing but the Barenblatt–Pattle steady state of a rescaled porous medium equation of exponent $m = 2$. In other words, equation (3.13) has the same steady state of the Fokker–Planck type equation

$$v_t = \left[v \left(\gamma v^{\frac{1}{2}} + \frac{x^2}{2} \right)_x \right]_x , \tag{3.15}$$

where γ is a given positive constant. Direct computations show that

$$u_t = -\frac{2}{\sqrt{6}} \left[v^{\frac{3}{2}} \left(\sqrt{6} v^{\frac{1}{2}} + \frac{x^2}{2} \right)_{xx} \right]_{xx} + \left[v \left(\sqrt{6} v^{\frac{1}{2}} + \frac{x^2}{2} \right)_x \right]_x . \tag{3.16}$$

Hence equation (3.13) is linked to equation (3.15) provided $\gamma = \sqrt{6}$. The steady states of both equations (3.15) and (3.13) are obtained by setting

$$v \left(\sqrt{6} v^{\frac{1}{2}} + \frac{x^2}{2} \right)_x = 0. \tag{3.17}$$

Thus, by studying entropies of the second order nonlinear degenerate diffusion equation (3.15) we obtain at once entropies for the fourth-order nonlinear diffusion equation (3.16). The exact form of the entropy associated to the steady state v_∞ given in (3.14) is given by:

$$H(f) = \int_{\mathbb{R}} \left(\frac{x^2}{2} f + \sqrt{\frac{8}{3}} f^{3/2} \right) dx. \tag{3.18}$$

Using entropy (3.18) one has (at least formally) integrating by parts in equation (3.16) that

$$\frac{d}{dt} H(v) = \int_{\mathbb{R}} v_t \left(\frac{x^2}{2} v + \sqrt{\frac{8}{3}} v^{3/2} \right) dx =$$

$$- \int_{\mathbb{R}} v \left(\frac{x^2}{2} + \sqrt{6} v^{1/2} \right)_x^2 dx - \frac{2}{\sqrt{6}} \int_{\mathbb{R}} v^{3/2} \left(\frac{x^2}{2} + \sqrt{6} v^{1/2} \right)_{xx}^2 dx. \tag{3.19}$$

Let $v_\infty(x)$ be the stationary solution defined by (3.14). As briefly discussed in the previous section, the relative entropy $H(v|v_\infty)$ satisfies the differential inequality

$$\frac{d}{dt} H(v|v_\infty) \le - \int_{\mathbb{R}} v \left(\frac{x^2}{2} + \sqrt{6} v^{1/2} \right)_x^2 dx = -D_p(v) \le 0. \tag{3.20}$$

We remark that $D_p(v)$ is the entropy production associated to the porous medium type equation. Lower bounds for the entropy production in terms of the relative entropy have been obtained, via the entropy method introduced in the previous Section[11,15,23,12]. These bounds assure that the following convex inequality of Sobolev type holds

$$H(v|v_\infty) \leq \frac{1}{2} D_p(v). \tag{3.21}$$

Applying (3.21) to $v(t)$ we finally deduce

$$\frac{d}{dt} H(v(t)|v_\infty) \leq -2H(v(t)|v_\infty),$$

which implies exponential convergence to equilibrium in relative entropy with an explicit rate. Going back to the original equation we conclude with the algebraic decay of the solution towards the similarity solution. All these computations can be rigorously justified[13].

4 The thin-film equation in a bounded domain

A similar analysis gives the rate of decay of the solution to equation (1.1),in a bounded domain, with periodic boundary conditions. For simplicity, we refer here to the range $0 < n < 2$. It is classical to regularize both the initial data and the equation to obtain an approximate solution that is a strong, smooth solution for all time[4,8]. The regularization, for any given $\epsilon > 0$, reads

$$u_t = -(P_\epsilon(u) u_{xxx})_x \qquad \text{in } Q_T = \Omega \times \mathbb{R}_+, \tag{4.22}$$

where

$$P_\epsilon(u) = \frac{u^{4+n}}{\epsilon u^n + u^4}. \tag{4.23}$$

Here the correct entropies are

$$H_{\epsilon,\beta}(u) = \int_\Omega [G_{\epsilon,\beta}(u(x,t)) - G_{\epsilon,\beta}(1)]\, dx, \tag{4.24}$$

where

$$G_{\epsilon,\beta}(u) = \frac{\epsilon u^{\beta-2}}{(3-\beta)(2-\beta)} + \frac{u^{2+\beta-n}}{(2+\beta-n)(1+\beta-n)}, \qquad 0 < \beta < 1. \tag{4.25}$$

These entropies satisfy, for $0 < \beta < 1$,

$$\frac{d}{dt} H_{\epsilon,\beta}(u)(t) = -\int_\Omega u^\beta u_{xx}^2\, dx - \frac{1}{3}\beta(1-\beta)\int_\Omega u^{\beta-2} u_x^4\, dx = -I_\beta(u), \tag{4.26}$$

where we have defined the entropy production associated with $G_{\epsilon,\beta}(u)$ as

$$I_\beta(u) = \int_\Omega u^\beta u_{xx}^2\, dx + \frac{1}{3}\beta(1-\beta)\int_\Omega u^{\beta-2} u_x^4\, dx. \tag{4.27}$$

A lower bound for $I_\beta(u)$ has been found by Lopez, Soler and the author[20]. Let $0 \le u \in H^2(\Omega)$, with $u_x(\pm\frac{1}{2}) = 0$. Then, if $I_\beta(u) < \infty$ for some $0 < \beta < 1$, $u^{1+\beta/2} \in H^2(\Omega)$ and, if

$$A_\beta = 16\frac{1-\beta}{(2+\beta)^2}\frac{1+2\beta}{4(1-\beta)^2+3\beta(4-\beta)}. \tag{4.28}$$

$$I_\beta(u) \ge A_\beta \int_\Omega \left[\left(u^{1+\frac{\beta}{2}}\right)_{xx}\right]^2 dx. \tag{4.29}$$

Various Poincaré inequalities, and the fact that, for all functions $0 \le u \in L^{2p}(\Omega) \cap L^\infty(\Omega)$ such that $\int_\Omega u\,dx = 1$,

$$D_p = \int_\Omega u^{2p}\,dx - \left(\int_\Omega u^p\,dx\right)^2 \ge \int_\Omega u^2\,dx - 1 \tag{4.30}$$

show that, for an explicitly computable constant c,

$$I_{\frac{n}{2}}(u) \ge c\left[\int_\Omega u^2\,dx - 1\right]. \tag{4.31}$$

This inequality is enough to guarantee exponential convergence in relative entropy[20], since, by Jensen's inequality, the right-hand side is un upper bound for the entropies $H_{0,\frac{n}{2}}(u)$.

5 A parabolic equation describing interface fluctuations

Entropy methods can also be applied to the nonlinear fourth-order equation

$$n_t = -(n(\log n)_{xx})_{xx}, \tag{5.32}$$

which arises in the study of interface fluctuations in spin systems[9] and in quantum semiconductor modeling[17]. Jüengel and the author[18] studied the asymptotic behavior of the solution to the problem

$$n_t = -(n(\log n)_{xx})_{xx} \quad \text{in } \Omega = (0,1), \tag{5.33}$$

$$n = 1,\ n_x = 0 \quad \text{on } \partial\Omega, \tag{5.34}$$

$$n(\cdot,0) = n_0 \quad \text{in } \Omega. \tag{5.35}$$

The existence of global-in-time non-negative weak solutions has been shown in[17] for initial data which are only measurable and satisfy

$$\int_\Omega (n_0 - \log n_0)dx < \infty.$$

As in the previous cases, one shows that the entropy production induced by the (semiclassical) relative entropy

$$H(n) = \int_\Omega \big(n(\log n - 1) + 1\big)\ dx, \tag{5.36}$$

is an upper bound for the entropy (5.36). This implies exponential convergence towards $n_\infty = 1$ with an explicit rate. A direct computation shows that the entropy production is

$$I(n) = \int_\Omega n(\log n)_{xx}^2\ dx = \int_\Omega \left(\frac{n_{xx}^2}{n} - \frac{1}{3}\frac{n_x^4}{n^3}\right)\ dx. \tag{5.37}$$

Integration by parts proves that

$$\int_\Omega (\sqrt{n})_{xx}^2\ dx = \frac{1}{4}\int_\Omega \left(\frac{n_{xx}^2}{n} - \frac{5}{12}\frac{n_x^4}{n^3}\right)\ dx \leq \frac{1}{4}\int_\Omega n(\log n)_{xx}^2\ dx. \tag{5.38}$$

¿From various Poincaré's inequalities one concludes with

$$\int_\Omega n(\log n)_{xx}^2\ dx \geq 4\|(\sqrt{n})_{xx}\|_{L^2(\Omega)}^2 \geq 8\|\sqrt{n} - 1\|_{L^\infty(\Omega)}^2. \tag{5.39}$$

The result follows considering that, if $n \in L^\infty(\Omega)$, $n \geq 0$ in Ω, then

$$\int_\Omega (n(\log n - 1) + 1)\ dx \leq (N+2)\|\sqrt{n} - 1\|_{L^\infty(\Omega)}^2, \tag{5.40}$$

where $N = \int_\Omega n\ dx$.

Acknowledgments

The author acknowledges supports both from the TMR Project "Asymptotic Methods in Kinetic Theory", grant number ERB-FMBX-CT97-0157, and from the Minister for Research of Italy (MIUR), Project "Asymptotic problems in Kinetic Theory".

References

1. A. Arnold *et al*, *Commun. PDE*, 43-100 (2001).
2. D. Bakry and M. Emery, *Sém. Proba.* **XIX**, **1123** Lecture Notes in Math. Springer, 177–206 (1985).
3. E. Beretta *et al*, *Arch. Rational Mech. Anal.* **129**, 175–200 (1995).
4. F. Bernis and A. Friedman, *J. Diff. Eqns.* **83**, 179–206 (1990).
5. F. Bernis, *Adv. in Diff. Eq.* **3**, 337–368 (1996).

6. F. Bernis *et al*, *Nonlinear Analysis* **18**, 217–234 (1992).
7. A. L. Bertozzi, *Notices of the AMS*, June–July 1998, 689–697 (1998).
8. A. L. Bertozzi and M. Pugh, *Comm. Pure Appl. Math.* **XLIX**, 85–123 (1996).
9. P. M. Bleher *et al*, *Commun. Pure Appl. Math* **XLIX**, 923–942 (1994).
10. J. A. Carrillo and G. Toscani, *Math. Meth. Appl. Sci.* **21** , 1269-1286 (1998).
11. J. A. Carrillo and G. Toscani, *Indiana Univ. Math.J.* **49**, 113–141 (2000).
12. J. A. Carrillo *et al*, *Monat. für Math.* **133**, 1–82 (2001).
13. J. A. Carrillo and G. Toscani. *Commun. Math. Phys.*, (2001) (to appear).
14. I. Csiszar, *Stud. Sci. Math. Hung.*, **2** 299–318 (1967).
15. J. Dolbeault and M. del Pino. *preprint CEREMADE 9905*, (1999).
16. L. Gross, *Amer. J. of Math.* **97** , 1061–1083 (1975).
17. A. Jüngel and R. Pinnau, *SIAM J. Math. Anal.*, 760–777, (2000).
18. A. Jüngel and G. Toscani, *ZAMP*, (2001) (to appear).
19. S. Kullback, *IEEE Trans. Inf. The.*, **4** 126–127, (1967).
20. J. L. López *et al*, *Comput. Math. Appl.* (2001), (to appear).
21. P.A. Markowich and C. Villani, *Revista Matematica Contemporanea* (SBM), (2001), (to appear).
22. T. G. Myers, *SIAM Reviews* **40**, 441–462 (1998).
23. F. Otto. *Commun. PDE* (2001).
24. A. Stam, *Inform. Control* **2** , 101–112 (1959).
25. G. Toscani, *C.R. Acad. Sc. Paris* , **A 324** série 1 689-694 (1997).
26. G. Toscani and C. Villani. *J. Statist. Phys.* **98** 1279–1309 (2000).

ON THE SYMMETRY CLASSIFICATION FOR A HEAT CONDUCTION MODEL

R. TRACINÀ
Dipartimento di Matematica
Università di Catania
E-mail: tracina@dmi.unict.it

In this paper we get the equivalence transformations for a special model for heat conduction. We show how it is possible to get symmetries from equivalence transformations by applying a projection theorem.

1 Intoduction

We consider a non linear system of partial differential equations describing the heat conduction in an isotropic rigid body [1] [2].

In one space dimension, this system can be written as:

$$z\left(\frac{q}{L}\right)_t + T_x = -\frac{q}{L}$$

$$\left(e_0 + \frac{z}{2}Aq^2\right)_t + q_x = 0, \tag{1}$$

where T is the absolute temperature, q the heat flux, z a positive relaxation constant, $L = L(T)$ the thermal conductivity, $e_0 = e_0(T)$ the equilibrium specific internal energy and $A := \frac{1}{L}\left(\frac{2}{T} - \frac{L'}{L}\right)$.

In our system the presence of two arbitrary functions, $e_0(T)$ and $L(T)$, suggests to obtain a symmetry classification by using the equivalence transformations.

The plane of the paper is the following. In Section 2 we give some elements concerned with the equivalence transformations for our system and we work in the space (x, t, T, q, C) (where $C = e_0'$) in order to obtain an equivalence classification with respect to L considering the arbitrary function C as a dependent variable. At this step of our procedure we consider the function C depending on t, x, T and q, i.e. we don't require the invariance of auxiliary conditions

$$C_x = C_t = C_q = 0$$

which characterize the functional dependence of specific heat at equilibrium. As a consequence we get only *weak equivalence transformations* [3]. In Section

3 we show how an equivalence transformation works for the system (1). Finally, in Section 4 we get an invariance classification starting from the weak equivalence classification.

2 Some elements on equivalence transformations and weak equivalence transformations

An equivalence transformation is a nondegenerate change of the variables t, x, T, q mapping any system of the form previous by considered into a system of the same form but, generally speaking, with different $e_0(T)$ and $L(T)$.

A weak equivalence transformation changes our system into a system having the same differential structure where the arbitrary function could depend not only on T but also on other independent or dependent variables.

In order to get the infinitesimal generators of this kind of transformations we use, as suggested by Ovsiannikov [4], the Lie infinitesimal criterion.

The infinitesimal equivalence operator Γ for the system (1) has the following form:

$$\Gamma = \xi^1\partial_x + \xi^2\partial_t + \eta^1\partial_T + \eta^2\partial_q + \mu\partial_C,$$

$$\xi^i = \xi^i(x,t,T,q), \quad \eta^i = \eta^i(x,t,T,q),$$

$$\mu = \mu(x,t,T,q,C).$$

The prolongation of Γ, which we need, is:

$$\tilde{\Gamma} = \Gamma + \zeta_1^1\partial_{T_x} + \zeta_2^1\partial_{T_t} + \zeta_1^2\partial_{q_x} + \zeta_2^2\partial_{q_t}$$

where the coefficients ζ_i^j, after putting

$$(x^1, x^2) \equiv (x,t), \quad (y^1, y^2) \equiv (T,q),$$

$$y_j^i = \frac{\partial y^i}{\partial x^j}, \quad y_{jk}^i = \frac{\partial^2 y^i}{\partial x^j \partial x^k},$$

$$D_j = \partial_{x^j} + y_j^i\partial_{y^i} + y_{jk}^i\partial_{y_k^i} + \ldots$$

are given by [3]

$$\zeta_j^i = D_j\eta^i - y_{x^1}^i D_j\xi^1 - y_{x^2}^i D_j\xi^2.$$

Requiring the invariance of our system with respect to the operator $\tilde{\Gamma}$ we get the determining system which leads to the following weak equivalence classification with respect to function L, where L_0 and k are constitutive constants, while c_0, c_1, c_2, c_3 and c_4 are arbitrary constants:

- L arbitrary:

$$\xi^1 = c_2 x + c_0, \quad \xi^2 = c_1,$$

$$\eta^1 = 0, \quad \eta^2 = -c_2 q,$$

$$\mu = -2c_2 C;$$

- $L = L_0 T^k, \quad k \neq 0, 2$:

$$\xi^1 = c_2 x + c_0, \quad \xi^2 = c_1,$$

$$\eta^1 = c_3 T, \quad \eta^2 = (-c_2 + kc_3 + c_3)q,$$

$$\mu = (-2c_2 + kc_3)C;$$

- $L = L_0 T^2$:

$$\xi^1 = c_2 x + c_0, \quad \xi^2 = -zc_4 e^{-t/z} + c_1,$$

$$\eta^1 = c_3 T, \quad \eta^2 = (c_4 e^{-t/z} - c_2 + 3c_3)q,$$

$$\mu = 2(c_4 e^{-t/z} + c_3 - c_2)C;$$

- $L = L_0$:

$$\xi^1 = c_4 x^2 + c_2 x + c_0, \quad \xi^2 = c_1 - c_4 z x,$$

$$\eta^1 = (c_4 x + c_3)T,$$

$$\eta^2 = (-c_4 x - c_2 + c_3)q - c_4 L_0 T,$$

$$\mu = -2(2c_4 x + c_2)C.$$

REMARK.

For $L = L_0 T^2$ we obtain a special case of the system for the heat propagation introduced by Ruggeri and Morro [6]. In fact, for this form of L, $A := \frac{1}{L}\left(\frac{2}{T} - \frac{L'}{L}\right)$ vanishes and the internal energy coincides with the internal energy at equilibrium $e_0(T)$.

3 Finite form for a equivalence transformations

In order to show how an equivalence transformation works, we consider the weak equivalence generator

$$\Gamma = (c_2 x + c_0)\partial_x + c_1\partial_t + c_3 T\partial_T + \\ +(3c_3 - c_2)q\partial_q + 2C(c_3 - c_2)\partial_C$$

associated with the following form of the system (1):

$$T_x + \frac{q}{L_0 T^2} - \frac{2zq}{L_0 T^3}T_t + \frac{z}{L_0 T^2}q_t = 0$$

$$q_x + CT_t = 0$$

assumed when $L = L_0 T^2$.

To get the finite form of equivalence transformation associated with the generator Γ, we must solve [4] the following Cauchy problem:

$$\frac{d\bar{x}}{d\epsilon} = \xi^1(\bar{x}, \bar{t}, \bar{T}, \bar{q}), \ \frac{d\bar{t}}{d\epsilon} = \xi^2(\bar{x}, \bar{t}, \bar{T}, \bar{q}),$$

$$\frac{d\bar{T}}{d\epsilon} = \eta^1(\bar{x}, \bar{t}, \bar{T}, \bar{q}), \ \frac{d\bar{q}}{d\epsilon} = \eta^2(\bar{x}, \bar{t}, \bar{T}, \bar{q}),$$

$$\frac{d\bar{C}}{d\epsilon} = \mu(\bar{x}, \bar{t}, \bar{T}, \bar{q}, \bar{C}),$$

$$\bar{x}(\epsilon = 0) = x, \ \bar{t}(\epsilon = 0) = t,$$

$$\bar{T}(\epsilon = 0) = T, \ \bar{q}(\epsilon = 0) = q,$$

$$\bar{C}(\epsilon = 0) = C$$

where ϵ is the small parameter.

After solving the previous Cauchy problem we get:

$$x = \frac{e^{-c_2\epsilon}(c_2\bar{x} + c_0) - c_0}{c_2}, \quad t = \bar{t} - c_1\epsilon, \quad T = e^{-c_3\epsilon}\bar{T},$$

$$q = e^{(c_2 - 3c_3)\epsilon}\bar{q}, \quad C = e^{2(c_2 - c_3)\epsilon}\bar{C}. \tag{2}$$

By trasforming our system with (2) we get:

$$L_0\bar{T}^3\bar{T}_{\bar{x}} + \bar{T}\bar{q} - 2z\bar{q}\bar{T}_{\bar{t}} + z\bar{T}\bar{q}_{\bar{t}} = 0$$

$$\bar{q}_{\bar{x}} + \bar{C}\bar{T}_{\bar{t}} = 0.$$

This system, as expected, has the same form but it is easy to see the new form of C, that is $\bar{C}$, is changed, for example if $C = C_0 T^3$ then $\bar{C} = C_0 e^{-3c_3\epsilon}\bar{T}^3$.

4 Symmetries from weak equivalence transformations

It is possible to get symmetries for the system (1) starting from the weak equivalence classification.

First of all we observe that the projection of an operator Γ in the space (t, x, T, q), i.e.

$$X = \xi^1 \partial_x + \xi^2 \partial_t + \eta^1 \partial_T + \eta^2 \partial_q,$$

is a symmetry operator.

Then, by using the operators obtained from our classification, we get some invariance algebras by applying the following theorem [5], [3]:

Theorem. *Let Γ be a weak equivalence operator for the system (1), the operator given by projection of Γ in the space $(t,\ x,\ T,\ q)$ is an symmetry operator if and only if the specializations of the function C are invariant with respect to Γ.*

As, in the search for weak equivalence operators, we considered C as if it was depending on (t, x, T, q), here now we take as a specialization: $C = C(T)$ **in agreement with the physics of the model.**

The invariance of the function C with respect to operator Γ of system (1), requested by the theorem, leads to the classifying equation for C:

$$\mu = C' \eta^1.$$

Specializing the form of η^1 and μ for each of the equivalence classes before obtained, we get a Lie invariance classification with respect to $C = C(T)$.

As an exemple, we consider the operator Γ admitted by our system for any form of L:

$$\Gamma = (c_2 x + c_0)\partial_x + c_1 \partial_t - c_2 q \partial_q - 2c_2 C \partial_C$$

then the condition $\mu = C'\eta^1$ becomes:

$$-2c_2 C = 0.$$

This equation, taking into account that the specific heat $C(T) > 0$, implies $c_2 = 0$, not only for any form of L but also for any form of $C(T)$ (physically suitable).

So, by substituting $c_2 = 0$ in Γ and by projecting in the space (t, x, T, q), we get this couple of generators:

$$X_1 = \partial_x, \quad X_2 = \partial_t.$$

These operators, because L is arbitrary, span the principal Lie algebras.

By applying in a similar way the previous theorem for the remaining generators of our weak equivalence classification with respect to L, we get additional generators and the corresponding form of specific heat (i.e. internal energy):

- $L = L_0 T^k$ and $C = C_0 T^h$:

$$X_3 = (k-h)x\partial_x + 2T\partial_T + (k+h+2)q\partial_q;$$

- $L = L_0$ and $C = C_0 T^{-4}$:

$$X_3 = 2x\partial_x + T\partial_T - q\partial_q,$$

$$X_4 = x^2\partial_x - zx\partial_t + xT\partial_T - (qx + L_0 T)\partial_q,$$

where C_0 is a constant constitutive constant.

Acknowledgments

This paper was supported by G.N.F.M. (Gruppo Nazionale per la Fisica Matematica).

References

1. G.Grioli, Rend.Acc.Naz.dei Lincei **67**, 332, (1979).
2. M. Torrisi and A. Valenti, Rend.Acc.Naz.dei Lincei **1**, 171,(1990).
3. M.Torrisi and R. Tracinà, Int. J. Non-Linear Mechanics **33, 3**, 473 (1998).
4. L. V. Ovsiannikov, *Group Analysis of Differential Equations* (Academic Press, New York 1982).
5. M.Torrisi, R. Tracinà and A. Valenti, J. Math. Phys. **37, 9**, 4758 (1996).
6. A. Morro and T. Ruggeri, J. Phys. C: Solid State Phys. **21**, 1743 (1988).

HYDRODYNAMIC ANALYSIS FOR HOT-CARRIERS TRANSPORT IN SEMICONDUCTORS

M. TROVATO

Dipartimento di Matematica e Fisica, Universitá di Sassari, via Vienna 2 - 07100 Sassari, Italy

e-mail: Trovato@ssmain.uniss.it, Trovato@dmi.unict.it

By considering the macroscopic variables of interest we develope the Maximum Entropy Principle (MEP) including the full-band effects with a total energy scheme. Furthemore, under spatialy homogeneous conditions, a closed set of balace equations for the fluctuations of these variables is costruct and a systematic study of small-signal analysis is provided. By generalizing the results obtained in previous papers we analyze quantitatively the different coupling processes, as functions of the electric field and we prove that, for n-type Si material, the coupling between the different moments can lead to a strongly no exponential decay of the correspondig *response functions*.

1 Introduction

Recentely [1,2,3,4,5,6,7,8] the maximum entropy principle has led to renewed interest in the construction of self-consistent closure relations for semiconductor transport equations using an arbitrary number of generalized kinetic fields

$$\psi_A(\vec{k}) = \{\varepsilon^p,\ \varepsilon^p u_{i_1},\ \varepsilon^p u_{i_1} u_{i_2} \ldots,\ \varepsilon^p u_{i_1} u_{i_2} \cdots u_{i_s},\ldots\} \tag{1}$$

where $p = 0, 1, ...N$, and $s = 1, 2, ...M$ (with arbitrary values for the integers N and M), u_i is the carrier group velocity and $\varepsilon(\vec{k})$ is the single particle band energy. In this context, it is further possible to reformulate in more general terms the theory for small-signal analysis for the hot carriers. As a matter of fact, by introducing a generalized *relaxation matrix* and generalized *fluctuating forces* we have succeeded in calculating the most relevant *Response Functions* of the electron system. We note that, the analysis of the decay in the time of the *Response functions* provides valuable information both on the *moments relaxation rate* and on overshoot phenomena of the system. Equation (1) is the key formula since it generalizes the kinetic fields to energy dispersions of general form. With this approach, we consider the expectation values

$$F_A = \int \psi_A(\vec{k})\, \mathcal{F}(\vec{k}, \vec{r}, t)\, d\vec{k} \tag{2}$$

The HD equations which are formally obtained within this scheme are [1,3,4,5,7]

$$\frac{\partial F_A}{\partial t} + \frac{\partial F_{Ak}}{\partial x_k} = -\frac{e}{\hbar} R_{Ai} E_i + P_A \quad , \qquad \text{with} \quad A = 1, \ldots \mathcal{N} \tag{3}$$

where $\mathcal{N}$ is the number of moments used, and F_{Ak}, P_{Ai}, R_A indicate, respectively, the fluxes, the external field productions, and the collisional productions. The structure of this equation shows that there are some unknown constitutive functions $H_A = \{F_{Ak}, P_{Ai}, R_A\}$ that must be determined, in a self-consistent way, in terms of the variables F_A. Following information theory, one can determine [1,3,5,7] sistematically the unknown constitutive functions, by introducing the MEP in terms of generalized kinetic fields (1).

2 Theory for the small-signal analysis

For space homogeneous and stationary conditions the solution [7] of the set of Eqs.(3) allows us to determine the *moments* of single-particle $\widetilde{F}_A = F_A/n$. By assuming that at the initial time the system of carriers is perturbed by an electric field $\delta E \xi(t)$ (where $|\xi(t)| \leq 1$), we will calculate the deviations from their average values of the moments denoted respectively with $\delta\widetilde{F}_A$. After linearizing Eqs.(3) around the stationary state, we obtain a system of equations which, under spatially homogenous conditions, can be written as

$$\frac{d\,\delta\widetilde{F}_\alpha(t)}{d\,t} = \Gamma_{\alpha\beta}\delta\widetilde{F}_\beta(t) - e\,\delta E\,\xi(t)\,\Gamma_\alpha^{(E)} \tag{4}$$

where the relaxation of the system to the stationary state is related to the *response matrix* $\Gamma_{\alpha\beta}$ which describes the time evolution of the moments after the perturbation of the electric field E and the $-e\delta E\xi(t)\ \Gamma_\alpha^{(E)}$ are the *fluctuating forces.* The equation (4) has the formal solution [9,10]

$$\delta\widetilde{\mathbf{F}}(t) = exp(\mathbf{\Gamma}t)\ \delta\widetilde{\mathbf{F}}(0) - e\,\delta E \int_0^t \mathbf{K}(s)\xi(t-s)\,ds \tag{5}$$

being

$$exp(\mathbf{\Gamma}t) = \mathbf{\Phi}\ diag\{exp(\lambda_1 t), \cdots, exp(\lambda_{\mathcal{N}-1}t)\}\ \mathbf{\Phi}^{-1} \tag{6}$$

$$\mathbf{K}(s) = exp(\mathbf{\Gamma}s)\ \mathbf{\Gamma}^{(E)} \tag{7}$$

with λ_α the eigenvalues of $\Gamma_{\alpha\beta}$ and $\mathbf{\Phi}$ the matrix of its eigenvectors.
The eigenvalues λ_α can be real or complex and they correspond to the *generalized relaxation rates* $\nu_\alpha = -\lambda_\alpha$, even if an exact correspondence between these rates and the respective relaxation processes exist only in the relaxation time approximation and in the absence of coupling between the variables $\widetilde{F}_\alpha$.
The response functions $\mathbf{K}(t)$ depends on the eigenvalues λ_α and it determines the linear response of the moments $\widetilde{F}_A$ to an arbitrary perturbation of the electric field. In particular, if we suppose that $\delta\widetilde{\mathbf{F}}(0) = 0$, the linear response

of hot carriers to a *step-like switching* of electric field and to a *small harmonic* electric field are of special interest. In the first case $\xi(t) = 1$ for all $t > 0$ and one obtains from Eq.(5) that

$$K_\alpha(t) = \frac{1}{e\delta E} \frac{d\, \delta \widetilde{F}_\alpha(t)}{d\, t} \tag{8}$$

Such a relation can be used for direct calculation of the *moments linear response functions* (see, for example for the *drift velocity* the Ref.[10]).
The advantages of the approach till here proposed, based on the MEP with a total energy scheme, are that: the formulation of a.c. and d.c. theory can be obtained, as at kinetic level, without the need to introducing external parameters and it can performed using an energy dispersion of general form (full-band approach).

3 Application of the total energy scheme

A simplified way to consider the total-energy scheme is to describe the *full complexity* of the band modeled in terms of a single particle with an effective mass which is a function of the average total energy $\widetilde{W}$ of the single carrier [4,7]. In this way the mass becomes a new constitutive function which should be independently determined from the fitting of experiments and/or MC calculations of the bulk material [4]. In this section the theory is applied to the case of n-Silicon using a *full set of scalar and vectorial moments* (i.e. M=1) and considering *scattering with phonons* of *f*- and *g*-type [7].

Figure 1 reports the generalized relaxation rates $\nu_\alpha = -\lambda_\alpha$ obtained using an increasing number of moments (i.e. $N = 2$, $N = 3$, $N = 5$), both in the parabolic (where m^* is constant) and nonparabolic (where $m^* = m^*(\widetilde{W})$) band models, respectively. The numerical results show that for small and intermediate values of the electric field there are same couple of complex conjugate eingenvalues due to the strong coupling between scalar and vectorial moments. In particular the velocity an energy relaxation rates are coupled by the electric field for all values of N. Besides, the width of the region with complex values of the eigenvalues depends both on the increasing number of moments used and on the nonparabolicity. A complex eigenvalue indicates the presence of some kind of *deterministic* relaxation [11] in the system. In the present case, the joint action of electric field and emission of optical phonons. In its extreme case this is well know as the condition of *streaming motion* [11]. The carrier is accelerated by the field up the energy of the optical phonon. From there, by emitting an optical phonon, it is scattered back to the bottom of the band and the cycle starts again. As a matter of fact, for the case of

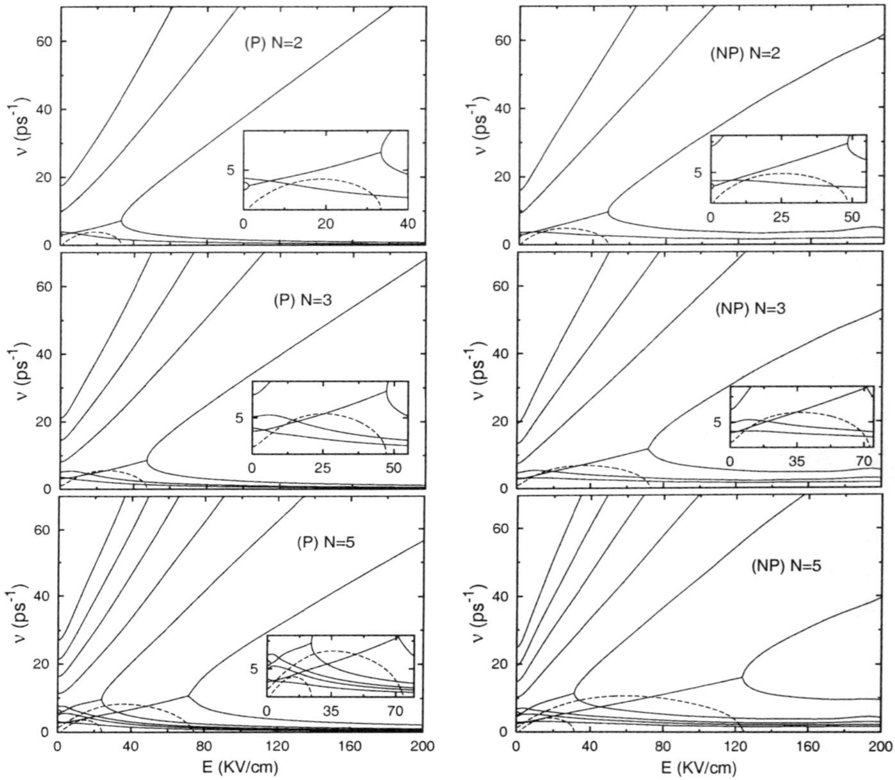

Figure 1. Eigenvalues of the realaxation matrix as a function of the electric field for electrons in Si in the case of Parabolic (P) and nonparabolic (NP) band models at $T_0 = 300K$. The continuous and the dashed lines (better evidenced inthe inserts) refer to the real part and the imaginary part to the eigenvalues evaluated in the linear approximation for a full set of scalar and vectorial moments with $N = 2$, $N = 3$ and $N = 5$ respectively.

electrons in Si at $T_0 = 300$ K carriers undergo many scattering events apart from optical phonon emissions, and therefore the streaming-motion regime is not fully achieved. By using many moments, the region with complex values of the eigenvalues is deeply enlarged towards higher fields (up to about 120 KV for $N = 5$). At these fields, the other scattering mechanisms are still efficent and the dissipation in the system is now well describe *only using many scalar and vectorial moments.* Figure 2 reports the time-dependence of the normalized linear response function respectively of velocity $K_v(t)/K_v(0)$,

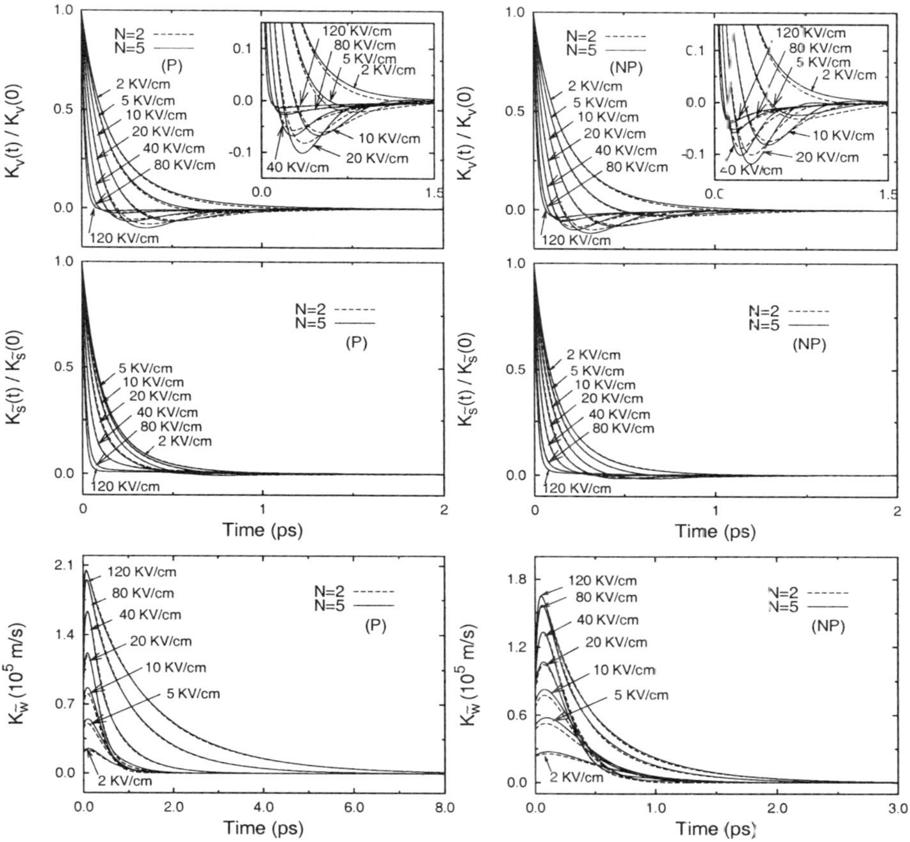

Figure 2. Normalized linear response of velocity $K_v(t)/K_v(0)$, energy flux $K_s(t)/K_s(0)$ and linear response of energy $K_w(t)$ vs time for electrons in Si in the parabolic (P) and nonparabolic (NP) band models at $T_0 = 300$ K and increasing electric fields. The dashed and the continuous lines refer to $N = 2$ and to $N = 5$ respectively.

energy flux $K_s(t)/K_s(0)$ and the linear response of energy $K_w(t)$ for $\{N = 2, N = 5\}$ at increasing electric fields. As a general trend, from Fig. 2 we notice that the response functions of vectorial moments exhibits a faster decay than that of energy, a well-know behavior associated to the property of the scattering mechanisms here considered which relaxes the velocity and the energy flux faster than the energy. The decay with time of the response functions is controlled essentially by the momentum and energy relaxation rates. At low electric fields, the shape of the velocity-response function is praticaly exponential, with a characteristic time constant which corresponds to momentum relaxation. The presence of higher electric fields couples the two relaxations processes, thus provoking a non-exponential shape of the decay. The shape of K_v becomes more complicated by exhibiting a negative part which is understood as follows [11]. The negative tail of $K_v(t)$ correspond to the velocity overshoot of carriers. In fact the velocity quickly increases with t when $K_v(t) > 0$ (since the initial momentum relaxation time of carriers is somewhat longer than that in the new steady state), reaches a maximum at time t corresponding to $K_v(t) = 0$ and then falls with t when $K_v(t) < 0$ (in this step the energy relaxation affects the momentum relaxation time that becomes shorter and the extra velocity of carriers is lost). At increasing fields, also the response function K_w clearly evidences the coupling between velocity and energy relaxation through a nonmonotonic behavior with a maximum which separates the velocity from the energy relaxation [12].

Acknowledgments

This research has been performed within the contract *Analisi qualitativa in problemi di meccanica dei continui e di biomatematica* and partially supported by the MURST and by the projct *Problemi matematici non lineari di propagazione e stabilità nei modelli del continuo.* Partial support from the GNFM and the INDAM is gratefully acknowledged.

References

1. I. Müller and T. Ruggeri, *Rational Extended Thermodynamics*, Springer, New York Berlin Heidelberg, (1998).
2. M. Trovato and P. Falsaperla, *Phys. Rev. B*, **57**, 4456 (1998), ibidem 57, 12617 (1998).
3. M. Trovato and L. Reggiani, *J. Appl. Phys.*, **85**, 4050 (1999).
4. M. Trovato, P. Falsaperla, L. Reggiani, *J.Appl. Phys.*, **86**, 5906 (1999).
5. H. Struchtrup, *Physica A*, **275**, 229 (2000).
6. F. Liotta, H. Struchtrup, *Solid-State Electron.*, **44**, 95 (2000).
7. M. Trovato and L. Reggiani, *Phys. Rev. B*, **61**, 16667 (2000).
8. M. Trovato and L. Reggiani, *Proceedings of IWCE, VLSI Design*, (2001).
9. P. Price, *J. Appl. Phys.*, **53**, 8805 (1982).
10. P. Price, *J. Appl. Phys.*, **54**, 3616 (1983).
11. T. Kuhn, L. Reggiani and L. Varani, *Phys. Rev. B*, **42**, 11133 (1990)
12. M. Nedjalkov, H. Kosina and S. Selberherr, *Proceedings of the SISPAD'99*, Eds. K. Taniguchi and N. Nakayama, p. 155 (Kyoto, 1999).

SMALL OSCILLATIONS OF A SPHERICAL LIQUID BRIDGE BETWEEN TWO EQUAL DISKS UNDER GRAVITY ZERO

D. VIVONA

Dipartimento di Metodi e Modelli Matematici
per le Scienze Applicate
Università La Sapienza, via A. Scarpa n.16-00161 ROMA (Italy)
e-mail: vivona@dmmm.uniroma1.it

In this paper, we consider the case of a spherical bridge built with an inviscid, incompressible liquid with density ρ. The study of the motion, which is assumed irrotational, is reduced to a variational equation on the function which gives the displacement of the free surface. We recognize that the stability of the equilibrium position depends essentially of the coerciveness of the bilinear form, which appears in this equation. The coerciveness can be studied by introducing an eigenvalue problem. We have shown in ([7]) that the smallest eigenvalue is always strictly greater than one, so that the bilinear form is coercive. Then, the existence of the eigenfrequencies of the problem is assured by a well-known method of functional analysis.

1 Position of the problem

We consider a liquid bridge between two equal coaxial (circular) disks D_0 and $D_0^{'}$ with radius R at distance $2h$, under gravity zero when the free surface in its equilibrium position is a part of a sphere S_0, with radius a, centre O on the axis of the disks. In equilibrium position and during the possible small oscillations, we suppose that the free surface comes to the boundaries C_0 and $C_0^{'}$ of the disks, it happens if the superficial tension is sufficiently great.

We introduce the orthogonal axes $Oxyz$, where Oz is the axis of the disks and the cylindrical coordinates r, θ, z. The volume τ_0 occupied by the liquid in the equilibrium position is determined by $z \in [-h, h]$, $r^2 + z^2 < a^2$. The equation of the free surface S_0 in the equilibrium position is $r^2 + z^2 = a^2 = R^2 + h^2$; the unit vector n normal to S_0 and directed to the exterior of the liquid is, in cylindrical coordinates, $(\frac{1}{a}\sqrt{a^2 - z^2}, 0, \frac{z}{a})$; the element of the area of S_0 is $dS_0 = a\ d\theta\ dz$.

The equation of the free surface during the oscillations is

$$r = \sqrt{a^2 - z^2} + \zeta(z, \theta, t) \qquad (-h \leq z \leq h);$$

and we suppose that the function ζ is "small" with its derivatives of the first and second order.

It must verify the following conditions:

$$\zeta(+h, \theta, t) = \zeta(-h, \theta, t) = 0 \tag{1}$$

$$\zeta \text{ periodic with respect to } \theta \text{ with period } 2\pi \tag{2}$$

$$\int_{S_0} \zeta\sqrt{a^2 - z^2} dS_0 = 0 \tag{3}$$

the last condition expresses the invariance of the total volume of the liquid.

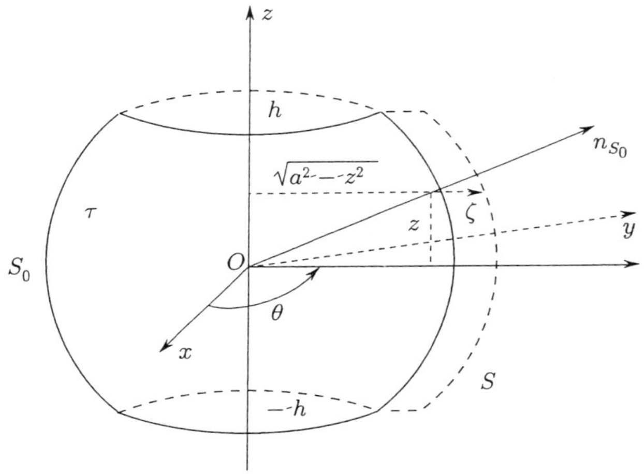

We will prove that the equilibrium position is stable and we study the small oscillations of the liquid about this position.

We suppose classically that the motion is irrotational, $\phi(r, z, \theta, t)$ is the velocity potential. The kinematic equations of the motion are

$$\begin{cases} \Delta\phi = 0 & \text{on } \tau_0 \\ \dfrac{\partial\phi}{\partial n} = \phi_z = 0 & \text{on } z = \pm h \\ \dfrac{\partial\phi}{\partial n} = \phi_z \dfrac{z}{a} + \phi_r \dfrac{\sqrt{a^2 - z^2}}{a} = \zeta_t \dfrac{\sqrt{a^2 - z^2}}{a} & \text{on } S_0 \, , \end{cases} \tag{4}$$

where $\frac{\partial\phi}{\partial n}$ is the external normal derivative and $\zeta_t = \frac{\partial\zeta}{\partial t}$.

The dynamical equation is reduced to the linearized Bernoulli's formula which expresses the pressure in the liquid:

$$p = -\rho\phi_t + c(t), \tag{5}$$

where c is an arbitrary function of the time linked to the indeterminateness of ϕ.

On the free surface, the dynamical condition (Laplace Law) is

$$p|_{S_0} - p_0 = -\alpha\left(\frac{1}{R_1} + \frac{1}{R_2}\right), \tag{6}$$

where p is the pressure of the liquid, p_0 the external constant pressure, α the constant superficial tension, R_1 and R_2 are the principal radii of the curvature of the free surface, regarded as negative when the principal centres of curvature lie in the same side of the surface as the liquid.

In (6) we can calculate the mean curvature by means of a classical formula [2] and from (6) we have

$$\phi_t|_{S_0} = \frac{\alpha}{\rho a}\frac{1}{\sqrt{a^2-z^2}}\left(\zeta + \zeta_{\theta\theta} + \frac{1}{a^2}\frac{\partial}{\partial z}[(a^2-z^2)^2\zeta_z]\right) + C(t) \tag{7}$$

where $C(t)$ is an arbitrary function of the time, too.

2 Functional equation of the problem

Now we consider the Neumann problem

$$\begin{cases} \Delta\phi = 0 & \text{on } \tau_0 \\ \dfrac{\partial\phi}{\partial n} = 0 & \text{on } z = \pm h \\ \dfrac{\partial\phi}{\partial n} = g\dfrac{\sqrt{a^2-z^2}}{a} & \text{on } S_0\ , \end{cases} \tag{8}$$

with the compatibility condition $\int_{S_0} g\sqrt{a^2-z^2}dS_0 = 0$.

We introduce the subspace $\widehat{H}$ of $L^2(S_0)$ defined by

$$\widehat{H} = \left\{g \in L^2(S_0)\Big/\int_{S_0} g\sqrt{a^2-z^2}\right\}.$$

It is well known that for every $g \in \widehat{H}$, the Neumann problem (8) has a unique weak solution

$$\phi \in \widehat{H}^1(\tau_0) = \left\{u \in L^2(S_0)\Big/\int_{S_0} u|_{S_0}\sqrt{a^2-z^2}dS_0 = 0\right\}$$

such that

$$\int_{\tau_0} \operatorname{grad}\phi \operatorname{grad}\psi \, d\tau = \int_{S_0} g\psi|_{S_0}\sqrt{a^2-z^2}dS_0 \quad \forall\psi \in \widehat{H}^1(\tau_0).$$

The trace $\phi|_{S_0}$ of ϕ on S_0 belongs to $\widehat{H}$. In this way we can introduce the operator $K : \widehat{H} \to \widehat{H}$ which associates to any function g the trace on S_0 of the corresponding solution of the Neumann problem

$$Kg = \phi|_{S_0}\,. \tag{9}$$

As in [3], we recognize that K is linear, symmetrical, positive definite and compact.

In our problem $g = \zeta_t$ and the equation (7) becomes

$$K\zeta_{tt} - \frac{\alpha}{\rho a}\frac{1}{\sqrt{a^2-z^2}}\left(\zeta + \zeta_{\theta\theta} + \frac{1}{a^2}\frac{\partial}{\partial z}[(a^2-z^2)^2\zeta_z]\right) = C(t).$$

Finally we have obtained the *functional equation* of the problem

$$K\zeta_{tt} + \frac{\alpha}{\rho a}M\zeta = 0, \tag{10}$$

where M is the operator given by

$$\begin{aligned} M\zeta = &-\frac{1}{\sqrt{a^2-z^2}}\Bigg[\zeta + \zeta_{\theta\theta} + \frac{1}{a^2}\frac{\partial}{\partial z}[(a^2-z^2)^2\zeta_z] + \\ &+\frac{1}{2\pi a^2\left(\operatorname{arcsen}\frac{h}{a} + \frac{h}{a}\sqrt{1-\frac{h^2}{a^2}}\right)}\cdot \\ &\cdot\left(\int_0^{2\pi}\int_{-h}^{h}\zeta d\theta dz + \frac{(a^2-h^2)^2}{a^2}\int_0^{2\pi}[\zeta_z(h,\theta,t)-\zeta_z(-h,\theta,t)]d\theta\right)\Bigg]. \end{aligned} \tag{11}$$

3 The bilinear form $m(\cdot,\cdot)$ associated to M and the mechanical meaning of K and $m(\cdot,\cdot)$

Calculating the scalar product $(Mu, v)_H$, with $u, v \in D$, we obtain, after integrating by:

$$(Mu, v)_H = \int_\Omega \left[u_\theta v_\theta + \frac{1}{a^2}(a^2-z^2)^2 u_z v_z - uv\right] d\theta dz.$$

The associated bilinear form $m(\cdot,\cdot) =: (Mu, v)_H$ is given by the same integral and its domain

$$V = \Big\{ u \in H^1(\Omega) \Big/ \int_\Omega u\sqrt{a^2 - z^2} d\theta dz = 0, \quad u|_{z=h} = u|_{z=-h} = 0;$$
$$\text{the traces of order 0 of on } \theta = 0, \theta = 2\pi \text{ are equal in } L^2(-h,h) \Big\}.$$

Now we give the mechanical meanings of K and $m(\cdot,\cdot)$. K is tied to the kinetic energy T of the liquid:

$$T = \frac{\rho}{2a}(K\zeta_t, \zeta_t)_H.$$

On the other hand, the potential energy Π of the surface tension forces is given by [1]:

$$\frac{d\Pi}{dt} = -\alpha \int_{S_0} \left(\frac{1}{R_1} + \frac{1}{R_2} \right) \zeta_t dS_0.$$

After integrating by parts, we obtain easily

$$\Pi = \frac{\alpha}{2a} m(\zeta, \zeta)$$

which gives the mechanical meaning of $m(\cdot,\cdot)$.

By Rumiantev's theorem [1], if $m(\cdot,\cdot)$ is coercive in V, the equilibrium position of the liquid is stable with respect to the norm in $L^2(S_0)$ of $\zeta, \zeta_\theta, \zeta_z, \zeta_t$.

4 Study of the coercivity of the bilinear form $m(\cdot,\cdot)$

In order to study the coercivity of

$$m(u,v) = \int_\Omega \left[u_\theta v_\theta + \frac{1}{a^2}(a^2 - z^2)^2 u_z v_z - uv \right] d\theta dz, \qquad (u, v \in V), \qquad (12)$$

we are going to use a method which can be found in [4], [5].

Let us set

$$\lambda = \inf_{u \in V} \frac{\int_\Omega [u_\theta^2 + \frac{1}{a^2}(a^2 - z^2)^2 u_z^2] d\theta dz}{\int_\Omega u^2 d\theta dz},$$

i.e. λ is the smallest eigenvalue of the following problem ([7]):

$$\begin{cases} u_{\theta\theta} + \frac{1}{a^2}\frac{\partial}{\partial z}[(a^2-z^2)^2 u_z] + \lambda u - \frac{\sqrt{a^2-z^2}}{4\pi a^2 h(a^2-\frac{h^2}{3})}\Big[(a^2-h^2)^{5\backslash 2}. \\ \quad \cdot \int_0^{2\pi}[u_z(h,z)-u_z(-h,z)]d\theta + \int_\Omega z(a^2-z^2)^{3\backslash 2} u_z d\theta dz\Big] = 0 \\ \int_\Omega u\sqrt{a^2-z^2}\,d\theta dz = 0 \\ u\ 2\pi - \text{periodic with respect to } \theta \\ u(\pm h,\theta) = 0. \end{cases}$$

We have proved that $\lambda > 1$; so we can choose ε such that $\lambda(1-\varepsilon)-1>0$, i.e.$0<\varepsilon<\frac{\lambda-1}{\lambda}$. Setting $\gamma = \min(\varepsilon, \lambda(1-\varepsilon)-1)$ we have

$$m(u,u) \geq \gamma ||u||^2_{H^2(\Omega)} \qquad \forall\, u \in V.$$

This means that $m(u,u)$ is *always coercive.*

Moreover $m(\cdot,\cdot)$, obviously symmetric, is also continuous.

5 Variational formulation of the problem and existence of the eigenfrequencies

Since $m(\cdot,\cdot)$ is always coercive in V, by virtue of Rumiantev's theorem [1], the spherical equilibrium position of the liquid bridge is always stable with respect to the norms in $L^2(S_0)$ of $\zeta, \zeta_\theta, \zeta_z, \zeta_t$.

Multiplying the functional equation (10)

$$K\zeta_{tt} + \frac{\alpha}{\rho a}M\zeta = 0.$$

by $\widetilde{\zeta} \in V$ and integrating in Ω, we obtain

$$(K\zeta_{tt}, \widetilde{\zeta})_{L^2(\Omega)} + \frac{\alpha}{\rho a} m(\zeta, \widetilde{\zeta})_{L^2(\Omega)} = 0 \qquad \forall \widetilde{\zeta} \in V.$$

We call $\widetilde{H}$ the completion of V with respect to the norm associated to the scalar product

$$(u,v)_{\widetilde{H}} = (Ku, v)_{L^2(\Omega)}.$$

The variational formulation of the problem is:

to find $\zeta(t) \in V$ such that

$$(\zeta_{tt}, \widetilde{\zeta})_{\widetilde{H}} + \frac{\alpha}{\rho a} m(\zeta, \widetilde{\zeta})_{L^2(\Omega)} = 0 \qquad \forall \widetilde{\zeta} \in V,$$

where $m(u,v)$ is symmetric, coercive and continuous in $V \times V$.

The embedding of V in $\widetilde{H}$ is obviously dense.
It is continuous, in fact for every $u \in V (\in \widetilde{H})$ it holds

$$||u||^2_{\widetilde{H}} = (Ku, u)_{L^2(\Omega)} \leq ||K|| \cdot ||u||^2_{L^2(\Omega)} \leq ||K|| \cdot ||u||^2_{H^1(\Omega)} \qquad \forall u \in V.$$

It is compact. Indeed, let $\{u_n\}$ a sequence which converges weakly to $u \in V \subset H^1(\Omega)$. By Rellich's theorem [6], it converges strongly to $u \in L^2(\Omega)$. As $u \in V(\in \widetilde{H})$ it is

$$||u_n - u||^2_{\widetilde{H}} \leq ||K|| \cdot ||u_n - u||^2_{L^2(\Omega)} \to 0,$$

so that $\{u_n\}$ converges strongly to $u \in \widetilde{H}$.

Therefore our problem is a *standard problem of vibration* [5]: there exists denumerable set of eigenvalues ω_n^2:

$$0 < \omega_1^2 \leq \omega_2^2 \leq \ldots \leq \omega_n^2 \leq \ldots, \qquad \omega_n^2 \to +\infty;$$

the corresponding eigenfunctions $\{\zeta_n\}$ form an orthogonal complete system in $\widetilde{H}$ and also in V equipped of the scalar product $m(u,v)$.
This proves the existence of the small oscillations of the spherical liquid bridge about its equilibrium position.

Acknowledgments

Research partially supported by GNFM, MURST (Italy)[a].

References

[1] N.N. Moiseyev and V.V. Rumiantsev, *Dynamic stability of bodies containing fluid*, (Springer-Verlag, Berlin, 1968).
[2] L. Bianchi, *Geometria differenziale* (Spoerri, Pisa, 1902).
[3] P. Capodanno and D. Vivona, Ricerche di Matematica **L**, 35 (2001).
[4] M. Roseau, *Vibrations in Mechanical Systems*, (Springer-Verlag, Berlin, 1987).

[a] This is an abbreviated paper: for details, see ([7]).

[5] E.W. Sanchez and E. Sanchez Palencia, *Vibration and coupling of continuous systems. Asymptotic methods*, (Spriger-Verlag, Berlin, 1989).

[6] H. Brezis, *Functional analysis. Theory and Applications* (Masson, Paris, 1983).

[7] D. Vivona, Bull.Polish Acad.Sc. **49**, 31 (2001).